2023 IEEE Radio Frequency Integrated Circuits Symposium (RFIC 2023)

San Diego, California, USA
11-13 June 2023

IEEE Catalog Number: CFP23MMW-POD
ISBN: 979-8-3503-2123-4

**Copyright © 2023 by the Institute of Electrical and Electronics Engineers, Inc.
All Rights Reserved**

Copyright and Reprint Permissions: Abstracting is permitted with credit to the source. Libraries are permitted to photocopy beyond the limit of U.S. copyright law for private use of patrons those articles in this volume that carry a code at the bottom of the first page, provided the per-copy fee indicated in the code is paid through Copyright Clearance Center, 222 Rosewood Drive, Danvers, MA 01923.

For other copying, reprint or republication permission, write to IEEE Copyrights Manager, IEEE Service Center, 445 Hoes Lane, Piscataway, NJ 08854. All rights reserved.

*** This is a print representation of what appears in the IEEE Digital Library. Some format issues inherent in the e-media version may also appear in this print version.

IEEE Catalog Number: CFP23MMW-POD
ISBN (Print-On-Demand):   979-8-3503-2123-4
ISBN (Online):   979-8-3503-2122-7
ISSN: 1529-2517

Additional Copies of This Publication Are Available From:

Curran Associates, Inc
57 Morehouse Lane
Red Hook, NY 12571 USA
Phone: (845) 758-0400
Fax: (845) 758-2633
E-mail: curran@proceedings.com
Web: www.proceedings.com

RFIC 2023 Session List

Sunday Sessions

PLENARY   RFIC Plenary

Monday Sessions

RMo1A	Circuits and Systems for High-Speed Optical and Wireline Communication
RMo1B	Silicon-Based Front-Ends and Building-Blocks
RMo1C	5G & mm-Wave Transceivers and Beamforming ICs
RMo2A	GaN Modeling, RFSOI Device and Chip Layout Automation
RMo2B	III/V Front-Ends and Building-Blocks
RMo2C	Systems and Applications at RF and mm-Wave
RMo3A	Reference Clock and Frequency Generation Techniques
RMo3B	High-Performance mm-Wave Low-Noise Amplifiers
RMo3C	THz & mm-Wave Communication Transceivers & Circuits
RMo4A	CMOS mm-Wave Frequency Multipliers
RMo4B	Advances in NB-IoT and WiFi Radios
RMo4C	High-Efficiency and Linear 5G mm-Wave Power Amplifiers

Tuesday Sessions

RTu1A	RF to THz LO Generation Solutions
RTu1B	Self-Interference Cancellation Techniques
RTu1C	mm-Wave & Sub-THz Circuits & Systems for Radar Sensing and Metrology
RTu2B	Emerging Circuits and Systems for Quantum Computing, Quantum Sensing, Photonics, and Built-In Self-Test (BIST) Applications
RTu2C	Systems for Applications: 5G and SATCOM
RTu3B	Advanced Building Blocks for mm-Wave & Beyond
RTu3C	IoT Transmitter and Sub-THz Power Amplifiers
RTu4C	Invited Industry Presentations

RFIC Plenary

Chair: Donald Y.C. Lie, Texas Tech University, USA — Co-Chair: Danilo Manstretta, Università di Pavia, Italy
20BCD, 17:30–19:00, Sunday, 11 June 2023

PAGE 1
PLENARY-1

The Roaring 20s: A Renaissance for the Semiconductor Industry?
Todd Younkin, SRC, USA

PAGE 3
PLENARY-2

Future System-on-Chip for Full Spectrum Utilization from RF to Optics
Mau-Chung Frank Chang, University of California, Los Angeles, USA

RMo1A: Circuits and Systems for High-Speed Optical and Wireline Communication

Chair: Bahar Jalali Farahani, Cisco, USA — Co-Chair: Antoine Frappé, Université de Lille, France
23ABC, 08:00–09:20, Monday, 12 June 2023

PAGE 5
RMo1A-1

A 112-Gbps, 0.73-pJ/Bit Fully-Integrated O-Band I-Q Optical Receiver in a 45-nm CMOS SOI-Photonic Process
Ghazal Movaghar, Viviana Arrunategui, Junqian Liu, Aaron Maharry, Clint Schow, James F. Buckwalter, University of California, Santa Barbara, USA

PAGE 9
RMo1A-2

A 42.7Gb/s Optical Receiver with Digital CDR in 28nm CMOS
Hyungryul Kang[1], Inhyun Kim[1], Ruida Liu[1], Ankur Kumar[1], Il-Min Yi[1], Yuan Yuan[2], Zhihong Huang[2], Samuel Palermo[1]
[1]Texas A&M University, USA ; [2]Hewlett Packard Enterprise, USA

PAGE 13
RMo1A-3

A 100-Gb/s 3-m Dual-Band PAM-4 Dielectric Waveguide Link with 1.9 pJ/Bit/m Efficiency in 28-nm CMOS
Kristof Dens[1], Joren Vaes[1], Christian Bluemm[2], Gabriel Guimaraes[1], Berke Gungor[1], Changsong Xie[2], Alexander Dyck[2], Patrick Reynaert[1]
[1]KU Leuven, Belgium ; [2]Huawei Technologies, Germany

PAGE 17
RMo1A-4

A 12-Bit 1.1GS/s Single-Channel Pipelined-SAR ADC with Adaptive Inter-Stage Redundancy
Xianshan Wen, Tao Fu, Ping Gui, Southern Methodist University, USA

RMo1B: Silicon-Based Front-Ends and Building-Blocks

Chair: Hao Gao, Technische Universiteit Eindhoven, The Netherlands
Co-Chair: Ramesh Harjani, University of Minnesota, USA
24ABC, 08:00–09:40, Monday, 12 June 2023

PAGE 21
RMo1B-1
A 65nm CMOS Current-Mode Receiver Frontend with Frequency-Translational Noise Cancelation and 425MHz IF Bandwidth
Benqing Guo[1], Haishi Wang[1], Lei Li[2], Wanting Zhou[2]
[1]CUIT, China ; [2]UESTC, China

PAGE 25
RMo1B-2
IIP2-Enhanced Receiver Front-End with Notch-Filtered Low-Noise Transconductance Amplifier for 5G New Radio Cellular Applications
Donggu Lee, Sukju Yun, Kuduck Kwon, Kangwon National University, Korea

PAGE 29
RMo1B-3
A Band-Shifting Millimeter-Wave T/R Front-End with Enhanced Imaging and Interference Rejection Covering 5G NR FR2 n257/n258/n259/n260/n261 Bands
Fuyuan Zhao, Wei Deng, Haikun Jia, Wenjing Ye, Ruichen Wan, Baoyong Chi, Tsinghua University, China

PAGE 33
RMo1B-4
A 6–22GHz CMOS Phase Shifter with Integrated mm-Wave LO
Natan Ershengoren, Eran Socher, Tel Aviv University, Israel

PAGE 37
RMo1B-5
A 300–320GHz Sliding-IF I/Q Receiver Front-End in 130nm SiGe Technology
Sumit Pratap Singh, Mostafa Jafari Nokandi, Mohammad Hassan Montaseri, Timo Rahkonen, Marko E. Leinonen, Aarno Pärssinen, University of Oulu, Finland

RMo1C: 5G & mm-Wave Transceivers and Beamforming ICs

Chair: Hongtao Xu, Fudan University, China — Co-Chair: Gernot Hueber, United Micro Technology, Austria
25ABC, 08:00–09:20, Monday, 12 June 2023

PAGE 41
RMo1C-1
A V-Band Four-Channel Phased Array Transmitter Beamforming IC with 0.7-Degree Phase Step in 20dB Dynamic Range
Cheol So, Eun-Taek Sung, Songcheol Hong, KAIST, Korea

PAGE 45
RMo1C-2
A 28/37GHz Frequency Reconfigurable Dual-Band Beamforming Front-End IC for 5G NR
Jaehun Lee[1], Hyoungkyu Jin[1], Gyuha Lee[1], Eun-Taek Sung[2], Songcheol Hong[1]
[1]KAIST, Korea ; [2]Samsung, Korea

PAGE 49
RMo1C-3
A 26.5-GHz 4×2 Array Switched Beam-Former Based on 2-D Butler Matrix for 5G Mobile Applications in 28-nm CMOS
Youngjoo Lee, Juwon Kim, Sungwon Kwon, Bosung Suh, Jun Hwang, Kyutae Park, Dohoon Chun, Kyujong Choi, Hongseok Choi, Dongho Yoo, Byung-Wook Min, Yonsei University, Korea

PAGE 53
RMo1C-4
A Phased-Array Receiver Front-End Using a Compact High Off-Impedance T/R Switch for n257/n258/n261 5G FR2 Cellular
Ying Chen, Xiaohua Yu, Samrat Dey, Venumadhav Bhagavatula, Chechun Kuo, Tienyu Chang, Ivan Siu-Chuang Lu, Sangwon Son, Samsung, USA

RFIC 2023 Table of Contents

RMo2A : GaN Modeling, RFSOI Device and Chip Layout Automation

Chair: Alvin Joseph, GlobalFoundries, USA — Co-Chair: Renyuan Wang, BAE Systems, USA
23ABC, 10:10–11:30, Monday, 12 June 2023

PAGE 57
RMo2A-1

 Exploration of Design/Layout Tradeoffs for RF Circuits Using ALIGN
Jitesh Poojary, Ramprasath S., Sachin S. Sapatnekar, Ramesh Harjani, University of Minnesota Twin Cities, USA Ⓐ

PAGE 61
RMo2A-2

 Optimizing RFSOI Performance Through a T-Shaped Gate and Nano-Second Laser Annealing Techniques
L. Lucci[1], S. Crémer[2], B. Duriez[1], T. Fache[1], S. Kerdiles[1], Y. Morand[1], J.-M. Hartmann[1], J. Azevedo-Goncalves[2], F. Gaillard[1], P. Chevalier[2]
[1]CEA-Leti, France Ⓐ *; [2]STMicroelectronics, France* Ⓐ

PAGE 65
RMo2A-4

 Artificial Neural Networks for GaN HEMT Model Extraction in D-Band Using Sparse Data
Andrea Arias-Purdue, Eythan Lam, Jonathan Tao, Everett O'Malley, James F. Buckwalter, University of California, Santa Barbara, USA Ⓐ

PAGE 69
RMo2A-5

 Benchmarking Measurement-Based Large-Signal FET Models for GaN HEMT Devices
Rafael Perez Martinez[1], Masaya Iwamoto[2], Jianjun Xu[2], Philipp Pahl[2], Srabanti Chowdhury[1]
[1]Stanford University, USA Ⓐ *; [2]Keysight Technologies, USA* Ⓐ

RFIC 2023 Table of Contents

RMo2B : III/V Front-Ends and Building-Blocks

Chair: Marcus Granger-Jones, Qorvo, USA — Co-Chair: Emanuel Cohen, Technion, Israel
24ABC, 10:10–11:30, Monday, 12 June 2023

PAGE 73
RMo2B-1

 A DC-to-12GHz 1.4–2.5dB IL 4×8 Switch Matrix with Three-Port Reconfigurable Inter-Stage Matching Network
Zhenyu Wang[1], Zhaowu Wang[1], Yicheng Wang[1], Xiaochen Tang[2], Yong Wang[1]
[1]UESTC, China Ⓐ *; [2]New Mexico State University, USA* Ⓐ

PAGE 77
RMo2B-2

 A DC-to-18GHz High Power and Low Loss Band-Divided SP3T Switch with Reconfigurable Pole-to-Throw Network in 0.25-μm GaN
Zhaowu Wang[1], Yicheng Wang[1], Zhenyu Wang[1], Xiaochen Tang[2], Yong Wang[1]
[1]UESTC, China Ⓐ *; [2]New Mexico State University, USA* Ⓐ

PAGE 81
RMo2B-3

 A 4.8–6.4-GHz GaN MMIC Front-End Module with Enhanced Back-Off Efficiency and Compact Size
Guansheng Lv, Wenhua Chen, Xiaofan Chen, Long Chen, Zhenghe Feng, Tsinghua University, China Ⓐ

PAGE 85
RMo2B-4

A 280GHz InP HBT Direct-Conversion Receiver with 10.8dB NF
Utku Soylu[1], Amirreza Alizadeh[1], Munkyo Seo[2], Mark J.W. Rodwell[1]
[1]University of California, Santa Barbara, USA Ⓐ *; [2]Sungkyunkwan University, Korea* Ⓐ

RMo2C: Systems and Applications at RF and mm-Wave

Chair: Mona Mostafa Hella, Rensselaer Polytechnic Institute, USA — Co-Chair: Rocco Tam, NXP Semiconductors, USA
25ABC, 10:10–11:30, Monday, 12 June 2023

PAGE 89
RMo2C-1
A CMOS 183GHz Millimeter-Wave Spectrometer for Exploring the Origins of Water and Evolution of the Solar System (INVITED PAPER)
Adrian Tang[1], Mau-Chung Frank Chang[2], Yanghyo Kim[3], Goutam Chattopadhyay[1]
[1]JPL, USA Ⓐ ; [2]University of California, Los Angeles, USA Ⓐ ; [3]Stevens Institute of Technology, USA Ⓐ

PAGE 93
RMo2C-2
A 140GHz Scalable On-Grid 8×8-Element Transmit-Receive Phased-Array with Up/Down Converters and 64QAM/24Gbps Data Rates
Amr Ahmed, Linjie Li, Minjae Jung, Gabriel M. Rebeiz, University of California, San Diego, USA Ⓐ

PAGE 97
RMo2C-3
A 57.6Gb/s Wireless Link Based on 26.4dBm EIRP D-Band Transmitter Module and a Channel Bonding Chipset on CMOS 45nm
Jose Luis Gonzalez-Jimenez, Alexandre Siligaris, Abdelaziz Hamani, Francesco Foglia-Manzillo, Pierre Courouve, Nicolas Cassiau, Cedric Dehos, Antonio Clemente, CEA-Leti, France Ⓐ

PAGE 101
RMo2C-4
A mm-Sized Implantable Glucose Sensor Using a Fluorescent Hydrogel
Hyeonkeon Lee[1], Honghyeon Park[2], Taein Kim[3], Mi Song Nam[3], Yun Jung Heo[3], Sanghoek Kim[3]
[1]LIG Nex1, Korea Ⓐ ; [2]Silicon Mitus, Korea Ⓐ ; [3]Kyung Hee University, Korea Ⓐ

RMo3A: Reference Clock and Frequency Generation Techniques

Chair: Salvatore Finocchiaro, Qorvo, USA — Co-Chair: Teerachot Siriburanon, University College Dublin, Ireland
23ABC, 13:30–14:50, Monday, 12 June 2023

PAGE 105
RMo3A-1
A 14.2mW 29–39.3-GHz Two-Stage PLL with a Current-Reuse Coupled Mixer Phase Detector
Yuan Liang[1], Chirn Chye Boon[2], Qian Chen[2]
[1]Guangzhou University, China Ⓐ ; [2]NTU, Singapore Ⓐ

PAGE 109
RMo3A-2
A Radiation-Hardened by Design 15–22GHz LC-VCO Charge-Pump PLL Achieving -240dB FoM in 22nm FinFET
David Dolt, Samuel Palermo, Texas A&M University, USA Ⓐ

PAGE 113
RMo3A-3
A Fast-Startup 80MHz Crystal Oscillator with 96×/368× Startup-Time Reductions for 3.0V/1.2V Swings Based on Un-Interrupted Phase-Aligned Injection
Chien-Wei Chen, Chao-Ching Hung, Yu-Li Hsueh, MediaTek, Taiwan Ⓐ

PAGE 117
RMo3A-4
Transformer-Coupled 2.5GHz BAW Oscillator with 12.5fs RMS-Jitter and 1-kHz Figure-of-Merit (FOM) of 210dB
Bichoy Bahr, Sachin Kalia, Baher Haroun, Swaminathan Sankaran, Texas Instruments, USA Ⓐ

RFIC 2023 Table of Contents

RMo3B: High-Performance mm-Wave Low-Noise Amplifiers

Chair: Vadim Issakov, Technische Universität Braunschweig, Germany
Co-Chair: Andrea Bevilacqua, Università di Padova, Italy
24ABC, 13:30–15:10, Monday, 12 June 2023

PAGE 121
RMo3B-1

A mm-Wave Wideband/Reconfigurable LNA Using a 3-Winding Transformer Load in 22-nm CMOS FDSOI
Mohammad Ghaedi Bardeh, Jierui Fu, Navid Naseh, Jeyanandh Paramesh, Kamran Entesari, Texas A&M University, USA

PAGE 125
RMo3B-2

High-Performance Broadband CMOS Low-Noise Amplifier with a Three-Winding Transformer for Broadband Matching
Joon-Hyung Kim[1], Jeong-Taek Son[1], Jung-Taek Lim[1], Jae-Eun Lee[1], Jae-Hyeok Song[1], Min-Seok Baek[1], Han-Woong Choi[1], Eun-Gyu Lee[1], Sunkyu Choi[1], Chong-Min Lee[2], Sung-Ku Yeo[2], Choul-Young Kim[1]
[1]Chungnam National University, Korea ; [2]Samsung, Korea

PAGE 129
RMo3B-3

A 28-GHz 12-dBm IIP3 Low-Noise Amplifier Using Source-Sensed Derivative Superposition of Cascode for Full-Duplex Receivers
Jonghoon Myeong, Byung-Wook Min, Yonsei University, Korea

PAGE 133
RMo3B-4

A SiGe BiCMOS D-Band LNA with Gain Boosted by Local Feedback in Common-Emitter Transistors
Guglielmo De Filippi, Lorenzo Piotto, Andrea Bilato, Andrea Mazzanti, Università di Pavia, Italy

PAGE 137
RMo3B-5

A D-Band to J-Band Low-Noise Amplifier with High Gain-Bandwidth Product in an Advanced 130nm SiGe BiCMOS Technology
Marcel Andree[1], Janusz Grzyb[1], Bernd Heinemann[2], Ullrich Pfeiffer[1]
[1]Bergische Universität Wuppertal, Germany ; [2]IHP, Germany

RFIC 2023 Table of Contents

RMo3C: THz & mm-Wave Communication Transceivers & Circuits

Chair: Omeed Momeni, University of California, Davis, USA
Co-Chair: Hossein Hashemi, University of Southern California, USA
25ABC, 13:30–14:50, Monday, 12 June 2023

PAGE 141
RMo3C-1

A 0.32-THz 6.6-dBm Single-Chain CW Transmitter Using On-Chip Antenna with 2.65% DC-to-THz Efficiency
Georg Zachl, Christoph Mangiavillano, Rohish Kumar Reddy Mitta, Tim Schumacher, Harald Pretl, Andreas Stelzer, Johannes Kepler Universität Linz, Austria

PAGE 145
RMo3C-2

A 26-Gb/s 140-GHz OOK CMOS Transmitter and Receiver Chipset for High-Speed Proximity Wireless Communication
Qiuyu Peng[1], Haikun Jia[1], Ran Fang[2], Pingda Guan[1], Mingxing Deng[1], Jiamin Xue[1], Wei Deng[1], Xin Liang[2], Baoyong Chi[1]
[1]Tsinghua University, China ; [2]BriRadio Technology, China

PAGE 149
RMo3C-3

A 189GHz Three-Stage Super-Gain-Boosted Amplifier with Power Gain of 10.7dB/Stage at Near-f_{max} Frequencies in 65nm CMOS
Fei He, Menghu Ni, Qian Xie, Zheng Wang, UESTC, China

PAGE 153
RMo3C-4

A Fully Integrated 400GHz OOK Transceiver with On-Chip Antenna in 90nm SiGe BiCMOS for Multi Gbps Wireless Communication
Sidharth Thomas, Sam Razavian, Aydin Babakhani, University of California, Los Angeles, USA

RMo4A: CMOS mm-Wave Frequency Multipliers

Chair: Andrea Mazzanti, Università di Pavia, Italy — Co-Chair: Foster Dai, Auburn University, USA
23ABC, 15:40–17:00, Monday, 12 June 2023

PAGE 157
RMo4A-1
A Double Balanced Frequency Doubler Achieving 70% Drain Efficiency and 25% Total Efficiency
Jesse Moody, Sandia National Laboratories, USA

PAGE 161
RMo4A-2
A 47GHz to 70GHz Frequency Doubler Exploiting 2nd-Harmonic Feedback with 10.1dBm P_{sat} and η_{total} of 22% in 65nm CMOS
Amin Aghighi, Mostafa Essawy, Arun Natarajan, Oregon State University, USA

PAGE 165
RMo4A-3
A 91.9–113.2GHz Compact Frequency Tripler with 44.6dBc Peak Fundamental Harmonic-Rejection-Ratio Using Embedded Notch-Filters and Area-Efficient Matching Network in 65nm CMOS
Xiangrong Huang, Haikun Jia, Wei Deng, Zhihua Wang, Baoyong Chi, Tsinghua University, China

PAGE 169
RMo4A-4
A Compact 70–86GHz Bandwidth Frequency Quadrupler with Transformer-Based Harmonic Reflectors in 28nm CMOS
Paolo Ricco[1], Gianfranco Avitabile[2], Danilo Manstretta[1]
[1]Università di Pavia, Italy ; [2]Politecnico di Bari, Italy

RMo4B: Advances in NB-IoT and WiFi Radios

Chair: Arun Paidimarri, IBM T.J. Watson Research Center, USA
Co-Chair: Roxann Broughton-Blanchard, Analog Devices, USA
24ABC, 15:40–16:40, Monday, 12 June 2023

PAGE 173
RMo4B-1
A 0.75mW Receiver Front-End for NB-IoT
Hossein Rahmanian Kooshkaki, Patrick P. Mercier, University of California, San Diego, USA

PAGE 177
RMo4B-2
A C-Band Compact High-Linearity Multibeam Phased-Array Receiver with Merged Gain-Programmable Phase Shifter Technique
Jingying Zhou[1], Nayu Li[1], Yuexiaozhou Yuan[1], Huiyan Gao[1], Shaogang Wang[1], Hang Lu[1], Chunyi Song[1], Yen-Cheng Kuan[2], Qun Jane Gu[3], Zhiwei Xu[1]
[1]Zhejiang University, China ; [2]NYCU, Taiwan ; [3]University of California, Davis, USA

PAGE 181
RMo4B-3
A Wi-Fi Tri-Band Switchable Transceiver with 57.9fs-RMS-Jitter Frequency Synthesizer, Achieving -42.6dB EVM Floor for EHT320 4096-QAM MCS13 Signal
Tsung-Ming Chen[1], Ming-Chung Liu[1], Pi-An Wu[1], Wei-Kai Hong[1], Ting-Wei Liang[1], Wei-Pang Chao[1], Po-Yu Chang[1], Yu-Ting Chou[1], Chien-Wei Chen[1], Sen-You Liu[1], Chang-Cheng Huang[1], Hsiu-Hsien Ting[1], Min-Shun Hsu[1], Yao-Chi Wang[1], Chao-Ching Hung[1], Yu-Li Hsueh[1], Eric Lu[2], Yuan-Hung Chung[1], Jing-Hong Conan Zhan[1]
[1]MediaTek, Taiwan ; [2]MediaTek, USA

RFIC 2023 Table of Contents

RMo4C: High-Efficiency and Linear 5G mm-Wave Power Amplifiers

Chair: Debopriyo Chowdhury, Broadcom, USA — Co-Chair: Patrick Reynaert, KU Leuven, Belgium
25ABC, 15:40–17:00, Monday, 12 June 2023

PAGE 185
RMo4C-1
A 26–40GHz 4-Way Hybrid Parallel-Series Role-Exchange Doherty PA with Broadband Deep Power Back-Off Efficiency Enhancement
Edward Liu, Hua Wang, ETH Zürich, Switzerland

PAGE 189
RMo4C-2
A 26GHz Balun-First Three-Way Doherty PA in 40nm CMOS with 20.7dBm Psat and 20dB Power Gain
Anil Kumar Kumaran[1], Masoud Pashaeifar[1], Hossein Mashad Nemati[2], Leo C.N. de Vreede[1], Morteza S. Alavi[1]
[1]Technische Universiteit Delft, The Netherlands ; [2]Huawei Technologies, Sweden

PAGE 193
RMo4C-3
A 26-GHz Linear Power Amplifier with 20.8-dBm OP1dB Supporting 256-QAM Wideband 5G NR OFDM for 5G Base Station Equipment
Zhilin Chen[1], Xiyu Wang[2], Xiaoxiao Ma[1], Min Lu[1], Jie Hu[1], Keqing Ouyang[1], Zhijun Long[1]
[1]Sanechips Technology, China ; [2]ZTE, China

PAGE 197
RMo4C-4
A 23–30GHz 4-Path Series-Parallel-Combined Class-AB Power Amplifier with 23dBm P_{sat}, 38.5% Peak PAE and 1.3° AM-PM Distortion in 40nm Bulk CMOS
Junjie Gu[1], Haoqi Qin[1], Hao Xu[1], Weitian Liu[1], Kefeng Han[2], Rui Yin[1], Lei Deng[3], Xiaoliang Shen[3], Zongming Duan[4], Hao Gao[5], Na Yan[1]
[1]Fudan University, China ; [2]Jiashan Fudan Institute, China ; [3]NICIC, China ; [4]ECRIEE, China ; [5]Technische Universiteit Eindhoven, The Netherlands

RFIC 2023 Table of Contents

RTu1A: RF to THz LO Generation Solutions

Chair: Andreia Cathelin, STMicroelectronics, France — Co-Chair: Wanghua Wu, Samsung, USA
23ABC, 08:00–09:20, Tuesday, 13 June 2023

PAGE 201
RTu1A-1

A 15.6-GHz Quad-Core VCO with Extended Circular Coil Topology for Both Main and Tail Inductors in 8-nm FinFET Process
Suoping Hu, Zhiyu Chen, Wanghua Wu, Pei-Yuan Chiang, Zhanjun Bai, Chih-Wei Yao, Sangwon Son, Samsung, USA

PAGE 205
RTu1A-2

A 10.8–14.5GHz 8-Phase 12.5%-Duty-Cycle Non-Overlapping LO Generator with Automatic Phase-and-Duty-Cycle Calibration for 60-GHz 8-Path-Filtering Sub-Sampling Receivers
Khoi T. Phan, Yang Gao, Howard C. Luong, HKUST, China

PAGE 209
RTu1A-3

A 4.4mW Inductorless 2–20GHz Single-Ended to Differential Frequency Doubler in 45nm RFSOI CMOS Technology
A. Meyer[1], M.L. Leyrer[2], C. Ziegler[1], M. Maier[1], V. Lammert[2], V. Issakov[1]
[1]Technische Universität Braunschweig, Germany ; [2]Infineon Technologies, Germany

PAGE 213
RTu1A-4

An Efficient 0.4THz Radiator with 20.6dBm EIRP and 0.2% DC-to-THz Efficiency in 90nm SiGe BiCMOS
Sidharth Thomas, Sam Razavian, Aydin Babakhani, University of California, Los Angeles, USA

RTu1B: Self-Interference Cancellation Techniques

Chair: Jin Zhou, MediaTek, USA — Co-Chair: Oren Eliezer, Samsung, USA
24ABC, 08:00–09:20, Tuesday, 13 June 2023

PAGE 217
RTu1B-1
A 28nm CMOS Dual-Band Concurrent WLAN and Narrow Band Transmitter with On-Chip Feedforward TX-to-TX Interference Cancellation Path for Low Antenna-to-Antenna Isolation in IoT Devices
Sai-Wang Tam, Alireza Razzaghi, Alden Wong, Sridhar Narravula, Weiwei Xu, Timothy Loo, Akash Kambale, Andrew Liu, Ovidiu Carnu, Yui Lin, Randy Tsang, NXP Semiconductors, USA

PAGE 221
RTu1B-2
A Distributed Cascode Power Amplifier with an Integrated Analog SIC Filter for Full-Duplex Wireless Operation in 65nm CMOS
Itamar Melamed, Nimrod Ginzberg, Omer Malka, Emanuel Cohen, Technion, Israel

PAGE 225
RTu1B-3
Frequency-Domain-Equalization-Based Full-Duplex Receiver with Passive-Frequency-Shifting N-Path Filters Achieving >53dB SI Suppression Across 160MHz BW
Sastry Garimella[1], Sasank Garikapati[1], Aravind Nagulu[2], Igor Kadota[1], Alfred Davidson[1], Gil Zussman[1], Harish Krishnaswamy[1]
[1]Columbia University, USA ; [2]Washington University in St. Louis, USA

PAGE 229
RTu1B-4
A Frequency-Tunable Dual-Path Frequency-Translated Noise-Cancelling Self-Interference Canceller RX with >16dBm SI Power-Handling in 65nm CMOS
Mostafa Essawy, Kareem Rashed, Amin Aghighi, Arun Natarajan, Oregon State University, USA

RTu1C: mm-Wave & Sub-THz Circuits & Systems for Radar Sensing and Metrology

Chair: Zeshan Ahmad, Texas Instruments, USA — Co-Chair: Ruonan Han, MIT, USA
25ABC, 08:00–09:40, Tuesday, 13 June 2023

PAGE 233
RTu1C-1
High-Linearity 76–81GHz Radar Receiver with an Intermodulation Distortion Cancellation and High-Power Limiter
N. Landsberg[1], M. Gordon[1], O. Asaf[1], N. Weisman[1], K. Ben-Atar[1], S. Levin[1], S. Pellerano[2], W. Shin[3], D. Nahmanny[1]
[1]Mobileye, Israel ; [2]Intel, USA ; [3]Apple, USA

PAGE 237
RTu1C-2
Mono/Multistatic Mode-Configurable E-Band FMCW Radar Transceiver Module for Drone-Borne Synthetic Aperture Radar
Kangseop Lee, Sirous Bahrami, Kyunghwan Kim, Jiseul Kim, Seung-Uk Choi, Ho-Jin Song, POSTECH, Korea

PAGE 241
RTu1C-3
A W-Band Spillover-Tolerant Mixer-First Receiver for FMCW Radars
Jingzhi Zhang, Sherif S. Ahmed, Amin Arbabian, Stanford University, USA

PAGE 245
RTu1C-4
A CMOS 160GHz Integrated Permittivity Sensor with Resolution of 0.05% $\Delta\varepsilon_r$
Hai Yu[1], Xuan Ding[1], Jingjun Chen[2], Sajjad Sabbaghi Saber[1], Qun Jane Gu[1]
[1]University of California, Davis, USA ; [2]Qualcomm, USA

PAGE 249
RTu1C-5
A 160-GHz FMCW Radar Transceiver with Slotline-Based High Isolation Full-Duplexer in 130nm SiGe BiCMOS Process
Xingcun Li, Huibo Wu, Shuyang Li, Wenhua Chen, Zhenghe Feng, Tsinghua University, China

RFIC 2023 Table of Contents

RTu2B: Emerging Circuits and Systems for Quantum Computing, Quantum Sensing, Photonics, and Built-In Self-Test (BIST) Applications

Chair: Fabio Sebastiano, Technische Universiteit Delft, The Netherlands — Co-Chair: Duane Howard, Amazon, USA
24ABC, 10:10–11:30, Tuesday, 13 June 2023

PAGE 253
RTu2B-1

 A Diamond Quantum Magnetometer Based on a Chip-Integrated 4-Way Transmitter in 130-nm SiGe BiCMOS
Hadi Lotfi[1], Michal Kern[1], Nico Striegler[2], Thomas Unden[2], Jochen Scharpf[2], Patrick Schalberger[1], Ilai Schwartz[2], Philipp Neumann[2], Jens Anders[1]
[1]Universität Stuttgart, Germany ; *[2]NVision Imaging Technologies, Germany*

PAGE 257
RTu2B-2

 A Cryo-CMOS DAC-Based 40Gb/s PAM4 Wireline Transmitter for Quantum Computing Applications
Niels Fakkel, Mohsen Mortazavi, Ramon Overwater, Fabio Sebastiano, Masoud Babaie, Technische Universiteit Delft, The Netherlands

PAGE 261
RTu2B-3

 A mm-Wave CMOS/Si-Photonics Hybrid-Integrated Software-Defined Radio Receiver Achieving >80-dB Blocker Rejection of <-10dBm In-Band Blockers
Ramy Rady, Yu-Lun Luo, Christi Madsen, Samuel Palermo, Kamran Entesari, Texas A&M University, USA

PAGE 265
RTu2B-4

 Mixer-Free Phase and Amplitude Comparison Method for Built-In Self-Test of Multiple Channel Beamforming IC
Seonjeong Park, Eun-Taek Sung, Seunghun Wang, Songcheol Hong, KAIST, Korea

---

RFIC 2023 Table of Contents

RTu2C: Systems for Applications: 5G and SATCOM

Chair: Bodhisatwa Sadhu, IBM T.J. Watson Research Center, USA — Co-Chair: Raja Pullela, MaxLinear, USA
25ABC, 10:10–11:30, Tuesday, 13 June 2023

PAGE 269
RTu2C-1

 A 24–30GHz 4-Stream CMOS Transceiver Based on Dual-LO Phase-Shifting Fully Connected Architecture
Qingfeng Zhang[1], Yiming Yu[1], Dongming Duan[1], Xin Xie[1], Shaoyu Meng[1], Haoran Wang[1], Chenxi Zhao[1], Huihua Liu[1], Yunqiu Wu[1], Wenquan Che[2], Quan Xue[2], Kai Kang[1]
[1]UESTC, China ; *[2]SCUT, China*

PAGE 273
RTu2C-2

 A 39GHz 2×16-Channel Phased-Array Transceiver IC with Compact, High-Efficiency Doherty Power Amplifiers
Joonho Jung, Jooseok Lee, Daehyun Kang, Jinhyun Kim, Woojae Lee, Hansik Oh, Jae-hong Park, Kihyun Kim, Dong-hyun Lee, Sangho Lee, Jeong Ho Lee, Ji Hoon Kim, Younghwan Kim, Taewan Kim, Sangyong Park, Seungwon Park, Seungjae Baek, Bohee Suh, Soyoung Oh, Dongsoo Lee, Juho Son, Sung-gi Yang, Samsung, Korea

PAGE 277
RTu2C-3

 A 14-nm Low-Cost IF Transceiver IC with Low-Jitter LO and Flexible Calibration Architecture for 5G FR2 Mobile Applications
Wanghua Wu[1], Jeiyoung Lee[2], Pak-Kim Lau[1], Taeyoung Kang[1], Kim Kiu Lau[1], Si-Wook Yoo[1], Xingliang Zhao[1], Ashutosh Verma[1], Ivan Siu-Chuang Lu[1], Chih-Wei Yao[1], Hou-Shin Chen[1], Gennady Feygin[1], Pranav Dayal[1], Kee-Bong Song[1], Sangwon Son[1]
[1]Samsung, USA ; *[2]Samsung, Korea*

PAGE 281
RTu2C-4

 A Quad-Band RX Phased-Array Receive Beamformer with Two Simultaneous Beams, Polarization Diversity, and 2.1–2.3dB NF for C/X/Ku/Ka-Band SATCOM
Zhaoxin Hu, Oguz Kazan, Gabriel M. Rebeiz, University of California, San Diego, USA

RFIC 2023 Table of Contents

RTu3B: Advanced Building Blocks for mm-Wave & Beyond

Chair: Yahya Tousi, University of Minnesota, USA — Co-Chair: Qun Jane Gu, University of California, Davis, USA
24ABC, 13:30–14:50, Tuesday, 13 June 2023

PAGE 285
RTu3B-1
An Ultra-Wideband and Compact Active Quasi-Circulator with Phase Alternated Differential Amplifier
Dongho Yoo, Jun Hwang, Byung-Wook Min, Yonsei University, Korea

PAGE 289
RTu3B-2
A D-Band Calibration-Free Passive 360° Phase Shifter with 1.2° RMS Phase Error in 45nm RFSOI
Mohammadreza Abbasi, Wooram Lee, Pennsylvania State University, USA

PAGE 293
RTu3B-3
A 140GHz RF Beamforming Phased-Array Receiver Supporting >20dB IRR with 8GHz Channel Bandwidth at Low IF in 22nm FDSOI CMOS
Shenggang Dong[1], Navneet Sharma[1], Sensen Li[1], Michael Chen[1], Xiaohan Zhang[2], Yaolong Hu[2], Jiantong Li[1], Yong Su[1], Xinguang Xu[1], Vitali Loseu[1], Eunyoung Seok[1], Taiyun Chi[2], Won-Suk Choi[1], Gary Xu[1]
[1]Samsung, USA ; [2]Rice University, USA

PAGE 297
RTu3B-4
A mm-Wave Blocker-Tolerant Receiver Achieving <4dB NF and -3.5dBm B1dB in 65-nm CMOS
Erez Zolkov, Nimrod Ginzberg, Emanuel Cohen, Technion, Israel

RFIC 2023 Table of Contents

RTu3C: IoT Transmitter and Sub-THz Power Amplifiers

Chair: Alexandre Giry, CEA-Leti, France — Co-Chair: Hyun-Chul Park, Samsung, Korea
25ABC, 13:30–14:50, Tuesday, 13 June 2023

PAGE 301
RTu3C-1

A Reactive Passive Mixer for 16-QAM Cartesian IoT Transmitters in 22nm FD-SOI CMOS
Lorenzo Tomasin, Daniele Vogrig, Andrea Neviani, Andrea Bevilacqua, Università di Padova, Italy

PAGE 305
RTu3C-2

A 110–170GHz Phase-Invariant Variable-Gain Power Amplifier Module with 20–22dBm P_{sat} and 30dBm OIP3 Utilizing SiGe HBT RFICs
Mustafa Sayginer, Michael Holyoak, Mike Zierdt, Mohamed Elkhouly, Joe Weiner, Yves Baeyens, Shahriar Shahramian, Nokia Bell Labs, USA

PAGE 309
RTu3C-3

A D-Band 20.4dBm OP_{1dB} Transformer-Based Power Amplifier with 23.6% PAE in a 250-nm InP HBT Technology
Senne Gielen[1], Yang Zhang[2], Mark Ingels[2], Patrick Reynaert[1]
[1]KU Leuven, Belgium ; [2]imec, Belgium

PAGE 313
RTu3C-4

305-GHz Cascode Power Amplifier Using Capacitive Feedback Fabricated Using SiGe HBT's with f_{max} of 450GHz
Suprovo Ghosh, Frank Zhang, Haidong Guo, Kenneth K. O, University of Texas at Dallas, USA

RTu4C : Invited Industry Presentations

Chair: Debopriyo Chowdhury, Broadcom, USA
25ABC, 15:40-16:40, Tuesday, 13 June 2023

(NA)
RTu4C-1
© D-Band Circuits and Systems Application in 55nm SiGe BiCMOS
Andrea Pallotta[1], Pascal Roux[2], David del Rio[3], Juan Francisco Sevillano[3], Mahmoud Pirbazari[1], Andrea Mazzanti[4], Vladimir Ermolov[5], Jussi Säily[5], Mario Giovanni Frecassetti[6], Maurizio Moretto[6]
[1]STMicroelectronics, Italy © ; [2]Nokia Bell Labs, France © ; [3]Ceit, Spain © ; [4]Università di Pavia, Italy © ; [5]VTT, Finland © ; [6]Nokia, Italy ©

(NA)
RTu4C-2
© Thermal Challenges in GaAs PA Design for 5G Applications
S.H. Tsai[1], C.S. Yeh[1], C. Potier[1], B. Thota[2], H. Andersen[1], B. François[2]
[1]iCana, Taiwan © ; [2]iCana, Belgium ©

(NA)
RTu4C-3
© 22FDX Technology Solutions for 5G mmWave
Shafiullah Syed[1], Zhixing Zhao[2], Shih Ni Ong[3], Lye Hock Kelvin Chan[3], Kirby Kheng Seong Tan[3], Chee Wai Wan[3], Wai Heng Chow[3], Koi Wai Chew[3], Amit Kumar Sahoo[3], Raghavendra Kammar Nagaraja[4], Andreas Knorr[1], Qiao Yang[5], Chris Boyer[1], Stephen Moss[1], Ming-Cheng Chang[2], Jen Shuang Wong[3], Dieter Lipp[2], Peter Javorka[2], Jan Hoentschel[2]
[1]GlobalFoundries, USA © ; [2]GlobalFoundries, Germany © ; [3]GlobalFoundries, Singapore © ; [4]GlobalFoundries, India © ; [5]GlobalFoundries, China ©

2023 IEEE Radio Frequency Integrated Circuits Symposium

San Diego, California, USA
11–13 June 2023

PROGRAM

San Diego Convention Center

Sponsored by

IEEE Microwave Theory and Technology Society
IEEE Electron Devices Society
and
IEEE Solid-State Circuits Society

RFIC Plenary, Reception, and Symposium Showcase
Sunday Evening, 11 June 2023
San Diego Convention Center

After a busy day immersed in RFIC'23 Workshops, enjoy a relaxing evening with your RFIC colleagues at these special Sunday night RFIC events, to be held in the 4th (Upper) Level of the San Diego Convention Center Ballroom 20.

17:30–19:00, Plenary Session, Ballroom 20BCD: The evening begins with the Student Paper Awards, Industry Paper Awards, and Tina Quach Service Award ceremony followed by two outstanding plenary speakers: Dr. Todd Younkin, President and CEO of the Semiconductor Research Corporation (SRC), USA, and Prof. Mau-Chung Frank Chang, the Wintek Chair in Electrical Engineering and Distinguished Professor of University of California, Los Angeles (UCLA) and the former President of the National Yang Ming Chiao Tung University (NYCU), Hsinchu, Taiwan.

19:00–21:00, RFIC Symposium Reception and Showcase, Sails Pavilion (across the hallway from Ballroom 20): Immediately following the Plenary Session is the RFIC'23 Symposium Reception and Showcase. Food and drinks will be provided while you connect with old friends, make new acquaintances, and catch up on the latest developments in the field.

The RFIC'23 Symposium Showcase is held concurrently with the reception and will feature our industry paper awards finalists, student paper awards finalists and the Systems & Applications Forum. The selected authors will be present to highlight their innovative work, summarized in electronic poster format, and some will also show a live demonstration. The media will cover this event, making it an excellent opportunity to announce the latest RFIC developments and breakthroughs.

Admittance to all RFIC Sunday evening events is included with the RFIC Symposium registration and the Super-pass registration. Additionally, Sunday-night-only tickets can be purchased for those who cannot attend the rest of the RFIC Symposium but don't want to miss the microwave week's opening event. Please see https://rfic-ieee.org/ for more details.

The RFIC Symposium is made possible through the generous support of our corporate sponsors:

RFIC 2023 Corporate Sponsors

Diamond

Platinum

Gold

Student Paper Contest

Student Programs

RFIC Symposium Schedule (11–13 June 2023)

Event	Location	Sat 10 June	Sun 11 June	Mon 12 June	Tue 13 June
Registration	Lobby D	08:00–17:00		07:00–18:00	
Speakers' Breakfast	Ballroom 20A			07:00–08:00	
Workshops	Rooms 23–25, 29–33		08:00–11:50 13:30–17:20		
Workshops Lunch	Meeting Room Foyer, Terrace Side		11:45–13:00		
Technical Lecture	Ballroom 20A		12:00–13:20		
Plenary Session	Ballroom 20BCD		17:30–19:00		
Reception and Symposium Showcase	Sails Pavilion		19:00–21:00		
Technical Sessions	Rooms 23–25			08:00–09:40 10:10–11:30 13:30–15:10 15:40–17:00	
Panel Sessions	Ballroom 20A, Room 32AB			12:00–13:30	
Student/Industry ChipChat	Room 32AB				17:00–19:00

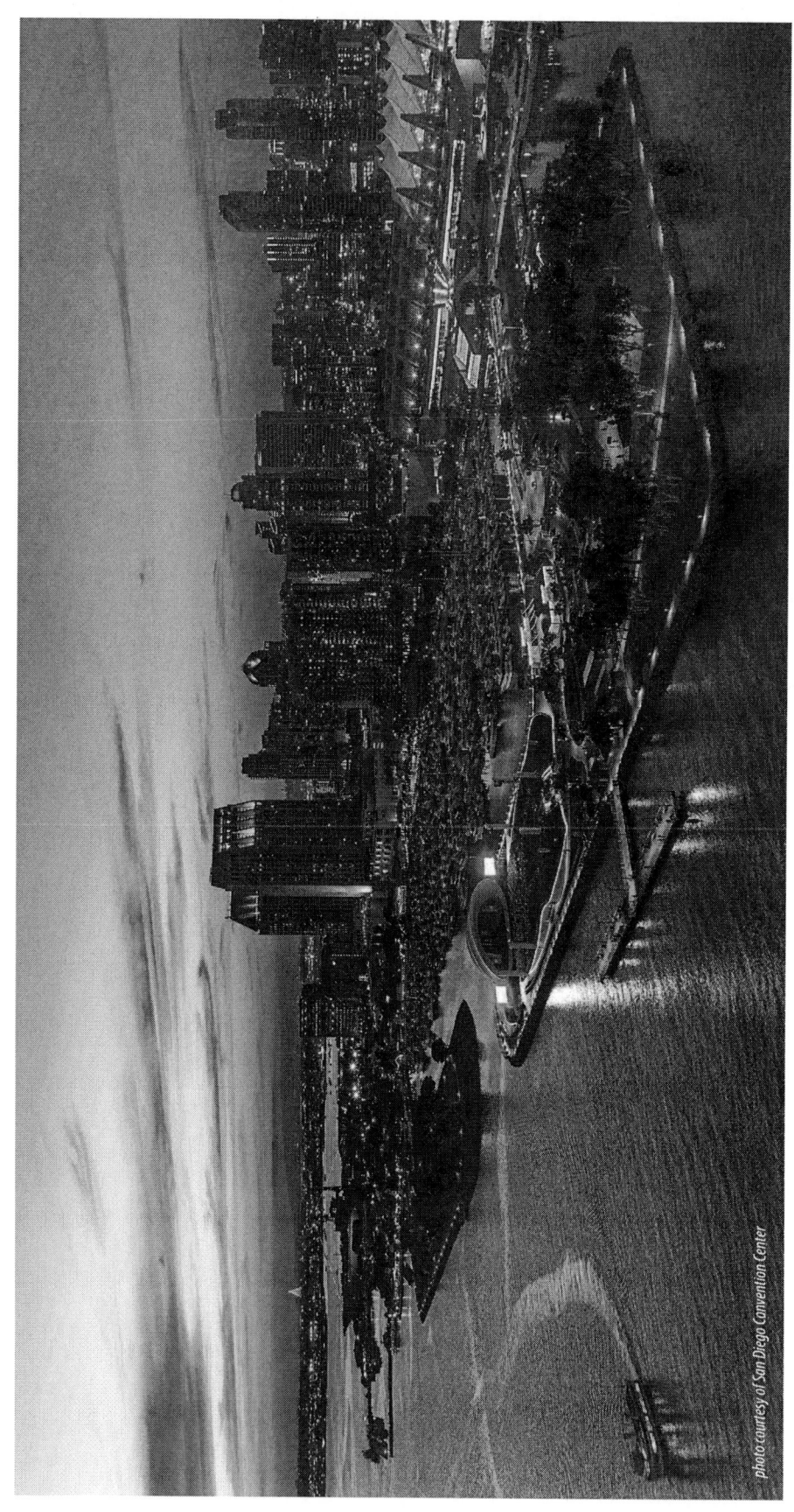

photo courtesy of San Diego Convention Center

Table of Contents

Table of Contents
Welcome Message from Chairs
Steering Committee
Support Staff
Executive Committee
Advisory Board
Technical Program Committee
RFIC 2023 Schedule
Plenary, Reception, and Symposium Showcase
Plenary Speakers
Student Paper Awards Finalists
Industry Paper Awards Finalists
Systems & Applications Forum and Best Student/Industry Paper Showcase
RFIC Panel Session
RFIC/IMS Joint Panel Session
Student/Industry ChipChat
Technical Lecture

MONDAY

AM1
RMo1A: Circuits and Systems for High-Speed Optical and Wireline Communication
RMo1B: Silicon-Based Front-Ends and Building-Blocks
RMo1C: 5G & mm-Wave Transceivers and Beamforming ICs

AM2
RMo2A: GaN Modeling, RFSOI Device and Chip Layout Automation
RMo2B: III/V Front-Ends and Building-Blocks
RMo2C: Systems and Applications at RF and mm-Wave

PM1
RMo3A: Reference Clock and Frequency Generation Techniques
RMo3B: High-Performance mm-Wave Low-Noise Amplifiers
RMo3C: THz & mm-Wave Communication Transceivers & Circuits

PM2
RMo4A: CMOS mm-Wave Frequency Multipliers
RMo4B: Advances in NB-IoT and WiFi Radios
RMo4C: High-Efficiency and Linear 5G mm-Wave Power Amplifiers

TUESDAY

AM1
RTu1A: RF to THz LO Generation Solutions
RTu1B: Self-Interference Cancellation Techniques
RTu1C: mm-Wave & Sub-THz Circuits & Systems for Radar Sensing and Metrology

AM2
RTu2B: Emerging Circuits and Systems for Quantum Computing, Quantum Sensing, Photonics, and Built-In Self-Test (BIST) Applications
RTu2C: Systems for Applications: 5G and SATCOM

PM1
RTu3B: Advanced Building Blocks for mm-Wave & Beyond
RTu3C: IoT Transmitter and Sub-THz Power Amplifiers
RTu4C: Invited Industry Presentations

RFIC Workshops
Social Events/Guest Program

Welcome Message from Chairs

On behalf of the Executive and Steering Committees, we would like to invite you to join us for the *2023 IEEE Radio Frequency Integrated Circuits (RFIC'23) Symposium*. The IEEE RFIC Symposium (RFIC) is the premier annual forum focused on presenting the latest breakthroughs and research results in all areas related to radio frequency (RF), millimeter-wave (mmWave), and wireless integrated circuits (ICs). The RFIC Symposium, combined with the International Microwave Symposium (IMS), ARFTG, and the Industry Exhibition, forms the "IMS Week", the world's largest RF and microwave technical meeting of the year. Fortunately, since the world has finally been recovering from the unprecedented COVID-19 pandemic, we are organizing the RFIC'23 Symposium as an in-person only event.

RFIC'23 will be held at the beautiful San Diego Convention Center, in the heart of the "America's Finest City", San Diego, California, from Sunday morning, 11 June, through Tuesday night, 13 June. The event starts with our workshops and technical lecture program on Sunday. The RFIC'23 Plenary Session is held on Sunday night, followed by a welcome reception and the Symposium Showcase. The RFIC technical sessions will be held on Monday and Tuesday in parallel tracks. The IMS technical exhibition will be held on Tuesday, Wednesday, and Thursday.

RFIC'23 technical papers will be presented through parallel sessions on Monday and Tuesday. Our sessions will include topics spanning from highly integrated wireless systems-on-chip and low-power radios to new power amplifiers (PAs), voltage-controlled oscillators (VCOs), and front-end circuitry designs. As mmWave 5G/6G research continues to gain attention, increasingly more mmWave and terahertz (THz) IC content is being published at RFIC. RFIC'23 also continues to expand its scope and has included a new RF Systems and Applications session dedicated to novel applications of RFICs at the systems level. This includes innovations in IC and system architectures, usage models, calibration techniques, and integration approaches. This systems initiative brings together researchers and practicing engineers at the boundary of RFICs and systems to the benefit of all. Additionally, RFIC'23 continues to cover emerging technologies in RF, such as on novel THz solutions, 3D ICs, silicon photonics, quantum computing ICs, hardware security, MEMS-based sensors and actuators, and AI/machine learning applied to RF circuits.

The 2023 RFIC Symposium will feature a rich educational program on Sunday 11 June with eleven RFIC focused workshops and one technical lecture. The RFIC workshops cover a wide range of advanced topics in RFIC technology as follows: mmWave, Advanced High-Speed Circuits and Systems, Low Power, Tutorial Style.

RFIC'23 will also feature an excellent 80 min short course, which we call a "Technical Lecture", delivered by world-renowned educator and author Prof. Behzad Razavi of UCLA, on "Modern Radio Receivers — From WiFi to 5G and Beyond". This lecture covers both RFIC and radio system design aspects, and would be instructive and beneficial for both students and newcomers as well as for senior practicing designers.

Following the full day of Sunday workshops, the RFIC Plenary Session will be held in the evening beginning with conference highlights, the presentation of the Student Paper Awards and the Industry Best Paper Awards. The RFIC'23 Plenary Session will conclude with two visionary plenary talks. In his talk "The Roaring 20s: A Renaissance for the Semiconductor Industry?", Dr. Todd Younkin, President and CEO of the Semiconductor Research Corporation (SRC), will share his vision for the future of global semiconductor technologies and design, especially those that will enable future RFIC breakthroughs. Prof. Mau-Chung Frank Chang, the Wintek Chair in Electrical Engineering and Distinguished Professor of University of California, Los Angeles (UCLA) and the former President of the National Yang Ming Chiao Tung University (NYCU), Hsinchu, Taiwan, will deliver his exciting vision on "Future System-on-Chip for Full Spectrum Utilization from RF to Optics".

Immediately after the plenary session, the RFIC Reception and Symposium Showcase will follow, with highlights from our industry showcase and student paper finalists in an engaging social and technical evening event supported by the RFIC Symposium 2023 corporate sponsors. The showcase will provide authors the opportunity to demo their work in a lab-like environment for more close-up discussion and interaction. You will not want to miss the 2023 RFIC Reception!

On Monday and Tuesday, RFIC will have multiple tracks of oral technical paper sessions and will offer panel sessions during the lunch breaks. Monday's lunchtime panel, titled "How Soon Will We Become Cyborgs?" will be dedicated to the debate on the expected impact of the increased use of various technologies, such as augmented reality and smart hearing aids, on our everyday lives. Tuesday's lunchtime panel, organized jointly with IMS'23, will discuss the topic of "AI/ML Based Wireless System Design and Operation — Hope or Hype?". This topic is interesting and controversial as the use of machine learning (ML), or more broadly, artificial intelligence (AI), has already been demonstrated in a wide range of applications, including even music composition and artistic design. This lunchtime panel, with both industry and academia experts, will explore how we may harness AI in wireless system design and operation, and will attempt to distinguish hope from hype.

RFIC'23 and the Microwave Week have many educational and professional development opportunities for students, all delivered at an exceptional value. Following its introduction in 2022, RFIC'23 will feature a new, dedicated Student Session, where students can meet, interact, and learn about exciting technology trends and their potential future careers from industry experts. This RFIC'23 Student Session includes an Industry ChipChat and reception on late Tuesday afternoon in Room 32AB, where future leaders meet prominent industry professionals to confess their secrets about their first years in their career. The industry executives/representatives will be talking about "3 Things to Know to Start your RFIC Design Career with a Bang". Bring your questions for an open discussion about the metamorphosis from student to professional RFIC designer, negotiate your salary and how to manage your talent to impact lives and more especially yours. Not enough time to extract all the secrets? Everyone is invited to stay and continue chip-chatting at the RFIC nacho station! Come and join this RFIC'23 special event customized by and for students and the RF industry! Free Food and drinks!

Last but not the least, the RFIC'23 will once again conduct a contest to select the top student papers from the symposium. The top student papers will also be featured at our Sunday's Symposium Showcase, providing an additional exposure opportunity. As part of IMS, students have the opportunity to participate in design competitions and an RF Bootcamp. Lastly, MTT-S offers a Ph.D. Student Sponsorship Initiative for new students to become engaged with Microwave Week, providing learning, networking, and volunteer experiences along with complimentary registration and accommodations to qualified and selected students. Students have the opportunity to purchase the Student Superpass, allowing them to experience every activity within Microwave Week, including a workshop, all three conferences (RFIC, IMS, and ARFTG), the Future Summit, a technical lecture, and much more, all at a deeply discounted price for IEEE student members

On behalf of the RFIC Steering and Executive Committees, we welcome you all to join us at the 2023 RFIC Symposium in beautiful San Diego, CA. Please visit the RFIC'23 website (https://rfic-ieee.org/) for more details and updates.

Donald Y.C. Lie
General Chair
Texas Tech University

Danilo Manstretta
TPC Chair
Università di Pavia

Hua Wang
TPC Co-Chair
ETH Zürich

Steering Committee

Donald Y.C. Lie, Texas Tech University, *General Chair*
Danilo Manstretta, Università di Pavia, *TPC Chair*
Hua Wang, ETH Zürich, *TPC Co-Chair*
François Rivet, University of Bordeaux, *Student Paper Chair*
Debopriyo Chowdhury, Broadcom, *Industry Paper Chair*
Steven Turner, BAE Systems, *Publications Chair*
Jennifer Kitchen, Arizona State University, *Publications Co-Chair*
Gernot Hueber, United Micro Technology, *Systems & Applications Forum Chair*
Hossein Hashemi, University of Southern California, *Transactions JSSC Guest Editor*
Jane Gu, University of California, Davis, *Transactions TMTT Guest Editor*
Bodhisatwa Sadhu, IBM T.J. Watson Research Center, *Workshops Chair / Technical Lecture Chair*
Amin Arbabian, Stanford University, *Workshops Co-Chair / Technical Lecture Co-Chair*
Oren Eliezer, Samsung, *Panel Sessions Chair*
Mona Hella, Rensselaer Polytechnic Institute, *Publicity Chair*
Kenichi Okada, Tokyo Tech, *Asia Pacific Liaison*
Yao-Hong Liu, imec, *European Liaison*
Margaret Szymanowski, Crane A&E, *Secretary*
Joseph Cali, Raytheon Technologies, *Session Organization Chair / Submission Website Chair*
Zaher Bardai, IMN Epiphany, *Visa Letters*
Michael Oakley, Raytheon Technologies, *Website Chair*

Support Staff

Elsie Vega, IEEE, *Conference Manager*
Robert Alongi, IEEE, *Finances*
Amanda Scacchitti, AAES, *IMS Publications / Plenary*
Stefanie Cuniffe, Horizon House, *Publicity / Exhibition / Registration*
Carl Sheffres, Horizon House, *Publicity / Exhibition / Registration*
George Vokalek, Causal Productions, *Publications*
Sandy Owens, EPMS, *Electronic Paper Management System*

Executive Committee

Stefano Pellerano, *Intel*
Waleed Khalil, *The Ohio State University*
Brian Floyd, *North Carolina State University*
Osama Shana'a, *MediaTek*
Donald Y.C. Lie, *Texas Tech University*

Advisory Board

Jenshan Lin, *University of Florida*
Stefan Heinen, *RWTH Aachen University*
Joseph Staudinger, *NXP Semiconductors*
Luciano Boglione, *U.S. Naval Research Laboratory*
Yann Deval, *University of Bordeaux*
Jacques C. Rudell, *University of Washington*
Lawrence Kushner, *Raytheon Technologies*
Bertan Bakkaloglu, *Arizona State University*
Kevin Kobayashi, *Qorvo*
Walid Y. Ali-Ahmad, *Apple*

Technical Program Committee

Ehsan Afshari, *University of Michigan*
Zeshan Ahmad, *Texas Instruments*
Alyssa Apsel, *Cornell University*
Amin Arbabian, *Stanford University*
Abdellatif Bellaouar, *Qualcomm*
Andrea Bevilacqua, *Università di Padova*
Roxann Broughton-Blanchard, *Analog Devices*
Joseph Cali, *Raytheon Technologies*
Steven Callender, *Intel*
Andreia Cathelin, *STMicroelectronics*
Debopriyo Chowdhury, *Broadcom*
SungWon Chung, *Neuralink*
Emanuel Cohen, *Technion*
Foster Dai, *Auburn University*
Zhiming Deng, *MediaTek*
Oren Eliezer, *Samsung*
Kamran Entesari, *Texas A&M University*
Salvatore Finocchiaro, *Qorvo*
Travis Forbes, *Sandia National Laboratories*
Antoine Frappé, *University of Lille*
Hao Gao, *Technische Universiteit Eindhoven*
Xiang Gao, *Zhejiang University*
Vito Giannini, *Uhnder*
Alexandre Giry, *CEA-Leti*
Marcus Granger-Jones, *Qorvo*
Jane Gu, *University of California, Davis*
Subhanshu Gupta, *Washington State University*
Ruonan Han, *MIT*
Hossein Hashemi, *University of Southern California*
Hiva Hedayati, *Broadcom*
Mona Hella, *Rensselaer Polytechnic Institute*
Chun-Huat Heng, *NUS*
Duane Howard, *Amazon Web Services*
Hsieh-Hung Hsieh, *TSMC*
Gernot Hueber, *United Micro Technology*

Vadim Issakov, *TU Braunschweig*
Bahar Jalali Farahani, *Cisco*
Alvin Joseph, *GlobalFoundries*
Tae Wook Kim, *Yonsei University*
Jennifer Kitchen, *Arizona State University*
Oleh Krutko, *AMD*
Timothy LaRocca, *Northrop Grumman*
Yao-Hong Liu, *imec*
Xun Luo, *UESTC*
Andrea Mazzanti, *Università di Pavia*
Omeed Momeni, *University of California, Davis*
Kenichi Okada, *Tokyo Tech*
Arun Paidimarri, *IBM T.J. Watson Research Center*
Hyun-Chul Park, *Samsung*
Edward Preisler, *Tower Semiconductor*
Raja Pullela, *MaxLinear*
Patrick Reynaert, *KU Leuven*
François Rivet, *University of Bordeaux*
Bodhisatwa Sadhu, *IBM T.J. Watson Research Center*
Fabio Sebastiano, *Technische Universiteit Delft*
Shahriar Shahramian, *Nokia Bell Labs*
Teerachot Siriburanon, *University College Dublin*
Margaret Szymanowski, *Crane A&E*
Rocco Tam, *NXP Semiconductors*
Yahya Tousi, *University of Minnesota*
Renyuan Wang, *BAE Systems*
Yanjie Wang, *SCUT*
David Wentzloff, *University of Michigan*
Magnus Wiklund, *NXP Semiconductors*
Justin ChiaHsin Wu, *Amlogic*
Wanghua Wu, *Samsung*
Hongtao Xu, *Fudan University*
Rabia Tugce Yazicigil, *Boston University*
Jin Zhou, *University of Illinois Urbana-Champaign*

RFIC 2023 Steering Committee

Donald Y.C. Lie
General Chair

Danilo Manstretta
TPC Chair

Hua Wang
TPC Co-Chair

François Rivet
Student Paper Chair

Debopriyo Chowdhury
Industry Paper Chair

Steven Turner
Publications Chair

Jennifer Kitchen
Publications Co-Chair

Gernot Hueber
Systems&ApplicationsForumChair

Hossein Hashemi
TransactionsJSSCGuestEditor

Jane Gu
TransactionsTMTTGuestEditor

Bodhisatwa Sadhu
Workshops Chair
Technical Lecture Chair

Amin Arbabian
Workshops Co-Chair
Technical Lecture Co-Chair

Oren Eliezer
Panel Sessions Chair

Mona Hella
Publicity Chair

Kenichi Okada
Asia Pacific Liaison

Yao-Hong Liu
European Liaison

Margaret Szymanowski
Secretary

Joseph Cali
Session Organization Chair
Submission Website Chair

Zaher Bardai
Visa Letters

Michael Oakley
Website Chair

continues ...

Support Staff

Elsie Vega
Conference Manager

Robert Alongi
Finances

Amanda Scacchitti
IMS Publications / Plenary

George Vokalek
Publications

Sandy Owens
ElectronicPaperManagementSystem

RFIC 2023 Executive Committee

Stefano Pellerano

Waleed Khalil

Brian Floyd

Osama Shana'a

Donald Y.C. Lie

RFIC 2023 Advisory Committee

Jenshan Lin

Stefan Heinen

Joseph Staudinger

Luciano Boglione

Yann Deval

Jacques C. Rudell

Lawrence Kushner

Bertan Bakkaloglu

Kevin Kobayashi

Walid Y. Ali-Ahmad

RFIC 2023 Technical Program Committee

Ehsan Afshari

Zeshan Ahmad

Alyssa Apsel

Amin Arbabian

Abdellatif Bellaouar

Andrea Bevilacqua

Roxann Broughton-Blanchard

Joseph Cali

Steven Callender

Andreia Cathelin

Debopriyo Chowdhury

SungWon Chung

Emanuel Cohen

Foster Dai

Zhiming Deng

Oren Eliezer

Kamran Entesari

Salvatore Finocchiaro

Travis Forbes

Antoine Frappé

Hao Gao

Xiang Gao

Vito Giannini

Alexandre Giry

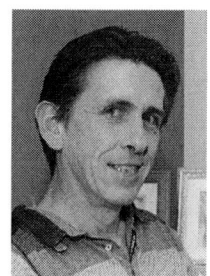
Marcus Granger-Jones

continues ...

RFIC 2023 Technical Program Committee

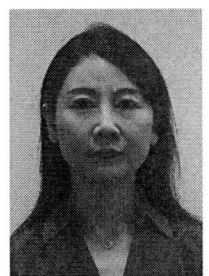
Jane Gu

Subhanshu Gupta

Ruonan Han

Hossein Hashemi

Hiva Hedayati

Mona Hella

Chun-Huat Heng

Duane Howard

Hsieh-Hung Hsieh

Gernot Hueber

Vadim Issakov

Bahar Jalali Farahani

Alvin Joseph

Tae Wook Kim

Jennifer Kitchen

Oleh Krutko

Timothy LaRocca

Yao-Hong Liu

Xun Luo

Andrea Mazzanti

Omeed Momeni

Kenichi Okada

Arun Paidimarri

Hyun-Chul Park

Edward Preisler

continues ...

RFIC 2023 Technical Program Committee

Raja Pullela

Patrick Reynaert

François Rivet

Bodhisatwa Sadhu

Fabio Sebastiano

Shahriar Shahramian

Teerachot Siriburanon

Margaret Szymanowski

Rocco Tam

Yahya Tousi

Renyuan Wang

Yanjie Wang

David Wentzloff

Magnus Wiklund

Justin ChiaHsin Wu

Wanghua Wu

Hongtao Xu

Rabia Tugce Yazicigil

Jin Zhou

RFIC 2023 Schedule
San Diego Convention Center

Saturday, 10 June 2023
08:00–17:00 Registration — Lobby D

Sunday, 11 June 2023
07:00–18:00 Registration — Lobby D
07:00–08:00 Speakers' Breakfast — 20A
08:00–11:50 Workshops — 23–25, 29–33
11:45–13:00 Workshops Lunch — Meeting Room Foyer, Terrace Side
12:00–13:20 Technical Lecture — 20A
             Modern Radio Receivers — From WiFi to 5G and Beyond
13:30–17:20 Workshops — 23–25, 29–33
17:30–19:00 RFIC Plenary — 20BCD
19:00–21:00 Welcoming Reception Featuring Symposium Showcase — Sails Pavilion

Monday, 12 June 2023
07:00–18:00 Registration — Lobby D
07:00–08:00 Speakers' Breakfast — 20A
08:00–09:20 RMo1A — 23ABC:
             Circuits and Systems for High-Speed Optical and Wireline Communication
08:00–09:40 RMo1B — 24ABC: *Silicon-Based Front-Ends and Building-Blocks*
08:00–09:20 RMo1C — 25ABC: *5G & mm-Wave Transceivers and Beamforming ICs*
09:40–10:10 Coffee Break — Foyer Outside Meeting Rooms
10:10–11:30 RMo2A — 23ABC: *GaN Modeling, RFSOI Device and Chip Layout Automation*
10:10–11:30 RMo2B — 24ABC: *III/V Front-Ends and Building-Blocks*
10:10–11:30 RMo2C — 25ABC: *Systems and Applications at RF and mm-Wave*
12:00–13:30 RFIC Panel Session — 20A: *How Soon Will We Become Cyborgs?*
13:30–14:50 RMo3A — 23ABC: *Reference Clock and Frequency Generation Techniques*
13:30–15:10 RMo3B — 24ABC: *High-Performance mm-Wave Low-Noise Amplifiers*
13:30–14:50 RMo3C — 25ABC: *THz & mm-Wave Communication Transceivers & Circuits*
15:10–15:40 Coffee Break — Foyer Outside Meeting Rooms
15:40–17:00 RMo4A — 23ABC: *CMOS mm-Wave Frequency Multipliers*
15:40–16:40 RMo4B — 24ABC: *Advances in NB-IoT and WiFi Radios*
15:40–17:00 RMo4C — 25ABC: *High-Efficiency and Linear 5G mm-Wave Power Amplifiers*

Tuesday, 13 June 2023
07:00–18:00 Registration — Lobby D
07:00–08:00 Speakers' Breakfast — Ballroom 20A
08:00–09:20 RTu1A — 23ABC: *RF to THz LO Generation Solutions*
08:00–09:20 RTu1B — 24ABC: *Self-Interference Cancellation Techniques*
08:00–09:40 RTu1C — 25ABC:
             mm-Wave & Sub-THz Circuits & Systems for Radar Sensing and Metrology
09:40–10:10 Coffee Break — Exhibition Floor
10:10–11:30 RTu2B — 24ABC: *Emerging Circuits and Systems for Quantum Computing, Quantum Sensing, Photonics, and Built-In Self-Test (BIST) Applications*
10:10–11:30 RTu2C — 25ABC: *Systems for Applications: 5G and SATCOM*
12:00–13:30 RFIC/IMS Joint Panel Session — 32AB:
             AI/ML Based Wireless System Design and Operation — Hope or Hype?
13:30–14:50 RTu3B — 24ABC: *Advanced Building Blocks for mm-Wave & Beyond*
13:30–14:50 RTu3C — 25ABC: *IoT Transmitter and Sub-THz Power Amplifiers*
15:10–15:40 Coffee Break — Exhibition Floor
15:40–16:40 RTu4C — 25ABC: *Invited Industry Presentations*

RFIC Plenary, Reception, and Symposium Showcase

Sunday Evening, 11 June 2023
San Diego Convention Center

17:30–19:00
RFIC Plenary
Ballroom 20BCD
Chair: Donald Y.C. Lie, Texas Tech University
Co-Chair: Danilo Manstretta, Università di Pavia

17:30   Welcome Message from General Chair and TPC Chairs
           Student Paper Awards, Industry Paper Awards, Tina Quach Service Award
18:00   *The Roaring 20s: A Renaissance for the Semiconductor Industry?*
           Todd Younkin, Semiconductor Research Corporation
18:30   *Future System-on-Chip for Full Spectrum Utilization from RF to Optics*
           Mau-Chung Frank Chang, University of California, Los Angeles

19:00–21:00
RFIC Welcoming Reception Featuring Symposium Showcase
Sails Pavilion

The RFIC Interactive Reception starts immediately after the Plenary Session and will highlight the Student Paper Awards finalists, the Industry Paper Awards finalists, and the Systems & Applications Forum in an engaging social and technical evening event with food and drinks. Authors of these showcase papers will present their innovative work, summarized in poster format. Some showcase papers will also offer live demonstrations. You will not want to miss the RFIC Reception! This event is supported by the RFIC Symposium corporate sponsors.

RFIC Plenary Speaker 1

Dr. Todd Younkin
President and CEO
Semiconductor Research Corporation

The Roaring 20s: A Renaissance for the Semiconductor Industry?

Abstract: Dr. Younkin will share his vision for the future of global semiconductor technologies and design, especially those that will enable future RFIC breakthroughs. Dr. Younkin will discuss the status of government investments and opportunities arising from the CHIPS and SCIENCE ACT of 2022, Korea's K-Belt strategy, Europe's CHIPS ACT, and more. Dr. Younkin leads a global research agenda of about $100M annually, supported by ~3k academic and industrial researchers, 27 international companies, and 3 U.S. government agencies (DARPA, NSF, and NIST). They have defined the opportunities for future compute and communication systems, as outlined by SRC's 2030 Decadal Plan for Semiconductors, and are now working with over 90 organizations to define the semiconductor hardware opportunities that will deliver that required system performance, via the NIST Microelectronic and Advanced Packaging Technologies (MAPT) Roadmap, awarded in April 2022 and scheduled for completion by September 2023.

Sunday, 11 June 2023   18:00–18:30   Ballroom 20BCD

RFIC Plenary Speaker 1 continued...

About Dr. Todd Younkin

Dr. Todd Younkin is a talented and seasoned executive with more than 20 years of experience in technology innovation. His extensive Research and Development experience spans Intel's 0.18µm to 5 nm nodes with technical contributions in novel materials, nanotechnology, integration, advanced lithography, and integrated photonics. Todd brings a wealth of expertise with strengths in areas such as cultivating relationships with strategic partners, entrepreneurship and investment strategies, technology innovation, operational excellence, and talent management. He has spent much of his career working alongside young minds that are aspiring to influence the ever-changing world of smart and autonomous electronics. He has built programs from the ground up, leveraging his entrepreneurial leadership to drive new business development that has generated multi-millions in funding. He has been a key contributor in introducing new technology advances and starting new global research in the U.S., Europe, and Asia. Dr. Younkin holds a Ph.D. from the California Institute of Technology in Pasadena, California. He completed his Bachelor of Science at the University of Florida in Gainesville, Florida. He aspires to continue to influence the next generation of technology and inventors, bringing ideas and investors together to drive heterogeneous electronic solutions that will deliver a smarter, shared future.

In August of 2020, Dr. Todd Younkin became the President & CEO of SRC. Recently, he engineered, launched, and led all programmatic aspects of the five-year, $240 million JUMP research initiative. It has six multi-university, multi-disciplinary innovation Centers with 133 faculty, 835 students, and 360 industrial engineering liaisons. It emphasizes the advancement of Computer Science, Electrical Engineering, and Materials to secure continued U.S. thought leadership. Following his appointment, SRC released its 2030 Decadal Plan for Semiconductors, where it identified the five "seismic shifts" shaping the future of information and communication technologies (ICT). Working closely with SIA, SRC called for greatly increased federal investments throughout the decade to establish a smarter pipeline for semiconductor R&D, aligned to SRC's Decadal Plan. This drove and resulted in the passage of the CHIPS and SCIENCE ACT of 2022 on 9 August 2022.

Dr. Todd Younkin is excited by the worldwide call for a renewed investment in semiconductor materials, hardware, and design, as well as the equally important calls for an emphasis on education and workforce development and our need for environmental sustainability. Only by investing in a bright, collective future, will we rise to the meet the opportunities presented by the next industrial revolution.

RFIC Plenary Speaker 2

Prof. Mau-Chung Frank Chang
Wintek Chair in Electrical Engineering and
Distinguished Professor of
University of California, Los Angeles

Future System-on-Chip for
Full Spectrum Utilization from RF to Optics

Abstract: The ever-increasing bandwidth requirement due to explosively growing 5/6G and AIoT data flows has compelled global commission authorities to release EM-spectra up to millimeterwave (30–300GHz) and even (sub)-millimeter-wave frequency regimes (>300GHz) for massively expanded sensing and network applications. In this talk, we will exemplify novel CMOS embedded technologies and methodologies developed at UCLA to enable System-on-Chip (SoC) realizations for multi-broadband radio, wideband radar, contactless/plastic interconnect, 3D-imaging and gas-phase rotational spectrometry at (sub)-mm-Wave frequencies, including:

- DiCAD (Digitally Controlled Artificial Dielectric), the only proven Digital-to-Permittivity Converter (DPC), embedded in CMOS-switched interconnects that can vary transmission-line permittivity in real-time (up to ×20 in practice) for (sub)-mm-wave frequency synthesis, direct-frequency modulation and reconfigurable (software defined) radio/radar/spectrometer implementations
- Self-Healing Radio (57–64GHz) with self-diagnosis and self-healing capabilities to secure high performance-yield and counter temperature/process variations & aging effects
- Multiband RF-Interconnect, beyond traditional baseband-only interconnect, to enable contactless and/or plastic waveguide communications up to Terahertz with unprecedented bandwidth, efficiency, dynamic re-configurability & multi-cast capabilities
- Fully integrated frequency synthesizer (PLL) at 560GHz and realized 1st active and passive CMOS imagers at 110GHz; 3-color (349/201/153GHz) and 3D imaging radars for sensing/ranging concealed objects
- Single-chip CMOS heterodyne H_2O-Detecting Spectrometer at 183 GHz to enable NASA's space exploration missions with reduced mass (6.5×) & power (5.5×) to meet strict payload and energy consumption requirements

We will also address challenges encountered in both design and implementations that may hinder further development of such systems, especially the major shortcomings in silicon technologies with limited dynamic range and power handling capabilities. We therefore propose replacing CMOS n-FET's drain with selectively grown wide bandgap cubic-phase GaN (c-GaN) for >10× improved breakdown voltages to secure desired sensing/communication range/coverage with cost-effectiveness.

We also elaborate on the possible growth of multi-wavelength light-emitting sources and detectors directly atop n-FET's c-GaN Drain with various indium contents of InGaN/GaN superlattice for RF-optical combined radio/radar/interconnect applications by creating unprecedented "Photonic System-on-Chip" with full EM-spectrum utilization from RF to optics.

Sunday, 11 June 2023   **18:30–19:00**   **Ballroom 20BCD**

RFIC Plenary Speaker 2 continued...

About Prof. Mau-Chung Frank Chang

Dr. Mau-Chung Frank Chang is the Wintek Chair in Electrical Engineering and Distinguished Professor at the University of California, Los Angeles. Prior to joining UCLA, he was the Assistant Director of the High Speed Electronics Laboratory of Rockwell Science Center (1983–1997), Thousand Oaks, California. In this tenure, he led the team to develop and transfer the MOCVD based AlGaAs/GaAs & InGaP/GaAs Heterojunction Bipolar Transistor (HBT) and BiFET (Planar HBT/MESFET) integrated circuit technologies from the research laboratory to production line (later became Skyworks Solutions). The HBT/BiFET productions have grown into multi-billion dollar businesses and dominated the cellphone power amplifier and front-end module markets for the past 30 years (currently exceeding 10 billion units/year and exceeding 50 billion units in the past decade). Throughout his career, his research has focused on the research & development of high-speed semiconductor devices and integrated circuits for radio, radar, imager, spectrometer and interconnect System-on-Chip applications. He invented the multiband, reconfigurable RF-Interconnects for Chip-Multi-Processor (CMP) inter-core communications and inter-chip CPU-to-Memory communications. He and his students were the 1st to demonstrate CMOS active and passive imagers at mm-Wave (100–180GHz) frequencies. His Lab also pioneered the development of self-healing 57–64GHz radio-on-a-chip (DARPA's HEALICS program) with embedded sensors, actuators and self-diagnosis/curing capabilities; and invented the Digitally Controlled Artificial Dielectric (DiCAD) embedded in CMOS technologies to vary transmission-line permittivity in real-time (up to 20× in practice) for realizing reconfigurable multiband/mode radios in (sub)-mm-Wave frequency bands. His UCLA Lab also realized the first CMOS Frequency Synthesizer for Terahertz operation (PLL at 560GHz) and devised the first tri-color CMOS active imager at 180–500GHz based on a Time-Encoded Digital Regenerative Receiver and the first 3-dimensional SAR imaging radar with sub-centimeter range resolution at 144GHz. More recently, his Lab has devised a Reconfigurable Convolution Neuron Network (RCNN) Accelerator for IoT applications, spun-off an Edge-AI startup company Kneron in San Diego, and won IEEE's 2021 Darlington Best Paper Award.

Prof. Chang is a Member of the US National Academy of Engineering, the European Academy of Sciences and Arts, the US National Academy of Inventors, the Academia Sinica of Taiwan, and a Fellow of the IEEE. He was also recognized with the IEEE David Sarnoff Award (2006), IET JJ Thomson Medal for Electronics (2017) and IEEE/RSE James Clerk Maxwell Medal (2023) for his seminal contributions to the heterojunction technology and realizations of (sub)-mm-Wave System-on-Chip with unprecedented bandwidth and re-configurability.

Prof. Chang has published more than 460 peer-reviewed technical papers and 60 US patents in various areas of high speed electronic devices and integrated circuits & systems. During his tenure with UCLA, he has graduated more than 50 Ph.D. students and 100 MS students. He also served as the President of the National Chiao Tung University, Hsinchu, Taiwan (2015–2019).

Sunday, 11 June 2023   **18:30–19:00**   **Ballroom 20BCD**

The Student Paper Awards Finalists

Chair: François Rivet, University of Bordeaux
Co-Chair: Debopriyo Chowdhury, Broadcom

The RFIC Symposium's Student Paper Award is devised to both encourage student paper submissions to the conference as well as give the authors of the finalists' papers a chance to promote their research work with the conference attendees after the plenary session during reception time. The outstanding student paper finalists listed below were nominated this year by the RFIC Technical Program Committee to enter the final contest. A committee of TPC judges selected the top three papers after rigorous reviews and discussions. All finalists benefit from a complimentary RFIC registration. The top-three Student Papers will be announced during the RFIC Plenary Session on 11 June 2023 in San Diego. Each winner will receive a plaque.

A 112-Gbps, 0.73-pJ/Bit Fully-Integrated O-Band I-Q Optical Receiver in a 45-nm CMOS SOI-Photonic Process
Ghazal Movaghar, Viviana Arrunategui, Junqian Liu, Aaron Maharry, Clint Schow, James F. Buckwalter
University of California, Santa Barbara, USA
RMo1A-1

A 100-Gb/s 3-m Dual-Band PAM-4 Dielectric Waveguide Link with 1.9 pJ/Bit/m Efficiency in 28-nm CMOS
Kristof Dens[1], Joren Vaes[1], Christian Bluemm[2], Gabriel Guimaraes[1], Berke Gungor[1], Changsong Xie[2], Alexander Dyck[2], Patrick Reynaert[1]
[1]KU Leuven, Belgium, [2]Huawei Technologies, Germany
RMo1A-3

A 140GHz Scalable On-Grid 8×8-Element Transmit-Receive Phased-Array with Up/Down Converters and 64QAM/24Gbps Data Rates
Amr Ahmed, Linjie Li, Minjae Jung, Gabriel M. Rebeiz
University of California, San Diego, USA
RMo2C-2

A 0.75mW Receiver Front-End for NB-IoT
Hossein Rahmanian Kooshkaki, Patrick P. Mercier
University of California, San Diego, USA
RMo4B-1

A 26–40GHz 4-Way Hybrid Parallel-Series Role-Exchange Doherty PA with Broadband Deep Power Back-Off Efficiency Enhancement
Edward Liu, Hua Wang
ETH Zürich, Switzerland
RMo4C-1

Mono/Multistatic Mode-Configurable E-Band FMCW Radar Transceiver Module for Drone-Borne Synthetic Aperture Radar
Kangseop Lee, Sirous Bahrami, Kyunghwan Kim, Jiseul Kim, Seung-Uk Choi, Ho-Jin Song
POSTECH, Korea
RTu1C-2

Sunday, 11 June 2023   **17:30–18:00**   **Ballroom 20BCD**

A Diamond Quantum Magnetometer Based on a Chip-Integrated 4-Way Transmitter in 130-nm SiGe BiCMOS

Hadi Lotfi[1], Michal Kern[1], Nico Striegler[2], Thomas Unden[2], Jochen Scharpf[2], Patrick Schalberger[1], Ilai Schwartz[2], Philipp Neumann[2], Jens Anders[1]

[1]Universität Stuttgart, Germany, [2]NVision Imaging Technologies, Germany

RTu2B-1

A D-Band Calibration-Free Passive 360° Phase Shifter with 1.2° RMS Phase Error in 45nm RFSOI

Mohammadreza Abbasi, Wooram Lee

Pennsylvania State University, USA

RTu3B-2

A D-Band 20.4dBm OP_{1dB} Transformer-Based Power Amplifier with 23.6% PAE in a 250-nm InP HBT Technology

Senne Gielen[1], Yang Zhang[2], Mark Ingels[2], Patrick Reynaert[1]

[1]KU Leuven, Belgium, [2]imec, Belgium

RTu3C-3

<u>Student Paper Contest Eligibility</u>: The student must have been a full-time student (9 hours/term graduate, 12 hours/term undergraduate) during the time the work was performed. The student must also be the lead author of the paper and must present the paper at the Symposium.

Sunday, 11 June 2023　　　　　**17:30–18:00**　　　　　**Ballroom 20BCD**

The Industry Paper Awards Finalists

Chair: Debopriyo Chowdhury, Broadcom
Co-Chair: François Rivet, University of Bordeaux

The RFIC Industry Showcase highlights the outstanding industry papers listed below. These papers received nominations for this recognition from the TPC sub-committees and godparents in a double-blind review. From these top papers, a two-stage double-blind review process was conducted by a committee of TPC judges. Finally, the Best Paper Chair and other key Steering Committee members finalize the top three winners after rigorous reviews and discussions. The top three will be displayed on the RFIC website and on a rolling slideshow prior to the RFIC Plenary Session on 11 June 2023 in San Diego. Each winner will receive a plaque and will be recognized in an upcoming Microwave Magazine article.

Optimizing RFSOI Performance Through a T-Shaped Gate and Nano-Second Laser Annealing Techniques
L. Lucci[1], S. Crémer[2], B. Duriez[1], T. Fache[1], S. Kerdiles[1], Y. Morand[1], J.-M. Hartmann[1], J. Azevedo-Goncalves[2], F. Gaillard[1], P. Chevalier[2]
[1]CEA-Leti, France, [2]STMicroelectronics, France
RMo2A-2

A Fast-Startup 80MHz Crystal Oscillator with 96×/368× Startup-Time Reductions for 3.0V/1.2V Swings Based on Un-Interrupted Phase-Aligned Injection
Chien-Wei Chen, Chao-Ching Hung, Yu-Li Hsueh
MediaTek, Taiwan
RMo3A-3

Transformer-Coupled 2.5GHz BAW Oscillator with 12.5fs RMS-Jitter and 1-kHz Figure-of-Merit (FOM) of 210dB
Bichoy Bahr, Sachin Kalia, Baher Haroun, Swaminathan Sankaran
Texas Instruments, USA
RMo3A-4

A Double Balanced Frequency Doubler Achieving 70% Drain Efficiency and 25% Total Efficiency
Jesse Moody
Sandia National Laboratories, USA
RMo4A-1

A Wi-Fi Tri-Band Switchable Transceiver with 57.9fs-RMS-Jitter Frequency Synthesizer, Achieving -42.6dB EVM Floor for EHT320 4096-QAM MCS13 Signal
Tsung-Ming Chen[1], Ming-Chung Liu[1], Pi-An Wu[1], Wei-Kai Hong[1], Ting-Wei Liang[1], Wei-Pang Chao[1], Po-Yu Chang[1], Yu-Ting Chou[1], Chien-Wei Chen[1], Sen-You Liu[1], Chang-Cheng Huang[1], Hsiu-Hsien Ting[1], Min-Shun Hsu[1], Yao-Chi Wang[1], Chao-Ching Hung[1], Yu-Li Hsueh[1], Eric Lu[2], Yuan-Hung Chung[1], Jing-Hong Conan Zhan[1]
[1]MediaTek, Taiwan, [2]MediaTek, USA
RMo4B-3

Sunday, 11 June 2023   **17:30–18:00**   **Ballroom 20BCD**

A 15.6-GHz Quad-Core VCO with Extended Circular Coil Topology for Both Main and Tail Inductors in 8-nm FinFET Process
Suoping Hu, Zhiyu Chen, Wanghua Wu, Pei-Yuan Chiang, Zhanjun Bai, Chih-Wei Yao, Sangwon Son
Samsung, USA
RTu1A-1

A 28nm CMOS Dual-Band Concurrent WLAN and Narrow Band Transmitter with On-Chip Feedforward TX-to-TX Interference Cancellation Path for Low Antenna-to-Antenna Isolation in IoT Devices
Sai-Wang Tam, Alireza Razzaghi, Alden Wong, Sridhar Narravula, Weiwei Xu, Timothy Loo, Akash Kambale, Andrew Liu, Ovidiu Carnu, Yui Lin, Randy Tsang
NXP Semiconductors, USA
RTu1B-1

A 14-nm Low-Cost IF Transceiver IC with Low-Jitter LO and Flexible Calibration Architecture for 5G FR2 Mobile Applications
Wanghua Wu[1], Jeiyoung Lee[2], Pak-Kim Lau[1], Taeyoung Kang[1], Kim Kiu Lau[1], Si-Wook Yoo[1], Xingliang Zhao[1], Ashutosh Verma[1], Ivan Siu-Chuang Lu[1], Chih-Wei Yao[1], Hou-Shin Chen[1], Gennady Feygin[1], Pranav Dayal[1], Kee-Bong Song[1], Sangwon Son[1]
[1]Samsung, USA, [2]Samsung, Korea
RTu2C-3

A 140GHz RF Beamforming Phased-Array Receiver Supporting >20dB IRR with 8GHz Channel Bandwidth at Low IF in 22nm FDSOI CMOS
Shenggang Dong[1], Navneet Sharma[1], Sensen Li[1], Michael Chen[1], Xiaohan Zhang[2], Yaolong Hu[2], Jiantong Li[1], Yong Su[1], Xinguang Xu[1], Vitali Loseu[1], Eunyoung Seok[1], Taiyun Chi[2], Won-Suk Choi[1], Gary Xu[1]
[1]Samsung, USA, [2]Rice University, USA
RTu3B-3

A 110–170GHz Phase-Invariant Variable-Gain Power Amplifier Module with 20–22dBm P_{sat} and 30dBm OIP3 Utilizing SiGe HBT RFICs
Mustafa Sayginer, Michael Holyoak, Mike Zierdt, Mohamed Elkhouly, Joe Weiner, Yves Baeyens, Shahriar Shahramian
Nokia Bell Labs, USA
RTu3C-2

<u>Industry Paper Contest Eligibility</u>: The first author must have an affiliation from industry. The first author must also be the lead author of the paper and must present the paper at the symposium.

Sunday, 11 June 2023   17:30–18:00   **Ballroom 20BCD**

RFIC Reception and Symposium Showcase
Featuring Systems & Applications Forum and Best Student/Industry Paper Showcase

Systems & Applications Forum Chair: Gernot Hueber, United Micro Technology
Student Paper Chair: François Rivet, University of Bordeaux
Industry Paper Chair: Debopriyo Chowdhury, Broadcom

The RFIC Interactive Reception, supported by the RFIC Symposium corporate sponsors, starts immediately after the plenary session and highlights the Student and Industry Paper Awards finalists in an engaging social and technical evening event with food and drinks. Furthermore, additional authors, both from academia and industry who choose to showcase/demo work focused on Systems and Applications, will be present. Authors will present their innovative work on large monitors as electronic posters and some will offer live demonstrations. Make sure to attend this event, where you will be able to network and see a preview of selected paper presentations that you can attend during the two days that follow. The following lists of participants were valid on 1 May 2023.

Student Paper Awards Finalists' Showcase/Demonstrations

A 112-Gbps, 0.73-pJ/Bit Fully-Integrated O-Band I-Q Optical Receiver in a 45-nm CMOS SOI-Photonic Process
Ghazal Movaghar, Viviana Arrunategui, Junqian Liu, Aaron Maharry, Clint Schow, James F. Buckwalter
University of California, Santa Barbara, USA
RMo1A-1

A 100-Gb/s 3-m Dual-Band PAM-4 Dielectric Waveguide Link with 1.9 pJ/Bit/m Efficiency in 28-nm CMOS
Kristof Dens[1], Joren Vaes[1], Christian Bluemm[2], Gabriel Guimaraes[1], Berke Gungor[1], Changsong Xie[2], Alexander Dyck[2], Patrick Reynaert[1]
[1]KU Leuven, Belgium, [2]Huawei Technologies, Germany
RMo1A-3

A 140GHz Scalable On-Grid 8×8-Element Transmit-Receive Phased-Array with Up/Down Converters and 64QAM/24Gbps Data Rates
Amr Ahmed, Linjie Li, Minjae Jung, Gabriel M. Rebeiz
University of California, San Diego, USA
RMo2C-2

A 0.75mW Receiver Front-End for NB-IoT
Hossein Rahmanian Kooshkaki, Patrick P. Mercier
University of California, San Diego, USA
RMo4B-1

Sunday, 11 June 2023   19:00–21:00   **Sails Pavilion**

A 26–40GHz 4-Way Hybrid Parallel-Series Role-Exchange Doherty PA with Broadband Deep Power Back-Off Efficiency Enhancement
Edward Liu, Hua Wang
ETH Zürich, Switzerland
RMo4C-1

Mono/Multistatic Mode-Configurable E-Band FMCW Radar Transceiver Module for Drone-Borne Synthetic Aperture Radar
Kangseop Lee, Sirous Bahrami, Kyunghwan Kim, Jiseul Kim, Seung-Uk Choi, Ho-Jin Song
POSTECH, Korea
RTu1C-2

DEMO *A Diamond Quantum Magnetometer Based on a Chip-Integrated 4-Way Transmitter in 130-nm SiGe BiCMOS*
Hadi Lotfi[1], Michal Kern[1], Nico Striegler[2], Thomas Unden[2], Jochen Scharpf[2], Patrick Schalberger[1], Ilai Schwartz[2], Philipp Neumann[2], Jens Anders[1]
[1]Universität Stuttgart, Germany, [2]NVision Imaging Technologies, Germany
RTu2B-1

DEMO *A D-Band Calibration-Free Passive 360° Phase Shifter with 1.2° RMS Phase Error in 45nm RFSOI*
Mohammadreza Abbasi, Wooram Lee
Pennsylvania State University, USA
RTu3B-2

A D-Band 20.4dBm OP_{1dB} Transformer-Based Power Amplifier with 23.6% PAE in a 250-nm InP HBT Technology
Senne Gielen[1], Yang Zhang[2], Mark Ingels[2], Patrick Reynaert[1]
[1]KU Leuven, Belgium, [2]imec, Belgium
RTu3C-3

Industry Paper Awards Finalists' Showcase/Demonstrations

Optimizing RFSOI Performance Through a T-Shaped Gate and Nano-Second Laser Annealing Techniques
L. Lucci[1], S. Crémer[2], B. Duriez[1], T. Fache[1], S. Kerdiles[1], Y. Morand[1], J.-M. Hartmann[1], J. Azevedo-Goncalves[2], F. Gaillard[1], P. Chevalier[2]
[1]CEA-Leti, France, [2]STMicroelectronics, France
RMo2A-2

A Fast-Startup 80MHz Crystal Oscillator with 96×/368× Startup-Time Reductions for 3.0V/1.2V Swings Based on Un-Interrupted Phase-Aligned Injection
Chien-Wei Chen, Chao-Ching Hung, Yu-Li Hsueh
MediaTek, Taiwan
RMo3A-3

Sunday, 11 June 2023   19:00–21:00   **Sails Pavilion**

Transformer-Coupled 2.5GHz BAW Oscillator with 12.5fs RMS-Jitter and 1-kHz Figure-of-Merit (FOM) of 210dB
Bichoy Bahr, Sachin Kalia, Baher Haroun, Swaminathan Sankaran
Texas Instruments, USA
RMo3A-4

A Double Balanced Frequency Doubler Achieving 70% Drain Efficiency and 25% Total Efficiency
Jesse Moody
Sandia National Laboratories, USA
RMo4A-1

A Wi-Fi Tri-Band Switchable Transceiver with 57.9fs-RMS-Jitter Frequency Synthesizer, Achieving -42.6dB EVM Floor for EHT320 4096-QAM MCS13 Signal
Tsung-Ming Chen[1], Ming-Chung Liu[1], Pi-An Wu[1], Wei-Kai Hong[1], Ting-Wei Liang[1], Wei-Pang Chao[1], Po-Yu Chang[1], Yu-Ting Chou[1], Chien-Wei Chen[1], Sen-You Liu[1], Chang-Cheng Huang[1], Hsiu-Hsien Ting[1], Min-Shun Hsu[1], Yao-Chi Wang[1], Chao-Ching Hung[1], Yu-Li Hsueh[1], Eric Lu[2], Yuan-Hung Chung[1], Jing-Hong Conan Zhan[1]
[1]MediaTek, Taiwan, [2]MediaTek, USA
RMo4B-3

A 28nm CMOS Dual-Band Concurrent WLAN and Narrow Band Transmitter with On-Chip Feedforward TX-to-TX Interference Cancellation Path for Low Antenna-to-Antenna Isolation in IoT Devices
Sai-Wang Tam, Alireza Razzaghi, Alden Wong, Sridhar Narravula, Weiwei Xu, Timothy Loo, Akash Kambale, Andrew Liu, Ovidiu Carnu, Yui Lin, Randy Tsang
NXP Semiconductors, USA
RTu1B-1

A 14-nm Low-Cost IF Transceiver IC with Low-Jitter LO and Flexible Calibration Architecture for 5G FR2 Mobile Applications
Wanghua Wu[1], Jeiyoung Lee[2], Pak-Kim Lau[1], Taeyoung Kang[1], Kim Kiu Lau[1], Si-Wook Yoo[1], Xingliang Zhao[1], Ashutosh Verma[1], Ivan Siu-Chuang Lu[1], Chih-Wei Yao[1], Hou-Shin Chen[1], Gennady Feygin[1], Pranav Dayal[1], Kee-Bong Song[1], Sangwon Son[1]
[1]Samsung, USA, [2]Samsung, Korea
RTu2C-3

A 140GHz RF Beamforming Phased-Array Receiver Supporting >20dB IRR with 8GHz Channel Bandwidth at Low IF in 22nm FDSOI CMOS
Shenggang Dong[1], Navneet Sharma[1], Sensen Li[1], Michael Chen[1], Xiaohan Zhang[2], Yaolong Hu[2], Jiantong Li[1], Yong Su[1], Xinguang Xu[1], Vitali Loseu[1], Eunyoung Seok[1], Taiyun Chi[2], Won-Suk Choi[1], Gary Xu[1]
[1]Samsung, USA, [2]Rice University, USA
RTu3B-3

A 110–170GHz Phase-Invariant Variable-Gain Power Amplifier Module with 20–22dBm P_{sat} and 30dBm OIP3 Utilizing SiGe HBT RFICs
Mustafa Sayginer, Michael Holyoak, Mike Zierdt, Mohamed Elkhouly, Joe Weiner, Yves Baeyens, Shahriar Shahramian
Nokia Bell Labs, USA
RTu3C-2

Sunday, 11 June 2023   **19:00–21:00**   **Sails Pavilion**

Systems & Applications Forum Showcase/Demonstrations

A CMOS 183GHz Millimeter-Wave Spectrometer for Exploring the Origins of Water and Evolution of the Solar System
Adrian Tang[1], Mau-Chung Frank Chang[2], Yanghyo Kim[3], Goutam Chattopadhyay[1]
[1]JPL, USA, [2]University of California, Los Angeles, USA, [3]Stevens Institute of Technology, USA
RMo2C-1

A 57.6Gb/s Wireless Link Based on 26.4dBm EIRP D-Band Transmitter Module and a Channel Bonding Chipset on CMOS 45nm
Jose Luis Gonzalez-Jimenez, Alexandre Siligaris, Abdelaziz Hamani,
Francesco Foglia-Manzillo, Pierre Courouve, Nicolas Cassiau, Cedric Dehos, Antonio Clemente
CEA-Leti, France
RMo2C-3

A 14.2mW 29–39.3-GHz Two-Stage PLL with a Current-Reuse Coupled Mixer Phase Detector
Yuan Liang[1], Chirn Chye Boon[2], Qian Chen[2]
[1]Guangzhou University, China, [2]NTU, Singapore
RMo3A-1

A C-Band Compact High-Linearity Multibeam Phased-Array Receiver with Merged Gain-Programmable Phase Shifter Technique
Jingying Zhou[1], Nayu Li[1], Yuexiaozhou Yuan[1], Huiyan Gao[1], Shaogang Wang[1], Hang Lu[1],
Chunyi Song[1], Yen-Cheng Kuan[2], Qun Jane Gu[3], Zhiwei Xu[1]
[1]Zhejiang University, China, [2]NYCU, Taiwan, [3]University of California, Davis, USA
RMo4B-2

A CMOS 160GHz Integrated Permittivity Sensor with Resolution of 0.05% $\Delta\varepsilon_r$
Hai Yu[1], Xuan Ding[1], Jingjun Chen[2], Sajjad Sabbaghi Saber[1], Qun Jane Gu[1]
[1]University of California, Davis, USA, [2]Qualcomm, USA
RTu1C-4

A mm-Wave CMOS/Si-Photonics Hybrid-Integrated Software-Defined Radio Receiver Achieving >80-dB Blocker Rejection of <-10dBm In-Band Blockers
Ramy Rady, Yu-Lun Luo, Christi Madsen, Samuel Palermo, Kamran Entesari
Texas A&M University, USA
RTu2B-3

A Quad-Band RX Phased-Array Receive Beamformer with Two Simultaneous Beams, Polarization Diversity, and 2.1–2.3dB NF for C/X/Ku/Ka-Band SATCOM
Zhaoxin Hu, Oguz Kazan, Gabriel M. Rebeiz
University of California, San Diego, USA
RTu2C-4

A mm-Wave Blocker-Tolerant Receiver Achieving <4dB NF and -3.5dBm B1dB in 65-nm CMOS
Erez Zolkov, Nimrod Ginzberg, Emanuel Cohen
Technion, Israel
RTu3B-4

Sunday, 11 June 2023　　　　　**19:00–21:00**　　　　　**Sails Pavilion**

RFIC Panel Session

Monday, 12 June 2023
12:00–13:30
Ballroom 20A

Panel Sessions Chair: Oren Eliezer, Samsung

How Soon Will We Become Cyborgs?

Panel Organizers and Moderators:
      Alyssa Apsel, *Cornell University*
      Travis Forbes, *Sandia National Laboratories*
      Oren Eliezer, *Samsung*

Panelists: **Renaldi Winoto**, *Mojo Vision*
            Carlos Morales, *Ambiq*
            Larry Larson, *Brown University*
            J.-C. Chiao, *Southern Methodist University*
            Gert Cauwenberghs, *University of California, San Diego*

Abstract: Augmented-reality contact lenses, cochlear implants, AI-aided earbuds, and thought-activated prosthetics have already demonstrated the restoration and enhancement of human capabilities, and the incorporation of artificial intelligence (AI) into these technologies can further increase their potential.

This lunchtime panel will host academia researchers and industry pioneers who are developing these technologies, and will debate how they will affect our near and long term lifestyles.

RFIC/IMS Joint Panel Session

Tuesday, 13 June 2023
12:00–13:30
Room 32AB

Panel Sessions Chairs
RFIC: Oren Eliezer, Samsung
IMS: Nuno Borges Carvalho, Universidade de Aveiro
Ke Wu, Polytechnique Montréal

AI/ML Based Wireless System Design and Operation — Hope or Hype?

Panel Organizers and Moderators:
Costas Sarris, *University of Toronto*
Qi-Jun Zhang, *Carleton University*
Bodhisatwa Sadhu, *IBM T.J. Watson Research Center*
Oren Eliezer, *Samsung*

Panelists: **Mike Shuo-Wei Chen**, *University of Southern California*
Anding Zhu, *University College Dublin*
Elyse Rosenbaum, *University of Illinois Urbana-Champaign*
Alberto Valdes-Garcia, *IBM T.J. Watson Research Center*
Sadasvan Shankar, *Stanford University*
Joonyoung Cho, *Samsung*

Abstract: The use of machine learning (ML), or more broadly, artificial intelligence (AI), has already been demonstrated in a wide range of applications, including even music composition and artistic design.

This lunchtime panel, with both industry and academia experts, will explore how we may harness AI in wireless system design and operation, and will attempt to distinguish hope from hype.

Student/Industry ChipChat

Tuesday, 13 June 2023
17:00–19:00
Room 32AB

Student Paper Chair: François Rivet, University of Bordeaux

The 3 Things to Start Your Career with a Bang

Moderators:
Travis Forbes, *Sandia National Laboratories*
Romane Dumont, *STMicroelectronics*

Panelists: **Osama Shana'a**, *MediaTek*
Domine Leenaerts, *NXP Semiconductors*
Stefano Pellerano, *Intel*
Kevin Tien, *IBM*
Ken Barnett, *GlobalFoundries*
S.C. Wong, *RichWave*

Abstract: Come and join the RFIC'23 special event customized by and for the students and the industry! A ChipChat and reception, where future leaders meet prominent industry professionals to confess their secrets about their first years in their career. Bring your questions for an open discussion on topics such as the metamorphosis from students to RFIC professionals, negotiation of your first salary and making use of your talent to impact lives, especially yours, and more.

Not enough time to extract all the secrets? Everyone is invited to continue chip-chatting at the RFIC nacho station.

RFIC Technical Lecture

Sunday, 11 June 2023
12:00–13:20
Ballroom 20A

Chair: Bodhisatwa Sadhu, IBM T.J. Watson Research Center
Co-Chair: Amin Arbabian, Stanford University

Modern Radio Receivers — From WiFi to 5G and Beyond

Speaker: **Behzad Razavi**, *Professor, University of California, Los Angeles*

Abstract: CMOS radios continue to evolve so as to satisfy the demands of new applications. Below 7 GHz, cellular and WiFi standards have been pushing the performance to support increasingly higher data rates while consuming less power. Such endeavors require novel architectures that also lend themselves to efficient circuit design. In addition, new radios have emerged around 30 GHz for 5G, around 60 GHz for WiGig, around 140 GHz for 6G, and around 300 GHz for sub-terahertz communications. Each of these frequency bands presents interesting and unique challenges, but a unifying principle among them is the need for beamforming.

This presentation deals with recent developments in receiver design for this broad range of applications. We examine the shortcomings of standard direct-conversion architectures and draw concepts from the state of the art to improve their performance. We also contend that heterodyne reception may outperform direct conversion in some cases. We then study beamforming techniques with emphasis on solutions that draw minimal power.

RFIC Technical Lecture continued...

About Prof. Behzad Razavi

Behzad Razavi (Fellow, IEEE) received the B.S.E.E. degree from the Sharif University of Technology, Tehran, Iran, in 1985, and the M.S.E.E. and Ph.D.E.E. degrees from Stanford University, Stanford, CA, USA, in 1988 and 1992, respectively. He was an Adjunct Professor at Princeton University, Princeton, NJ, USA, from 1992 to 1994, and Stanford University in 1995. He was with AT&T Bell Laboratories, Holmdel, NJ, USA, and Hewlett-Packard Laboratories, Palo Alto, CA, USA, until 1996. Since 1996, he has been an Associate Professor and a Professor of electrical engineering at the University of California at Los Angeles, Los Angeles, CA, USA. He is the author of the book Principles of Data Conversion System Design (IEEE Press, 1995), RF Microelectronics (Prentice Hall, 1998, 2012) (translated to Chinese, Japanese, and Korean), Design of Analog CMOS Integrated Circuits (McGraw-Hill, 2001 and 2016) (translated to Chinese, Japanese, and Korean), Design of Integrated Circuits for Optical Communications (McGraw-Hill, 2003, and Wiley, 2012), Design of CMOS Phase-Locked Loops (Cambridge University Press, 2020), and Fundamentals of Microelectronics (Wiley, 2006, 2014, and 2021) (translated to Korean, Portuguese, and Turkish); and an editor of Monolithic Phase-Locked Loops and Clock Recovery Circuits (IEEE Press, 1996) and Phase-Locking in High-Performance Systems (IEEE Press, 2003). His current research interests include wireless and wireline transceivers and data converters. Dr. Razavi is a member of the U.S. National Academy of Engineering and a fellow of the U.S. National Academy of Inventors. He received the Beatrice Winner Award for Editorial Excellence at the 1994 ISSCC, the Best Paper Award at the 1994 European Solid-State Circuits Conference, the Best Panel Award at the 1995 and 1997 ISSCC, the TRW Innovative Teaching Award in 1997, the Best Paper Award at the IEEE Custom Integrated Circuits Conference in 1998, the McGraw-Hill First Edition of the Year Award in 2001, the 2012 Donald Pederson Award in Solid-State Circuits, the Lockheed Martin Excellence in Teaching Award in 2006, the UCLA Faculty Senate Teaching Award in 2007, and the CICC Best Invited Paper Award in 2009 and 2012. He was a co-recipient of the Jack Kilby Outstanding Student Paper Award and the Beatrice Winner Award for Editorial Excellence at the 2001 ISSCC, the 2012 and the 2015 VLSI Circuits Symposium Best Student Paper Awards, and the 2013 CICC Best Paper Award. He was also recognized as one of the top 10 authors in the 50-year history of ISSCC. He was a recipient of the American Society for Engineering Education PSW Teaching Award in 2014 and the 2017 IEEE CAS John Choma Education Award. He served on the Technical Program Committee of the International Solid-State Circuits Conference (ISSCC) from 1993 to 2002 and the VLSI Circuits Symposium from 1998 to 2002. He has also served as a Guest Editor and an Associate Editor for the IEEE JOURNAL OF SOLID-STATE CIRCUITS, IEEE TRANSACTIONS ON CIRCUITS AND SYSTEMS, and International Journal of High Speed Electronics. He served as the founding Editor-in-Chief for the IEEE SOLID-STATE CIRCUITS LETTERS. He has served as an IEEE Distinguished Lecturer.

WORKSHOPS

Sunday, 11 June 2023

Workshops and Short Courses are offered on Sunday, Monday and Friday at the San Diego Convention Center. Please see daily handout on Sunday, Monday, and Friday in the registration area and from volunteers throughout the meeting floors to confirm room location.

WSB (half-day): 08:00–11:50
A Deep Dive into Circuit Design for Wireline/Optical and Wireless Transceivers: Commonalities and Differences

Sponsor: **RFIC**

Organizers: **Mahdi Parvizi**, *Cisco*
Bahar Jalali Farahani, *Cisco*

Abstract: This workshop presents the similarities and differences between wireless and wireline/optical communication along with circuit design innovations that enable the next generation of these systems. There are undeniable similarities between the systems and electronic building blocks in wireline/optical and wireless transceivers. In this event, first commonalities and differences of wireline/optical system versus an advanced wireless link will be discussed, next advanced modulation schemes to close the gap with Shannon limit in wireline links will be reviewed. Next, advanced circuit design techniques for wireless and optical transmitters, which is power amplifiers and modulator drivers will be presented. The last talk covers the optical and wireless receiver front-ends where novel circuit design techniques for low-noise, low-power LNAs and TIAs will be highlighted.

Speakers:

1. "Wireless-Inspired Wireline Communication Systems", **James F. Buckwalter**, *University of California, Santa Barbara*

2. "How Close are we to the Shannon Limit? The Role of Modulation Schemes to Close the Gap", **Ali Sheikholeslami**, *University of Toronto*

3. "Energy Efficiency and Linearity Improvements for Next Generation of Power Amplifiers and Modulator Drivers", **Munehiko Nagatani**, *NTT*

4. "Design of Low-Power, Low-Noise Receiver Front-Ends: Wireless vs. Wireline", **Behzad Razavi**, *University of California, Los Angeles*

WSC (full-day): 08:00–17:20
6G Challenge:
Overpass RF Bandwidth Limitation to Reach 100Gbs to 1Tbs

Sponsor: **RFIC**

Organizers: **Didier Belot**, *CEA-Leti*
Hao Gao, *Technische Universiteit Eindhoven*
Pierre Busson, *STMicroelectronics*

Abstract: Wireless systems with small RF bandwidths, high-order modulations, and advanced signal-processing techniques have reached a saturation point. They run into spectrum saturation and interference troubles under the sub-6GHz frequency band. International Telecommunication Union (ITU) announced the opening of 275GHz to 450GHz for super high data-rate communication applications. 5G is becoming a reality worldwide, and 6G is in a championship worldwide. The complete paradigm change of this new generation implies the evolution from today, and one of the elements to be defined will be the revolution in the transceiver functions: The data-rate is targeted beyond 100Gbps, and the carrier frequency to support such data transfer will be in the combination of mm-wave and sub-THz. In the 6G, the mm-wave/sub-THz front-end has challenges on bandwidth, power consumption, antenna coupling, array integration, etc. In this workshop, we also dedicate attention to silicon-based building blocks' present realizations targeting 5G to 6G evolution.

Speakers:

1. "Revolutionary Ideas for 6G (and Next-G) Transceivers Overcoming Fundamental Limitations of Conventional Architectures", **Payam Heydari**, *University of California, Irvine*

2. "140GHz Two-Dimensional 8×8 Phased Arrays for 6G MIMO Systems", **Amr Ahmed, Gabriel M. Rebeiz**, *University of California, San Diego*

3. "A Channel Aggregation Architecture TX-RX in D-Band with 84Gbps Data-Rate in RFSOI Process", **Jose Luis Gonzalez-Jimenez**, *CEA-Leti*

4. "Sub-THz Base Station Radio Architecture", **Rui Hou**, *Ericsson*

5. "THz Waveforms for Communication and Sensing", **Sofie Pollin**, *KU Leuven*

6. "Silicon-Based mm-Wave Broadband RF Front-End", **Hao Gao**, *Technische Universiteit Eindhoven*

WSD (half-day): 08:00–11:50
EM Circuit Co-Design and Conflation of
Passive/Active Circuits at mm-Wave Frequency

Sponsor: **RFIC**

Organizers: **Vadim Issakov**, *Technische Universität Braunschweig*
Ruonan Han, *MIT*

Abstract: Integration of passive electromagnetic structures and particularly integration of antennas on silicon becomes feasible at frequencies above 100GHz due to wavelength-related size reduction. The goal of this workshop is to give inspiration on the various novel circuit techniques relying on conflation of passive and active devices. Furthermore, this workshop discusses potential emerging applications towards THz and presents the latest developments on integrated EM devices and co-design with active circuits at high mm-wave frequencies. We discuss how to realize passive on-chip components, such as transformers, coupler baluns and antennas and how to combine them with the active circuitry. Furhermore, novel techniques involving antennas to realize certain functions are discussed. Antennas can be co-designed synergistically with active circuits to realize novel hybrid antenna-electronics with "on-radiator" and near-field functions, such as power combining/splitting, impedance scaling/filtering, active load modulation, noise cancellation and reconfigurability. A significant research challenge in hybrid active circuit/electromagnetic electronics is the application of suitable multi-physics simulation tools and co-design/co-optimization methodologies. This requires 3D full-physics solutions for electromagnetic simulation. Several world renowned speakers will provide an overview on the techniques, applications and the practical design considerations on realization of these approaches. In this half-day workshop we will discuss emerging techniques for on-chip mm-wave active/passive circuit co-design and applications of these new techniques. Distinguished speakers from leading companies and academia will present a wide range of topics to cover various aspects of EM-circuit co-design. A brief concluding discussion will round-off the workshop to summarize the key learnings of aspects presented during the day.

Speakers:

1. "Co-Design Techniques of mm-Wave Circuits with Electromagnetics and Radiation", **Hua Wang**, *ETH Zürich*

2. "EM-to-Information Approach for Reconfigurable THz Sensors and Surfaces", **Kaushik Sengupta**, *Princeton University*

3. "Embedding Networks and Automation to Enhance Power Gain and Compression in Amplifiers Above 100GHz", **James F. Buckwalter**, *University of California, Santa Barbara*

4. "Advanced Circuit Techniques Using Integrated Transformers", **Andrea Bevilacqua**, *Università di Padova*

5. "Co-Design and Coupling Effect in Highly Integrated mm-Wave Systems on Chip", **Fabio Padovan**, *Infineon Technologies*

WSF (full-day): 08:00–17:20
Enabling Quantum Computing: A Survey of Readout Technologies

Sponsor: **RFIC**

Organizers: **Duane Howard**, *Amazon*
Fabio Sebastiano, *Technische Universiteit Delft*
Kevin Tien, *IBM*

Abstract: The continued prevalence of microwave system techniques for interacting with superconducting transmon qubits and spin qubits have driven a resurgence of interest in cryogenic circuit and systems for quantum computing. Moreover, quantum computing applications demand low power, high scalability, and high precision in control signal generation and readout signal processing, which has led to several recent demonstrations of innovative system building blocks, as well as end-to-end control and readout chains. In this workshop, we introduce the state-of-the-art in system architectures for qubit control and readout, and then focus on the recent developments in technologies related to qubit readout. We will examine current building blocks found in high-end systems, then look at the next generation of high performance cryo-LNA technologies. Finally, we conclude with deep dives into full readout chain construction, and test and metrology for this very challenging ecosystem of components.

Speakers:

1. "Workshop Introduction", **Duane Howard[1], Fabio Sebastiano[2], Kevin Tien[3]**, [1]*Amazon*, [2]*Technische Universiteit Delft*, [3]*IBM*

2. "Probing Spin Qubits with Radiofrequency Reflectometry", **M. Fernando González-Zalba**, *Quantum Motion*

3. "Readout Chains for Transmon Qubits in Production Scaled Quantum Computers", **David Lokken-Toyli**, *IBM*

4. "State-of-the-Art Cryo-LNAs in III-V Technology for Scalable Quantum Computing", **Arsalan Pourkabirian**, *Low Noise Factory*

5. "Scaling Considerations for Superconducting Quantum-Limited Amplifiers", **José Aumentado**, *NIST*

6. "SiGe and CMOS Cryogenic Amplifiers for Superconducting Qubit Readout", **Joseph Bardin**, *UMass Amherst*

7. "Wideband-Noise-Matching Considerations for Cryo-CMOS LNAs", **Leonid Belostotski**, *University of Calgary*

8. "Panel Discussion: the Ecosystem for Cryo-LNAs — What's Next?"

9. "A Cryogenic CMOS RF Receiver for Multiple Spin Qubit Readout: From Specifications to Implementation and Qubit Testing", **Masoud Babaie**, *Technische Universiteit Delft*

10. "Scaling Measurement Methodologies using Cryogenic TaaS Framework for Higher Quality Cryo-LNAs and Reliable Qubit Readout Chains", **Brandon Boiko**, *FormFactor*

WSG (full-day): 08:00–17:20
Fundamentals of RF Power Amplifiers: From the Basics to Advanced PA Architectures, Practical Design Aspects, and Process Technologies

Sponsor: **RFIC/IMS**

Organizers: **Debopriyo Chowdhury**, *Broadcom*
Jennifer Kitchen, *Arizona State University*

Abstract: The RF Power Amplifier (PA) is a performance bottleneck of most RF wireless transmit systems and a critical design component of any RF system. Fundamental PA design knowledge and realization expertise are highly desired and regarded skills in the RF community. With their numerous process technologies, architectures, and implementation "tricks", the design of RF PAs may quickly become overwhelming. Moreover, the knowledge is typically acquired through years of design experience and multiple failed design attempts. This workshop jump-starts you into the world of PA design by walking you through the various aspects of RF PA design, starting from the basics and then introducing the most popular forms of advanced PA architectures. The various tutorials within the workshop will categorize the different PA design methodologies to give you a better understanding behind their motivations. Experts from industry and academia will also summarize the strengths of various process technologies, enabling you to better select processes depending on your target application. Finally, PA designers with decades of experience will provide insight into successfully implementing RF PAs, including practical design aspects ("tricks of the trade"), accounting for PA memory and thermal effects (the big "gotcha"), and effectively simulating PA designs to closely predict performance. This workshop will provide design insights not obtained from textbook reading, thus benefiting those who are new to the RF PA design field and seasoned warriors who would like a rapid refresher.

Speakers:
1. "Introduction to the Workshop", **Jennifer Kitchen[1], Debopriyo Chowdhury[2]**, *[1]Arizona State University, [2]Broadcom*

2. "Foundations of RF Power Amplifiers", **Joseph Staudinger**, *NXP Semiconductors*

3. "Comparison of Efficiency Enhancement Techniques for RF PAs", **Matthew Heins**, *University of Texas at Dallas*

4. "Digital Power Amplifiers and Transmitters Based on RF Digital-to-Analog Converters", **Sangmin Yoo**, *Samsung*

5. "Envelope Tracking Systems for RF PA Efficiency Enhancement", **Peter Asbeck**, *University of California, San Diego*

6. "The Promise of Load Modulation and Doherty Power Amplifiers", **Hua Wang**, *ETH Zürich*

7. "CMOS vs SOI vs GaAs — What is the Best Technology for RF and mm-Wave PA Design?", **Ali M. Darwish**, *Army Research Laboratory*

8. "Practical Challenges in Integrated CMOS Power Amplifier Design", **Ali Afsahi**, *Broadcom*

9. "Concluding Remarks", **Jennifer Kitchen[1], Debopriyo Chowdhury[2]**, *[1]Arizona State University, [2]Broadcom*

WSH (full-day): 08:00–17:20
Integration of 6G Systems from BB to Antenna for 6G Phased Arrays

Sponsor: **RFIC**

Organizers: **Gernot Hueber**, *United Micro Technology*
Shahriar Shahramian, *Nokia Bell Labs*

Abstract: Wireless networks have enabled socio-economic growth worldwide and are expected to further advance to foster new applications such as autonomous vehicles, virtual/augmented-reality, and smart cities. Due to limitations of further growth in capacity in the sub-6GHz spectrum, mm-wave and sub-Thz frequencies are gaining an important role in the emerging 6G and the communication-on-the-move applications. In 6G, RF/mm-wave/sub-THz front-ends have challenges on bandwidth, power consumption, antenna coupling, array integration, etc. We examine the integration technologies and packaging challenges. 6G covering from sub-10GHz to high frequency as well the complexity of systems is increasing, which demands implementations in the right technology (CMOS, SiGe, ...) and integration of chipsets heterogeneously from basedband, transceiver to the antenna. The heterogeneous integration will be important with the multitude of frequency bands covered, e.g. 7–14GHz bands up to frequencies >100GHz.

Speakers:
1. "The Challenges of Integration in 6G Transceiver Systems", **Gernot Hueber[1], Shahriar Shahramian[2]**, *[1]United Micro Technology, [2]Nokia Bell Labs*

2. "Toward 6G: From New Hardware Design to Wireless Semantic and Goal-Oriented Communication Paradigms", **Emilio Calvanese-Strinati**, *CEA-Leti*

3. "6G from System Architectures Multi-Band Transceivers and Integration", **Harish Viswanathan**, *Nokia Bell Labs*

4. "THz CMOS Phased Array Transceiver for 6G", **Kenichi Okada**, *Tokyo Tech*

5. "Advances in Packaging and Integration for 6G Phased Array Transceiver Systems", **Hsin-Chia Lu**, *National Taiwan University*

6. "Architectures, Algorithms, and Vertical Integration for Advanced Beamforming in Next Generation 5G and 6G Systems", **Alberto Valdes-Garcia**, *IBM T.J. Watson Research Center*

7. "Architectures, Technology Partitioning, and Challenges for 6G Beamformers", **Giuseppe Gramegna**, *imec*

8. "Techniques for MIMO and Extreme Data-Rates at mm-Wave/Sub-THz", **Harish Krishnaswamy**, *Columbia University*

WSI (full-day): 08:00–17:20
mm-Wave Integrated Radars: Opportunities and Challenges

Sponsor: **RFIC**

Organizers: **Yahya Tousi**, *University of Minnesota*
Vito Giannini, *Uhnder*

Abstract: The unique sensing capabilities of mm-wave radars bolstered by modern nano-scale silicon technology and advanced image processing has created the opportunity for integrated radar technology to create substantially improved image perception at a considerably lower size and cost compared to the radars of the 20th century. There is a growing effort in both academia and industry to bring this technology to fruition. In this workshop, we overview the existing opportunities in this field and the challenges that need to be overcome in order to standardize and commercialize integrated radar technology. The workshop brings together a complementary mix of top academic and industry speakers with a breadth of expertise and experience in this field ranging from the fundamental aspects of circuit design, system integration to sensor fusion, product design and testing.

Speakers:

1. "Introduction to the mm-Wave Radar Workshop", **Yahya Tousi**, *University of Minnesota*

2. "Automotive Radar — Applications and Technology Trends", **Juergen Hasch**, *Robert Bosch*

3. "Imaging Radars at Scale — From Automotive to Security Applications", **Sherif Ahmed**, *Stanford University*

4. "High-Performance and Low-Power mm-Wave Radar Systems: Requirements and Challenges", **Krishnanshu Dandu**, *Texas Instruments*

5. "Phased-Array-based Real-Time 3D Radar for AI-Based Event Classification", **Alberto Valdes-Garcia**, *IBM T.J. Watson Research Center*

6. "A Digital-Perception Radar Platform for Automotive Safety", **James Maligeorgos**, *Uhnder*

7. "140GHz Automotive Radar — Sense and Nonsense", **Ilja Ocket**, *imec*

8. "THz and mm-Wave High-Resolution Imaging and Radar Sensing for Low-Power and Short-Range Applications", **Omeed Momeni**, *University of California, Davis*

9. "Soli: Radar for Intelligent Human-Computer Interactions", **Jaime Lien**, *Google*

WSJ (full-day): 08:00–17:20
mm-Wave and Sub-THz PA Design for
Next-Gen Wireless and Sensing Applications

Sponsor: **RFIC/IMS**

Organizers: **Steven Callender**, *Intel*
Sungwon Chung, *Neuralink*

Abstract: There is no silver bullet power amplifier (PA) design that provides a one-size-fits-all solution for next-gen communication and sensing systems due to the diversity of applications and their associated PA specs (eg output power, linearity, bandwidth, and back-off efficiency). The goal of this workshop is to explore leading mm-wave and sub-THz applications and the associated PA specs for these systems. The applications of focus are massive MIMO and large-scale phased-arrays, sub-orbital satellite communication (SATCOM), and mm-wave radar. A balanced mix of both industry and academic perspectives will be provided, offering both a high-level familiarization of the application and associated specifications, along with deeper technical dives into PA design techniques in modern process nodes.

Speakers:
1. "Design of D-Band PAs in Bulk-CMOS and FinFET", **Patrick Reynaert**, *KU Leuven*
2. "Generating "Efficient" D-Band Power Using Nanoscale CMOS Technology", **Ali M. Niknejad**, *University of California, Berkeley*
3. "Reliable mm-Wave and Sub-THz PA Design", **Jefy Jayamon**, *Qualcomm*
4. "Power Amplifiers for Large-Scale SATCOM Phased Arrays", **Kaushik Dasgupta**, *Amazon*
5. "GaN and GaAs Power Amplifier Design for Arrays", **Taylor Barton**, *University of Colorado Boulder*
6. "High-Efficiency Silicon PAs for mm-Wave Radar Sensors", **Tolga Dinc, Krishnanshu Dandu, Swaminathan Sankaran, Brian Ginsburg**, *Texas Instruments*

WSK (full-day): 08:00–17:20
To 100Gb/s and Beyond: High-Data-Rate Interconnect Technologies, Who Will Win at Which Scenario?

Sponsor: **RFIC**

Organizers: **Jane Gu**, *University of California, Davis*
Wooram Lee, *Penn State University*

Abstract: Interconnect bottlenecks have been a long-standing grand challenge over decades, caused by the increasing gap between exponentially growing data generation and transmission demand, and slowly-increasing supporting data bandwidth supply. Both Electrical Interconnect (EI) and Optical Interconnect (OI) have been investigated extensively to try to combat the challenge, however, both of them face their own inherent constraints. The newly emerging sub-THz/THz Interconnect (TI) aims to complement the existing EI and OI to close the interconnect gap. This workshop plans to bring experts from different domains, OI, EI, and emerging TI, to discuss the challenges, opportunities and best use scenarios of each interconnect scheme.

Speakers:
1. "Analog and Digital Optical Interconnects", **Vladimir Stojanovic**, *University of California, Berkeley*
2. "High-Density, Low-Power Optical Communications for AI, Data Center, and More", **Jonathan Proesel**, *Nubis Communications*
3. "Silicon Photonics-Based Optical I/O for Next-Gen XPUs", **Ganesh Balamurugan**, *Intel*
4. "Waveguide Interconnects — D/F-Band Systems for >100Gbps Medium Reach Links", **Thomas W. Brown**, *Intel*
5. "High-Speed Short-Reach Interconnects Using Dielectric Waveguide", **Hyeon-Min Bae**, *KAIST*
6. "Going Beyond 100Gbps with Polymer Microwave Fibers", **Patrick Reynaert**, *KU Leuven*
7. "A Path to 200+Gb/s Transceiver Design for Electrical Interconnects", **Jihwan Kim**, *Intel*
8. "Next-Generation Electrical Interconnects: Chips and Chiplets", **Tod Dickson**, *IBM T.J. Watson Research Center*

WSL (full-day): 08:00–17:20
State-of-the-Art mm-Wave GaN Transistor and MMIC Technologies and Future Perspective

Sponsor: **IMS/RFIC**

Organizers: **Farid Medjdoub**, *IEMN (UMR 8520)*
Keisuke Shinohara, *Teledyne Scientific & Imaging*

Abstract: Owing to superior electrical and thermal properties of GaN-on-SiC material systems, tremendous progress has been made on GaN-based transistor and MMIC technologies. Advanced heterostructure material designs, epitaxial growth techniques, and transistor scaling processes enabled GaN MMICs to extend their applications from microwave to mm-wave frequencies (up to W-band). Next-generation RF systems require high efficiency and high linearity for more complex modulation schemes to support very high data-rates. The traditional trade-off among efficiency, linearity, and power density imposes performance limitations on GaN MMICs, which become more pronounced at mm-wave frequencies. In this workshop, world-leading experts will discuss the present status, challenges, and future perspective of mm-wave GaN transistor and MMIC technologies, covering emerging materials and devices, device modeling, thermal management, reliability, and circuit designs.

Speakers:

1. "N-Polar GaN Devices for Efficiency and Linearity", **Matthew Guidry, Umesh Mishra**, *University of California, Santa Barbara*

2. "High-Efficiency High-Robustness mm-Wave AlN/GaN Transistors", **Farid Medjdoub**, *IEMN (UMR 8520)*

3. "Progress in Highly Linear and Efficient mm-Wave GaN HEMTs and MMICs", **Jeong-sun Moon**, *HRL Laboratories*

4. "Polarization-Engineered III-N mm-Wave Transistors for Linearity, Efficiency, and Reconfigurability", **Patrick Fay**, *University of Notre Dame*

5. "Broadband mm-Wave GaN MMICs: Technology Aspects and Design Examples", **Fabian Thome**, *Fraunhofer IAF*

6. "GaN Transistor Reliability Drivers — Temperature and Electric Fields", **Martin Kuball**, *University of Bristol*

7. "The Interplay Between Deep Level Effects and Reliability in Deep Submicron-Gate GaN HEMTs for RF Applications", **Enrico Zanoni**, *Università di Padova*

8. "N-Polar GaN HEMT Technology for mm-Wave Amplifiers using Commercial 4-inch Wafer Process Facilities", **Kozo Makiyama**, *Sumitomo Electric*

9. "ScAlN/GaN Heterostructure Field Effect Transistors for Ultra-High-Power and Wide-Band MMICs", **Eduardo Chumbes**, *Raytheon*

10. "GaN Transistor Designs for mm-Wave Applications", **Keisuke Shinohara**, *Teledyne Scientific & Imaging*

WSA (half-day): 13:30–17:20
Recent Advances in
Ultra-Low-Power Wireless Communication Technology

Sponsor: **RFIC**

Organizers: **Sai-Wang Rocco Tam**, *NXP Semiconductors*
Yao-Hong Liu, *imec*
Oren Eliezer, *Samsung*
Minyoung Song, *imec*

Abstract: Ultra-Low-Power (ULP) wireless communication technology provides many unique features over conventional wireless communication such as high energy efficiency, low cost, small form factor, large scale deployments, reconfigurability and simple architecture. This workshop will bring together experts from academia and industry to highlight recent works and applications in this exciting technology. In the first topic, we are going to review the industry impacts on the most successful and large-scale commercialization using ULP wireless communication technologies such as RFID and Near-Field Communication (NFC). After that, we are going to shift our focus to recent research advances in using RF backscattering techniques in Reconfigurable Intelligent Surface (RIS) and WLAN/BT connectivity solutions. In the last topic of this workshop, we will discuss recent advances from medical, industrial and academic fields in biomedical implants with technologies such as co-optimizing antenna and RFIC to miniaturize radio module volume. Unconventional wireless propagation methods are also introduced, such as body channel communication, Magnetoelectric, ultrasound, etc.

Speakers:

1. "Recent Circuit and System Architecture Design Advances in RFID/NFC Products", **Peter Thüringer**, *NXP Semiconductors*

2. "Recent Advances in Reconfigurable Intelligent Surfaces (RIS) and Backscatter Communication", **Manos Tentzeris**, *Georgia Tech*

3. "Enabling Low-Power yet Standards Compatible Wireless Communication via Wake-Up Receivers and Backscatter Circuits", **Patrick Mercier**, *University of California, San Diego*

4. "Circuit Techniques for Wirelessly Powered Sensors and Actuators", **Aydin Babakhani**, *University of California, Los Angeles*

5. "Magneto-Electric Power and Data Transfers to Millimetric Bioelectronic Implants", **Kaiyuan Yang**, *Rice University*

WSE (half-day): 13:30–17:20
FDSOI CMOS Energy Efficient 5G and IoT Design Techniques and Related Technology

Sponsor: **RFIC**

Organizers: **Wanghua Wu**, *Samsung*
Andreia Cathelin, *STMicroelectronics*

Abstract: Thanks to the extended body biasing feature, FDSOI process has enabled new system and circuit design techniques to drastically improve the RF and mm-wave system performance. Tremendous industry collaboration efforts have committed to bring up the FDSOI to higher volumes of production to serve the wireless, IoT, and automotive market in the near future. This workshop includes an overview introductory presentation followed by 4 talks on FDSOI technology and its industry design examples for RF and mm-wave applications. The introduction provides the overview on FDSOI technology and its benefits for analog/RF/mm-wave circuit design, focusing on a technology perspective. The following three talks demonstrate RF and mm-wave system design examples using FDSOI technology, for 5G infrastructure and user terminal as well as for ULP IoT. The last talk reveals the advanced FDSOI process design roadmap and what is expected in the near future.

Speakers:
1. "Benefits of FD-SOI Technology for Analog/RF/mm-Wave Circuits", **Andreia Cathelin**, *STMicroelectronics*

2. "5G FR2 UE Phased Array Transceiver Solution Using 28nm CMOS FD-SOI", **Xiaohua Yu**, *Samsung*

3. "Ultra-Low-Power IoT Frequency Synthesis Solutions Based on FD-SOI Technology", **Yann Deval[1], David Gaidioz[2], Andres Mauricio Asprilla Valdes[2], Denis Michael Flores Pazos[2], Andreia Cathelin[2]**, *[1]IMS (UMR 5218), [2]STMicroelectronics*

4. "mm-Wave Front-End Circuits for 5G and 6G Applications Using FD-SOI", **Yang Zhang**, *imec*

5. "22FDX Platform and Features Optimized for Demanding RF Applications Ranging from WiFi Connectivity and mm-Wave Cellular to Auto Radar", **Andreas Knorr, Tianbing Chen, Shafi Syed, Randy Wolf, Zhixing Zhao, Mingcheng Chang, Steffen Lehmann, Peter Javorka, Shih Ni Ong, Amit Kumar Sahoo, Jen Shuang Wong, Wai Heng Chow, Kok Wai Johnny Chew, Nicholas Comfoltey, Farzad Michael David Inanlou, Stephen Moss, Julio Costa**, *GlobalFoundries*

WSO (half-day): 13:30–17:20
Advanced Wafer-Level Heterogeneous Integration and Packaging for mm-Wave 5G and 6G Applications

Sponsor: **IMS/RFIC**

Organizers: **Kamal Samanta**, *Sony*
Kevin Xiaoxiong Gu, *Metawave*

Abstract: This workshop will cover various recently developed technologies and the state-of-the-art performance in wafer-level integration and packaging technologies and manufacturing techniques with challenges and possible future directions and solutions. In particular, it will highlight the latest advances in the areas such as embedded wafer-level ball grid array (eWLB) technology for system integration with high Q interconnects and passives in thin-film Re-Distribution Layers (RDL), wafer-level heterogeneous integration of different substrates, BiCMOS embedded TSVs, sub-THz on-chip antenna integration, innovative Fan-Out technologies for wafer-level package, RF IPD, and FOSiP, and embedding various chips within the silicon Metal-Embedded Chip/Chiplet Assembly. Further, the workshop will present the practical realization of highly integrated systems, including 60GHz and 77GHz eWLB transceiver modules with integrated antennas, 3D wafer-level packaging for mm-wave and sub-mm-wave space systems, and hetero-integration technology solutions to enable a full 2D array of phased array systems above 120GHz.

Speakers:

1. "Developments in Wafer-Level Packaging for mm-Wave Communication and Radar System", **Maciej Wojnowski, Klaus Pressel**, *Infineon Technologies*

2. "Advanced Packaging and Heterogeneous Integration Technologies for mm-Wave and THz Applications", **Mehmet Kaynak**, *IHP*

3. "Fan-Out Packages Enabling Pivotal mm-Wave Performance", **CP Hung**, *ASE Group*

4. "Metal-Embedded Chip/Chiplet Assembly (MECA) Platform for High-Frequency RF Subsystems", **Souheil Nadri**, *HRL Laboratories*

5. "Wafer-Level Packaging for High Frequency Applications at Northrop Grumman Space Systems", **Matthew Laurent**, *Northrop Grumman*

IEEE
445 Hoes Lane
Piscataway, NJ 08854, USA

2023 RFIC Symposium
San Diego, California, USA
11–13 June 2023

RFIC Plenary Speaker 1

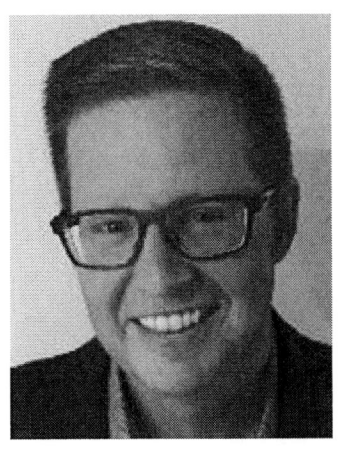

Dr. Todd Younkin
President and CEO
Semiconductor Research Corporation

The Roaring 20s: A Renaissance for the Semiconductor Industry?

Abstract: Dr. Younkin will share his vision for the future of global semiconductor technologies and design, especially those that will enable future RFIC breakthroughs. Dr. Younkin will discuss the status of government investments and opportunities arising from the CHIPS and SCIENCE ACT of 2022, Korea's K-Belt strategy, Europe's CHIPS ACT, and more. Dr. Younkin leads a global research agenda of about $100M annually, supported by ~3k academic and industrial researchers, 27 international companies, and 3 U.S. government agencies (DARPA, NSF, and NIST). They have defined the opportunities for future compute and communication systems, as outlined by SRC's 2030 Decadal Plan for Semiconductors, and are now working with over 90 organizations to define the semiconductor hardware opportunities that will deliver that required system performance, via the NIST Microelectronic and Advanced Packaging Technologies (MAPT) Roadmap, awarded in April 2022 and scheduled for completion by September 2023.

About Dr. Todd Younkin

Dr. Todd Younkin is a talented and seasoned executive with more than 20 years of experience in technology innovation. His extensive Research and Development experience spans Intel's 0.18μm to 5 nm nodes with technical contributions in novel materials, nanotechnology, integration, advanced lithography, and integrated photonics. Todd brings a wealth of expertise with strengths in areas such as cultivating relationships with strategic partners, entrepreneurship and investment strategies, technology innovation, operational excellence, and talent management. He has spent much of his career working alongside young minds that are aspiring to influence the ever-changing world of smart and autonomous electronics. He has built programs from the ground up, leveraging his entrepreneurial leadership to drive new business development that has generated multi-millions in funding. He has been a key contributor in introducing new technology advances and starting new global research in the U.S., Europe, and Asia. Dr. Younkin holds a Ph.D. from the California Institute of Technology in Pasadena, California. He completed his Bachelor of Science at the University of Florida in Gainesville, Florida. He aspires to continue to influence the next generation of technology and inventors, bringing ideas and investors together to drive heterogeneous electronic solutions that will deliver a smarter, shared future.

In August of 2020, Dr. Todd Younkin became the President & CEO of SRC. Recently, he engineered, launched, and led all programmatic aspects of the five-year, $240 million JUMP research initiative. It has six multi-university, multi-disciplinary innovation Centers with 133 faculty, 835 students, and 360 industrial engineering liaisons. It emphasizes the advancement of Computer Science, Electrical Engineering, and Materials to secure continued U.S. thought leadership. Following his appointment, SRC released its 2030 Decadal Plan for Semiconductors, where it identified the five "seismic shifts" shaping the future of information and communication technologies (ICT). Working closely with SIA, SRC called for greatly increased federal investments throughout the decade to establish a smarter pipeline for semiconductor R&D,

aligned to SRC's Decadal Plan. This drove and resulted in the passage of the CHIPS and SCIENCE ACT of 2022 on 9 August 2022.

Dr. Todd Younkin is excited by the worldwide call for a renewed investment in semiconductor materials, hardware, and design, as well as the equally important calls for an emphasis on education and workforce development and our need for environmental sustainability. Only by investing in a bright, collective future, will we rise to the meet the opportunities presented by the next industrial revolution.

RFIC Plenary Speaker 2

Prof. Mau-Chung Frank Chang
Wintek Chair in Electrical Engineering and
Distinguished Professor of
University of California, Los Angeles

Future System-on-Chip for
Full Spectrum Utilization from RF to Optics

Abstract: The ever-increasing bandwidth requirement due to explosively growing 5/6G and AIoT data flows has compelled global commission authorities to release EM-spectra up to millimeterwave (30–300GHz) and even (sub)-millimeter-wave frequency regimes (>300GHz) for massively expanded sensing and network applications. In this talk, we will exemplify novel CMOS embedded technologies and methodologies developed at UCLA to enable System-on-Chip (SoC) realizations for multi-broadband radio, wideband radar, contactless/plastic interconnect, 3D-imaging and gas-phase rotational spectrometry at (sub)-mm-Wave frequencies, including:

- DiCAD (Digitally Controlled Artificial Dielectric), the only proven Digital-to-Permittivity Converter (DPC), embedded in CMOS-switched interconnects that can vary transmission-line permittivity in real-time (up to ×20 in practice) for (sub)-mm-wave frequency synthesis, direct-frequency modulation and reconfigurable (software defined) radio/radar/spectrometer implementations
- Self-Healing Radio (57–64GHz) with self-diagnosis and self-healing capabilities to secure high performance-yield and counter temperature/process variations & aging effects
- Multiband RF-Interconnect, beyond traditional baseband-only interconnect, to enable contactless and/or plastic waveguide communications up to Terahertz with unprecedented bandwidth, efficiency, dynamic re-configurability & multi-cast capabilities
- Fully integrated frequency synthesizer (PLL) at 560GHz and realized 1st active and passive CMOS imagers at 110GHz; 3-color (349/201/153GHz) and 3D imaging radars for sensing/ranging concealed objects
- Single-chip CMOS heterodyne H_2O-Detecting Spectrometer at 183 GHz to enable NASA's space exploration missions with reduced mass (6.5×) & power (5.5×) to meet strict payload and energy consumption requirements

We will also address challenges encountered in both design and implementations that may hinder further development of such systems, especially the major shortcomings in silicon technologies with limited dynamic range and power handling capabilities. We therefore propose replacing CMOS n-FET's drain with selectively grown wide bandgap cubic-phase GaN (c-GaN) for >10× improved breakdown voltages to secure desired sensing/communication range/coverage with cost-effectiveness.

We also elaborate on the possible growth of multi-wavelength light-emitting sources and detectors directly atop n-FET's c-GaN Drain with various indium contents of InGaN/GaN superlattice for RF-optical combined radio/radar/interconnect applications by creating unprecedented "Photonic System-on-Chip" with full EM-spectrum utilization from RF to optics.

About Prof. Mau-Chung Frank Chang

Dr. Mau-Chung Frank Chang is the Wintek Chair in Electrical Engineering and Distinguished Professor at the University of California, Los Angeles. Prior to joining UCLA, he was the Assistant Director of the High Speed Electronics Laboratory of Rockwell Science Center (1983–1997), Thousand Oaks, California. In this tenure, he led the team to develop and transfer the MOCVD based AlGaAs/GaAs & InGaP/GaAs Heterojunction Bipolar Transistor (HBT) and BiFET (Planar HBT/MESFET) integrated circuit technologies from the research laboratory to production line (later became Skyworks Solutions). The HBT/BiFET productions have grown into multi-billion dollar businesses and dominated the cellphone power amplifier and front-end module markets for the past 30 years (currently exceeding 10 billion units/year and exceeding 50 billion units in the past decade). Throughout his career, his research has focused on the research & development of high-speed semiconductor devices and integrated circuits for radio, radar, imager, spectrometer and interconnect System-on-Chip applications. He invented the multiband, reconfigurable RF-Interconnects for Chip-Multi-Processor (CMP) inter-core communications and inter-chip CPU-to-Memory communications. He and his students were the 1st to demonstrate CMOS active and passive imagers at mm-Wave (100–180GHz) frequencies. His Lab also pioneered the development of self-healing 57–64GHz radio-on-a-chip (DARPA's HEALICS program) with embedded sensors, actuators and self-diagnosis/curing capabilities; and invented the Digitally Controlled Artificial Dielectric (DiCAD) embedded in CMOS technologies to vary transmission-line permittivity in real-time (up to 20× in practice) for realizing reconfigurable multiband/mode radios in (sub)-mm-Wave frequency bands. His UCLA Lab also realized the first CMOS Frequency Synthesizer for Terahertz operation (PLL at 560GHz) and devised the first tri-color CMOS active imager at 180–500GHz based on a Time-Encoded Digital Regenerative Receiver and the first 3-dimensional SAR imaging radar with sub-centimeter range resolution at 144GHz. More recently, his Lab has devised a Reconfigurable Convolution Neuron Network (RCNN) Accelerator for IoT applications, spun-off an Edge-AI startup company Kneron in San Diego, and won IEEE's 2021 Darlington Best Paper Award.

Prof. Chang is a Member of the US National Academy of Engineering, the European Academy of Sciences and Arts, the US National Academy of Inventors, the Academia Sinica of Taiwan, and a Fellow of the IEEE. He was also recognized with the IEEE David Sarnoff Award (2006), IET JJ Thomson Medal for Electronics (2017) and IEEE/RSE James Clerk Maxwell Medal (2023) for his seminal contributions to the heterojunction technology and realizations of (sub)-mm-Wave System-on-Chip with unprecedented bandwidth and re-configurability.

Prof. Chang has published more than 460 peer-reviewed technical papers and 60 US patents in various areas of high speed electronic devices and integrated circuits & systems. During his tenure with UCLA, he has graduated more than 50 Ph.D. students and 100 MS students. He also served as the President of the National Chiao Tung University, Hsinchu, Taiwan (2015–2019).

979-8-3503-2123-4/23 $31.00 © 2023 IEEE

RMo1A-1

A 112-Gbps, 0.73-pJ/bit Fully-Integrated O-band I-Q Optical Receiver in a 45-nm CMOS SOI-Photonic Process

Ghazal Movaghar, Viviana Arrunategui, Junqian Liu, Aaron Maharry, Clint Schow, James Buckwalter

Dept. of Elec. and Comp. Eng, University of California, Santa Barbara, USA

(ghazalmovaghar, arrunategui-norvick, junqian, amaharry, schow, buckwalter)@ucsb.edu

Abstract—A 1310-nm (O-band) coherent optical receiver (CORX) is demonstrated with a monolithic electronic/photonic integrated circuit (MEPIC) process for short range optical interconnects. The CORX is designed using a 45-nm CMOS SOI photonic process for quadrature phase shift keying (QPSK) and includes I/Q channels that are characterized up to 56 GBaud (112 Gbps) with FEC-acceptable BER. The receiver achieves 0.73 pJ/bit energy efficiency and, to our knowledge, is the best energy efficiency reported for a coherent optical receiver.

Keywords—optical receiver, coherent demodulation, energy efficiency, CMOS SOI, silicon photonics

I. INTRODUCTION

Intra-data center traffic interconnects aim for data rates above 200 Gbps per wavelength while reducing power consumption. Intensity-modulation direct detection (IMDD) with high-order pulse amplitude modulation (PAM) is an alternative to on-off keying as it improves spectral efficiency for a given bandwidth. However, scaling beyond PAM-4 demands linearity and power consumption in the transmitter and receiver. Another approach to increasing spectral efficiency is to use quadrature phase shift keying (QPSK) coherent modulation as a scalable alternative to IMDD links if energy-efficient demodulation is possible using a coherent optical receiver (CORX) [1].

Recent work has investigated low-power coherent optical links above 100 Gbps/wavelength with heterogeneous integration between electronic and photonic ICs [2]. Nevertheless, the complexity of coherent optical signal processing places requirements on both photonic and electronic circuits, as well as low latency for phase locking. In particular, the long loop delay of an optical phase locked loop results in unstable locking. Monolithic optical transceivers benefit from reducing parasitics between the photonic and RF integrated circuit components as well as satisfying some of the transient requirements in locking behavior. The first fully-integrated coherent receiver at C-band was demonstrated with a photonic BiCMOS 0.25 μm SiGe technology [3].

To improve energy efficiency, RF CMOS circuit techniques complement silicon photonic devices. A high-performance 45-nm CMOS SOI technology ($f_T = 290 GHz$) supports a process development kit that includes complete optical structures for waveguides, photodetectors, fiber coupling, polarization control structures, as well as ring and Mach-Zehnder modulators [4].

A self-homodyne receiver architecture further simplifies the use of coherent optical architecture in low-power

environments. The local oscillator (LO) laser is forwarded from the transmit side and used as a reference for the receiver to lock the phase of the LO laser to an incoming wavelength channel with an optical delay lock loop (ODLL) [5]. Fig. 1 illustrates the self-homodyne QPSK coherent link where the CORX includes an optical hybrid to produce quadrature versions of the LO and received data and also includes an optical phase shifter. The electronic circuits capture differential I/Q signals and amplify these as well as driving a Costas phase/frequency detector to lock the phase of LO laser [6].

In this paper, an O-band QPSK CORX is presented that operates up to 56 Gbaud with record energy efficiency of 0.73 pJ/bit, including photonic tuning elements. In Section II, we present energy-efficient circuit design techniques that complement photonic devices. Additionally, we describe the operation of the self-homodyne architecture. In Section III, we present measurement results of the QPSK constellation at different data rates and the BER curves indicating error free operation using forward error correction (FEC).

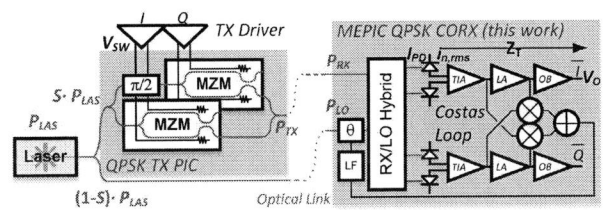

Fig. 1. Self-homodyne coherent optical data link with photonic hybrid transmitter IC and monolithic electronic/photonic integrated circuit for the QPSK coherent receiver with an optical delay locked loop.

II. CIRCUIT DESIGN AND SIMULATION

A. Self-Homodyne Detection

To avoid the regeneration of the LO at the receiver, a self-homodyne link architecture is proposed for energy-efficient coherent operation. The self-homodyne link splits off a portion of the laser power to use as the LO, P_{LO}, and forwards this to the RX on a separate fiber or polarization. Coherent detection can, in theory, improve the sensitivity of the receiver and support lower input power. In the RX, the received photodetector (PD) current is

$$I_{PD} = R_{PD}\sqrt{P_{LO}P_{RX}} = R_{PD}P_{LAS}\sqrt{\frac{S(1-S)}{L}}, \quad (1)$$

where R_{PD} is PD responsivity and P_{RX} is the received optical power as illustrated in Fig. 1. Clearly, a low P_{RX} can be offset

979-8-3503-2123-4/23 $31.00 © 2023 IEEE

by higher P_{LO}. Since the total transmitted laser power is split between the TX PIC with ratio S and LO with ratio $1-S$, we reach the second equality in (1) where L accounts for the transmitter and receiver PIC losses in the signal path. The losses, particularly through a silicon-based TX PIC might be 10-20 dB and the PIC losses are assumed to dominate the total link losses over a short reach (< 1km). The PD current depends on $\sqrt{L}$ as opposed to L as in direct detection, making coherent more tolerant to loss.

Based on a desired output voltage, V_O, the transimpedance is

$$Z_T = \frac{V_O}{R_{PD}P_{LAS}}\sqrt{\frac{L}{S(1-S)}}. \qquad (2)$$

The *minimum* transimpedance occurs for an equal power split. In this case, if V_O is 0.1 V, R_{PD} is 1 A/W, P_{LAS} is 10 mW, then $Z_T = 20\Omega \times \sqrt{L}$. For L = 30 dB, the transimpedance should be at least 56 dBΩ.

Higher P_{LAS} reduces the Z_T and receiver power consumption. On the other hand, higher P_{LAS} increases to TX power consumption, $P_{TX_{DC}}$, at the expense of the RX power consumption. If the $P_{TX_{DC}}$ is dominated by the laser DC power operating at efficiency, η_{LAS}, $P_{TX_{DC}} = P_{LAS}/\eta_{LAS}$. The total power consumption consists of the laser power and the receiver power, e.g. $P_{DC} = P_{DC,TX} + P_{DC,RX}$. Consequently, the minimum DC power of the receiver for self-homodyne coherent detection is

$$P_{DC,RX,OPT} = \sqrt{\frac{K_Z V_O}{\eta_{LAS}R_{PD}}\sqrt{\frac{L}{S(1-S)}}}. \qquad (3)$$

where K_Z is a technology-dependent coefficient that relates the desired Z_T to power consumption, i.e. $P_{DC,RX} = K_Z \times Z_T$. Building upon earlier assumptions and considering that $\eta = 10\%$ and $K_Z = 0.01mW/\Omega$, then $P_{DC,RX,OPT} = 4.5mW\sqrt[4]{L}$ or 25 mW per channel for L = 30 dB. For an I/Q receiver operating at 100 Gb/s, the energy efficiency of the receiver is hypothetically 0.5 pJ/b.

These assumptions can be tested against the implementation in a 45-nm CMOS SOI process. Fig. 2 shows the schematic of the high-speed I/Q receiver data

path with an on-chip 90° optical hybrid to assess the potential power consumption and losses in a silicon photonic technology.

B. Optical Front-end

In Fig. 2, a PN-type phase shifter is introduced to allow a large tuning range as well as high speed operation to track LO phase through an ODLL. Slower, thermal phase shifters are placed in each quadrature path of the hybrid to allow tuning of the I/Q balance and directional couplers used as splitters/combiners form the optical hybrid. The loss of the optical hybrid is estimated to be 0.4 dB.

The optical hybrid produces differential optical signals that drive integrated Germanium PD pairs with approximately 50-fF capacitance and more than 40 GHz of electro-optical bandwidth; measurements of a PD test structure yielded 60 GHz of bandwidth.

C. Electronic Front-end

The differential PD current is amplified through a low-power pseudo-differential push-pull shunt-feedback transimpedance amplifier (TIA) followed by 4 limiting amplifier (LA) stages to increase gain. The pseudo-differential eliminates the tail current source; however, it reduces common mode rejection. The TIA uses resistive feedback to produce high gain from the composite g_m of both nmos and pmos devices while operating at low dc currents to increase the g_{ds}. This produces high intrinsic gain while minimizing power consumption for a given bandwidth [7]. The shunt feedback TIA can be designed to maximize Z_T for a given power consumption; however, achieving desired bandwidth requires inductive peaking [8].

The output of the I TIA and Q LA shown in Fig. 2 also drive a passive ring mixer to form the Costas loop shown in Fig. 1. Simulation shows 47-dBΩ transimpedance at TIA, 58.7 dBΩ at the LA and 61.2 dBΩ at the receiver output. The TIA stage consumes around 2.4 mW, suggesting a transimpedance power efficiency of $K_Z = 0.01mW/\Omega$. Fig. 3 plots input referred noise current (IRNC) as well as the receiver frequency response assuming 400-pH output wirebond inductance indicating 40-GHz 3dB bandwidth. The IRNC

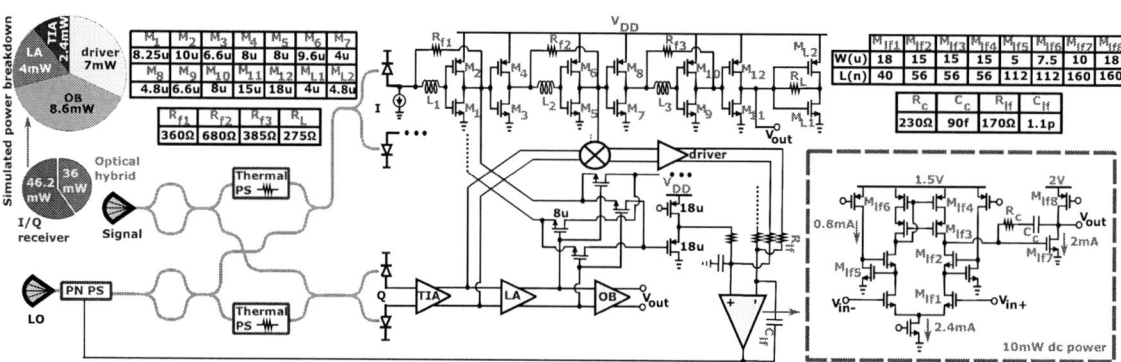

Fig. 2. Optical receiver implemented in 45-nm RF/photonic integrated circuit process consisting of an optical hybrid, TIA, LA, 50Ω output buffer, passive ring mixers, and a loop filter.

979-8-3503-2123-4/23 $31.00 © 2023 IEEE

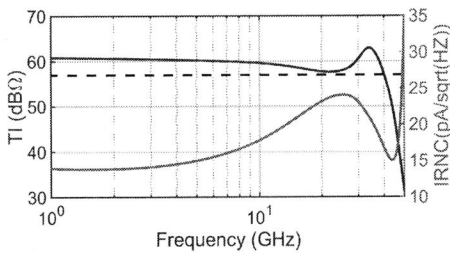

Fig. 3. Transimpedance and input-referred noise current spectral density for the RX channel

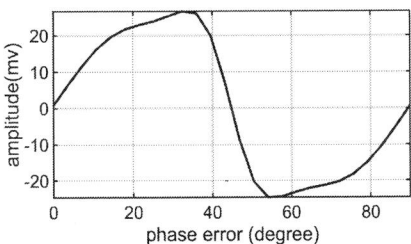

Fig. 4. PFD output as a function of phase error.

integrated over twice the 3dB bandwidth results in 3.9-μA total integrated input noise current. Hence, to achieve a BER below FEC limit of 3.8×10^{-3}, receiver sensitivity is 20.3 μA.

Fig. 4 plots the simulated post-layout Costas loop PFD response. This design also includes an on-chip integrating op-amp to form the loop filter in Fig. 1. The loop filter generates an infinite loop gain at dc to drive a 3mm long PN optical phase shifter to change LO phase until the PFD output is set to 0 V and LO laser is locked to the correct phase.

III. MEASUREMENT RESULTS

The chip micrograph, chip-on-board assembly, and measurement setup is shown in Fig. 5. The entire MEPIC is contained within 2.6 mm by 1.1 mm, where a significant area is required for the LO phase shifter. The optical hybrid and electronics have relatively equal area. The die is wirebonded to a high-speed test PCB and the outputs are connected through high speed connectors to a realtime oscilloscope (RTO). Probes are used only to introduce the optical fibers to the waveguide

grating couplers on the chip. For testing, a 1310-nm external cavity laser (ECL) splits into LO and signal paths where 25% of ECL power goes to LO and 75% goes to the transmitter. In the signal path, a reference transmitter is driven with a 500-mV PRBS-15 signal from a bit pattern generator (BPG) (SHF 12105A) [2]. The signal path also includes an O-band fiber amplifier (PDFA) to compensate for high transmitter loss and an attenuator for sensitivity measurements. The receiver I/Q channels are connected to a 70 GHz real-time oscilloscope (RTO) (Keysight UXR0702A) with a 0.875 μs acquisition time at 256 GSa/s to capture the received QPSK signal. The ECL power is set to 20 dBm providing 14-dBm LO and 18.7-dBm input power to the TX. The signal power at the output of the attenuator with minimum attenuation of 0.6dB is 2.8dBm. The 14-dBm LO power and 2.8-dBm signal power correspond to 0.3-mA LO and 4-μA peak signal current per PD indicating coupling loss of 12.2 dB for LO and 19.8 dB for signal. Manual alignment and sensitivity to mechanical perturbations of couplers contributed to the high coupling loss. The differential dual-channel electrical circuit draws 42-mA current from a 1.1-V supply, close to the predicted 50 mW of RX power consumption, and the thermal phase shifter inside the optical hybrid consumes 36 mW for quadrature bias corresponding to 82.2-mW dc power consumption for the entire CORX. A significant portion of the total receiver power was therefore consumed in optical tuning elements.

Fig. 6a and Fig. 6b plot the measured constellations at 40 and 56 Gbaud based on the I / Q electrical outputs of the receiver. The bit error rate (BER) as a function of signal power incident at each PD, is shown in Fig. 7. At lower data rates the error rate is mainly due to noise while as data rate increases intersymbol interference degrades the error rate and sensitivity. As a result we can find receiver input referred noise current at 28 Gbps, where the minimum signal power to achieve BER below FEC limit of 3.8×10^{-3} is -35 dBm. -4.2 dBm of LO power incident at each PD results in a sensitivity of 17.4 μA, also close to the predicted receiver sensitivty calculated from the IRNC simulation. The single-ended output voltage peak swing is 30 mV for 34-μA input current per PD driving the TIA showing 867-Ω (58.7 dBΩ) transimpedance. Future measurements will explore the PFD closed loop behavior of the ODLL, which utilizes the phase shifter in LO path for phase recovery. This phase shifter was left unbiased for current measurements.

A performance summary for this design is provided in Table 1 with comparison against recent work at similar data rates. Notably, this result is fully integrated and was tested on a PCB assembly and not probed electrically. While the finFET CMOS has indicated excellent power, this process does not support silicon photonic integration and the measured results are not for a full link optical assembly. When compared to prior monolithic coherent design in O-band, we achieved a 6-fold improvement in energy efficiency for similar data rates.

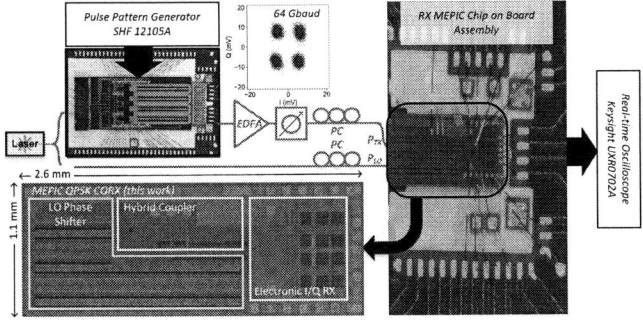

Fig. 5. Chip microphotograph of the 45-CLO self-homodyne coherent optical receiver and chip-on-board receiver assembly with link measurement setup with TX constellation as generated in [2].

Table 1. State-of-the-Art Comparison

Ref	Process	Modulation	Speed (Gb/s)	TI (dBΩ)	Energy efficiency (pJ/bit)	Chip area (mm^2)
[3]	0.25 μm SiGe-Photonic	QPSK*	128[1]	77	3.2	2.5×1.1
[9]	0.25 μm SiGe-Photonic	QPSK	112[1]		4.3	3.65×1.45
[10]	130-nm SiGe BiCMOS EIC 90-nm Silicon Photonics PIC	QPSK	100	67.2	3.32[2]	1.475×1.9 EIC
[11]	45-nm CMOS SOI EIC 90-nm Silicon Photonics PIC	QPSK	100[1]	53.6	0.93[2]	1.885×1.28 EIC
[8]	22 nm FinFET CMOS	NRZ, PAM4	80, 128[1]	59.3	0.098[3]	0.23×0.11[4]
This work	45-nm CMOS SOI-Photonic	QPSK	112	58.7	0.73, 0.41[2]	2.63×1.1

* C-band, [1] With post-processing equalization, [2] EIC only, [3] No integration with PIC, [4] EIC active area without pads

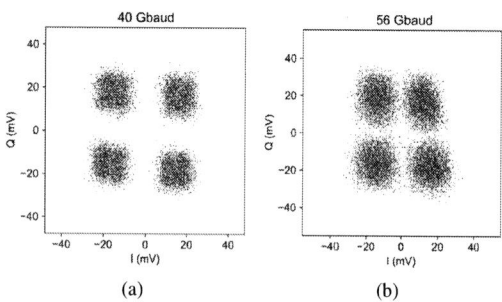

Fig. 6. QPSK constellations at (a) 40 Gbaud and (b) 56 Gbaud.

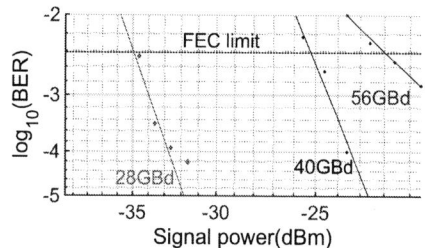

Fig. 7. BER curves at different bit rates indicating the power penalty to higher data rates as referenced to the FEC limit.

IV. CONCLUSION

This paper describes the performance of a 0.73-pJ/bit coherent optical receiver fabricated in the 45CLO technology that supports both silicon photonic components as well as high speed electronics. Measured constellations and sensitivity curves show performance up to 56 Gbaud below FEC BER limit of 3.8×10^{-3}. Future work will include PFD and ODLL verification.

ACKNOWLEDGMENT

The information, data, or work presented herein was funded in part by the Advanced Research Projects Agency-Energy (ARPA-E), U.S. Department of Energy, under Award Number DE-AR0000848. The views and opinions of authors expressed herein do not necessarily state or reflect those of the United States Government or any agency thereof. The authors would like to thank GlobalFoundries for providing silicon fabrication through the 45SPCLO university program. We would also like to thank Ted Letavic, Rod Augur, Ken Giewont, Takako Hirokawa, and Kevin Dezfulian at GlobalFoundries for technical discussions.

REFERENCES

[1] T. Hirokawa et al., "Analog Coherent Detection for Energy Efficient Intra-Data Center Links at 200 Gbps Per Wavelength," in Journal of Lightwave Technology, vol. 39, no. 2, pp. 520-531, 15 Jan.15, 2021

[2] A. Maharry et al., "First Demonstration of an O-Band Coherent Link for Intra-Data Center Applications," 2022 European Conference on Optical Communication (ECOC), 2022,

[3] C. Kress et al., "64 GBd Monolithically Integrated Coherent QPSK Single Polarization Receiver in 0.25 μ m SiGe-Photonic Technology," 2018 Optical Fiber Communications Conference and Exposition (OFC), 2018, pp. 1-3.

[4] M. Rakowski, et al. "45nm CMOS - Silicon Photonics Monolithic Technology (45CLO) for next-generation, low power and high speed optical interconnects," in Optical Fiber Communication Conference (OFC) 2020, OSA Technical Digest (Optica Publishing Group, 2020), paper T3H.3.

[5] R. Ashok, S. Naaz, R. Kamran and S. Gupta, "Analog Domain Carrier Phase Synchronization in Coherent Homodyne Data Center Interconnects," in Journal of Lightwave Technology, vol. 39, no. 19, pp. 6204-6214, Oct.1, 2021

[6] M. Lu et al., "An Integrated 40 Gbit/s Optical Costas Receiver," in Journal of Lightwave Technology, vol. 31, no. 13, pp. 2244-2253, July1, 2013

[7] S. Saeedi, S. Menezo, G. Pares and A. Emami, "A 25 Gb/s 3D-Integrated CMOS/Silicon-Photonic Receiver for Low-Power High-Sensitivity Optical Communication," in Journal of Lightwave Technology, vol. 34, no. 12, pp. 2924-2933, 15 June15, 2016

[8] S. Daneshgar, H. Li, T. Kim and G. Balamurugan, "A 128 Gb/s, 11.2 mW Single-Ended PAM4 Linear TIA With 2.7 Arms Input Noise in 22 nm FinFET CMOS," in IEEE Journal of Solid-State Circuits, vol. 57, no. 5, pp. 1397-1408, May 2022, doi: 10.1109/JSSC.2022.3147467

[9] P. M. Seiler et al., "56 GBaud O-Band Transmission using a Photonic BiCMOS Coherent Receiver," 2020 European Conference on Optical Communications (ECOC), 2020, pp. 1-4, doi: 10.1109/ECOC48923.2020.9333218.

[10] L. A. Valenzuela, Y. Xia, A. Maharry, H. Andrade, C. L. Schow and J. F. Buckwalter, "A 50-GBaud QPSK Optical Receiver With a Phase/Frequency Detector for Energy-Efficient Intra-Data Center Interconnects," in IEEE Open Journal of the Solid-State Circuits Society, vol. 2, pp. 50-60, 2022

[11] H. Andrade, Y. Xia, A. Maharry, L. Valenzuela, J. F. Buckwalter and C. L. Schow, "50 GBaud QPSK 0.98 pJ/bit Receiver in 45 nm CMOS and 90 nm Silicon Photonics," 2021 European Conference on Optical Communication (ECOC), 2021, pp. 1-4.

A 42.7Gb/s Optical Receiver with Digital CDR in 28nm CMOS

Hyungryul Kang[1], Inhyun Kim[1], Ruida Liu[1], Ankur Kumar[1], Il-Min Yi[1], Yuan Yuan[2], Zhihong Huang[2], Samuel Palermo[1]

[1]Electrical and Computer Engineering, Texas A&M University, USA

[2]Hewlett-Packard Laboratories, Hewlett-Packard Enterprise, USA

{hrkang, spalermo}@tamu.edu

Abstract — **This paper presents a broadband optical receiver that employs multiple bandwidth extension techniques in the analog front-end (AFE) and has efficient digital clock and data recovery (CDR). Total AFE bandwidth is extended by 5.5X with continuous-time linear-equalizer (CTLE) peaking, series inductances between each AFE stage, and active inductors in the CTLE output and variable gain amplifier (VGA) stages. The resolution of a digitally controlled oscillator (DCO) is optimized at 9-bit to balance quantization and random noise-induced jitter from the CDR. Fabricated in 28nm CMOS, the 42.7Gb/s optical receiver achieves an optical modulation amplitude (OMA) sensitivity of -3.6dBm at a bit error rate (BER)<10⁻¹², 10MHz CDR bandwidth, and 3.4pJ/bit energy efficiency.**

Keywords — **Bandwidth extension techniques, DCO resolution, digital CDR, optical receiver.**

I. INTRODUCTION

Electrical link systems have been historically used to support chip-to-chip communication in applications such as data centers and high-performance computing systems. However, achievable electrical line rates cannot support bandwidth demands as these systems scale in size and performance [1]. Excessive high-frequency electrical channel loss motivates the use of optical interconnects systems that take advantage of low-loss optical channels. This necessitates efficient wideband optical receivers. However, trade-offs between gain, bandwidth, and noise make the design of these optical analog front-end (AFE) receivers challenging.

Another key receiver challenge is the design of clock-and-data recovery systems with efficient power and area utilization. Digital CDR loop filter implementations, enabled through synthesis flows in nanometer CMOS technologies, provides significant area savings and the flexibility to program filter parameters to optimize loop dynamics. However, due to the digital implementation, quantization noise is added and must be considered in the digital CDR design procedure.

This paper presents a 42.7 Gb/s optical receiver that employs broadband AFE design techniques and includes a complete digital CDR system. Section II gives an overview of the optical receiver architecture, discusses key AFE design details, and details how the CDR's digitally-controlled oscillator (DCO) resolution is set to minimize clock jitter. Receiver measurement results from a 28nm CMOS prototype are presented in Section III. Finally, Section IV concludes this paper.

II. PROPOSED OPTICAL RECEIVER

Fig. 1 shows the block diagram of the proposed optical receiver, which consists of the AFE, slicers, deserializers, and CDR. The AFE block is composed of a transimpedance amplifier (TIA), CTLE, and VGA. A shunt feedback TIA, consisting of a simple inverter with a feedback resistor (R_F), converts the photodiode (PD) current to a voltage signal. This is followed by a tunable CTLE that provides bandwidth extension to enable a larger input TIA gain and reduced noise. Further bandwidth extension is achieved with series passive peaking employed at the TIA input and between the TIA, CTLE, and VGA blocks and with active inductors at the CTLE output and in each VGA stage. The AFE output is sampled by data and edge slicer banks that consist of double-tail sense amplifiers [2]. After being demultiplexed from quarter-rate to $1/16^{th}$ rate, each data pattern is buffered out and fed to a bit error rate tester (BERT). The eight quarter-rate clock phases are generated from the CDR that consists of a bang-bang phase detector (BBPD), digital loop filter (DLF), LC-DCO, and a delay-locked loop (DLL).

A. AFE

The AFE is implemented exclusively with inverter-based stages. The transimpedance of the shunt feedback TIA can be represented by

$$Z_T(s) = -R_F \frac{1}{1 + \frac{1}{A(s)} + \frac{sR_FC_T}{A(s)}}, \qquad (1)$$

where C_T denotes total capacitance seen at an input node of TIA block. By representing an inverter as an analog amplifier with a single pole system, $A(s) = A/(1 + (s/\omega_A))$, the transimpedance $Z_T(s)$ can be represented by as a second order system [3]. Assuming a Butterworth response for maximally flat frequency response yields the following TIA bandwidth.

$$\omega_{3dB} = \omega_0 = \frac{\sqrt{(A+1)2A}}{R_FC_T} \qquad (2)$$

The input-referred current noise power spectral density is also expressed in [3].

$$\overline{i_n^2}(f) = \frac{4kT}{R_F} + \frac{4kT\gamma}{g_mR_F^2} + 4kT\gamma\frac{(2\pi C_T)^2}{g_m}f^2 \qquad (3)$$

Thus, R_F has to be optimized considering these trade-offs between gain, noise, and bandwidth. Iterative post-layout simulations are employed to set the 400Ω value to support

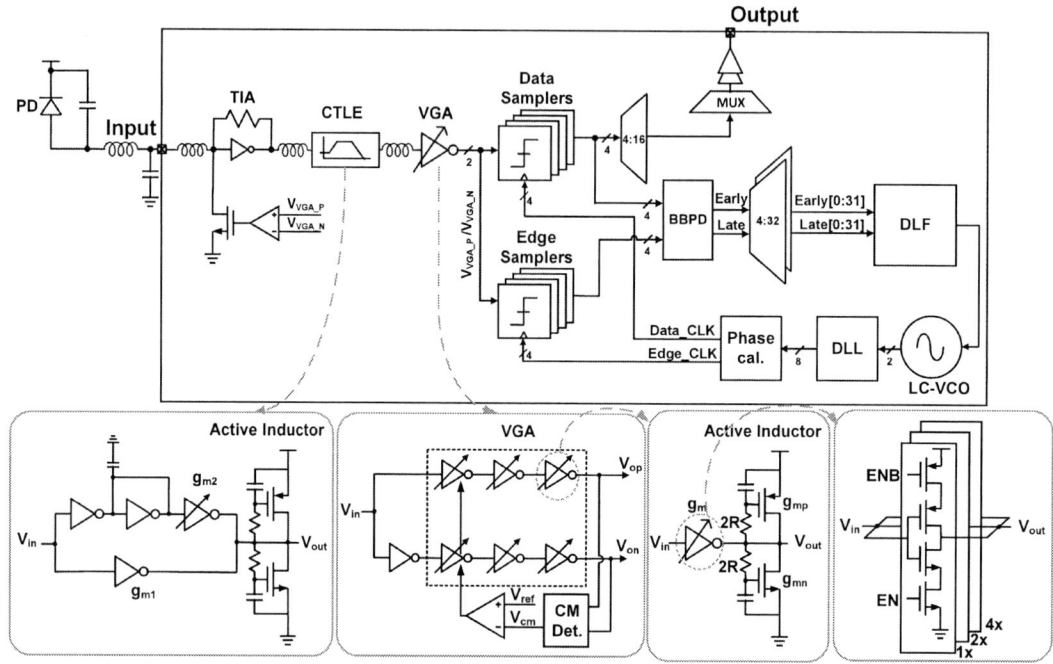

Fig. 1. Optical receiver block diagram.

data rates over 40 Gb/s with a simulated $3\mu A_{rms}$ input-referred noise current. Series interstage inductances are added between the AFE blocks to isolate the capacitance between stages and provide bandwidth extension [4], [5]. The inverter-based CTLE provides 7 dB adjustable low-frequency gain to compensate for PVT variations. Inverter-based active inductor loads also extend the bandwidth in the CTLE and VGA stages by adding an additional zero

$$Z_L = \frac{1}{G_m}\frac{1+sRC_{gs}}{1+sCgs/G_m} = \frac{1}{G_m}\frac{1+s/\omega_z}{1+s/\omega_T}, \quad (4)$$

where $G_m = g_{mp} + g_{mn}$, C_{gs} is total inverter gate capacitance, and ω_T is the transit frequency. For frequency range $\omega << \omega_T$, this impedance can be approximated as

$$Z_L \approx \frac{1}{G_m}(1+s/\omega_z) = \frac{1}{G_m} + s\frac{R}{\omega_T}. \quad (5)$$

The overall simulated frequency response in Fig. 2 shows that the CTLE provides 5dB peaking by subtracting the low-frequency signal component by utilizing a top-path low-pass filter [6]. An additional 5.5dB peaking is provided by the four-stage VGA block. Overall, these bandwidth extension techniques extend the AFE bandwidth by a factor of 5.5X.

B. CDR

A block diagram of the proposed digital PLL-based CDR is shown in Fig. 3. The digital implementation allows flexibility in the loop dynamics, as the proportional (K_p) and integral gain (K_i) factors can be adjusted digitally. In order to support high data rates with low jitter, an LC-VCO is chosen due to its lower phase noise relative to a ring-based VCO. In addition to VCO phase noise, quantization noise due to the

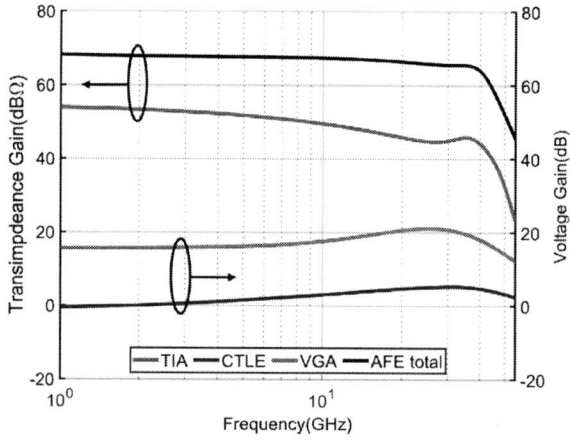

Fig. 2. Simulated AFE frequency response.

digital loop filter implementation affects jitter. As mentioned in [7], total jitter is the sum of random-noise-induced jitter and quantization-induced jitter.

$$J_{tot} = J_{rn} + J_q \quad (6)$$

There exists an optimum point where the total jitter of random noise and quantization noise is minimized. The DCO resolution is set through first performing simulations that sweep the DCO resolution and collect total jitter values including DCO noise. Identical simulations are then conducted without DCO noise to acquire the jitter related to quantization noise only. Since total jitter is the sum of random-noise-induced jitter and quantization-induced jitter [7], random noise can be extracted.

979-8-3503-2123-4/23 $31.00 © 2023 IEEE

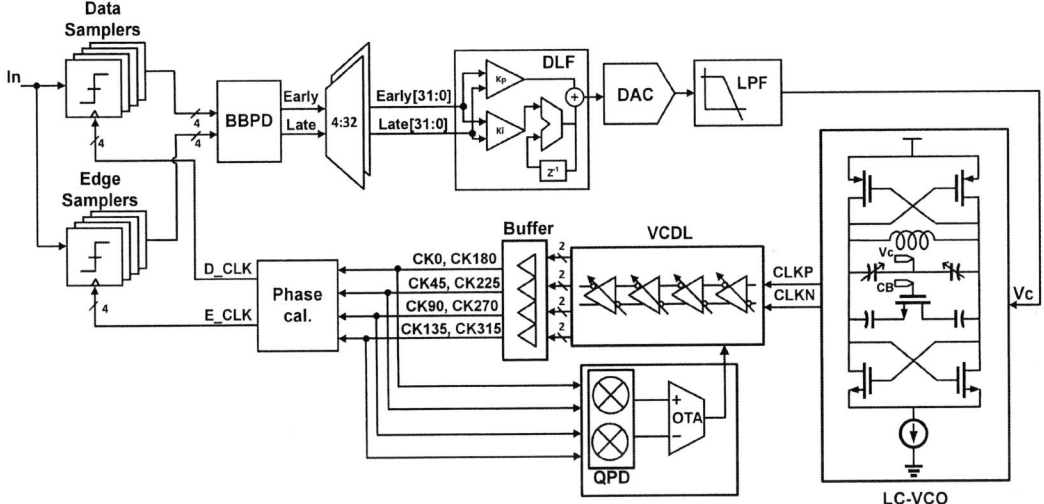

Fig. 3. Digital CDR block diagram.

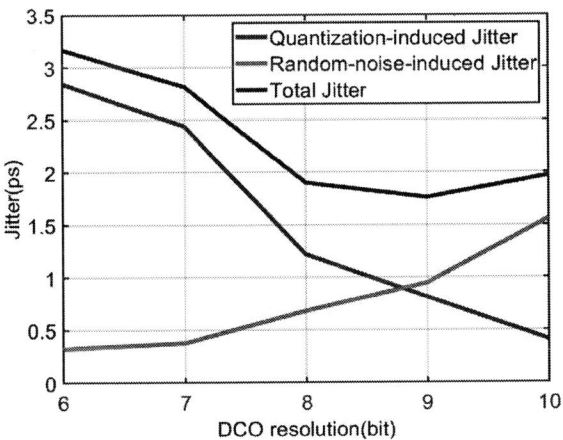

Fig. 4. Jitter as a function of DCO resolution.

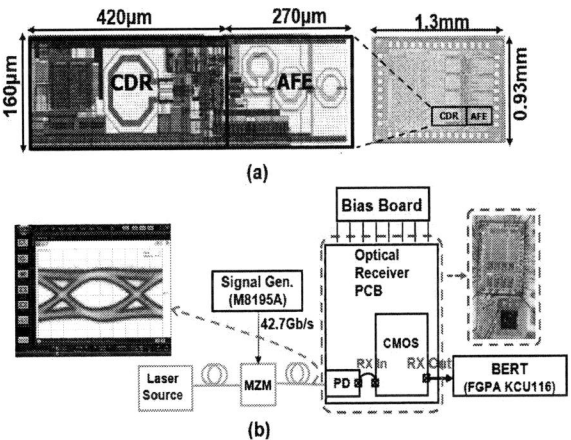

Fig. 5. (a) Chip micrograph with layout details and (b) optical test setup.

Fig. 4 provides the jitter values at a BER of 10^{-12} obtained from simulation for different DCO resolutions. As the DCO resolution increases, the CDR bandwidth decreases and leads to an increase in VCO random jitter. Simulation results show that 9-bit is an optimum resolution for minimum jitter. A DAC with this resolution is used to convert the loop filter digital output to an analog voltage that controls the LC-VCO. The VCO output feeds a DLL, consisting of a voltage-controlled-delay-line (VCDL), phase detector, and operational transconductance amplifier (OTA), that generates 8-phase clock signals. The DLL works by adjusting the VCDL delay cell load capacitance in a negative feedback manner to generate 8 equally spaced phases. The OTA bandwidth and load capacitance, which affects the overall DLL bandwidth, are also carefully designed to not affect CDR loop dynamics.

III. EXPERIMENTAL RESULTS

The multi-channel optical receiver prototype was fabricated in a 28nm CMOS process. Fig. 5 shows the chip micrograph, the layout details of a single RX channel, and its optical test setup. The RX chip was wire-bonded to a GaAs/InP PIN-PD. An M8195A generates a pseudo-random binary sequence (PRBS) pattern to drive a Mach-Zehnder Modulator (MZM) fed with a DC laser source. The output modulated signal is then coupled to the PD. By sweeping the laser power, the OMA sensitivity is acquired at a certain BER. The measured BER versus input optical power curve of Fig. 6 shows that at a 42.7 Gb/s data rate, the receiver achieves an OMA sensitivity of -3.6dBm at a BER of 10^{-12}. Jitter tolerance is also measured with the M8195A that applies sinusoidal jitter with different frequency/amplitude values, limited by the equipment to 1UI jitter amplitude. Fig. 7 provides a jitter tolerance plot measured at BER=10^{-12} with 42.7 Gb/s PRBS data. It shows that CDR has 10MHz of bandwidth with 0.1UI high-frequency jitter tolerance. The performance summary and the comparison with other optical receivers are shown in Table I. Except [8], which uses a more advanced technology, this paper provides an

979-8-3503-2123-4/23 $31.00 © 2023 IEEE

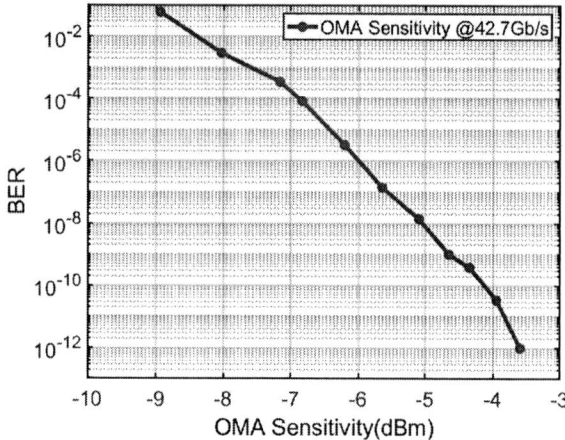

Fig. 6. Measured OMA sensitivity.

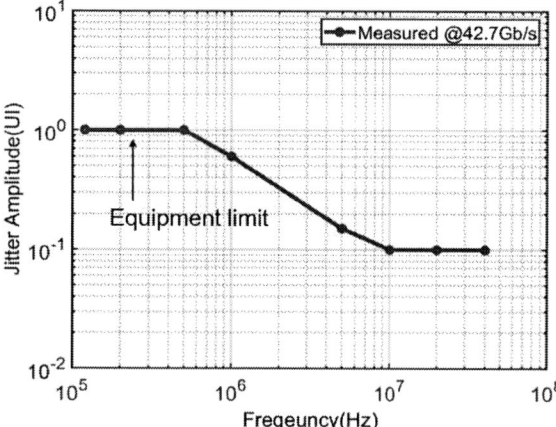

Fig. 7. Measured jitter tolerance.

CDR-based optical receiver prototype with the highest data rate. The presented AFE design with the described bandwidth extension techniques allows the prototype receiver to achieve high bandwidth with comparable gain.

IV. CONCLUSION

This paper presented a 42.7 Gb/s CDR-based optical receiver fabricated in a 28nm CMOS process. In addition to a CTLE, series peaking and active inductor loads were used to proved bandwidth extension. The AFE has a transimpedance of 67.5dBΩ over a 38.7GHz bandwidth to allow for wideband applications. The digital CDR employs an optimum 9-bit digital loop filter to minimize clock jitter. Measurements show that the digital CDR has a 10MHz bandwidth with $0.1UI_{pp}$ jitter tolerance. The receiver is wire-bonded with a PIN-PD and the measured OMA sensitivity was -3.6dBm at 3.4pJ/bit energy efficiency.

ACKNOWLEDGMENT

This work was supported by Hewlett Packard Labs.

Table 1. Comparison Table

	[8]	[9]	[10]	[11]	[12]	**This Work**
Technology	14nm CMOS	28nm CMOS	28nm CMOS	32nm SOI	40nm CMOS	28nm CMOS
Data rate (Gb/s)	56	25	40	25	25.78	42.7
PD resp. (A/W)	-	0.8	0.6	0.5	0.5	0.8
OMA Sens. (dBm) @BER 10^{-12}	-4*	-6.8	-1.9	-10.9	-10.6**	-3.6
Efficiency (pJ/b)	2.2	0.17	3.4	4.4	9.8	3.4
Gain (dBΩ)	-	-	75	-	-	67.5
BW (GHz)	-	-	27	-	-	38.7
Area (mm²)	0.06	0.0018	0.009	0.06♯	3.72	0.11

*Measured @BER 10^{-11} **44μApp converted to -10.6 ♯RX core area

REFERENCES

[1] S. Fathololoumi et al., "1.6 tbps silicon photonics integrated circuit and 800 gbps photonic engine for switch co-packaging demonstration," Journal of Lightwave Technology, vol. 39, no. 4, pp. 1155–1161, 2021.

[2] D. Schinkel, E. Mensink, E. Klumperink, E. van Tuijl, and B. Nauta, "A double-tail latch-type voltage sense amplifier with 18ps setup+hold time," in 2007 IEEE International Solid-State Circuits Conference. Digest of Technical Papers, 2007, pp. 314–605.

[3] E. Säckinger, Broadband circuits for optical fiber communication. John Wiley & Sons, 2005.

[4] J. Kim and J. F. Buckwalter, "A 40-gb/s optical transceiver front-end in 45 nm soi cmos," IEEE Journal of Solid-State Circuits, vol. 47, no. 3, pp. 615–626, 2012.

[5] S. Shekhar, J. Walling, and D. Allstot, "Bandwidth extension techniques for cmos amplifiers," IEEE Journal of Solid-State Circuits, vol. 41, no. 11, pp. 2424–2439, 2006.

[6] K. Zheng, Y. Frans, S. L. Ambatipudi, S. Asuncion, H. T. Reddy, K. Chang, and B. Murmann, "An inverter-based analog front-end for a 56-gb/s pam-4 wireline transceiver in 16-nm cmos," IEEE Solid-State Circuits Letters, vol. 1, no. 12, pp. 249–252, 2018.

[7] G. Marucci, S. Levantino, P. Maffezzoni, and C. Samori, "Analysis and design of low-jitter digital bang-bang phase-locked loops," IEEE Transactions on Circuits and Systems I: Regular Papers, vol. 61, no. 1, pp. 26–36, 2014.

[8] I. Ozkaya et al., "A 56gb/s burst-mode nrz optical receiver with 6.8ns power-on and cdr-lock time for adaptive optical links in 14nm finfet cmos," in 2018 IEEE International Solid - State Circuits Conference - (ISSCC), 2018, pp. 266–268.

[9] S. Saeedi and A. Emami, "A 25gb/s 170μw/gb/s optical receiver in 28nm cmos for chip-to-chip optical communication," in 2014 IEEE Radio Frequency Integrated Circuits Symposium, 2014, pp. 283–286.

[10] L. Szilagyi, M. Khafaji, J. Pliva, R. Henker, and F. Ellinger, "40-gbit/s 850-nm vcsel-based full-cmos optical link with power-data rate adaptivity," IEEE Photonics Technology Letters, vol. 30, no. 7, pp. 611–613, 2018.

[11] A. Rylyakov et al., "A 25 gb/s burst-mode receiver for low latency photonic switch networks," IEEE Journal of Solid-State Circuits, vol. 50, no. 12, pp. 3120–3132, 2015.

[12] J. Wang et al., "A fully integrated 25 gb/s low-noise tia+cdr optical receiver designed in 40-nm-cmos," IEEE Transactions on Circuits and Systems II: Express Briefs, vol. 66, no. 10, pp. 1698–1702, 2019.

A 100-Gb/s 3-m Dual-Band PAM-4 Dielectric Waveguide Link with 1.9 pJ/bit/m Efficiency in 28-nm CMOS

Kristof Dens[#], Joren Vaes[#], Christian Bluemm[$], Gabriel Guimaraes[#], Berke Gungor[#], Changsong Xie[$], Alexander Dyck[$], Patrick Reynaert[#]

[#]ESAT-MICAS, KU Leuven, Belgium
[$]Huawei Technologies Dusseldorf GmbH, Germany

{kristof.dens; joren.vaes; patrick.reynaert}@kuleuven.be

Abstract — This work presents a plastic fiber link in 28-nm CMOS, operating in two adjacent bands centered on 117.5 and 152.5 GHz. Each band supports multi-level (PAM-4) intensity-modulated signaling, which can be detected non-coherently, obviating the need for carrier synchronization. Data rates up to 100 Gb/s are reported for fiber lengths up to 3 m and links up to 11 m are demonstrated at a reduced data rate. A rectification-based detector is proposed to support linear non-coherent demodulation.

Keywords — CMOS, millimeter wave communication, dielectric waveguide, dual band, diplexer, transceiver, pulse amplitude modulation (PAM).

I. INTRODUCTION

As the demand for high-speed communication continues to evolve, interconnect technologies must navigate a complex trade-off between achievable data rate, distance, power consumption and cost. In recent years, there has been an increased interest in the use of dielectric waveguide interconnects to cover an application space where conventional electrical and optical links struggle to deliver [1]–[6]. This technology boasts low-cost plastic waveguides and power-efficient mm-Wave transceivers, which are key qualities for target applications such as data center communication. In addition, the galvanic isolation of the fiber and its robustness in the presence of dirt, mechanical vibrations and large temperature differences make the technology attractive for automotive and harsh industrial environments.

Recent works increasingly utilize coherent modulation schemes, allowing for an increased spectral efficiency and efficient equalization. However, this comes at the cost of higher system complexity, power consumption and chip area. For lab-type proof of concept, a global oscillator signal is often shared between transmitter (Tx) and receiver (Rx) with manual carrier phase tuning [3]–[5]. This is not feasible for real-world systems, where Tx and Rx are spatially separated and operate stand-alone.

In this work, we present a 28-nm CMOS dual-band plastic waveguide link, operating from 100 to 170 GHz. An amplitude-shift keying (ASK) modulation format obviates the need for carrier recovery at the receiver. Broadband linear modulator and detector circuits are introduced, supporting both NRZ and PAM-4 signaling. The link achieves a data rate of 100 Gb/s at fiber lengths up to 3 m and

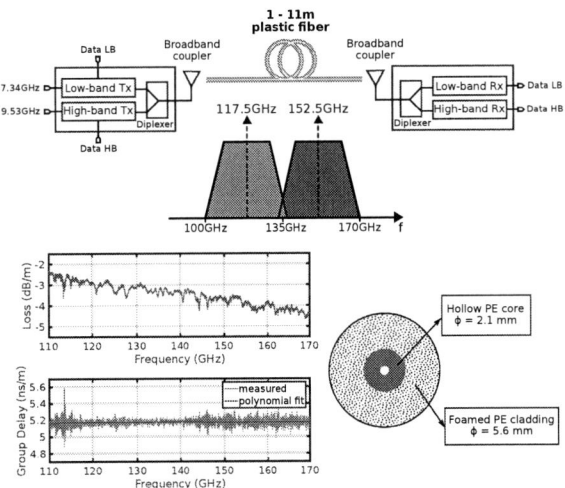

Fig. 1. System overview (top) and measured plastic fiber characteristics (bottom) of the presented link

covers distances up to 11 m at reduced data rates. The state-of-the-art performance, coupled with fully stand-alone operation, advances the technology towards market adoption.

II. DUAL-BAND PAM-4 COMMUNICATION LINK

Fig. 1 shows a system-level overview of the proposed link. A foam-cladded plastic fiber serves as the dielectric channel. The fiber consists of a hollow 2.1-mm diameter high-density polyethylene (HDPE) core, surrounded by a foamed PE cladding to avoid unwanted electromagnetic interaction. The dimensions of the core and cladding are optimized to yield a low dispersion and to minimize loss in the 100-170 GHz range while still being flexible. Covering the full 70-GHz bandwidth with a single channel leads to stringent requirements on the baseband interfaces. Furthermore, the passband design becomes inefficient due to the large fractional bandwidth required while operating close to $f_{max}/2$ of the technology. To alleviate these challenges, the transmitter and receiver utilize two adjacent frequency bands, with the low band (LB) and high band (HB) spanning from 100 to 135 GHz and from 135 to 170 GHz, respectively. This multi-band operation also decreases the dispersion in each band, thereby diminishing its detrimental effects.

979-8-3503-2123-4/23 $31.00 © 2023 IEEE

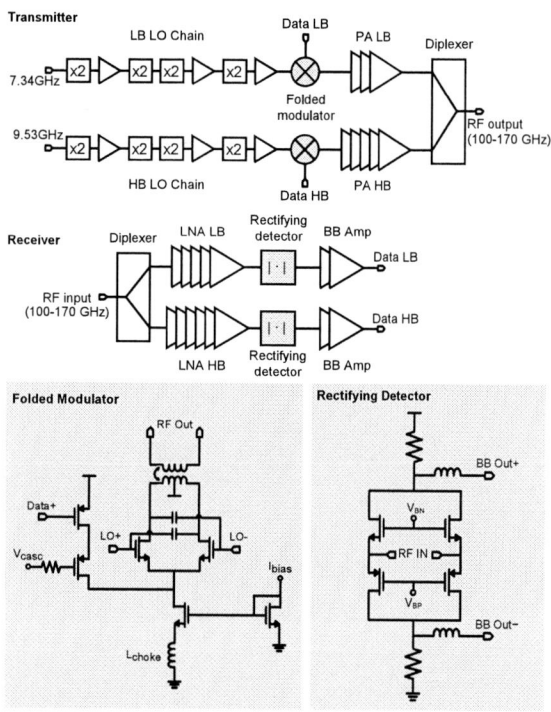

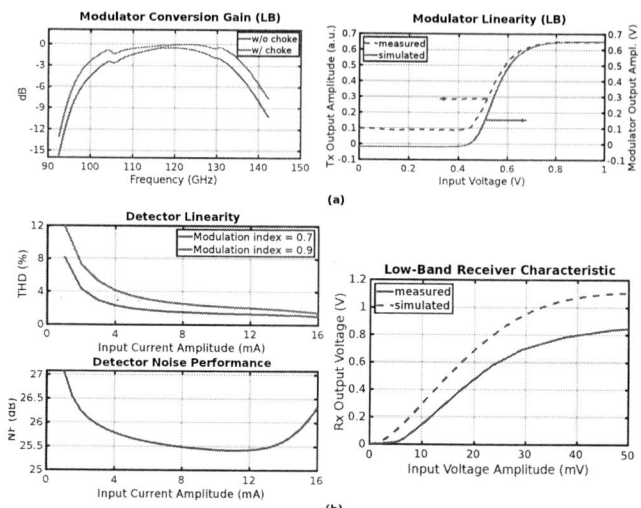

Fig. 3. (a) Simulation results of the low-band modulator (b) Simulation results of the low-band band detector and Rx characteristics

Fig. 2. Block diagrams of Tx and Rx (top) Schematic of the folded modulator and the rectifying detector (bottom)

A. Transmitter

The block diagram of the transmitter is shown in Fig. 2. Each sub-band transmitter has an LO multiplier chain, a linear amplitude modulator for upconversion, and a power amplifier (PA) to boost the output power to around 0 dBm. The low-band PA contains three gain stages, while the high-band PA requires five, due to both the reduction in available transistor gain closer to f_{max} and a lower conversion gain in the high-band modulator. Staggered matching networks are utilized to achieve a flat response over the targeted bandwidth, despite a strong roll-off of the transistor gain. Both LO chains use four doubler stages for a multiplication factor of 16, bringing the required input frequencies down to approximately 7.34 GHz and 9.53 GHz. The external references are kept below 10 GHz, to support a broad range of commercial off-the-shelf (COTS) phase-locked loop sources. Each doubler consists of a push-push stage followed by a transformer-based balun. Neutralized common-source buffers are inserted where necessary to increase the drive strength and to improve harmonic rejection. The amplitude modulator is based on a folded-cascode mixer topology. The PMOS cascode device isolates the input device from the voltage swing at the source of the neutralized switching pair, improving the linearity of the modulator. In order to provide a differential input, the modulator is duplicated (not shown) for the negative data input and connected to a dummy load. A choke at the source of the bias transistor improves the conversion-gain bandwidth of the modulator by increasing the impedance of the current source at

higher frequencies, aiding in the folding of the signal current. Simulation results of the low-band modulator are presented in Fig. 3 (a).

B. Receiver

Fig. 2 shows the architecture of the Rx. A diplexer separates the incoming signal into two sub-bands, with each band containing a low-noise amplifier (LNA), a linear detector and a baseband amplifier. The diplexer layout is shown in Fig. 4. Outside of the passband the LNA matching networks present a short circuit at their input, making a series connection of their transformer primaries a natural choice for the diplexer topology. The series connection is made through two $\lambda/2$ transmission lines, allowing the bands to be spaced further apart, thereby easing layout constraints and further reducing parasitic coupling between them. A shunt inductor connected between the differential input pads resonates out the pad capacitance, significantly broadening the input match. The transmitter uses a similar diplexer structure to combine both bands. The additional filtering provided by the multi-stage LNAs is sufficient to allow the diplexer to be fully integrated on-chip, avoiding multi-layer packaging as proposed previously [5].

The low-band LNA contains five neutralized common-source stages, whereas the transistor gain roll-off necessitates a six-stage LNA in the high band. Traditional mm-Wave envelope detectors have a square-law characteristic, which is not sufficiently linear to support PAM-4 signaling. The detector topology proposed in this work contains a tuned implementation of a CMOS bridge rectifier. The circuit rectifies the mm-Wave input current, resulting in an approximately linear relationship between the output voltage and the amplitude of the received mm-Wave signal. Furthermore, both linear and non-linear components of the transistor response contribute to the rectification operation, thereby improving the conversion gain and decreasing the

979-8-3503-2123-4/23 $31.00 © 2023 IEEE   14

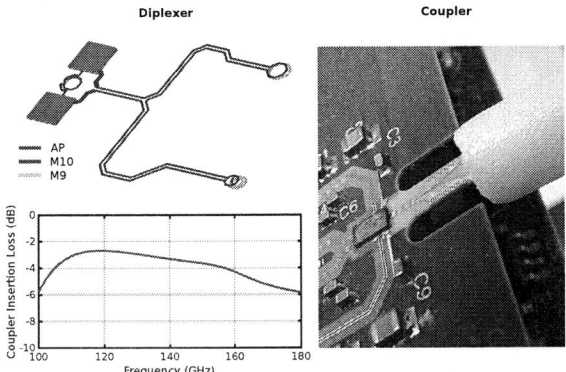

Fig. 4. Diplexer layout and the fabricated coupler with simulated insertion loss

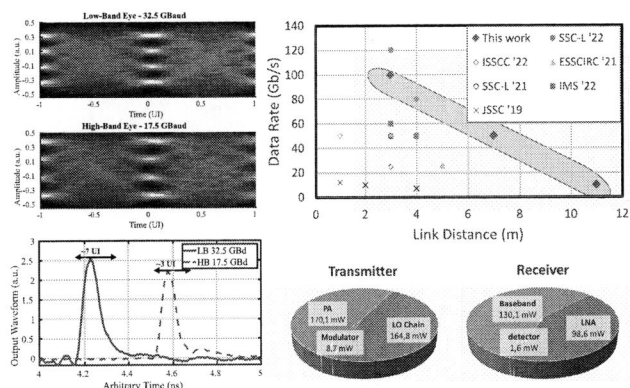

Fig. 5. Measurement results, including power breakdown and measured pulse responses of a full 3-m link

noise figure. Series peaking inductors are used to improve the bandwidth of the detector. The detector is AC-coupled to the two-stage baseband amplifier, which consists of resistively-loaded differential pairs employing bridged T-coil peaking to minimize bandwidth degradation in the presence of pads and ESD protection.

C. Dielectric Waveguide Interface

A Vivaldi-style coupler is used as an interface to the waveguide, since the non-resonant topology is well suited for covering the wide 100-170 GHz frequency range. Differential 100-ohm mm-Wave interfaces obviate the need for a balun on either IC or PCB. The coupler is implemented on a single metal layer, thus avoiding vias in the mm-Wave signal path. Slots in the side of the coupler reduce standing waves orthogonal to the coupling direction, increasing the bandwidth. The coupler is fabricated on an Isola Astra MT77 substrate and the ICs are connected to the PCB using a flip-chip process. Fig. 4 shows the fabricated coupler, together with its simulated performance.

III. MEASUREMENT RESULTS

Fig. 5 presents the measured eye diagrams, pulse response of the 3m link as well as a comparsion of the measured The measurement setup is shown in Fig. 6 (c). The data inputs are generated by either a Keysight M8194A arbitrary waveform generator or an Anritsu M1900a with MU196020A PAM-4 generator. At the receiver-side, a Keysight four-channel, 70-GHz UXR oscilloscope captures the output signals, which are then equalized offline using MATLAB.

The link is characterized in three different application scenarios. In a first use case, both bands operate simultaneously in an unbalanced fashion, using 32.5 GBd PAM-4 in the low band and 17.5 GBd PAM-4 in the high band. In this way, a data rate of 100 Gb/s is measured over a 3 meter link. Each band is pulse-shaped with a root-raised-cosine filter at both Tx and Rx, together forming a raised cosine filter. Furthermore, a feed-forward equalizer (FFE) is applied at the Rx with a combination of 21 linear taps and 7 non-linear 2nd order Volterra taps. The

taps are spaced at the baudrate to avoid the excessive power consumption associated with oversampled equalizers. The bit-error rate (BER) is better than 2.2e-4, corresponding to threshold of the "KP4" Reed Solomon (544, 514) forward error correction (FEC), which is standardized for a variety of 100 Gb/s IEEE 802.3 Ethernet scenarios. The superior performance in the low band is due to the significantly lower loss of the fiber in addition to better performance of the CMOS technology in this frequency range. Furthermore, using the same equalization but disabling the high band to conserve power, the low band can operate at 25 GBd PAM-4 for a data rate of 50 Gb/s over a distance of 7 m, with a degraded BER of 2.2e-3. Finally, the low band can operate at a BER below 1e-6 at distances up to 11 m with a data rate of 10 Gb/s, using NRZ modulation. Here, equalization resembles a low-cost and low-power COTS clock and data recovery chip with a combination of continuous-time linear equalizer (CTLE) and a 3-tap linear FFE. When disabling the equalization, BERs of 1.7e-3 are still achieved for this case. The dual-band link consumes 574 mW, which includes the LO chains on the Tx as well as the 50-ohm buffers on the Rx.

Table 1 compares this work to the state-of-the-art. The designed non-coherent dual-band link achieves a data rate of 100 Gb/s over a single fiber. The improved receiver sensitivity enables a 50 Gb/s data rate over links as long as 7 m, while consuming only 0.8 pJ/bit/m. The presented link can also operate up to a distance of 11m with a data rate of 10 Gb/s. Fig. 6 shows die photographs of the manufactured chips. The sizes of Tx and Rx are 2.64 mm by 1.36 mm and 1.99 mm by 1.34 mm respectively.

IV. CONCLUSION

This paper has presented the design and measurement results of a dual-band dielectric waveguide link. The presented non-coherent dual-band implementation is the first stand-alone link to achieve 100 Gb/s over a single fiber and does not require a shared LO between Tx and Rx. A link distance of 11m with a data rate of 10 Gb/s is also demonstrated, which

979-8-3503-2123-4/23 $31.00 © 2023 IEEE

Table 1. Performance summary and comparison to prior works

	JSSC 2019 [6]		SSC-L 2021 [3]		ESSCIRC 2021 [1]		ISSCC 2022 [2]		IMS 2022 [4]		SSC-L 2022 [5]		This Work		
Technology	28nm CMOS		22nm FinFET		28nm CMOS		28nm CMOS		16nm FinFET		16nm FinFET		28nm CMOS		
Frequency (GHz)	140		134		115		70		108		109 + 135		117.5 + 152.5		
Modulation	FSK		16QAM		ASK (PAM4)		ASK (SSB)		16QAM		16QAM		ASK (PAM4)		
Data Rate (Gbps)	12	7	40	56	50	24	50	25	60	50	120	80	100	50	10
Link Length (m)	1	4	1.2	3	3	5	1	3	3	4	3	4	3	7	11
BER	1e-12		1.5e-7	7.5e-5	1e-6		1e-9	1e-12	5.8e-4	3e-4	1e-3	1e-4	2.2e-4 (KP4 FEC)	2.2e-3	1e-6
P$_{DC}$ (mW)	230		494*†		130*†		212*		636*†		1116*†		574*	271*‡	
Energy Efficiency (pJ/bit/m)	19.2	8.2	10.3	2.9	0.9	1.1	4.2	2.8	3.5	3.2	3.1	3.5	1.9	0.8	2.5
Shared LO between TX & RX?	No		Yes		No		No		Yes		Yes		No		
Equalization	No		20-tap fractionally-spaced matrix FFE		141-tap Volterra equalizer		3-tap Tx-side§ FFE		20-tap fractionally-spaced matrix FFE		20-tap fractionally-spaced matrix FFE		28-tap Volterra equalizer	CTLE + 3-tap FFE	

*Not including DSP power †Not including 50-ohm output buffers ‡Only low band active §Not mentioned in proceedings

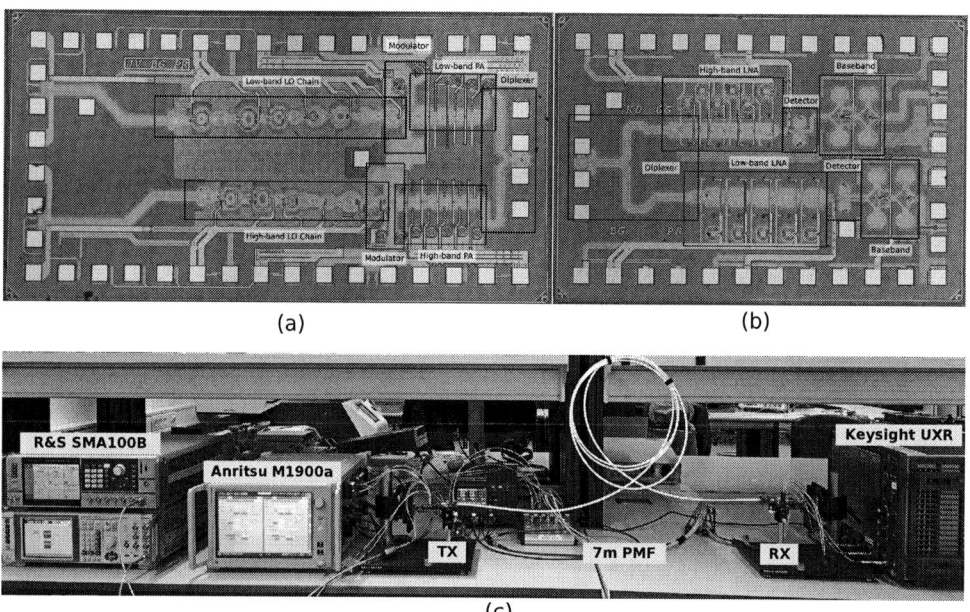

Fig. 6. Chip micrograph of (a) Transmitter, (b) Receiver (c) The measurement setup

is the highest reported data rate for this link distance. The results demonstrate the potential of dielectric waveguides for high-speed data transmission over short and medium distances.

ACKNOWLEDGMENT

The authors would like thank KU Leuven for the support of this research (C2 Project).

REFERENCES

[1] K. Dens, J. Vaes, S. Ooms, M. Wagner, and P. Reynaert, "A pam4 dielectric waveguide link in 28 nm cmos," in *ESSCIRC 2021 - IEEE 47th European Solid State Circuits Conference (ESSCIRC)*, 2021, pp. 479–482.

[2] H.-I. Song, H. Choi, J. y. Yoo, H.-S. Won, C. M. Lee, H. Jin, T. y. Kim, W. Kwon, K. Lim, K. Kwon, C.-A. Kim, T. Kim, J. G. Jo, J. Eu, S. Park, and H.-M. Bae, "A 50gb/s pam-4 bi-directional plastic waveguide link with carrier synchronization using pi-based costas loop," in *2022 IEEE International Solid- State Circuits Conference (ISSCC)*, vol. 65, 2022, pp. 1–3.

[3] T. W. Brown, G. C. Dogiamis, Y.-S. Yeh, D. Correas-Serrano, T. S. Rane, S. Ravikumar, J. C. Chou, V. B. Neeli, J. Koo, M. Marulanda, N. P. Gaunkar, I. Huang, H. Chandrakumar, J. W. Bates, Z. Tuli, Q. Yu, M. Weiss, J. Rangaswamy, C. F. Nieva, D. Frolov, T. Kamgaing, Y. S. Nam, H. Braunisch, and S. Rami, "A 50-gb/s 134-ghz 16-qam 3-m dielectric waveguide transceiver system implemented in 22-nm finfet cmos," *IEEE Solid-State Circuits Letters*, vol. 4, pp. 206–209, 2021.

[4] G. C. Dogiamis, T. W. Brown, N. P. Gaunkar, Y. S. Nam, T. S. Rane, S. Ravikumar, V. B. Neeli, J. C. Chou, and S. Rami, "A 60-gbps 108-ghz 16-qam dielectric waveguide interconnect with package integrated filters," in *2022 IEEE/MTT-S International Microwave Symposium - IMS 2022*, 2022, pp. 556–559.

[5] G. C. Dogiamis, T. W. Brown, N. Prabhu Gaunkar, Y. S. Nam, T. S. Rane, S. Ravikumar, V. B. Neeli, J. C. Chou, S. Rami, and J. Swan, "A 120-gb/s 100–145-ghz 16-qam dual-band dielectric waveguide interconnect with package integrated diplexers in intel 16," *IEEE Solid-State Circuits Letters*, vol. 5, pp. 178–181, 2022.

[6] M. De Wit, Y. Zhang, and P. Reynaert, "Analysis and design of a foam-cladded pmf link with phase tuning in 28-nm cmos," *IEEE Journal of Solid-State Circuits*, vol. 54, no. 7, pp. 1960–1969, 2019.

979-8-3503-2123-4/23 $31.00 © 2023 IEEE

RMo1A-4

A 12-bit 1.1GS/s Single-Channel Pipelined-SAR ADC with Adaptive Inter-stage Redundancy

Xianshan Wen, Tao Fu, Ping Gui
Southern Methodist University, USA
xianshanw@smu.edu, pgui@smu.edu

Abstract—This paper presents a 12-bit single-channel Pipelined-SAR ADC capable of operating at 1.1GS/s. An adaptive inter-stage redundancy scheme is proposed to mitigate the speed overhead caused by inter-stage redundancy bit. A new switching scheme is proposed in the first stage that largely reduces the switching power of the capacitive DAC. Implemented in a 28nm CMOS process, it achieves an SNDR of 60.1dB with power consumption of 8.5mW, corresponding to a Walden FOM of 9.3fJ/conv.-step and a Schreier FOM of 168.2dB.

Keywords—ADC, RF sampling, single-channel, SAR, pipeline ADC, inter-stage redundancy.

I. INTRODUCTION

As the CMOS process advances, processing RF signals in digital domain becomes advantageous in terms of both the speed and energy efficiency. Hence, medium-resolution ($\geq$ 10bit) RF sampling ADCs with Gigahertz sampling rate are desired in wireless receivers to digitize the input signals. The Pipeline-SAR architecture makes an attractive candidate because it is capable of achieving the speed and resolution requirements for such applications with high power efficiency. Three-stage Pipelined-SAR ADCs have been shown capable of operating at 1GS/s with 12-bit resolution [1][2]. This paper presents a 1.1GS/s 12-bit single-channel Pipelined-SAR ADC with SNDR of 61.3dB and 60.1dB at low input frequency and Nyquist input, respectively. It consumes 8.5mW when operating at 1.1GS/s, which achieves a Walden FOM (FOM$_{Walden}$) of 9.3fJ/conv.-step and a Schreier FOM (FOM$_{Schreier}$) of 168.2dB.

II. CIRCUIT

A. Architecture of the Proposed ADC

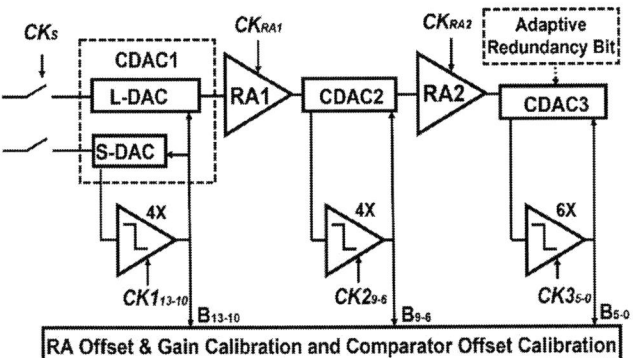

Fig. 1. Block diagram of the proposed ADC.

As shown in Fig. 1, the proposed ADC has a three-stage structure with two residue amplifiers (RA1 and RA2) in between to provide 8x voltage amplification. The three stages resolve 4b, 4b and 6b respectively, with 1-bit inter-stage redundancy in both the 2nd and 3rd stage. The RAs are implemented with open-loop Gm-R structure with Harmonic-Injecting Cross-Coupled Pair (HXCP) [1] to improve the gain and linearity.

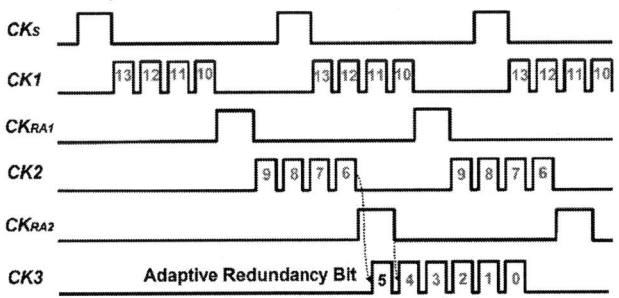

Fig. 2. Timing diagram of the proposed ADC.

The timing diagram of the proposed ADC is shown in Fig. 2. The operation of the comparators and RAs are timed asynchronously to enhance the speed and reduce power consumption associated with high speed clocks. The conversion of the 3rd stage's MSB (Bit 5) is implemented as the adaptive inter-stage redundancy bit, which is explained in section C.

B. The Proposed CDAC Switching Scheme

As shown in the timing diagram of Fig. 2, the 1st and 2nd stages need to complete the signal sampling, 4-bit conversions and the residue amplification within one period, while the 3rd stage needs to complete the sampling and 6-bit conversions within the same time. The overall speed of the Pipelined-SAR ADC is limited by the slowest stage. Normally, the 1st stage is the speed bottleneck because of the longer settling time caused by the large CDAC size (~700fF) imposed by the noise requirement. This issue can be resolved by the Large-DAC (L-DAC)/Small-DAC (S-DAC) approach in [2], with the S-DAC performing the bit conversions and transferring the bit decisions to the L-DAC to generate a precise residue voltage for RA1. Hence, the speed of the bit conversions is only limited by the settling time of S-DAC. However, the switching of S-DAC during bit conversions has power overhead, which increases the overall switching power of the 1st stage.

To alleviate the power overhead issue and further reduce the overall switching power, we proposed a DAC switching

979-8-3503-2123-4/23 $31.00 © 2023 IEEE   17   2023 IEEE Radio Frequency Integrated Circuits Symposium

scheme that eliminates the unnecessary switching in the L-DAC based on the S-DAC's bit decisions, as shown in Fig. 3. When the S-DAC generates different bit decisions for its 1st-MSB (B_{13}) and 2nd-MSB (B_{12}) i.e. the input is within half of the full-scale range (i.e. from -Vref1/2 to Vref1/2, where Vref1 is the reference voltage of the 1st stage), the L-DAC skips the switching of its 1st-MSB (B'_{13}) and only switches its 2nd-MSB (B'_{12}) based on B_{13} to generate the correct residue voltage. When B_{13} and B_{12} have the same value, these two bits are directly transferred to B'_{13} and B'_{12}. Overall, the proposed switching scheme saves about 25% of the switching power in the L-DAC compared to the conventional switching scheme, as shown in Fig. 4.

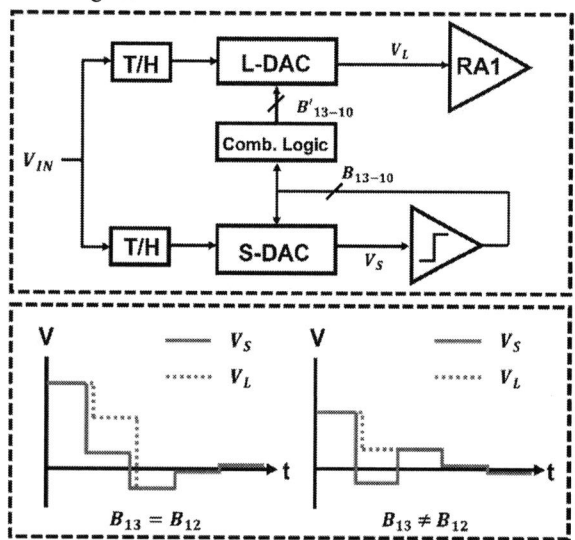

Fig. 3. Propose switching scheme of the CDAC1.

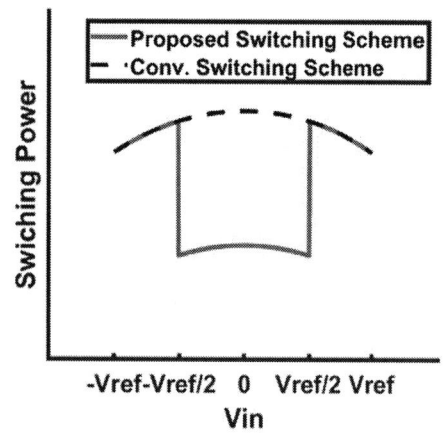

Fig. 4. Switching power on L-DAC.

Similar topology can be extended to the 3rd MSB and the 4th MSB, but with limited power improvement and reduced settling time for the 4-bit L-DAC in our 12-bit ADC. The proposed switching scheme is more beneficial if applied to SAR-assisted ADCs with higher resolution where the switching power plays a larger role in the overall power consumption because of the larger CDAC.

C. Adaptive Inter-Stage Redundancy

With the L-DAC/S-DAC approach, the speed of the 1st stage is comparable to that of the 2nd stage. The 3rd stage does not seem to limit the overall speed because it does not need an RA amplification time after the bit conversions. However, the speed of the comparators has a negative correlation to the magnitude of its input as shown in Fig. 5. The averaged SAR-conversion time of each bit in the 3rd stage is longer than that of the 1st or the 2nd stage as the averaged input of the comparators in the 3rd stage is smaller due to the 6-bit conversions including one inter-stage redundancy bit.

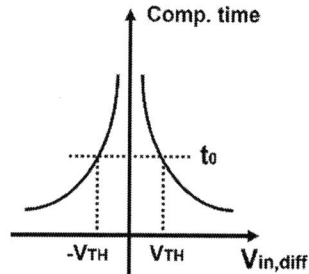

Fig. 5. Comparator resolving time versus input voltage.

To reduce the 3rd stage conversion time and mitigate the performance degradation caused by potential metastability, we proposed an adaptive inter-stage redundancy scheme which eliminates the speed overhead of the conventional inter-stage redundancy as shown in Fig. 6.

Ideally, the input range of the 3rd stage is 8·LSB2, where LSB2 is the LSB of the 2nd stage, which is the same as half of the 3rd stage full-scale range and thus the 1-bit redundancy (B_5) is not necessary. With nonidealities such as the DAC settling error and the gain/offset error in the 2nd stage and RA2, the input of 3rd stage occasionally exceeds $8 \cdot LSB2$, hence the redundancy bit is needed to ensure the output of RA2 does not saturate the 3rd stage.

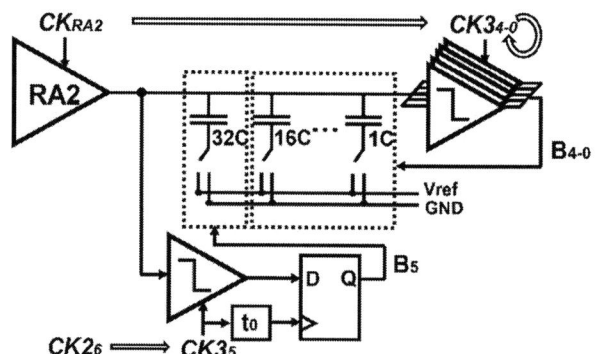

Fig. 6. Simplified block diagram of the proposed adaptive inter-stage redundancy bit.

In our proposed adaptive inter-stage redundancy as shown in Fig. 6, the comparator clock of the redundancy bit ($CK3_5$) is asynchronously generated by the last bit of the 2nd stage ($CK2_6$), such that the comparator is enabled during the RA2's settling

979-8-3503-2123-4/23 $31.00 © 2023 IEEE

phase. When the 3rd stage input is larger than a certain threshold voltage V_{TH}, the comparison result can be resolved within the time frame t_0, the redundancy bit decision is made, and B_5 is switched right after the RA2 amplification. When the 3rd stage input is smaller than V_{TH} such that the bit decision cannot be made during t_0, the next bit (B_4) starts right after the RA2 amplification and the redundancy bit is not used. The operation example is shown in Fig. 7. In either case, the time allocated for the redundancy bit overlaps with the RA2 amplification. The time required for the 3rd stage is therefore reduced by one bit of the comparison time and the associated SAR logic delay. In addition, the switching power of CDAC3 is largely reduced because its MSB is not switched for a large number of samples. Moreover, since the redundancy bit is only resolved when the input is larger than V_{TH}, the settling error $e_{settling}$ of RA2 during the redundancy bit's comparison does not affect the comparison results as long as

$$\left| e_{settling} \right| < V_{TH}. \tag{1}$$

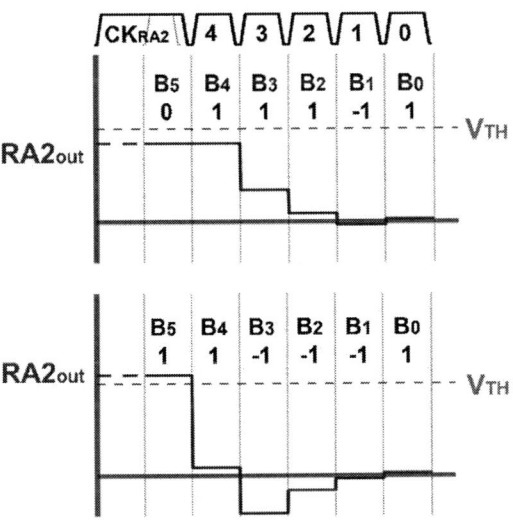

Fig. 7. Operational example of the proposed adaptive inter-stage redundancy bit scheme.

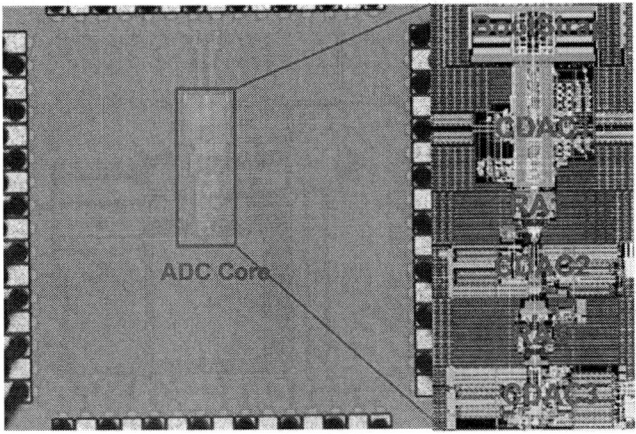

Fig. 8. Chip photo.

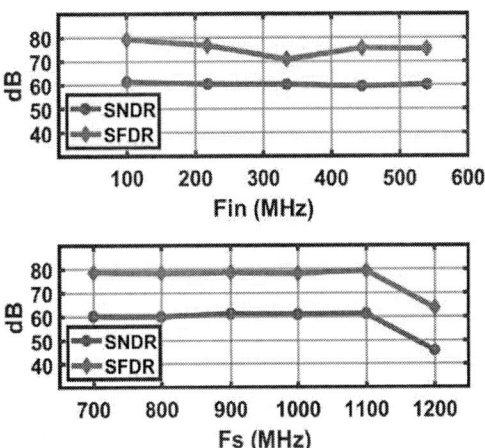

Fig. 9. Measured SNDR/SFDR with input frequency and sampling frequency sweep.

III. MEASUREMENT RESULTS

The presented ADC is fabricated in 28nm CMOS process. The chip photo is shown in Fig. 8. The measured SNDR/SFDR versus input frequency (Fin) at 1.1GS/s (top) and the SNDR/SFDR versus sampling frequency (Fs) with input frequency of 100MHz (bottom) are shown in Fig. 9. The SNDR and SFDR are maintained above 59dB and 70dB respectively when sweeping the input frequency at 1.1GS/s. When sweeping the sampling frequency with 100MHz input frequency, the SNDR is maintained above 60dB and the SFDR is maintained above 78dB up to 1.1GS/s. Fig. 10 shows the FFT spectrum measured at 1.1GS/s with low input frequency (top) and Nyquist input (bottom). It achieves 61.3dB at low input frequency and 60.1dB SNDR at Nyquist input. The measured DNL and INL is +0.41/-0.4LSB and +1.65/-1.17LSB, respectively, as shown in Fig. 11.

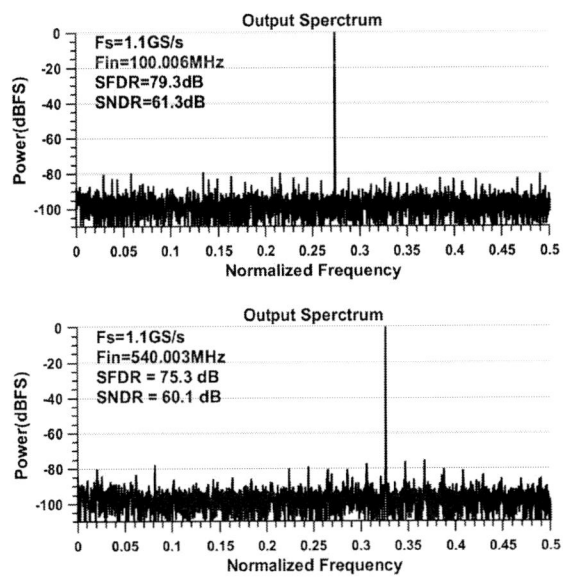

Fig. 10. Measured output spectrum (8192 FFT points, 256x decimated).

Table 1. Comparison table.

	This work	[6] SSCL'22 L. Fang	[5] VLSI'20 B.Hershberg	[4] CICC'18 J. Largos	[3] ISSCC'17 L. Kull	[2] ISSCC'19 W. Jiang	[1] CICC'21 L. Fang
Process	**28nm**	28nm	16nm	28nm	14nm	28nm	28nm
Architecture	**Pipelined SAR**	Pipelined SAR	Pipeline	Pipeline	Pipelined SAR	Pipelined SAR	Pipelined SAR
Resolution (bits)	**12**	12	11	12	10	12	12
Sample Rate (MS/s)	**1100**	1000	1000	1000	1500	1000	1000
Supply Voltage (V)	**0.9**	0.9	0.9	0.9	0.95	1	0.9
SFDR @Nyq. (dB)	**75.3**	76*	75.9	73.1	58.39	74.56	73.4
SNDR @Nyq. (dB)	**60.1**	61*	59.5	56.6	50.1	60.02	60.7
Power (mW)	**8.5**	5.8	10.9	24.8	6.92	7.6	6.8
FoM$_{Walden}$@Nyq. (fj/conv-step)	**9.3**	6.3*	14.1	45	17.7	9.28	7.76
FoM$_{Schreier}$@Nyq. (dB)	**168.2**	168.1*	166.1	159.6	160.5	168.2	169
Active Area (mm²)	**0.0096**	0.015	0.095	0.54	0.0016	0.0091	0.007

*Measured with near Nyquist input (~0.6·f$_{Nyq}$)

The ADC consumes 8.5mW at 1.1GS/s with VDD of 0.9V, corresponding to a Walden FOM (FOM$_{Walden}$) of 9.3fJ/conv.-step and a Schreier FOM (FOM$_{Schreier}$) of 168.2dB. The power consumption of the reference voltages is only 0.68mW, which demonstrates the effectiveness of the proposed switching scheme in the first stage.

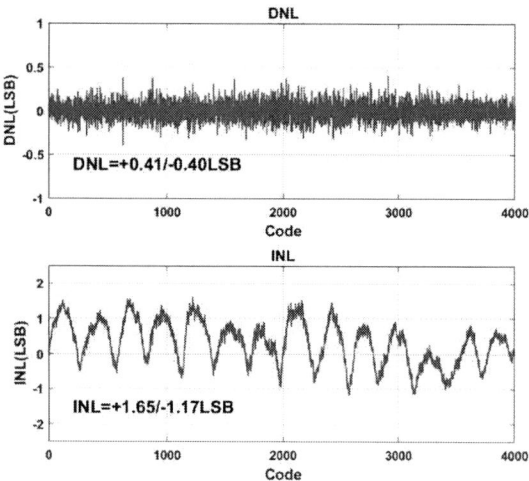

Fig. 11. Measured DNL and INL.

The comparison of this ADC to the prior similar works is summarized in Table 1. Compared to the published state-of-the-arts, this ADC improves the sampling rate of the 12-bit single-channel ADCs while achieving comparable FOM$_{Walden}$ and FOM$_{Schreier}$, validating the proposed adaptive inter-stage redundancy bit.

IV. SUMMARY

With the assistance of the new switching scheme in the 1st stage and the adaptive inter-stage redundancy bit in the 3rd stage, the presented ADC can operate at 1.1GS/s and achieve competitive SNDR, SFDR as well as the Figure of Merits at Nyquist input. The new switching scheme provides a solution to suppress the switching power in a large CDAC. The proposed adaptive inter-stage redundancy mitigates the time overhead of the conventional inter-stage redundancy bits without affecting the precision of the ADC.

REFERENCES

[1] L. Fang *et al.*, "A 1GS/s 82dB Peak-SFDR 12b Single-Channel Pipe-SAR ADC with Harmonic-Injecting Cross-Coupled-Pair and Fast N-replica Bootstrap Switch Achieving 7.5fj/convstep," *IEEE Custom Integrated Circuits Conf. (CICC)*, Apr. 2021.

[2] W. Jiang *et al.*, "A 7.6mW 1GS/s 60dB SNDR single-channel SAR-assisted pipelined ADC with temperature-compensated dynamic GM-R-based amplifier," *IEEE Int. Solid-State Circuits Conf. (ISSCC)*, Feb. 2019.

[3] L. Kull *et al.*, "A 10b 1.5GS/s Pipelined-SAR ADC with Background Second-Stage Common-Mode Regulation and Offset Calibration in 14nm CMOS FinFET," *IEEE Int. Solid-State Circuits Conf. (ISSCC)*, Feb. 2017.

[4] J. Lagos *et al.*, "A 1Gsps, 12-bit, single-channel pipelined ADC with dead-zone degenerated ring amplifiers," in *IEEE Custom Integrated Circuits Conf. (CICC)*, Apr. 2018.

[5] B. Hershberg *et al.*, "A 1MS/s to 1GS/s Ringamp-Based Pipelined ADC with Fully Dynamic Reference Regulation and Stochastic Scope-on-Chip Background Monitoring in 16nm," in *Proc. IEEE Symp. VLSI Circuits (VLSIC)*, June 2020

[6] L. Fang et al., "A 12-b 1-GS/s 61-dB SNDR Pipelined-SAR ADC With Inverter-Based Residual Amplifier and Tunable Harmonic-Injecting Cross-Coupled-Pair for Distortion Cancelation Achieving 6.3 fJ/conv-step," in IEEE Solid-State Circuits Letters, vol. 5, pp. 194-197, 2022

979-8-3503-2123-4/23 $31.00 © 2023 IEEE

RMo1B-1

A 65 nm CMOS Current-Mode Receiver Frontend with Frequency-translational Noise Cancelation and 425 MHz IF Bandwidth

Benqing Guo[#1], Haishi Wang[#2], Lei Li[*], Wanting Zhou[*]

[#]Chengdu university of information technology, P. R. China

[*]University of electronic science and technology, P. R. China

[1]rficgbq@gmail.com, [2]whs@cuit.edu.cn

Abstract—An LNA-first receiver frontend including the main/auxiliary paths is proposed. A frequency-translational noise cancellation is attained at the input of a transimpedance amplifier (TIA) shared by the dual paths. The RF input matching is examined, based on an up-conversion of baseband (BB) parasitic. The TIA is customized to provide wideband coverage and high-order filtering. With current mirrors applied to the RF/BB, the presented current-mode frontend has been fabricated in a 65 nm CMOS. Measurements show that the BB bandwidth covers 425 MHz while the S_{11} <-10 dB is maintained up to ~4.7 GHz. At typical 2 GHz LO, a 40 dB conversion gain and a 2.2 dB NF are obtained within the passband. The in-band and out-of-band IIP3 linearity are measured with -13 dBm and 14.5 dBm, respectively. The presented circuit core dissipates 44 mW in the signal path and occupies a chip area of 0.52 mm^2.

Keywords—LNA-first receiver, transimpedance amplifier, noise cancellation, wideband IF, input matching.

I. INTRODUCTION

Nowadays, fifth-generation (5G) communications are advancing continuously around the world. At the subordinated sub-6 GHz band, to render high data rate transmission, a solution compatible with previous-generation communications is a carrier aggregation (CA) technique. It typically uses multiple sub-receiver chains in parallel to combine a larger RF channel bandwidth (BW). The adverse influence of the solution is an increased budget of power consumption and hardware cost in proportion to sub-receivers' number, even less to say complicated phase lock loops to support couples of LO frequency sources. In contrast, an agile way is to increase the baseband (BB) BW, for example, covering >100 MHz, and resulting in enlarged RF bandwidth, equivalently. Consequently, the aforementioned budget is effectively saved. On the other hand, removing the discrete SAW filter inevitably imposes a heavy burden on the filtering of the integrated receiver to save cost/area. Thus, the BW and blocker resilience has become a primary focus for 5G wireless applications.

The CMOS-integrated Receiver (RX) is divided into the Mixer-first (MF) and LNA-first (LF) architectures[1,2]. The MF architecture appears to be an attractive candidate for the 5G RX due to its simplicity and high out-of-band (OOB) linearity. Nevertheless, since the input impedance (Z_{in}) of the baseband transimpedance amplifier (TIA) needs to satisfy the input matching, the in-band (IB) linearity is often limited by signal swings at the TIA input. And its NF is once poor, reducing the receiver link budget. Recently, to alleviate passive mixers' loss/noise, its local oscillator absorbs considerable power to drive large mixer switches and enable low on-resistance (Rsw). Moreover, its input matching has shown an overall narrow band profile due to the mixer's reciprocity and parasitic effect, causing unwanted signal reflections. In contrast, the LF RX provides a better compromise among gain, noise, input matching, and power consumption. But additional considerations should be applied to the linearity. Especially, compared to the traditional single-path LF RX structure, additional auxiliary paths have been innovatively proposed to alleviate the noise issue [3] and even blocker interference[1]. Thereafter, an auxiliary BB path was further applied for the MF RX to attain a BB noise cancellation (NC) while an N-path filter via the BB capacitor C_N suppresses RF blocker interferences [2]. Generally, narrow IF BWs have been commonly obtained based on the traditional feedbacked OpAmps structure [1,3], which typically cannot fulfill high data rate applications. More recently, the report [4] presents an MF RX loaded by a current-mode 3rd order filter TIA with 130 MHz BW, but yielding higher NF. Meanwhile, an invertor-based Miller-compensated structure in [2] reports a 175 MHz BW with large dc power.

In the paper, we propose an LF RX frontend, attaining a frequency-translational NC at the baseband TIA input. Based on the current-mode structure, the open-loop TIA provides a wideband coverage and 4th order low pass filtering profile. With moderate power dissipation, the presented frontend features large BW, low noise, good input matching, and blocker interference filtering. The coming section details the front-end circuit, followed by measurements on a prototype chip.

II. PROPOSED CIRCUIT

Shown in Fig. 1(a) is the simplified diagram of the presented LF RX with dual paths. Based on the noise-canceling LNA with a current mirror (CM) load[5], inserting passive mixers at dual paths constructs an NC after frequency translation. Specifically, the main path consists of a common source (CS) input stage (M_1) with active shunt feedback (M_5), and a CM (M_3 and M_4) where a voltage commutating mixer is inserted. Meanwhile, along the auxiliary path, a noise-canceling stage (M_2) driving a current commutating mixer is constructed. The noise canceling stage enables noise and distortion cancellation while increasing the transconductance of the circuit, too. The RF input impedance matching is provided by an active feedback sub-amplifier($M_{1,3,5}$). The CM network is adopted here to render the linear current amplification, which effectively avoids voltage-current conversion distortions in

979-8-3503-2123-4/23 $31.00 © 2023 IEEE

conventional voltage mode RXs. The presented RX using the CMs at the RF and BB is shown as a current-mode receiver.

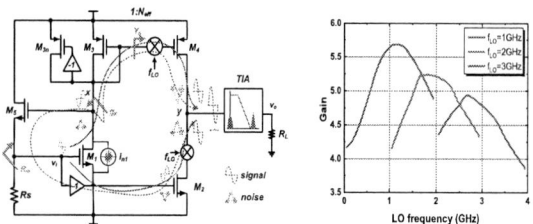

Fig. 1. (a)Simplified schematic of presented receiver front-end with frequency-translational noise cancellation. (b)Simulated sub-amplifier gain under different LO frequencies.

A. Input impedance matching

The single-ended input resistance of the presented circuit in Fig. 1, R_{in} can be simply expressed as

$$R_{in} = \frac{1}{g_{m5}(1+A_0)} \quad (1)$$

where a CS stage gain A_0 simply takes $g_{m1}/(g_{m3}-g_{m3n}+g_x)$. The g_{m1}, g_{m3}, g_{m3n}, and g_{m5} are the small signal transconductance of transistors M_1, M_3, M_{3n}, and M_5. The parameter g_x stands for the conductance observed towards the CS stage and takes

$$g_x = \frac{g_{m1}g_{m5}R_s}{1+g_{m5}R_s} + g_{ds1} \cong g_{m1}g_{m5}R_s + g_{ds1} \quad (2)$$

where g_{ds1} is the M_1's output conductance. The output conductance of smaller-sized M_3 and M_5 is neglected in (2) and below for simplicity. The first item in (2) is contributed by the mapped source resistance Rs via the active feedback path. Inserting the mixer into the CM brings about a BB resistance up-conversion due to the mixer's reciprocity. It implicitly affects the input matching via the frequency-dependent A_0. Then, the CS stage gain considering the effect is

$$A_0(\omega_{LO}+\omega_{IF}) \cong \frac{g_{m1}}{g_{m3}-g_{m3n}+g_x+j(\omega_{LO}+\omega_{IF})C_0+\frac{1}{\gamma_0}j\omega_{IF}C_{gsM4}}$$

$$= \frac{g_{m1}}{g_{m3}-g_{m3n}+g_x+j(\omega_{LO}+\omega_{IF})(C_0+C_{eq})} \quad (3)$$

where the coefficient $\gamma_0(\sim2/\pi^2)$ comes from a 25% duty-cycle LO driving. The ω_{IF} can take positive/negative values depending on the high/low sideband chosen. Then, the equivalent parasitic C_0 and C_{eq} at the RF side are estimated by

$$C_0 = C_{gsM3} + C_{gsM3n} + C_{gdM5} + C_{mix} \quad (4)$$

$$C_{eq} = \frac{\omega_{IF}}{\omega_{LO}+\omega_{IF}} \frac{C_{gsM4}}{\gamma_0} \quad (5)$$

Note that for a small BB BW case featured in traditional RXs, the BB parasitic up-converted to the RF side, C_{eq} can be neglected. In the paper, the large BW makes the item prominent. Take ω_{IF}:400 MHz and ω_{LO}:1 GHz as an example, it reaches $C_{eq}=1.4C_{gsM4}$. Namely, the CM merged with mixers introduces more BB parasitic up-conversion for the wide IF case. With additional mixer parasitic, C_{mix}, the presented RX potentially faces a matching difficulty, especially at high sideband.

To avoid the process complexity of using small threshold voltage transistors in[5], standard devices for M_3/M_4 are utilized in the paper. Particularly, a negative resistance cell (M_{3n}) is nested here to trade-off the gain $A_0(\omega)$ and its load voltage headroom. On the other hand, to alleviate the $A_0(\omega)$'s limitation by the imaginary part, the M_4 with a lower g_m value is selected. Particularly, the gain of the sub-amplifier, $A_0(\omega)$, in simulation is displayed in Fig.1(b) under different LO frequencies, where a wide IF of >400 MHz is assumed. Note that at the $\omega_{IF}>0$ side, the $A_0(\omega)$ gradually decreases as expected. The gain curve for 1 GHz LO, however, doesn't peak at the negative ω_{IF} band edge because smaller series coupling capacitors in the circuit attenuate the gain appreciably. The ~5x gain is roughly kept across the interested LO frequency range.

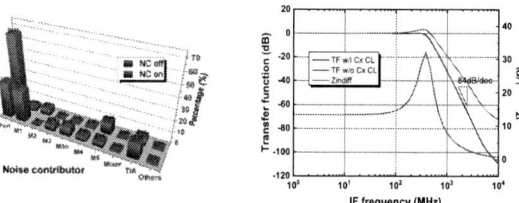

Fig. 2. (a) Noise contribution percentage. (b) Transfer function and input resistance of TIA.

B. Transconductance gain and noise cancellation

Due to the boost effect by the negative cell, the overall current ratio N_{eff} is, however, maintained as large enough, (Neglecting the parasitic) and takes the form below

$$N_{eff} = \frac{g_{m4}}{g_{m3}-g_{m3n}+g_x} \quad (6)$$

With the NC constraint shown below soon, the overall transconductance gain between vi and vy, incorporating the frequency down-conversion, is given by

$$G_m = \sqrt{\gamma_0}\left(N_{eff}g_{m1}+g_{m2}\right)$$
$$= 2\sqrt{\gamma_0}N_{eff}\left(g_{m1}+g_{m3}-g_{m3n}+g_x\right) \quad (7)$$

According to Fig.1(a), the thermal noise current, i_{n1} from M_1, firstly creates noise voltage at net x, then down-converted and reaches net y along the main path. In parallel with it, the noise voltage at net x is also down-converted to a noise current by the auxiliary path. The two noise currents with opposite phases cancel each other once upon meeting the NC condition:

$$g_{m2} = \sqrt{\gamma_0}N_{eff}\left[g_{m1}+2\left(g_{m3}-g_{m3n}+g_x\right)\right] \quad (8)$$

The attractive point is that the diode(cross)-connected M_3/M_{3n} obeys the same noise-canceling principle, and contributes no thermal noise in principle. As in Fig.1(a), compared to the LNA in [5], the noise current i_{n1} experiences a more different time delay between the two paths inserted with voltage/current-mode mixers, yielding additional phase error that may potentially degrade the noise cancellation property. Specifically, the LO divider closer in layout to the mixer of the auxiliary path than to that of the main path beneficially compensates for the phase error, guaranteeing noise cancellation. The noise percentage contribution is reported in Fig.2(a).

979-8-3503-2123-4/23 $31.00 © 2023 IEEE

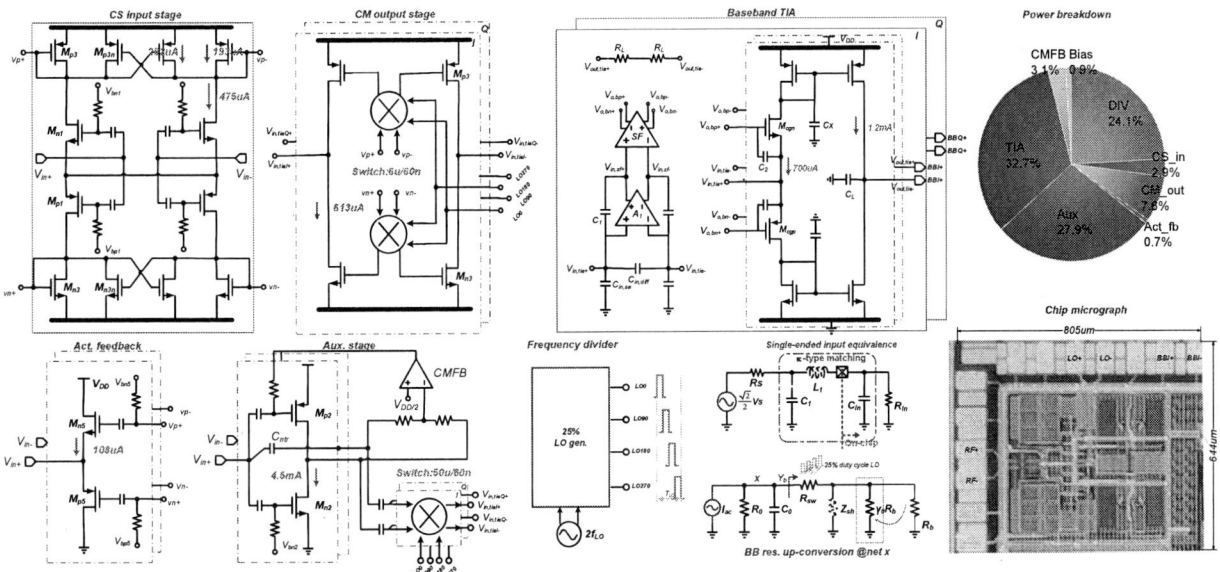

Fig. 3. Diagram of presented receiver frontend and individual circuits, and power breakdown and micrograph of fabricated chip.

C. Baseband TIA

Although the common-gate(CG)-based TIA appears noisier than the OpAmp-based counterpart, the preceding large transconductance gain G_m, makes the TIA's noise requirement fairly loosened. The CG current-mode TIA in [4,6] is customized for further increased BW and filtering order here. Namely, the related poles/zeros of the TIA need to move toward higher frequencies. The frequency upper line of g_{mCG}/C_{gsCG} can be increased firstly by moderately increasing the bias current of the CG transistor, which also is more gainful for approaching an input virtual ground. Secondly, by properly scaling down these related capacitors of C_{in}, C_1, and C_2, poles of ω_{pa}/γ and ω_{pb} defined in [4] are pushed towards higher frequencies. Thirdly, two real poles related to the capacitors C_x and C_L are added to enhance the filtering profile. Typically, one pole closes to the passband edge to reinforce the TIA as a 4th order filter. The other pole compensates for the high frequency zero, ω_{zh}. The TIA thus keeps the 4th order filtering up to the zero contributed by the source follower (SF) stage[6]. Once the frequency goes beyond the zero point (~5 GHz), the 3rd order filtering profile appears. With optimized parameters, the normalized transfer function simulations of the TIA are depicted in Fig.2(b). It has shown a steep 84 dB/dec roll-off, thanks to the addition of C_x and C_L. Quantitively, the BW covers >400 MHz, while ~14 Ω in-band differential input resistance is maintained. Such a good virtual ground across a wideband range is fairly difficult for conventional OpAmp-based BB implementations.

III. EXPERIMENTAL RESULTS

The presented frontend circuit prototype as in Fig. 3, has been fabricated in standard 65 nm CMOS technology, including individual sub-circuits. The stacked n/pMOS structure is applied to enable a power efficiency design. An off-chip π-type

matching network is used to extend input matching bandwidth besides the utility of off-chip input balun of 1:$2^{0.5}$ turn ratio. It occupies an area of 805 um×644 um including pads, as displayed in Fig. 3. The dc power of the signal path is 44 mW under a 1.8 V supply. And the dynamic power of the logic circuit is 14 mW/GHz with a 1.2 V supply. The effect of input balun and output buffer were de-embedded from measurements

The input reflection coefficient, S_{11} results with varying LO frequency is shown in Fig.4(a). Note that the curve valley appears not exactly at the LO frequency point but at a lower frequency offset to that, because the imaginary item related to the low sideband compensates that related to the RF parasitics. In contrast, the peak S_{11} at the high sideband worsens the input matching, which validates the aforementioned analysis of BB parasitic up-conversion. Fig. 4(b) shows the simulated and measured gain with discrete 1, 2, and 3 GHz LO frequencies. It displays that, at typical 2 GHz f_{LO}, the in-band gain of the receiver is 40 dB. And the -3 dB gain BW is found at ~ 425 MHz at the BB side. As shown in Fig. 5(a), the NF result with respect to IF variations is provided, again at 2 GHz f_{LO}, where the minimum measured NF value takes 2.2 dB located at f_{IF}=30 MHz. The NF simulated with NC off is also shown in comparisons. There is ΔNF=3.8 dB difference between the NC enabled/disenabled, which verifies the NC effectiveness.

The block interference is injected to check the performance degeneration of noise and gain, by fixing the IB signal at f_{IF}=50 MHz. As displayed in Fig.5(b), when the blocker offset frequency over the BB BW (Δf/BW) takes 3, the measured NF is 3.6 dB, under the 0 dBm blocker level. The good TIA's virtual ground and high voltage headroom of the auxiliary stage guarantee the decent noise-canceling operation of the auxiliary stage upon the blocker injection. The simulation also indicates a B_{1dB}=-13.7 dBm with the same Δf/BW setup. For the OOB IIP3 measurement, two test tones of f_1 and f_2 are selected at Δf

979-8-3503-2123-4/23 $31.00 © 2023 IEEE

and $2\Delta f$-40*MHz*. The resulting IM3 product always falls at 40 MHz. The measured result is shown in Fig. 6(a). Quantitatively, the OOB IIP3 of 8 and 14.5 dBm is obtained by setting $\Delta f/BW$=5 and 7 with the same f_{LO}=2 GHz. The enhanced OOB IP3 benefits from the linear current-mode RX structure and high-order filtered TIA. The IB IP3 of -13 dBm is measured with a similar two tones setup. For LO variations, the gain, noise, and in-band IP3 are shown in Fig. 6(b). A higher frequency operation can be restricted by the frequency divider. And the π-type input matching also has upper limit of ~4.7 GHz.

Table 1 benchmarks the performances of the proposed circuit against recently published reports. After applying the open-loop TIA with moderate power consumption, the presented RX frontend has achieved a BW of 425 MHz, over all the other reports. In contrast, large BW in [2] is obtained but at cost of large power consumption to support a high OpAmp GBW. The TIA with high order filter also assists the presented RX with better OOB linearity than other LF RX reports. The input matching of MF RXs generally behaves as a narrow band characteristic, where their S_{11} valley points show a frequency offset to central LO frequencies. The presented circuit like other LF RX reports has displayed a good wideband input matching.

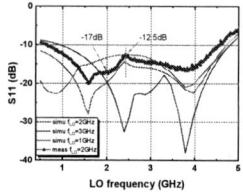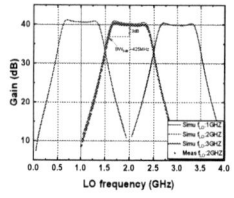

Fig. 4. Simulated and measured (a) S_{11}, and (b)gain.

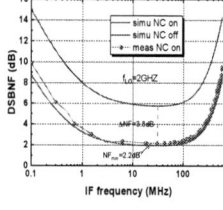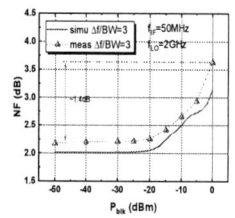

Fig. 5. (a)Simulated/measured NFs with respect to IF variations. (b)Simulated and measured NF degeneration under blocker interferences.

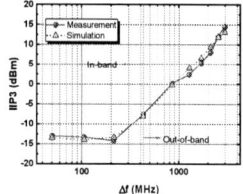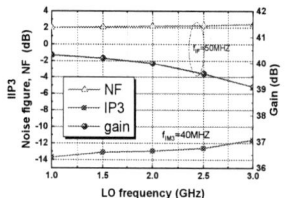

Fig. 6. (a)Out-of-band linearity with varying frequency offsets of blocker interference. (b) Measured gain, noise, and in-band IP3 with LO variations.

IV. CONCLUSION

In the paper, we proposed an LNA-first analog frontend incorporating frequency-translational noise cancellation and a wideband high-order TIA filter. In terms of competitive bandwidth, input matching, and low noise, the presented current-mode receiver is eligible for high data rate applications.

ACKNOWLEDGMENT

This work was supported by the National Natural Science Foundation of China (61871073), and the Natural Science Foundation of Sichuan province (2022NSFSC0522).

Table 1. Performance summary and comparisons

Para.	Gain	BW	NF	BNF$^\$$	OB-IIP3	Power	CMOS
Unit	dB	MHz	dB	dB	dBm	mW	nm
T.W.[L]	40	425	2.2	3.6(3)	14.5(7)	44°+14•	65
[1][L]	50	10	1.8	14(5)	5(5)	27-40	40
[3][L]	72	2	1.9	4.1(40)	13.5(10)	35.1-78•	40
[7][L]	48.2	40	3.4	NA	-3.7(10)	22.2	28
[8][L]	36	80	2.7	8.4(6.3)	13(5)	58.5+17.6•	40
[2][M]	22	175	2.5	NA	13(3)	172	22♦
[4][M]	32.4	130	5.5	9.7(8)	21(3)	21.6+7.8	28
[6][M]	33.5	200	2.3	3.5(3)	19(3)	34+26	65

[L]LNA-first, [M]Mixer-first, °single path, •LO generation, $^\$$0 dBm blocker, ♦FDSOI, •8 phase clock for harmonic rejections.

REFERENCES

[1] H. Hedayati, W. A. Lau, N. Kim, V. Aparin and K. Entesari, "A 1.8 dB NF Blocker-Filtering Noise-Canceling Wideband Receiver With Shared TIA in 40 nm CMOS," in IEEE Journal of Solid-State Circuits, vol. 50, no. 5, pp. 1148-1164, May 2015, doi: 10.1109/JSSC.2015.2403324.

[2] A. N. Bhat, R. van der Zee, et al., "A Baseband-Matching-Resistor Noise-Canceling Receiver Architecture to Increase In-Band Linearity Achieving 175MHz TIA Bandwidth with a 3-Stage Inverter-Only OpAmp," in 2019 IEEE Radio Frequency Integrated Circuits Symposium (RFIC), Boston, MA, USA, Jun. 2019, pp. 155–158.

[3] D. Murphy et al., "A Blocker-Tolerant, Noise-Cancelling Receiver Suitable for Wideband Wireless Applications," IEEE J. Solid-State Circuits, vol. 47, no. 12, pp. 2943–2963, Dec. 2012.

[4] G. Pini, D. Manstretta, and R. Castello, "Analysis and Design of a 260-MHz RF Bandwidth +22-dBm OOB-IIP3 Mixer-First Receiver With Third-Order Current-Mode Filtering TIA," IEEE J. Solid-State Circuits, vol. 55, no. 7, pp. 1819–1829, Jul. 2020.

[5] B. Guo, J. Chen, L. Li, H. Jin, and G. Yang, "A wideband noise-canceling CMOS LNA with enhanced linearity by using complementary nMOS and pMOS configurations," IEEE J. Solid-State Circuits, vol. 52, no. 5, pp. 1331–1344, 2017.

[6] B. Guo, H. Wang, et al., "A Mixer-First Receiver Frontend with Resistive-Feedback Baseband Achieving 200 MHz IF Bandwidth in 65 nm CMOS," in 2022 IEEE Radio Frequency Integrated Circuits Symposium (RFIC), Jun. 2022, pp. 31–34.

[7] B. Guo, D. Prevedelli, et al., "A 0.08 mm 2 1-6.2 GHz receiver front-end with inverter-based shunt-feedback balun-LNA," in 2020 IEEE Radio Frequency Integrated Circuits Symposium (RFIC), 2020, pp. 379–382.

[8] M. A. Montazerolghaem, S. Pires, L. C. N. de Vreede, and M. Babaie, "6.5 A 3dB-NF 160MHz-RF-BW Blocker-Tolerant Receiver with Third-Order Filtering for 5G NR Applications," in 2021 IEEE International Solid-State Circuits Conference (ISSCC), Feb. 2021, vol. 64, pp. 98–100. doi: 10.1109/ISSCC42613.2021.9365849.

RMo1B-2

IIP2-Enhanced Receiver Front-End with Notch-Filtered Low-Noise Transconductance Amplifier for 5G New Radio Cellular Applications

Donggu Lee[1], Sukju Yun, Kuduck Kwon[2]

RF/Analog Circuits and Systems (RACAS) Lab., Kangwon National University, South Korea

[1]ldgldg209209@kangwon.ac.kr, [2]kdkwon@kangwon.ac.kr

Abstract—This study presents an IIP2-enhanced receiver front-end with a notch-filtered low-noise transconductance amplifier (LNTA) for 5G new radio (NR) cellular applications. The LNTA employs a dual-band third-order *LC* notch filter with a band-switchable differential inductor to reject TX leakages and out-of-band blockers. Consequently, the receiver front-end has enhanced blocker tolerance and satisfies the IIP2 specification with no IIP2 calibration. Fabricated through a 65-nm CMOS process, the receiver front-end was primarily characterized in the low band and mid band of 5G NR. It achieved a noise figure of 3.5 dB, conversion gain of 41.5 dB, out-of-band IIP3 of 2.1 dBm, and calibration-free IIP2 of more than 59 dBm.

Keywords—5G, IIP2 calibration, band-switchable differential inductor, blocker-tolerant, dual-band *LC* notch filter, transmitter leakage rejection.

I. INTRODUCTION

The development of a 5G new radio (NR) to support ultra-high data rate, ultra-low latency, and hyper connectivity for the fourth industrial revolution is underway. The cellular transceiver should support the 5G NR and legacy modes. Blocker-tolerant and highly linear broadband transceivers are crucial to efficiently handle many blockers and interfering signals. Particularly, in frequency-division duplexing (FDD) systems, the transmitter (TX) leakage enters a low-noise transconductance amplifier (LNTA) input in the receiver (RX) as a strong blocker owing to the limited TX-RX isolation of surface acoustic wave (SAW) duplexers. Second-order intermodulation distortions (IMD2), which are generated by the mixer with a large TX leakage, substantially degrade the signal-to-noise ratio of the overall RX. Thus, most cellular RXs use a SAW-less RX architecture with input-referred second-order intercept point (IIP2) calibration for FDD systems to address this IMD2 issue [1]–[3]. However, most IIP2 calibration methods increase the design complexity owing to high-resolution digital-to-analog adjustment. Additionally, multiple calibrations must be performed separately for each operating frequency band owing to frequency-dependent mismatched components. Consequently, IIP2 calibration increases the overall test time and cost of the products. As shown in Fig. 1, investigations on a novel blocker-tolerant and linear RX architecture to satisfy the RX IIP2 specification for 5G NR sub-6-GHz cellular applications without complicated IIP2 calibration are underway [4]. An RF filtering LNTA can reduce IMD2 generated by the mixer by $2A_{Rej}$ dB by filtering the TX leakage and out-of-band (OB) blockers by A_{Rej} dB before the mixer. That is, the RF filtering effectively enhances the mixer IIP2 by $2A_{Rej}$ dB.

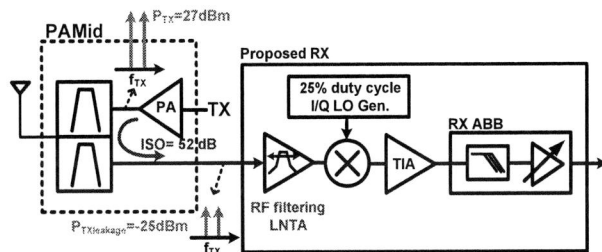

Fig. 1. A new blocker-tolerant and IIP2-enhanced RX architecture with an RF filtering LNTA

This study proposes a blocker-tolerant and IIP2-enhanced RX front-end with a notch-filtered LNTA for 5G NR sub-6 GHz cellular applications. The LNTA employs a dual-band third-order *LC* notch filter with a band-switchable differential inductor to reject TX leakage and OB blockers and improves the overall IIP2 performance of the RX.

II. PROPOSED IIP2-ENHANCED RX FRONT-END WITH NOTCH-FILTERED LNTA

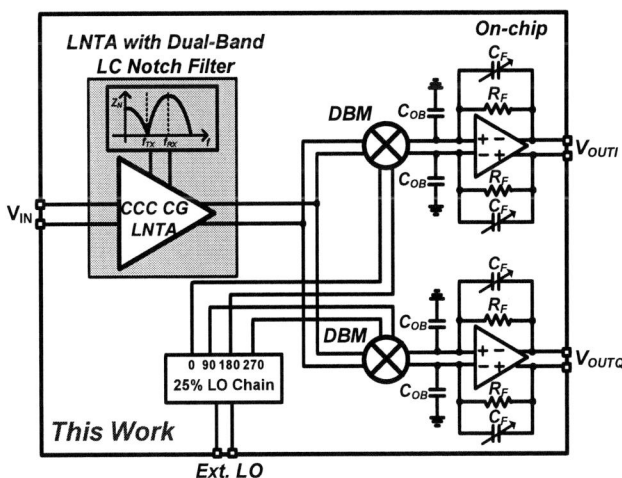

Fig. 2. Block diagram of the proposed IIP2-enhanced RX front-end with a notch-filtered LNTA

A simplified block diagram of the proposed blocker-tolerant and IIP2-enhanced RX front-end for 5G NR sub-6 GHz cellular applications is shown in Fig. 2. This includes an LNTA with dual-band third-order *LC* notch filter, in-phase/quadrature (*I/Q*) double-balanced current-mode passive mixers with non-overlapping 25% duty-cycle *I/Q* LO signals, and *I/Q* transimpedance amplifiers (TIAs). Improved IIP2 performance

979-8-3503-2123-4/23 $31.00 © 2023 IEEE   25   2023 IEEE Radio Frequency Integrated Circuits Symposium

can be achieved with no IIP2 calibration because RF notch filtering effectively improves the IIP2 of the mixer and the RX blocker tolerance. This study uses an *LC* notch filter instead of an N-path filter for RF filtering. This is because the power dissipation of a 25% duty-cycle LO chain required to drive the N-path filter is significant and the N-path filter cannot sufficiently reject the TX leakage signal at small RX-TX separation frequencies. Furthermore, in front of the TIA, C_{OB} was added to maintain a low TIA input impedance in the OB frequency range. Consequently, the RX linearity can be further enhanced.

A. Capacitor Cross-Coupled Common-Gate LNTA with Dual-Band Third-Order LC Notch Filter

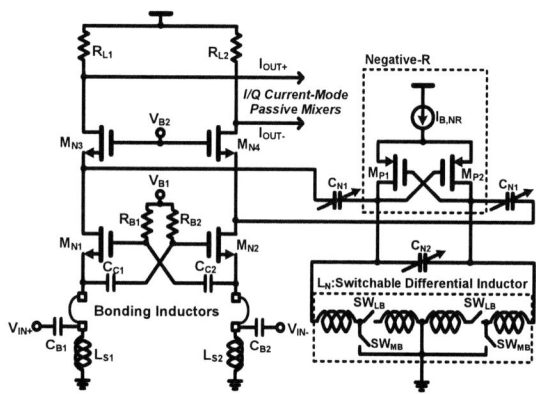

Fig. 3. Proposed LNTA with a dual-band third-order LC notch filter

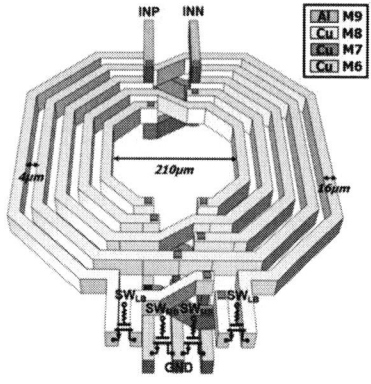

Fig. 4. Layout of the band-switchable differential inductor with a center tap

The proposed LNTA, comprising a capacitor cross-coupled (CCC) common-gate (CG) input stage (M_{N1} and M_{N2}), a dual-band third-order Q-boosted *LC* notch filter, cascode transistors (M_{N3} and M_{N4}), and load resistors (R_{L1} and R_{L2}), is shown in Fig. 3. The embedded notch filter was located at the sources of M_3 and M_4. The effective transconductance (G_m) from the LNTA differential input voltage to the differential output current can be simply expressed as follows:

$$G_m = \frac{I_{OUT+} - I_{OUT-}}{V_{IN+} - V_{IN-}} = g_m \left(\frac{2C_C + C_{gs}}{C_C + C_{gs}} \right) \frac{g_{mC} Z_N}{(2 + g_{mC} Z_N)}, \quad (1)$$

where g_m and g_{mC} are the transconductances of $M_{N1,2}$ and $M_{N3,4}$, C_C is the capacitance of $C_{C1,2}$, C_{gs} is the gate-to-source capacitance of $M_{N1,2}$, and Z_N is the equivalent impedance seen to the *LC* notch filter, assuming an infinite Q-factor of the filter, expressed as follows:

$$Z_N = \frac{1 + s^2 L_N (\frac{C_{N1}}{2} + C_{N2})}{\frac{sC_{N1}}{2}(1 + s^2 L_N C_{N2})} = \frac{1 + s^2/\omega_N^2}{\frac{sC_{N1}}{2}(1 + s^2/\omega_P^2)}. \quad (2)$$

Z_N has three poles and two zeros. When $C_C \gg C_{gs}$, G_m becomes $2g_m$ in the passband frequencies and zero in the notch frequency of f_N. The effective G_m of the CCC CG LNTA is approximately twice than that of conventional CG LNTA. Consequently, a higher G_m and better noise figure (NF) can be achieved with the same current consumption. Moreover, the CCC CG LNTA has a better common-mode rejection ratio. The source impedance of the conventional CG input stage is $2/g_m$ in the differential and common modes. Conversely, the source impedance in the differential mode is $1/g_m$ in the CCC CG input stage; however, that of the common-mode becomes ideally infinite when $C_C \gg C_{gs}$. Therefore, the CCC CG LNTA has better common-mode rejection capability and g_m-boosting effect. The Q-factor of the *LC* notch network is influenced by the finite Q-factor of the L_N and the on-resistance of the switches in capacitor arrays of C_{N1} and C_{N2}. As shown in Fig. 3, a Q-enhancement circuit is utilized to enhance notch selectivity [5]. Suppose channel-length modulation is ignored. In that case, the negative resistance can be approximated as $Z_{NR} = -2/g_{mp}$, where g_{mp} is the transconductance of $M_{P1,2}$. $I_{B,NR}$ can adjust g_{mp}. Z_{NR} can be set to an appropriate value to cancel the parasitic resistance of the *LC* notch network with $I_{B,NR}$ and $M_{P1,2}$. Additionally, C_{N1} and C_{N2} can be tuned using digitally controlled 5-bit signals to adjust the notch and passband frequencies of the third-order *LC* notch filter according to the operating band. L_N employs a band-switchable differential inductor with a center tap to support the low band (LB) and mid band (MB) of 5G NR, as shown in Fig. 4 [6]. The differential inductor has a width of 16 μm, outer radius of 210 μm, and metal spacing of 4 μm. When the LB switch is turned on and the MB switch is turned off, the entire differential inductor is used for LB operation; however, it is partly disconnected for the MB operation when the MB switch is turned on and the LB switch is turned off. The inductance and Q-factor of the proposed band-switchable differential inductor were simulated using the EMX electromagnetic (EM) simulation tool. The inductances and Q-factors for LB and MB operations were approximately 9.2 nH/11 and 3.4 nH/13, respectively. The simulated frequency response of the notch-filtered LNTA at Band 5/n5, Band 2/n2, and Band 1/n1 is shown in Fig. 5. In Band 2/n2, the TX leakage can be rejected by 16.4 dB. RF filtering can improve the effective IIP2 performance of the RX.

B. Current-Mode Passive Mixer with TIA

The proposed RX employs a double-balanced current-mode passive mixer with a subsequent TIA, which achieves sufficient

979-8-3503-2123-4/23 $31.00 © 2023 IEEE

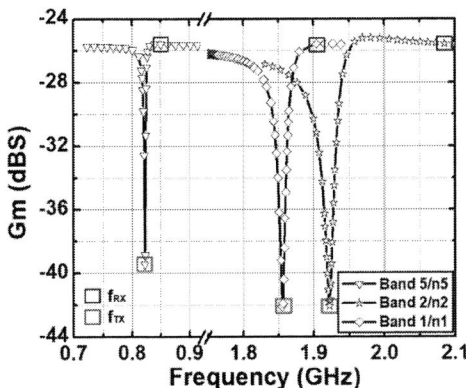

Fig. 5. Simulated frequency response of the notch-filtered LNTA

linearity because low impedances at the input and output of the mixer is maintained. A nonexistent DC current also enhances the flicker noise performance. Furthermore, AC-coupling capacitors are included in front of the mixer to perform high-pass filtering and reject low-frequency IMD2 signals generated by the preceding LNTA. The nonoverlapping 25% duty-cycle I/Q LO signals drive the mixer to improve the conversion gain, noise, and linearity performances. Operational amplifier (OPAMP)-based TIAs were adopted to perform current-to-voltage conversion and first-order low-pass filtering and to maintain a low input impedance. The TIA utilizes a modified feed-forward (FF) OPAMP, which has a large DC gain and high unity-gain frequency for a given current consumption [7].

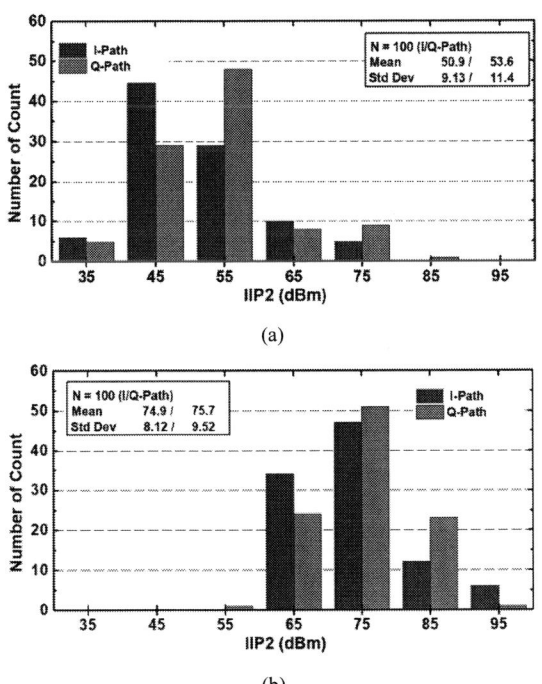

Fig. 6. Monte Carlo IIP2 simulation results of the two RX front-end for Band 2/n2 with random device mismatches and process variations (N=100): (a) without RF notch filter (b) with RF notch filter

C. IIP2 Enhancement

The Monte Carlo IIP2 simulation results of two RX front-end designs were performed with random device mismatches and process variations, as shown in Fig. 6. The IIP2 of the RX front-end that employed the proposed notch-filtered LNTA was compared to the IIP2 of the RX front-end that utilized a conventional LNTA. The IIP2 comparison was performed in Band 2/n2, an RX-TX separation (f_{TX-RX}) of which is the smallest (80 MHz). The two-tone test conditions for the IIP2 were $f_{LO} = 1.96$ GHz, $f_1 = 1.88$ GHz, $f_2 = 1.881$ GHz, and $p_1 = p_2 = -28$ dBm. As shown in Fig. 6, the simulated mean IIP2 values of the proposed RX were approximately 24 dB higher than those of the RX with the conventional LNTA. Thus, RF notch filtering of the proposed LNTA improves the overall IIP2 performance of the RX and satisfies the IIP2 requirement for 5G NR cellular applications without IIP2 calibration.

III. MEASUREMENT RESULTS

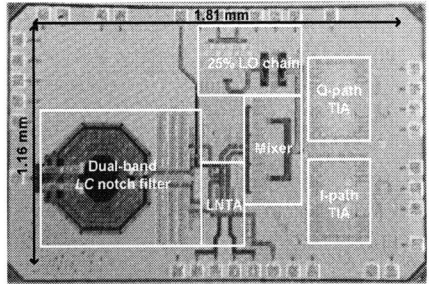

Fig. 7. Chip microphotograph

The blocker-tolerant and IIP2-enhanced RX front-end with the proposed dual-band notch-filtered LNTA was fabricated using a 65-nm CMOS process and the die was assembled with a chip-on-board package. A chip microphotograph is shown in Fig. 7. The active area was 2.1 mm². Excluding the LO chain, the RX front-end consumes 11.7 mA from a nominal supply voltage of 1 V. All measurements were taken for Band 28/n28, Band 5/n5, Band 8/n8, Band 3/n3, Band 2/n2, and Band 1/n1 for the LTE/5G NR FDD frequency bands with channel bandwidths of 10 and 20 MHz. These conditions are rigorous for measuring satisfactory TX leakage-induced IIP2 performances. The measured S_{11} is less than −10 dB in the LB and MB of 5G NR cellular applications. The measured conversion gains are shown in Fig. 8. The measured conversion gains for LB and MB operations were approximately 38–41.5 dB, respectively. In Band 2/n2 ($f_{TX-RX} = 80$ MHz), the TX leakage and blocker rejection ratio of the RX front-end was 28 dB for f_{TX} (= f_{RX}–80 MHz). Fig. 9 depicts the measured double side-band (DSB) NFs and in-band(IB)/OB input-referred third-order intercept points (IIP3s) of the RX front-end. The minimum DSB NF of 3.5 dB was obtained. The measured maximum IB/OB IIP3s were −14.5/2.1 dBm, respectively. The measured OB IIP2 performances is shown in Fig 10. The two-tone test conditions for the OB IIP2 were $f_1 = f_{TX}$, $f_2 = f_{TX} - 1$ MHz and $p_{f1} = p_{f2} = -28$ dBm. Measured I/Q OB IIP2s of over 59 dBm were obtained with no IIP2 calibration methods in entire operating band for LTE/5G NR cellular applications.

979-8-3503-2123-4/23 $31.00 © 2023 IEEE

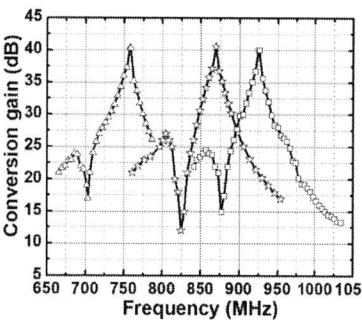

(a)

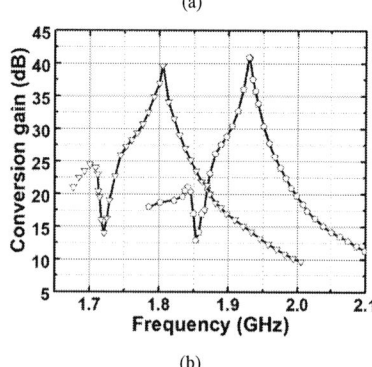

(b)

Fig. 8. Measured conversion gains: (a) LB (b) MB

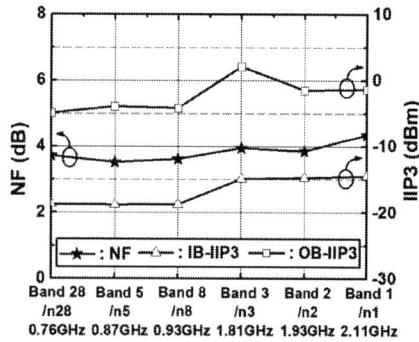

Fig. 9. Measured NF and IIP3 versus operating bands

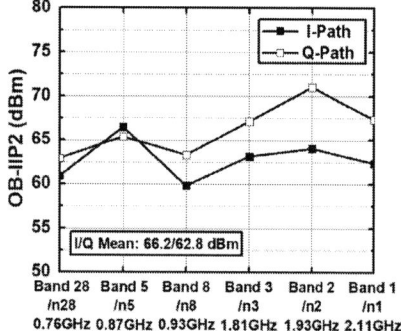

Fig. 10. Measured OB IIP2 versus operating bands

The summary and comparison of the measured performance are listed in Table I. An IIP2 performance better than 59 dBm can be achieved without additional IIP2 calibration through RF notch filtering of the LNTA.

Table 1. Performance Summaries and Comparison

	[1]	[2]	[4]	This work
Design	Full RX	Full RX	RX FE	RX FE
Process	14-nm FinFET CMOS	12-nm FinFET CMOS	65 nm CMOS	65 nm CMOS
Frequency (GHz)	0.575–5.925	0.6–6	0.7-2.2	0.7-2.2
Conversion gain (dB)	NA*	NA	43.5	41.5
NF (dB)	NA	1.5 with eLNA‡	4.5-4.9	3.5-4.4
IB IIP3 (dBm)	NA	NA	−12.2	−14.5
OB IIP3 (dBm)	NA	NA	2.33	2.1
OB IIP2 (dBm)	> 70	> 55	>55	>59
IIP2 Calibration	YES	YES	NO	NO
RF Filtering	NO	NO	N-path BPF	LC notch
Power (mW)	114**	142**	15.1	11.7
Supply Voltage (V)	NA	0.9/1.3/1.8	1	1
Area (mm²)	25.7†	39.84†	2.24	2.1

*NA: not available **transceiver power consumption per 1 carrier component
†total area of MIMO transceiver ‡eLNA: external LNA

IV. CONCLUSION

An IIP2-enhanced RX front-end with an RF notch-filtered LNTA for 5G NR cellular applications was implemented in a 65-nm CMOS process. The proposed RF filtering LNTA could enhance the IIP2 performance of the overall RX without IIP2 calibration. Further, a band-switchable differential inductor was proposed to support the LB and MB of the 5G NR. The implemented RX front-end achieved a maximum conversion gain of 41.5 dB, minimum NF of 3.5 dB, maximum OB IIP3 of 2.1 dBm, and minimum OB IIP2 of over 59 dBm.

ACKNOWLEDGMENT

This work was supported by the National Research Foundation of Korea(NRF) grant funded by the Korea government(MSIT) (No. 2023R1A2C1003227).

REFERENCES

[1] J. Lee *et al.*, "6.1 A Low-Power and Low-Cost 14nm FinFET RFIC Supporting Legacy Cellular and 5G FR1," *2021 IEEE International Solid- State Circuits Conference (ISSCC)*, San Francisco, CA, USA, 2021.

[2] M. -D. Tsai *et al.*, "10.3 A 12nm CMOS RF Transceiver Supporting 4G/5G UL MIMO," *2020 IEEE International Solid- State Circuits Conference - (ISSCC)*, San Francisco, CA, USA, 2020.

[3] J. Lee *et al.*, "21.6 A Sub-6GHz 5G New Radio RF Transceiver Supporting EN-DC with 3.15Gb/s DL and 1.27Gb/s UL in 14nm FinFET CMOS," *2019 IEEE International Solid- State Circuits Conference - (ISSCC)*, San Francisco, CA, USA, 2019, pp. 354-356.

[4] D. Shin, K. Lee and K. Kwon, "A Blocker-Tolerant Receiver Front End Employing Dual-Band N-Path Balun-LNA for 5G New Radio Cellular Applications," in *IEEE Transactions on Microwave Theory and Techniques*, vol. 70, no. 3, pp. 1715-1724, March 2022.

[5] K. Kwon, S. Kim and K. Y. Son, "A Hybrid Transformer-Based CMOS Duplexer With a Single-Ended Notch-Filtered LNA for Highly Integrated Tunable RF Front-Ends," in *IEEE Microwave and Wireless Components Letters*, vol. 28, no. 11, pp. 1032-1034, Nov. 2018.

[6] D. Lee and K. Kwon, "Sub-6 GHz noise-cancelling balun-LNTA with dual-band Q-enhanced LC notch filter for 5G new radio cellular applications," *J. Semiconductor Technology and Science*, vol. 22, no. 3, pp. 161-167, June 2022.

[7] K. Kwon and J. Han, "A 2G/3G/4G SAW-Less Receiver Front-End Adopting Switchable Front-End Architecture," in *IEEE Transactions on Microwave Theory and Techniques*, vol. 62, no. 8, pp. 1716-1723, Aug. 2014.

RMo1B-3

A Band-Shifting Millimeter-Wave T/R Front-End with Enhanced Imaging and Interference Rejection Covering 5G NR FR2 n257/n258/n259/n260/n261 Bands

Fuyuan Zhao[#], Wei Deng[#$], Haikun Jia[#], Wenjing Ye[#], Ruichen Wan[#], Baoyong Chi[#]

[#]Tsinghua University, China

[$]Research Institute of Tsinghua University in Shenzhen, China

wdeng@tsinghua.edu.cn

Abstract—For achieving multi-band fusion within the fifth-generation (5G) new radio (NR) frequency range 2 (FR2), a band-shifting transmit/receive (T/R) front-ends (FEs) with the reconfigurable auxiliary paths, covering n257/n258/n259/n260/n261, is proposed in this paper. To overcome the deterioration from capacitance-based frequency tuning, an inductance-mutation transformer technique is proposed in the interstage matching of both TX and RX FEs, to enable operation band shifting between low frequency (LF) and high frequency (HF). The proposed transceiver FE chip, composed of a differential power amplifier (PA) with cascade output stage, a two-stage cascade low-noise amplifier (LNA), and a compact dual-resonance T/R switch at antenna interface, is designed and fabricated in standard 28-nm CMOS process. The measured results show the FE can cover 24.25-29.5/37-43.5GHz at LF/HF with enhanced interference and imaging rejection, and realizes <−25 dB EVM for 400 Msym/s 64-QAM wireless communication with >9 dBm average output power in the target band.

Keywords—5G NR, multi-band fusion, band-shifting, inductance-mutation, front-end.

I. INTRODUCTION

The fifth-generation (5G) new radio (NR) frequency range 2 (FR2) standard has evolved to support a broad millimeter-wave (mm-wave) spectrum. The main operating bands of 5G NR FR2-1, including n257, n258, n259, n260, and n261, occupy two continuous ranges at mm-wave frequencies, i.e., the low frequency (LF) range (24.25-to-29.5 GHz) and the high frequency (HF) range (37-to-43.5 GHz). A multi-band multi-standard 5G system is desired to support extensive wireless application prospects. The reported multi-band 5G transceivers are enabled either by separated 28/39 GHz transmit/receive (T/R) front-ends (FEs) or continuous ultra-wide band T/R FEs [1-3]. Contrary to the former category with large silicon area, the latter realizing band fusion in a single FE is considered as a compact scheme for large-scale 5G phased-array systems. However, an ultra-wide band T/R FE suffers from cross-band interference issue, as well as the image signal issue depending on the intermediate-frequency (IF). One solution is to adopt sharp image-rejection filters (IRFs) or quadrature mixers. Nevertheless, it significantly increases the overall system complexity. Besides continuous ultra-wide band structure, another effective approach for a single channel to cover multiple bands is a tunable RF FE. As shown in Fig. 1, apparently, this structure offers a relative narrower instantaneous bandwidth, hence leads to an inherent out-of-

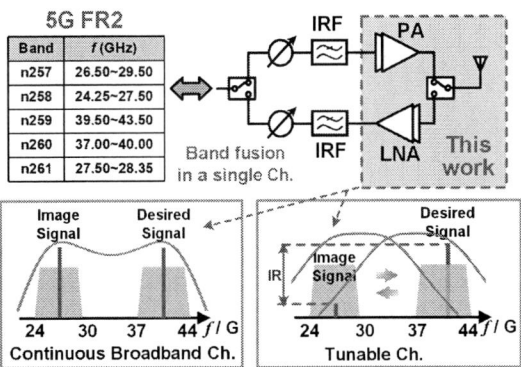

Fig. 1. Conceptual graph—multi-band fusion of 5G FR2 in a single RF channel and two different fusion methods: continuous broadband channel and tunable channel.

band attenuation. This can provide better coexistence performance and stronger image rejection, which relaxes the requirement on sharp IRFs to some extent.

Conventionally, a frequency tunable FE can be realized by controlling the switched-capacitor array in the matching networks. However, this method suffers from reduced bandwidth, poor ripple, large switch loss, and deteriorated gain at its low frequencies, as portrayed in Fig. 2a. To overcome this issue, a band-shifting T/R FE based on the proposed inductance-mutation transformer (LM-XFMR) technique is presented in this paper. The LM-XFMR operated as the matching networks can provide variable equivalent inductance to change the resonance frequency. Taking the merit of this technique, the proposed FE features reconfigurable architecture and can shift the operation band between low frequency (LF) and high frequency (HF), to cover 5G FR2 bands.

This paper is organized as follows. Section II analyses the and the LM-XFMR circuit and explains the operation principle of the band-shifting function. Section III illustrates the complete circuit design of the 5G RF FE in details. Section IV demonstrates experimental results of the FE chip. Finally, the conclusions are presented in Section V.

II. OPERATION ANALYSIS OF BAND-SHIFTING

In consideration of the large span of 5G FR2, simply using switched-capacitor arrays only to change the operation band in such range will worsen the circuit performance significantly. The reasons are: 1) the parasitic resistance of the matching

979-8-3503-2123-4/23 $31.00 © 2023 IEEE   29   2023 IEEE Radio Frequency Integrated Circuits Symposium

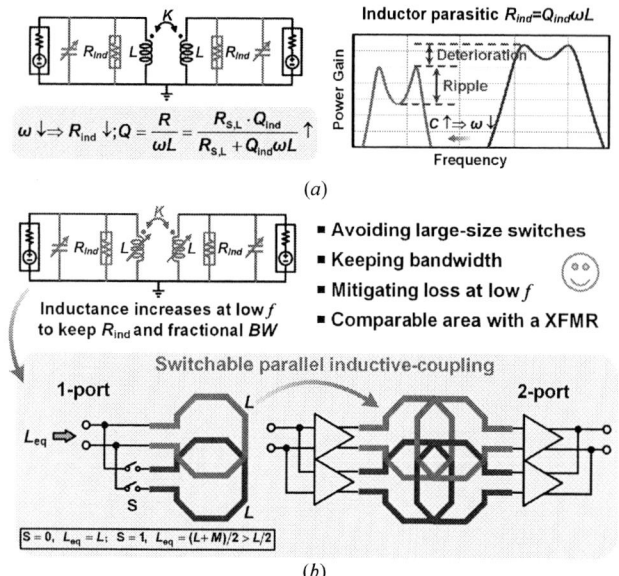

(a)

(b)

Fig. 2. Comparison of different methods of frequency band tuning: *(a)* switched-capacitor-based tuning; *(b)* proposed LM-XFMR tuning.

inductor leads to more insertion loss at lower frequencies; 2) the Q-factor of the network increases when the frequency decreases, and result in a decline in the relative bandwidth; 3) large tuning range means large-size capacitors, and much parasitic loss is introduced as well. These problems above can be solved if the matching inductors are also changed along with the switched capacitors. Based on this point, the LM-XFMR is proposed and employed in this FE design.

The formation procedure is illustrated in Fig. 2*b*. Firstly, in a two-coil-coupling structure, when switch S is turned on, the two coils with positive mutuality leads the equivalent inductance $L_{eq} > L/2$. Considering the equivalent capacitance is doubled to $2C$, resonant frequency f_0 is less than $1/(2\pi\sqrt{LC})$. Therefore, with mutable inductance, the operating band can be shifted by setting S on/off. In the actual implementation, the switch S could be replaced by gain stages for mitigating loss, and an LM-XFMR matching network for band-shifting is obtained.

The proposed LM-XFMR can be applied for the interstage matching in the mm-wave power amplifier (PA) and low-noise amplifier (LNA). As shown in Fig. 3, an amplifier with the LM-

XFMR is composed of a main path and an auxiliary path, with the same gain stage sizes. When both of the two paths are activated, the interstage network formed by the four coupled inductors can be equivalent to a transformer. If there are $L_1 = L_3$, $L_2 = L_4$, $k_{12} = k_{34}$, the equivalent inductances are obtained as follow,

$$L_{1e} = \frac{L_1}{2} + \frac{M_{13}}{2}, \quad L_{2e} = \frac{L_2}{2} + \frac{M_{13}}{2}, \quad M_{eq} = \frac{M_{12}}{2} + \frac{M_{14}}{2}. \quad (1)$$

Here, the equivalent primary and secondary windings are greater than $L_1/2$, $L_2/2$, which shifts the circuit to operate at the LF. Likewise, when the auxiliary path is powered down and only the main path works, L_1, L_2 constitute an equivalent transformer affected by $L_{3,4}/C_{3,4}$. Thus, the expression of the equivalent inductor (L_{1e}) is given as:

$$L_{1e} = L_1 + \frac{\omega^2 M_{1e}^2}{1/C_e - \omega^2 L_e} = \begin{cases} L_1 - k_{1e}L_1, & C_e \gg 1/\omega^2 L_e \\ L_1, & C_e \to 0 \end{cases} \quad (2)$$

where L_e is the equivalence of L_3, L_4, C_e is the equivalence of C_3, C_4 and M_{ie} is the mutual inductance of L_i and L_e. It is seen that only $C_{3,4}$ is substantially large or small enough, the equivalent inductors L_{1e}, L_{2e}, will drive the circuit operate at the HF. Considering the existing parasitic capacitance of the transistors, $C_{3,4}$ with high capacitance are selected. In a word, choosing proper inductances, capacitances and couple coefficients, the amplifier is shifted between LF and HF effectively by enabling/disabling the auxiliary path.

III. CIRCUIT DESIGN

The detailed circuit of the proposed mm-wave T/R FE is shown in Fig. 4*a*. Both of the PA and LNA consist of a main path and an auxiliary path, with an LM-XFMR interstage network and fixed input/output matching networks. The structure and component parameters in the auxiliary path are exactly the same with that in the main path. The PA consists of a common-source driver stage and a cascode output stage, and the LNA has two cascode stages with a shared CS transistor at the input. The inductors in the LM-XFMR are designed as the rhombus shape (Fig. 4*b*) to increase the overlap area thus the coupling coefficient between the main and auxiliary paths could be enhanced. The mutual inductances of L_1, L_3 and L_2, L_4 are 48 pH and 30 pH, respectively, leading to 52.4% and 25.9% increasing in equivalent inductances when the auxiliary path turns on. Thanks to the parallel coupling, the area sizes of the

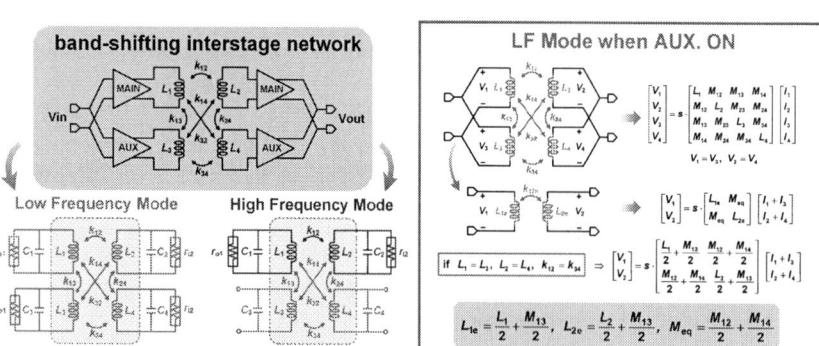

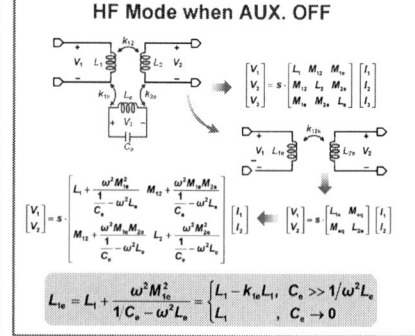

Fig. 3. Operation principle and equivalent modelling analysis of the band-shifting interstage network.

979-8-3503-2123-4/23 $31.00 © 2023 IEEE

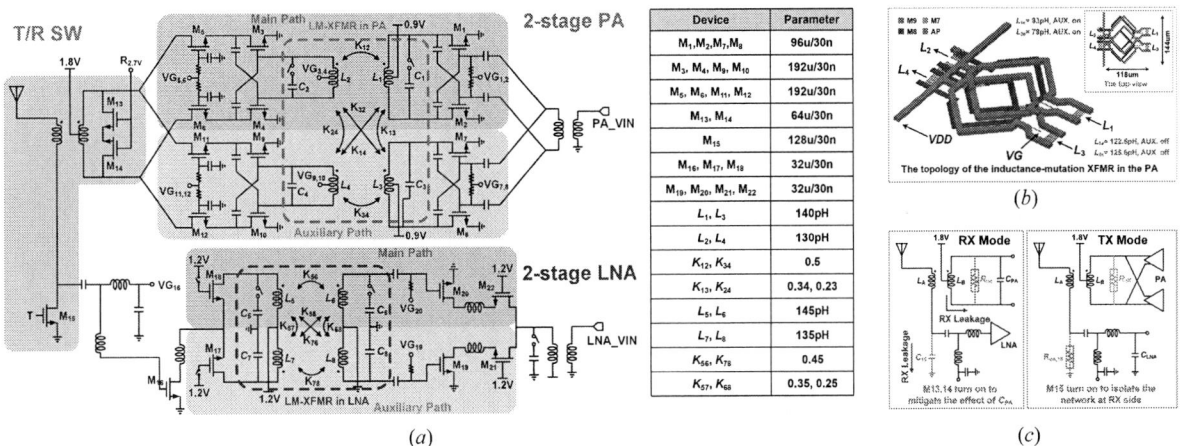

Fig. 4. Circuit design of the proposed FE: (*a*) full schematic of the T/R FE; (*b*) layout of the LM-XFMR in the PA; (*c*) equivalent circuits at the antenna interface in TX/RX mode.

LM-XFMRs are approximately equal to that of a typical transformer, such as the LM-XFMR of the PA with an area of 118×144 μm². Therefore, despite two paths employed, the proposed FE still features a compact area size compared to designs with two separate paths.

The wideband T/R switch is realized by a switched-inductor at the antenna interface, which reduces insertion loss effectively with compact topology and enables dual-resonance matching for RX. As illustrated in Fig. 4*c*, in the TX mode, the RX path is shorted with M_{15} conducted, and the LNA is powered down. In the RX mode, M_{15} is open so that the switched-inductor is involved in the RX network. At the same time, with VG_{1-4}, VG_{7-10} set to ground, $M_{13,14}$ are turned on and short the PA output to mitigate the RX leakage.

IV. MEASUREMENT RESULTS

The proposed band-shifting T/R FE is implemented in a standard 28nm CMOS process, and the occupied core area is 0.26 mm², as shown in Fig.5. The continuous-wave measurement results are plotted in Fig. 6. Under the LF mode

for the TX, the two paths are conducted, and the measured peak gain is 19.1 dB with a 3-dB bandwidth of 24.1-to-38.8 GHz. Under the HF mode, the auxiliary path is shut down, and the corresponding TX peak gain is 17.2 dB with a 3-dB bandwidth of 35.4-to-43.5 GHz. For the large-signal TX measurement, the peak P_{1dB} of 17.8 dBm at the LF and the peak P_{1dB} of 15.0 dBm at the HF are obtained. The PA under the LF/HF mode achieves

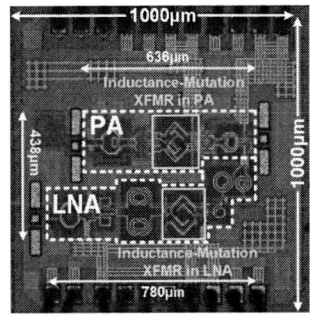

Fig. 5. Chip micrograph.

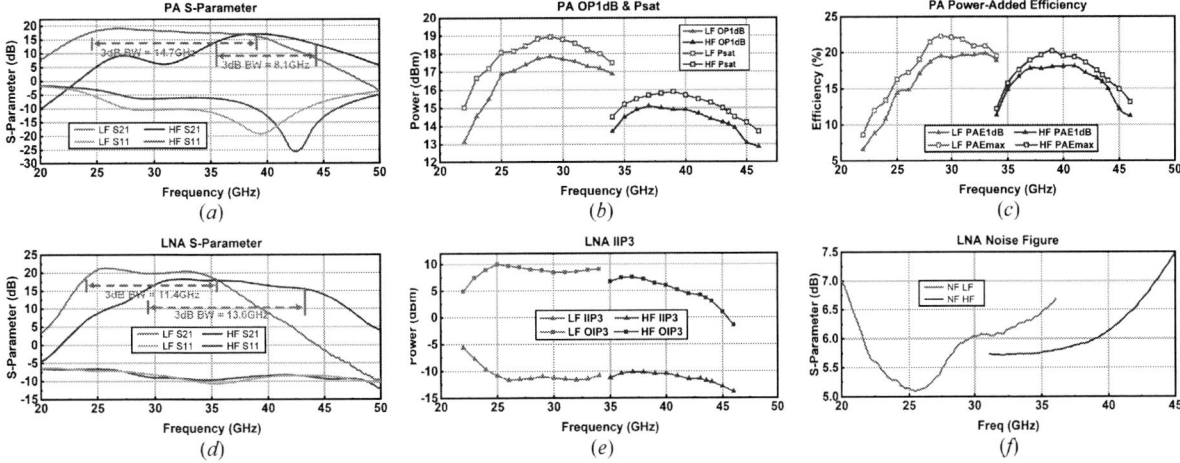

Fig. 6. Experimental results of the TX/RX for small & large-signal measurements over the band at LF/HF: (*a*) PA S-parameter; (*b*) PA OP$_{1dB}$ and P$_{SAT}$; (*c*) PA PAE; (*d*) LNA S-parameter; (*e*) LNA IIP3 and OIP3; (*f*) LNA Noise figure.

979-8-3503-2123-4/23 $31.00 © 2023 IEEE

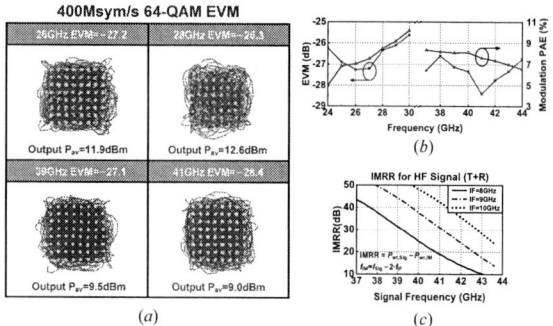

Fig. 7. Communication performances: (a) modulation measurement results of the PA under 400Msym/s 64-QAM; (b) EVM & PAE curves under 64-QAM modulation. (c) IMRR for the signal at HF from 8-to-10GHz IF band.

Table 1. Performance summary and comparison with the prior bulk CMOS FE designs for 5G NR FR2.

Performance		This work		[3] ISSCC 2019		[4] ESSCIRC2021	[5] MWCL2020	[6] IMS2021
		LF	HF	28G Band	37G Band			
Technology		28nm Bulk CMOS		65nm Bulk CMOS		28nm Bulk CMOS	65nm Bulk CMOS	90nm Bulk CMOS
Spectrum Mask		Switchable band		Continous Wideband		Continous band	Continous band	Continous band
PA	Peak Gain (dB)	19.1	17.2	30	26.2	19	21	18.3
	Frequency (GHz)	24.1–43.5		27.5–30.5*	35–38.5*	22~31	24.9–30.0	30~39.2
		24.1–38.8	35.4–43.5					
	BW$_{3dB}$ (GHz)	14.7	8.1	3*	3.5*	9	5.1	9.2
	OP1dB (dBm)	17.8	15.0	14.1	15.2	14	13.3	13.3
	Psat (dBm)	19.0	15.8	15.8	16.8	15*	15.3	15.1
	PAE@1dB (%)	19.4	18.1	15.0	18.8	18.7	23.2	24
	PAEmax (%)	22.1	20.2	20.0	22.2	21.0	25.1	29
	EVM (dB)	400M 64QAM −27.2	400M 64QAM −28.4	250M 64QAM −26.3		400M 64QAM −26.0	44M 64QAM −21*	--
LNA	Peak Gain (dB)	21.3	18.3	16.1	11	17	19.6	17.6
	Frequency (GHz)	23.9–43.7		26.5–29.5*	31.5–39*	22~30.6	26~32.2*	34~38.1
		23.9–35.3	29.2–43.7					
	BW$_{3dB}$ (GHz)	11.4	13.6	3*	7.5*	8.6	6.2	4.1
	NF (dB)	5.1	5.8	6.2	7.0	4.9	4.9	4.7
	IIP3 (dBm)	−10.8	−10.1	−6.1	−3.2	−9.2	—	−6.4
	P$_{dc}$ (mW)	19.7	19.8	37.6		35	28.8	25
Covered 5G Bands		n257/n258/n259/n260/n261		n257/n260/261		n257/n258/n261	n258/n261	
Area (mm²)		0.26		0.48		0.25	0.22	0.21

* Estimated from literature

P_{SAT} of 19.0/15.8 dBm at 29/39 GHz, as well as 22.1%/20.2% PAE_{MAX} and 19.4%/18.1% PAE at P_{1dB}. In the RX measurement, the LNA consumes less than 20 mW under both LF and HF modes. A 21.3-dB peak gain with 23.9-to-35.3 GHz 3-dB bandwidth and an 18.3-dB peak gain with 29.2-to-43.7 GHz 3-dB bandwidth are measured at the LF and HF, respectively. For the two-tone test, the measured IIP3 is >−12.5/−14 dBm within LF/HF.

Fig. 7 demonstrates the PA performance for the modulation measurement at 26/28/39/41GHz. The TX EVM for the single-carrier signal under 400 Msym/s 64QAM are −27.2 dB, −26.3 dB, −27.1 dB, −28.4 dB, respectively, with the corresponding

back-off P_{av}/PAE of 11.9 dBm/9.1%, 12.6 dBm/10.7%, 9.6 dBm/9.3%, 9.0 dBm/8.7%. With the measured HF power gain of the PA and LNA combined, the image rejection ratio (IMRR) is > 18 dB for a 39/41 GHz signal from 8-to-10GHz IF band (Fig. 5c), and a marked interference suppression below 25 GHz is also achieved.

Table 1 gives the performance summary and comparison with the prior bulk CMOS T/R FEs for 5G FR2. The proposed band-shifting FE with enhanced imaging and interference rejection can fully support 5G bands: n257/n258/ n259/n260/n261. The FoM against fractional bandwidth charts of the PA and LNA are plotted in Fig. 8 as well. The proposed T/R FEs features comparable FoMs with the largest fractional bandwidth compared to the previous work.

V. Conclusion

This paper proposes a band-shifting mm-wave T/R FE fabricated in a standard 28nm CMOS process. Exploiting the LM-XFMRs as the matching networks, the operation bands of both the PA and LNA can be shifted between the LF and the HF, which fully covers n257/n258/n259/n260/n261 5G FR2 bands. With the excellent effective RF bandwidth, the proposed FE design realizes appropriate gain power, linearity, noise figure, and features interference and imaging rejection to relax the demand on high-Q filters. Lower than −25dB EVM is demonstrated under 400 Msym/s 64QAM within the target bands. In addition, the LNA consumes a relatively low P_{DC} of 19.8 mW, which is suitable for mobile devices with an asymmetric uplink-to-downlink channel ratio (e.g., RX: TX=4:1).

Acknowledgment

This work was partly supported by supported by the National Natural Science Foundation of China under Grant 62131013, the Shenzhen Science and Technology Program (No. JCYJ20200109113601723 and No. JCYJ20210324115610028), the Tsinghua-Samsung Joint Research Project, and Beijing Advanced Innovation Center for Integrated Circuits.

References

[1] A. Verma et al., "A 16-channel, 28/39GHz dual-polarized 5G FR2 phased-array transceiver IC with a quad-stream IF transceiver supporting non-contiguous carrier aggregation up to 1.6GHz BW," *IEEE Int. Solid-State Circuits Conf. (ISSCC)*, pp. 1-3, Feb. 2022.

[2] J. Park and H. Wang, "A 26-to-39GHz broadband ultra-compact high-linearity switchless hybrid N/PMOS bi-directional PA/LNA front-end for multi-band 5G large-scaled MIMO system," *IEEE Int. Solid-State Circuits Conf. (ISSCC)*, pp. 322-324, Feb. 2022.

[3] S. Mondal et al., "A reconfigurable bidirectional 28/37/39GHz front-end supporting MIMO-TDD, carrier aggregation TDD and FDD/full-duplex with self-interference cancellation in digital and fully connected hybrid beamformers," *IEEE Int. Solid-State Circuits Conf. (ISSCC)*, pp. 348-350, Feb. 2019.

[4] D. Manente et al., "A 22–31 GHz bidirectional 5G transceiver front-end in 28 nm CMOS," *IEEE 47th Eur. Solid State Circuits Conf. (ESSCIRC)*, pp. 283-286, Sept. 2021.

[5] W. Lee and S. Hong, "28 GHz RF front-end structure using CG LNA as a switch," *IEEE Microw. Wireless Compon. Lett.*, vol. 30, no. 1, pp. 94-97, Jan. 2020.

[6] T.-Y. Chiu et al, "A Ka-band transformer-based switchless bidirectional PA-LNA in 90-nm CMOS process," *IEEE MTT-S Int. Microw. Symp. (IMS)*, pp. 450-453, June 2021.

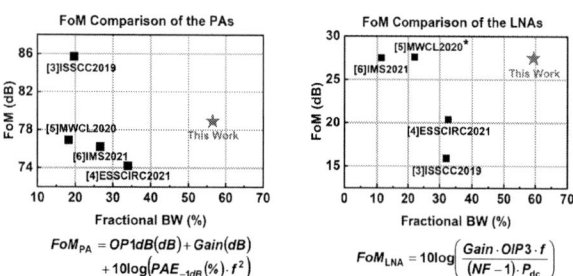

Note: For designs with multi-band, FoM is weighted average based on BW.

Fig. 8. FoM against fractional bandwidth charts of the PAs and LNAs.

A 6-22 GHz CMOS Phase Shifter with Integrated mm-Wave LO

Natan Ershengoren[#1], Eran Socher[#2]

[#]High Frequency Integrated Circuits Lab, Tel Aviv University, Israel

[1]gershengorn@mail.tau.ac.il, [2]socher@tauex.tau.ac.il

Abstract—In this work we propose a new architecture for a wideband phase shifter, based on upconversion and downconversion of the wideband input signal using phase-shifted mm-wave local oscillator (LO) signals. Phase shifting of the LO at a single frequency is achieved using an on-chip quadrature voltage-controlled oscillator (QVCO) and vector modulation (VM). The issue of mixer image band phase shift interference is mitigated using filtering at the mm-wave intermediate frequency (IF). The concept is demonstrated in 65nm CMOS in the design of a 6-22 GHz full 360° phase shifter using a 84 GHz LO and 62-78 GHz IF range. The phase shifter achieves phase error <1.05° and amplitude error <0.2 dB for 7-bit resolution with power consumption of 137 mW (including buffers) and core area of 0.26 mm^2.

Keywords—active phase shifter, vector modulator, QVCO, wideband system.

I. Introduction

Wideband RF frontends are needed for high-data-rate broadband communication, high-resolution radar, and precise positioning. This type of technology finds application in weather monitoring, air traffic control, marine vessel traffic management, the defence and military sectors, and vehicle speed detection for law enforcement.

All of these applications require the use of phased array systems, in which the transmitter/receiver module must be small in size and multi-functional, and the phase shifter (PS) is one of the most important and, at the same time, one of the most area-consuming elements of the phased array, depending on whether the PS topology is active or passive.

In comparison to the active, the passive PS has advantages of certain advantages of power consumption and better dynamic range and linearity. However, passive phase shifters have a certain trade-off between the operation bandwidth and insertion losses. Moreover, a scatter of the losses between the phase states is usually significant and degrades with the quantization of the phase step. Moreover, all phase-shifting elements must be designed for wideband operation to perform accurate phase shifts across a bandwidth of interest. Such design is a difficult task in the lower frequency bands, significantly below the X band, due to the nature of the devices. Special techniques for the bandwidth extension consume more chip area, which is limited by the cost-effectiveness of the final product.

Active PS can offer, of course, active gain and fewer limitations on phase shift quantization, which gives an opportunity to extend bandwidth without overrated area consumption. Active topology typically use a passive quadrature signal synthesizer and a vector modulator (vector-sum circuit) to achieve the desired phase shift by weighting four quadrature signals. The generation of quadrature signals needs a passive hybrid, with the significant challenge of accurate broadband operation and area. Variations across frequency typically result in resolution limitations and a need to assign different phase states at different frequencies.

This work proposes a new PS architecture for covering the 6-22 GHz with an active topology that employs phase shifting at a fixed 84 GHz LO frequency and using it to upconvert and downconvert back the wide 6-22 GHz band to apply that phase shift simultaneously across the band.

II. Idea and Circuit Architecture

The proposed PS architecture is based on the basic principle of the mixer operation and the well-known LO phase-shifting approach.

As shown in Fig.1, an LO signal is used first to upconvert the RF input signal into a high IF band, and then the same LO frequency is used again, though with a phase shift, to downconvert the signal back into the original RF band at the output.

The same RF frequency is achieved for two up/down conversion processes. The first is $(f_{RF}+f_{LO})\text{-}f_{LO}$ and the second is $f_{LO}\text{-}(f_{LO}\text{-}f_{RF})$. Thus, the phase shift in each term has an opposite sign. Therefore, the total phase shift at the output is ideally completely rejected except for singular cases. For the phase-shifting concept to work, one of the contributions has to be filtered out at the IF domain. This function is performed in the active filter stage, as depicted in Fig. 1. To be able to amplify the lower IF band ($f_{LO}\text{-}f_{RF}$), and reject the higher IF band ($f_{LO}+f_{RF}$), the RF band must have a minimum RF frequency, which would be half the two IF band separation.

The topology, Fig. 2, represents a new principle of the operation and makes the design process easier because only the input stage, upconversion mixer, and output stage, downconversion mixer, must be designed for the wideband operation on the RF frequencies.

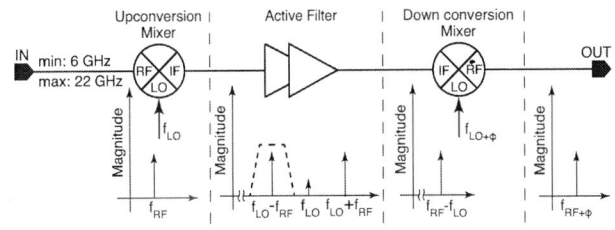

Fig. 1. Proposed circuit concept.

979-8-3503-2123-4/23 $31.00 © 2023 IEEE

The active filter operates at higher frequencies, where the fractional bandwidth is a more feasible task. One more advantage of the proposed topology is that the phase/magnitude control circuit translates the changes indirectly to the RF signal and sets the LO of the mixers, which must be constant in frequency. It means that a modulation circuit can be designed for narrowband operation, which relaxes the design procedure and also improves the phase and magnitude errors of the resulting output signal.

III. CIRCUIT IMPLEMENTATION

A. Upconvertion and Downconversiton Mixers

Both mixers are implemented by commonly used active Gilbert cell. This choice is also made for compensation of interconnect and matching network losses and adjusting LO and signal levels.

A few techniques were applied to extend the operation bandwidth of the conventional Gilbert cell.

Instead of the regular CS RF input stage, the CG transistors with shunt inductors are used in the upconversion mixer design. At the resonance frequency, the input impedance will be approximately 1/gm. The shunt inductor was realized using the second coil of the input balun, which is also used to perform the single-ended measurement.

To neutralize parasitic capacitance between the RF stage and LO quad, which degrades the performance bandwidth of the upconversion mixer, serial peaking inductors were used.

Due to the output LO transistor capacitance, the interstage transformer was designed using the low-k technique to cancel it at the frequencies of interest to improve the mixer performance. This transformer performs a couple of missions: relaxes voltage headroom issues, raises the conversion gain and operation bandwidth by nulling the capacitance at the output node, and is designed to load the input of the amplifier stage with optimal input impedance.

The downconversion mixer utilizes the same topology and techniques for improvement as the up-conversion mixer, with a difference in the load stage. Resistors were used

for wideband operation, as well as peaking inductors, to mitigate the parasitic drain capacitance of the switching quad. Measurement setup limitations forced the use of a single-ended output, and thus an active balun was added as a final stage with a differential input and a PMOS active load despite its linearity limitations.

B. Active Filter

The active filter circuit has two amplifier stages, one of which is a common source cross-coupled, and the second is cascode amplifier. They were developed using the conjugate matching technique to suppress the image signal after the upconverting mixer while amplifying the desired signal.

The transformers were used for interstage matching because of their compact layout. They also have a good common-mode rejection ratio.

Because the tightly coupled windings provide significant parasitic capacitive coupling, a high-k transformer in the matching network between stages frequently results in narrowband efficiency. The low-k transformer technique is employed in this research to overcome this problem at IF frequencies.

C. LO Generation Circuit and Phase Shifting

Both LO signals must be coherent for the correct operation of the circuit. That is why both mixers use a single QVCO as an LO source. However, using a single source in proximity allows the simple use of a free-running QVCO without a phase-locking loop.

In order to minimize loading influence and frequency pulling of the VCO, buffers were added to the two sides at its outputs for isolation, Fig. 2. They were built using regular differential pair with inductive load. The gates obtain bias directly from the outputs of the QVCO core.

A QVCO is used instead of a simple VCO LO to allow controllable 360° LO phase shift through the use of vector modulation by combining two variable gain amplifiers (VGA), designed using Gilbert cell, driven by quadrature signals from QVCO.

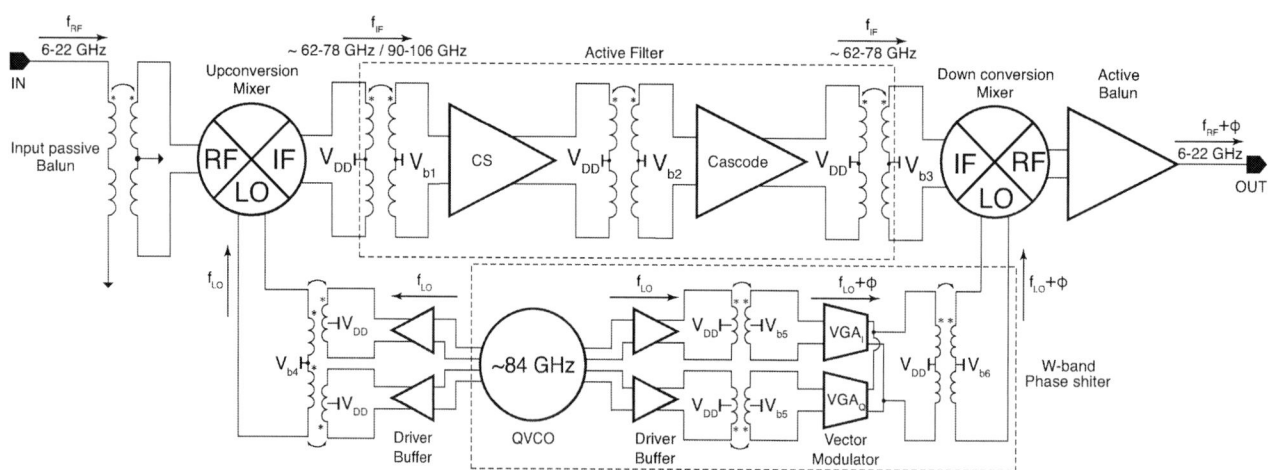

Fig. 2. Topology of the designed phase shifter.

A magnetically coupled QVCO is used to reduce power consumption and degradation of the oscillation performance due to the additional interconnections needed for the traditional actively coupled QVCO.

IV. MEASUREMENTS RESULTS

The chip was fabricated using TSMC 65nm CMOS technology with one poly and 9 metal layers (1P9M). The fabricated circuit is shown in Fig. 3.

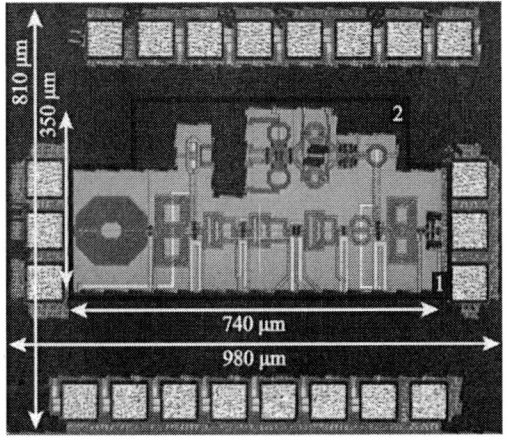

Fig. 3. Fabricated die microphotograph, showing LO area (2) and RF path area (1).

The size of the chip, excluding DC and RF pads, is 0.35x0.740 mm^2. As can be seen from the die micrograph, the layout has a symmetric structure. It is done to minimize mismatches and imbalances in the differential signals. Fig.3 visually can be divided into two parts. The upper part, enclosed by half rectangle 2, is responsible for the LO generation and distribution, while the bottom part, enclosed by half rectangle 1, from GSG RF pads from the left to the GSG RF pads from the right, handle the RF signal flow.

The die was measured using a probe station and network analyzer. A Matlab script and GPIB connection were used for the measurement automatization by controlling individual transistor biasing in the vector modulator.

The circuit was characterized for 7-bit resolution. The measured return losses at the ports show a nice correlation with the simulations, as shown in Fig. 4, considering process variation and the broadband target.

The measurements of the reflection coefficient hardly change over the phase state. It is reasonable as S_{11} mainly depends on the up-conversion mixer, and S_{22} depends on the output balun. Some variation might result from temperature and drift throughout the measurements over states. In the operation band, S_{11} kept less than -10 dB, and S_{22} was less than -6.5 dB.

The overall circuit gain, shown in Fig. 5, has the same correlation pattern as return losses. A 3dB bandwidth magnitude performance was observed between 7.2-19 GHz.

The normalized 7-bit phase response is shown in Fig. 6. The measured state control settings are not

frequency-dependent, which is a significant advantage, considering future use and integration of the circuit.

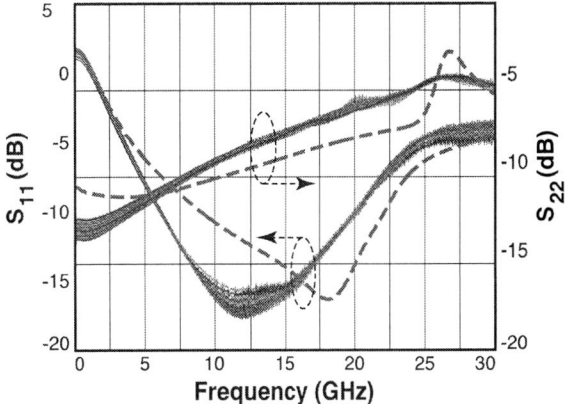

Fig. 4. Measured and simulated (dashed) return losses of the input and output ports for all 7-bit states.

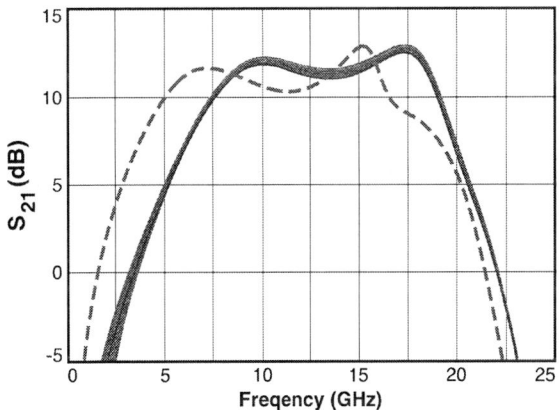

Fig. 5. Measured and simulated (dashed) gain for all 7-bit states.

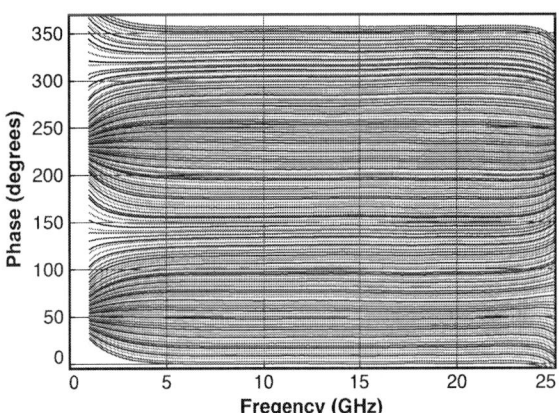

Fig. 6. Measured 7-bit phase response relative to 0° state.

A VM phase shifters have good performance in terms of the quantization level and phase/magnitude errors. Because the phase shift of the RF signal is generated with the help of the VM circuit, the phase shift performance of the overall

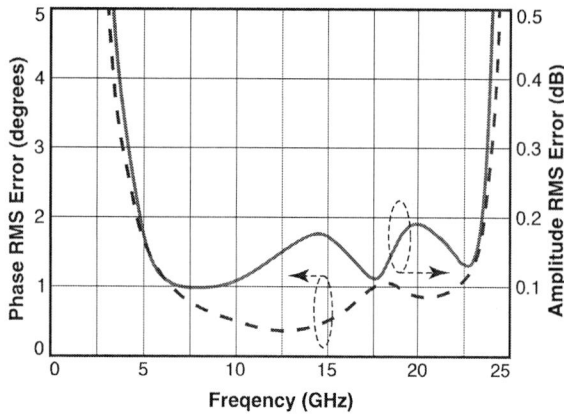

Fig. 7. Amplitude RMS error and Phase RMS error.

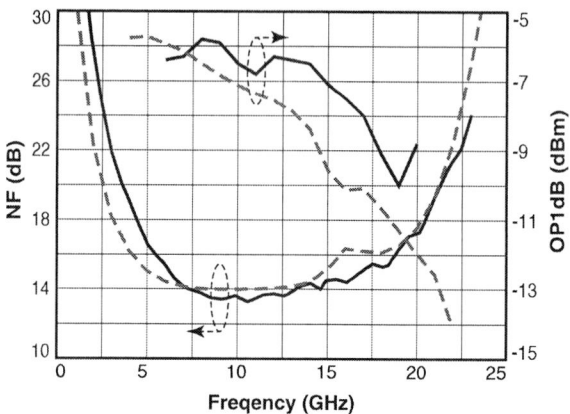

Fig. 8. Measured and simulated (dashed) NF and OP1dB.

sample will be determined by it. Therefore, the results in Fig. 7 demonstrate the expected and typical results for the VM PS.

In the 6-22 GHz bandwidth, amplitude error is less than 0.2 dB, while phase error is less than 1.1°.

The NF and OP_1dB also was measured. Results well correlate with simulations, Fig. 8. As was written previously, P_1dB performance degrades mainly because of the active balun, used to measure the circuit using a single-ended approach.

The measured and processed data were compared to the state-of-the-art circuits and represented in Table 1.

This comparison shows that compared to the state-of-the-art, this work measured results offer the highest phase resolution and bandwidth, lowest phase and gain errors, and a compact area. Power consumption is higher in order to have positive gain across the wide bandwidth driving 50 Ohm load.

V. CONCLUSION

This work proposes a new architecture for wideband phase shifters. The technique, which uses vector modulation at a mm-wave LO frequency driving up- and down-conversion mixers, provides wideband RF operation and high-resolution control of phase and amplitude across frequency. Applicable to X-, Ku-, and K-band phased arrays, the new concept is demonstrated in 65nm CMOS with a silicon area of 0.35x0.74

mm², which is competitive or smaller than the existing state of the art topologies. The fundamental operation of the proposed active phase shifters is to translate the phase shift generated by the interpolation circuit from IQ signals of the QVCO thru the LO port of the mixer to the RF input signal. The analysis shows the necessity and impact of image band rejection in the circuit active filter on the resulting phase error. The measurements conform well with simulated results and show a 7-bit phase shifting operation with a 6-22 GHz range, less than 1.05° phase error, and 0.2 dB gain error. The proven new concept could further be extended to wider bandwidth and higher resolution.

REFERENCES

[1] Y. Gong, M. K. Cho, and J. D. Cressler, "A bi-directional, X-band 6-Bit phase shifter for phased array antennas using an active DPDT switch," *IEEE Radio Frequency Integrated Circuits Symposium (RFIC)*, pp. 288–291, 2017.

[2] K. Kibaroglu, M. Sayginer, and G. M. Rebeiz, "A Low-Cost Scalable 32-Element 28-GHz Phased Array Transceiver for 5G Communication Links Based on a 2x2 Beamformer Flip-Chip Unit Cell," *IEEE Journal of Solid-State Circuits*, vol. 53, no. 5, pp. 1260–1274, May 2018.

[3] K. J. Koh and G. M. Rebeiz, "A 6–18 GHz 5-bit active phase shifter," *IEEE MTT-S International Microwave Symposium*, 2010.

[4] F. Qiu, H. Zhu, W. Che, and Q. Xue, "A Simplified Vector-Sum Phase Shifter Topology With Low Noise Figure and High Voltage Gain," *IEEE Transactions on Very Large Scale Integration (VLSI) Systems*, vol. 30, no. 7, pp. 966–974, July 2022.

[5] O. Kazan, Z. Hu, L. Li, A. Alhamed, and G. M. Rebeiz, "An 8-Channel 5–33-GHz Transmit Phased Array Beamforming IC With 10.8–14.7-dBm Psat for C -, X -, Ku -, and Ka -Band SATCOM," *IEEE Transactions on Microwave Theory and Techniques(Early Access)*, December 2022.

Table 1. Comparison with state-of-the-art phase shifters.

	Frequency GHz	Technology	Resolution bit	Gain dB	RMS Phase Error degree	RMS Amplitude Error dB	Power mW	Area mm²
[1]	8-12	130-nm SiGe BiCMOS	6	11.5	<2.2	<0.9	195	3.9
[2]	28	180-nm SiGe BiCMOS	6	9	<7*	<1*	-	-
[3]	6-18	180-nm SiGe BiCMOS	5	16.5-19.5	<5.6	<1.1	61	0.9
[4]	24-30	65-nm CMOS	5	6.5-10	<3.5	<0.9	28	0.34
[5]	5-33	90-nm SiGe HBT	5	1.5-4.5	<8.3	<2.2	52	-
This work	6-20	65-nm CMOS	7	<13	<1.05	<0.2	137	0.79 (0.26)**

*Approximated from the plot on the Fig. 14 (j) [4]. **Excluding pads, core area.

979-8-3503-2123-4/23 $31.00 © 2023 IEEE

RMo1B-5

A 300-320 GHz Sliding-IF I/Q Receiver Front-End in 130 nm SiGe Technology

Sumit Pratap Singh[1], Mostafa Jafari Nokandi[2], Mohammad Hassan Montaseri[1], Timo Rahkonen[2], Marko E. Leinonen[1], Aarno Pärssinen[1]

[1]Centre for Wireless Communication (CWC), University of Oulu, Finland
[2]Circuits and Systems (CAS) Research Unit, University of Oulu, Finland
FirstName.LastName@oulu.fi

Abstract — This paper presents a 300-320 GHz sliding-IF I/Q receiver front-end in 130 nm SiGe BiCMOS technology with f_t/f_{max} of 300 GHz/450 GHz. The architecture, unlike direct conversion receiver at sub-THz/THz frequency range, removes the need of LO frequency same as carrier frequency. Consequently, power consumption of the LO chain is significantly reduced. Signal amplification is performed at IF stage. LO frequency at two-third and one-third of carrier frequency is generated, from external 50 GHz LO signal, using on-chip frequency doublers for RF and I/Q mixers, respectively. The receiver provides 15.2 dB of conversion gain at 310 GHz. The 3-dB RF and BB bandwidths are measured to be 26 GHz and 8 GHz, respectively. Input referred compression point (ICP) and SSB noise figure of the receiver are measured to be -17 dBm and 29.5 dB, respectively. RF and LO chain of the receiver consume 296 mW and 110 mW, respectively.

Keywords — 6G, Receivers, Sliding-IF, multipliers, I/Q, 300 GHz, sub-THz and THz Integrated circuits, SiGe BiCMOS, Maximum frequency of oscillation (f_{max}).

I. INTRODUCTION

Future vision of high speed wireless data communication is going to rely on bandwidth provided by sub-THz and THz bands of electromagnetic spectrum, where both industry and academia are using their research capabilities [1]. SiGe BiCMOS technologies provide a viable platform to implement radio front-end to achieve extremely high data rate communication towards 6G radio [2]. Recently, 220 GHz - 330 GHz band has been utilised to implement receiver front-ends using both SiGe and CMOS technologies [3]–[9].

RF gain and power are drastically reducing when carrier frequency approaches f_{max} or even exceeds the $f_{max}/2$ frequency of the IC technology. Direct down-conversion receiver at sub-THz frequency range, with LO frequency similar to the carrier frequency, suffers from high DC power consumption to drive the mixers [4], [8]. This work presents a 310 GHz sliding IF I/Q receiver front-end implemented using IHP SG13G2 SiGe BiCMOS process with f_t/f_{max} of 300 GHz/450 GHz. This sliding-IF based receiver requires the highest LO frequency at two-third of carrier frequency and it consumes relatively low DC power in LO chain to generate the sufficient LO power to drive the mixers. Low gain and high noise figure at frequencies near f_{max} demotivates to use low noise amplifier (LNA) before mixer, and instead, signal amplification can be done at IF stages to improve overall gain and noise performance of the receiver. Section II discusses the architecture of the receiver and Section III discusses the circuit

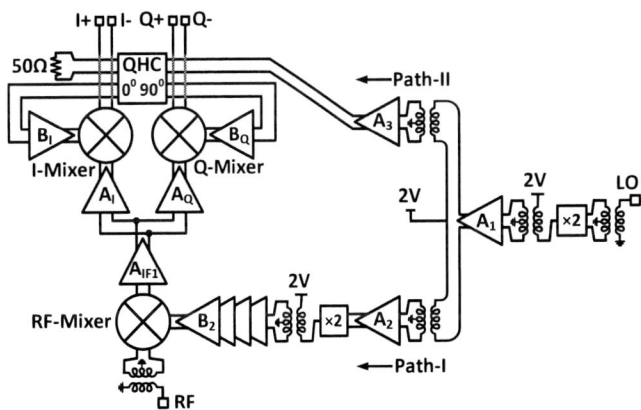

Fig. 1. Architecture of the proposed sliding IF I/Q receiver front-end.

design for both RF and LO chain. Measurement results and conclusion are described in Sections IV and V, respectively.

II. RECEIVER ARCHITECTURE

The block diagram of the mixer-first sliding-IF receiver front-end is shown in Fig 1. LNA is avoided due to its lower gain at frequencies close to f_{max}. Here, sliding-IF facilitates double downconversion of RF signal, first by two-third and then by one-third of carrier frequency, respectively. The selection of RF, LO and IF frequencies causes image frequencies to fall at one-third of carrier frequency which is very far from the passband of the first mixing stage. Due to higher gain at one-third of carrier frequency, signal amplification is done at IF stages followed by its downconversion by I/Q mixers. Entire RF chain is fully differential.

LO signal at two-third and one-third of carrier frequency is generated by the multiplication of external LO signal of 50 GHz. Here external LO input signal frequency is multiplied by 2 by a frequency doubler which generates LO signal at 100 GHz. This signal is amplified and then divided into two paths. Path-I of the LO chain generates the LO signal for the RF mixer using one additional frequency doubler and LO signal amplifiers. Path-II generates quadrature LO signal for I/Q mixer stages by using signal amplifiers and a quadrature hybrid coupler (QHC) for 90 degree phases shift.

979-8-3503-2123-4/23 $31.00 © 2023 IEEE   37   2023 IEEE Radio Frequency Integrated Circuits Symposium

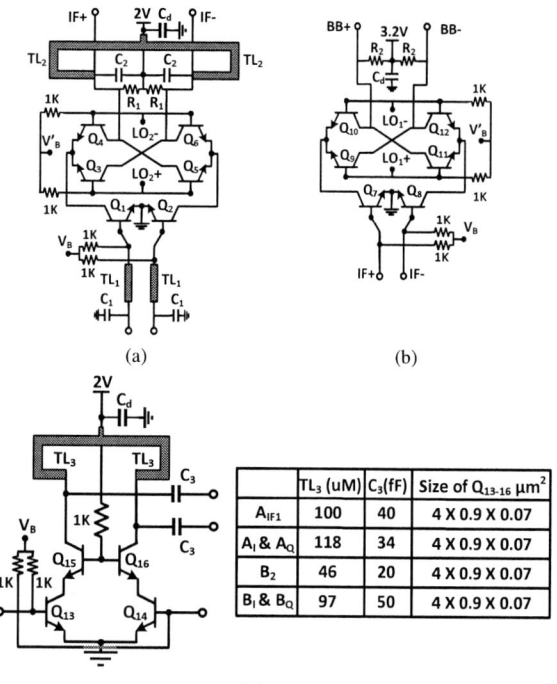

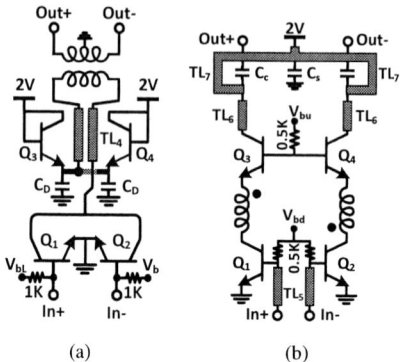

Fig. 3. Schematic of (a) frequency doublers (b) 100 GHz amplifiers (A_2 and A_3 are shown in the receiver's block diagram in Fig. 1).

	TL_3 (uM)	C_3(fF)	Size of Q_{13-16} μm^2
A_{IF1}	100	40	4 X 0.9 X 0.07
A_I & A_Q	118	34	4 X 0.9 X 0.07
B_2	46	20	4 X 0.9 X 0.07
B_I & B_Q	97	50	4 X 0.9 X 0.07

(c)

Fig. 2. Schematic of (a) differential double balanced Gilbert cell of RF mixer (b) differential double balanced Gilbert cell of I/Q mixers and (C) differential cascode gain amplifiers units for A_{IF1}, A_I, A_Q, B_2, B_I and B_Q as shown in Fig 1.

A. Signal Chain

As shown in Figs. 2a and 2b, active double-balanced Gilbert cell based mixer is used for both RF and I/Q downconversion because of its lower LO power level requirements, better port to port isolation and common mode noise immunity [4], [6], [7]. For high conversion gain and low noise contribution, transconductance stages of both mixers are sized using HBTs (Q_{1-2} and Q_{7-8}) with emitter area of $8 \times 0.9 \times 0.07$ um^2. Switching quad of Gilbert cells is designed using HBTs (Q_{3-6} and Q_{9-12}) with emitter area of $4 \times 0.9 \times 0.07$ um^2. All HBTs are biased near their peak f_{max} and serially programmable 5 bit on-chip digital-to-analog converters (DACs) are used in each bias node to allow some individual circuit-level tuning. RF mixer and I/Q mixer are biased with 2V and 3.2V voltage supplies. De-coupling capacitor (C_d) of $100fF$ is used at the closest common node of power supplies. Differential input of the RF mixer is connected to GSG pads through a rectangular shaped Marchand balun and a LC matching network. LC matching network is implemented with 4 μm wide MS lines (TL_1 = 40 μm) and metal-insulator-metal (MIM) capacitors (C_1 = 7fF). RLC load of the RF mixer is implemented using 4 μm wide MS line (TL_2 = 47 μm), resistors (R_1 = 200 ohm), and capacitors ($C_2 = 70fF$). On the other hand, load of the I/Q mixer is implemented using resistors (R_2 = 200 ohm). As shown in Fig 2c, IF and LO amplifiers (A_{IF1}, A_I, A_Q, B_2, B_I and B_Q) are designed using differential cascode

circuit topology. Their loads are designed using 2 μm wide transmission lines (TL_3) and coupling capacitors (C_3). Length of TL_3, capacitance value of C_3 and sizes of transistors are mentioned in the table shown in Fig 2c.

B. LO chain

The frequency multipliers are based on push-push common-emitter doubler (Q_1 and Q_2), generating the second harmonic and canceling the fundamental frequency. They are followed by a balun, which makes a differential output and filters the fundamental frequency, as shown in Fig. 3a. The second-order nonlinearity is maximized by sizing $Q_{1\&2}$ $10 \times 0.9 \times 0.07$ um^2 and biasing them in class-B region. The cancellation of the fundamental in the output is sensitive to the device mismatches and the imbalance of the input signals. So, two independently adjustable biasing branches (V_{bL} and V_{bR}) adjust the linear gains separately to minimize the leaking of the fundamental frequency, as in [10]. Q_3 and Q_4 serve to lower the collector voltage of $Q_{1\&2}$ to avoid breakdown. $Q_{3\&4}$ are sized $10 \times 0.9 \times 0.07$ um^2 to provide more headroom. Two MIM capacitors of $C_D = 831fF$ are added in parallel to bypass $Q_{3\&4}$ from the signal path. The single-ended output is connected to the balun by a 4 μm wide transmission line (TL_4) whose length is 86 μm and 56 μm for the first and second frequency multipliers, respectively.

Besides the independent biasing and output filtering in the multipliers, for spurious suppression of the LO path, it is also important to improve the differential symmetry by enhancing the common mode rejection of the amplifying stages. This is achieved by a cascode amplifier employing two coupled inductors between the cascoded devices, as shown in Fig. 3b. The chosen size of $4 \times 0.9 \times 0.07 um^2$ for Q_{1-4} provides a reasonable compromise between the voltage gain, consumed current, and reactance of the input impedance. TL_{5-7} and C_c are utilized to implement matching networks at the input and output. $TL_{5\&6}$ are 4 μm wide and TL_5 has 56 μm length, while TL_3 is 161 μm, 25 μm, and 71 μm long for amplifiers A_{1-3}, respectively. In the case of TL_7, a width of 2 μm is chosen and the length is 102 μm, 93 μm, and 83 μm for A_{1-3}. C_c is used for both coupling and matching, while C_s works

979-8-3503-2123-4/23 $31.00 © 2023 IEEE

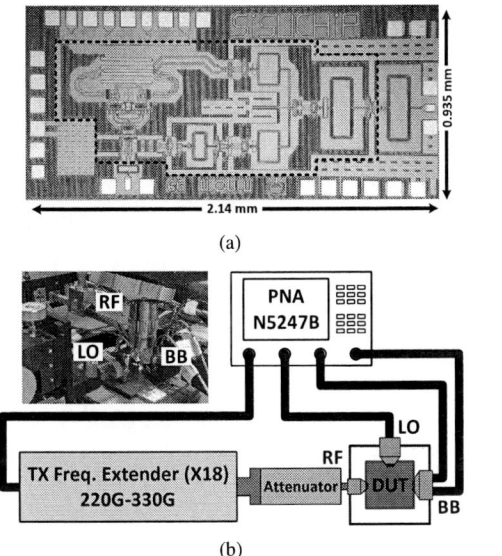

(a)

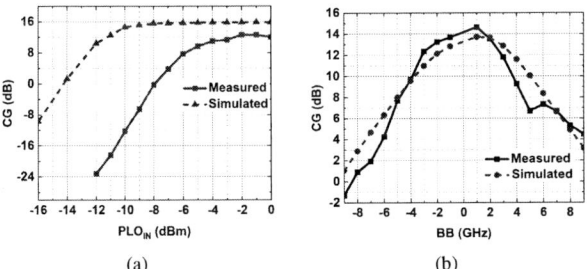

(b)

Fig. 4. (a) Die micrograph of the proposed receiver, where black dotted rectangular encloses the active area ($1.02\ mm^2$) without pads and balun. (b) Measurement set-up.

Fig. 5. Measured vs simulated curves of (a) CG (dB) vs input LO power (dBm) (b) CG (dB) vs baseband frequency.

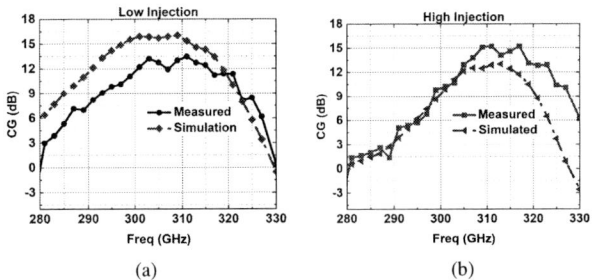

Fig. 6. Measured vs simulated curves of CG (dB) vs RF frequency (GHz) for (a) low injection (b) high injection.

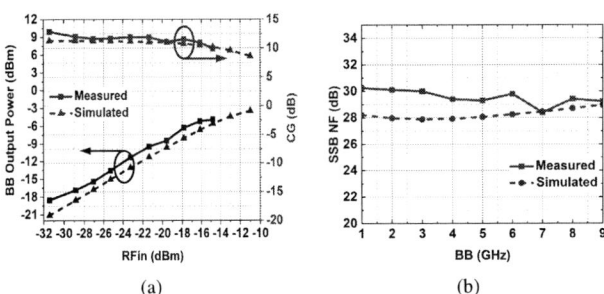

Fig. 7. Measured vs simulated curves of (a) CG (dB) and output power (dBm) vs input power (dBm) (b) single-sideband noise figure.

as a supply filter. Their values are $70\ fF$ and $76\ fF$ for A_2, and $80\ fF$ and $149\ fF$ for A_3, respectively.

III. MEASUREMENTS

The micrograph of the receiver die is shown in Fig. 4a with chip area of $2.14 \times 0.935\ mm^2$. As shown in Fig. 4b, the on-chip measurements of the receiver are performed using spectrum analyzer mode of Keysight's PNA-X N5247B, VDI's WR3.4 CW Tx frequency extender and an external waveguide attenuator. Both RF and LO signals are fed through GSG probes and the differential I/Q baseband outputs are measured using GSGSG probe. The receiver on-chip conversion gain (CG) is calculated based on input RF power and output baseband power including input balun, input and output probe pads.

Fig. 5a compares the simulated vs measured CG by sweeping the input LO power at 50 GHz. RF frequency is fixed at 302 GHz. It is observed that -3 dBm LO power is sufficient for maximizing the mixer gains. The saturation point in the conversion gain vs. LO power sweep also shows the largest discrepancy between the measured and simulated responses we have. Obviously the simulation of the LO multiplier

part is not as accurate as the mixer branch conversion gain simulations. Fig. 5b is plotted by sweeping the RF frequency form 300-318 GHz and fixed LO at 51.5 GHz. It can be seen that receiver measured peak CG is 17 dB with 3-dB and 6-dB BB bandwidth of 8 GHz and 12 GHz, respectively. Fig. 6a and 6b depicts the RF response of the receiver for low injection and high injection case, respectively. For RF response, both RF and LO frequencies are swept together by keeping fixed baseband frequency at 2 GHz. Measured 3-dB and 6-dB RF bandwidths of the receiver are 26 GHz and 36 GHz, respectively.

Fig. 7a plots the measured and the simulated input-referred compression (P_{1dB}) of the receiver. Here, RF is fixed at 309 GHz and LO is kept at 52 GHz. Measured input-referred compression point is -17 dBm. The discrepancy between measured and simulated results is probably due to block-wise EM simulation of the whole receiver which is inevitable due large electrical dimension of receiver. This doesn't capture all interstage coupling effects. Additionally, HBT models are characterized below 100 GHz frequency range which also adds to discrepancies between simulated and measured results. Noise figure measurement is done using method mentioned in [11]. Output noise power (N_o) of the receiver is measured using Keysight's UXA N9040B signal analyzer. Fig. 7b compares the simulated vs measured single-sideband noise figure. With input LO signal at 51.5 GHz, input thermal noise floor (N_i) of -174 dBm/Hz from 300 GHz to 318 GHz is downconverted to baseband frequencies 1GHz-9GHz. With known value of N_i, measured N_o and CG, NF is calculated. Average noise figure is measured to be 29.5 dB.

Table 1 shows the comparison with state-of-the-art sub-THz receiver front-ends in both SiGe and CMOS

979-8-3503-2123-4/23 $31.00 © 2023 IEEE

Table 1. Summary of the important performance parameters and comparison with related works

	This work	[11]	[4]	[5]	[3]	[6]	[7]	[12]	[8]	[9]
Technology	130 nm SiGe	130 nm SiGe	130 nm SiGe	130 nm SiGe	130 nm SiGe	55 nm SiGe	130 nm SiGe	130 nm SiGe	40 nm CMOS	28 nm CMOS
f_t/f_{max} (GHz)	300/450	280/435	300/450	300/500	300/500	320/370	300/500	300/500	-/280	-/-
Frequency (GHz)	310	320	240	240	240	240	240	190	265	293
Topology	Mixer first Sliding IF	Mixer first DCR	LNA first DCR	Mixer first DCR	LNA first DCR	Mixer first DCR	LNA first DCR	LNA first DCR	Mixer first DCR	Mixer first Sliding IF
LO	X2, X4	X9	X16	X8	LO driver	VCO, X2, /128	Ext. LO	X12	X6	PLL
Con. Gain (dB)	15.2	-14	11	13	18	23	20.6	49	26	20
3-dB/6-dB RFBW (GHz)	26/36	13/-	18/-	-	-	74/-	-	-	-	-
3-dB/6-dB BBBW (GHz)	8/12	-	-	55/-	25/-	59/-	40	-/23	-	8/-
SSB NF (dB)	29.5	36	16	18	21[a]	27.5	13.2	16.5	17.4	20
Input-Referred P_{1dB} (dBm)	-17.0	-	-18	-12	-27	-27.5	-22	-48	-34	-17.3
P_{diss} (mW)	296,[b] 110[c]	72,[b] 3000[c]	146,[b] 720[c]	500	512	293,[b] 566[c]	134	29,[b] 171[c]	467	52
Area (mm^2)	2	0.92	1.56	1.25	2.1	1.837	0.43	3.6	4.55	0.72

[a] Simulated. [b] Signal chain. [c] LO chain.

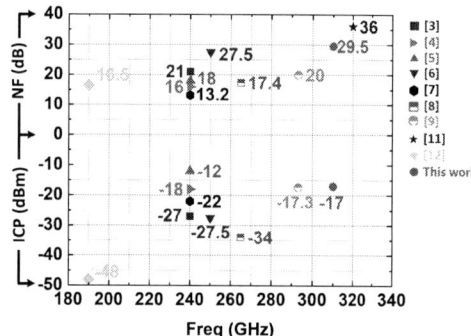

Fig. 8. NF and ICP comparison with state-of-the-art. ICP of [11] is not available.

technologies. It can be seen that, due to the sliding-IF architecture, DC power consumption in this work is lowest among the SiGe BiCMOS based receivers. The difference between NF and ICP values signifies the dynamic range of the receiver. Smaller gap between NF and ICP provides better dynamic range. Fig. 8 compares NF and ICP of this work with the previous arts. It can be seen that this receiver offers one of the best dynamic range in this category.

IV. CONCLUSION

A 300-320 GHz sliding-IF I/Q receiver front-end in 130 nm SiGe BiCMOS technology is presented. Due to the paucity of the power and the gain at sub-THz/THz frequency range, the proposed receiver with sliding-IF architecture, operating near f_{max}, removes the need of high frequency LO same as carrier frequency. Consequently, this noticeably reduces the power consumption of the LO chain and also provides an opportunity to amplify the signal at the IF stage. In LO chain, the major effort was in reducing spurious tones in the multipliers and maximizing the symmetry of differential LO signals. The receiver indicates its potential use in phased array receivers at sub-THz and THz regions for future ultra high speed communication system like 6G.

ACKNOWLEDGMENT

This research work has been financially supported by the Academy of Finland 6G Flagship (grant 346208). The authors would like to thank Keysight Technologies for measurement equipment donation and Klaus Nevala for his help during measurements.

REFERENCES

[1] M. A. Uusitalo et al., "6G Vision, Value, Use Cases and Technologies from European 6G Flagship Project Hexa-X," *IEEE Access*, vol. 9, 2021.

[2] H. Rücker and B. Heinemann, "High-performance SiGe HBTs for next generation BiCMOS technology," *Semiconductor Science and Technology*, vol. 33, no. 11, oct 2018.

[3] M. Elkhouly et al., "A 240 GHz Direct Conversion IQ Receiver in 0.13μm SiGe BiCMOS Technology," *2013 IEEE Radio Frequency Integrated Circuits Symposium (RFIC)*, p. 305–308, 2013.

[4] N. Sarmah et al., "A Fully Integrated 240-GHz Direct-Conversion Quadrature Transmitter and Receiver Chipset in SiGe Technology," *IEEE Trans. Microw. Theory Techn.*, vol. 64, no. 2, p. 562–574, 2016.

[5] M. H. Eissa et al., "A 220-275 GHz Direct-Conversion Receiver in 130-nm SiGe:C BiCMOS Technology," *IEEE Microwave and Wireless Components Letters*, vol. 27, no. 7, p. 675–677, 2017.

[6] U. Alakusu et al., "A 210–284-GHz I–Q Receiver With On-Chip VCO and Divider Chain," *IEEE Microwave and Wireless Components Letters*, vol. 30, no. 1, p. 50–53, 2020.

[7] E. Turkmen et al., "A 225-265 dB I-Q Receiver in 130-nm SiGe BiCMOS for FMCW Radar Applications," *IEEE Microwave and Wireless Components Letters*, vol. PP, no. 99, p. 1–4, 2022.

[8] S. Hara et al., "A 76-Gbit/s 265-GHz CMOS Receiver With WR-3.4 Waveguide Interface," *IEEE Journal of Solid-State Circuits*, vol. PP, no. 99, p. 1–11, 2022.

[9] O. Memioglu, Y. Zhao, and B. Razavi, "A 300GHz 52mW CMOS Receiver with On-Chip LO Generation," *2022 IEEE International Solid-State Circuits Conference (ISSCC)*, vol. 65, p. 1–3, 2022.

[10] M. J. Nokandi, A. Pärssinen, and T. Rahkonen, "A Coherent Spurious Analysis for Sub-THz Frequency Multiplier Chains," in *2020 IEEE International Symposium on Circuits and Systems (ISCAS)*, 2020, pp. 1–5.

[11] E. Öjefors, B. Heinemann, and U. R. Pfeiffer, "Subharmonic 220- and 320-GHz SiGe HBT Receiver Front-Ends," *IEEE Transactions on Microwave Theory and Techniques*, vol. 60, no. 5, p. 1397–1404, 2012.

[12] Y. Zhang et al., "LO Chain (×12) Integrated 190-GHz Low-Power SiGe Receiver with 49-dB Conversion Gain and 171-mw DC Power Consumption," *IEEE Transactions on Microwave Theory and Techniques*, vol. 69, no. 3, p. 1943–1954, 2021.

RMo1C-1

A V-band Four-Channel Phased Array Transmitter Beamforming IC With 0.7-Degree Phase Step in 20 dB Dynamic Range

Cheol So[#1], Eun-Taek Sung[#], and Songcheol Hong[#]

[#]KAIST, Republic of Korea

[1]cheolso93@gmail.com

Abstract—This paper presents a V-band 4-channel transmitter beamforming IC fabricated in 28-nm CMOS process, which includes power amplifiers (PA), active vector modulators, and power dividers. The PA is linearized by a cold-FET controlled coupled inductor, improving the AM-AM and AM-PM. The vector modulator consists of a summing amplifier that combines I and Q vectors to determine the amplitude and phase of a synthesized signal, along with an active I/Q generator. The nonlinearity of the vector summing amplifier is overcome by introducing a CLC network of a 45° phase shifter. It shows the 0.1 dB RMS gain and 2° RMS phase error at 60 GHz, with 3-bit gain and 9-bit phase control. An OP1dB of 10.8 dBm and a peak gain of 14.3 dB are achieved in the 3-dB frequency bandwidth of 57 GHz to 63.3 GHz. The chip area is 4.4 mm^2, and the power consumption is 295 mW.

Keywords— Millimeter-wave (mm-wave), CMOS, V-band, phased-array, transmitter, linearizer, cold-FET, CLC network, coupled inductor, active R-L poly-phase filter.

I. INTRODUCTION

A wide frequency bandwidth of unlicensed bands around 60 GHz offers high data rates in communications and high range resolutions in radars. However, the frequency bands suffer from high atmospheric attenuation and path loss. To overcome these, beamforming techniques can be used to improve the signal-to-noise ratio(SNR) of a system by aligning transmitting and receiving beams with phased antenna arrays[1], [2], [3]. As the number of array channels is increased by N times, an array transmitter's total power consumption is also increased by N times, while the EIRP is increased by N^2 times. This results in an overall system efficiency improvement by N times[4]. The improved angular resolution of a system increases the detection accuracies of targets and allows it to be a robust system from interference signals. In addition, as the output power required in each channel is reduced, a beamforming front-end IC including power amplifiers can be integrated in a CMOS process.

However, as the EIRP of a main lobe beam becomes very large in large-scale antenna arrays, the power of an interfering sidelobe beam is also not negligible. Therefore, a high phase resolution of a phased array antenna system is necessary to keep the sidelobe level low and to reduce interference signals in undesired directions. Since the sidelobe level can be lowered by amplitude tapering in the antenna array, a gain control function in each channel is required[1], [5].

This paper presents a V-band 4-channel transmitter beamforming IC for a large-scale antenna array. A cascode PA is proposed to improve the AM to AM and AM to

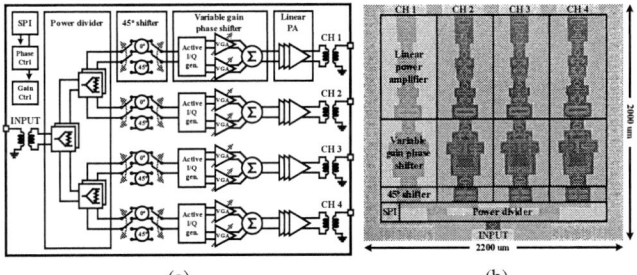

(a)   (b)

Fig. 1. The V-band 4-channel transmitter beamforming IC. (a) Block diagram, and (b) microphotograph.

PM, in which cold-FET controlled inductors are introduced in the inter-stage nodes for linearization. The phase and gain of a signal are synthesized by using an active vector modulator[6], [7]. The intrinsic arctangent non-linearity of the vector modulator is addressed, which is simply avoided by introducing a 45° phase shifter made of a CLC network.

II. THE TRANSMITTER BEAMFORMER IC ARCHITECTURE

The proposed V-band 4-channel phased array transmitter beamforming IC is illustrated in Fig. 1, which includes a block diagram and a microphotograph of the implemented chip. The input signal is distributed to the 4 channels via power dividers. Each channel comprises of a 45° phase shifter, a variable gain phase shifter, and a linear PA.

A. Power amplifier linearized by cold-FET controlled coupled inductor

The scaling in an advanced RF CMOS transistor leads to a low breakdown voltage, making it difficult to achieve a high output power from a PA. A stacked structure for a high supply voltage is widely used to get a high output power. However, the gain and efficiency are reduced due to the large parasitic capacitances at the inter-stage nodes. Thus, series or parallel inductors at the node are used to cancel out the parasitic capacitances, which improve the gain and efficiency of a stacked amplifier[8].

In addition, a millimeter wave (mm-wave) CMOS PA often operates at a class-AB bias rather than class-A to achieve high efficiency and linearity[9], [10]. However, there is a gain compression problem due to the harmonics. In addition, as the input power increases, the average gate-source capacitance of the CS amplifier decreases. This makes phase leads[11].

979-8-3503-2123-4/23 $31.00 © 2023 IEEE   41   2023 IEEE Radio Frequency Integrated Circuits Symposium

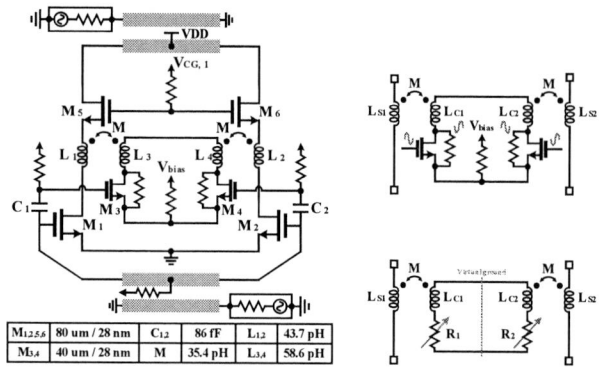

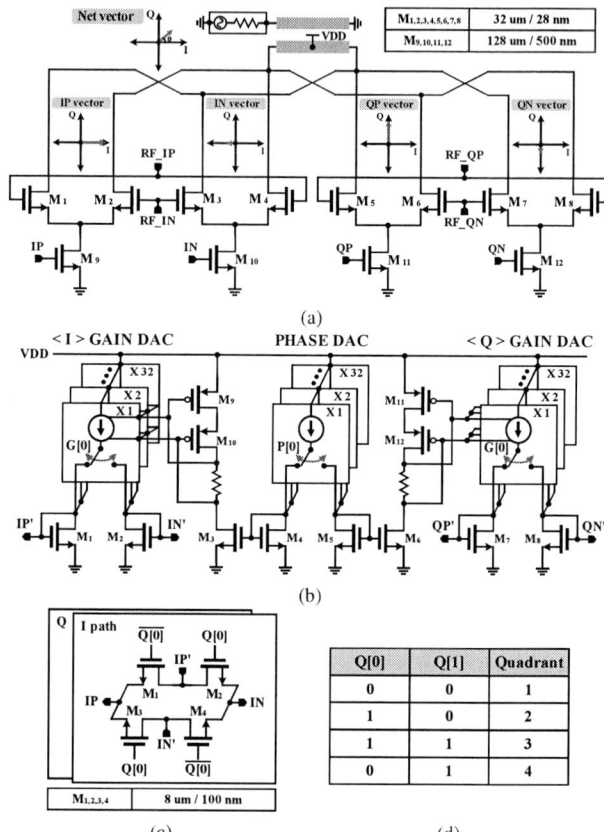

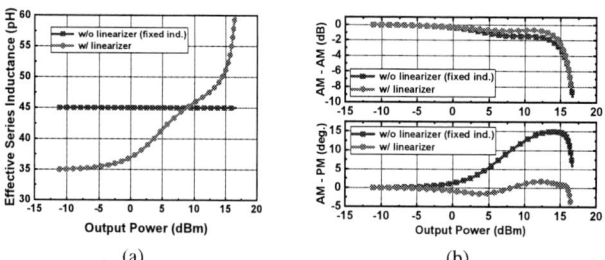

| $M_{1,2,5,6}$ | 80 um / 28 nm | $C_{1,2}$ | 86 fF | $L_{1,2}$ | 43.7 pH |
| $M_{3,4}$ | 40 um / 28 nm | M | 35.4 pH | $L_{3,4}$ | 58.6 pH |

(a)   (b)

Fig. 2. Proposed linear power amplifier. (a) Schematics, and (b) cold-FET controlled coupled inductor linearizer at inter-stage node, and its equivalent circuits.

(a)   (b)

Fig. 3. Simulation results. (a) The effective inter-stage series inductance, (b) AM to AM, AM to PM with the linearizer on and off.

To address the gain compression and the phase distortion of the cascode PA, a linearization technique with a cold-FET controlled coupled inductor at the inter-stage node is proposed.

The PA consists of a two-stage driver amplifier and a linear power amplifier. The two-stage driver amplifiers have differential cascode configurations with cross-coupled capacitors to achieve sufficient gains[12]. Fig. 2 shows the schematics of the proposed linear power amplifier, cold-FET controlled coupled inductor linearizer, and its equivalent circuits. In Fig. 2 (b), the proposed linearizer consists of the interstage series inductor (Ls), coupled inductor (Lc), and a cold-FET transistor which is connected to Lc. The effective resistance of the cold-FET transistor increases as the amplitude of the voltage swing at the drain increases with the increase of the power when the same DC voltage is applied to the drain and source, and a specific bias voltage is applied to the gate[13]. In the proposed structure, the opposite phase voltage swing is also applied to the drain and gate, resulting in more resistance changes when the same voltage swing is applied. By expressing the cold-FET as a variable resistance R, the effective inter-stage series inductor Ls can be expressed as

$$L_{S,eff} = L_S - \frac{w^2 M^2 L_C}{R^2 + w^2 L_C^2}. \quad (1)$$

As the input power increase, R also increases, leading to the larger inductance Ls. It helps to cancel out parasitic capacitances at the inter-stage node, resulting in improved AM to AM. Additionally, as the series inductance of the signal path

| M$_{1,2,3,4,5,6,7,8}$ | 32 um / 28 nm |
| M$_{9,10,11,12}$ | 128 um / 500 nm |

(a)

(b)

M$_{1,2,3,4}$	8 um / 100 nm	

Q[0]	Q[1]	Quadrant
0	0	1
1	0	2
1	1	3
0	1	4

(c)   (d)

Fig. 4. Schematics of the vector modulator. (a) Vector summing amplifier, (b) current steering DACs for phase and gain selections, (c) DPDT switches for quadrant selection, and (d) the quadrant selection logic.

increases, it causes a phase lag. This compensates for AM to PM distortion.

B. Vector modulator as a variable gain phase shifter

Fig. 4 shows the schematic of the proposed variable gain active phase shifter. The active I/Q generator is used, which provides a finite gain in the I/Q generator by placing an R-L poly-phase filter in the inter-stage node of a cascode amplifier without an additional impedance matching network[7]. The I and Q vectors are fed into a vector summing amplifier, which generates four output vectors: I positive, I negative, Q positive, and Q negative. The sizes of the vectors are determined by the respective current sources, of which currents are controlled by current steering DACs. The DAC uses binary weighted cells. It is specially designed to keep the same total DC current of the vector summing amplifier regardless of all gain and phase states, which is achieved through complementary operations. The ratio of I and Q vectors is determined by the phase control DAC, while the ratio of positive and negative vectors is determined by the gain control DAC for each I and Q vector. By changing the polarities of the positive and negative vectors through the DPDT switches between the DAC and the vector summing amplifier, the gain and phase of the other three quadrants can also be expressed[6], [7].

979-8-3503-2123-4/23 $31.00 © 2023 IEEE   42

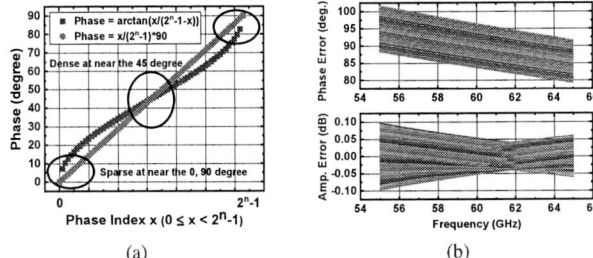

(a)   (b)

Fig. 5. Non-ideal effects of the vector modulator. (a) Arctangent non-linearity of phase, and (b) I/Q errors due to load impedance variations of the I/Q generator.

Fig. 5 shows the non-ideal effects of the vector modulator. The phase is determined through the DAC, which can be described with an arctangent function in the proposed phase shifter. As a result, the phase step increases near $0°$ and $90°$. For example, when a 6-bit resolution represents a quadrant, the phase step is $1.4°$ in a linear step, while the maximum phase step is $7.2°$ in the vector modulator-based phase shifter.

In addition, the current flowing through the current source changes depending on the gain and phase state in the vector summing amplifier. In Fig. 4 (a), the input impedance changes as the operating region of M_{1-8} changes from the saturation to the triode region, which causes the load impedance variations of the I/Q generator, resulting in I/Q errors. When the real and imaginary parts of the load impedance vary by 20 %, the amplitude and phase errors of the I/Q generator become 0.12 dB and $12.8°$, respectively. Therefore, there is a limitation in expressing a phase with a high phase resolution using the vector modulator.

C. 45° phase shifter for high phase resolution

Due to the abovementioned phase nonlinearity of the vector modulator, the constellations near $0°$ and $90°$ show large gain and phase errors. Additionally, I/Q errors due to input impedance variations of the vector summing amplifier also cause large gain and phase errors near $0°$ and $90°$. Therefore, when an additional $45°$ shifter has been introduced, the constellations near $0°$ and $90°$ are also expressed by the more linear region of the vector modulator. This reduces the gain and phase errors near $0°$ and $90°$. The phase shifter is implemented with a π-type CLC artificial transmission line for a compact size and a low return loss[14]. Fig. 6 shows the schematic and simulation results of the implemented CLC based $22.5°$ phase shifter. Considering the cut-off frequency of the CLC network, the $45°$ phase shifter is implemented using two stages of $22.5°$ phase shifters. To have the phase shift of an angle ϕ, the capacitance(C_t) and inductance(L_t) of the CLC network can be expressed as

$$C_t = \frac{1-\cos(\phi)}{wZ_o\sin(\phi)}, \quad L_t = \frac{Z_o\sin(\phi)}{w}. \quad (2)$$

A switched capacitor and coupled inductor are used to implement an effective transmission line with two different

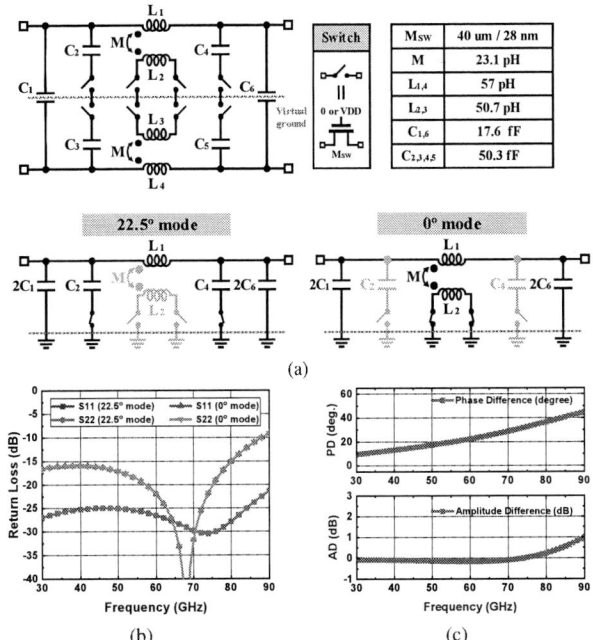

(a)

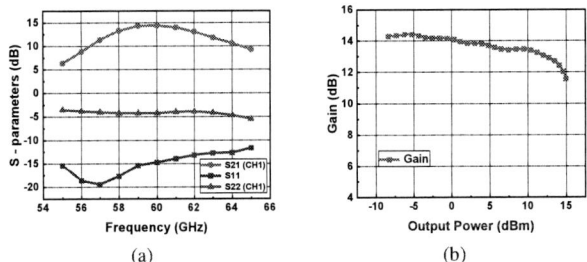

(b)   (c)

Fig. 6. 2-stage CLC $45°$ phase shifter. (a) Schematic of the CLC network based $22.5°$ phase shifter, (b) simulated S-parameters, and (c) amplitude and phase difference of the $22.5°$ phase shifter.

Fig. 7. Measured results. (a) Average S-parameters at maximum gain state, and (b) gain versus output power.

phase modes. C and L in each phase mode can be expressed as

$$C_{t,22.5°} = 2C_1 + C_2, \quad L_{t,22.5°} = L_1, \quad (3)$$

$$C_{t,0°} = 2C_1, \quad L_{t,0°} = L_1 - \frac{M^2}{L_2}. \quad (4)$$

C_1, C_2, L_1, L_2, and M are set to make $\phi1$-$\phi2$ to be $22.5°$. The return loss of the $22.5°$ phase shifter is less than -15 dB, while the phase and amplitude differences are $22.3°$ and 0.16 dB at 60 GHz, respectively.

III. MEASUREMENT RESULTS

Fig. 1 (b) presents a microphotograph of the proposed 4-channel transmitter beamforming IC, which is implemented using a 28-nm bulk CMOS process. It has a total chip size of 2200 um x 2000 um, including the I/O pads. The measured DC power consumption is 295 mW. The S-parameter and P1dB are measured by using an Anritsu 37397D vector

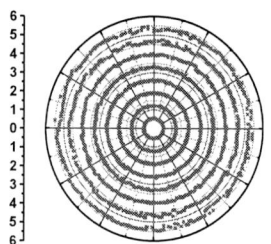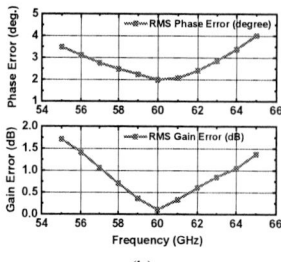

(a)   (b)

Fig. 8. Measured results. (a) Constellation for 3-bit gain and 9-bit phase states at 60 GHz, and (b) RMS gain and phase errors.

Table 1. Performance summary and comparison

	This work	[1]	[2]	[3]
Process	28 nm CMOS BULK	65 nm CMOS BULK	65 nm CMOS BULK	45 nm CMOS SOI
Frequency (GHz)	60	60	60	60
Number of Channel	4	4	4	1
3-dB Bandwidth (GHz)	6.3	11.5	N/A	8
Gain / Channel (dB)	20.3*	15	0	22
Gain Dynamic Range (dB)	20	5.2	N/A	14
Phase Resolution (°)	0.7	22.5	22.5	11
RMS Gain Error (dB)	1.06	0.53	2.5	0.7
RMS Phase Error (°)	2.76	8.8	18	5
OP1dB / Channel (dBm)	10.8	7.1	5	10
Pdc / Channel (mW)	73.8	101	100	67
Size (mm^2)	4.4	2.88	3.74	0.45

* Additional 6 dB due to 4 channel distribution.

network analyzer. All gain and phase states of the channels are controlled through a Total Phase Cheetah SPI host adaptor, and the data is captured through GPIB.

Fig. 7 shows the measured S-parameters and OP1dB of channel 1 at the maximum gain state. It achieves a peak gain of 14.3 dB in the 3-dB frequency range of 57 GHz to 63.3 GHz and the OP1dB of 10.8 dBm. Fig. 8 (a) shows the constellation of 9-bit phase states in 360° and 3-bit gain states in a 20 dB dynamic range. Fig. 8 (b) shows the RMS gain and phase errors of the 4-channel transmitter beamforming IC. The RMS gain and phase errors show 1.06 dB and 2.76° in the 3-dB frequency range, respectively.

Table 1 summarizes the performance of the proposed 4-channel phased array transmitter beamforming IC and compares them with the other state-of-the-art designs.

IV. CONCLUSION

A V-band 4-channel phased array transmitter beamforming IC is presented, which is implemented using a 28-nm bulk CMOS process. It utilizes a cold-FET controlled coupled inductor linearizer at the inter-stage node to improve the AM to AM and AM to PM in the cascode power amplifier. Additionally, a 45° phase shifter is introduced to obviate the non-ideal effects of the vector summing phase shifter, which results in a high phase resolution performance of 0.7° phase step.

ACKNOWLEDGMENT

This work was supported by the Institute of Information Communications Technology Planning and Evaluation (IITP) funded by the Korea Government, Ministry of Science and ICT (MIST) under Grant 2019-0-00826. The authors also would like to thank the Integrated circuit Design Education Center (IDEC) for their support in CAD tools. Dr. Eun-Taek Sung is currently working at Samsung Electronics, Republic of Korea.

REFERENCES

[1] J. G. Lee, T. H. Jang, G. H. Park, H. S. Lee, C. W. Byeon, and C. S. Park, "A 60-GHz Four-Element Beam-Tapering Phased-Array Transmitter With a Phase-Compensated VGA in 65-nm CMOS," *IEEE Transactions on Microwave Theory and Techniques*, vol. 67, no. 7, pp. 2998–3009, 2019.

[2] J.-L. Kuo, Y.-F. Lu, T.-Y. Huang, Y.-L. Chang, Y.-K. Hsieh, P.-J. Peng, I.-C. Chang, T.-C. Tsai, K.-Y. Kao, W.-Y. Hsiung, J. Wang, Y. A. Hsu, K.-Y. Lin, H.-C. Lu, Y.-C. Lin, L.-H. Lu, T.-W. Huang, R.-B. Wu, and H. Wang, "60-GHz Four-Element Phased-Array Transmit/Receive System-in-Package Using Phase Compensation Techniques in 65-nm Flip-Chip CMOS Process," *IEEE Transactions on Microwave Theory and Techniques*, vol. 60, no. 3, pp. 743–756, 2012.

[3] Y. Wang, H. Chung, Q. Ma, and G. M. Rebeiz, "A 57.5–65.5 GHz Phased-Array Transmit Beamformer in 45 nm CMOS SOI With 5 dBm and 6.1% Linear PAE for 400 MBaud 64-QAM Waveforms," *IEEE Transactions on Microwave Theory and Techniques*, vol. 69, no. 3, pp. 1772–1779, 2021.

[4] B. Sadhu, X. Gu, and A. Valdes-Garcia, "The More (Antennas), the Merrier: A Survey of Silicon-Based mm-Wave Phased Arrays Using Multi-IC Scaling," *IEEE Microwave Magazine*, vol. 20, no. 12, pp. 32–50, 2019.

[5] K. Kibaroglu, M. Sayginer, T. Phelps, and G. M. Rebeiz, "A 64-Element 28-GHz Phased-Array Transceiver With 52-dBm EIRP and 8–12-Gb/s 5G Link at 300 Meters Without Any Calibration," *IEEE Transactions on Microwave Theory and Techniques*, vol. 66, no. 12, pp. 5796–5811, 2018.

[6] K.-J. Koh and G. M. Rebeiz, "0.13-μm CMOS Phase Shifters for X-, Ku-, and K-Band Phased Arrays," *IEEE Journal of Solid-State Circuits*, vol. 42, no. 11, pp. 2535–2546, 2007.

[7] C. So, E.-T. Sung, and S. Hong, "A 60-GHz Variable-Gain Phase Shifter With an Active RL Poly-Phase Filter," *IEEE Transactions on Microwave Theory and Techniques*, vol. 71, no. 2, pp. 593–601, 2023.

[8] J. Oh, B. Ku, and S. Hong, "A 77-GHz CMOS Power Amplifier With a Parallel Power Combiner Based on Transmission-Line Transformer," *IEEE Transactions on Microwave Theory and Techniques*, vol. 61, no. 7, pp. 2662–2669, 2013.

[9] B. Park, S. Jin, D. Jeong, J. Kim, Y. Cho, K. Moon, and B. Kim, "Highly Linear mm-Wave CMOS Power Amplifier," *IEEE Transactions on Microwave Theory and Techniques*, vol. 64, no. 12, pp. 4535–4544, 2016.

[10] D. Kang, D. Yu, K. Min, K. Han, J. Choi, D. Kim, B. Jin, M. Jun, and B. Kim, "A Highly Efficient and Linear Class-AB/F Power Amplifier for Multimode Operation," *IEEE Transactions on Microwave Theory and Techniques*, vol. 56, no. 1, pp. 77–87, 2008.

[11] S. Kang, D. Baek, and S. Hong, "A 5-GHz WLAN RF CMOS Power Amplifier With a Parallel-Cascoded Configuration and an Active Feedback Linearizer," *IEEE Transactions on Microwave Theory and Techniques*, vol. 65, no. 9, pp. 3230–3244, 2017.

[12] C. So and S. Hong, "A V-Band Differential Push–Push Frequency Doubler With a Current-Reuse gm-Boosted Buffer," *IEEE Microwave and Wireless Technology Letters*, vol. 33, no. 3, pp. 299–302, 2023.

[13] S.-M. Weng, Y.-C. Lee, T.-H. Chen, and J. Y.-C. Liu, "A 60-GHz Adaptively Biased Power Amplifier with Predistortion Linearizer in 90-nm CMOS," in *2018 IEEE/MTT-S International Microwave Symposium - IMS*, 2018, pp. 651–654.

[14] T.-S. Chu, J. Roderick, and H. Hashemi, "An Integrated Ultra-Wideband Timed Array Receiver in 0.13 μm CMOS Using a Path-Sharing True Time Delay Architecture," *IEEE Journal of Solid-State Circuits*, vol. 42, no. 12, pp. 2834–2850, 2007.

979-8-3503-2123-4/23 $31.00 © 2023 IEEE

RMo1C-2

A 28/37 GHz Frequency Reconfigurable Dual-Band Beamforming Front-End IC for 5G NR

Jaehun Lee[#$1*], Hyoungkyu Jin[#*], Gyuha Lee[#*], Eun-Taek Sung[$], Songcheol Hong[#]

[#]Wave Embedded Integrated Systems Lab, KAIST, Republic of Korea

[$]Samsung Electronics, Republic of Korea

[1]jaehun0919@kaist.ac.kr

Abstract — A 28/37 GHz frequency reconfigurable dual-band beamforming front-end IC is presented. It integrates frequency reconfigurable structures in a power amplifier and a vector-summing type phase shifter to cover widely separated dual frequency bands with a single front-end IC. A wideband variable-gain low-noise amplifier and switches are also included to cover the dual bands. It has a 6-bit phase control resolution and a 16 dB gain control range with 0.9° and 0.62° RMS phase errors at 28 and 37 GHz, respectively. It achieves 13.5 and 13.5 dBm 1-dB compressed output power, 15.5% and 14.1% transmitter efficiency, and a noise figure of 4.4 and 4.9 dB at 28 and 37 GHz, respectively. The front-end IC delivers linear output power of 9.3 and 8.6 dBm with a 64-quadrature amplitude modulation 200 MHz signal at 28 and 37 GHz, respectively.

Keywords — Dual band, Reconfigurable architectures, Power amplifiers, Phase shifters, Low-noise amplifiers, Switches, 5G mobile communication, Phased arrays.

I. INTRODUCTION

During the evolution of the fifth-generation (5G) communication standard with the frequency range 2 (FR2) bands, multiple noncontiguous FR2 bands, including those at 28 GHz and 37 GHz, have been allocated in different countries [1], [2]. Therefore, to support worldwide use or international roaming of 5G communication, a beamforming front-end IC that supports multiple bands is crucial.

Many building blocks, such as power amplifiers (PAs), low-noise amplifiers (LNAs), phase shifters, switches, and variable-gain amplifiers, are required to implement a beamforming front-end IC to support phased array antennas. If two front-end ICs operating on different frequencies are configured in parallel for multi-band operation, the number of building blocks will be doubled, as shown in Fig. 1(a). This increases the form factor and the cost of the RF front-end.

In this paper, a dual-band frequency reconfigurable beamforming front-end IC that covers both the 28 GHz and 37 GHz bands with a single front-end IC is presented (Fig. 1(b)). The proposed beamforming IC not only operates on dual bands but can also control the beam precisely by integrating phase shifters that can control both the gain and phase with high-bit resolution.

II. CIRCUIT DESIGN

A. Frequency Reconfigurable Front-End IC Architecture

Fig. 2 shows the proposed frequency reconfigurable front-end IC architecture, which consists of a frequency

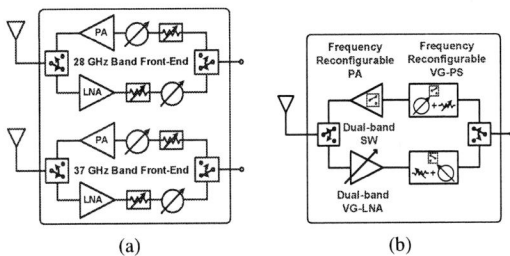

Fig. 1. Block diagram of the front-end ICs: (a) paralleled two front-ends for dual bands, (b) the proposed frequency reconfigurable dual-band front-end IC.

reconfigurable variable-gain phase shifter (VGPS), a frequency reconfigurable PA, a dual-band switch, and a dual-band variable-gain LNA (VGLNA). A vector-summing type active phase shifter is employed to implement the frequency reconfigurable VGPS, which provides both gain and phase control with fine resolution. Here, a frequency reconfigurable I/Q generator is used to produce dual-band I/Q signals. The PA utilizes reconfigurable transmission line transformer (TLT) structures for the output and inter-stage matching networks, allowing it to operate on either the 28 GHz or the 37 GHz band [3]. The respective optimal impedance matching on 28 GHz and 37 GHz bands is achieved by reconfiguring the matching networks. The dual-band switch has an intrinsic dual-band matching property shared by the transmitter (TX) and receiver (RX). It resonates out the off-capacitances (C_{OFF}) of turned-off transistors at both bands. As the leakage due to C_{OFF} is eliminated, the switch shows high isolation and low noise figure (NF) degradation on both bands. All switch transistors are located within the matching network layout to reduce the chip size compared to that of parallel-configured 28 GHz and 37 GHz switches. The dual-band VGLNA has a differential common-gate (CG) structure with a wideband NF input-matching characteristic. A CG cross-summing structure is introduced at the second stage, where the parasitic C_{gd}'s of the amplifiers are neutralized to have a phase-invariant property during the gain control process. The input and output matching networks use broad-side coupled transformers with high k values, and the inter-stage matching network is implemented with a strongly coupled resonator structure to use frequency peak splitting for the dual-band operation. In particular, the combination of the CG structure and the high-k input transformer enables dual-band operation in terms of both the NF and the gain. The proposed beamforming front-end IC

---

*equal contribution

979-8-3503-2123-4/23 $31.00 © 2023 IEEE   45   2023 IEEE Radio Frequency Integrated Circuits Symposium

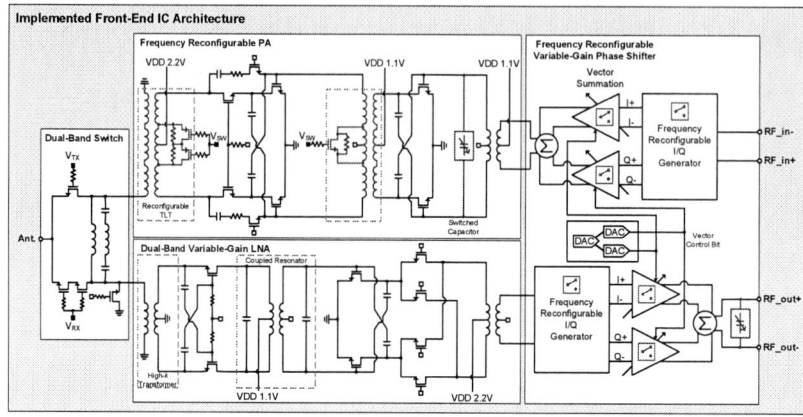

Fig. 2. The proposed dual-band front-end IC architecture with frequency reconfigurable VGPS and PA, and dual-band switch and VGLNA.

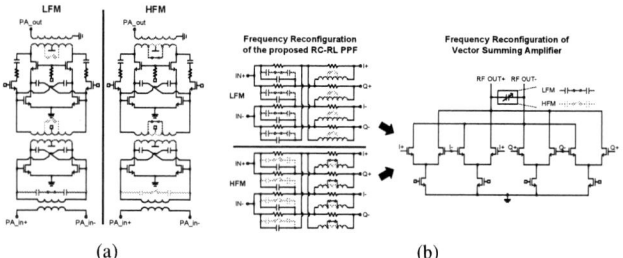

(a)   (b)

Fig. 3. Frequency reconfigurable dual-band operation of (a) PA and (b) VGPS.

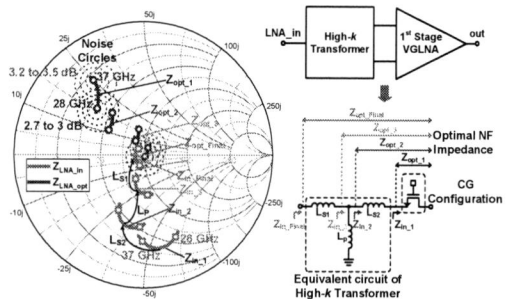

Fig. 4. Simultaneous noise and input matching of the VGLNA with CG configuration and high-k input transformer.

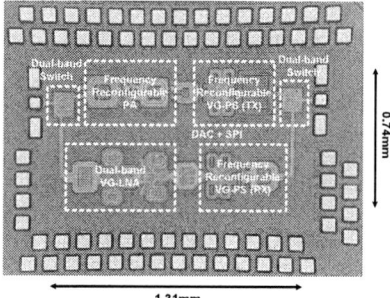

Fig. 5. Die microphotograph of the fabricated front-end IC.

with these blocks provides dual-band operation on the 28 GHz and 37 GHz bands due to the use of frequency reconfigurable circuits as well as wideband circuits. In addition, the TX path achieves a gain control function with the VGPS, and the RX path enables gain control with both the VGPS and VGLNA.

B. Dual-Band Operation of the Frequency Reconfigurable Front-End IC

Fig. 3 shows the frequency reconfigurable dual-band operation of the proposed front-end IC based on the dual-band operation of the PA and VGPS. The PA and VGPS can be reconfigured for low-frequency-mode (LFM) and high-frequency-mode (HFM) by the mode selection switches, which are integrated into the reconfigurable TLTs of the PA and the reconfigurable RC-RL poly-phase filter (PPF) of the VGPS. In LFM for 28 GHz operation, the reconfigurable TLTs

of the PA are switched off to realize higher inductances of the output and inter-stage matching networks, and the switched capacitor is turned on to realize higher capacitance of the input matching network. The PPF switches are controlled so that the PPF has higher capacitance and inductance, generating I/Q signals on the 28 GHz band. The generated I/Q signal is summed using a vector-summing amplifier, which is set to operate at a lower frequency by turning on a switched capacitor at the output. The gain and phase of the signal are determined by controlling the DC biases of the vector-summing amplifier, and the PA in the LFM amplifies the synthesized signal. In HFM for 37 GHz operation, the mode-selection switches of the PA operate in a direction opposite to that of LFM to realize lower inductances of the output and inter-stage matching networks and lower capacitance of the input matching network. The mode-selection switches of the VGPS also operate in a direction opposite to that of LFM so that the PPF has lower capacitance and inductance, and the output of the vector-summing amplifier is tuned to the higher frequency. The inductances of the matching networks that determine the load and source impedances of the PA are reconfigured to the optimal inductance for each frequency band, enabling optimal TX performance for each band.

Fig. 4 shows the simultaneous noise and input matching of the VGLNA. For the VGLNA with a CG configuration and a high-k input transformer, the trajectories for the input impedance (Z_{LNA_in}) and the optimal NF impedance (Z_{LNA_opt})

979-8-3503-2123-4/23 $31.00 © 2023 IEEE

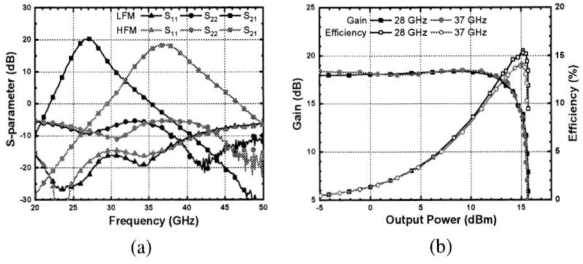

Fig. 6. Measured TX performance: (a) S-parameters, (b) gain and efficiency.

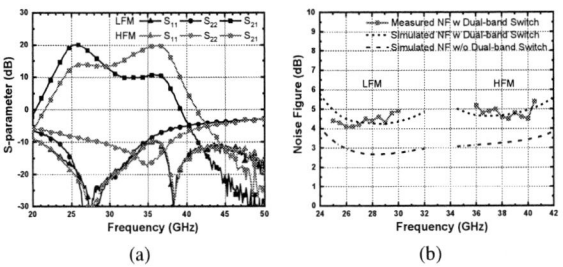

Fig. 7. Measured RX performance: (a) S-parameters, (b) noise figure.

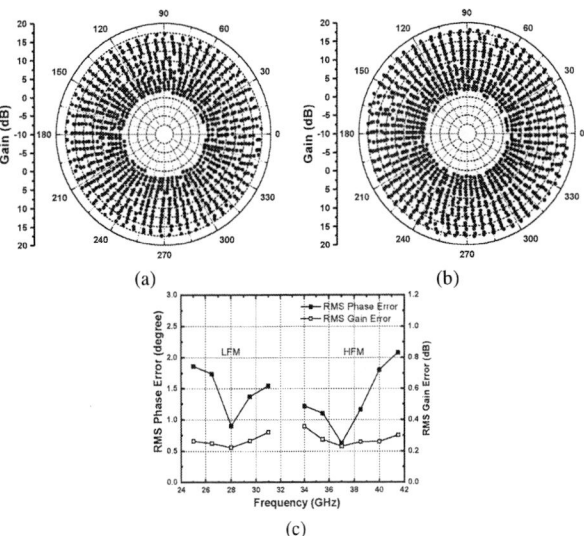

Fig. 9. Measured phase/gain control performance: TX vector constellation at (a) 28 GHz, (b) 37 GHz, and (c) TX RMS phase/gain errors.

with noise circles are plotted on a Smith chart from 28 to 37 GHz. Z_{LNA_in} and Z_{LNA_opt} move along the impedance path caused by L_{S2}, L_P, and L_{S1}, which are components of the input transformer equivalent circuit. As a result, Z_{LNA_opt} and Z_{LNA_in} can be simultaneously matched at the dual bands, allowing the VGLNA to operate on dual bands. A dual-band RX path is readily implemented by integrating the VGLNA and the VGPS. Since they operate as wideband and frequency reconfigurable components, the beamforming front-end IC allows dual-band operations at 28 GHz and 37 GHz with a small chip size.

III. IMPLEMENTATION AND MEASUREMENTS

The proposed frequency reconfigurable dual-band beamforming front-end IC was fabricated using a 28-nm FDSOI process. A microphotograph of the IC is shown in Fig. 5, and the core area is 0.97 mm².

Fig. 6 shows the measured TX performance. The TX gains are 18.5 and 18.4 dB at 28 and 37 GHz, respectively. The TX shows output 1-dB compression points (OP$_{1dB}$'s) of 13.5 and 13.5 dBm with peak efficiencies of 15.5% and 14.1% at 28 and 37 GHz, respectively. Fig. 7 shows the measured RX

performance. The RX gains are 16.4 and 19.1 dB at 28 and 37 GHz, respectively. The NF of the RX is 4.4 and 4.9 dB at 28 and 37 GHz, respectively. The simulated NF degradations due to the dual-band switch are 1.6 and 1.4 dB at 28 and 37 GHz, respectively, which are the minimum values in each band. When the VGLNA controls the gain, the NF of the receiver is degraded, but the input 1-dB compression point (IP$_{1dB}$) is not greatly affected. In contrast, when the VGPS controls the gain, the IP$_{1dB}$ of the receiver is degraded but the NF is not. Therefore, a proper distribution of the gain control function to the LNA and VGPS can minimize the respective degradation of the NF and IP$_{1dB}$ during the gain control. To distribute the gain control function to the RX chain on the 28/37 GHz bands, the gains of 9.3/9.8 dB are controlled by the VGPS and the gains of 6.7/7.2 dB are controlled by the VGLNA. The variations of the NF and IP$_{1dB}$ during 14 dB gain control are 4.4 to 6.2 dB and -23.8 to -17.3 dBm at 28 GHz, and 4.9 to 5.7 dB and -29.1 to -21.8 dBm at 37 GHz, as shown in Fig. 8(a). The peak phase variations of the RX are 1.8° and 1.1° for 14 dB gain control at 28 and 37 GHz, respectively, as shown in Fig. 8(b).

Fig. 9 shows the measured phase and gain control performance. At 28 GHz and 37 GHz, the frequency reconfigurable VGPS has a phase control resolution of 5.625° and a gain control range of 16 dB with 1 dB steps. Fig. 9(a) and Fig. 9(b) show the TX vector constellations at 28 GHz and 37 GHz, respectively. The vector constellations show that the phase and gain can be controlled at both frequency bands. The measured RMS gain and phase errors are 0.22/0.23 dB and 0.9/0.62° at 28/37 GHz, respectively. The RMS gain and phase errors are less than 0.26 dB and 1.74° from 26.5 to 29.5 GHz and less than 0.26 dB and 1.81° from 37 to 40 GHz, as shown in Fig. 9(c).

Fig. 10 shows the error vector magnitudes (EVMs) and adjacent channel leakage ratios (ACLRs) measured with a 5G

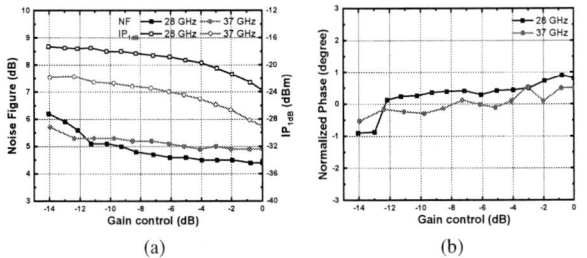

Fig. 8. Measured RX performance: (a) NF/IP$_{1dB}$ variation with gain control, (b) phase variation with gain control.

979-8-3503-2123-4/23 $31.00 © 2023 IEEE   47

Table 1. Comparison with state-of-the-art multiple-band front-end ICs.

Parameter	This work		[1] VLSI22		[2] RFIC21	[4] ISSCC20	
Process	28nm FDSOI		65nm CMOS		90nm SiGe	65nm CMOS	
Freq. (GHz)	28	37	25.8-39.2		17.1-52.4	28	37
Phase Control Resolution (bit)	6		6.5		5	-	-
Gain Control Resolution (bit)	4		6		-	-	-
Gain Control Range (dB)	16	16	8 at 28 GHz 6 at 39 GHz		15	-	-
RMS Phase Error (°)	0.9	0.62	<1.9		<5.6	-	-
RMS Gain Error (dB)	0.22	0.23	<0.38		-	-	-
RX Gain (dB)	16.4	19.1	<19.2		-	44	37
RX NF (dB)	4.4	4.9	>3.7		-	8.5*	9.5*
RX IP$_{1dB}$ (dBm)	>-23.8	>-29.1	-22.8		-	-29	-22
TX Gain (dB)	18.5	18.4	<22.1		<22	43.5	40
TX OP$_{1dB}$ (dBm)	13.5	13.5	>12.8		12-14.8	14	14.2
TX Peak Efficiency (%)	15.5	14.1	>14.5		11.2 at 28 GHz 12.4 at 35 GHz	21	21.5
Modulation Results	64-QAM 200 MHz BW 9.3 dBm Pout 7.8% PAE -25.1 dB EVM	64-QAM 200 MHz BW 8.6 dBm Pout 7.1% PAE -25 dB EVM	28 GHz 64-QAM 600 MHz BW 10.5 dBm Pout 7.5% PAE -31.2 dB EVM	39 GHz 64-QAM 600 MHz BW 7.7 dBm Pout 4.9% PAE -31 dB EVM	64-QAM 1 GHz BW 9.5 dBm Pout 4.6% PAE -25 dB EVM	-	-
1 Channel Core Area (mm^2)	0.97		0.195		0.765	1.05	
Integration Level	PA, LNA, PS, SW		PA, LNA, PS, SW		PA, PS, VGA	Transceiver, mixer, LO	

*Graphically estimated.

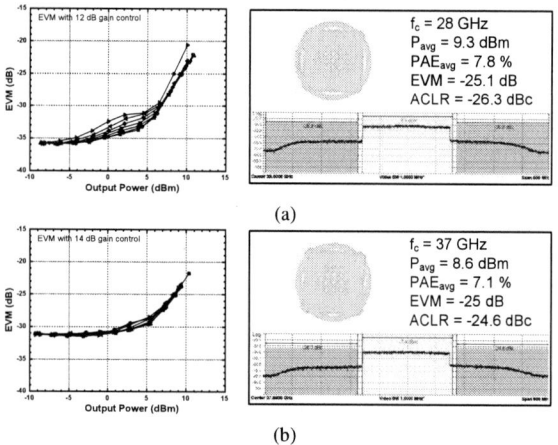

Fig. 10. Measured EVMs and ACLRs at (a) 28 GHz and (b) 37 GHz.

new radio (NR) 64-quadrature amplitude modulation (QAM) (7.62 dB peak-to-average power ratio) 200 MHz signal. The front-end IC achieves average output power of 9.3 and 8.6 dBm and efficiency of 7.8% and 7.1% with EVMs of -25.1 and -25 dB at 28 and 37 GHz, respectively. The average output power variations are 9.3 to 8.4 dBm with 12 dB gain control at 28 GHz and 8.6 to 8.3 dBm with 14 dB gain control at 37 GHz.

Table 1 compares the results of this work with those of reported state-of-the-art multiple-band front-end ICs. Owing to the proposed frequency reconfigurable structure, this front-end IC supports dual-band operation with high linear output powers at both 28 GHz and 37 GHz under a 64-QAM modulated signal. It also has low NFs of the RX on both bands. This work has achieved the widest gain-control range with the lowest RMS phase error reported to date and a low RMS gain error

at both bands.

IV. CONCLUSION

This paper presented a frequency reconfigurable dual-band beamforming front-end IC for 5G NR that has linear output power of 9.3 dBm and 8.6 dBm and noise figures of 4.4 dB and 4.9 dB at 28 GHz and 37 GHz, respectively. It integrates a frequency reconfigurable VGPS, a frequency reconfigurable PA, a dual-band switch, and a dual-band VGLNA to cover the dual bands. The transmitter path uses VGPS for the gain control while the receiver path uses both VGPS and VGLNA. It has a 6-bit phase control resolution and a 16 dB gain control range with 0.9° and 0.62° RMS phase errors at 28 and 37 GHz, respectively.

ACKNOWLEDGMENT

This work was supported by Samsung Electronics Co., Ltd.

REFERENCES

[1] W. Zhu, R. Wang, J. Zhang, J. Wang, C. Li, and Y. Wang, "An ultra-compact bidirectional T/R folded 25.8-39.2GHz phased-array transceiver front-end with embedded TX power detection/self-calibration path supporting 64-/256-/512-QAM at 28-/39-GHz band for 5G in 65nm CMOS technology," in *2022 IEEE Symposium on VLSI Technology and Circuits (VLSI Technology and Circuits)*, 2022, pp. 102–103.

[2] A. A. Alhamed and G. M. Rebeiz, "A global multi-standard/multi-band 17.1-52.4 GHz Tx phased array beamformer with 14.8 dBm OP1dB supporting 5G NR FR2 bands with multi-Gb/s 64-QAM for massive MIMO arrays," in *2021 IEEE Radio Frequency Integrated Circuits Symposium (RFIC)*, 2021, pp. 99–102.

[3] J. Lee, J.-S. Paek, and S. Hong, "Millimeter-wave frequency reconfigurable dual-band CMOS power amplifier for 5G communication radios," *IEEE Transactions on Microwave Theory and Techniques*, vol. 70, no. 1, pp. 801–812, 2022.

[4] S. Mondal, L. R. Carley, and J. Paramesh, "4.4 a 28/37GHz scalable, reconfigurable multi-layer hybrid/digital MIMO transceiver for TDD/FDD and full-duplex communication," in *2020 IEEE International Solid-State Circuits Conference - (ISSCC)*, 2020, pp. 82–84.

979-8-3503-2123-4/23 $31.00 © 2023 IEEE

RMo1C-3

A 26.5-GHz 4x2 Array Switched Beam-Former Based on 2-D Butler Matrix for 5G Mobile Applications in 28-nm CMOS

Youngjoo Lee, Juwon Kim, Sungwon Kwon, Bosung Suh, Jun Hwang, Kyutae Park, Dohoon Chun, Kyujong Choi, Hongseok Choi, Dongho Yoo, Byung-Wook Min

School of Electrical and Electronic Engineering, Yonsei University, Korea

{yjlee94, bmin}@yonsei.ac.kr

Abstract — This paper demonstrates a 26.5-GHz 2-D Butler matrix based 4×2 array switched beam-former. By using a 2-D Butler matrix, feed lines between IC and antennas are uniform, which is a critical problem in an integrated Butler matrix for a 1-D array since it requires complex phase matched routing on PCB. A proposed switched beam-former consists of a signal distribution IC and two switched beam-former ICs. Reconfigurable switches with a function of power divider/combiner are integrated for additional beam patterns. The proposed switched beam-former can generate total 22 beams, which cover a whole scan angle with a low gain variation. Measured beam patterns show that the proposed switched beam-former can cover any 3-D spatial angle of ±44° in azimuth and ±43° in elevation even with a low spatial beam resolution. To our knowledge, this is the first 2-D array switched beam-former based on the Butler matrix in millimeter wave bands.

Keywords — Butler matrix, dual-port excitation, switched beam-former, reconfigurable switch, fast beam-switching, mobile applications

I. INTRODUCTION

Millimeter wave bands give tremendous amount of spectrum to 5G communications. However, due to a high path loss of the millimeter wave signal, a complex beam-forming system based on lots of lossy phase shifters is required. On the other hand, Butler matrix based switched beam systems are very simple and offer a fast beam-switching since multiple beams are always available without an array calibration and can be selected at the input of the Butler matrix. Although these switched beam systems have a low spatial resolution of available beams, they are suitable for mobile devices with a small number of array elements and a wide beamwidth.

Several switched beam-formers with an integrated Butler matrix have been reported, but they are limited for a 1-D antenna array [1]–[10]. The integrated Butler matrix for the 1-D antenna array has a critical problem that feed lines between IC and antennas cannot be uniform and therefore, complex phase matched routing on PCB is required, which is very difficult at the millimeter wave band. However, the feed lines can be uniform by using a 2-D Butler matrix for a 2×2 antenna array. In this paper, we present the 2-D Butler matrix based switched beam-former for a 4×2 antenna array for 5G mobile devices. Compared to a conventional 8×8 Butler matrix for a 8-element array that can generate 8 beams, the proposed switched beam-former can generate 22 beams with dual-port excitation modes, by using a reconfigurable switch network to relieve limited beam patterns [7]–[12].

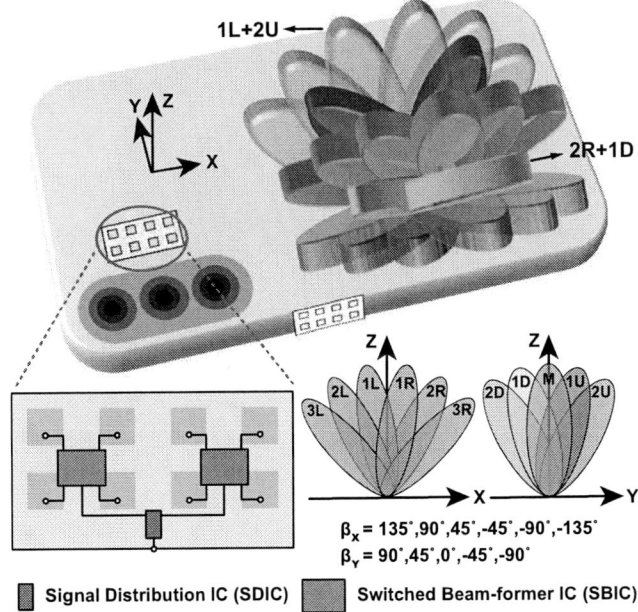

Fig. 1. A 26.5-GHz 4×2 switched beam array for 5G mobile applications.

II. DESIGN AND CONFIGURATION

A. Concept of Proposed Switched Beam-Former

The proposed 4×2 array switched beam-former in Fig. 1 can generate total 22 beams and consists of a signal distribution IC (SDIC), two switched beam-former ICs (SBICs) and antennas. Signal routes between the SDIC and two SBICs have same length and direction, and therefore SBICs for 2×2 sub-arrays can be in a same control state. Since 4 elements are connected per one SBIC, feed lines between the SBIC and antennas can be uniform with a minimized routing distance, resulting in improved system noise figure (NF) and transmitted power and eliminating a requirement of any array calibration [13].

The SBIC shown in Fig. 2(a) includes four TRX channels, 4×4 Butler matrix for 2-D array, and SP4T switch network consisting of a SPDT switch and two reconfigurable SPDT switches (RSPDTs) which can operate as a SPDT switch or a power divider/combiner. The TRX channel consists of front-end circuits, bi-directional amplifier, 1-bit 45° phase shifter to eliminate a crossover of the Butler matrix. The SDIC shown in Fig. 2(b) includes the RSPDT, hybrid coupler, and

979-8-3503-2123-4/23 $31.00 © 2023 IEEE   49   2023 IEEE Radio Frequency Integrated Circuits Symposium

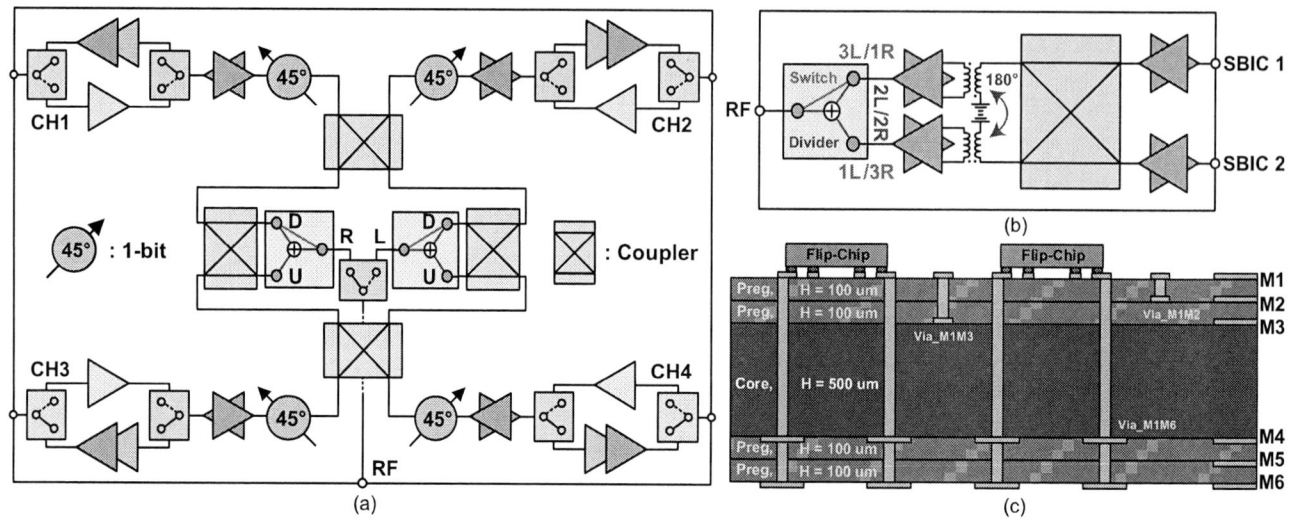

Fig. 2. Block diagrams of (a) switched beam-former IC (SBIC) and (b) signal distribution IC (SDIC), and (c) PCB metal stack-up.

bi-directional amplifiers of which output baluns are inverted with each other for 180° phase shift [10].

A PCB metal stack-up is shown in Fig. 2(c), the SDIC and SBICs are flipped on M1 of the PCB, and antennas are placed on M4 and M6. M2 is used as the bottom ground of the grounded CPW lines with M1 signal lines. The antenna is designed as a stacked patch antenna, and M3 is used as the ground plane of the antennas. The patch antenna and the parasitic patch are placed on M4 and M6, respectively. There is no metal in the M5, and this layer is used to increase a bandwidth of the antenna. Bias lines of the ICs are routed with M1 and M2.

B. Switched Beam-Forming with SDIC and SBICs

States of the RSPDT in the SDIC, SP4T switches and 45° phase shifters in the SBICs for all available beams are summarized in Fig. 3(a). When the 3L/1R path is activated in the SDIC, 3L or 1R (β_x=135° or -45°) beams in the x-z azimuth plane can be generated depending on the SPDT state (L or R) in SBIC. For this condition, the 45° phase shifters of CH1 and CH3 in the SBIC are in delay state. The elevation angle can be selected out of 2U/M/2D (β_y=-90°/0°/90°) through the RSPDT states in the SBIC. M beams can be generated by exciting two input ports of the Butler matrix with the divider mode of the RSPDT. An example of SDIC and SBIC states for 1R+2D beam is shown in Fig. 3(b). 1L or 3R beams (β_x=45° or -135°) can also be generated in the similar way by selecting 1L/3R path in the SDIC and with the 45° delay state of CH2 and CH4 in the SBIC. Therefore, total 12 beams can be generated when the RSPDT in the SDIC operates as the switch mode.

2L or 2R beams (β_x=90° or -90°) can be generated when the RSPDT in the SDIC operates as the divider mode. Since the output baluns of the bi-directional amplifiers between the RSPDT and coupler are inverted with each other, the two SBICs are excited with 180° phase difference. Since the 45°

SDIC	SP4T State	45° Delay State	Beam
3L/1R	R(L) - U	CH1, CH3	1R(3L)+2U
	R(L) - D	CH1, CH3	1R(3L)+2D
	R(L) - UD	CH1, CH3	1R(3L)+M
1L/3R	R(L) - U	CH2, CH4	3R(1L)+2U
	R(L) - D	CH2, CH4	3R(1L)+2D
	R(L) - UD	CH2, CH4	3R(1L)+M
2L/2R	R(L) - U	-	2R(2L)+2U
	R(L) - D	-	2R(2L)+2D
	R(L) - UD	- / CH1,2 / CH3,4	2R(2L)+ M / 1U / 1D

(a)

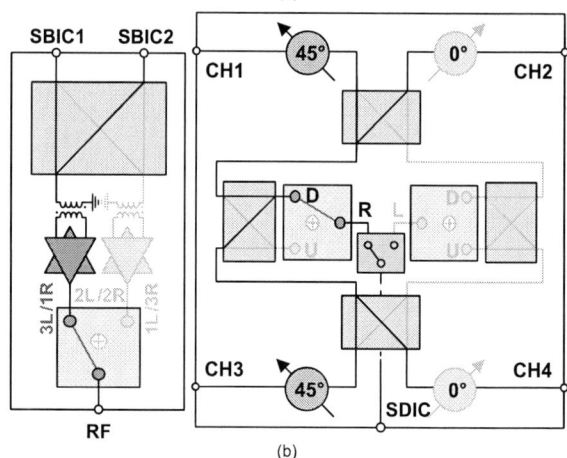

(b)

Fig. 3. Switch network and 1-bit phase shifter states of the proposed switched beam-former for total 22 beams (a) and an example of SDIC and SBIC states for 1R+2D beam (b).

phase shifter is not used for beam-forming in the azimuth plane, the phase shifter can be used in the elevation plane, resulting in 1U and 1D beams. Therefore, total 10 beams can be generated when the RSPDT in the SDIC operates as the divider mode.

979-8-3503-2123-4/23 $31.00 © 2023 IEEE

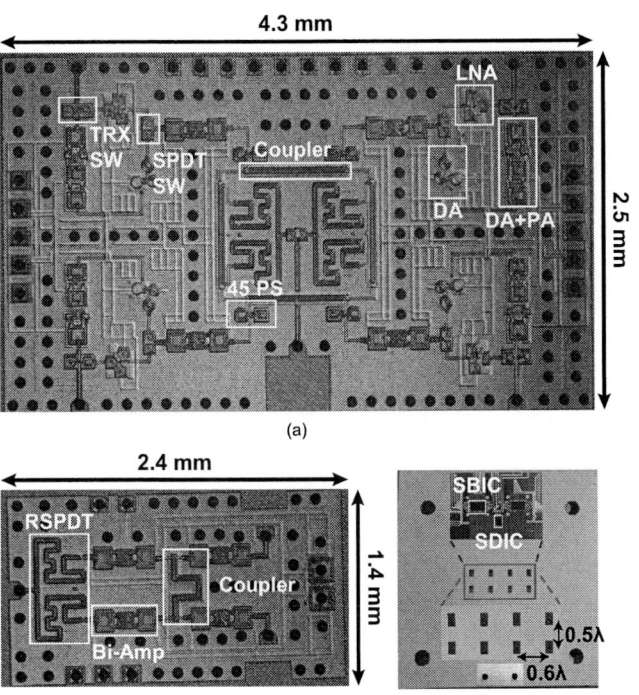

Fig. 4. Die micrograph of the (a) SBIC, (b) SDIC and (c) 4×2 array switched beam-former PCB photograph.

So, total 22 beams can be generated with the proposed switched beam-former by using the RSPDTs, additional 14 beams are generated compared to the conventional 8×8 Butler matrix for the 8-element array that can generate 8 beams.

III. MEASUREMENT

The 26.5-GHz 4×2 array switched beam-former was fabricated in Samsung 28-nm CMOS process and the die-micrograph of the SBIC and SDIC is shown in Fig. 4(a) and Fig. 4(b), respectively. The overall chip area of the SBIC and SDIC is $4.3 \times 2.5~mm^2$ and $2.4 \times 1.4~mm^2$, respectively. And the array PCB photograph is shown in Fig. 4(c). An antenna spacing is 0.6λ in the horizontal direction and 0.5λ in the vertical direction. The stacked patch antennas are vertically polarized, so the E-plane and H-plane scans are in the elevation and azimuth planes, respectively. The array antenna patterns were measured in an anechoic chamber in the RX mode at 27 GHz without any array calibration and are shown in Fig. 5.

The 3-dB beam coverages of the proposed switched beam-former are $\pm 44°$ in the azimuth and $\pm 43°$ in the elevation plane. Measured beam patterns show that the gains of 2L and 2R beams, which are generated when the RSDPT in the SDIC operates as the divider mode, are higher than the gains of 3L,1L,1R and 3R beams, which are generated when the RSPDT in the SDIC operates as the switch mode. This is because the divider mode of the RSPDT has lower ohmic loss than the switch mode. In the azimuth plane, four beams (2L/1L/1R/2R) are steered with < −10 dB side lobes. Due to the limited single-element antenna beamwidth, 3L/3R beams

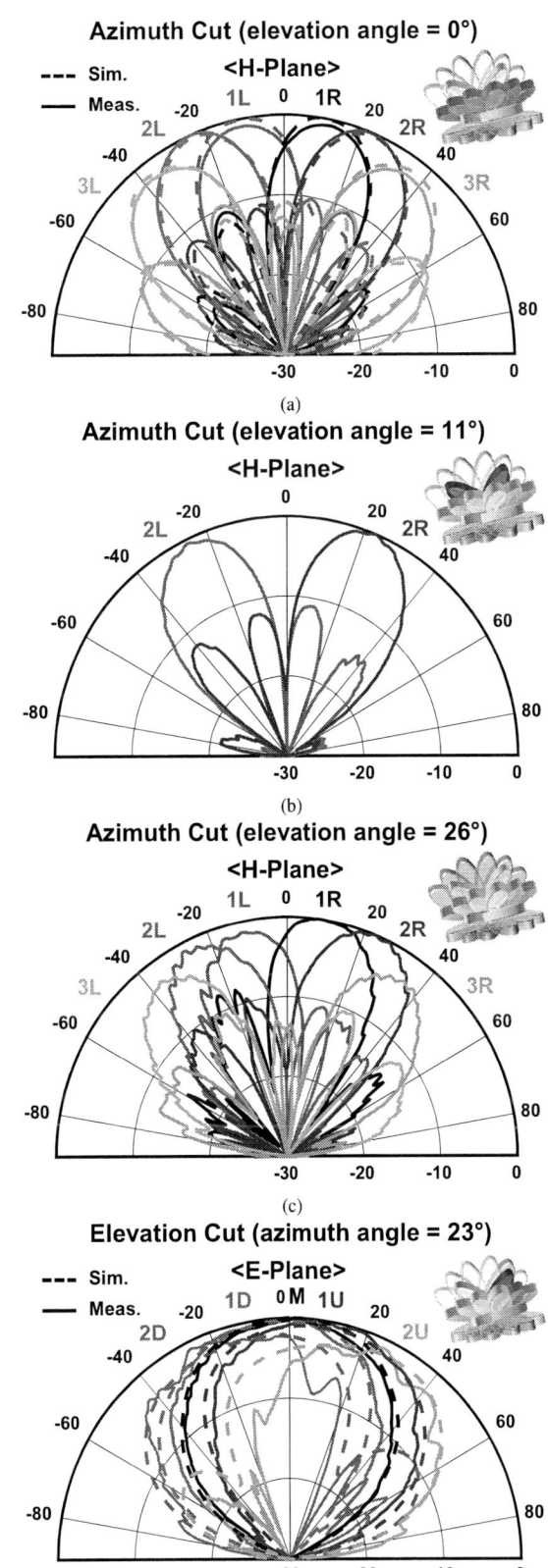

Fig. 5. Measured beam patterns of the 4×2 array switched beam-former in the RX mode at 27 GHz without an array calibration, (a) M beams, (b) 1U beams and (c) 2U beams in the azimuth cut and (d) 2R beams in the elevation cut.

Table 1. Comparison with previous fully integrated switched beam-formers

Reference	Process	Center Freq. (GHz)	Array Dimension	Number of Beams	IC–Ant. Distribution	Additional Beams	Bi-dir.	Ant.	Including Front-End
This Work	28-nm CMOS	26.5	4×2	22	Uniform	O	O	O	O
[4] TMTT17	0.15-μm GaAs	2.4	4×4	16	Non-uniform	X	O	O	X
[7] MOTL19	65-nm CMOS	28	8×1	15	Non-uniform	O	O	O	X
[5] TMTT11	0.13-μm CMOS	5.5	8×1	8	Non-uniform	X	O	O	X
[9] IMS22	28-nm CMOS	28	4×1	7	Non-uniform	O	O	X	X
[10] TMTT23	28-nm CMOS	28	4×1	7	Non-uniform	O	O	X	X
[2] JSSC14	0.13-μm SiGe	230	4×1	4	Non-uniform	X	X	X	O
[1] TMTT10	0.13-μm CMOS	60	4×1	4	Non-uniform	X	O	O	X
[3] IMS21	0.15-μm GaAs	24	4×1	4	Non-uniform	X	O	X	X

are steered with side lobe levels about –6.8 dB. In the RX array, the measured electronic RX gain is 25 dB at 26.5 GHz. Based on the measured NF of 2 channels of SBIC, the full system RX NF is calculated to be 4.4 dB at 26.5 GHz. In the TX array, the measured peak EIRP is 41 dBm at the saturated output power and 37.7 dBm at P1dB and the electronic TX gain is 34.7 dB at 26.5 GHz. DC power consumptions in the TX and RX modes are 4.36 W and 410 mW, respectively, when all the bi-directional amplifiers in the SDIC are turned on.

Table 1 summarizes the comparison with previous fully integrated switched beam-formers. All of the previous works require complex phase matched routing because distances between IC and antennas are non-uniform, which deteriorates the system NF and transmitted power. However, the proposed switched beam-former has the uniform IC–antennas distribution and also includes the front-end circuits and antennas. Furthermore, it is possible to produce the most number of beam patterns.

IV. CONCLUSION

This paper presented a 26.5-GHz 4×2 array switched beam-former based on a 2-D Butler matrix for 5G mobile applications in 28-nm CMOS. By using a 2-D Butler matrix, feed lines between IC and antennas are uniform with a minimized routing distance, resulting in improved system NF and transmitted power and eliminating a requirement of any array calibration. Although the number of beams that can be steered is limited in the switched beam system, the proposed switched beam-former with RSPDTs can generate total 22 beams, which cover a whole scan angle with a low gain variation. The proposed switched beam-former can support any 3-D spatial angle of ±44° in azimuth and ±43° in elevation even with a low spatial beam resolution. Therefore, a mobile device mounted with the proposed switched beam-formers in 4 different directions can cover an entire hemisphere with fast beam-switching and without any calibration. To the best our knowledge, this is the first 2-D array switched beam-former based on the Butler matrix in millimeter wave bands.

ACKNOWLEDGMENT

This work was supported by Samsung Electronics Co., LTD (IO201209-07875-01), and IITP grant funded by the Korea government(MSIT) (No. 2020000218). The EDA tool was supported by the IC Design Education Center(IDEC), Korea.

REFERENCES

[1] W. Choi, K. Park, Y. Kim, K. Kim, and Y. Kwon, "A V-band switched beam-forming antenna module using absorptive switch integrated with 4×4 Butler Matrix in 0.13-μm CMOS," *IEEE Trans. Microw. Theory Tech.*, vol. 58, no. 12, pp. 4052–4059, Dec. 2010.

[2] Elkhouly, Mohamed, et al. "A G-Band Four-Element Butler Matrix in 0.13 μm SiGe BiCMOS Technology." *IEEE J. Solid-State Circuits*, vol. 49, no. 9, pp. 1916–1926, Sep. 2014.

[3] Q. -Y. Jiang and Y. -S. Lin, "A 24-GHz Butler-Matrix-Based Switched Beamformer in GaAs," *IEEE MTT-S. Int. Microw. Symp. (IMS)*, 2021, pp. 534-537.

[4] W.-T. Fang, C.-H. Chen, and Y.-S. Lin, "2.4-GHz absorptive MMIC switch for switched beamformer application," *IEEE Trans. Microw. Theory Tech.*, vol. 65, no. 10, pp. 3950–3961, Dec. 2017.

[5] B. Cetinoneri, Y. A. Atesal, and G. M. Rebeiz, "An 8×8 Butler matrix in 0.13-μm CMOS for 5-6-GHz multibeam applications," *IEEE Trans. Microw. Theory Tech.*, vol. 59, no. 2, pp. 295–301, Feb. 2011.

[6] T.-Y. Chin, S.-F. Chang, J.-C. Wu, and C.-C. Chang, "A 25-GHz compact low-power phased-array receiver with continuous beam steering in CMOS technology," *IEEE J. Solid-State Circuits*, vol. 45, no. 11, pp. 2273–2282, Nov. 2010.

[7] J. Park, and J.-G. Kim, "A 28 GHz CMOS Butler matrix for 5G mm-wave beamforming systems," *Microw. Opt. Tech. Lett.*, vol. 62, no. 7, pp. 2499-2505, March 2020.

[8] Y. Lee, B. Suh and B.-W. Min, "A 28-GHz Butler Matrix based Switched Beam-Forming Network with Phase Inverting Switch for Dual-Port Excitation in 28-nm CMOS," *IEEE MTT-S. Int. Microw. Symp. (IMS)*, 2022, pp. 1002-1005.

[9] Y. Lee, B. Suh and B.-W. Min, "A Ka-Band Bi-Directional Reconfigurable Switched Beam-Forming Network Based on 4×4 Butler Matrix in 28-nm CMOS," *IEEE Trans. Microw. Theory Tech.*, doi:10.1109/TMTT.2023.3235999.

[10] B. Suh and B.-W. Min, "A 28-GHz Reconfigurable SP4T Switch Network for a Switched Beam System in 65-nm CMOS," *IEEE Trans. Microw. Theory Tech.*, vol. 68, no. 6, pp. 2057-2064, June 2020.

[11] C. Chang, R. Lee and T. Chin, "Design of 2.4-GHz CMOS reconfigurable dual-function switch network and its application on array beamforming," in *Proc. 39th Eur. Microw. Conf. (EuMC)*, Sep./Oct. 2009, pp. 1018-1021.

[12] Kerim Kibaroglu, Mustafa Sayginer and Gabriel M. Rebeiz. "A low-cost scalable 32-element 28-GHz phased array transceiver for 5G communication links based on a 2 × 2 beamformer flip-chip unit cell," *IEEE J. Solid-State Circuits*, 53.5 (2018): 1260-1274.

RMo1C-4

A Phased-Array Receiver Front-End Using a Compact High Off-Impedance T/R Switch for n257/n258/n261 5G FR2 Cellular

Ying Chen , Xiaohua Yu, Samrat Dey, Venumadhav Bhagavatula, Chechun Kuo, Tienyu Chang,
Ivan Siu-Chuang Lu, Sangwon Son
Samsung Semiconductor, Inc., USA
chen.ying@samsung.com

Abstract— This paper presents a low-power and compact phased-array receiver (RX) front-end (FE) in 28nm CMOS FD-SOI for n257/n258/n261 5G FR2 cellular applications. A compact high off-impedance SPST T/R switch with an embedded 2-bit 18dB variable attenuator is proposed to minimize TX degradation while maintaining a low insertion loss with additional gain control for RX. With the T/R switch and PA output matching network (OMN), the TX/RX interface is co-optimized for TX output power and RX NF. Over the frequency band of 24.25-29.5GHz the RXFE achieves measured RX NF of 4.4-5.6dB with ~41dB gain control range and ~35dB IP1dB range. The RXFE occupies a small die area of 0.14mm². The RX operation consumes only 13.4mW and 4.7mW DC power in sensitivity mode and low-gain/high-linearity mode respectively.

Keywords—5G, FR2, phased array, receiver, front-end, switch.

I. INTRODUCTION

Cellular handsets supporting 5G FR2 with frequency bands of n257 (26.5–29.5GHz), n258 (24.25–27.5GHz), n261 (27.5–28.35GHz) requires the millimeter-wave (mm-wave) transceiver chip to be low power, low cost, and with a small form factor, along with meeting stringent performance specifications. Phased arrays using multiple FE elements have been widely adopted for the mm-wave transceiver. As the number of FE elements increases, the power consumption and the die size of each FE element becomes critical to the total power and the form factor of the chip. In particular, the low power RX is essential to achieve low overall transceiver chip power consumption due to the duty cycle of RX mode being as high as 80% of the transceiver operating time.

Recent publications on 5G FR2 RXFE have been focusing on advancing mm-wave performance [1]-[7]. However, they have a higher power consumption and a larger die size, making them less favorable for mobile handset applications. Wafer-level chip-scale package (WLCSP) is a popular option for mm-wave applications as it offers low packaging cost with reasonable mm-wave performances [1], [2], [4], [5]. However, the minimum pitch between adjacent bump pads is limited to be ~300μm by the board manufacturing technology, resulting in a significant ground inductance for the on-die Ground-Signal-Ground (GSG) structure, which requires careful considerations to achieve optimal performance. Unfortunately, few literatures discussed this.

This paper presents a low-power and compact phased-array RXFE in WLCSP that supports 5G FR2 n257/n258/n261 bands covering 24.25–29.5GHz. To achieve low power consumption, passive phase shifter consuming zero power is selected, whose insertion lose can be compensated by a higher

intermediate-frequency (IF) stage gain. The power saving for the whole receiver can be achieved since the power saving from each FE is multiplied by the number of array elements. The slicing technique is used in low-noise amplifier (LNA) for additional power saving during gain back off. A compact-size T/R switch is inserted in the RX path achieving a low insertion loss for RX mode, and a high off-impedance for TX mode that minimizes degradation on TX output power and efficiency. Variable attenuations are embedded into the switch design without degrading the performance or increasing the size. The TX/RX interface is co-optimized using the T/R switch and PA OMN to minimize degradation on TX output power and RX NF due to inter-dependent TX/RX loading. Also, the on-die GSG structure design consideration is discussed and the structure is co-optimized with RXFE floorplan and grounding strategies to minimize GSG loss.

Fig. 1 shows the block diagram of the proposed phased-array receiver. TX and RX share the same antenna with TDD operation. The RXFE consists of a T/R switch, a two-stage LNA, and followed by a phase shifter. While loaded with TX, the RXFE implemented in a 28nm CMOS FD-SOI process achieves measured 4.4–5.6dB NF, ~41dB gain control range with ~35dB IP1dB range over 24.25-29.5GHz. The RXFE consumes only 4.7-13.4mW power from 1V supply and occupies 0.14mm², achieving the lowest power consumption and the smallest die area among recently published 5G FR2 RXFEs.

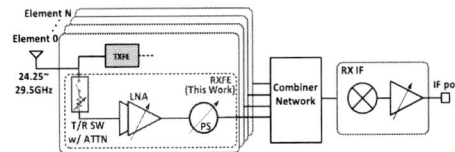

Fig. 1. Block diagram of a 5G FR2 phased-array receiver

II. PHASED ARRAY RECEIVER FRONT-END DESIGN

A. High Off-Impedance T/R Switch Design Embedded with a 2-Bit Variable Attenuator

Inductor-based T/R switch topology is commonly used at mm-wave, where an additional series inductor or quarter-wavelength transmission-line is used in combination with a shunt switch [1], [2], [5], [6], [7]. However, it occupies a large die area due to the passives and has a finite off-impedance (<~300Ω typically) mainly limited by the Q-factor of on-chip passives. As shown in Fig. 2(a) the proposed T/R switch is a single-pole single-throw (SPST) switch implemented in CMOS SOI. In order to minimize degradation on TX, the

979-8-3503-2123-4/23 $31.00 © 2023 IEEE   53   2023 IEEE Radio Frequency Integrated Circuits Symposium

switch is inserted only in the RX path. To achieve a high off-impedance with a compact-size, no inductors are used. Also, to meet the linearity and reliability requirements during TX, two series switches are stacked. To protect ESD at RF port, the secondary winding of PA output transformer acts as the ESD discharge path from the RF port to the ground. The switch is designed to also provide variable attenuations to meet linearity requirements under RX blocker interference. Multiple switch branches are connected in parallel for both series and shunt switches with each separately controlled. Higher attenuations can be obtained with fewer series branches and more shunt branches. The total attenuation range of the switch is 18dB with a step of 6dB.

1) Design considerations for IL, R_{off}, and linearity/reliability

During RX the back-gates of the series switch transistors are positively biased to reduce R_{ON} and the IL. To withstand the peak voltage swing of 4.5V with 17dBm TX output power under VSWR=2:1, multiple stacks of series switches must be used. To minimize the number of series switches, G-to-S/D of all series switch transistors are reverse biased to the maximum DC limit in order to deeply turn off series switches for better linearity and power handling capability. Neglecting the small voltage swing across the shunt switch and assuming evenly divided voltage swing across G-to-S/D of the series switch transistors, the maximum allowable peak voltage swing can be approximated as $V_{p\_rf\_max} = 2 \cdot (V_{SG\_DC\_max} + V_{th}) \cdot N_{stack}$ where N_{stack} is the number of stacks, and $V_{SG\_DC\_max}$ is limited by gate oxide breakdown. Beyond that the series switches will start to turn on during positive swing cycles causing non-linearity. For cellular applications, there is often a requirement for antenna port VSWR conditions, under which the voltage standing wave can be higher than under 50-Ω condition, posing more stringent requirement on reliability. RF stress reliability is checked with targeted average output power under VSWR=2:1 to ensure the peak-to-peak voltage swings across G-to-S/D are less than $2 \times V_{SG\_DC\_max}$.

Another consideration is the switch off-impedance. As shown in Fig. 3, R_{off} and C_{off} represent the parallel equivalent resistance and capacitance looking into the switch in TX mode. C_{off} can be absorbed into the design of PA output matching network, but R_{off} loads the TX output leading to output power losses. It is critical to have a sufficiently high R_{off}. Simulation shows TX output power degrades by about 0.7–0.8dB when loaded with R_{off} of 300Ω. As R_{off} increases to 1kΩ, the output power degradation becomes only 0.1–0.2dB. To favor a low IL in RX, series switches need to be large in size and shunt switches need to be small. However, the large series switches would have more couplings to the lossy substrate, while the higher R_{ON} of small shunt switches would reduce quality factor of the switch off-capacitance, both of which lead to a lower R_{off}.

2) Switch design variants comparisons

Fig. 2(b), (c), (d) show three switch design variants with optimal voltage bias settings in the format of Vbias_RX/Vbias_TX. Since RF port is connected to ground in DC for ESD protection, it is always 0V. To bias the LNA

input transistor (e.g. 0.35V) a series AC coupling capacitor C_1 needs to be inserted in between. A large value of C_1 is necessary for a sufficiently low impedance to the signal path but with a drawback of higher parasitic capacitance to lossy substrate resulting in signal losses. The inserted location of C_1 has an impact on optimal bias settings and the switch performance. The shown three design variants provide insights on trade-offs between key performance parameters and architecture choices. In *Design A*, C_1 is inserted at switch output and negative voltages are needed to deeply turn off series switches. The benefit of such is a larger R_{off} without seeing the parasitics of C_1 at switch input. However, the negative voltage generator occupies a significant die area with increased complexity. In *Design B*, C_1 is inserted at the input of the switch so that only positive voltages are used to turn on and off the switches. The shunt switch is removed for better IL, but with a drawback of a lower R_{off} and higher RF stress to LNA input transistor during TX. In *Design C*, a shunt switch is added back to increase R_{off} in favor of TX performance while trading-off the RX IL. Table 1 shows the simulated results of the three variants. *Design C* is selected for achieving R_{off}>1kΩ without using negative voltage generator.

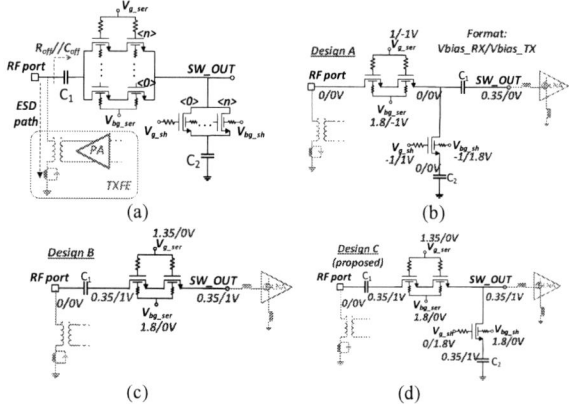

Fig. 2. (a) Proposed switch with embedded variable attenuations (b) *Design A* (c) *Design B* (d) *Design C*

Table 1. Simulated performances of the three switch design variants.

	Design A	*Design B*	*Design C*
IL (dB)	0.85~0.95	0.75~0.85	0.85~0.95
R_{off} (Ω)	1.2k	700	1k
IP1dB (dBm) VSWR=1:1	+27	+29.5	+29
Vp-p G-to-S/D VSWR=2:1 $P_{out\ TX}$=17dBm	2V	1.9V	1.9V
Negative voltage	Yes	No	No

B. LNA Design

The LNA is a single-ended two-stage design to achieve more than 25dB gain. For both amplifier stages, the gain back off is implemented with slicing technique. Compared to the cascode current bleeding technique, of which current consumption remains the same during back off, the slicing technique saves power consumption during gain back-offs. To provide additional gain back off range, a small shunt switch connected to the gate of the CS transistor turns on to further attenuate the input signal. The total gain back off range of LNA alone is 21dB with a step of 3dB.

979-8-3503-2123-4/23 $31.00 © 2023 IEEE

C. PA Output Matching Network Co-optimization

Fig. 3 shows the PA OMN which is co-designed with the GSG structure and RX-antenna interface. There is no switch added in the OMN to avoid additional loss during TX operation. In TX mode, the OMN performs impedance transformation from 50Ω antenna impedance to the optimal load impedance for the required output power and efficiency. In RX mode, the OMN transforms the off-mode output impedance from PA core into parallel equivalent resistance (R_{off_PA}) and inductance (L_{off_PA}) seen by the RXFE. R_{off_PA} needs to be large enough to minimize the loading effect on RX, while L_{off_PA} tunes out the capacitance at the signal I/O pad.

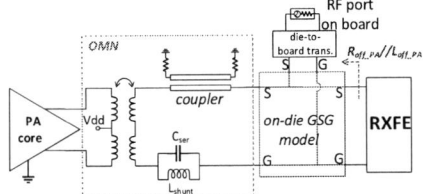

Fig. 3. Circuit diagram of the PA OMN and interfacing circuit blocks

To achieve a high R_{off_PA}, a series capacitor (C_{ser}) is added on the ground side of the secondary of the PA balun [8]. The use of C_{ser} also serves impedance transformation and improving signal balance at PA drain during TX mode. The simulated R_{off_PA} is 170–200Ω across the band. To provide ESD protection at RF port, a shunt inductor (L_{shunt}) is added in parallel with C_{ser} to provide a low impedance path to ground for ESD currents to discharge. Compared with adding another shunt ESD inductor at the signal pad, this approach offers a lower OMN loss during TX operation. Fig. 4 shows the impact of inter-dependent TX/RX loading effects on PA OP5dB and RX NF in simulation. Across 24.25–29.5GHz band the PA OP5dB degrades by 0.05–0.25dB due to R_{off} loss from RX loading, while the RX NF degrades by 0.6–1.1dB due to R_{off_PA} loss.

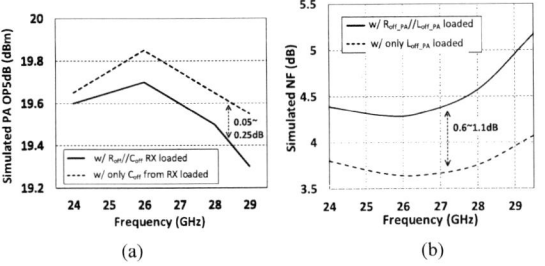

Fig. 4. (a) RX loading effect on PA OP5dB (b) TX loading effect on RX NF

D. Passive Phase Shifter

Fig. 5 shows the circuit diagram of the passive phase shifter. The input interface with TX input and RX output is in differential, while the output interface with the combiner network is in single-ended. The phase shift is achieved by I/Q vector summation. Both input and output are matched to 50Ω. The I and Q variable attenuators adjust the magnitudes of the I/Q signals before passing through a 90° coupler that performs the quadrature summation. The phase shifter achieves 4-bit phase resolution with an average IL of 7dB.

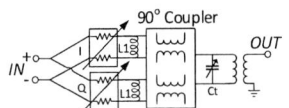

Fig. 5. Circuit diagram of phase shifter

III. ON-DIE GSG AND EFFECT OF SUBSTRATE COUPLINGS

Fig. 6(a) illustrates the floorplan with on-die GSG configuration. The signal pad is shared between TXFE and RXFE. The TXFE and the RXFE are both surrounded by ground planes made of a low ohmic metal layer, to which the local ground of individual circuit block along the line-up is connected. The T/R switch being the first circuit block of RXFE line-up, is placed as close as possible to the signal pad so as to minimize the length of the interconnect trace. This helps to maintain a high R_{off} of T/R switch seen by TX. Ground bridges are added along the line-up to reduce the loss incurred by the GSG structure. Fig. 6(b) shows the simplified equivalent circuit of the GSG interface including the substrate couplings between ground planes. With the capacitive substrate couplings, the ground network between *gnd_rf* and *G_pad* can have a parallel resonance that would cause a higher loss in the GSG structure. With the shunt inductance L_gnd_bridge from ground bridges, the parallel resonant frequency can be shifted to much higher than the frequency band of operation, so that the in-band loss is minimized. The simulated loss of the on-die GSG structure improves from 0.7dB to 0.2dB at 28GHz with the ground bridges.

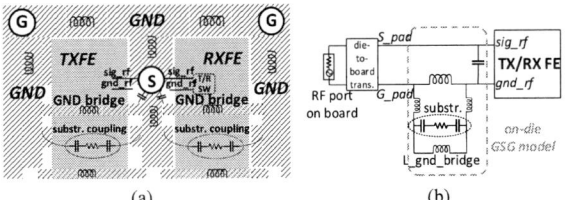

(a)   (b)

Fig. 6. (a) Floorplan with on-die GSG configuration (b) Simplified equivalent circuit of the GSG interface with effect of substrate couplings

IV. MEASURED RESULTS

Fig. 7 shows the WLCSP packaged die photo of the RXFE fabricated in 28nm CMOS FD-SOI occupying a compact area of 0.14mm² per element. The performance was characterized through conducted measurement using a soldered down board with fixed IF output frequency at 9GHz. The cable and board trace losses were de-embedded from the results. Fig. 8 shows the measured gain and phase at different phase states with gain and phase RMS errors of <0.9dB and <6 degrees respectively. Fig. 9 shows the measured RX gain of ~40dB control range and IP1dB of ~35dB range by varying RXFE gains in a single channel configuration with gains from subsequent stages set to minimum. The IP1dB is >-1dBm and >-33dBm with RXFE gains set to minimum and maximum respectively. Fig. 10(a) shows the measured and simulated S11 and NF with bumps included and also loaded by TX. The measured NF is 4.4–5.6dB over 24.25–29.5GHz. Fig. 10(b) shows the measured EVM in 5×1 channel configuration with 5G NR FR2 400MHz 64QAM CP-OFDM modulated signal at 28GHz with varying

979-8-3503-2123-4/23 $31.00 © 2023 IEEE

Table 2. Comparison with state-of-the-art 5G FR2 phased-array RXFE

	This work	JSSC'22 [1]	JSSC'21 [2]	TMTT'19 [3]	TMTT'21 [4]	ISSCC'18 [5]	ASSCC'21 [6]	RFIC'21 [7]
Tech.	28nm SOI	65nm CMOS	65nm CMOS	40nm CMOS	65nm CMOS	28nm CMOS	65nm CMOS	28nm SOI
Package	WLCSP	WLCSP	WLCSP	On-wafer probe	WLCSP	WLCSP	On-wafer probe	WLCSP
Freq. (GHz)	24.25~29.5	24~29.5	26.5~29.5	27~30	24~28	26.5~29.5	24~30	24.25~29.5
RX Gain/ Range (dB)	28.5&/39	14.2*/31.5	18/26	16.8/8	23.2/24	34&/9	25.8/30.7	32&/35
NF (dB)	4.4~5.6&	4.3~6.0*	4.9	5.5~6	>4.4	4.4~4.7&	5.1~6*	4.3~6.4&
PS resol. (bit)	4	6	5	3	>8.8	3	6	4
Phase/Gain RMS error (deg./dB)	6/0.9	1.8/0.22	2/0.4	5/-	0.47/0.4	-	2/0.24	6/0.9
RX EVM (dB)	-35.5	-25	-29.4	-	-	-36.5	-24.7	-33.1
RX IP1dB (dBm)	>-33/-1 (max/min gain)	-23.7~-22.2	-	-16	-16.5	-	-29	>-37.8/-6.6 (max/min gain)
P_{DC} per RXFE (mW)	4.7~13.4*	82*	80*	32*	45	41.8**	80*	17.3*
Die area per RXFE (mm²)	0.14	7.5#	0.48	0.54#	0.4#	1.16	1.5#	0.46
TX P_{SAT}/ PAE max (dBm/%)	19.4/26.9%	18/20.5	16.1/22	16/27.5	18.2/21.1	14/20	16.5/-	-

*: RXFE only, **: including LO and IF power, &: including IF contribution, #: estimated from die dimension

RXFE gains. An RF input power range of >44dB is achieved for EVM<30dB. In TX mode, the TXFE achieves measured OP5dB of 19.4dBm and peak PAE of 26.9% while loaded by the RXFE with the high off-impedance T/R switch. Table 2 summarizes the measured results in comparison with state-of-the-art 5G FR2 phased array RXFEs. This work has achieved the lowest power consumption (i.e. 13.4mW in sensitivity mode and 4.7mW in low-gain/high-linearity mode) and the smallest die area of all.

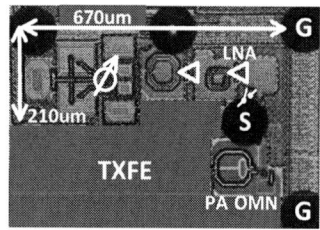

Fig. 7. Die photo of the RXFE

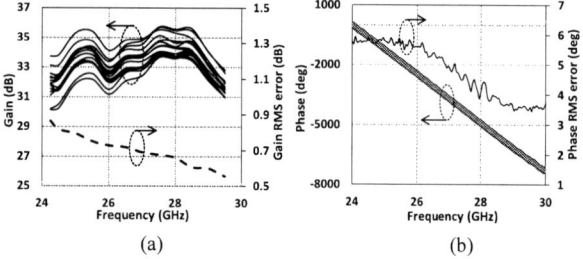

Fig. 8. Measured (a) gain at different phase states and respective gain RMS error (b) phase at different phase states and respective phase RMS error

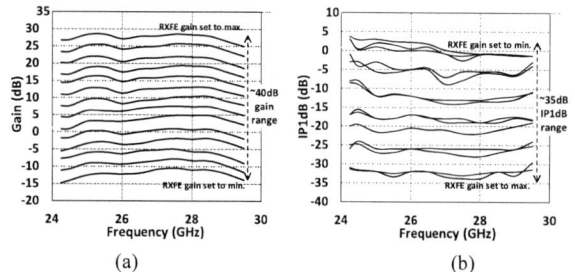

Fig. 9. Measured (a) RX gain and (b) IP1dB with varying RXFE gains in single channel configuration with subsequent gains set to minimum

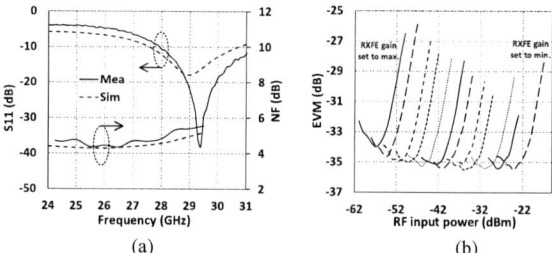

Fig. 10. (a) Measured and simulated S11 and NF (bump inclusive, loaded by TX) (b) Measured RX EVM at 28GHz in 5×1 channel configuration

V. CONCLUSION

A 24.25–29.5GHz phased-array RXFE has been presented. A compact high off-impedance T/R switch is proposed to minimize TX degradation while maintaining a low insertion loss with additional gain control for RX. The RXFE has a compact die area of 0.14mm² and demonstrates a low power consumption of 4.7–13.4mW, making it a great candidate for 5G FR2 cellular handset applications.

REFERENCES

[1] Y. Yi, D. Zhao et al., "A 24-29.5-GHz Highly Linear Phased-Array Transceiver Front-End in 65-nm CMOS Supporting 800-MHz 64-QAM and 400-MHz 256-QAM for 5G New Radio," IEEE JSSC, vol. 57, no. 9, pp. 2702-2718, Sep. 2022.

[2] J. Pang et al., "A CMOS Dual-Polarized Phased-Array Beamformer Utilizing Cross-Polarization Leakage Cancellation for 5G MIMO Systems," IEEE JSSC, vol. 56, no. 4, Apr. 2021.

[3] S. Shakib et al., "A Wideband 28-GHz Transmit-Receive Front-End for 5G Handset Phased Arrays in 40-nm CMOS," IEEE TMTT, vol. 67, no. 7, Jul. 2019.

[4] W. Zhu et al., "A 24-28-GHz Four-Element Phased-Array Transceiver Front End With 21.1%/16.6% Transmitter Peak/OP1dB PAE and Subdegree Phase Resolution Supporting 2.4 Gb/s in 256-QAM for 5-G Communications," IEEE TMTT, vol. 69, no.6, Jun. 2021.

[5] J. D. Dunworth et al., "A 28GHz Bulk-CMOS Dual-Polarization Phased-Array Transceiver with 24 Channels for 5G User and Baseband Equipment," in ISSCC, 2018, pp. 70-72.

[6] X. Huang et al., "A 24-30GHz 4-Element Phased Array Transceiver with Low Insertion Loss Compact T/R Switch and Bidirectional Phase Shifter in 65 nm CMOS Technology," in A-SSCC, 2021, pp. 1-3.

[7] X. Yu et al., "A 17.3-mW 0.46-mm² 26/28/39GHz Phased-Array Receiver Front-End with an I/Q-Current-Shared Active Phase Shifter for 5G User Equipment," in proc. RFIC Symp., 2021, pp. 107-110.

[8] C. Kuo et al., "A 5G FR2 (n257/n258/n261) transmitter front-end with a temperature-invariant integrated power detector for closed-loop EIRP control," in IEEE Radio Freq. Integr. Circuits Symp. (RFIC) Dig. Papers, Jun. 2021, pp. 175–178.

RMo2A-1

Exploration of Design / Layout Tradeoffs for RF Circuits using ALIGN

Jitesh Poojary, Ramprasath S, Sachin S. Sapatnekar, Ramesh Harjani

University of Minnesota, Twin Cities, USA

Abstract— **The extended manual layout process for RF and analog/mixed-signal design restricts design space exploration and limits design productivity. This work demonstrates the efficacy of an automated layout generator versus a manual approach using a state-of-the-art MIMO receiver. Multiple smaller floorplans of the layout are generated automatically in hours compared to weeks for a single manual layout. Measured results from an automatically generated layout fabricated in TSMC 65nm CMOS show performance numbers comparable to the manual design. Measured in-band/in-notch IIP_3 and out-of-band/in-notch IIP_3 are 18.3dBm and 23.64dBm, respectively.**

Keywords— **Automated RF-AMS layout synthesis, Design space exploration, Design productivity, MIMO receiver**

I. INTRODUCTION

RF/analog/mixed-signal (RF-AMS) circuits require careful design to avoid expensive respins and failures in the field. To build robust, high-performance RF-AMS blocks, it is desirable to evaluate a number of design options. However, conventional design methods are mostly manual, time-consuming and do not allow the designer to fully explore the design space. A typical RF-AMS design flow involves: (a) architecture design, (b) sub-block design, (c) device sizing, (d) layout, typically done manually, (e) final verification with post-layout parasitics. In particular, for RF circuits the layout step is a critical determinant of circuit performance. During design iterations, devices may be resized in step (c) based on post-layout parasitics, but such sizing operations perturb the layout, leading to further changes in the parasitics, leading to long iterations between steps (c)–(e) till the design specifications are met. The layout step (d) is a tedious manual process, requiring expert human layout/mask designers. Advanced process technologies involve complex design rule checks (DRCs), which further slow the layout process. Typical design/layout iterations run into multiple weeks, limiting the number of designs that can be evaluated before tape-out. To reduce the design/layout iterations, designers resort to conservative parasitic estimates, resulting in designs with sub-optimal power and/or performance [1].

An automated RF-AMS layout synthesis flow can be useful to a designer as it addresses the critical bottleneck of long layout times while simultaneously handling complex DRCs. An automated tool can help explore the design space by generating multiple designs/layouts in the same amount of time it takes for a single manual layout. There have been multiple recent open-source approaches to achieve this goal including [2]–[4], and ALIGN [5]. In this work, we compare the efficacy of automatic layout generation using ALIGN versus a manual layout for a state-of-the-art MIMO receiver [6]. Multiple automated layouts are generated simultaneously, all of which satisfy the required layout

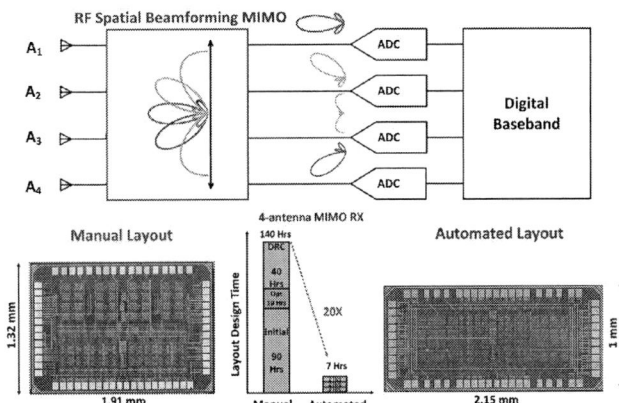

Fig. 1. MIMO design used to illustrate the process. Chip micrographs for manual and automatic layouts and productivity improvement using the tools.

constraints such as symmetry, ordering, common-centroid and matching. With rapid layout synthesis, ALIGN quickly estimates parasitics during the design phase, reducing the number of design/layout iterations. The designer's intent for a floorplan can be specified in ALIGN in the form of user-defined constraints provided to the layout generator [7]. The layout is generated hierarchically and the designer can pick the best-performing layout for each hierarchy using post-layout extracted simulations. Fig. 1 shows the manual and automated layouts (microphotos) of the MIMO RX and shows a measure of how design productivity is enhanced by the latter. The manual approach took weeks for a single layout, as against a few hours required by the automated process to generate multiple complete chip-level MIMO layouts.

We describe the ALIGN flow in Section II. Section III overviews the MIMO design and compares a manual layout against multiple automated layouts. One of the automated layouts was fabricated and tested. Section IV compares its measured performance against the fabricated manual design. Section V discusses the productivity gain from automation and Section VI concludes the paper.

II. DESIGN FLOW

An overview of the steps involved in the ALIGN flow is shown in Fig. 2. At the very basic level, the input can be a netlist and the output is a hierarchical layout in GDSII format. There are four major steps which are briefly described: (a) Constraint generation identifies known sub-circuits in the netlist and layout constraints such as symmetry, common-centroid, ordering, and matching. ALIGN uses graph convolutional network (GCN) to identify hierarchies like OTA, LNA, etc. The designer may examine these identified hierarchies and constraints and augment them

979-8-3503-2123-4/23 $31.00 © 2023 IEEE   57   2023 IEEE Radio Frequency Integrated Circuits Symposium

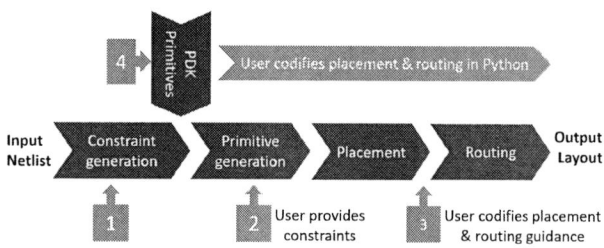

Fig. 2. Overview of the ALIGN flow.

to reflect designer intent. Primitives are one or more devices that are typically laid out as a single layout entity such as resistors, capacitors, current mirrors, and differential pairs. (b) The second step generates layouts for each of the primitives in the first step. (c) The third step assembles these primitive layouts into a legal layout that meets layout constraints. (d) The last step is routing which connects various nets with wires of appropriate widths. This step also generates power grids for the supply/ground nets and connects them to the devices. Apart from the auto-generated constraints in the first step, users can input placement constraints such as maximum width/height, aspect ratio, and spacing between any pair of blocks, and routing constraints such as shielding for critical nets, clock nets, and matched routing for symmetric nets. As shown in Fig. 2, the user can intervene in the ALIGN flow at multiple points and add/delete constraints within the flow. There is also support to code the entire placement and routing using relative positions of blocks. To ensure that the layouts generated are compatible with foundry-specified PDKs, an abstract set of rules are honoured by all the layout generators. These rules are chosen to be broad enough to work for all tested foundries with minimal changes to the flow. The arithmetic values for the layout rules change for different foundries and technologies.

Black-box methodology: Designers can reuse layouts of sub-circuits whose performance is verified either in silicon or via simulation. ALIGN supports the inclusion of such layouts through a black box methodology. In this methodology, the user-input layouts are abstracted into the library exchange format (LEF) with defined pins, ports and obstacles. The abstraction step is automated for the input layouts in GDSII format. These layouts are instantiated in the placement step and appropriate connections are made during routing.

Engineering change order (ECO): Design/layout iterations are performed to subsume the impact of layout parasitics. In each iteration, the layout is perturbed due to one of the following: alteration of device sizes, the spacing between devices, or inclusion of new placement/routing constraints. Depending on the hierarchy at which such a change is made, the impact on the layout could be localized or span the entire design. ALIGN handles such a change using an ECO methodology. As an example, we may add space between blocks to reduce coupling, which could perturb the corresponding hierarchy, its parents and neighbours. ALIGN automatically identifies such a perturbation and rapidly performs incremental placement and routing on those blocks.

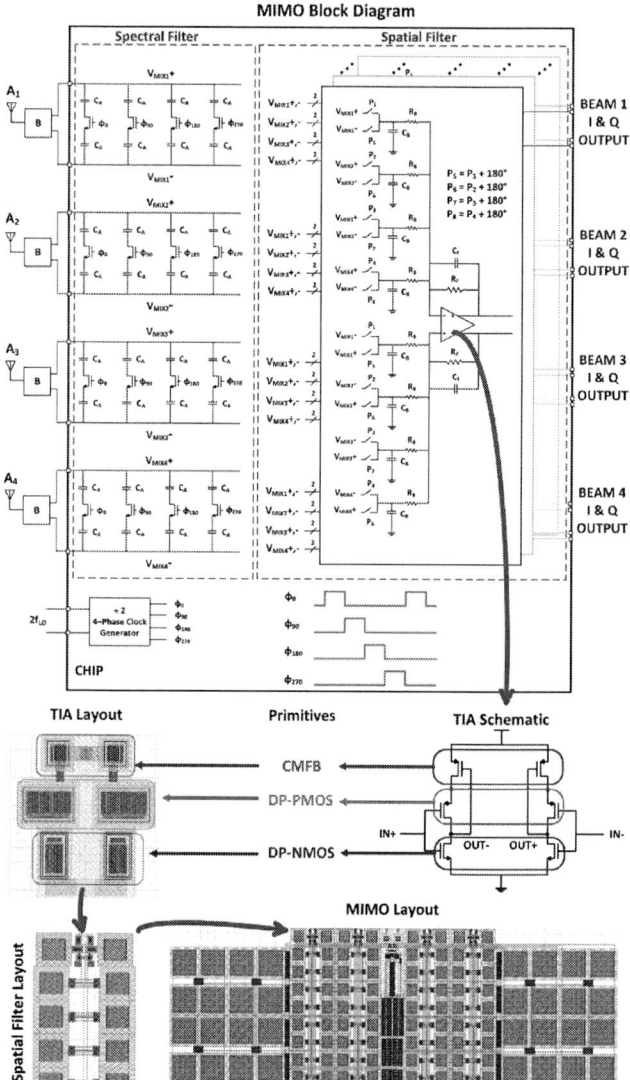

Fig. 3. Auto-annotation and hierarchical layout generation of the spatial filter.

III. MIMO

Fig. 3 shows the MIMO architecture with four spectral filters, eight spatial filters and the clock generation block. The spectral filter consists of a differential bottom-plate mixer architecture for improved IIP_3. The spatial filter consists of a differential summing amplifier with capacitor C_B acting as the voltage source. Spatial beamforming is performed by combining different antenna inputs with phase shifts. Fig. 3 shows some blocks recognized by ALIGN: the transimpedance amplifier (TIA) with common primitives such as common mode feedback (CMFB) transistor pairs, differential NMOS and PMOS pairs. After identifying these primitives associated with the amplifier, ALIGN automatically creates a symmetrical layout based on the device sizes. Internal routing widths can be user-defined, based on performance needs.

Fig. 4 compares various MIMO layouts generated using ALIGN with the aforementioned placement and routing constraints against a manual layout. For a fair comparison

979-8-3503-2123-4/23 $31.00 © 2023 IEEE

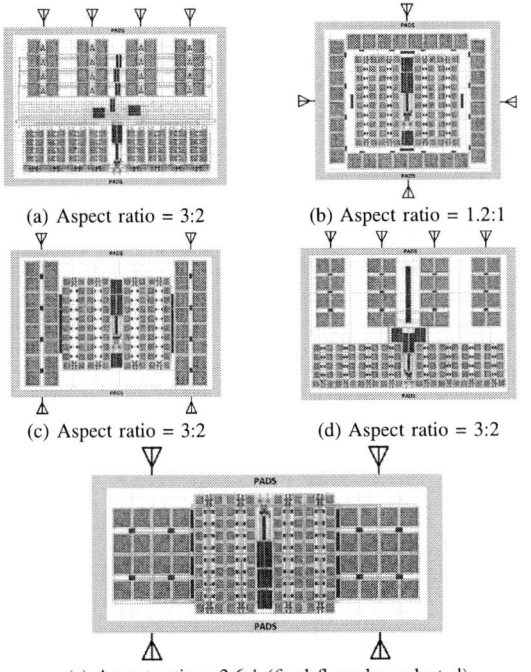

(a) Aspect ratio = 3:2

(b) Aspect ratio = 1.2:1

(c) Aspect ratio = 3:2

(d) Aspect ratio = 3:2

(e) Aspect ratio = 2.6:1 (final floorplan selected)

Fig. 4. Various MIMO layouts (a) Manual layout, (b) and (c) automatically generated layouts, (d) ALIGN mimicking manual layout through constraints, and (e) ALIGN layout with user-specified maximum height constraint.

between manual and ALIGN-generated layouts, the layouts of primitive cells such as MIM capacitors and special RF transistors used in the manual layout were reused in ALIGN layouts using the black-box methodology. Fig. 4(a) shows the manual layout and Fig. 4(b) and (c) show two ALIGN-generated layouts with just the clock net constraint. The layout in Fig. 4(b) is the most compact of all variants, and its square aspect ratio of the layout makes it easy to match routing parasitics using an H-tree structure.

In each iteration, simulations with post-layout extracted parasitics were used to identify the performance-critical nets and blocks. The following changes were made in successive iterations based on the simulations: (a) improving the resistance of critical nets by widening wires using the net-specific routing width constraint, (b) reducing coupling by (i) increasing the spacing between blocks, and (ii) adding shielding between adjacent signal nets. These changes involved perturbation to both placement and routing and were implemented automatically using the ECO mode described in Section II. The entire placement and routing in ECO mode took only tens of minutes in each iteration. Fig. 4(d) shows the layout generated by ALIGN mimicking the manual layout. This ALIGN layout was achieved by manually specifying constraints for all the hierarchies. An external limitation on the die size constrained the maximum height of the MIMO layout to be $600\mu m$ which when input to ALIGN generated the layout in Fig. 4(e). This layout was selected for the tapeout.

IV. MEASUREMENT

A prototype of a four antenna MIMO system was implemented in TSMC's 65nm CMOS process. The die photo

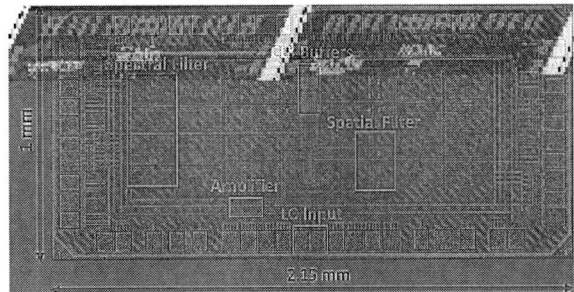

Fig. 5. Die Photo

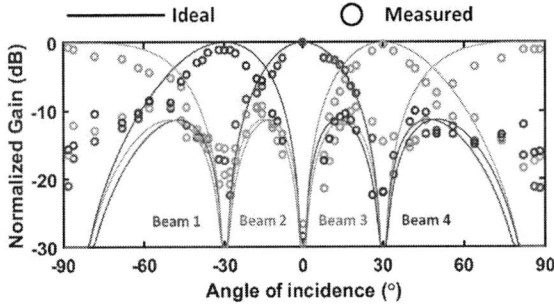

Fig. 6. Spatial gain across four beams.

is shown in Fig. 5. The dies were wire bonded to a 60-pin QFN and then mounted on a two-layer PCB. Four BALUNs were placed on the PCB to create differential RF signals.

Spatial Gain: For gain measurement, the RF output from a signal generator is split into four via a power divider. These four outputs are then passed through different PCB traces to create different phase shifts for a beam. The spatial gain for all four beams were measured from 1GHz to 2.5GHz and exploits the phase versus delay relationship. Fig. 6 shows the spatial gain for all four beams. A maximum spatial suppression of 28.4dB was measured between the broadside (Beam 1) beam and +30° (Beam 4) beam at 0° angle of incidence.

Gain and Bandwidth: Fig. 7 shows the measured gain of four output beams (Beam 1-Beam 4) for a broadside input beam at 1GHz. The measured low frequency gain and 3dB bandwidth for output Beam1 were 13dB and 30MHz respectively. As seen, we observe additional parasitic poles around 80MHz IF. The gain was measured for operating range of 1–2.5GHz. The measured gain of 13dB dropped after 2.3GHz. Hence, the measured operating range for this design is 1–2.3GHz.

IIP_3 **and** B_{1dB}**:** Figure 8 shows the measured IIP_3 and

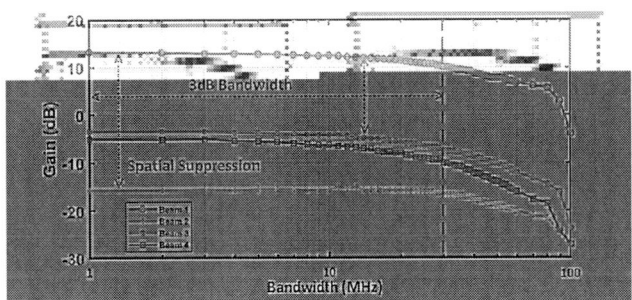

Fig. 7. Measured gain versus IF bandwidth for all four beams.

979-8-3503-2123-4/23 $31.00 © 2023 IEEE

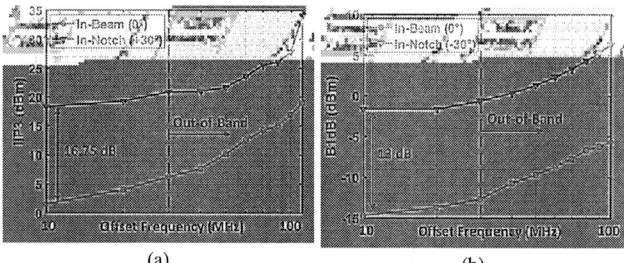

(a)   (b)

Fig. 8. (a) IIP_3 versus offset frequency. (b) B_{1dB} versus offset frequency.

Table 1. Measured performance comparison for manual vs ALIGN layouts.

	ISSCC'21[6]	This work
Technology	65nm CMOS	65nm CMOS
Layout type	Manual	Automated
Operating frequency range (GHz)	1-3	1-2.3
Single element conversion gain (dB)	15	13
Max spatial suppression (dB)	27	28.4
In-band/In-beam OIP_3 (dBm)	17.4	14.6
Out-of-band/In-beam IIP_3 (dBm)	18.8 $\Delta f/BW$=4.6	19.3 $\Delta f/BW$=4.6
In-band/In-notch IIP_3 (dBm)	17.8	18.3
Out-of-band/In-notch IIP_3 (dBm)	19.3 $\Delta f/BW$=2	23.64 $\Delta f/BW$=2
In-band/In-beam B_{1dB} (dBm)	-11.97	-14.66
In-band/In-notch B_{1dB} (dBm)	1	-1.66
Area (sq.mm)	2.53	2.15
Power (mW)	130-242	130-175

B_{1dB} values for different offset frequencies. The two-tones in the IIP_3 measurement were kept at broadside. Thus, in-beam IIP_3 measurement were taken at Beam 1 (0°) beam and in-notch IIP_3 measurement were taken at Beam 2 (+30°) beam. The measured in-band/in-beam IIP_3 and in-band/in-notch IIP_3 at 10MHz offset were 1.58dBm and 18.33dBm respectively. In the B_{1dB} measurement, the signal was kept at 0° angle and blocker was kept at either 0° angle (in-beam case) or -30° angle (in-notch case). The measured in-band/in-beam B_{1dB} and in-band/in-notch B_{1dB} at 10MHz offset were -14.66dBm and -1.66dBm respectively.

Performance Comparison: A comparison of the measured performance for the manual and automated layouts is shown in Table 1. The performance parameters of the manual layout [6] has been included with a loss calibration of 5dB. As can be seen, spatial suppression, IIP_3 and B_{1dB} of the automated layout are close to/exceeds manual layout's performance except for RF frequency range. We suspect this is because all the clock buffers were placed in the center for the automated placement, resulting in an operating frequency of 1-2.3GHz as opposed to manual layout's 1-3GHz range. This parameter can be improved with few additional iterations in ALIGN.

V. Productivity Improvement

Fig. 9 compares the time required to generate the layout of a MIMO receiver using manual and automated approaches. FP 1 and FP 2 correspond to the floorplans shown in Fig. 4(b) and (e) respectively. FP 2.1 refers to the default layout generated by ALIGN with the maximum height constraint of 600μm. Post-layout extracted simulations on this layout identified critical nets whose resistance needed to be improved. Resistance parasitics were improved using net-specific routing

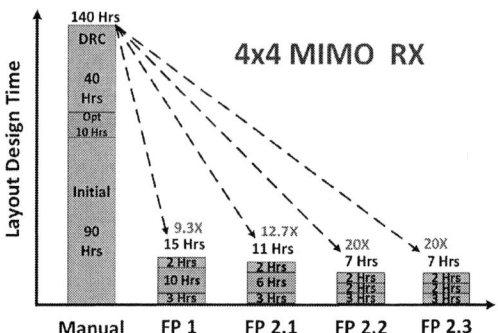

Fig. 9. Productivity gain: manual vs automated layout (four iterations)

width constraints and ECO mode described in Section II was used to realize these constraints. Simulations based on this layout identified nets whose coupling capacitance needed improvement. Using this feedback, shielding and increased spacing constraints were added and a second iteration of ECO was used to arrive at the final layout. As seen in Fig. 9, each of the iterations took hours to generate the layout and cleanup DRCs against the manual approach that took days for a single layout. The resultant automated layout has a similar performance to the manual layout as shown in Section IV. The productivity gain chart shows that within the same amount of time spent in generating a single manual layout, multiple automated layouts can be explored. As demonstrated, each layout can also be iteratively improved in a short time using performance evaluated with post-layout simulations.

VI. Conclusion

The efficacy of an automated RF-AMS layout synthesis flow has been demonstrated using productivity improvement on a state-of-the-art MIMO design. The layout is iteratively improved using an ECO mode with feedback from post-layout simulations. The automated flow generates a layout with performance similar to the manual layout with an order of magnitude smaller overall time. The time saved helps explore the design space and other architectures for the same design.

Acknowledgements

This work was supported in part by the DARPA IDEA program (contract N660011824048).

References

[1] J. Liu, S. Su *et al.*, "From Specification to Silicon: Towards Analog/Mixed-Signal Design Automation using Surrogate NN Models with Transfer Learning," in *Proc. ICCAD*, 2021, pp. 1–9.

[2] E. Chang, J. Han *et al.*, "BAG2: A process-portable framework for generator-based AMS circuit design," in *Proc. CICC*, 2018, pp. 1–8.

[3] T. Ajayi, S. Kamineni *et al.*, "An Open-source Framework for Autonomous SoC Design with Analog Block Generation," in *Proc. IEEE Int. Conf. VLSI & SoC*, 2020, pp. 141–146.

[4] B. Xu, K. Zhu *et al.*, "MAGICAL: Toward Fully Automated Analog IC Layout Leveraging Human and Machine Intelligence," in *Proc. ICCAD*, 2019, pp. 1–8.

[5] K. Kunal, M. Madhusudan *et al.*, "ALIGN – Open-Source Analog Layout Automation from the Ground Up," in *Proc. DAC*, 2019, pp. 1–4.

[6] J. Poojary and R. Harjani, "A 1-to-3GHz Co-Channel Blocker Resistant, Spatially and Spectrally Passive MIMO Receiver in 65nm CMOS with +6dBm In-Band/In-Notch B1dB," in *Proc. ISSCC*, 2021, pp. 96–98.

[7] "ALIGN: Analog layout, intelligently generated from netlists," https://github.com/ALIGN-analoglayout/ALIGN-public.

979-8-3503-2123-4/23 $31.00 © 2023 IEEE

RMo2A-2

Optimizing RFSOI Performance through a T-shaped Gate and Nano-Second Laser Annealing Techniques

L. Lucci[#], S. Crémer[$], B. Duriez[#], T. Fache[#], S. Kerdiles[#], Y. Morand[#],
J.-M. Hartmann[#], J. Azevedo-Goncalves[$], F. Gaillard[#], P. Chevalier[$]

[#]CEA-Leti, Université Grenoble Alpes, France
[$]STMicroelectronics, France

luca.lucci@cea.fr

Abstract — **We report on two experiments that were carried out in order to boost the RF performances of PD-SOI devices. In the first experiment we implemented a T-shaped gate on a nominally 40 nm long device to mimic on an advanced RFSOI platform the mushroom gate shape that is usually found in III-V devices. T-shaped gate more than halved the longitudinal gate resistance improving RF figure-of-merits for long finger devices. In a second experiment we used a nano-second laser anneal of the gate poly-Si layer. The reduction of the vertical component of the gate resistance helped to improve the performances of short finger devices.**

Keywords — **RFSOI, CMOS, HF noise, NLA, T-shaped.**

I. INTRODUCTION

Today's advanced CMOS silicon devices are challenging III-V devices for applications in the mmW and sub-THz spectra, paving the way towards cost-effective solutions required by novel market applications.

In the wide array of CMOS options, Partially-Depleted Silicon-on-Insulator (PD–SOI)— a mature technology which development started long ago [1]—is the technology of choice for a range of RF products. When market opportunities are identified, there is still plenty of opportunities for technology upgrades or developments [2], [3]. PD–SOI will definitely have its place among the technologies adopted for emerging applications in 6G telecommunications, SATCOM, and the internet-of-things (IoT) [4].

In this contribution we present the outcome of two innovative process modules implemented during the fabrication of state-of-the-art PD–SOI devices [5]. The objective being to boost the RF performances, notably by reducing the effective RF gate resistance, R_G, a well-known RF weakness of advanced CMOS devices.

More precisely, in the first experiment we modified the gate-stack process for a 40 nm channel length poly gate to obtain a mushroom-like structure or T-shaped gate and have a morphology similar to to those routinely found in III-V. The aim is to minimize the longitudinal resistance component of the top silicided layer. In the second experiment we used nanosecond laser annealing to locally reduce the vertical component of the gate resistance that is dominant in short finger devices.

The manuscript is organized as follows: in the first part we will quickly outline the currently known models for gate resistance in CMOS RF devices, quickly clarifying the

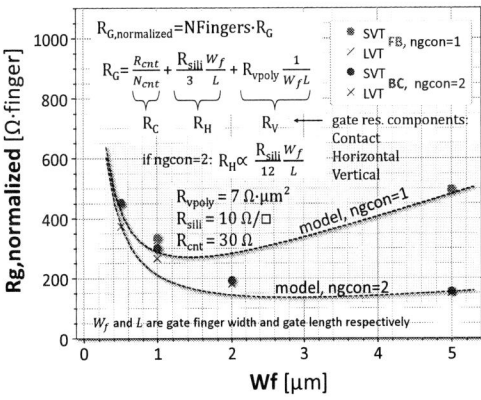

Fig. 1. Data and model comparison for minimum channel length devices of gate resistance vs. gate finger width (Wf). Symbols are data for different device flavours, *i.e.* floating-body (FB) devices in red, or body-contacted (BC) devices in blue. Different threshold voltage (V_T) devices are also reported: standard-V_T (SVT) with circles and low-V_T (LVT) with crosses. A simple three parameter model is still capable of reproducing with good accuracy the extracted RF gate resistance on a reference Process-Of-Record (POR) wafer.

measurement procedure and the extraction methodology. In the second section we will illustrate the impact of a T-shaped gate on RF small-signal properties of the device, R_G and f_{max}. In the third part we will address the laser annealing and its impact on small-signal properties, mostly regarding the high-frequency noise.

All experiments were carried out on 300mm wafers in our facility. Processed wafers went through standard DC and AC characterization at M2-metal layer. RF measurements were then carried out on fully processed wafers at Mtop. Keysight or R&S VNAs and 300-mm probe stations were used for S-parameter measurements up to 67 GHz. When appropriate, high-frequency noise was characterized up to 50 GHz with a Keysight PNAX setup. Unless explicitly indicated, all S-parameters are assumed to be de-embedded down to Metal 1 with all BEOL contributions removed by a standard Open-Short procedure.

II. SMALL-SIGNAL AND GATE RESISTANCE MODELING

In the early days of RFCMOS, the horizontal/longitudinal component, R_H, and the vertical component, R_V, of the gate resistance R_G were identified. The R_H is dominated by the top silicide resistance R_{sili} in $\Omega/\square$. R_V is dominated by the

979-8-3503-2123-4/23 $31.00 © 2023 IEEE   61   2023 IEEE Radio Frequency Integrated Circuits Symposium

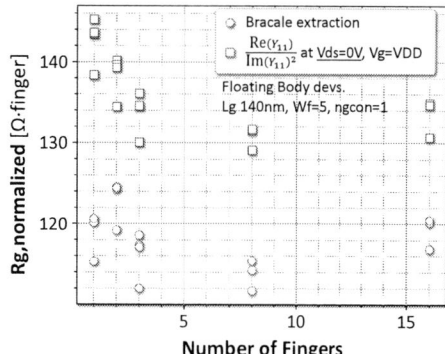

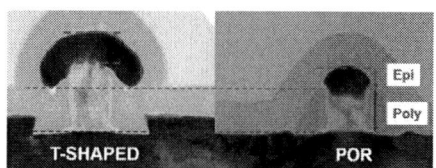

Fig. 2. The cold-FET R_G extraction used in this contribution [16] is more immune to systematic/bias errors w.r.t. simpler approaches like extracting $\mathrm{Re}\,[Y_{22}]\,/\mathrm{Im}\,[(Y_{22})]^2$. Between the two extractions there is correlation, but a small 10% offset is present, due to the uncompensated channel resistance with the simple extraction, which is also showing a un-physical dependence on the number of fingers in the device.

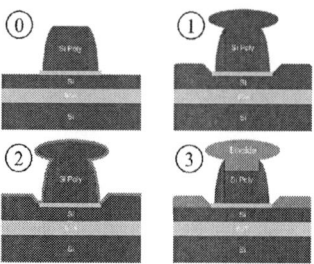

Fig. 4. Illustration of the main steps introduced or modified for a T-shaped gate: steps (0) and (2) are in common with the nominal device. The epitaxial growth (1) is added and has an impact both on the polysilicon gate and S/D regions. The silicidation (3) is present both in the nominal device and in the T-shaped flow, but lasts longer in the latter device.

Fig. 3. TEM images of a T-shaped gate (left) and of a standard device (right). For a minimum channel length device, the top silicide layer is almost 3x in width and also thicker.

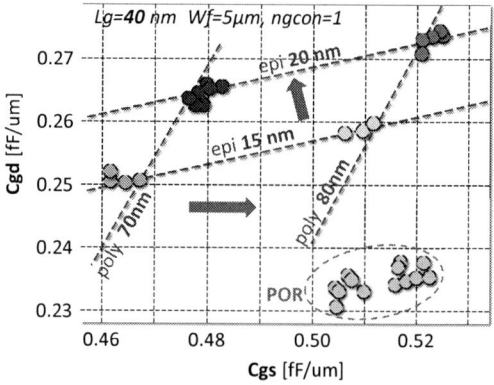

Fig. 5. Effect on C_{GS} and C_{GD} of the process split on polysilicon deposition height, and epi-layer thickness ('Poly' and 'Epi' in Fig. 3). Care was taken for the T-shaped devices to have the same gate length and same spacer geometry for a fair comparison. C_{GS} and C_{GD} are extracted at $V_{DS}{=}1$ V and V_{GS} bias where peak-f_t value is obtained for the device-under-test.

poly-silicon resistance R_{vpoly} in $\Omega \cdot \mathrm{cm}^2$. It was demonstrated that R_H and R_V distributed nature could be accounted for using a simple factor 1/3 [6]. The model was updated with a factor 1/12 in the case of double gate connection [7]. The silicided-poly to contact resistance was then added [8]. The R_G model as such was integrated in CMOS spice models like BSIM [7] and PSP [9] and shown to be valid for metal gates and in new advanced devices [10]. The simple model is capable of reproducing extracted R_G values in PD–SOI technology as shown in Fig 1.

Current 'effective' RF resistance extraction is discussed in [11] and known to be challenging in recent technologies [12], [13] because the resistance is strongly reduced. Previous similar experiments to reduce the gate resistance made use of gate pre-doping. Results were reported in [14] for a 28 nm bulk technology, and in [15] for PD–SOI.

The methodology used for effective R_G extraction is detailed in [16] and is giving, in our experience, more consistent results than standard textbooks extraction [17] as shown in Fig. 2.

III. T-SHAPED GATE

The idea of this structure (see Fig. 3) is to increase the width of the top conductive layer of the gate finger, which is made of highly conductive silicide, while keeping it narrow at its base, and consequently preserve a short channel at the interface between the poly and the gate oxide [18], [19]. The longitudinal resistance, R_H, should therefore decrease, especially for minimum channel length devices. The exact

process changes to turn a regular gate into a T-shaped gate in a nutshell are (Fig. 4): Starting wafers are stopped after S/D implant (0); some Silicon selective epitaxial growth is performed (1); a longer silicidation process is used to fully silicide the top of the gate (3).

The drawback that could be expected is the increase of the parasitic capacitance between the gate and the source/drain regions. The C_{GD} degradation was limited in the design phase by process simulations. Then process splits on poly deposition and epitaxial growth time were engineered to preserve the overall gate width to height aspect ratio while minimizing the added parasitic capacitances (Fig. 5).

Fig. 6 shows that the extracted resistance is unchanged for short finger widths but strongly reduced for longer fingers, a clear signature of a reduced longitudinal sheet resistivity. Its main impact is a strong increase in the f_{max} performance at constant f_t. A perfect correlation could be found between the R_G and the f_{max}, as shown in Fig. 7. A smaller boost (but at higher f_{max} values) could be measured on Wf=2 μm devices.

IV. NANO-SECOND LASER ANNEAL

Laser annealing, high temperature treatment using a laser source, is a well-know technique that has found extensive

979-8-3503-2123-4/23 $31.00 © 2023 IEEE

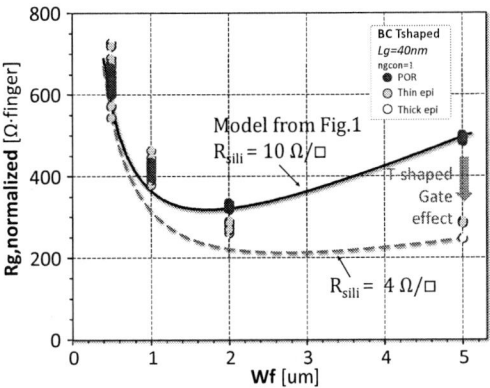

Fig. 6. Impact of a T-shaped gate on the extracted gate resistance. In the picture we show the biggest improvement (reduction), *i.e.* single side contacted (ngcon=1) min gate length devices. Above 2μm, finger gate resistance is halved. T-shaped gate data can be fitted simply reducing of the vertical resistance physical parameter R_{sili}. The impact does not depends on the device flavour (SVT or LVT) and is equivalent between body-contacted and floating-body devices.

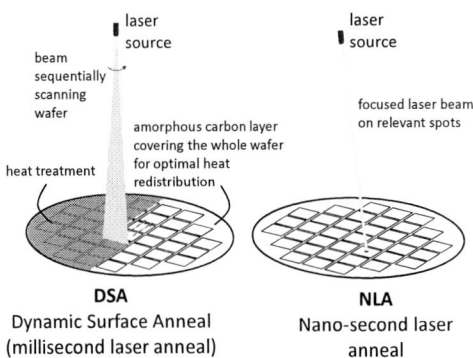

Fig. 8. Illustrative comparison between DSA and NLA anneals. Both techniques are used in the present experiment. DSA is a millisecond laser anneal implemented with a redistribution layer that spreads the heat budget homogeneously and allows a consistent annealing reaching some microns deep into the substrate. Meanwhile, NLA is a much shallower, higher energy annealing technique which is pattern-dependent and which enables to anneal only the poly-Si/salicide part of the gate stack.

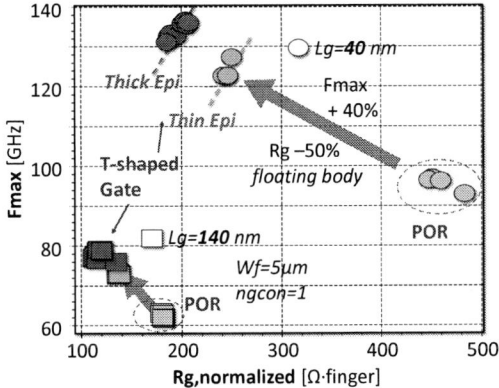

Fig. 7. Correlation between f_{max} and R_G for long finger (Wf=5μm) devices at Lmin and at Lg=140 nm. The boost is unaffected by the device flavour.

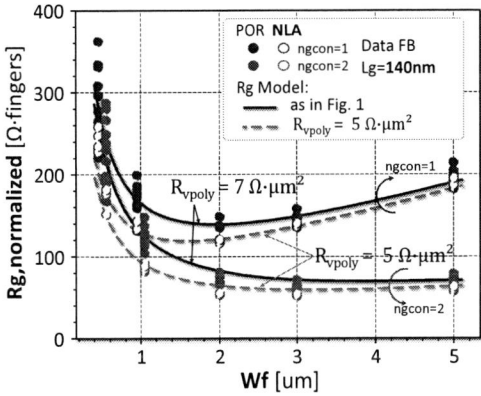

Fig. 9. Extracted gate resistance versus finger width Wf for devices with gate length Lg=140 nm. Both single contact (ngcon=1) and double contact (ngcon=2) devices are reported. Filled symbols are standard devices and empty symbols devices that received a nano-second laser anneal. Data at Wf=0.5 and 1.0 μm are plotted slightly offset to distinguish the two sets. The impact of NLA was fitted simply with a 30% reduction of the polysilicon resistance R_{vpoly}.

use in 65 nm and lower technology node devices. The dynamic-surface anneal (DSA) or milli-second laser anneal, as illustrated in Fig. 8, is routinely used in combination with a moderate temperature RTA (rapid thermal annealing) [20]. Using a millisecond-laser anneal after a moderate temperature RTA yields better dopant activation while limiting undesired diffusion, if compared to a high temperature RTA alone. In this experiment a nano-second laser anneal (NLA), using much short pulses at higher energy/frequency [21], was used to maximize dopant activation in short-channel narrow finger width poly-gates. The expected results is an improved vertical gate resistance. Thanks to the negligible thermal budget, other device parameters should not be impacted.

Effort was spent to identify the key process parameters, in this particular case the energy density of the UV-laser beam and the maximum number of pulses at the same location. Experiments were carried on test wafers with different energy densities and various number of pulses till damages were spotted on the poly and silicided areas. In our specific case we

found a maximum applicable energy density of 300 mJ/cm² (ranges up to 600 mJ/cm² were tested) with no further gain obtained after 10x repetitions (up to 300x repetition tested).

In Fig. 9 the impact of NLA is shown on long channel devices, Lg=140 nm, because test structure with a variety of finger widths were available for measurements. Analogous R_G reductions could be measured at Lmin. It is clearly shown that NLA reduces the normalized finger resistance for shorter fingers, were the vertical component is predominant. It also reduces the significant spread in the extracted values.

No device parameter other than R_G significantly changed before/after NLA. The R_G reduction clearly boosted f_{max}, although the very small R_G values involved caused f_{max} extraction issues. The most interesting results were obtained for the four high-frequency noise parameters: NF_{min}, Γ_{opt} and

979-8-3503-2123-4/23 $31.00 © 2023 IEEE

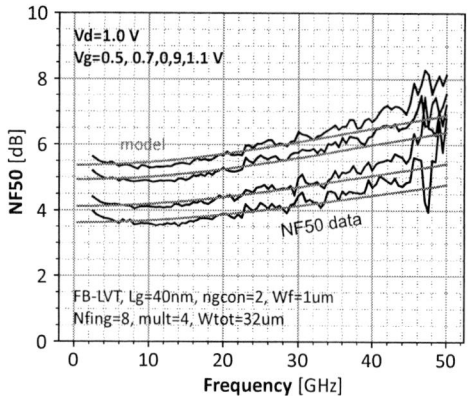

Fig. 10. Extraction of high-frequency noise parameters from NF50 measurements. A small-signal model is extracted from the S-parameters, then a T_{out} value is calibrated for a Pospiezalski noise model to reproduce the NF50 data. The methodology follows [22].

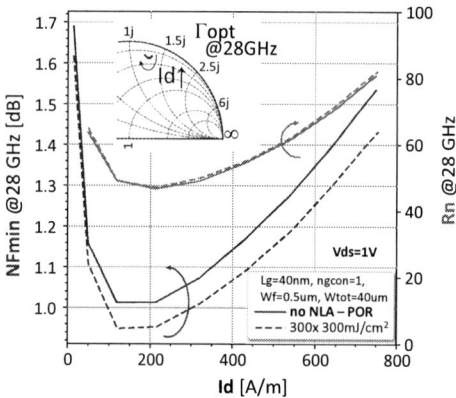

Fig. 11. Extracted NF_{min} values vs drain current for a nominal device and NLA: reducing R_G results in lower high-frequency noise.

R_n. Noise measurements were performed concurrently with the S-parameters in selected devices. Small-signal and noise parameter extraction was carried out using the methodology described in [23]. Results are provided in Fig. 10.

In Fig. 11 it is shown that the decrease in gate resistance measured on the selected device is at the origin of a small (0.1 dB) reduction of the NF_{min} at 28 GHz. Same reduction could be measured on all narrow-finger short-channel devices.

V. Conclusion

Process experiments were implemented during fabrication of state-of-the-art PD–SOI devices to assess the impact of two technological options on the RF figures of merit.

Implementing a T-shaped gate reduced R_G and boosted f_{max} in long finger devices with introduction of just one extra epitaxial growth step.

A nano-second laser anneal was capable of selectively bringing a poly finger near its melting temperature, so reducing its vertical resistance component by maximizing dopant activation, without any trade-off on other device parameters. Improvements in f_{max} and a 0.1 dB reduction in noise figure were measured in very short finger devices.

These results and assessment pave the way to the perspective of using both techniques concurrently to simultaneously reduce both the silicide and the poly-silicon components of the gate resistance.

Acknowledgment

The experiments in this contribution were funded by the French authorities within the framework of the NANO2022 and BEYOND5 European projects.

References

[1] S. Lee *et al.*, "Record RF performance of 45-nm SOI CMOS technology," in *IEEE IEDM Tech. Digest*, 2007.
[2] S. H. Jain *et al.*, "Novel mmWave NMOS device for high Pout mmWave power amplifiers in 45RFSOI," in *IEEE ESSDERC Proc.*, Sep. 2021.
[3] S. Khokale *et al.*, "LNFET device with 325/475 THz f_T/f_{max} and 0.47 dB NF$_{min}$ at 20 GHz for SATCOM applications in 45nm PDSOI CMOS," in *IEEE RFIC Symp.*, 2022.
[4] B. Martineau *et al.*, "Si and SOI CMOS technologies for millimeter wave wireless applications," in *IEEE IEDM Tech. Digest*, 2020.
[5] P. Chevalier *et al.*, "PD-SOI CMOS and SiGe BiCMOS technologies for 5G and 6G communications," in *IEEE IEDM Tech. Digest*, 2020.
[6] B. Razavi *et al.*, "Impact of distributed gate resistance on the performance of MOS devices," *IEEE Trans. Circuits and Syst. I, Fundam. Theory Appl.*, vol. 41, no. 11, pp. 750–754, 1994.
[7] X. Jin *et al.*, "An effective gate resistance model for CMOS RF and noise modeling," in *IEEE IEDM Tech. Digest*, 1998.
[8] A. Litwin, "Overlooked interfacial silicide-polysilicon gate resistance in MOS transistors," *IEEE Trans. Electron Devices*, vol. 48, no. 9, pp. 2179–2181, 2001.
[9] A. J. Scholten *et al.*, "The new CMC standard compact MOS model PSP: Advantages for RF applications," *IEEE J. Solid-State Circuits*, vol. 44, no. 5, pp. 1415–1424, 2009.
[10] B. Dormieu *et al.*, "Revisited RF compact model of gate resistance suitable for high-k/metal gate technology," *IEEE Trans. Electron Devices*, vol. 60, no. 1, pp. 13–19, 2013.
[11] I. Jo *et al.*, "Accurate extraction of effective gate resistance in RF MOSFET," *Circuits and Systems*, vol. 06, no. 05, pp. 143–151, 2015.
[12] R. Torres-Torres *et al.*, "MOSFET gate resistance determination," *Electronics Letters*, vol. 39, no. 2, p. 248, 2003.
[13] ——, "Impact of technology scaling on the input and output features of RF-MOSFETs: effects and modeling," in *IEEE ESSDERC Proc.*, 2003.
[14] C. Schwan *et al.*, "CMOS RF performance gain by gate resistance optimization," in *IEEE RFIC Symp.*, 2016.
[15] D. Lederer *et al.*, "45nm PD-SOI FET gate resistance optimization for mmW applications," in *IEEE SOI-3D-Subthreshold Microelect. Tech. Unified Conf. (S3S)*, Oct. 2018.
[16] A. Bracale *et al.*, "A new approach for SOI device small-signal parameters extraction," *Analog Integr. Circuits and Signal Process.*, vol. 25, no. 2, pp. 157–169, 2000.
[17] C. Enz *et al.*, "MOS transistor modeling for RF IC design," *IEEE J. of Solid-State Circuits*, vol. 35, no. 2, pp. 186–201, 2000.
[18] D. Hisamoto *et al.*, "A low-resistance self-aligned T-shaped gate for high-performance sub-0.1-μm CMOS," *IEEE Trans. Electron Devices*, vol. 44, no. 6, pp. 951–956, Jun. 1997.
[19] H.-C. Lin *et al.*, "A novel self-aligned T-shaped gate process for deep submicron Si MOSFETs fabrication," *IEEE Electron Device Lett.*, vol. 19, no. 1, pp. 26–28, Jan. 1998.
[20] O. Gluschenkov *et al.*, "Laser annealing in CMOS manufacturing," *ECS Transactions*, vol. 85, no. 6, p. 11, 2018.
[21] K. Huet *et al.*, "Doping of semiconductor devices by laser thermal annealing," *Materials Science in Semiconductor Processing*, vol. 62, pp. 92–102, 2017.
[22] F. Danneville *et al.*, "RF and broadband noise investigation in high-k/metal gate 28-nm CMOS bulk transistor," *Int. J. Numerical Modelling: Electronic Networks, Devices and Fields*, vol. 27, no. 5-6, pp. 736–747, Jan. 2014.
[23] O. M. Kane *et al.*, "22nm ultra-thin body and buried oxide FDSOI RF noise performance," in *IEEE RFIC Symp.*, 2019.

RMo2A-4

Artificial Neural Networks for GaN HEMT Model Extraction in D-band Using Sparse Data

Andrea Arias-Purdue, Eythan Lam, Jonathan Tao, Everett O'Malley, James F. Buckwalter

University of California, Santa Barbara, USA

andrea_arias@ucsb.edu

Abstract— We describe the application of Artificial Neural Networks (ANNs) for Gallium Nitride (GaN) High-Electron Mobility Transistor (HEMT) model parameter extraction to improve the model accuracy between 110 and 170 GHz. Fully-connected ANNs trained by backpropagation relate the physics-based ASM-HEMT model parameters to RF transistor measurements. The effects of ANN activation function, number of layers, number of nodes and number of training set data points on training accuracy are studied. For the 12 model parameters that dominate the 40-nm GaN HEMT RF characterization, we obtained a combined root-mean-squared (RMS) error of 2.5% between the ANN prediction and the training set, which is acceptable for most design tasks.

Keywords— GaN, HEMT, ASM-HEMT, machine learning, artificial neural network, backpropagation, millimeter-wave.

I. INTRODUCTION

GaN HEMTs are amongst the most promising devices to substantially increase power density in millimeter-wave bands for communications and radar [1]. Scaled GaN HEMTs exhibit high transistor figures of merit such as transit frequency (f_T) and maximum frequency of oscillation (f_{MAX}) while simultaneously achieving excellent breakdown voltage. In [2], the 40-nm T-gate process offers f_T/f_{MAX} = 200/400 GHz with a breakdown exceeding 40 V and has been shown for a 140-GHz, 23-dBm power amplifier (PA) [3]. Due to the limited available gain (below 6 dB) at D-band (110-170 GHz), a careful performance trade-off study must be undertaken to achieve the optimal loadline for power and/or power added efficiency (PAE). The design of a D-band PA requires precise tuning with the transistor internal parasitics to achieve desired performance and typically requires several iterations in III-V processes as illustrated in the conventional approach in Fig. 1.

Most III-V device model development is typically constrained to frequency bands under 60 GHz. Current control (I-V) and small-signal (S-parameters) inform model parameters in physics-based models while large-signal loadpull (LP) is often used for model validation. Hence, the device physics is fully characterized by the small-signal and DC data alone. Standard model extraction techniques (either physics-based or phenomenological) are linear in nature: the model parameters are extracted either through DC or AC tests, however most parameters are affected by both conditions.

Moreover, these models oftentimes become inaccurate at higher frequency bands where non-linear (power sweep) device data may not be available. In D-band, a comparison of simulation and measurement of the 40-nm GaN HEMT is shown in Fig. 2 and indicates model accuracy of around 8-15%, which is a significant uncertainty to design circuits that might be tuned over a 10% bandwidth.

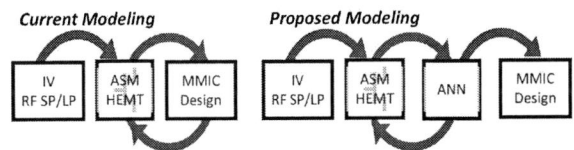

Fig. 1. Proposed ANN model modifies foundry models to fit measurement characteristics at higher frequency bands.

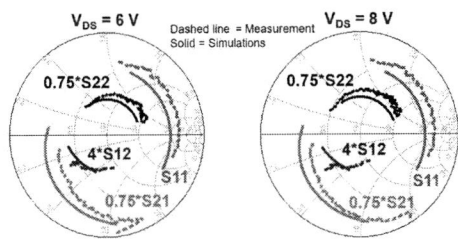

Fig. 2. Measured transistor S-parameters (dashed line) and simulations (solid line) from 110-170 GHz of a 40-nm GaN device (gate width W_g= 4x37.5 μm) at 0.2 A/mm. The accuracy of the model is around 8-15%.

This work proposes Artificial Neural Networks (ANNs) as part of a design methodology for rapid and accurate physics-based device model parameter extraction as illustrated in Fig. 1. THe ASM-HEMT model is emerging as a standard physics-based model, compatible with most circuit simulators [4]. Our approach leverages the ANN to map the non-linear interrelations between model parameters with DC and RF performance. Prior work has successfully used deep learning for DC-parameter model extraction, training to a total of 6 parameters [5].

The proposed methodology significantly extends the limitation of the prior work to develop an ANN model from sparse data in the form of basic DC and S-parameter characterization to complete a set of ASM-HEMT parameters. We identify 12 model parameters that most affect the model and demonstrate an RMS training and validation error of 2.3-2.7%. In section II, we review the appropriate choice of machine learning approaches for device modeling. We demonstrate the connection of the ANN to the ASM-HEMT model and the selection of hidden layers for an optimally-trained ANN. Section III describes the results of the training and improvements in the S-parameters measurement agreement between 110 and 170 GHz.

979-8-3503-2123-4/23 $31.00 © 2023 IEEE

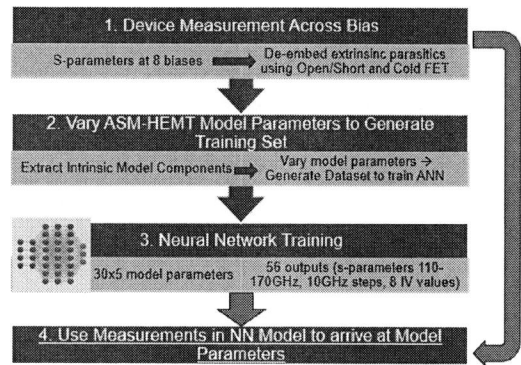

Fig. 3. Training methodology for ANN of the ASM HEMT model.

II. ANN-BASED MODELING METHODOLOGY

Neural networks have been considered for transistor modeling since the early days of computer-based semiconductor device modeling [6] including small-signal [7]-[8], and large-signal modeling [9]-[10]. Various neural network architectures have been proposed for transistor modeling, including fully-connected convolutional networks (CNNs) [11], and recurrent networks (RNN) [12]. Here, we use the ASM-HEMT model, a potential-based physical model that has been shown to offer good agreement for GaN HEMTs [13] and that can be parameterized through the ANN training methodology summarized in Fig. 3.

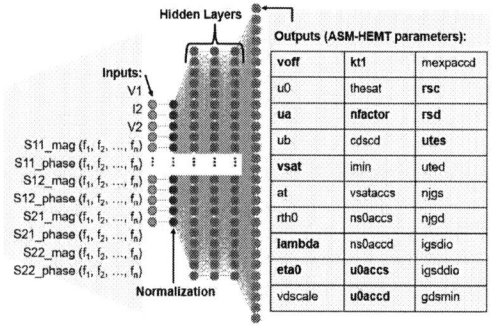

Fig. 4. ANN model for ASM-HEMT parameter extraction.

First, representative S-parameters measurements from 110-170 GHz are made to arrive at the intrinsic model parameters. These intrinsic parameters are used in the ASM-HEMT model and a total of 30 model parameters are varied (5 variations per parameter), with each simulation generating a set of outputs (8 IV values and S-parameters at 7 frequency steps from 110 - 170 GHz). To train the ANN, a dataset containing the model parameter inputs and corresponding IV and S-parameters at each of the 7 frequencies is generated while the ANN outputs are the 30 model parameters. The training datasets contain a varying amount of entries in multiples of 12k points (110-170 GHz in 10 GHz steps). The ASM-HEMT model simulations were performed using Keysight ADS, varying each of the 30 model parameters identified in Fig. 4 arranged in 6 sub-groups per

bias point (V_D, I_D). Each simulation generated up to 115k points, which was found optimal in that it kept the simulation time under 2 minutes.

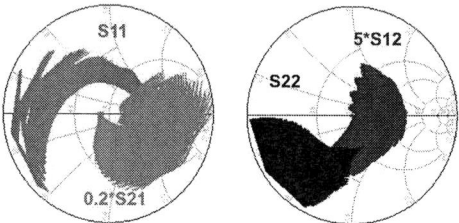

Fig. 5. Composite S-parameter plot of partial training set varying 6 model parameters at V_{DS} = 8 V and I_D = 20 mA.

The fully-connected neural network contains a normalization layer and a varying number of nodes and hidden layers (Fig. 4). Each node is indexed and updated independently from its layer, so that it can be easily de-activated without incurring additional computational losses and without having to modify the layer. The ANN inputs are trained using the backpropagation algorithm [14].

In the forward direction, the jth node output, Output$_j$, is calculated through an activation function using

$$\text{Output}_j = f_{\text{Activation}}(\text{Bias} + \Sigma(\text{Input}_i \times W_i)), \quad (1)$$

where W_i is the weight for the ith input. The weights are updated by applying a correction (ΔW_i) for output nodes:

$$\Delta W_i = \eta \times \Delta N_{\text{OUT}} \times f'_{\text{Activation}} \times \text{Input}_i + \alpha \times \Delta W_{i(n-1)} \quad (2)$$

where η is the training coefficient and

$$\Delta N_{\text{OUT}} = \text{Training Data - Output (Output Node)} \quad (3)$$

The derivative of the activation function ($f'_{\text{Activation}}$) is evaluated at the current training instance, Input$_i$ is the ith input, α is the momentum coefficient, and $\Delta W_{i(n-1)}$ is the delta weight from the previous training instance.

For hidden nodes, output errors need to be backpropagated through the neural net. Eq. (2) still applies, however, ΔN_{OUT} is now

$$\Delta N_{\text{OUT}} = \Sigma(W_k \times \Delta N_{\text{OUT},k}) \quad \text{(Hidden Node)} \quad (4)$$

where k is the index of the kth output node connected to the hidden node.

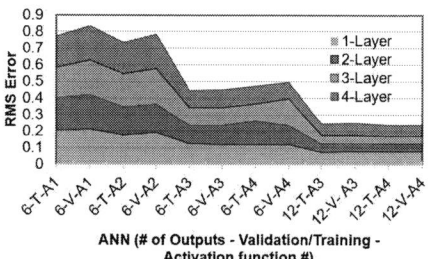

Fig. 6. Stack graph of RMS error performance of training (T) and validation sets (V) 1- through 4-layer ANNs, with activation functions 1 through 4, with 6 and 12 outputs. The activation functions used are: Tanh (A1), sigmoid + Gaussian (A2), sigmoid (A3), and ReLU + sigmoid (A4).

III. ANN TRAINING MATRIX AND RESULTS

The ANN described in the last section provides foundational insight into ANN structure, such as the number of hidden nodes and layers, activation functions and training coefficients. Our assumption is that semiconductor systems exhibit universal IV behavior signatures, manifested as saturation and pinch-off. The RF behavior is generally influenced by dynamic effects such as feedback capacitance and charge modulation that are also present in all devices to varying degrees. These effects can be globally and accurately captured by the appropriate activation functions, weights, bias and other ANN parameters.

Our goal is to optimize the ANN design in terms of its size (layer and node count), activation function, number of training set data points and training coefficient. The ANN performance parameters are the average RMS error relative to the training set and the training time. Commonly used activation functions were included in this study, including sigmoid, tanh, ReLU, Gaussian, arc(cos) and Shockley equation. Each activation function bias and leakage is randomized prior to training. The backpropagation training coefficient (η) was also varied from 0.2 to 0.0001 (where the baseline value is 0.002). The ANN design matrices are constructed by, first, completing a coarse version, followed by a finer set of ANNs with varying number of nodes, training set size and backpropagation coefficient. In addition, we also repeated some of the first matrix variations (e.g. number of layers, number of outputs and activation function) to ensure that the chosen design space was optimal.

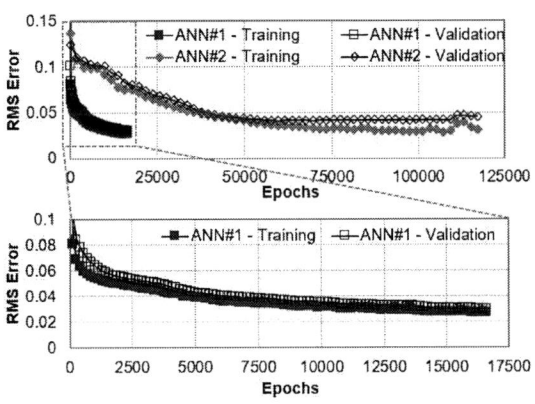

Fig. 7. RMS training error of downselected ANNs. ANN#1 trains significantly faster than ANN#2, while reaching the same overall RMS error of 2.3-2.7%.

Fig. 6 summarizes the results of the first ANN design matrix, showing a stack graph of each ANN version simulated. Each ANN has 16 nodes per layer. The same training data set comprised of 115k entries was used to understand the effect of the number of layers (fixed number of nodes per layer), the activation function and the number of outputs being learned relative to the number of inputs. Fig. 6 includes the training error and the validation set error for each ANN. The validation set is generated by selecting 10% of the training set data at random and removing it from the training/backpropagation parts of the algorithm. The simulations indicate that 2 to 3

layers appear to perform better as compared to 1- or 4-layer ANNs. In all cases, the RMS error of the validation set is very similar to that of the training set. We are able to calculate the ANN overall RMS error since the ANN numerical outputs are all normalized to the same limits. From Fig. 6, the choice of activation function has a noticeable effect on the RMS error, with functions A3 and A4 performing the best.

Next, we varied the number of nodes followed by the number of training set data points, from 115k to 128k and 140k. Varying the number of nodes per layer from 12 to 24 in steps of 4 revealed that RMS performance was maintained for the 12 - 20 nodes case while it deteriorated for the 24 node ANN for activation functions A2 and A3 (from 5% to 7%). Increasing the number of training data points from 115k to 140k did not improve RMS error, with the best performance being about 5% RMS error for both the training and the validation sets (same as baseline ANN obtained with the first design matrix). Likewise, varying the backpropagation coefficient η did not signifcantly change the error, although we note that if the parameter η is too large (>0.1) the ANN weights overflow to very large figures and the net cannot recover.

A 5% overall error is generally not sufficient to generate an accurate ANN model, hence we revised the ANN input structure next. The new structure key difference is in the frequency dependence of the data; instead of frequency being a free parameter with 7 entries, the ANN inputs were specified for each frequency trained, so the number of input nodes was increased from 12 (current, voltage, frequency and 8 S-parameters) to 59 (8 S-parameters per frequency point). The modified data array improved the RMS error by two-fold, from 5% (Fig. 6, best case) to 2.3-2.7% (Fig. 7). Further improvement can likely be effected by increasing the dataset bandwidth, although we will show that we can obtain robust model predictions based on our sparse model.

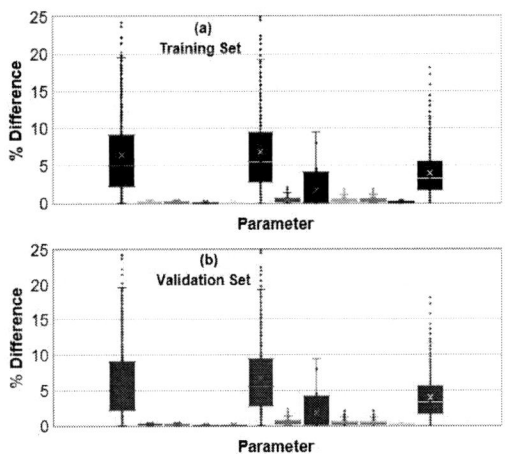

Fig. 8. Box plot graph of percent error of each of the 12 parameters tracked for the training (a) and validation sets (b).

A similar study varying activation function, layer and node count, training set size and backpropagation coefficient was performed on our revised ANN design (59 inputs) with the

---

979-8-3503-2123-4/23 $31.00 © 2023 IEEE   67

resulting RMS averages mostly around 3%, confirming our findings during our first set of ANN simulations.

While non-optimal activation functions can increase RMS error, it appears that the appropriate ANN design space allows for a relaxed array of functions. We note that when attempting to achieve errors below 0.5%, activation function choice optimization likely becomes important. In our case, where the target is <2.5% RMS, ultimately, the RMS error lower bound appears to be limited by dataset structure. ANN design parameters such as activation function, backpropagation rate and other factors can improve RMS error but they mostly affect training time, an important efficiency metric.

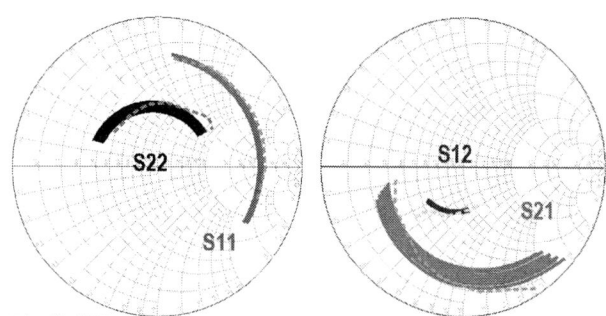

Fig. 9. Composite graph showing the ANN-generated model (solid) alongside measured S-parameters (dashed line) from 110-170 GHz at 0.2 A/mm (30 mA) V_D= 8 V bias.

The lowest error ANNs are demonstrated in Fig. 7. We highlight that an optimal activation function choice can significantly improve training time (under 2 minutes for ANN#1). ANN#1 uses a combination of ReLU and sigmoid function nodes, while ANN#2 uses sigmoid only. For ANN#1, the individual parameter RMS error averages are <1% for 6 parameters, between 1 and 5% for 2 parameters, between 5 and 8% for 2 parameters and 8-15% for the last 2 parameters (Fig. 8). Using the ANN-predicted model parameters and associated error, we simulated the device S-parameters and compared it against our D-band measurements. A composite plot is shown in Fig. 9 demonstrating reasonably good agreement compared to the Foundry model in Fig. 2.

IV. CONCLUSION

Through optimization of dataset structure, activation function, number of layers and nodes and backpropagation coefficient, we demonstrated an ANN model that successfully predicts the ASM-HEMT model parameters in the 110-170 GHz range at an overall RMS error of 2.3-2.7%. The ANN model training time was only 2 minutes, which compares very favorably to hours/days typically spent using standard fitting methods that often include manual model parameter optimization. The results highlight the importance of dataset structure and identifies various ANN design tradeoffs such as activation function and backpropagation rate as secondary contributors that can improve RMS error but mostly affect training time for the ANN design space analyzed in this work.

ACKNOWLEDGMENT

This work was supported by Defense Advanced Research Projects Agency (DARPA) through a DSSP under the SRC ComSenTer Program. The authors gratefully acknowledge DARPA program manager Dr. Tom Kazior's guidance and support through the grant that funded this effort. The authors thank GradientN for making their ANN software NNetDesigner© available for this work. The authors also thank HRL technical staff for valuable discussions and guidance and Dr. Tim Hancock from DARPA for facilitating access to the GaN foundry.

REFERENCES

[1] Z. Popovic, "Amping up the pa for 5g: Efficient gan power amplifiers with dynamic supplies," *IEEE Microwave Magazine*, vol. 18, no. 3, pp. 137–149, 2017.

[2] K. Shinohara, A. Corrion, D. Regan, I. Milosavljevic, D. Brown, S. Burnham, P. J. Willadsen, C. Butler, A. Schmitz, D. Wheeler, A. Fung, and M. Micovic, "220ghz f_t and 400ghz f_{max} in 40-nm gan dh-hemts with re-grown ohmic," in *2010 International Electron Devices Meeting*, 2010, pp. 30.1.1–30.1.4.

[3] E. Lam, A. Arias-Purdue, E. O'Malley, and J. F. Buckwalter, "A 23.5-dbm, 7.9%-pae pseudo-differential power amplifier at 136 ghz in 40-nm gan," in *2022 17th European Microwave Integrated Circuits Conference (EuMIC)*, 2022, pp. 119–122.

[4] S. Ghosh, S. A. Ahsan, A. Dasgupta, S. Khandelwal, and Y. S. Chauhan, "Gan hemt modeling for power and rf applications using asm-hemt," in *2016 3rd International Conference on Emerging Electronics (ICEE)*, 2016, pp. 1–4.

[5] F. Chavez, D. T. Davis, N. C. Miller, and S. Khandelwal, "Deep learning-based asm-hemt i-v parameter extraction," *IEEE Electron Device Letters*, vol. 43, no. 10, pp. 1633–1636, 2022.

[6] J. J. Hopfield, "Neural networks and physical systems with emergent collective computational abilities," *Proceedings of the National Academy of Sciences*, vol. 79, no. 8, pp. 2554–2558, 1982, 1982.

[7] H. Habal, D. Tsonev, and M. Schweikardt, "Compact models for initial mosfet sizing based on higher-order artificial neural networks," in *2020 ACM/IEEE 2nd Workshop on Machine Learning for CAD (MLCAD)*, 2020, pp. 111–116.

[8] P. Watson, M. Weatherspoon, L. Dunleavy, and G. Creech, "Accurate and efficient small-signal modeling of active devices using artificial neural networks," in *GaAs IC Symposium. IEEE Gallium Arsenide Integrated Circuit Symposium. 20th Annual. Technical Digest 1998 (Cat. No.98CH36260)*, 1998, pp. 95–98.

[9] A. Jarndal, S. Husain, M. Hashmi, and F. M. Ghannouchi, "Large-signal modeling of gan hemts using hybrid ga-ann, pso-svr, and gpr-based approaches," *IEEE Journal of the Electron Devices Society*, vol. 9, pp. 195–208, 2021.

[10] J. Xu and D. E. Root, "Artificial neural networks for compound semiconductor device modeling and characterization," in *2017 IEEE Compound Semiconductor Integrated Circuit Symposium (CSICS)*, 2017, pp. 1–4.

[11] S.-C. Han, J. Choi, and S.-M. Hong, "Acceleration of semiconductor device simulation with approximate solutions predicted by trained neural networks," *IEEE Transactions on Electron Devices*, vol. 68, no. 11, pp. 5483–5489, 2021.

[12] B. S. Paskaleva, V. Shitole, and E. Rosenbaum, "Data-driven compact modeling of bipolar junction transistors with recurrent neural networks." 8 2021. [Online]. Available: https://www.osti.gov/biblio/1884897

[13] A. Dasgupta, S. Ghosh, Y. S. Chauhan, and S. Khandelwal, "Asm-hemt: Compact model for gan hemts," in *2015 IEEE International Conference on Electron Devices and Solid-State Circuits (EDSSC)*, 2015, pp. 495–498.

[14] D. E. Rumelhart, G. E. Hinton, and R. J. Williams, "Learning representations by back-propagating errors," *Nature*, vol. 323, no. 6088, pp. 533–536, 1986.

RMo2A-5

Benchmarking Measurement-Based Large-Signal FET Models for GaN HEMT Devices

Rafael Perez Martinez[*#], Masaya Iwamoto[#], Jianjun Xu[#], Philipp Pahl[#], Srabanti Chowdhury[*]

[*]Department of Electrical Engineering, Stanford University, USA

[#]Keysight Laboratories, Keysight Technologies Inc., USA

{rafapm, srabanti}@stanford.edu

Abstract — This paper compares the accuracy and attributes of measurement-based large-signal FET models in the context of GaN HEMT modeling. We compare three FET models that have been implemented within PathWave Advanced Design System. In particular, the benefits and drawbacks of using neural networks to model the I-V and Q-V relations in a general way are analyzed. This is done by characterizing a 150 nm gate length 8x50 μm GaN-on-SiC HEMT and extracting the respective FET models based on DC-IV, small-signal, and large-signal data in the device's operating range. The three models are validated and benchmarked at different operating conditions and higher frequencies than their extraction frequency to show how neural network technology can serve as a powerful tool for the accurate modeling of thermal and trapping effects of GaN HEMTs.

Keywords — GaN HEMT, device modeling, linearity, neural networks, thermal effects, trapping effects.

I. INTRODUCTION

In recent years, Gallium Nitride (GaN) technology has become an appealing solution for various RF/mm-Wave applications due to its ability to provide higher output power than conventional III-V (e.g., Gallium Arsenide) and Silicon semiconductors [1]. Nonetheless, GaN high-electron-mobility transistors (HEMTs) are far from reaching their theoretically achievable performance due to thermal and trapping effects, which limit the output power, efficiency, and linearity, all of which are critical performance metrics in power amplifiers (PAs). The nonidealities found in existing commercial GaN HEMT foundry processes make the design of GaN MMICs a challenging task. As such, the ability to accurately model these nonidealities has become increasingly important in the design of high-power GaN MMICs at mm-wave frequencies to reduce the number of design iterations.

For GaN technology to be readily adopted, an accurate nonlinear large-signal model that captures thermal and trapping effects is necessary to simulate DC, S-parameters, and nonlinear figures of merit (FoMs) such as AM-AM/AM-PM distortion, intermodulation distortion, EVM, and ACPR. Measurement-based large-signal FET models are a particularly attractive solution as they can capture the nonidealities (e.g., thermal and trapping effects) present in the device without relying on physics-based equations. They are particularly useful when good measurement data is available and it is not possible to obtain a physics-based model of the device in use [2]. In the present work, three measurement-based FET models are validated and benchmarked in the context of GaN HEMT modeling. These three models have been natively implemented

within PathWave Advanced Design System (ADS). All three models are extracted and validated using the same data set, which includes DC, S-parameter, and large-signal NVNA data (although only two of the models require DC and S-parameter data to be fully extracted).

The flow of this paper is organized as follows: Section II introduces the three models used in this work. Section III shows and describes the measurement data used to extract and construct all three GaN HEMT models. The validation data to benchmark all three models are shown in Section IV. Lastly, Section V summarizes the paper.

II. MEASUREMENT-BASED MODELS

Several types of measurement-based models have been implemented in the literature. These include empirical models (e.g., fitting closed-form solutions to measurement data), table-based models with spline interpolation, and neural network-based models. In the present work, we evaluate three models implemented natively in PathWave ADS. The first model is a table-based model that is constructed from DC and S-parameters [3]. The second model is based on neural networks – it uses the non-quasi-static dynamics framework of the first model along with neural networks to construct a model with smoother constitutive relations (e.g., I-V and Q-V relations) [4]. The third model is a neural network-based model that is based on DC, small-signal, and large-signal NVNA data taken at two different temperatures [5], [6]. All three models are described in more detail below.

A. The Root Model

The Root model is a measurement-based table-driven model that computes the state functions of measured raw data and stores the values of these functions in tables, which are then interpolated using built-in spline functions during simulation [3]. A key advantage of the Root model is the simplicity of the model extraction and generation procedure since it only requires DC, S-parameters, and parasitic elements to generate the model. However, a limitation of this model is the interpolation from spline functions when generated from noisy data. In such a scenario, the spline functions will interpolate the noise in the data, leading to reduced accuracy at lower amplitude signal levels in nonlinear simulations [4].

B. The NeuroFET Model

To solve the limitations of the Root Model, NeuroFET replaces the spline functions with neural networks to

979-8-3503-2123-4/23 $31.00 © 2023 IEEE

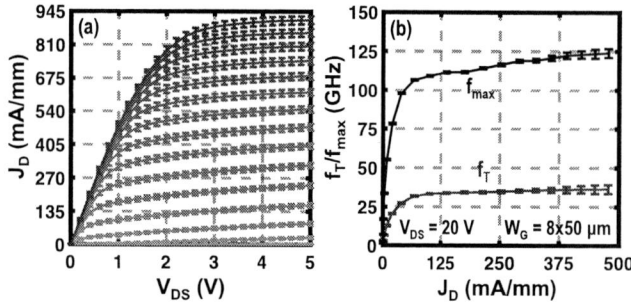

Fig. 1. Measured (a) I_D-V_D characteristics and (b) f_T/f_{max} of seven different 8x50 μm GaN-on-SiC devices from the same die. The error bars denote the standard deviation and the lines show the average values.

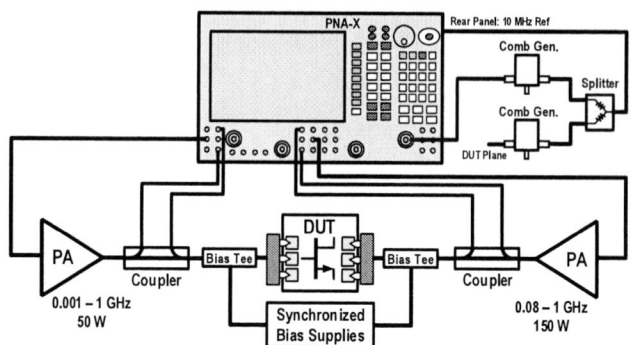

Fig. 2. Measurement setup for active source injection characterization with a temperature-controlled wafer chuck and driving amplifiers to boost the internal sources of NVNA.

generate constitutive relations that are smooth, infinitely differentiable, and uniform approximations of the desired device characteristics [4]. Similar to the Root Model, NeuroFET employs the same framework/concepts, data, and model generation infrastructure [7]. The main difference between these models is that the model's constitutive relations are expressed analytically using bias-dependent functions obtained from neural networks instead of using a table-based approach along with interpolation. Therefore, the simulation accuracy is improved at lower power levels, which is vital when trying to simulate linearity metrics such as the output third-order intercept point (OIP_3).

C. The DynaFET Model

The third model explored is the DynaFET model, which incorporates large-signal nonlinear vector network analyzer (NVNA) data to account for nonlinear dynamic effects such as thermal and trapping effects. When paired with DC and S-parameter data, NVNA data provides valuable information such as detailed large-signal waveforms which aid nonlinear circuit modeling compared to models with data limited only to DC and S-parameters [5], [6]. To fully extract this model, measurement data at two or more different temperatures are required. Similar to NeuroFET, DynaFET is a neural network-based model, where the objective of the training procedure is to identify constitutive relations such as the gate/drain currents and charges. As such, DynaFET can be thought of as a global solution that predicts DC, S-parameters, large-signal nonlinearities (e.g., distortion, load-pull, and PAE), and long-term memory effects without the need for application-specific model tuning. One limitation of this model is that it is inaccurate for FET switch designs since the model's formulation (e.g., trap definitions/calculations) does not apply to switch applications. As such, the model primarily targets PA applications [5], [6].

III. GaN HEMT Characterization

The device chosen for the extraction of the three models is a 150 nm gate length 8x50 μm GaN-on-SiC HEMT. Seven different devices with the same geometry were measured and characterized at room temperature (i.e., T = 25°C) to account for any variability in drain current (I_D) and unity

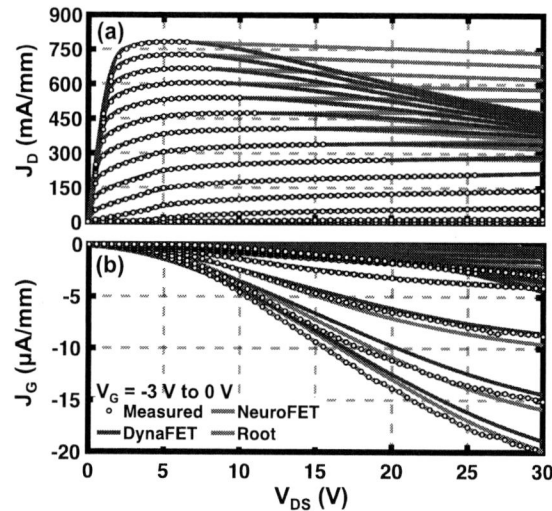

Fig. 3. Measured and simulated (a) I_D and (b) I_G as a function of V_{DS} at room temperature for DynaFET, NeuroFET, and Root models. V_{GS} is swept from -3 V to 0 V.

current (f_T) and unity power gain frequencies (f_{max}) as seen in Fig. 1. The error bars denote the standard deviation at each bias, while the lines show the average values. The device with the average performance was then chosen as the "golden device" for model extraction. Large-signal NVNA measurements were taken at two different temperatures (T = 25°C and 55°C), various input power levels (ranging from small-signal to large-signal conditions), and load impedances (using active load pull) for a fundamental frequency of 100 MHz. This frequency was chosen as it is fast enough to capture the transistor response in a low nanosecond regime (i.e., 10-100x faster than pulsed IV measurements) while at the same time supporting data collection up to the 20th harmonic. The data acquisition was done using an automated, adaptive, large-signal data acquisition software application, which is part of the NVNA nonlinear measurement system [8]. The hardware configuration for this setup is shown in Fig. 2.

Following the data acquisition of the large-signal NVNA data, broadband S-parameters were measured from 50 MHz to 25 GHz for various bias conditions for model validation and

979-8-3503-2123-4/23 $31.00 © 2023 IEEE

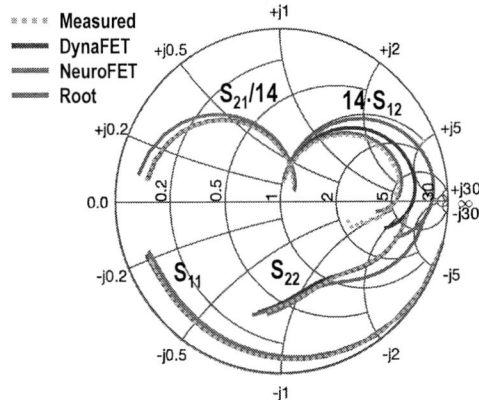

Fig. 4. Measured and simulated broadband S-parameters from 50 MHz to 25 GHz at V_{DS} = 20 V and V_{GS} = -2 V.

parasitic extraction. In particular, S-parameters were measured under pinched-off and Cold FET conditions for parasitic extraction (e.g., R/L's) as well as an array of HEMTs with different geometries for parasitic capacitance identification via linear regression using the method described in [2]. The thermal resistance of the GaN HEMT was also measured using the method described in [9] as this parameter is needed for the extraction of the DynaFET model.

Lastly, DC and S-parameter data for a fixed frequency of 3.7 GHz with a DC power compliance of 5.5 W/mm were taken as the last step for two different temperatures (T = 25°C and 55°C) since any extreme operating conditions can stress the device and change its characteristics. The DC validation results for all three models are shown in Fig. 3. Excellent agreement between measured and simulated I_D and gate current (I_G) is observed for all three models. It is important to mention that the DynaFET extraction package has a feature to add extrapolated data to account for self-heating effects in the IV characteristics. Such a feature was applied to the training data for extraction of both the NeuroFET and DynaFET models, whereas the Root Model only used the raw measured data. All three models were constructed using the same data set, e.g., Root Model and NeuroFET using DC and S-parameter data at room temperature, and DynaFET using DC, S-parameter data, and large-signal data at 25°C and 55°C.

IV. BENCHMARKING AND MODEL VALIDATION

The measured and simulated S-parameters for the three models at T = 25°C are in good agreement over the entire frequency and bias range where the model was extracted. For the intended application bias point, the Root model was able to achieve the best result, followed by the DynaFET model which shows a few discrepancies for S_{12}, and then NeuroFET, which was slightly inaccurate for S_{12} and S_{21}. This is shown in Fig. 4 for one bias point of interest (i.e., V_{DS} = 20 V and V_{GS} = -2 V). However, outside the intended application point, the Root model was less accurate, while DynaFET showed better accuracy over the entire bias range.

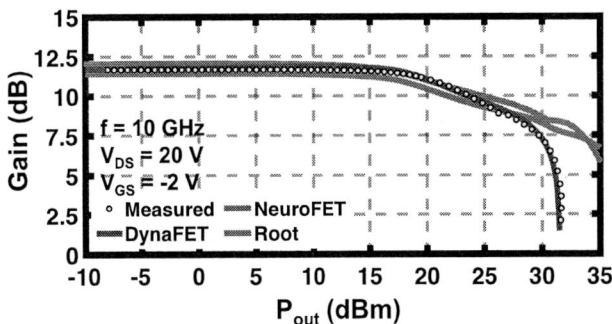

Fig. 5. Measured and simulated single-tone CW sweep gain versus output power at 10 GHz under a 50 Ω environment. The device was biased at V_{DS} = 20 V and V_{GS} = -2 V.

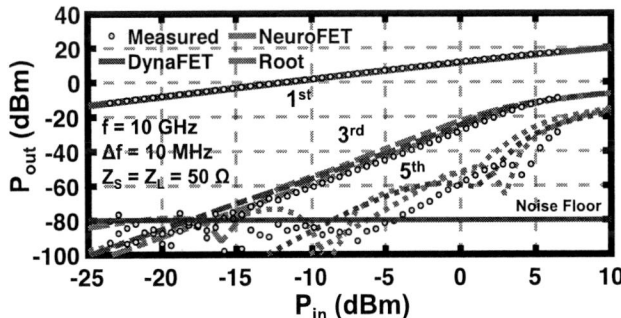

Fig. 6. Measured and simulated two-tone intermodulation nonlinear results showcasing fundamental, third, and fifth-order intermodulation products for f = 10 GHz and Δf = 10 MHz under a 50 Ω environment. The device was biased at V_{DS} = 20 V and V_{GS} = -2 V.

Most importantly, large-signal nonlinear measurements were performed to validate and benchmark the accuracy of the three models. Fig. 5 shows the measured and simulated continuous wave (CW) large-signal single-tone stimulus for a frequency of 10 GHz under a 50 Ω environment. All three models predict the gain with good accuracy at low to medium power levels. It is noted that the DynaFET model does a better job of predicting the high-power-level regime of the device (gain compression), whereas the Root and NeuroFET models struggle to predict the compression characteristics of the device at high power levels.

Two-tone intermodulation nonlinear measurements were also performed to validate and benchmark each model's capacity to simulate and quantify intercept points (e.g, OIP_3). This measurement was done for a center frequency (f_c) of 10 GHz with a frequency spacing (Δf) of 10 MHz under a 50 Ω environment and is shown in Fig. 6. It is observed that the NeuroFET model predicts the device's large-signal intermodulation characteristics with better accuracy at all power levels followed by DynaFET, and lastly the Root model. The Root model shows poor accuracy at low power levels as expected due to the interpolation issue from the spline functions. Moreover, OIP_3 was measured as a function of current density as shown in Fig. 7. The NeuroFET model shows good agreement between measured and simulation data. It is observed that the DynaFET model over-predicts OIP_3 at

979-8-3503-2123-4/23 $31.00 © 2023 IEEE

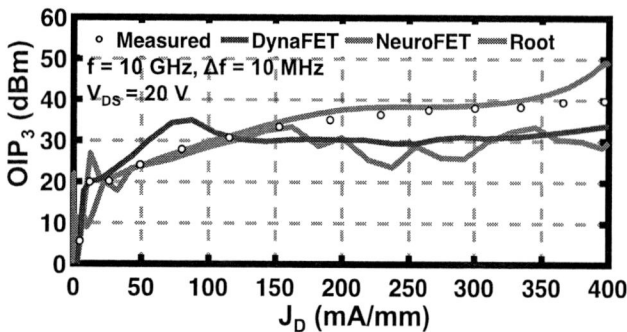

Fig. 7. Measured and simulated OIP_3 as a function of current density for $f = 10$ GHz and $\Delta f = 10$ MHz under a 50 Ω environment.

low current density and underpredicts OIP_3 at high current densities. Lastly, the Root Model predicts accurately OIP_3 at low current densities. It is worth noting that the input power level for which OIP_3 was extracted for the Root model was much higher due to the inaccuracies at low power levels.

In terms of model accuracy, there exists a trade-off between data acquisition complexity, extraction/training time, and intended circuit applications. While all three models were extracted using the same data set, both the Root and NeuroFET models only require DC and S-parameter data at room temperature as opposed to DC, S-parameter, and NVNA data at two different temperatures for the DynaFET model. This implies that it is not strictly necessary to have a temperature-controlled chuck or an NVNA to extract the Root and NeuroFET models. Most importantly, for the DynaFET model to be extracted successfully, the device under test (DUT) should be able to withstand (i.e., not degrade significantly) strenuous large-signal stimuli under different loads as this data acquisition process can take several hours or days. As such, devices that have not been qualified to be "reliable" are not good candidates for the DynaFET model. The training time needed to extract the neural network-based models varies for the NeuroFET and DynaFET models. With modern computing resources, the NeuroFET model takes 2-3 hours to fully train with I_D taking the longest. DynaFET takes a much longer time (12-24 hours) due to the immense amount of NVNA data. On the other hand, a Root model can be extracted in a matter of seconds.

Lastly, amplifier applications that strictly require nonlinear simulation results at high power levels will benefit greatly from the DynaFET model. In addition, DynaFET is expected to perform better under different load impedances as the model was extracted using large-signal NVNA for a variety of load impedances based on previous work shown [6]. However, a good alternative for applications that require low-noise amplifiers (LNAs), or switching applications are well-suited using NeuroFET since the model's formulation covers the operating region for FET switch designs, and it also has it captures with good accuracy the intermodulation products under a 50 Ω environment, which is needed for intercept point simulations. The Root model can be used to check the accuracy

of the extrinsic parasitics for NeuroFET or DynaFET due to its fast extraction time.

V. CONCLUSION

This works provides a comparison of the accuracy and attributes of three measurement-based large-signal FET models natively implemented in ADS, specifically in relation to GaN HEMT modeling. We test this by characterizing a 150 nm gate length 8x50 µm GaN-on-SiC HEMT and extracting three FET models from DC-IV, small-signal, and large-signal data under the device's operating range. The benefits and drawbacks of each model are discussed and it is found that DynaFET provides the best solution to large-signal GaN PA design as it is able to capture trapping/thermal effects in mature/reliably-qualified devices. NeuroFET is shown to be a good alternative for LNAs and switches since the model's formulation also applies to FET switching applications and it captures with good accuracy two-tone intermodulation simulations. Both neural network-based models demonstrate the effectiveness of neural network technology in the modeling of GaN MMICs.

ACKNOWLEDGMENT

This work was supported in part by the Stanford Graduate Fellowship (SGF). The authors are grateful to S. Cochran, E. Schmidt, D. Root, and Keysight Technologies management for their support.

REFERENCES

[1] U. K. Mishra, P. Parikh, and Yi-Feng Wu, "AlGaN/GaN HEMTs-an overview of device operation and applications," *Proceedings of the IEEE*, vol. 90, no. 6, pp. 1022-1031, June 2002.

[2] J. C. Pedro, D. E. Root, J. Xu, and L. C. Nunes, *Nonlinear Circuit Simulation and Modeling: Fundamentals for Microwave Design.* Cambridge, U.K.: Cambridge Univ. Press, 2018.

[3] D. E. Root, S. Fan, and J. Meyer, "Technology Independent Large Signal Non Quasi-Static FET Models by Direct Construction from Automatically Characterized Device Data," *1991 21st European Microwave Conference*, 1991, pp. 927-932.

[4] J. Xu, D. Gunyan, M. Iwamoto, A. Cognata, and D. E. Root, "Measurement-Based Non-Quasi-Static Large-Signal FET Model Using Artificial Neural Networks," *2006 IEEE MTT-S International Microwave Symposium Digest*, 2006, pp. 469-472.

[5] J. Xu, J. Horn, M. Iwamoto, and D. E. Root, "Large-signal FET model with multiple time scale dynamics from nonlinear vector network analyzer data," *2010 IEEE MTT-S International Microwave Symposium*, 2010, pp. 417-420.

[6] J. Xu, R. Jones, S. A. Harris, T. Nielsen, and D. E. Root, "Dynamic FET model - DynaFET - for GaN transistors from NVNA active source injection measurements," *2014 IEEE MTT-S International Microwave Symposium (IMS2014)*, 2014, pp. 1-3.

[7] J. Xu, M. C. E. Yagoub, R. Dingm, and Q. J. Zhang, "Exact adjoint sensitivity analysis for neural-based microwave modeling and design," *IEEE Transactions on Microwave Theory and Techniques*, vol. 51, no. 1, pp. 226-237, Jan. 2003.

[8] "Nonlinear Vector Network Analyzer (NVNA) - Breakthrough Technology for Nonlinear Vector Network Analysis from 10 MHz to 67 GHz." Accessed: Jan. 10, 2023. [Online]. Available: https://www.keysight.com/zz/en/assets/7018-01822/brochures/5989-8575.pdf

[9] D. E. Root, J. Xu, and M. Iwamoto, "Thermal Resistance Formulation and Analysis of III-V FETs Based on DC Electrical Data," *2021 IEEE BiCMOS and Compound Semiconductor Integrated Circuits and Technology Symposium (BCICTS)*, 2021, pp. 1-4.

979-8-3503-2123-4/23 $31.00 © 2023 IEEE

RMo2B-1

A DC-to-12 GHz 1.4–2.5 dB IL 4×8 Switch Matrix with Three-Port Reconfigurable Inter-Stage Matching Network

Zhenyu Wang[#], Zhaowu Wang[#], Yicheng Wang[#], Xiaochen Tang[$], Yong Wang[#]

[#]University of Electronic Science and Technology of China, China

[$]New Mexico State University, USA

yongwang@uestc.edu.cn

Abstract—An ultra-wideband low insertion loss (IL) reconfigurable 4×8 switch matrix is proposed in this paper. With the proposed three-port reconfigurable inter-stage matching network and symmetrically-routed structure, the switch matrix breaks the limits between bandwidth (BW), IL, and the number of branches. The proposed switch matrix is fabricated in a commercial 0.25-μm GaN HEMT process and achieves a favourable IL of 1.4–2.5 dB over DC-to-12 GHz. The switch matrix is operated in dual-band mode. In low-band mode (DC-to-5 GHz), isolation is higher than 35 dB and input 1-dB compression is higher than 40.3 dBm; in high-band mode (4-to-12 GHz), isolation is higher than 28 dB and input 1-dB compression is higher than 35 dBm. The core area is 1.5×1.6 mm².

Keywords—reconfigurable architectures, switches, ultra-wideband technology.

I. INTRODUCTION

Modern complex communication systems have an increasing demand for bandwidth, flexibility, and multi-mode/multi-band capabilities. Switch matrix with the capability of multi-channel signal routing facilitates more flexible configurations of reconfigurable RF Transmit/Receive systems, satellite communications, and test equipment. Two main switch matrix types are blocking and common highway (CH). Blocking type allows multiple signal paths to operate simultaneously by employing multiple switches, it suffers from a large area and high cost (e.g., a 4×4 blocking switch matrix generally needs eight SP4T switches). CH type is small-sized and supports one signal path at a time, which is suitable for broadband, miniaturized, highly reconfigurable RF front ends or devices (e.g., 4×4 CH switch matrix generally needs two SP4T switches). A reconfigurable RF front-end with CH switch matrix is shown in Fig. 1a. Switch matrices with wider bandwidth (BW) and more signal routing paths (i.e., more branches) are preferred for the RF front-end to be more flexible and compatible with more standards. However, the number of branches, BW, and insertion loss (IL) limit each other in the switch matrix circuit design. For instance, an $N×M$ CH switch matrix has $N+M-2$ leakage paths, and these paths deteriorate IL performance as frequency increases. Therefore, it is difficult for a large-scale switch matrix to maintain a small IL over an ultra-wide BW.

Switch matrix is usually constructed by several single-pole single-throw (SPST) or single-pole multi-throws (SPMT) switches. A CMOS 4×4 switch matrix is reported in [1], it achieves a 2.0–3.3±0.3 dB IL over 0.01-to-8 GHz BW. Employing sixteen SPST switches and four SP4T switches, [2]

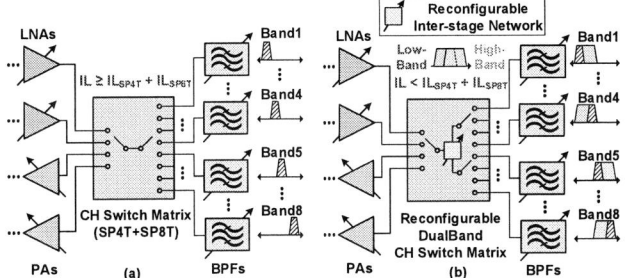

Fig. 1. (a) A reconfigurable RF front-end with conventional CH switch matrix. (b) Reconfigurable RF front-end with the proposed dual-band mode reconfigurable CH switch matrix.

achieves a 27.5-to-30 GHz blocking-type GaAs 4×4 switch matrix with an IL of 1–6 dB and a chip size of 8.16 mm². Reference [3] presents a blocking-type GaN 2×4 switch matrix with two SPDT switches and two DPDT switches, it achieves 2–2.5 dB IL over 27-to-30.5 GHz BW and occupies 16 mm² chip area. Recently, [4] proposes a reconfigurable matching network technique, and achieves an SP10T switch with 1.1–3.2 dB IL and 18-GHz BW.

This work proposes a reconfigurable structure for CH switch matrix to achieve a large scale and an ultra-wide BW without sacrificing IL performance. Instead of simple switch splicing in the conventional approach, the proposed 4×8 switch matrix combines three symmetrically-routed SP4T switches through a three-port reconfigurable inter-stage network (RSIN), as shown in Fig. 1. Thanks to the symmetrical SP4T architectures, a sole inter-stage network enables input and output matching for the whole 4×8 matrix. Different from the reconfigurable input matching network in [4], the proposed RSIN is three-port, and it handles all the impedance matching between the input and output switches and reduces signal leakage to decrease IL. In addition, the RSIN configures in dual-band operation modes, which increases the isolation between the output switches. Therefore, the switches can be optimized for IL and compression point of each operation mode separately. A prototype of the proposed switch matrix is fabricated in a 0.25 μm GaN HEMT process.

II. CIRCUIT DESIGN AND IMPLEMENTATION DETAILS

A. Architecture Design

A conventional CH-type 4×8 switch matrix is shown in Fig. 2. When the switch matrix is operating, the signal leaks to the closed branches during transmission as shown in Fig. 2a. In the

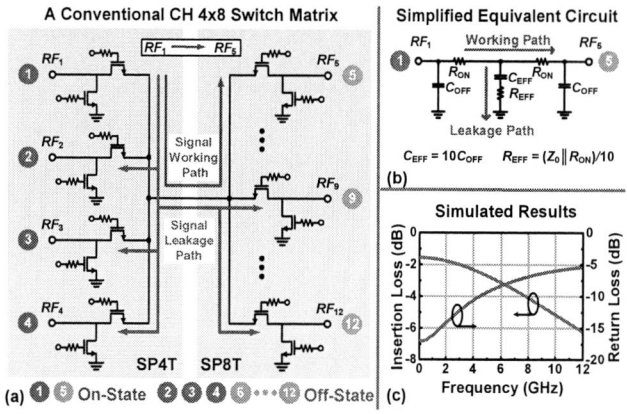

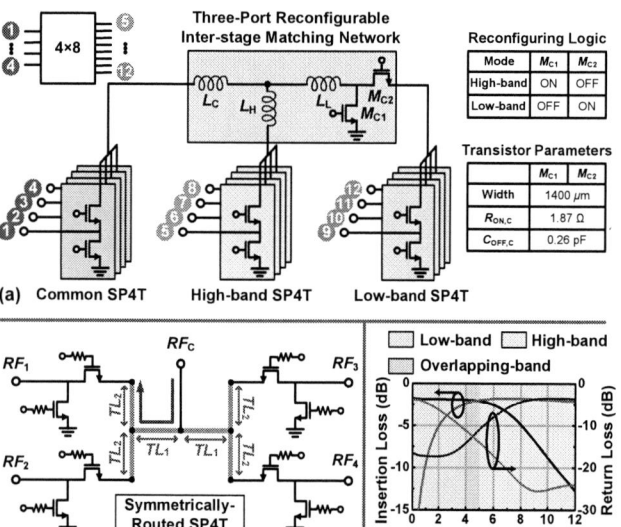

Fig. 2. (a) A conventional CH-type 4×8 switch matrix, which composes of an SP4T and an SP8T. (b) Simplified equivalent circuit of the switch matrix. (c) Simulated results of IL and return loss (RL).

Fig. 3. (a) The proposed reconfigurable dual-band 4×8 switch matrix. (b) The symmetrically routed SP4T switch. (c) Simulation results of IL and RL.

simplified equivalent circuit of the switch matrix, these closed branches present as a large effective capacitor C_{EFF} and a small effective resistor R_{EFF}, as shown in Fig. 2b. The IL of the equivalent circuit is expressed as

$$IL_{\text{CONV}} = 20\log\left(\frac{1}{2Z_0}|j\omega C_{EFF}(R_{ON} + Z_0 + j\omega C_{OFF}R_{ON}Z_0)^2 + \right.$$
$$\left. (2 + 2j\omega C_{OFF}Z_0)(R_{ON} + Z_0 + j\omega C_{OFF}R_{ON}Z_0)|\right) \quad (1)$$

where Z_0 is port impedance, C_{OFF} and R_{ON} are the off-state parasitic capacitor and on-state parasitic resistor of a transistor, respectively, and the small R_{EFF} is ignored. From (1), the C_{EFF} significantly impacts the IL, especially at high frequencies. Simulation results also verify this observation. As shown in Fig. 2c, the IL of the switch matrix (red line) reaches 6 dB at 12 GHz. Furthermore, the large C_{EFF} deteriorates the impedance matching, as shown by the blue line in Fig. 2c. In all, signal leakage and matching deterioration bring significant challenges to the design of ultra-wideband large-scale switch matrix.

The proposed 4×8 switch matrix addresses the conventional design challenges with a symmetrically-routed structure and

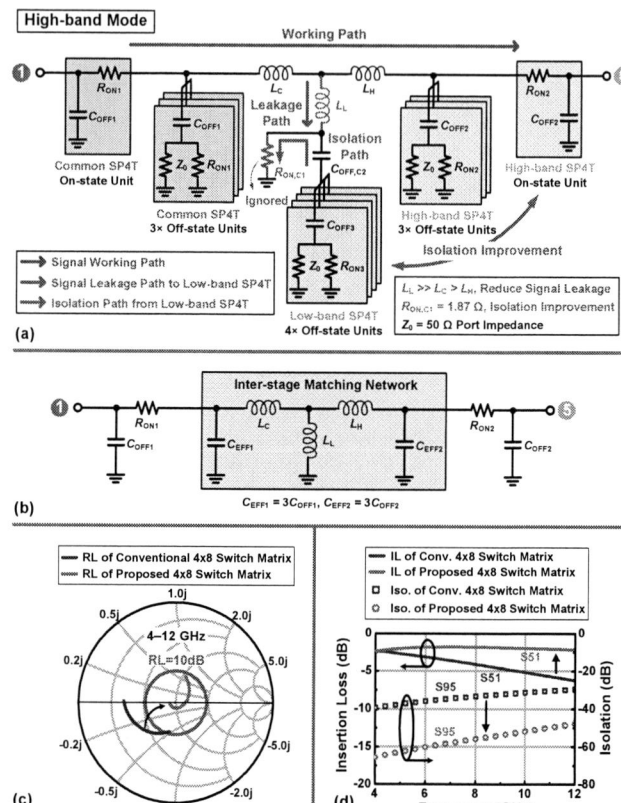

Fig. 4. (a) Equivalent circuit of the proposed 4×8 switch matrix at high-band mode. (b) Simplified equivalent circuit of the switch matrix at high-band mode. (c) Simulated RL of the conventional and proposed switch matrix. (d) Simulated IL and isolation of the conventional and proposed switch matrix.

reconfigurable inter-stage matching network, as shown in Fig. 3. Three symmetrically-routed SP4T switches are employed in the design. A common SP4T switch operates in the full-frequency band (DC-to-12 GHz), and the other two SP4T switches operate in the high-frequency (4-to-12 GHz) and low-frequency (DC-to-5 GHz) bands, respectively. In the symmetric routing structure, the paths from the common port (RF_C) to each branch have equal lengths and feature the same parasitic network, as shown in Fig. 3b. It indicates that a single matching network at the common port can cover the impedance matching of all branches. In the switch matrix, all the common ports of the three SP4T switches are connected to the RISN. Thus, the impedance matchings of all ports are achieved by a single RISN. Simulation results of IL and RL are shown in Fig. 3c, the IL is less than 2dB and RL is higher than 10 dB.

B. Reconfigurable Inter-stage Matching Network

The proposed RISN and its control logic table are presented in Fig. 3a. Large gate width is adopted for the control transistors (i.e., M_{C1} and M_{C2}) to achieve low IL and high isolation, and the transistor parameters are listed in Fig. 3a. The inter-stage network configuration is dual-band modes, i.e., high-band and low-band modes.

The equivalent circuit of the switch matrix at high-band mode is illustrated in Fig. 4. A leakage path from the common SP4T switch to the low-band SP4T switch exits, as shown in

979-8-3503-2123-4/23 $31.00 © 2023 IEEE

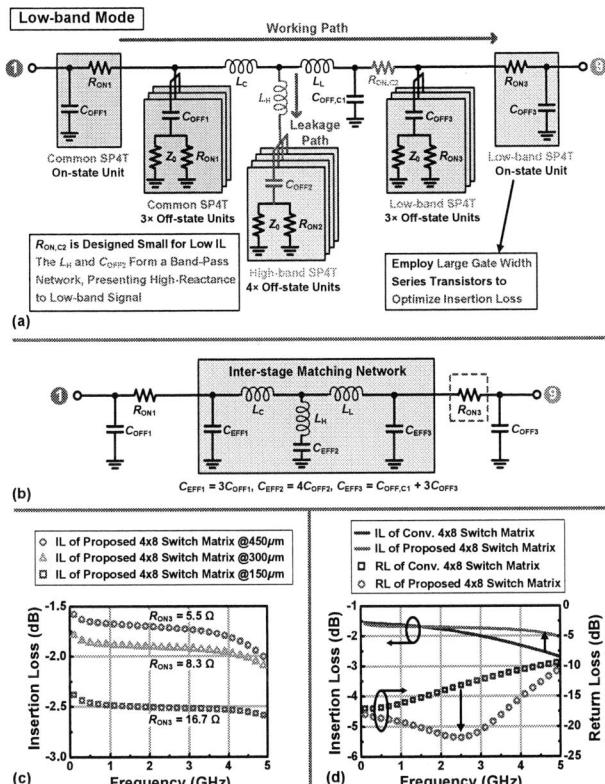

Fig. 5. (a) Equivalent circuit of the proposed 4×8 switch matrix at low-band mode. (b) Simplified equivalent circuit of the switch matrix at low-band mode. (c) Simulated IL of the conventional and proposed switch matrix. (d) Simulated IL and RL of the conventional and proposed switch matrix.

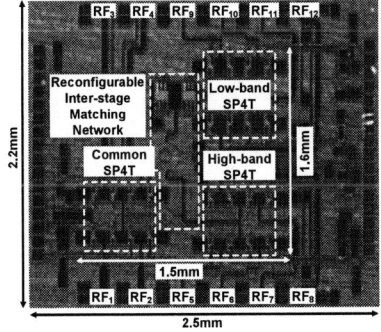

Fig. 6. Die micrograph

Fig. 4a. Thus, a large inductor L_L is introduced into the leakage path, which presents a high reactance to high-frequency signal. As a result, the signal leakage is reduced, and IL is also decreased. In addition, the common port of the low-band switch is nearly short-circuited to the ground due to the small on-state resistor of M_{C1} (i.e., $R_{ON,C1}$). Therefore, unwanted signals are difficult to leak from the off-state low-band switch to the high-band SP4T switch, so that the isolation between the two switches is improved. A simplified equivalent circuit is shown in Fig. 4b. As observed, the effective parasitic capacitors (i.e., C_{EFF1} and C_{EFF2}) of the off-state branches of the common and high-band SP4T switches are involved in the matching circuit, forming a double-π matching network with inductors L_C, L_L, L_H.

Simulated S-parameters are shown in Fig. 4c and 4d. The results show that the proposed 4×8 switch matrix has better impedance matching and features IL of 2-dB at 12 GHz, which is 4-dB less than the conventional 4×8 switch matrix. Moreover, the isolation from the low-band switch to the high-band switch (i.e., S95) is more than 50 dB in the proposed switch matrix, which is 20 dB higher than the conventional design.

The equivalent circuit at low-band mode is illustrated in Fig. 5a. M_{C1} is turned off, while M_{C2} is on and connected in series in the signal transmission line. To reduce IL, a large gate width is chosen for M_{C2}. Additionally, the series resonance frequency of L_H and the parasitic capacitors (C_{OFF2}) of the high-band switch is higher than 32 GHz, thereby this band-pass network consisting of L_H and C_{OFF2} has high reactance to the low-band signals, reducing the signal leakage. Due to the high isolation between the low-band and high-band SP4T switches, the low-band switch can be optimized independently, without sacrificing low-band IL to improve high-band IL as in conventional design. The simplified equivalent circuit is shown in Fig. 5b, where R_{ON3} is the parasitic resistor of the series transistor of the low-band switch. Large R_{ON3} in series in the signal path harms the IL, thereby large-size series transistors with small parasitic resistance are preferred, as demonstrated by the simulation results in Fig. 5c. Additionally, large transistors are beneficial for linearity improvement. Figure 5d presents simulated IL and return loss (RL) of conventional and proposed switch matrix. It reveals that the proposed matrix has better impedance matching and features 2-dB IL at 5 GHz.

III. MEASUREMENT RESULTS

A prototype of the proposed 4×8 switch matrix is fabricated in a 0.25 μm GaN HEMT technology and occupies a core area of 1.5×1.6 mm^2 (total chip area of 2.2×2.5 mm^2 including the pad frame). The die micrograph is shown in Fig. 6.

S-parameters measurements on the prototype are shown in Fig. 7. Two reconfigured frequency bands count as DC-to-5 GHz and 4-to-12 GHz. The two bands have a 1 GHz overlap to guarantee practical usage at frequencies around 4.5 GHz. As shown in Fig. 7a and 7b, curves of IL of all channels overlap well. This proves the conformance of the symmetrically-routed structure in three SP4T switches. The measured ILs are 1.4–2.3 dB and 1.9–2.5 dB for low and high frequency bands, respectively. In low-band mode, the isolation from the low-band switch to the common switch (e.g., S101) is higher than 39 dB, and the isolation from the high-band switch to the low-band switch (e.g., S59) is higher than 35 dB. In high-band mode, the isolation from the high-band switch to the common switch (e.g., S63) is higher than 28 dB, and the isolation from the low-band switch to the high-band switch (e.g., S95) is higher than 41 dB. All the inputs and outputs are matched. The input P_{1dB} are 40.3–40.9 dBm and 35.0–36.1 dBm for low-band mode and high-band mode, respectively.

Table 1 compares the work with the prior arts. As a rule of thumb, IL is inversely correlated to the number of $N \times M$ and the bandwidth of a switch matrix. Even so, this 4×8 switch matrix outperforms the commercial 4×2 switch matrix in [5] which operates at DC-to-3 GHz, and it exhibits 1-dB less IL than the

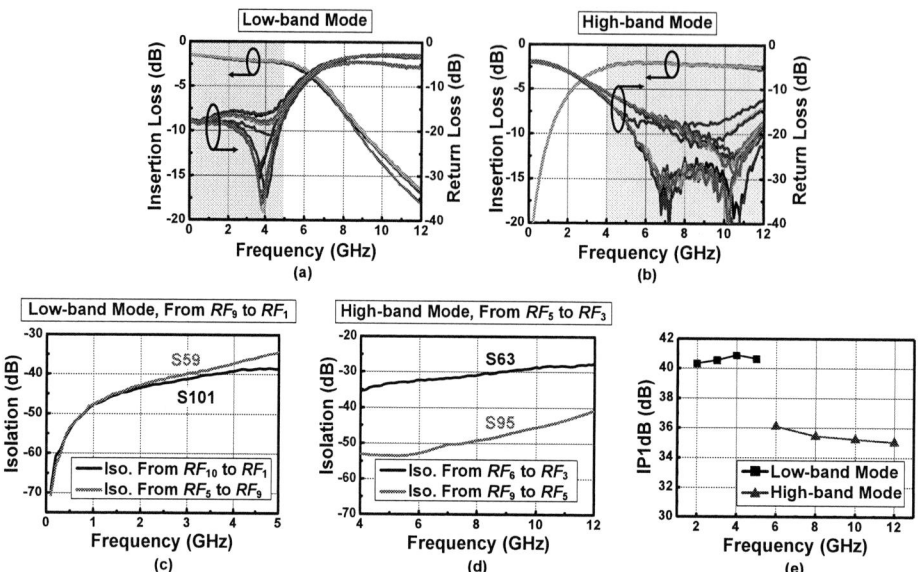

Fig. 7. Measured IL and RL at (a) low-band mode and (b) high-band mode. (c) Measured isolation at low-band mode, where S59 indicates the isolation from high-band SP4T switch to low-band SP4T switch. (d) Measured isolation at high-band mode, where S95 indicates the isolation from low-band SP4T switch to high-band SP4T switch. (e) Measured 1-dB compression points

Table 1. Performance Summary and Comparison

	Topology	Architecture	Freq. (GHz)	Insertion Loss (dB)	Isolation (dB)	IP$_{1dB}$ (dBm)	Return Loss (dB)	Tech.	Area (mm^2)
This Work	4×8	1. Symmetrically Routing 2. Reconfigurable Inter-stage Matching Network	DC–5 (Band-1)	1.4–2 (DC–3 GHz)	>40 (DC–3 GHz)	40.3–40.5	>13	0.25 μm GaN	2.2×2.5 (1.5×1.6**)
				1.4–2.3 (DC–5 GHz)	>35 (DC–5 GHz)	40.3–40.9			
			4–12 (Band-2)	1.9–2.4 (4–8 GHz)	>31, >49* (DC–8 GHz)	35.4–36.1	>11		
				1.9–2.5 (4–12 GHz)	>28, >41* (DC–12 GHz)	35.0–36.1			
[5] Analog Devices (HMC596LP4)	4×2	Four SPDT Switches	0.2–3	6–8	>34	22	>10	CMOS	NA
[1] TMTT (2012)	4×4	Crossbar	0.01–8	2–3.6***	>44	8–10	>10	0.13 μm CMOS	1.4×1.5
[3] EuMIC (2022)	2×4	SPDT+DPDT Switches	27–30.5	2–2.5	>27	32	>15	0.15 μm GaN	4×4
[4] RFIC (2022)	SP10T	1. Symmetrically Routing 2. Reconfigurable Matching Network	DC–18	2.2–2.9 (8–14.5 GHz)	>26	23.2–23.4	>10	0.15 μm GaAs	1.7×2.4
				1.1–3.2 (DC–18 GHz)	>24	21.2–24.0			

*Isolation from low-band SP4T switch to high-band SP4T switch. **Core area. *** Read from figure.

4×4 switch matrix in [1] which operates at DC-to-8 GHz. Moreover, this work exhibits low IL that is comparable to the SP10T switch in [4] at 4-to-12 GHz.

IV. CONCLUSION

This paper presents a high-power low-IL ultra-wideband 4×8 CH-type switch matrix, which is achieved using the proposed symmetrically-routed SP4T structure and reconfigurable inter-stage network. A prototype is fabricated in a 0.25 μm GaN HEMT, and the measurement results show a favourable IL over DC-to-12 GHz, which will be attractive for broadband, miniaturized, highly reconfigurable RF systems.

REFERENCES

[1] D. Shin, D. -W. Kang and G. M. Rebeiz, "A 0.01–8-GHz (12.5 Gb/s) 4×4 CMOS switch matrix," *IEEE Trans. Microw. Theory Tech.*, vol. 60, no. 2, pp. 381-386, Feb. 2012.

[2] A. Barigelli, *et al.*, "Scalable Ka band switch matrix in compact LTCC package for satellite communication application," in *Proc. 47th Eur. Microw. Conf. (EuMC)*, Nuremberg, Germany, Oct. 2017, pp. 1073-1076.

[3] A. Biondi, *et al.*, "A Ka-Band GaN 2x4 MMIC Switch for compact and scalable Switch Matrices," in *Proc. 17th Eur. Microw. Integr. Circuits Conf. (EuMIC)*, Milan, Italy, Sep. 2022, pp. 25-28.

[4] Z. Wang, Z. Wang, T. Yang, and Y. Wang, "A DC-to-18 GHz SP10T RF switch using symmetrically-routed series-TL-shunt and reconfigurable single-pole network topologies presenting 1.1-to-3.2 dB IL in 0.15 μm GaAs pHEMT," in *Proc. IEEE Radio Freq. Integr. Circuits Symp. (RFIC)*, Jun. 2022, pp. 79-82.

[5] "HMC596LP4 data sheet," Analog Devices, Wilmington, Massachusetts, United States.

RMo2B-2

A DC-to-18 GHz High Power and Low Loss Band-Divided SP3T Switch with Reconfigurable Pole-to-Throw Network in 0.25-μm GaN

Zhaowu Wang[#], Yicheng Wang[#], Zhenyu Wang[#], Xiaochen Tang[$], Yong Wang[#]

[#]University of Electronic Science and Technology of China, China

[$]New Mexico State University, USA

yongwang@uestc.edu.cn

Abstract—**A high power and low loss band-divided single-pole three-throw (SP3T) switch is presented in this work. A reconfigurable pole-to-throw network (RPTN) is proposed to minimize insertion loss (IL). The RPTN enables the three throws to operate in different bands, with overlays to each other. Thanks to the RPTN, parasitic of off-state elements can be either bypassed or adsorbed into the filtering networks; transistor sizes in the series-shunt units can be increased to reduce resistive losses since large parasitic capacitances can be either bypassed or absorbed. Fabricated in a 0.25-μm GaN HEMT process, the SP3T shows better than 1.47 dB IL, over 14.6 dB RL, over 20 dB isolation, and 35.2-to-39.8 dBm input 0.1 dB compression point (IP$_{0.1dB}$) in an ultra-wideband from DC to 18 GHz.**

Keywords—**gallium nitride (GaN), multi-band, reconfigurable, single-pole three-throw (SP3T), switches, ultra-wideband.**

I. INTRODUCTION

Single-pole multi-throw (SPMT) switches are key components in radio frequency (RF) systems. Conventional SPMT switches serve as transmit/receive (T/R) switches in time-division duplex (TDD) systems. In these applications, insertion loss (IL), isolation, and linearity of a switch become main considerations. In recent years, broadband or multi-band RF systems have started to emerge [1], [2]. It makes the switch design even more difficult since bandwidth, insertion loss (IL), isolation, and linearity always limit each other.

A typical trend in broadband or multi-band RF systems is that SPMT switches are used for band dividing and combining. For example, a pair of SPMT switches are always used to build filter/amplifier arrays, where each filter/amplifier module is narrow band, as shown in Fig. 1(a). Figure 1(b) shows another example where an SPMT switch is used for band-divided up/down-conversion transceivers. Different from the requirements on T/R switches, requirement on isolation is less stringent. This is because the unselected branches can be turned off (e.g., active modules), or the unselected branches themselves have good out-off band rejections (e.g., filters). However, IL and linearity deserve more attention, since they dominate noise figures, power added efficiencies, gain requirements, etc. Another different point on such switch designs is that each throw does not need to cover full band; it means that the throws can be band divided, and the passbands of throws should have certain overlays to ensure practical usage.

Recently, reconfigurable technology is proven to be efficient to break through trade-offs of switch designs. Several remarkable works on reconfigurable/multi-band SPMT switches can be found in literature. A tri-band SP3T switch in

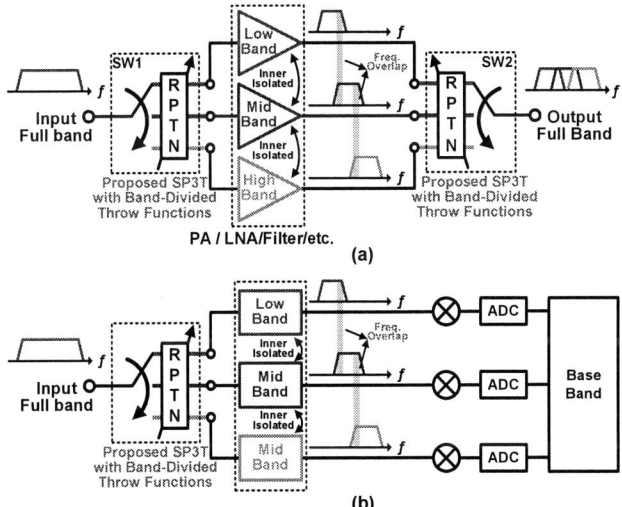

Fig. 1. Application examples of SP3Ts in broadband/multi-band systems.

mm-wave is realized by reconfiguring the length of the $\lambda/4$ transmission-line (TL) [3]. However, the IL of this SP3T is up to 3 dB, which is unacceptable for many applications. A dual-band SP4T, reported in [4], achieves favourable IL but exhibits a narrow bandwidth. An SP10T with reconfigurable input matching network and symmetric structure is reported in [5]. Each branch of this SP10T sees the same performance in a frequency range from DC to 18 GHz, but its IL is up to 3 dB.

This work proposes a reconfigurable pole-to-throw network (RPTN) to achieve the high power and low loss SP3T design for the abovementioned applications. It consists of two transistors, two inductors, and two capacitors. The RPTN is a four-port network, where one port works as a pole and the rest ports connect to three series-shunt units. With proper reconfiguring logic, the three throw paths present low-pass, band-pass, and high-pass functions respectively. The beauty of the RPTN is that once one throw path is turned on, the parasitic parameters of the rest two off-state series-shunt units will be either leveraged or bypassed. This leads to less parasitic leakage. What's more, large gate widths of the series transistors in the series-shunt units can be selected to reduce the resistive loss of R_{ON} and increase linearity, because their off-state parasitic capacitances are either bypassed or absorbed into the RPTN. Therefore, a great improvement in the IL can be attained. To reach the power handling limit, a 0.25-μm GaN HEMT process is chosen to prototype the concept. The fabricated chip presents

979-8-3503-2123-4/23 $31.00 © 2023 IEEE 77 2023 IEEE Radio Frequency Integrated Circuits Symposium

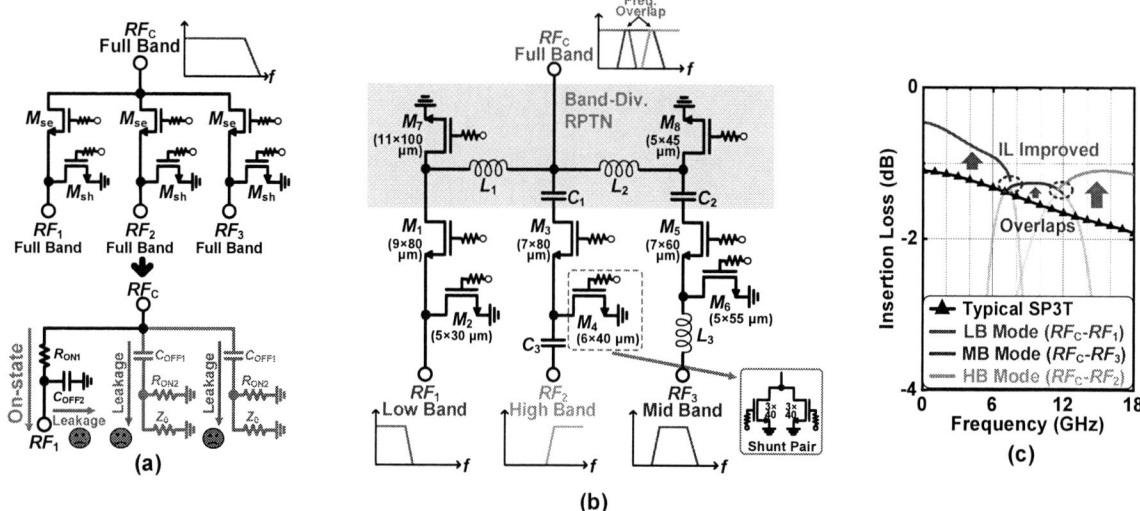

Fig. 2. (a) Typical SP3T structure and its equivalent network. (b) The proposed SP3T with reconfigurable pole-to-throw network (RPTN). (c) Simulated IL for the typical SP3T and the proposed SP3T.

Table 1. Figure of merits of switch technologies.

Reference	[6],[7]	[8]	[4],[7],[8]	[8]	This Work
Technology	SOI	GaAs pHEMT	CMOS	GaN HEMT	
Linearity	Medium	Medium	Low	High	
$R_{ON} \times C_{OFF}$ (fs)	50~200	100~300	150~500	400	500*

* Simulation result of the 0.25 μm GaN HEMT in this work

Table 2. Gate widths versus performance of switches.

		IL @ low freq.	IL @ high freq.	Isolation	Linearity
Series Width (major)	large	🙂	🙁	🙁	🙂
	small	🙁	🙂	🙂	🙁
Shunt Width (minor)	large	😐	🙁	🙂	🙂
	small	😐	🙂	🙁	🙁

1.20-to-1.47 dB IL, 35.2-to-39.0 dBm $IP_{0.1dB}$, and DC-to-18 GHz bandwidth.

II. CIRCUIT ANALYSIS AND DESIGN DETAILS

A. Analysis of Issues in Traditional SP3T Structure

Figure of merit (FoM) of $R_{ON} \times C_{OFF}$ is an efficient tool to evaluate the technologies for switch designs. Smaller R_{ON} and C_{OFF} are desired to maintain low loss and high isolation. Since a smaller R_{ON} always leads to a larger C_{OFF}, the FoM often stays nearly as a constant for a certain technology. Therefore, trade-off should be made between IL, isolation, and frequency. Table 1 shows the typical FoM and linearity of four mainstream technologies [4], [6], [7], [8]. As can be seen, the GaN HEMT has the highest power handling capability, but its FoM is the worst. This poses great difficulties for a GaN switch design.

Fig. 2(a) shows structure and equivalent circuit of a typical series-shunt SP3T. There are three factors deteriorating IL: 1) the resistive loss of R_{ON1}; 2) the leakage through C_{OFF1}; 3) the leakage via C_{OFF2}. Wherein, R_{ON1} determines the IL of low frequencies and C_{OFF1} dominates the IL in higher frequencies. Both R_{ON1} and C_{OFF1} are determined by the gate width of series

transistors (M_{se}). As known to all, pursuing a transistor with small R_{ON1} and small C_{OFF1} is impossible, therefore the IL for a broadband SPMT is hard to optimize. In addition, gate width of the shunt transistor (M_{sh}) is another factor that needs to be compromised. A large M_{sh} result in better isolation due to the small R_{ON2}, but leads to worse IL due to the larger C_{OFF2}. Moreover, the gate widths of both series and shunt transistors are positively correlated with the power handling capability. Table 2 summarizes the above trade-offs. Considering that the GaN HEMT has the worst FoM, it is a fine platform to prototype the efficiency of the proposed topology in this work.

B. Proposed RPTN SP3T and Its Special Features

Figure 2(b) shows the schematic of the proposed SP3T. It consists of the RPTN and three series-shunt units. The RPTN is composed of two transistors (M_7 and M_8), two inductors (L_1 and L_2), and two capacitors (C_1 and C_2). M_7 and M_8 are used to configure the three operation modes. C_3 and L_3 are adopted to improve output matching.

Impressive features of the proposed RPTN are: 1) With proper switching of M_7 and M_8, parasitic elements in the off-state branches can be either bypassed or adsorbed into the filtering network; 2) Large gate widths of the series transistors in the series-shunt units can be selected to reduce the resistive loss and improve the linearity, because their off-state parasitic capacitances are either absorbed into the RPTN or bypassed, hereby tradeoff on the transistor sizes in lower and higher frequencies can be broken; 3) Three throw ports presents band-divided features, with low-pass, band-pass, and high-pass functions respectively; 4) Passbands of the three throws can be overlapped.

Simulations in figure 2(c) prove the above features, and compare this work to the typical SP3T design. Both results are obtained from layout simulation and optimized for low IL in the whole band. Compared to the typical SP3T design, the IL is greatly improved due to features 1-to-3. Band overlaps of feature-4 facilitate practical usage.

979-8-3503-2123-4/23 $31.00 © 2023 IEEE

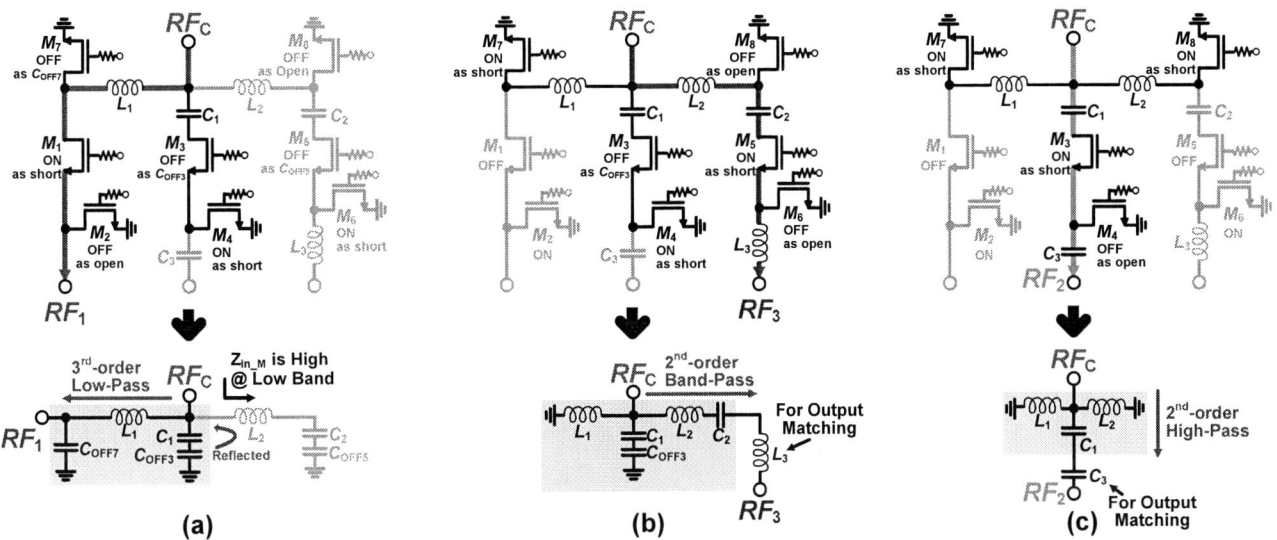

Fig.3. Schematic analysis and simplified equivalent circuits of the (a) low band mode (RF_1), (b) mid band mode (RF_3), and (c) high band mode (RF_2).

C. Operation Analysis of the Proposed RPTN SP3T

Fig. 3 shows the schematic analysis and simplified equivalent circuits of the three switching modes. All the on-state transistors are treated as short circuit to simplify the analysis. The off-state transistors with small widths are treated as open circuit and the off-state transistors with large widths are treated as capacitors (C_{OFF}). On and off statuses are marked beside the transistors. Elements that can be ignored are colored in grey.

Operation of the low-band (LB) throw branch is illustrated in Fig. 3(a), where RF_C connects to RF_1. Seeing from RF_C, the turned-off mid-band branch (i.e., the right branch RF_C-to-RF_3) forms a series LC resonance where the resonant frequency is far out of the low band, therefore it presents high impedance and can be ignored to simplify the analysis. Parasitic of all the other elements are ignored because they are shorted by the on-state transistors. Parasitic capacitances of C_1 and C_{OFF3} in the high-band branch (i.e., RF_C-to-RF_2) are absorbed into the low-band filtering network. M_7 is with a large gate width, contributing another capacitor to the filter network. Thus, the low-band mode can be simplified into a 3rd-order low-pass network, as shown at the bottom of Fig. 3(a).

Operation of the mid-band (MB) branch is shown in Fig. 3(b), where RF_C connects to RF_3. In this mode, M_7 turns on to insert L_1 into the mid-band filtering network, meanwhile parasitic of series-shunt unit in low-band branch is bypassed. Small gate width is selected for M_8 to reduce the impact of the parasitic capacitance. Parasitic capacitances of C_1 and C_{OFF3} in the high-band branch are absorbed into the filtering network. At RF_3 port, an inductor L_3 is employed to improve the output matching. Therefore, the mid-band branch can be simplified as a 2nd-order band-pass network.

In the high band (HB) mode, RF_C connects to RF_2, as shown in Fig. 3(c). Both M_7 and M_8 are turned on to push L_1 and L_2 into the high-band filtering network. Meanwhile, the parasitic of series-shunt units in both low-band and mid-band branches

is bypassed. Thus, L_1, L_2, and C_1 form a 2nd-order high-pass network. Here, C_3 is to improve the output matching.

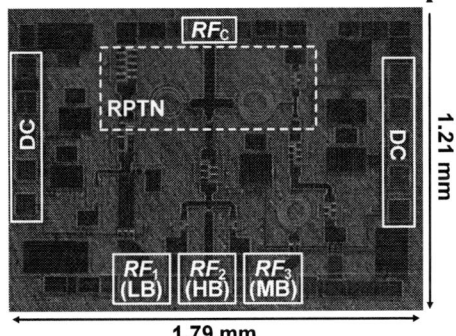

Fig. 4. Die micro-photograph.

III. MEASUREMENT

The proposed SP3T with RPTN topology is fabricated in a 0.25-μm GaN HEMT technology. The die occupies an area of 1.79×1.21 mm² including pads. The micro-photograph is shown in Fig. 4.

On chip measurements are conducted for obtaining both small signal and large signal results. The measured results are presented in Fig. 5. The operation bands of the three throws are DC-to-7.5 GHz, 7-to-12 GHz, and 11.5-to-18 GHz. The three bands have more than 0.5 GHz overlaps to each other. The measured ILs are 0.42–1.21 dB for LB mode, 1.30–1.47 dB for MB mode, and 1.05–1.20 dB for HB mode, as shown in Fig. 5(a). The input RL and output RL are shown in Fig. 5(b) and Fig. 5(c). All RLs are larger than 14.6 dB, indicating that all ports are well matched. As mentioned before, isolations within the passband of each mode are interested. All the measured isolations for three modes are better than 20 dB, as shown in Fig. 5(d). In the large signal measurement, a 2–18 GHz high-power amplifier with 40-dBm output limit is used to boost the output of signal source. A 30 dB attenuator is connected to the output port of the proposed SP3T, so as to protect the power

979-8-3503-2123-4/23 $31.00 © 2023 IEEE

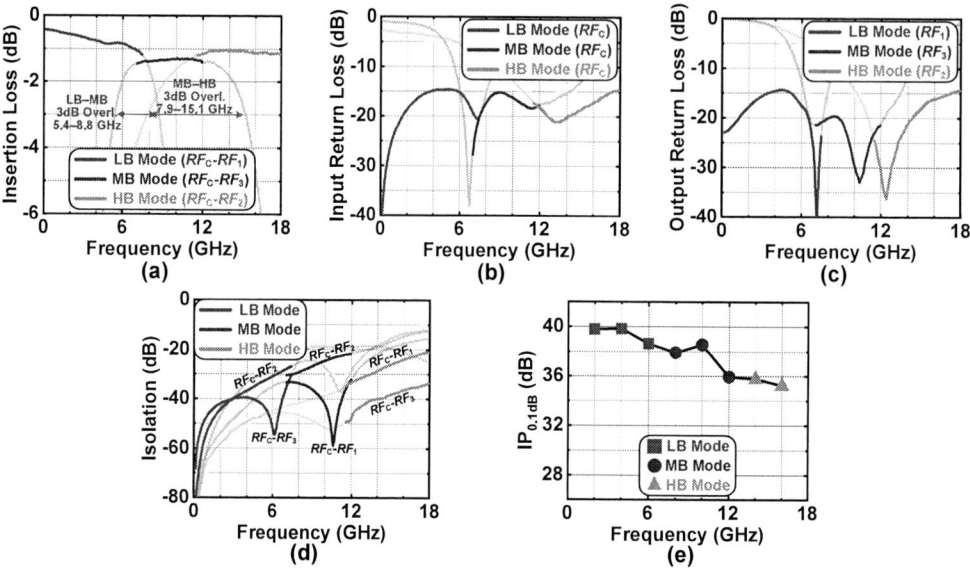

Fig. 5. Measured (a) insertion loss, (b) input return loss, (c) output return loss, (d) isolation, and (e) input 0.1 dB compression point.

Table 3. Performance summary and comparison.

Ref.	[9]ADI HMC641A	[3]TMTT (2020)	[4]TCAS II (2022)	[10]EuMC (2019)	This Work		
Multi-band	NO	**YES** Tri-band	**YES** Dual-band	NO	**YES** Tri-band		
SPMT	SP4T	SP3T	SP4T	**SPDT**	SP3T		
Freq. (GHz)	DC-18	14.7-44.3(SW1)/ 16.4-41(SW2)	DC-5.3	2-12	DC-7.5	7-12	11.5-18
Insertion Loss (dB)	<2.3	<3.7/ <3.9	<2	<1.5	<1.21	<1.47	<1.20
Return Loss (dB)	>15	NA	NA	>10	>14.6	>15.3	>14.6
Isolation (dB)	>36	>18.2/ >18.8	>16.8	>35	>26.8	>21.5	>20.9
$IP_{0.1dB}$ (dBm)	22 (IP_{1dB})	12.8/12.8 (IP_{1dB})	7 (IP_{1dB})	40 ($IP_{0.3dB}$)	>39 ($IP_{0.1dB}$)	>36 ($IP_{0.1dB}$)	>35.2 ($IP_{0.1dB}$)
Tech.	GaAs	65 nm CMOS	28 nm FDSOI	0.25 μm GaN	0.25 μm GaN HEMT		
Area (mm²)	1.92 × 1.60	1.05 × 0.79	0.453 × 0.443*	4 × 2	1.78 × 1.21		

* Core area

and operation principles of the proposed RPTN are illustrated in detail. The work is prototyped in a 0.25-μm GaN HEMT process, showing promising IL and linearity performance. The results reveal that the proposed topology will be attractive for broadband and multi-band RF systems.

meter. Because the input 1 dB compression point (IP_{1dB}) is larger than the power limit of the high-power amplifier, $IP_{0.1dB}$ is measured alternately, which is from 35.2 to 39.8 dBm, as shown in Fig 5(e).

Table 3 summarizes and compares the work with the prior-arts. Even though GaN presents the worst FoM compared to GaAs, CMOS, and SOI technologies (see Table 1), the proposed SP3T presents significant advantages in IL and bandwidth compared to [3], [4], and [9]. In contrast to the GaN SPDT switch in [10], the proposed SP3T achieves favourable IL and linearity; moreover, it has much larger bandwidth, one more throw, and smaller chip area.

IV. CONCLUSION

This paper presents a high power and low loss band-divided SP3T design with the proposed RPTN topology. The benefits

REFERENCES

[1] K. Choi, H. Park, J. Kim and Y. Kwon, "6–18 GHz Watt-Level Switchless Dual-Band PA MMIC using Coupled- Line-Based Diplexer," *2018 IEEE/MTT-S International Microwave Symposium (IMS)*, 2018, pp. 1352-1355.

[2] K. Lim et al., "A 65-nm CMOS 2×2 MIMO Multi-Band LTE RF Transceiver for Small Cell Base Stations," *IEEE J. Solid-State Circuits*, vol. 53, no. 7, pp. 1960-1976, July 2018.

[3] Y. Kim, I. Lee and S. Jeon, "A New mm-Wave Multiple-Band Single-Pole Multiple-Throw Switch with Variable Transmission Lines," *IEEE Trans. Microw. Theory Tech.*, vol. 68, no. 7, pp. 2551-2561, July 2020.

[4] W. Lee and S. Hong, "Frequency-Reconfigurable SP4T Switch with Plaid Metal Transistors and Forward Body Biasing for Enhanced RON × COFF Characteristics," *IEEE Trans. Circuits Syst. II: Exp. Briefs*, vol. 69, no. 2, pp. 399-403, Feb. 2022.

[5] Z. Wang, Z. Wang, T. Yang and Y. Wang, "A DC-to-18 GHz SP10T RF Switch Using Symmetrically-Routed Series-TL-Shunt and Reconfigurable Single-Pole Network Topologies Presenting 1.1-to-3.2 dB IL in 0.15 μm GaAs pHEMT," *IEEE Radio Frequency Integrated Circuits Symposium (RFIC)*, 2022, pp. 79-82.

[6] B. Yu et al., "Ultra-Wideband Low-Loss Switch Design in High-Resistivity Trap-Rich SOI with Enhanced Channel Mobility," *IEEE Trans. Microw. Theory Tech.*, vol. 65, no. 10, pp. 3937-3949, Oct. 2017.

[7] F. Meng, K. Ma and K. S. Yeo, "A 130-to-180GHz 0.0035mm² SPDT Switch with 3.3dB Loss and 23.7dB Isolation in 65nm Bulk CMOS," *2015 IEEE ISSCC Dig. of Tech. Papers*, 2015, pp. 1-3.

[8] T. V. Dinh, P. Descamps, D. Pasquet, D. Lesénéchal and S. Wane, "Experimental Characterization of Packaged Switch Devices for RF and Millimeter-Wave Applications," *IEEE Radio Frequency Integrated Circuits Symposium (RFIC)*, 2016, pp. 47-50.

[9] "HMC641A data sheet," Analog Devices, Wilmington, United States.

[10] M. Hangai, R. Komaru, S. Miwa, Y. Kamo and S. Shinjo, "2–12 GHz High-Power GaN MMIC Switch Utilizing Stacked-FET Circuits," *2019 49th European Microwave Conference (EuMC)*, 2019, pp. 840-843.

RMo2B-3

A 4.8-6.4-GHz GaN MMIC Front-End Module with Enhanced Back-off Efficiency and Compact Size

Guansheng Lv, Wenhua Chen, Xiaofan Chen, Long Chen, Zhenghe Feng
Tsinghua University, China

Abstract—A back-off efficient transmit/receive (T/R) front-end module (FEM) architecture is presented in this paper. On one hand, switchless class-G (SLCG) topology is adopted for power amplifier (PA) to improve the back-off efficiency in TX mode. On the other hand, co-design asymmetric T/R switch scheme is applied to reduce the switch loss in TX path. A 4.8-6.4-GHz FEM is implemented in a 0.15-μm GaN-HEMT process for validation, and the chip size is only 1.45 mm × 1.6 mm. The TX mode realizes a saturated power of 37.2–38.9 dBm and a 6-dB back-off drain efficiency (DE) of 40.2%–43.4%. Applying a 160-MHz LTE signal with 8.5-dB PAPR, an average DE of 33.3%–37% at an average power of 28.3–30 dBm is measured, and the ACPR is better than -46 dBc after digital predistortion. The RX mode achieves a noise figure of 1.8–2.2 dB and an IIP3 of 20.8–25 dBm.

Keywords—Front-end module (FEM), gallium nitride (GaN), monolithic microwave integrated circuit (MMIC), high efficiency, class-G, compact, transmit/receive (T/R) switch, power amplifier (PA), low-noise amplifier (LNA).

I. INTRODUCTION

Spectrally efficient modulation schemes are widely used in modern wireless communication systems, which leads to a large peak-to-average ratio (PAPR). Therefore, it is critical for power amplifier (PA) to exhibit high efficiency at power back-off (PBO) region. However, most reported front-end modules (FEMs) adopt inefficient PA topologies, such as class-A/AB, resulting in a quite low average efficiency [1]–[4]. Doherty PA topology is employed in some FEMs to improve efficiency [5]–[7], but the required λ/4 transmission line (TL) will limit the overall bandwidth and occupy a large chip area. Recently, chen *et al.* propose a novel switchless class-G (SLCG) PA to overcome the abruptly dropped efficiency and non-continuous nonlinearity of conventional class-G technique [8]. SLCG PA shows compact structure and no explicit bandwidth limitation and is very attractive for FEM design.

In addition to PA, transmit/receive (T/R) switch also has a great influence on the TX efficiency. Conventionally, a single-pole double-throw (SPDT) switch is used as T/R switch. Although the design of FEMs is simplified, the insertion loss (IL) of SPDT switch degrades the TX efficiency significantly. To alleviate this problem, asymmetric topology of T/R switch and co-design technique have been presented previously [1]–[4]. These two methods are usually applied simultaneously in practical design to achieve a better performance.

In this paper, a novel back-off efficient FEM (BE-FEM) based on SLCG PA topology and co-design asymmetric T/R switch scheme is proposed to enhance the average efficiency under high-PAPR modulated signals. A 4.8–6.4-GHz monolithic microwave integrated circuit (MMIC) FEM is implemented in a 0.15-μm GaN-HEMT process for validation,

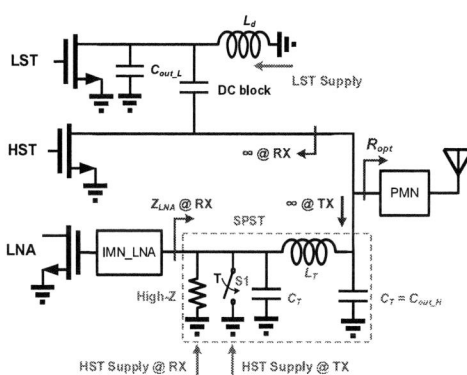

Fig. 1. Architecture of the proposed back-off efficient FEM.

and a 6-dB PBO drain efficiency (DE) of 40.2%–43.4% is realized with a compact chip size of 1.45 mm × 1.6 mm.

II. PROPOSED BACK-OFF EFFICIENT FEM

Fig. 1 shows the architecture of the proposed BE-FEM. As can be seen, SLCG scheme is used for PA, and asymmetric T/R switch topology is adopted to reduce the insertion loss (IL) in TX path. A post-matching network (PMN) is inserted between the conjunction port and antenna port for further impedance transforming. SLCG PA is composed of a smaller low-side transistor (LST) with lower voltage supply and a larger high-side transistor (HST) with higher voltage supply, the intrinsic optimal load impedances of which are both R_{opt}. LST is biased at class-AB state while HST is biased at class-C state, and the handover required by class-G operation is realized by two-quadrant modulation (TQM) mechanism. To support TQM, the adopted technology should be able to operate in the 3rd quadrant, i.e. both the drain voltage and current are below zero. In the back-off region, the output power is provided by LST, and an efficiency peak can be observed prior to the HST turns on. In the saturation region, the output power is provided by HST, and LST works as a rectifier [8]. Since a small portion of RF power is converted into DC current by LST, the output power of SLCG PA is a little lower than that of HST. The power back-off (PBO) range can be estimated as

$$PBO \ (dB) = 20\lg(n) \qquad (1)$$

where n is the size ratio of HST to LST.

In the adopted T/R switch topology, the switch in TX path is eliminated directly, and a λ/4-TL based single-pole single-throw (SPST) switch is employed in RX path. The shunt switch device S1 is turned on in TX mode and turned off in

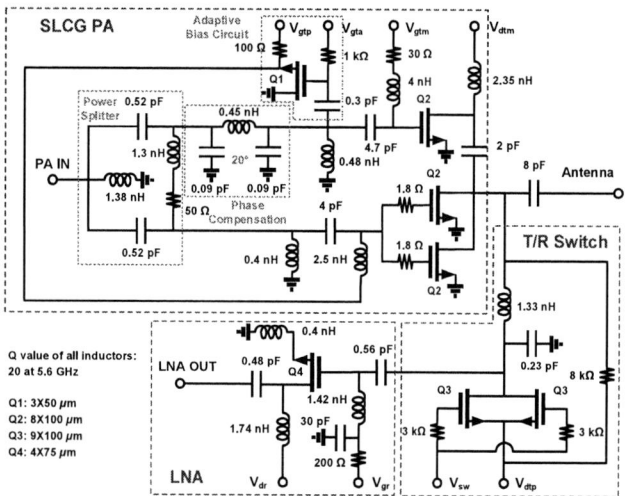

Fig. 2. Complete schematic of the implemented BE-FEM.

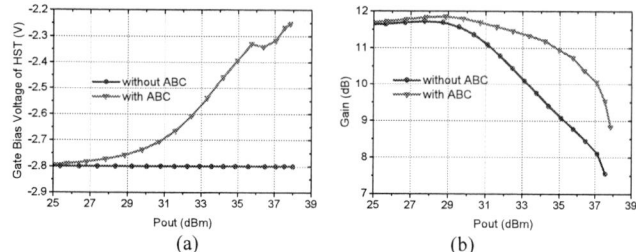

Fig. 3. (a) Gate bias voltage of HST versus output power with and without ABC, (b) gain versus output power with and without ABC.

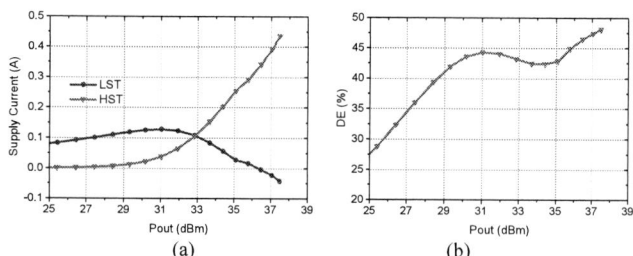

Fig. 4. Simulated (a) supply currents and (b) DE versus output power.

RX mode. LST's output capacitance C_{out_L} is neutralized by the drain bias inductor L_d, while HST's output capacitance C_{out_H} is absorbed into the SPST. The value of L_d can be expressed as

$$L_d = 1/(\omega_0^2 C_{out\_L}) \qquad (2)$$

where ω_0 is the operating angular frequency. Rather than using an additional choke inductor, the drain voltage of HST is supplied via the SPST. As the shunt switch device S1 will be turned off in RX mode, a high-impedance shunt resistor is employed to ensure the drain supply of HST in RX mode.

The $\lambda/4$ TL of the SPST is realized by a π-type network for a compact size. In order to absorb C_{out_H} completely, the shunt capacitance C_T of the π-type network should be not less than C_{out_H}. Assuming $C_T = C_{out\_H}$, the characteristic impedance Z_T of the $\lambda/4$ TL and series inductance L_T can be calculated as

$$Z_T = 1/(\omega_0 C_{out\_H}) \qquad (3)$$

$$L_T = 1/(\omega_0^2 C_{out\_H}) \qquad (4)$$

Z_{LNA} in Fig. 1 denotes the source impedance of the LNA at ω_0, and it can be calculated as

$$Z_{LNA} = Z_T^2/R_{opt} \qquad (5)$$

The input matching network (IMN) of LNA can be designed individually after the determination of Z_{LNA}. Alternatively, the IMN can also be designed with the T/R switch incorporated.

III. CIRCUIT IMPLEMENTATION

The proposed BE-FEM is implemented in a 0.15-μm GaN-HEMT process from WIN Semiconductor for validation, and the center frequency is chosen to be 6 GHz. Fig. 2 presents the complete schematic. The detailed design procedure will be presented in the following.

A. SLCG PA

A 8×100-μm transistor with 14-V drain supply and a $2 \times 8 \times 100$-μm transistor with 28-V drain supply are used as LST and HST, respectively. R_{opt} is extracted to be 50 Ω. C_{out_L} and C_{out_H} are extracted to be 0.3 pF and 0.53 pF, respectively. Since R_{opt} is already 50 Ω, PMN can be a simple dc-block capacitor, as shown in Fig. 2. Using (2), L_d is calculated to be 2.35 nH. High-pass LC networks are employed for the input matching of LST and HST. To ensure a good performance of SLCG PA in the load-modulation region, the output currents of the LST and HST should be phase-aligned. Therefore, a 20° phase compensation network is inserted in front of LST. A LC-ladder based lumped Wilkinson power splitter is adopted to allocate the input power into LST and HST equally.

Owing to the loading effect of LST in the saturation region, SLCG PA exhibits relatively large gain compression. To alleviate this problem, an adaptive bias circuit (ABC) is used in our design. As presented in Fig. 2, ABC samples the input voltage of LST through a 0.3-pF capacitor, and then converts the sampled voltage into current through transistor Q1. RC circuit composed of a 100-Ω resistor and the equivalent input capacitor of HST works as a low-pass filter. Since V_{gtp} is a negative voltage, the actual gate bias voltage of HST will become less negative with the increase of the input power. Fig. 3(a) plots the gate bias voltage of HST versus output power. Notably, the gate bias voltage increases from -2.8 V in the back-off region to -2.25 V in the saturation region. Fig. 3(b) presents the gain versus output power. As can be seen, the gain compression is reduced significantly after using ABC.

Fig. 4(a) shows the simulated supply currents of LST and HST. The supply current of LST reaches its maximum around 6-dB PBO point, and then decreases gradually due to the load modulation effect. Finally, the LST is driven into 3rd quadrant, leading to a negative supply current. Fig. 4(b) presents the simulated DE, and a 6-dB PBO DE of 44% is observed.

B. T/R Switch

With $C_{out\_H} = 0.53$ pF, L_T is calculated to be 1.33 nH using (4). The required inductance is realized by a customized spiral inductor with a quality factor around 20. A transistor size of

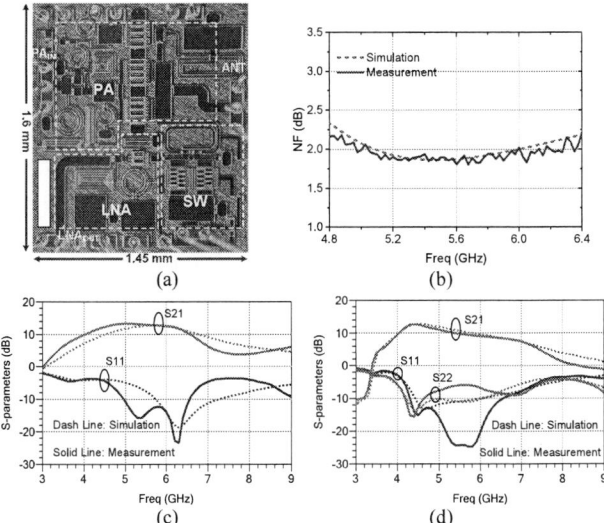

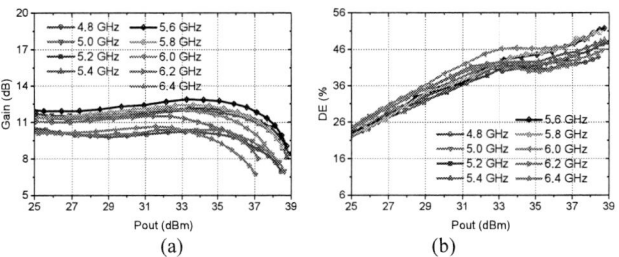

Fig. 6. Measured (a) gain and (b) DE versus output power in TX mode.

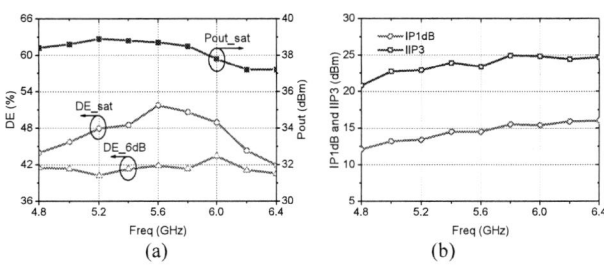

Fig. 7. (a) TX-mode and (b) RX-mode large-signal performance summary.

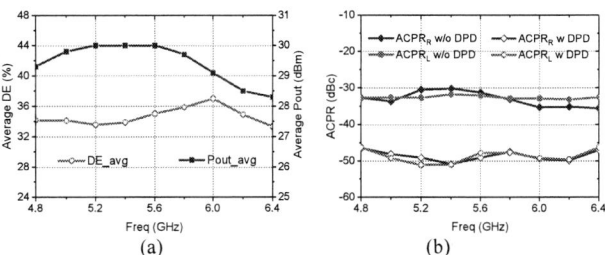

Fig. 8. Measured modulated performance under a 160-MHz LTE signal with 8.5-dB PAPR: (a) average power and DE, (b) ACPR with and without DPD.

Fig. 5. (a) Chip photo of the fabricated BE-FEM, (b) measured NF, (c) measured S-parameters in TX mode, (d) measured S-parameters in RX mode.

$2 \times 9 \times 100$ μm is used for the shunt switch device S1 to ensure enough drain supply current for HST in TX mode. The drain supply of HST is maintained in RX mode by an 8-kΩ shunt resistor. The off-state capacitance C_{off} of the switch device are extracted to be 0.3 pF, and an extra 0.23-pF shunt capacitor is paralleled with the switch to obtain a total 0.53-pF capacitance. The off-state resistance R_{off} of the switch device is strongly related to the gate resistance. A larger gate resistance leads to a higher R_{off}, but a slower switch speed. In our design, a 1.5-kΩ gate bias resistor is used, and the value of R_{off} is up to 5500 Ω. According to electromagnetic (EM) simulation results, the ILs of the T/R switch in TX and RX mode are 0.4 dB and 0.75 dB, respectively. The isolation of the T/R switch in TX mode is around -23 dB, which is good enough to protect LNA.

C. LNA

The design of LNA is based on classical inductive source-degenerated topology, as shown in Fig. 2. A 4×75-μm transistor with 10-V drain supply is adopted, and the quiescent current is selected to be 15 mA for a tradeoff between gain and noise performance. The source degenerated inductance is chosen as 0.4 nH to align the optimal source impedances for NF and input return loss with the minimum gain degradation. Simple high-pass LC networks are used for both input and output matchings, and the T/R switch is incorporated into the input matching of the LNA to achieve a larger bandwidth. A 200-Ω resistor is inserted at the gate bias path for robustness and low-frequency stability improvement.

IV. MEASUREMENT RESULTS

The fabricated BE-FEM is shown Fig. 5(a), and the chip size is only 1.45×1.6 mm^2. Both RF and DC pads are wire bonded to an external printed circuit board (PCB) for the convenience of measurement. The IL of RF-in and RF-out TLs on PCB is de-embedded from measurement results. The performance of the BE-FEM is characterized by small-signal,

large-signal and modulated-signal measurements.

A. Small-signal Measurement

Fig. 5(c) and Fig. 5(d) present the measured S-parameters from 3 to 9 GHz in TX and RX mode, where the simulated S-parameters are also depicted for comparison. A rather good agreement is observed, and the small-signal gain in 4.8-6.4 GHz is 11.2–13.1 dB in TX mode and 8.5–11.5 dB in RX mode. Fig. 15(b) plots both the measured and simulated NF performance. As can be seen, the measured NF is close to the simulation, and its value is 1.8–2.2 dB in 4.8–6.4 GHz.

B. Large-signal Measurement

The large-signal performance in TX mode is evaluated by pulsed continuous-wave (CW) with 10-μs pulse width and 10% duty cycle. Fig. 6 presents the measured gain and DE versus output power from 4.8 to 6.4 GHz in TX mode. As can be seen, a saturated output power of 37.2–38.9 dBm, a saturated DE of 41.9%–51.7% and a 6-dB PBO DE of 40.2%–43.4% are achieved. The power gain around 31 dBm is 10–12.5 dB. Fig. 7(a) summarizes the saturated and PBO performances at different frequencies. Fig. 7(b) shows the measured input 1-dB compression point (IP1dB) and input referred third-order intercept point (IIP3) in RX mode. IP1dB is measured under a pulsed CW signal while IIP3 is acquired under a two-tone signal with 10-MHz spacing and 17-dBm total output power.

979-8-3503-2123-4/23 $31.00 © 2023 IEEE

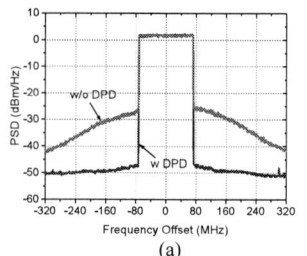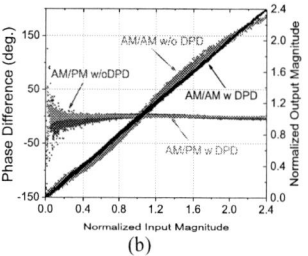

(a)   (b)

Fig. 9. Linearity performance with and without DPD at 5.4 GHz: (a) output power spectrum density, (b) AM-AM and AM-PM.

As can be seen, an IP1dB of 12.1–16 dBm and an IIP3 of 20.8–25 dBm are realized.

C. Modulated-signal Measurement

The modulated performance of the BE-FEM is evaluated by a 160-MHz LTE signal with 8.5-dB PAPR. The input signal is generated by a R&S SMW200A signal generator and the output modulated signal is captured by a R&S FSW43 spectrum analyzer. Fig. 8 shows the measured modulated performance at different frequencies. As can be noted, an average output power of 28.3–30 dBm and an average DE of 33.3%–37% are realized in 4.8–6.4 GHz. Measured adjacent channel power ratio (ACPR) is better than -30 dBc across the operating band, and is further improved to better than -46 dBc after applying digital pre-distortion (DPD). Fig. 9(a) presents the output power spectrum density (PSD) at 5.4 GHz. ACPR with and without DPD are -31.8/-30.2 dBc and -51/-50.9 dBc, respectively. The AM–AM and AM–PM characteristics at 5.4 GHz are also given in Fig. 9(b). It can be seen that the linearity of the BE-FEM is markedly improved after DPD.

D. Performance Comparison

The performance of the proposed BE-FEM is summarized and compared with other GaN FEMs or PAs in Table 1. Remarkably, the back-off DE of the BE-FEM shows a great improvement over previous reported GaN FEMs, while an ultra-compact size and a large fractional bandwidth (FBW) are also achieved. Moreover, the 6-dB PBO DE of the BE-FEM is comparable with that of the standalone Doherty PA in [9] thanks to the co-design asymmetric T/R switch scheme.

V. Conclusion

In this paper, a novel BE-FEM architecture is proposed by adopting SLCG PA topology and co-design asymmetric T/R switch scheme. To verify the proposed architecture, a 4.8–6.4-GHz FEM is implemented in a 0.15-μm GaN-HEMT process. A saturated output power of 37.2–38.9 dBm and a 6-dB PBO DE of 40.2%–43.4% are measured in TX mode. According to the comparison in Table 1, the realized PBO efficiency of the BE-FEM is apparently higher than other reported GaN FEMs.

Acknowledgment

This work was supported in part by National Key R&D Program of China (Grant No. 2019YFB1804100) and Beijing Science and Technology Plan (Grant No. Z211100004821010).

Table 1. Comparison with other GaN MMIC FEMs or PAs

Ref.	This Work	[1]	[2]	[3]	[9]
GaN Process	0.15-μm	0.25-μm	0.25-Mm	0.25-μm	0.25-μm
Freq. (GHz)	4.8–6.4	5.7–6.7	5.2–5.6	S band	4.1–5.6
FBW (%)	28.6	16.1	7.4	13	30.9
Integration	**FEM**	FEM	FEM	FEM	PA
PA Topology	**SLCG**	Class-AB	Class-AB	Class-AB	Doherty
Area (mm^2)	**2.3**	2.4	36	49	7.8
TX Performance					
Gain (dB)	**10–12.5**	10	18.5	35	8.3–11.2
Psat (dBm)	**37.2–38.9**	33.9	46	46	38.4–39.5
DE$_{sat}$ (%)	**41.9–51.7**	48.5	36	42	51.7–60.8
DE$_{6dB}$ (%)	**40.2–43.4**	22[a]	NA	21–30[a]	38.5–46.5
RX Performance					
Gain (dB)	**8.5–11.5**	8	31.5	30	NA
NF (dB)	**1.8–2.2**	3.7	2.4	1.75	NA
IP$_{1dB}$ (dBm)	**12.1–16**	6[a]	-14[a]	-13[a]	NA
P$_{DC}$ (mW)	**150**	66	700	1350	NA

a: Read from curves.

References

[1] P. Choi et al., "A 5.9-GHz Fully Integrated GaN Frontend Design With Physics-Based RF Compact Model," *IEEE Trans. Microw. Theory Techn.*, vol. 63, no. 4, pp. 1163-1173, Apr. 2015.

[2] M. van Heijningen et al., "C-Band Single-Chip Radar Front-End in AlGaN/GaN Technology," *IEEE Trans. Microw. Theory Techn.*, vol. 65, no. 11, pp. 4428-4437, Nov. 2017.

[3] R. Giofrè et al., "S-Band GaN Single-Chip Front End for Active Electronically Scanned Array With 40-W Output Power and 1.75-dB Noise Figure," *IEEE Trans. Microw. Theory Techn.*, vol. 66, no. 12, pp. 5696-5707, Dec. 2018.

[4] Y. Yi et al., "A 24–29.5-GHz Highly Linear Phased-Array Transceiver Front-End in 65-nm CMOS Supporting 800-MHz 64-QAM and 400-MHz 256-QAM for 5G New Radio," *IEEE J. Solid-State Circuits*, vol. 57, no. 9, pp. 2702-2718, Sept. 2022.

[5] Y. H. Chee, F. Golcuk, T. Matsuura, C. Beale, J. F. Wang and O. Shanaa, "A digitally assisted CMOS WiFi 802.11ac/11ax front-end module achieving 12% PA efficiency at 20dBm output power with 160MHz 256-QAM OFDM signal," in *IEEE Int. Solid-State Circuits Conf. (ISSCC) Dig. Tech. Papers*, Feb. 2017, pp. 292-293.

[6] K. Nakatani et al., "A highly integrated RF frontend module including Doherty PA, LNA and switch for high SHF wide-band massive MIMO in 5G," *IEEE Topical Conference on RF/Microwave Power Amplifiers for Radio and Wireless Applications (PAWR)*, Jan. 2017, pp. 37-39.

[7] D. Parat, A. Serhan, P. Reynier, R. Mourot and A. Giry, "A Linear High-Power Reconfigurable SOI-CMOS Front-End Module for WI-FI 6/6E Applications," in *Proc. IEEE Radio Freq. Integr. Circuits Symp. (RFIC)*, Jun. 2022, pp. 39-42.

[8] X. Chen, M. Zhao, W. Chen and Z. Feng, "A 700-2800MHz Switchless Class-G Power Amplifier with Two-Quadrant Modulation for Back-off Efficiency Improvement," in *IEEE MTT-S Int. Microw. Symp. Dig.*, Jun. 2022, pp. 410-413.

[9] J. Pang et al., "Broadband GaN MMIC Doherty Power Amplifier Using Continuous-Mode Combining for 5G Sub-6 GHz Applications," *IEEE J. Solid-State Circuits*, vol. 57, no. 7, pp. 2143-2154, Jul. 2022.

RMo2B-4

A 280 GHz InP HBT Direct-Conversion Receiver with 10.8 dB NF

Utku Soylu[#1], Amirreza Alizadeh[#], Munkyo Seo[*], and Mark J. W. Rodwell[#]

[#]Department of Electrical and Computer Engineering, University of California, Santa Barbara, USA
[*]Department of Electrical and Computer Engineering, Sungkyunkwan University, South Korea
[1]utkusoylu@ucsb.edu

Abstract— We report a fully integrated 280 GHz direct-conversion receiver in 250 nm InP HBT technology. The receiver has > 17 dB conversion gain over 264-297 GHz, consumes 455 mW, and has 10.8-11.6 dB DSB noise figure over 261-300 GHz. Its −3 dB bandwidth is 16.5 GHz, while its −6 dB bandwidth is 34.5 GHz. The local oscillator is generated by an internal 8:1 active frequency multiplier. To the authors' knowledge, the IC demonstrates record noise performance for an integrated receiver operating near 280 GHz.

Keywords—H-band, millimeter wave, direct conversion, noise figure (NF), indium phosphide (InP), double heterojunction bipolar transistor (DHBT).

I. INTRODUCTION

The 200-300 GHz frequency range can support wireless backhaul links with capacities well above 100 Gb/s, serving future communications networks. Integrated receivers operating near 280 GHz have been demonstrated [1-9] in several technologies. Key design objectives for such receivers are low noise figure for increased link range and high bandwidth for high symbol rate hence high data rate. Here we report an integrated receiver with a 10.8-11.6 dB double sideband (DSB) noise figure over 261-300 GHz, record for receivers operating near 280 GHz. The receiver −3 dB bandwidth is 16.5 GHz, while its −6 dB bandwidth is 34.5 GHz. With moderate post-equalization, this bandwidth should be sufficient to support 50 GSymbol/second data transmission.

II. RECEIVER DESIGN

The receiver was fabricated in Teledyne 250 nm InP HBT technology [10], this having a maximum 650 GHz power gain cut-off frequency (f_{max}) at top metal, a maximum 3 mA/μm current density, and 4.5 V BV_{CEO}. The technology provides four Au interconnect layers, 50 Ω/square thin film resistors, and 0.3 fF/μm^2 MIM capacitors.

The receiver has a broadband mm-wave low-noise amplifier, followed by down conversion mixers and broadband 50 Ω baseband output amplifiers for the in-phase (I) and quadrature-phase (Q) signals (Fig. 1). The mixers' local oscillators are generated by an 8:1 LO frequency multiplier, an LO buffer amplifier, and a passive LO quadrature phase generation circuit. In the LNA and frequency multiplier, IC interconnects are microstrip lines (MSL's) using the lowest

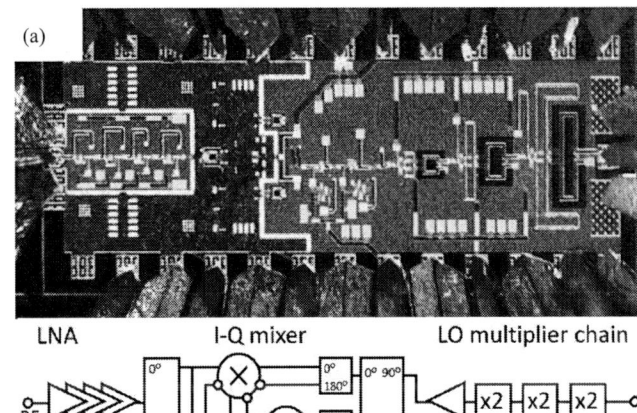

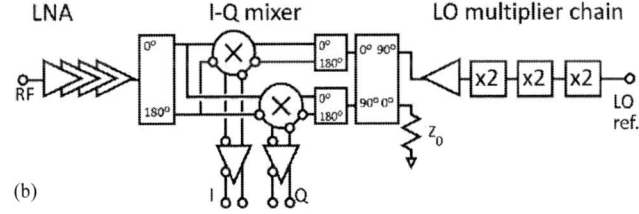

Fig. 1. 280 GHz direct-conversion receiver: integrated circuit photograph (a), and block diagram (b). The IC is 2.0 mm × 0.8 mm.

metal plane (M1) for ground. This provides lowest loss, hence best LNA noise figure. For the mixer and baseband amplifiers, the interconnects are inverted microstrip lines (IMSL's), using the upper (M4) metal plane for ground. IMSL does not require holes in the ground plane where each transistor is placed. This minimizes ground-return parasitics in dense circuits having many transistors, such as (I, Q) Gilbert-cell mixers. At 280 GHz, 50 Ω MSL loss is 1 dB/mm while IMSL loss is 1.9 dB/mm. The simulated MSL to IMSL transition loss is < 0.15 dB.

A. LO Multiplier (x8)

The 280 GHz LO is generated from a 35 GHz input using three cascaded frequency doublers (Fig. 2a); the multiplier chain was earlier reported in [11]. Each balanced frequency doubler cell uses a push-push emitter-coupled pair. The chain has an input transformer balun. At 280 GHz, the measured 8th harmonic power is −0.6 dBm. With −2 dBm input power, the multiplier 3-dB bandwidth is 48 GHz. Adjacent harmonics are suppressed by better than 28 dBc; given the LNA's finite RF bandwidth, other LO harmonics will not result in significant receiver spurious responses.

979-8-3503-2123-4/23 $31.00 © 2023 IEEE   85   2023 IEEE Radio Frequency Integrated Circuits Symposium

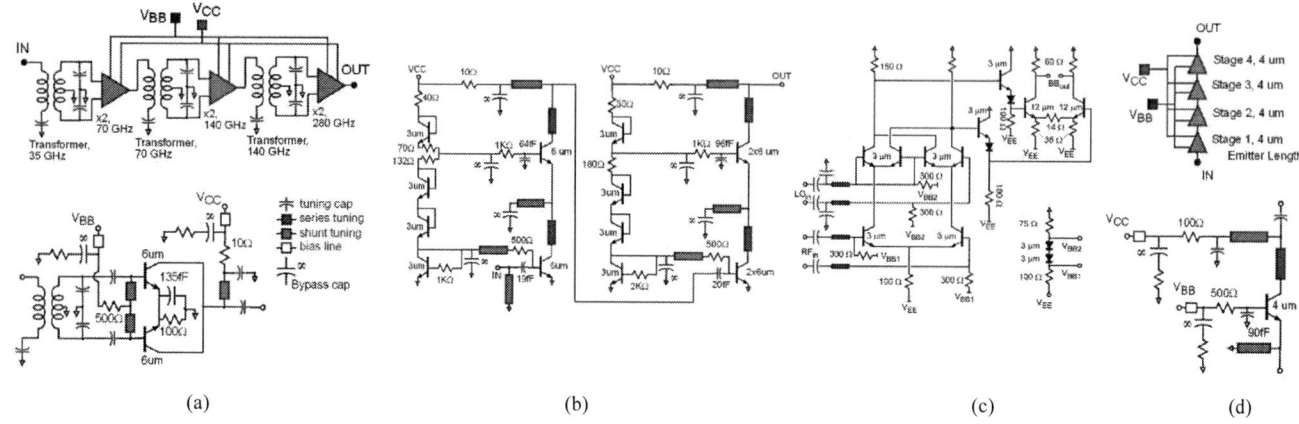

Fig. 2. Circuit schematics: (a) LO Multiplier (×8), (b) LO Driver, (c) Mixer and 50-ohm Baseband Driver, (d) LNA

B. LO Driver

The driver amplifier increases the LO power to that necessary to drive the mixers. Though the DC bias circuit (Fig. 2b) is that of a pair of cascode stages, to provide highest output power and PAE there is impedance tuning between the outputs of the common-emitter (CE) and the inputs of the common-base (CB) transistors. The LO buffer amplifier is therefore best viewed as a four-stage (CE, CB, CE, CB) cascade. All inter-stage impedance tuning is for maximum saturated output power, given 0 dBm input power, not maximum small-signal gain. CE and CB transistors are individually stabilized, with the 500 Ω shunt resistors stabilizing the CE transistors and the finite base capacitance stabilizing the CB transistors. The finite base capacitance also provides significant cascade power-combining between the CB and CE stages [12, 13]. In simulation, given 0 dBm input power, the output power is > 7 dBm over 240-340 GHz. The simulated DC power is 160 mW.

C. I-Q Mixer and 50-Ohm Baseband Driver

The LO driver amplifier output feeds a Lange coupler and a pair of transformer baluns, generating the four (0°, 90°, 180°, 270°) LO phases required for the I and Q mixers. Each mixer is double-balanced in the Gilbert configuration (Fig. 2c), with impedance-matching on the LO and RF ports. There is no impedance matching between the input RF and LO switching transistors. Increasing the baseband load resistance from 50 Ω to 150 Ω reduces the noise contributions of the resistor itself and of the output buffer. The differential output amplifiers, with emitter followers as level-shifters and emitter degeneration for linearity, directly drives an external 50 Ω load. Simulated I-Q imbalance is < 0.4 dB in amplitude and < 7.8° in phase for DC-20 GHz output frequencies, including layout parasitics of Lange coupler, LO/RF transformers and the baseband output amplifier drivers.

D. LNA

The LNA (Fig. 2d) is a four-stage CB amplifier with base capacitive degeneration [12, 13, 14]. The base capacitance provides unconditional stability from DC to f_{max} and adjusts the source impedance for minimum noise measure to be the complex conjugate of the input impedance, allowing simultaneous input matching for reflection coefficient and for minimum cascaded noise figure. The HBT junction area is scaled, together with the DC current and the base capacitance, so that the source conductance for minimum noise measure is 20 mS; this permits the input stage to be noise-matched to 50 Ω with a single inductive shunt element, avoiding the added attenuation, hence the added noise, of a series matching element [14]. Center frequencies of the four stages are stagger-tuned to provide wider bandwidth.

III. MEASUREMENT RESULTS

Measurements (Fig. 3) were performed on 3-mil thinned die (Fig. 1). A synthesizer and cascaded 12:1 and 2:1 VDI frequency multipliers generated the input signal, which passed through a directional coupler and a 220-330 GHz GGB wafer probe. The coupler −20 dB port, connected to a harmonic mixer and spectrum analyzer, monitored the input power; before IC testing, this input power monitor was calibrated against a VDI/Erickson THz power meter. The receiver output power was measured by a power meter and its spectrum monitored by a spectrum analyzer. Measurements are corrected for the input probe loss using the loss stated on the probe data sheet.

In operation, the 8:1 multiplier was biased with V_{CC}=2.9 V and V_{BB}=1.9 V, drawing 29.8 mA and 1.2 mA respectively. The 4-stage LNA was biased with V_{CC} =1.6 V and V_{BB} =0.85 V, drawing 11 mA and 0.51 mA respectively. The LO driver was biased with V_{CC} =4.7 V, and drew 31 mA from this supply. The mixers and output buffers drew 81 mA from V_{EE} = −2.5 V. The DC power dissipation is 455 mW.

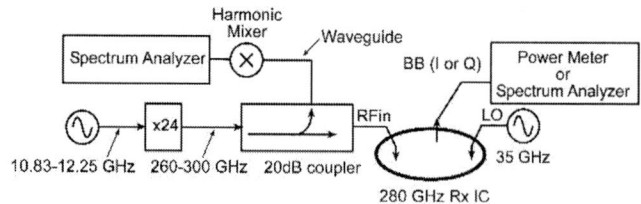

Fig. 3. Receiver continuous wave (CW) testing setup.

979-8-3503-2123-4/23 $31.00 © 2023 IEEE

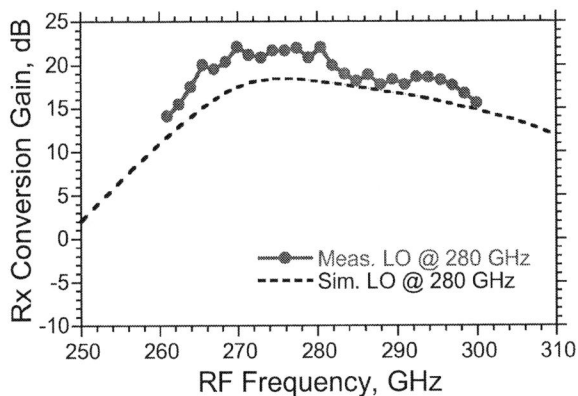

Fig. 4. Conversion gain vs. RF frequency at a fixed 280 GHz LO frequency.

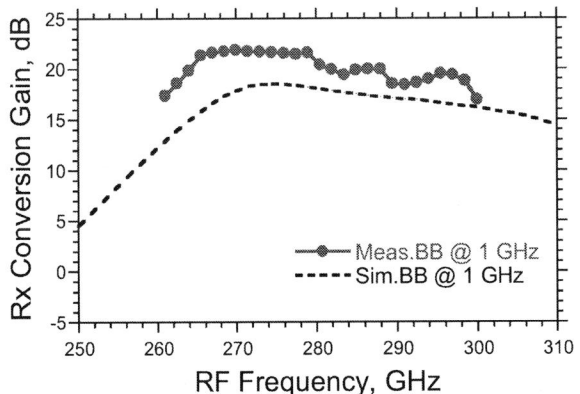

Fig. 5. Conversion gain vs. RF frequency at a fixed 1 GHz output frequency.

The input RF frequency f_{RF} was first swept with a fixed 280 GHz LO frequency (Fig. 4). The peak measured conversion gain, to a single-ended output, was 22 dB; gains using the differential outputs would be 3 dB greater. Over 260-280 GHz, the measured gain was ~2-4 dB greater than simulated; over 280-300 GHz it is ~1-2 dB greater. The measured –3 dB bandwidth is 16.5 GHz, while the measured –6 dB bandwidth is 34.5 GHz. CB amplifiers in this technology often show higher gain than simulated [10].

The RF and LO frequencies were then swept with a fixed 1 GHz baseband frequency (Fig. 5). This showed 20.4 dB measured conversion gain at 280.5 GHz and ~40 GHz LO-tuning bandwidth at the –6 dB points.

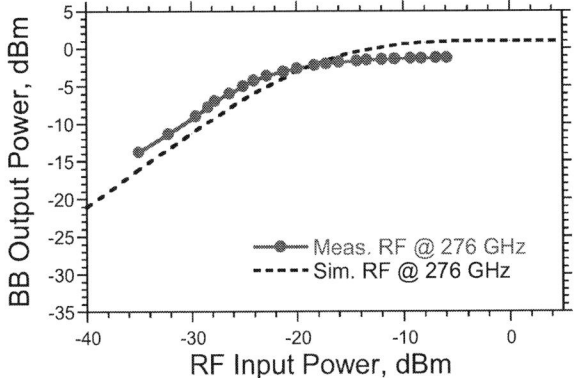

Fig. 6. Baseband output power vs. RF input power at f_{LO} = 280 GHz.

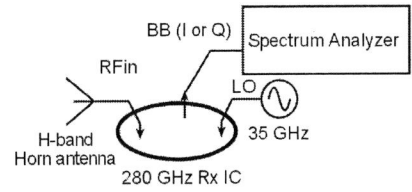

Fig. 7. Receiver noise figure measurement by the gain technique.

To measure the dynamic range, the input RF power at 276 GHz was swept while applying a 280 GHz local oscillator. P_{in1dB} is -23.6 dBm, with 455 mW dissipation (Fig. 6).

Because there is no commercially available 280 GHz hot/cold noise source, noise was instead measured by the less accurate gain method; even with the hot/cold method, measurements are difficult unless (T_{hot}–T_{cold}) is comparable to or larger than T_{DUT}. Receiver gain was first measured using the configuration of Fig. 3, with input and output powers measured using a power meter. Output noise power spectral density was then measured, at 200 MHz, using a spectrum analyzer (Fig. 7), with the RF input connected through a 220-330 GHz wafer probe to an H-band horn antenna, providing a 50 Ω, 300 K source termination. The input referred noise was determined by dividing the measured output noise by the measured receiver gain, dividing by kT, and subtracting 3 dB to convert from single sideband (SSB) to DSB noise figure. The measured receiver DSB noise figure is 10-11.6 dB over 261-300 GHz (Fig. 8). Measured noise is 1.5-2.5 dB below original design simulations. However, over 260-300 GHz, the measured receiver gain is 1-4 dB greater than that simulated. If this discrepancy between measured and simulated receiver gain were due to greater LNA gain than simulated, then the contributions to the receiver noise of the mixer and of the 2nd, 3rd and 4th LNA stages would be reduced. A simple calculation shows that the reduction of the mixer noise alone would reduce the disparity between simulation and measurement to 1-1.5 dB, even neglecting the reduced noise contributions of the 2nd-4th LNA stages. The measured DSB receiver noise (Fig. 8) is very similar to that of [15], for an LNA at the same frequency in the same technology, and similar to [16], an LNA in a similar technology, though [15,16] are LNA, not receiver results.

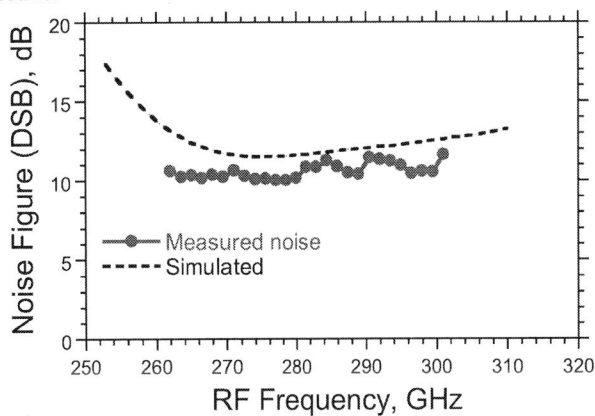

Fig. 8. Measured vs simulated receiver DSB noise figure.

Table 1. State-of-the-art receivers at above 200 GHz.

Ref	[1]	[2]	[3]	[4]	[5]	[6]	[7]	[8]	[9]	**This work**
Freq, GHz	240	220	220	240	240	240	200	300	340	**280**
Conversion Gain, dB	25	1.5	3.5	18	32	10.5	25	26	14	**22**
P_{DC}, mW	260	-	110	512	575	866	825	482	472	**455**
Chip Size, mm²	2	1.5	4.8	2.1	4.5	1.56	2	1.32	3	**1.6**
DSB NF, dB	15[a]	7.5[a]	7.4	18[b]	10.4	12	8	12-16.3	17	**10.8**
Technology	65 nm CMOS	50 nm GaAs mHEMT	100 nm GaAs mHEMT	130 nm SiGe HBT	130 nm SiGe HBT	130 nm SiGe HBT	250 nm InP HBT	250 nm InP HBT	250 nm InP HBT	**250 nm InP HBT**

[a]calculated, [b]simulated

We do not believe that an on-wafer 220-330 GHz receiver noise measurement, by the gain method, can be accurate to better than approximately +/-1 dB.

IV. CONCLUSION

In this paper, a fully integrated 280 GHz receiver with record noise figure is presented. The peak conversion gain is 22 dB with 34.5 GHz of -6 dB modulation bandwidth with 455 mW dissipation. The measured DSB noise figure is 10.8 dB at 281.5 GHz.

ACKNOWLEDGMENT

This work was supported by ComSenTer, a JUMP program sponsored by the Semiconductor Research Corporation. The authors thank Teledyne Scientific & Imaging for the IC fabrication.

REFERENCES

[1] S. V. Thyagarajan, S. Kang and A. M. Niknejad, "A 240 GHz Fully Integrated Wideband QPSK Receiver in 65 nm CMOS," in IEEE Journal of Solid-State Circuits, vol. 50, no. 10, pp. 2268-2280, Oct. 2015.

[2] I. Kallfass et al., "All Active MMIC-Based Wireless Communication at 220 GHz," in IEEE Transactions on Terahertz Science and Technology, vol. 1, no. 2, pp. 477-487, Nov. 2011.

[3] M. Abbasi et al., "Single-Chip 220-GHz Active Heterodyne Receiver and Transmitter MMICs With On-Chip Integrated Antenna," in IEEE Transactions on Microwave Theory and Techniques, vol. 59, no. 2, pp. 466-478, Feb. 2011.

[4] M. Elkhouly, Y. Mao, S. Glisic, C. Meliani, F. Ellinger and J. C. Scheytt, "A 240 GHz direct conversion IQ receiver in 0.13 μm SiGe BiCMOS technology," 2013 IEEE Radio Frequency Integrated Circuits Symposium (RFIC), Seattle, WA, USA, 2013, pp. 305-308.

[5] M. H. Eissa et al., "Wideband 240-GHz Transmitter and Receiver in BiCMOS Technology With 25-Gbit/s Data Rate," in IEEE Journal of Solid-State Circuits, vol. 53, no. 9, pp. 2532-2542, Sept. 2018.

[6] N. Sarmah et al., "A Fully Integrated 240-GHz Direct-Conversion Quadrature Transmitter and Receiver Chipset in SiGe Technology," in IEEE Transactions on Microwave Theory and Techniques, vol. 64, no. 2, pp. 562-574, Feb. 2016.

[7] M. Seo, A. S. H. Ahmed, U. Soylu, A. Farid, Y. Na and M. Rodwell, "A 200 GHz InP HBT Direct-Conversion LO-Phase-Shifted Transmitter/Receiver with 15 dBm Output Power," 2021 IEEE MTT-S International Microwave Symposium (IMS), Atlanta, GA, USA, 2021, pp. 378-381.

[8] S. Kim et al., "300 GHz Integrated Heterodyne Receiver and Transmitter With On-Chip Fundamental Local Oscillator and Mixers," in IEEE Transactions on Terahertz Science and Technology, vol. 5, no. 1, pp. 92-101, Jan. 2015.

[9] Y. Yan, Y. B. Karandikar, S. E. Gunnarsson, M. Urteaga, R. Pierson and H. Zirath, "340 GHz Integrated Receiver in 250 nm InP DHBT Technology," in IEEE Transactions on Terahertz Science and Technology, vol. 2, no. 3, pp. 306-314, May 2012.

[10] M. Urteaga, Z. Griffith, M. Seo, J. Hacker and M. J. W. Rodwell, "InP HBT Technologies for THz Integrated Circuits," in Proceedings of the IEEE, vol. 105, no. 6, pp. 1051-1067, June 2017.

[11] U. Soylu, A. Alizadeh, M. Seo and M. J. W. Rodwell, "A 280 GHz (x8) Frequency Multiplier Chain in 250 nm InP HBT Technology," 2022 IEEE BiCMOS and Compound Semiconductor Integrated Circuits and Technology Symposium (BCICTS), Phoenix, AZ, USA, 2022, pp. 191-194.

[12] A. S. H. Ahmed, A. A. Farid, M. Urteaga and M. J. W. Rodwell, "204GHz Stacked-Power Amplifiers Designed by a Novel Two-Port Technique," 2018 European Microwave Integrated Circuits Conference (EuMIC), Madrid, 2018, 23-25 Sept. 2018

[13] A. S. H. Amed, M. Seo, A. Farid, M. Urteaga, J. Buckwalter, M. Rodwell, "A 140GHz power amplifier with 20.5dBm output power and 20.8% PAE in 250-nm InP HBT technology" 2020 IEEE International Microwave Symposium, June 21-26, Los Angeles, CA

[14] U. Soylu, A. S. H. Ahmed, M. Seo, A. Farid and M. Rodwell, "200 GHz Low Noise Amplifiers in 250 nm InP HBT Technology," 2021 16th European Microwave Integrated Circuits Conference (EuMIC), London, United Kingdom, 2022, pp. 129-132.

[15] K. Eriksson, S. E. Gunnarsson, V. Vassilev and H. Zirath, "Design and Characterization of H-Band (220–325 GHz) Amplifiers in a 250-nm InP DHBT Technology," in IEEE Transactions on Terahertz Science and Technology, vol. 4, no. 1, pp. 56-64, Jan. 2014.

[16] H. -J. Song and M. Yaita, "On-Wafer Noise Measurement at 300 GHz Using UTC-PD as Noise Source," in IEEE Microwave and Wireless Components Letters, vol. 24, no. 8, pp. 578-580, Aug. 2014.

RMo2C-1

A CMOS 183 GHz Millimeter-Wave Spectrometer for Exploring the Origins of Water and Evolution of the Solar System

Adrian Tang[1,2], Mau-Chung Frank Chang[2], Yanghyo Kim[3], Goutam Chattopadhyay[1]

[1]Jet Propulsion Laboratory, USA
[2]University of California, Los Angeles, USA
[3]Stevens Institute of Technology, USA
ajtang@jpl.nasa.gov

Abstract— This paper discusses advanced CMOS-based remote sensing emission spectrometers towards measuring the D/H (Deuterium/Hydrogen) ratio throughout the solar system. These isotopic measurements are critical to gain a clearer understanding of the water on Earth and its origins. We briefly review the MIRO instrument aboard the ESA/Rosetta mission, the isotopic measurements performed at a Jovian comet 67P and the challenges they pose to the cometary hypothesis for the origin of water on Earth. We then present the next generation low-cost CMOS-based water sensing spectrometer and describe how it can help address these critical science questions. We discuss design and development of the CMOS spectrometer and its first space test flight aboard the NASA ReckTangLE sub-orbital mission in 2019.

I. INTRODUCTION

The origins of water on Earth, the processes by which it was deposited, and how those processes relate to the evolutionary model of our solar system remain a contentious issue [1,2,3] in planetary science. While the events in question occurred 4.5 billion years ago, several observable markers or indicators exist which provide us evidence towards the origins of water on our planet. Of these markers, the deuterium to hydrogen ratio or "D/H ratio" [4] is a readily accessible observable in the study of water's origin. D/H ratio looks at the ratio of abundance of heavy water molecules to more common hydrogen-16 water molecules in an observed sample. These two isotopes have different molecular resonances in the 183 and 500-600 GHz bands that can be observed through remote sensing emission spectroscopy [5]. Ground based radio telescopes have performed these emission measurements for many decades focusing on large planetary bodies (large size necessary for detectability) and found that indeed the D/H ratio is extremely consistent. Jovian objects, Oort cloud objects and chondrites (primitive meteorites) observed by radio astronomers were measured to have D/H ratios around 2×10^{-4}, extremely similar to the D/H ratios found on Earth. This implies a common origin and gives rise to the hypothesis that Earth's oceans were delivered by Jovian or Oort family comets. Also, it was observed that the four outer planets in our solar system have a considerably different D/H ratio, closer to the values suspected of the proto-solar medium that existed prior to the solar system's formation, again supporting the hypothesis that Earth's water arrived after the planet's initial formation. Fig. 1 provides a comparison of several families of planetary objects and shows key ground and spaceborne measurements made of D/H ratio over the last three decades. In 2004 ESA launched the Rosetta planetary mission, that carried on it an instrument called the Microwave Instrument for the Rosetta Orbiter (MIRO) [6], which was a combination 180 GHz and 600 GHz emission spectrometer instrument specifically for investigating D/H ratios. In Aug 2014 Rosetta rendezvoused with Churyumov–Gerasimenko (67P as the short identifier) which is a Jovian family comet of a somewhat unusual shape.

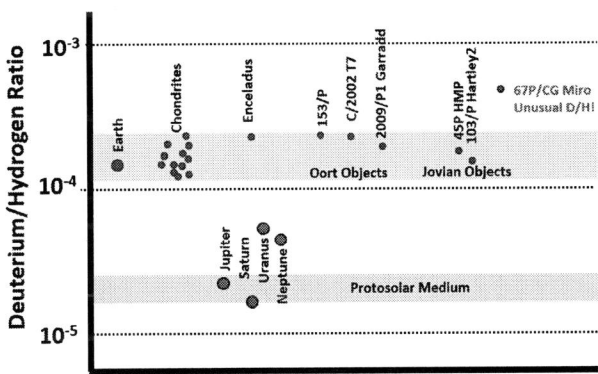

Fig. 1. Measured D/H ratio of key planetary bodies and the families of solar system objects they belong to.

As seen in Fig. 1 the D/H ratio measured by MIRO of comet 67P is vastly different [7], almost 2 times the value of other Jovian and Oort objects, a highly unexpected result, which has now unsettled the current model of the cometary deposit of water here on Earth. To further investigate this fundamental question of water's origins on Earth requires us to conduct many more detailed D/H measurements on a wide range of Oort, Jovian and other planetary objects so that we can understand how statistically rare or common the MIRO-67P result is, and to place it in the context of the other solar system regions (Inner planets and the Kuipler belt). The challenge of conducting these measurements is that many of the Oort and Jovian objects that would make ideal targets to explore the cometary hypothesis are too small for ground-based radio telescopes to observe and demand a probe mission. MIRO on Rosetta (shown in Fig. 2) was a complex instrument on a flagship mission that cost over a billion Euros and took over 5 years to build. The MIRO instrument itself demanded considerable payload resources, with a system mass of over 20

979-8-3503-2123-4/23 $31.00 © 2023 IEEE 89 2023 IEEE Radio Frequency Integrated Circuits Symposium

kg, and power consumption of almost 70W during observations. Clearly with this scale and scope of science instrument, the cost and effort needed to visit the 10s or even 100s of planetary objects required to settle this hypothesis is not a realistic endeavour.

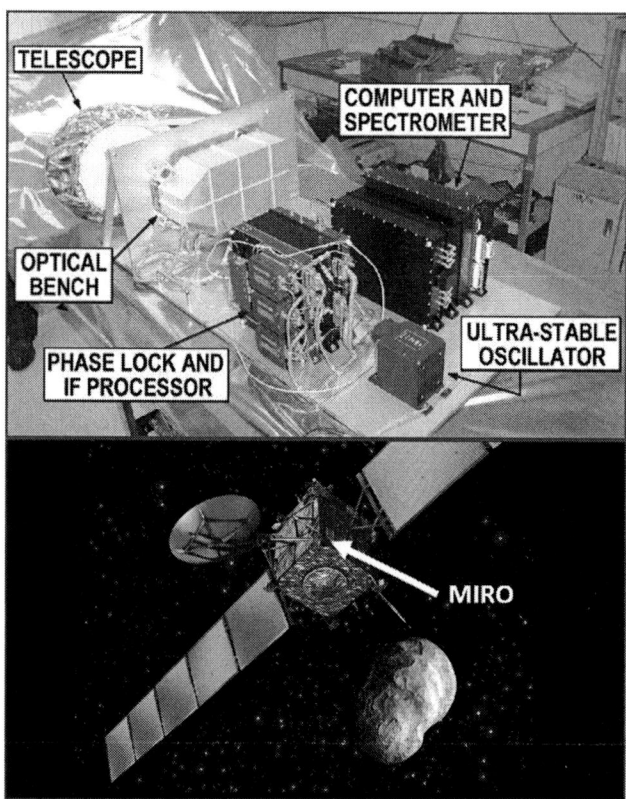

Fig. 2. The MIRO spectrometer [6] aboard the Rosetta Orbiter mission launched by ESA in 2004 which observed D/H ratios on Jovian family comet 67P in August 2014.

II. CMOS-Based Emission Spectrometers

For this debate about the origins of Earth's water to be settled a much smaller class of mission and D/H observing instrument is required, one that can be deployed to many objects across the solar system at a far lower cost than the original MIRO instrument. To that end, NASA is supporting development of CMOS based D/H spectrometers that can be carried on interplanetary CubeSats. These are small spacecraft that can visit other planets and were recently demonstrated by the JPL/NASA MARCO missions [8] in 2018. While the 500-600 GHz band of MIRO remains beyond silicon's ability to provide the receiver sensitivities needed for emission spectroscopy, considerable efforts have been made on CMOS implementations of emission spectrometers for the 183 GHz band [9,10]. Figure 3 shows the block diagram of a recently developed 183 GHz CMOS-based spectrometer system implemented in a 28nm TSMC process. The single-chip spectrometer is a heterodyne architecture with a double-sideband down conversion mixer. The chip contains a phase-locked synthesizer operating with a 91.5 GHz fundamental VCO that's then doubled to provide the LO to a 183 GHz

mixer stage. Phase noise for spectroscopy is far less critical than for communication systems as 1) the measurement is averaged many times and phase noise is a zero-mean process, 2) the emission signals involved in molecular spectroscopy are incoherent noise and broadband relative to the LO linewidth (1-10 MHz range at typical atmospheric pressures), and 3) no other 183 GHz signals exist on other planets so blocking and reciprocal mixing effects are not of concern. The silicon receiver's 183 GHz amplifier chain is preceded by a two stage MMIC low noise amplifier implemented in a 35nm InP-HEMT technology allowing the receiver to achieve the system temperatures necessary for sensing weak molecular emission signals at room-to-spaceborne temperatures (10K to 300K). The CMOS spectrometer chip is packaged on a PCB which supports the low frequency digital control and DC connections. The receiver input port is coupled into a differential quartz probe mounted on top of the PCB substrate to carry the high frequency input signal while avoiding substrate losses of the PCB material. The quartz probe then couples the receiver to a WR-5 waveguide flange mechanically fastened to the edge of the PCB module for interfacing to the MMIC LNAs. As reported in [9], this configuration provides a receiver that has InP level noise performance but takes advantage of CMOS integration for the synthesizer, IF stages, calibration circuits, and digital control, drastically reducing form factor and power consumption vs. a traditional III-V implementation like MIRO.

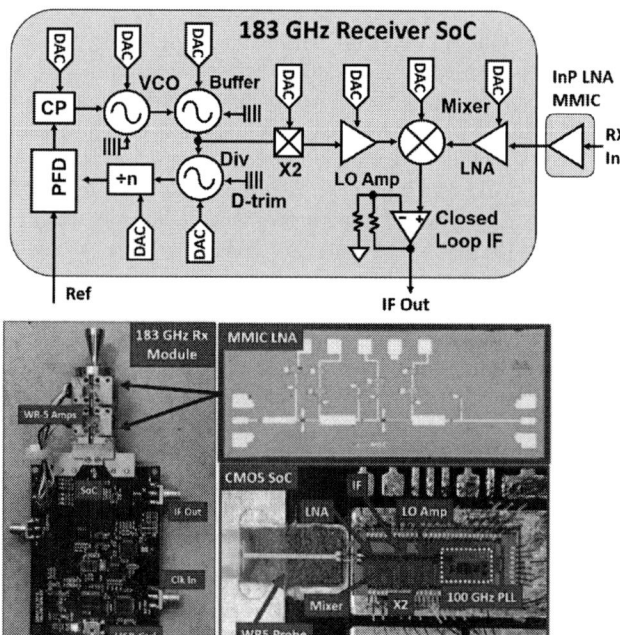

Fig. 2. Block diagram of the JPL developed 183 GHz spectrometer chip and photograph of packaged spectrometer assembly, CMOS chip and MMIC LNA [9,10].

Beyond the receiver noise performance of the spectrometer which is well supressed by the InP LNA (achieving under 400K at 183 GHz as reported in [9,10]) a second major consideration for spectroscopic remote sensing is that of deep stability quantified by the system's Allan time. Allan time

979-8-3503-2123-4/23 $31.00 © 2023 IEEE

describes the timescale over which the benefit of time averaging to a power measurement's statistical variance at the receiver output σ_{prx} is overtaken by the deep non-zero mean drift occurring from changing aging, supply, radiation, and thermal conditions. This is elegantly expressed as two competing terms in the radiometer equation [9] and shown visually in Fig 4.

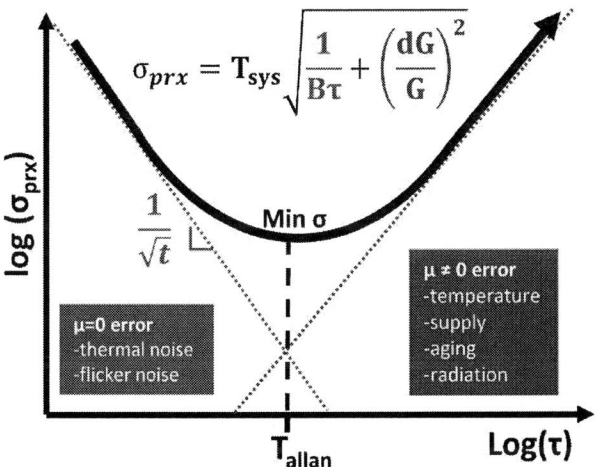

$$\sigma_{prx} = T_{sys}\sqrt{\frac{1}{B\tau} + \left(\frac{dG}{G}\right)^2}$$

Fig. 4. The radiometer equation showing competing terms of deep-drift and noise and their optimal cross-over point at T_{allan}.

The practical significance of Allan time is that it describes how frequently the spectrometer system must be calibrated, at the cost of observation time. Note this is similar but not the same to Allan deviation which describes frequency stability as oppose to amplitude stability. Interplanetary CubeSat missions are too small to carry the necessary fuel for orbital injection, and so they are restricted to flyby missions, making total observation windows limited, and placing a high importance on long Allan times to increase observational efficiency. While the mitigation of receiver thermal and flicker noise is well understood by the RFIC community and can be addressed through a mixture of RF circuit design techniques, receiver topologies, and architectures, solutions to mitigate deep-drift effects are addressed through several circuit approaches or "circuit philosophies" that place importance on maintaining a static system. First and most effective is to use a closed-loop circuit topology wherever possible, especially for high-gain stages like in IF amplification. Closed loop stages are less dependent on transistor parameters and typically depend on ratios of parameters (like the OTA example of Fig. 5a). This makes them more robust to temperature, aging, radiation and other non-zero mean changes encountered during spaceflight. Second, when open-loop structures cannot be avoided (for example the 183 GHz stages at the RF front-end) use PTAT and active feedback around those circuits to stabilize the die temperature like shown in Fig. 5b. Although operating the receiver circuits at a elevated temperature increases the receiver NF/Tsys, the benefits to the deep-drift suppression easily outweigh the added noise in the overall σ_{prx} error and final Tallan. Third, for critical circuits (references & biasing) place PTAT monitors for temperatures, process monitors for

aging and total ionized dose radiation monitors [11] nearby as shown in Fig 5c. When the chip is millions of miles away in deep space, having these sensors through telemetry provides added insight into the overall RFIC conditions that would otherwise leave operators or scientists making blind decisions on if the systematics have changed between measurements. Finally, where possible operate open-loop circuits in a well saturated condition as shown in Fig 5d. to minimize power variation vs time. This is especially true in the LO amplification chain as conversion loss for most mixer topologies is highly sensitive to the LO drive power.

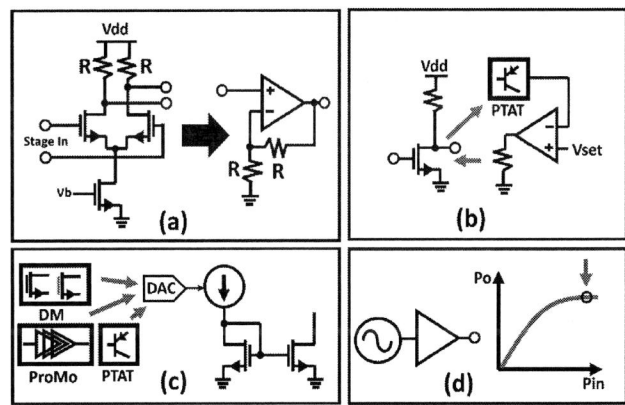

Fig. 5. Key circuit approaches for mitigating deep-drift. (a) Employing closed-loop analog circuit topologies where possible. (b) Stabilizing temperatures near sensitive circuits. (c) Monitoring aging temperature and radiation near key circuits. (d) Operating LO amplifiers in a saturated condition.

To provide an idea of the advantages these approaches to deep-drift mitigation provide, in [9] an open and closed loop IF was implemented, where the open-loop variant exhibited a T_{allan} of only 2.1 seconds compared with a closed-loop T_{allan} of 19.3 seconds. With the thermal stabilization off, this T_{allan} was found to degrade to only 14.5 seconds providing a sense of the contribution from each of those techniques. Even with the improved Allan time the spectrometer's output power spectrum must still be calibrated, which is accomplished by pointing the instrument to known directions in space and observing the Cosmic Microwave Background or "CMB". The CMB is the static broadband microwave energy that remains from the big-bang and inflationary period of the universe, and has been surveyed by the Planck telescope, BiCEP telescope and other observatories to milli-kelvin radiometric accuracy and sub-arcsecond spatial resolution. Using the CMB as a calibration reference is a standard approach in astronomy and planetary science, and not only allows the spectrometer instrument vs time, but also vs other instruments on-board the same spacecraft (in our case a second receiver observing other spectral lines of water like HDO and $H2O_{18}$ in the 600 GHz band).

III. Sub Orbital Demonstration

The 183 GHz CMOS-based spectrometer was integrated into a sub-orbital space-balloon payload as shown and combined with a baseband processor [12] that digitizes the resulting IF signal and computes the FFT to extract the spectral features. The payload construction is shown in Fig. 6a where the spectrometer was integrated with an optical chain and scanning primary mirror that can steer the beam to various points in the sky to perform calibration. The mission, called the NASA ReckTangLE (Reck-Tang Limbsounder Experiment) also carried out a separate 600 GHz experiment (not the focus of this paper) and used a wire grid polarizer to diplex the optical beam between the two experiments. The mission was launched from, New Mexico on Oct 17, 2019 and flew at an altitude of 124,500 ft with the mission terminating in the Texas panhandle approximately 4 hours and 40 minutes later. Fig. 6b also shows an example spectral sounding of the $H2O_{16}$ isotope taken at an altitude of 124,500 ft with a one-second integration time for pointing angles from $-45°$ below the horizon (called the limb) to $+45°$ above the horizon. At angles near $-45°$ the warm Earth in the background, and losses of the thick atmosphere at mm-wavelengths overwhelm the $H2O$ emission and begin to saturate the spectrometer. Alternatively, near $+45°$ observation angles, the beam path through the atmosphere is much shorter than the limb view, so the emission signal is weaker as less $H2O$ molecules are present in the beam path between the antenna and space.

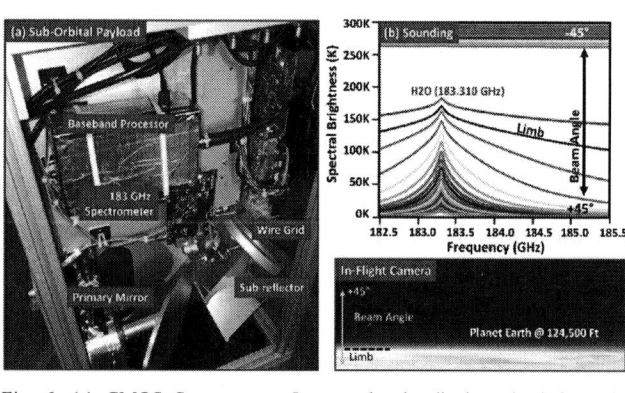

Fig. 6. (a) CMOS Spectrometer Integrated sub-oribtal payload from the ReckTangLE mission [9]. (b) Spectral sounding at various pointing angles revealing the $H2O_{16}$ line at 183.310 GHz. Also shown in a camera image indicating the orientation of the pointing angles.

IV. CONCLUSIONS

To the author's knowledge the 2019 ReckTangLE result represents the first demonstration of a millimeter-wave CMOS device performing science in space. More importantly, this sub-orbital fight demonstration represents the first step towards realizing a D/H spectrometer instrument with a low enough mass, power consumption, and suitable system cost to be accommodated on an interplanetary cubesat. This work is not yet complete, as similar advancements are required for improving the mass and power of the 500-600 GHz channel MIRO also carried out to 67P during the 2004 mission. An effort is also underway to develop such an instrument [13] based on hybrid CMOS+GaAs Schottky receivers, which has its own sub-orbital test-flight scheduled for June 2023. Assuming these tests are also successful, the resulting dual band instrument provides a realizable path to measuring the large number of Oort and Jovian family objects in the solar

system required to finally settle the hypothesis that Earth's water originated from comets.

ACKNOWLEDGMENT

Part of this research was carried out at the Jet Propulsion Laboratory, California Institute of Technology, under a contract with the National Aeronautics and Space Administration. The Authors are grateful to TSMC for their excellent 28nm manufacturing, NGC for their InP 35nm fabrication support, and the NASA Columbia Scientific Ballooning Facility for their mission support, launch, fight, and recovery operations.

REFERENCES

[1] Albertsson, T., D. Semenov, and Th Henning. "Chemodynamical deuterium fractionation in the early solar nebula: The origin of water on earth and in asteroids and comets." The Astrophysical Journal 784.1 (2014): 39.

[2] Morbidelli, Alessandro, et al. "Source regions and timescales for the delivery of water to the Earth." Meteoritics & Planetary Science 35.6 (2000): 1309-1320.

[3] Hartogh, Paul, et al. "Ocean-like water in the Jupiter-family comet 103P/Hartley 2." Nature 478.7368 (2011): 218-220.

[4] Mousis, Olivier, et al. "Constraints on the formation of comets from D/H ratios measured in H2O and HCN." Icarus 148.2 (2000): 513-525.

[5] JPL Spectroscopy Catalog (spec.jpl.nasa.gov)

[6] Gulkis, S., et al. "MIRO: Microwave instrument for Rosetta orbiter." Space Science Reviews 128.1 (2007): 561-597.

[7] Marshall, D. W., et al. "Spatially resolved evolution of the local H2O production rates of comet 67P/Churyumov-Gerasimenko from the MIRO instrument on Rosetta." Astronomy & Astrophysics 603 (2017): A87

[8] Klesh, Andrew, et al. "MarCO: Early operations of the first CubeSats to Mars." (2018).

[9] Adrian Tang, et al. "Sub-Orbital Flight Demonstration of a 183 / 540-600 GHz Hybrid CMOS-InP and CMOS-Schottky-MEMS Limb-Sounder", IEEE JMW, vol. 1, no. 2, pp. 560-573, Spring 2021.

[10] Y. Kim, Y. Zhang, T. Reck, D. Nemchick, G. Chattopadhyay, B. Drouin, M-C F. Chang and A. Tang, "A 183-GHz InP/CMOS-Hybrid Heterodyne-Spectrometer for Spaceborne Atmospheric Remote Sensing," in IEEE Trans. on Terahertz S&T, vol. 9, no. 3, pp. 313-334, May 2019.

[11] A. Tang, Y. Kim, MC-Chang, "Logic-I/O Threshold Comparing Gamma Radiation Dosimeter in Insensitive Deep-Sub-Micron CMOS", IEEE Transactions on Nuclear Science, vol. 63, no. 2, pp. 1247-1250, April 2016.

[12] A. Tang et al., "CMOS System-on-Chip Spectrometer Processors for Spaceborne Microwave-to-THz Earth and Planetary Science and Radioastronomy," in IEEE JMW, vol. 2, no. 4, pp. 599-613, Oct. 2022.

[13] Goutam Chattopadhyay, Theodore Reck, Adrian Tang, Cecile Jung-Kubiak, Choonsup Lee, Jose Siles, Erich Schlecht, Yanghyo M. Kim, M-C F. Chang, and Imran Mehdi, "Compact Terahertz Instruments for Planetary Missions", IEEE European Antennas and Propagation, 2015.

RMo2C-2

A 140 GHz Scalable On-Grid 8x8-Element Transmit-Receive Phased-Array with Up/Down Converters and 64QAM/24 Gbps Data Rates

Amr Ahmed, Linjie Li, Minjae Jung, Gabriel M. Rebeiz

University of California San Diego, USA

(aaahmed, lil030, mijung, grebeiz)@ucsd.edu

Abstract—**This paper presents a scalable wafer-scale transmit-receive phased array at 140 GHz. The chip is composed of 2-D 8x8 140 GHz channels employing radio-frequency (RF) beamforming with 4-bit phase and gain controls at 135-145 GHz. An up/down-converter (UDC) channel with a x6 LO chain and an IF of 9-14 GHz is also integrated on the chip. The on-chip RF distribution network is composed of Wilkinson divider/combiner networks, along with line amplifiers (LAs) to provide signal amplification. The chip is fabricated in GlobalFoundries CMOS 45RFSOI technology with an area of 9.84*8.27 mm² and is flipped on a low-cost organic RF PCB containing 8x8 patch antenna array placed at 1.07x1.22 mm grid (0.5λ x 0.57λ at 140 GHz). The array can scan ±60° for both transmit and receive operations. The measured TX EIRP is 37.5 dBm at 140 GHz. The measured RX array input 1-dB compression point is -11 dBm with an electronic gain of 20 dB at 140 GHz. The array supports 16/64 QAM operation with up to 24 Gb/s with <4% EVM_rms for both transmit and receive operations. This work presents the first phased-array at 140 GHz with wide-angle scanning and full scalability in the X and Y directions.**

Keywords—**140 GHz, beamforming, CMOS SOI, D-band, phased-arrays, transmitter, receiver, wafer-scale.**

I. INTRODUCTION

D-band communication systems (110-170 GHz) can be a key to fulfill the increasing demand on low latency and high data rate links; thanks to the wide available unallocated frequency spectrum of up to 60 GHz. Hence, D-band wireless systems have been an emerging topic recently in industry and academia [1]-[4]. However, at such high frequencies, phased arrays are essential to overcome the high space loss factor to allow for larger communication distances.

In order to build 2-D scalable arrays at such high frequencies, the RF phase shifting architecture is adopted in order to remove the mixer and associated drivers at each antenna element (IF and LO phase shifting topologies). Also, for wide scan angles, the elements must be spaced as close to 0.5λ as possible. This is achieved using an 8x8 silicon RFSOI chip with on-grid antenna ports, and coupled to a low-cost organic interposer. The chip design and layout is done for unlimited scalability and in both X and Y directions. A hybrid beamforming architecture is adopted with an up/down-converter every 64-elements to either add the sub-arrays in the IF domain, or to allow for a different digital stream every 64-elements for NxN MIMO systems.

II. ARRAY ARCHITECTURE

The array is composed of 8x8 RF beamforming transmit-receive RF channels, RF distribution networks and a UDC

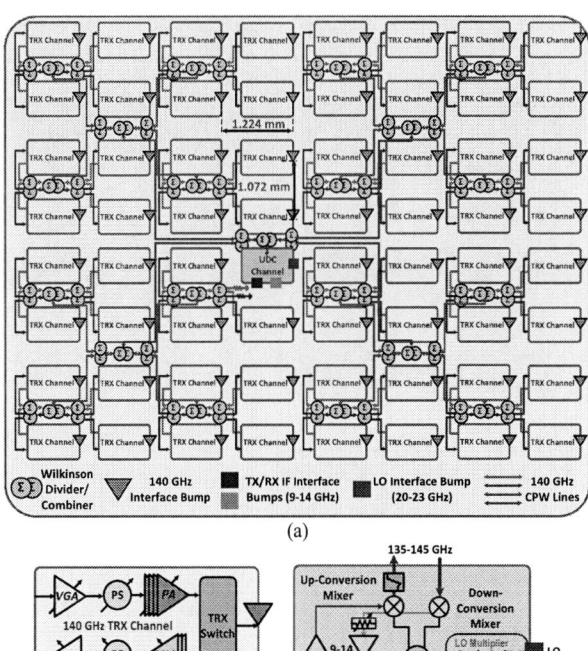

Figure 1. (a) Simplified block diagram of the 2D 140 GHz 8x8 transmit-receive array with up/down converter to IF, (b) TRX and UDC channels' details.

channel at its centre (Fig. 1(a)). Each RF channel (Fig. 1(b)) consists of a VGA (3-bit gain control), vector modulator (4-bit phase control) and PA (3-bit gain control) in the transmit path and an LNA (3-bit gain control), vector modulator (4-bit phase control) and VGA (3-bit gain control) in the receive path along with a single-shunt TRX SPDT switch at the antenna interface port. The simulated output power of the TX channel is 4-5 dBm and the simulated noise figure of the RX channel is 6.2 dB including the TRX switch loss.

One of the 64 elements is replaced with a UDC channel (Fig. 1(b)). This allows the chip input signal interface to be at the IF (9-14 GHz) and greatly relaxes the complexity of the array feeding network for scalability. The input LO frequency is 20-23 GHz (nominally 21.5 GHz) which is multiplied by a chain of on-chip frequency multipliers (doubler and tripler). The multiplied frequency passes by a Wilkinson divider and feeds both up-conversion and down-conversion channels in the UDC (based on active Gilbert-cell mixers). For RX operation, the mixer is followed by a variable attenuator and

979-8-3503-2123-4/23 $31.00 © 2023 IEEE   93   2023 IEEE Radio Frequency Integrated Circuits Symposium

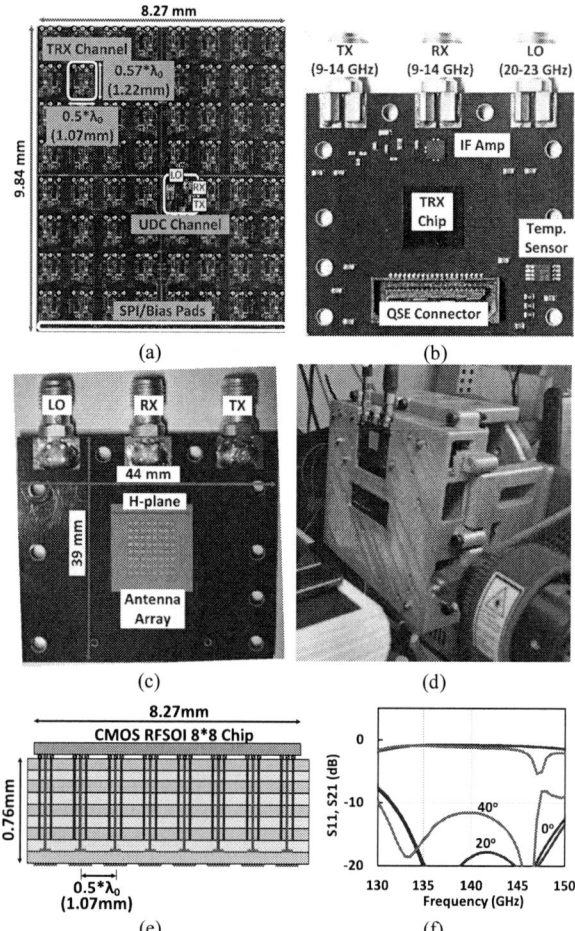

(a)

(b)

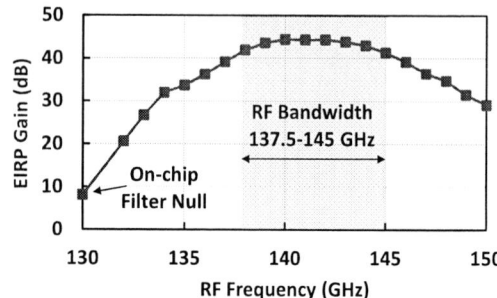

Figure 3. Measured TX array EIRP gain with a fixed IF frequency (11 GHz).

(c)

(d)

(e)

(f)

Figure 2. (a) Array layout rotated 90° when compared to Fig.1, (b) RF PCB front side with the array flipped on it and showing all the components, (c) RF PCB back side showing the antenna array, (d) RF PCB mounted on antenna positioner, (e) Cross section showing direct feed from chip to antennas, (f) Simulated H-plane transmission and reflection coefficients over different scan angles.

an IF amplifier. For TX operation, the IF signal is first amplified before being fed to the mixer. The mixer is followed by an image rejection high pass elliptical filter and a driver amplifier [3].

Two active Wilkinson distribution networks are integrated on the chip: one for TX and one for RX, and this is done to eliminate the use of two SPDT switches every time an amplifier is placed in the distribution network. The distribution networks consist of CPW transmission lines, line amplifiers (LAs) and $\lambda/4$ Wilkinson dividers/combiners (0.4 dB loss at 140 GHz). The overall CPW lines length from the UDC channel to the input of each RF channel is around 11 mm resulting in 22 dB of loss (2 dB/mm at 140 GHz ohmic loss [5]). Therefore, 7 and 4 stages of single-stage line amplifiers, each with 8 dB gain are placed along the TX and RX distribution networks respectively in order to compensate for the transmission-line and Wilkinson losses.

The chip also has an SPI module to control every register in the array. In addition, the reference current can be either an external current or and on-chip constant-Gm current

generation circuit (with PTAT operation). Finally, ESD protection circuits are included at all bias and digital pads. All pads are located within the chip grid for scalability.

III. PACKAGING AND ANTENNA ARRAY

The array is fabricated in GlobalFoundries CMOS 45RFSOI technology occupying a total area of $9.84*8.27mm^2$ and is flipped on an RF PCB (Fig. 2(a) and (b)). An 8x8 microstrip antenna array is fabricated on the RF PCB (equivalent to Tachyon 100G or Megtron with ε_r=3.4, tans=0.004 at 140 GHz) and is also placed at 0.5λ x 0.57λ grid (at 140 GHz) in the X and Y directions (Fig. 2(c)). The antenna is a probe-fed stub-coupled design, built for wideband operation, with the patch on the bottom M10 layer and the stub on the M9 layer. M8 layer is the antenna ground, and the total thickness of the antenna substrate is 0.11mm to mitigate the scan blindness. Figure 2(e) presents the cross section (not to scale) of the PCB showing the RF ports feeding directly the antennas for minimal transition loss. The antennas are co-designed with the RF chip bumps and are well matched with S_{11}<-20 dB (at 0° scanning). The antenna transmission coefficient (includes ohmic and mismatch losses) remains <-1 dB on axis at 135-147 GHz. Figure 2(f) presents the simulated HFSS scan results in the H-plane.

Figure 2(b) and (c) show the top and bottom sides of the PCB. The top side has the chip, external RX IF amplifier, temperature sensor IC, connectors for TX, RX and LO, and for supply, bias and digital controls. The bottom side only has the antenna array. The organic PCB material is no different than what is already used in 5G (28 GHz, 39 GHz) or 60 GHz. The laminate used is a standard copper clade laminate.

The RF PCB is connected to a DC/Control PCB and mounted to a Millibox GMI03 antenna positioner for over-the-air (OTA) measurements (Fig. 2(d)).

IV. MEASUREMENTS

A. Transmit OTA Measurements

The TX array is supplied with an IF signal (9-14 GHz) using a Keysight PNA (N5340C), and the RF radiated signal from the array (at 130-150 GHz) is received with a WR-6 standard horn antenna placed at 0.74m from the array. The RF signal is then down-converted back to IF using a WR6.5 VDI unit and fed into the PNA. After de-embedding all the setup losses, the gain, EIRP and patterns are measured.

979-8-3503-2123-4/23 $31.00 © 2023 IEEE

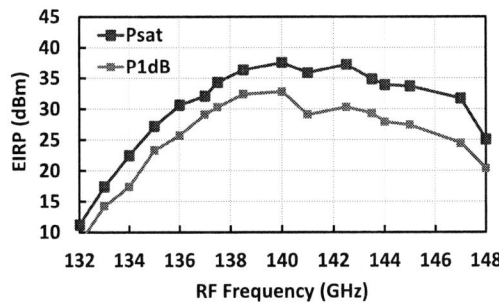

Figure 4. Measured TX array EIRP at Psat and P1dB across frequency.

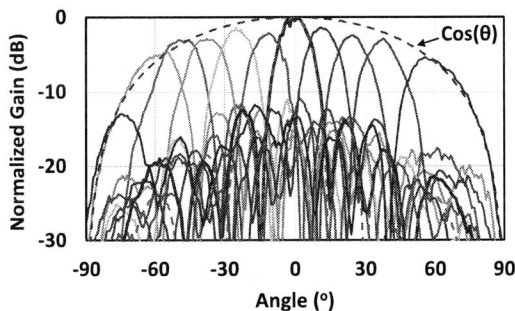

Figure 5. Measured TX array ±60° scanning patterns at 140 GHz.

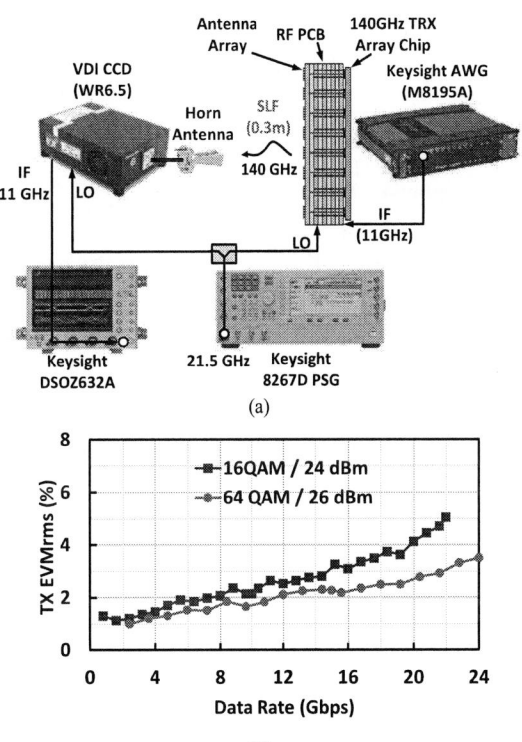

(a)

(b)

Figure 6. TX OTA (a) modulation measurement setup, (b) measured EVM$_{rms}$ versus data rate for 16 and 64 QAM signals and EIRP of 24 and 26 dBm, respectively.

Figure 3 presents the measured array EIRP gain with a fixed IF frequency (11 GHz) and swept LO and RF frequencies. The EIRP gain is defined as the radiated EIRP at the RF frequency divided by the input IF power at the SMA connector port. This measurement shows the TX RF bandwidth of the chip and antennas with a peak gain of 44 dB and a 3-dB bandwidth of 137.5-145 GHz. The measured instantaneous bandwidth is 3-5 GHz in the 136-146 GHz range (fixed LO and swept IF).

Figure 4 presents the measured EIRP across frequency with a peak of 37.5 dBm Psat. To our knowledge, this is the highest demonstrated EIRP from a wafer-scale phased array at 140 GHz in silicon technology. The simulated EIRPsat is 45 dBm (20logN+Gel+Pel=36+4+5), but the T/R modules on each element and the line amplifiers in the active Wilkinson combiner had lower gains than simulated, and the UDC did not have the power to drive the array into saturation. The EIRP is therefore limited by the UDC OPsat in this design.

The measured H-plane (x-direction) scan patterns at 140 GHz are presented in Fig. 5. The phased array is electronically scanned from -60° to 60° with the pattern gain drop of 5.3 dB at ±60° scan angles. The E-plane (Y-direction) scan patterns are up to ±40° and not included for brevity. To the author knowledge, this is the largest electronic scan range reported at 140 GHz for an active phased-array.

Finally, a Keysight AWG (M8195A) is used to generate modulation waveforms at IF (11 GHz) and the signal is directly fed to the TX array. The far-field signal is down-converted to an IF and fed to a Keysight digital oscilloscope (DSOZ632A) for demodulation (Fig. 6(a)). The measured

EVM$_{rms}$ versus data rate for 16 and 64 QAM modulated signals are shown in Fig. 6(b) (at 24/26 dBm EIRP for 16/64QAM). Measurements of up to 22 and 24 Gbps for 16 and 64QAM waveforms, respectively, are shown with EVM$_{rms}$ of 5% and 3.5%, respectively.

B. Receive OTA Measurements

The Rx test setup is similar to Fig. 6(a), but with a WR6.5 VDI up-converter. The electronic gain of the RX array is measured at a fixed IF frequency (11 GHz) while sweeping both the LO and RF frequencies (Fig. 7). The electronic receive gain is defined as the output IF signal power from the RF PCB divided by the total incidence power (P$_{inc}$) on the array aperture (Fig. 7). A gain of 20 dB is measured with a 3-dB bandwidth of 135-142.5 GHz.

The measured RX patterns are shown in Fig. 8. The RX array can also be scanned from -60° to 60° with the pattern gain drop of 5.5 dB at ±60° scan angles.

Fig. 9 presents the measured EVM$_{rms}$ for 16 and 64QAM modulated signals versus data rate at P$_{inc}$ of -32 dBm (on the total array aperture). Up to 22 and 24 Gbps for 16 and 64QAM with EVM$_{rms}$ of 4.7% and 3.65%, respectively, is obtained.

The measured array P$_{inc}$ 1-dB compression point at 135-142 GHz is -9 to -11 dBm. The input P1dB per channel is therefore -10 dBm - 18 dB (64 elements) = -28 dBm, which is high (simulated is -35 dBm). Again, this is due to the lower gain per channel and in the line amplifiers.

979-8-3503-2123-4/23 $31.00 © 2023 IEEE

Table 1 Comparison with the state-of-the-art D-Band Phased Arrays

	This Work		[3] JSSC 2022	[4] JSSC 2022	[2] RFIC 2020		[1] TMTT 2022
Technology	45nm RFSOI		45nm RFSOI	45nm RFSOI	130-nm SiGe		22FDX + InP
# of Elements	64 (8x8)		8 (4x2)	8 (4x2)	8	8	8 (8x1)
Transmit/Receive	Transmit + Receive		Transmit Only	Receive Only	Transmit Only	Receive Only	Transmit Only
Beamforming Arch.	RF		IF	IF	RF	RF	Digital
Frequency (GHz)	137.5-145[a]	135-142.5[a]	136-147[a]	139-155[a]	130-170[b]	130-170[b]	135
Peak TX EIRP (dBm)	37.5	N/A	32	N/A	N/A	N/A	38.5
TX Peak Gain (dB)	44[c]	N/A	21[d]	N/A	16[e]	N/A	N/R
RX Peak Gain (dB)	N/A	20[d]	N/A	27.5[d]	N/A	22[e]	N/A
Scan Range (°)	±60		±30	±35	Not Specified	Not Specified	±5
Chip-Antenna Package	Antenna In Package		Quartz Superstrate	Quartz Superstrate	Radio-on-glass	Radio-on-glass	LTCC Interposer
OTA Communication (Data Rate (Gb/s))	22 (16-QAM) 24 (64-QAM)		16 (16/64-QAM)	10 (16-QAM) 9 (64-QAM)	N/A	N/A	3 (16-QAM)

N/A: Not Applicable, [a]RF BW, [b]Chip BW, [c]EIRP Gain, [d]Electronic Gain, [e]Chip Gain.

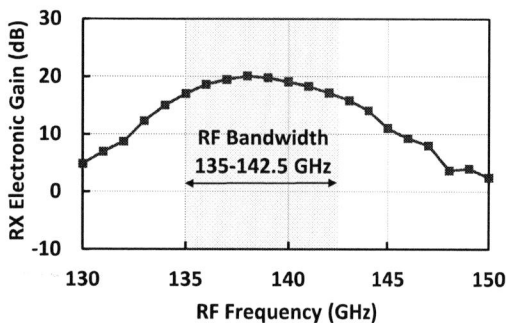

Figure 7. Measured RX array electronic gain with a fixed IF frequency.

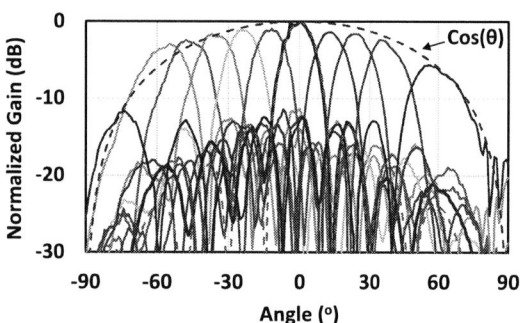

Figure 8. Measured RX array ±60° scanning patterns at 138 GHz.

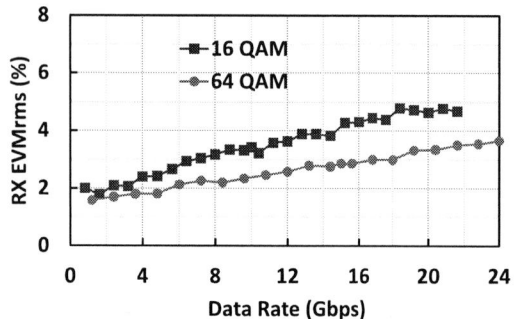

Figure 9. RX OTA measured EVM$_{rms}$ versus data rate for 16 and 64 QAM signals with P$_{inc}$ of -32 dBm respectively.

Table 1 compares this work with the state-of-the-art D-band phased arrays. This work achieves the widest scanning range, the highest OTA data rate and the highest EIRP for silicon technologies to-date.

V. CONCLUSION

This work demonstrated a scalable 2-D wafer-scale 8x8 transmit-receive phased array at 140 GHz. The work shows that low-cost silicon and, most important, low-cost packaging, can be used to build scalable phased-arrays at D-band. Also, while this array is built using an 8x8 chip, scalable low-cost topologies could be done using 4x4 or 16x16 chips.

ACKNOWLEDGMENT

This work is supported by the Semiconductor Research Corporation (SRC) under the JUMP program. The authors would also like to thank Cadence for EMX, Ansys for HFSS, GlobalFoundries for chip fabrication and Keysight for the measurement software. The authors thank Qualcomm for help and access to the PCB technology.

REFERENCES

[1] A. A. Farid, A. S. H. Ahmed, A. Dhananjay and M. J. W. Rodwell, "A Fully Packaged 135-GHz Multiuser MIMO Transmitter Array Tile for Wireless Communications," in IEEE Transactions on Microwave Theory and Techniques, vol. 70, no. 7, pp. 3396-3405, July 2022, doi: 10.1109/TMTT.2022.3161972.

[2] M. Elkhouly et al., "D-band Phased-Array TX and RX Front Ends Utilizing Radio-on-Glass Technology," 2020 IEEE Radio Frequency Integrated Circuits Symposium (RFIC), Los Angeles, CA, USA, 2020, pp. 91-94.

[3] S. Li, Z. Zhang and G. M. Rebeiz, "An Eight-Element 136–147 GHz Wafer-Scale Phased-Array Transmitter With 32 dBm Peak EIRP and >16 Gbps 16QAM and 64QAM Operation," in IEEE Journal of Solid-State Circuits, vol. 57, no. 6, pp. 1635-1648, June 2022, doi: 10.1109/JSSC.2022.3148385.

[4] S. Li, Z. Zhang, B. Rupakula and G. M. Rebeiz, "An Eight-Element 140-GHz Wafer-Scale IF Beamforming Phased-Array Receiver With 64-QAM Operation in CMOS RFSOI," in IEEE Journal of Solid-State Circuits, vol. 57, no. 2, pp. 385-399, Feb. 2022, doi: 10.1109/JSSC.2021.3102876.

[5] S. Li and G. M. Rebeiz, "High Efficiency D-Band Multiway Power Combined Amplifiers With 17.5–19-dBm Psat and 14.2–12.1% Peak PAE in 45-nm CMOS RFSOI," in IEEE Journal of Solid-State Circuits, vol. 57, no. 5, pp. 1332-1343, May 2022, doi: 10.1109/JSSC.2022.3145394.

A 57.6 Gb/s Wireless Link based on 26.4 dBm EIRP D-band Transmitter Module and a Channel Bonding Chipset on CMOS 45nm

Jose Luis Gonzalez-Jimenez, Alexandre Siligaris, Abdelaziz Hamani, Francesco Foglia-Manzillo,
Pierre Courouve, Nicolas Cassiau, Cedric Dehos, Antonio Clemente
Université de Grenoble-Alpes, CEA-Leti, France
joseluis.gonzalezjimenez@cea.fr

Abstract—This paper presents a baseband to D-band wireless link based on a transmitter module integrating a 45-nm CMOS channel bonding chipset and a high-gain antenna in PCB technology. The link realized using a commercial receiver at 42 cm achieves a data rate of 57.6 Gb/s using a multi-channel 16-QAM with a transmitting energy efficiency of 27.4 pJ/b.

Keywords—D-band, wireless-link, 16-QAM, CMOS 45nm.

I. INTRODUCTION

Wireless transmitters for high-data-rate links are receiving increasing interest as building blocks for the next-generation communication networks [1]-[5]. Most of them target the sub-THz spectrum due to the large available bandwidth (BW). However, these architectures have to face challenging trade-offs among power consumption, bandwidth, and transmitted power due to the high path loss and the feasible modulation schemes that can be used for large-band channels, which is usually limited to 4 or 6 bits/s/Hz.

This work presents a D-band wireless link system demonstration comprising a wideband channel-bonding transmitter (TX) chipset and a high-gain antenna. A channel-bonding architecture [1] is selected for the up-converter from baseband (BB) to the D-band, because it allows to optimize the signal fractional bandwidth at each frequency conversion step, without limiting the total bandwidth of the emitted signal. With this approach, the power consumption at all system levels, including the digital interfaces, can be reduced [6].

The developed D-band module covers a band of 17.3 GHz by up-converting 8 baseband channels of 2.16 GHz each. It comprises two different 45-nm CMOS integrated circuits (ICs). The first one performs a four-channel up-conversion and a channel aggregation from I/Q BB to an intermediate frequency (IF) band, centered at 61.56 GHz. The second IC is a two-channel transmitter, similar to that in [6]. The channel-bonding operation at IF is implemented on-chip using a passive hybrid 4-way power combiner, whereas at D-band it is realized over the air, using a high-gain flat lens antenna. This antenna is illuminated by two printed antennas, each fed by one of the two output signals of the TX IC, realized in the low-cost printed-circuit board (PCB) where the TX IC is assembled. The overall system enables a data rate (DR) of 57.6 Gb/s wireless point-to-point link over 42 cm using a 16 QAM with 1.8 Gbaud per baseband channel, with a low energy consumption (27.4 pJ/b) at 42 cm using and overall ante gain of 25 dBi. The link is demonstrated using a commercial receiver (RX) in a hardware-in-the-loop system-level validation platform. The link range is greatly enhanced with respect to state-of-the-art D-band

modules based on low-gain antennas [3], [7] without resorting to energy-hungry phased array systems [4], [5].

II. TRANSMITTER ARCHITECTURE AND BUILDING BLOCKS

The module is composed of two units of BB- to-IF up-converters and a two-channel TX IC, as shown in Fig. 1.

A. Baseband-to-IF channel bonding IC

The BB-to-IF up-converter and channel-bonding IC (see Fig. 2) is composed of four similar lanes, each for a different I/Q differential BB input. They include a 2nd order filter, an I/Q up-conversion mixer and an IF amplifier, as well as a dedicated local oscillator (LO) generator each. The four LO frequencies are set to 58.32 GHz, 60.48 GHz, 62.64 GHz and 64.8 GHz, so that the four BB channels are up-converted to adjacent IF channels centered at 61.56 GHz. The outputs of the four lanes are combined using an on-chip passive 4-way hybrid. Its average insertion loss is 8 dB across the IF band, with a ±1 dB variation. The IC is flip-chipped on a connectorized board, as shown in Fig. 2. The BB-to-IF IC is fabricated in 45-nm CMOS silicon-on-insulator (SOI) technology and occupies an area of 2×3.5 mm^2. Its power consumption is 495 mW for a 1 V supply.

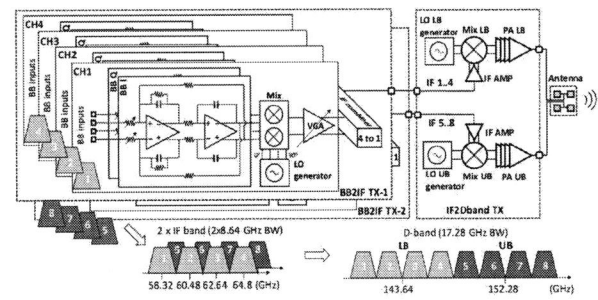

Fig. 1. Block diagram of the TX module, comprising two 2 units of four BB-to-IF up-converters, and an IF-to-D-band TX IC, which excites the antenna feeds of the flat lens.

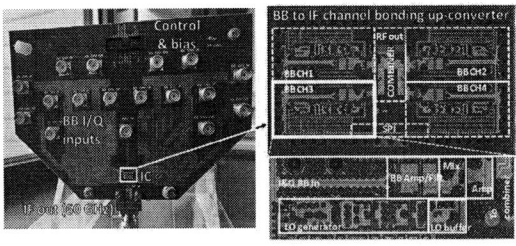

Fig. 2. BB-to-IF up-converter board and IC photograph.

The maximum gain from any of the BB inputs to the combined output is -26.5 dB. It includes a 1.5-dB loss due to the input BB connectors board traces, and a 4-dB loss on the IF output trace and connector. Each lane has 9 dB and 6 dB of programmable gain variation at the BB filters and IF amplifier, respectively. The 1-dB output compression point for the maximum gain is -22 dBm per channel. It is worth to note that, when all channel are combined, the total output power increases by 6 dB. The gain value is chosen to account for large dynamic ranges at the BB input (e.g. a few hundreds of mV) and a low enough IF power is chosen for not saturating the D-band transmitter.

One of the main limitations of channel-bonding architectures is the channel to channel interferences. They are caused basically by two mechanisms. First, the LO spurs at frequencies in adjacent channels produce some leakage from channel to channel. The spurs are due to the LO generator architecture that is based on frequency multiplication, as in [7]. The measured LO leakage of the proposed transmitter is less than -30 dBc in every channel, as shown in Fig. 3a. Secondly, the Nyquist replicas of the BB signals may fall in adjacent channels. Fig. 3b shows the IF output spectrum for 1.8 Gbauds modulated BB signal inputs. They BB signals come from a multi-channel AWG using a sampling rate of 2.5 GS/s. Please, note that each I and Q BB signal has a BW of 900 MHz. The 2nd order filter of each lane reduces out of band Nyquist replicas, but does not completely eliminate them, as shown in Fig. 3c. The measured signal-to-interferer ratio (SIR) between the wanted channel and the replicas is 20 dB in the worst case, obtained by comparing Fig. 2b and Fig. 2c for CH1 and CH2. This SIR is high enough for a 16-QAM but unsuited for 64-QAM. This limitation can be overcome in future versions of the

IC by further increasing the filter order.

B. IF to D-band channel bonding TX

The IF-to-D-band up-conversion and transmission is implemented by a two-channel TX module similar to that proposed in [6]. Each of the two lanes is composed of an IF amplifier, an up-conversion mixer and a power amplifier (PA). In the fabricated IC prototype, both IF signals are taken from the same input connector and then split on-chip for simplicity. The LO signals for each lane are generated on-chip using the frequency multiplication technique presented in [7], with frequencies 82.08 GHz and 90.72 GHz, respectively. They are generate from a common input reference (REF IN) signal of 4.32 GHz generated using a commercial PLL. The D-band TX brings each of the IF input signals to adjacent sub-bands (LB and UB) centered at 143.34 GHz and 152.28 GHz, respectively, covering a frequency range of 17.28 GHz. These two signals are provided at separated outputs and connected to two adjacent pairs of patch antennas (see Fig. 5), realized in the PCB board where the IC is flip-chipped. These antennas serve as the focal-plane feeds for a flat lens antenna, also called transmitarray (TA). The TA design is similar to that in [8] and comprises three metal layers in a PCB. The distance between feed and lens is 8.3 mm. The TA phase profile is optimized so that directive beams pointing to directions very close to boresight are formed for the overall TX signal band, i.e. in the frequency spectra of the two signals emitted by the TX IC. Thus, the antenna performs over-the-air power combining. The measured antenna gain is 25 dBi and varies less than 1 dB from 139 GHz to 159 GHz. The measured effective isotropic radiated power (EIRP) of the D-band TX module at the 1-dB compression point (1dBCP) is shown in Fig. 6a. An average of 26.5 dBm is radiated across the full band.

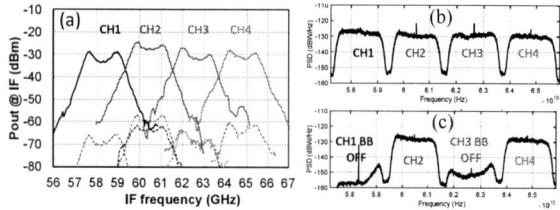

Fig. 3. (a) Single-tone characterization of the BB-to-IF TX. The main responses for each BB channel input are shown in solid lines and the leakage to adjacent channels due to LO spurs in dashed lines. (b) Typical IF output signal for modulated BB inputs, when all BB inputs are on and (c) when only CH2 and CH4 BB inputs are on.

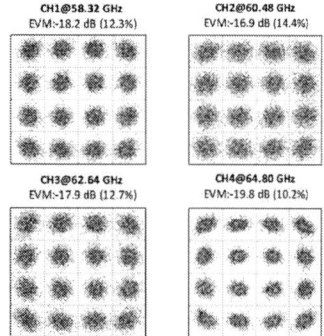

Fig. 4. IF output constellations for 16-QAM, 1.8 Gbauds BB inputs. The IF signal power is -24 dBm.

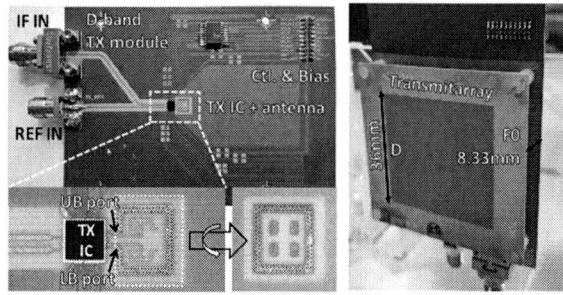

Fig. 5. Pictures of the D-band TX module: details of the board including TX IC and feed antennas, and of the full system with flat lens antenna.

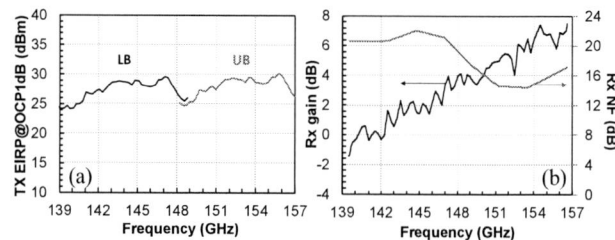

Fig. 6. (a) EIRP at the 1-dB compression point of the D-band TX module. (b) Measured gain and noise figure of the commercial Rx.

III. D-BAND LINK AND OVER-THE-AIR TESTS

A point-to-point link is implemented using a commercial receiver (RX) with 4 dB of average gain and 8 dB and noise figure (NF), as shown in Fig. 6b, equipped with a 20-dBi horn antenna. The maximum link distance is calculated taking into account the 1dBCP EIRP than can be radiated by the TX and a back-off of 10 dB. Such a large back-off is necessary for multi-channel modulated signals, since the combined time-domain multi-channel signal has larger peak-to-average ratio than the individual channel signals. The target BER is 10^{-2}, considering that acceptable packet error rates can be achieved by applying a suitable forward error correction (FEC) code [9]. Table 1 shows the link budget, resulting in an RX SNR of 21 dB for 42-cm link range. Some margin is left for other contributions to the error vector magnitude (EVM) that should be of -14 dB for the target BER and 16 QAM modulation.

The TX and RX were mounted on plastic carriers and connected to laboratory equipment in order to implement a full BB-to-BB link, as shown in Fig. 7. The digital baseband (DBB) section of the TX is implemented using a multi-channel arbitrary waveform generator (AWG). Each of the four BB I and Q differential input signals is generated using a sampling frequency of 2.5 GS/s. The baud rate is 1.8 Gbauds and a 0.18 Root Raised Cosine digital filter is used at the last stage of the DBB TX. Each BB channel signal is organized in a frame with 50 symbols, each one composed of 512 samples (that corresponds to the FFT length) and preceded by a preamble with QPSK modulation used on the RX DBB for channel,

Table 1. Link budget analysis.

Parameter	Per channel	Full band
BW (GHz)	2.16	IF: 2×4×2.16 RF 8×2.16
BB I or Q diff. peak amp. (mV @ 100 Ω)	750	
BB I or Q rms amp. (mV @ 100 Ω)	270	
BB to IF up-converter gain (dB)	-29	
IF signal power (dBm)	-30.4	-24.4 (×2)
D-band TX gain (dB)	37	
EIRP@1dBCP (dBm)	17.4	26.4
Back-off (dB)	10	
EIRP for IF power (dBm)	7.4	16.4
Path Loss @ 42 cm (dB)	68.4	
Rx antenna gain (dBi)	20	
Rx NF (dB)	18	
Rx Signal @ antenna aperture (dBm)	-41	-32
Rx Noise @ antenna aperture (dBm)	-62	-53
SNR	21	
Margin for BER 10^{-2} on 16-QAM (dB)	7	

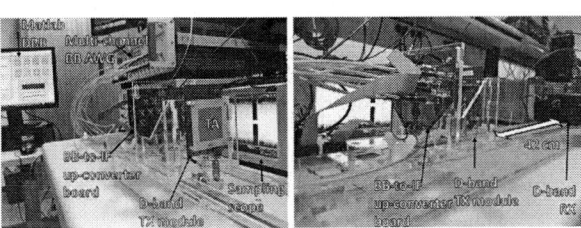

Fig. 7. Point-to-point D-band link setup.

carrier frequency offset (CFO), and I/Q imbalance estimation and correction. The payload includes BPSK modulated scattered pilots (not shown in the constellation results) used for slow phase variations tracking at the RX DBB. The two channel-bonding TX boards were connected using V-band connectors and a waveguide to minimize the losses.

The RX, which consists in a wideband down-converter mixer, is placed at 42 cm from the flat lens of the TX. The LO used for the RX and the reference generator used for the internal LO generators of the TX are not synchronized. The RX provides a down-converted signal from DC to 20 GHz. This signal is sent to a 50 GS/s sampling scope for digitization and, next, read by the RX DBB. Both the TX and RX DBB were implemented in Matlab. The AWG, controlled by the TX DBB, continuously generates a train of frames of 18 µs of duration and spaced by 2 µs, at each BB output. The scope triggers on the rising section of the frames preamble received at the RX and sends the data to the Matlab DBB RX, that performs off-line demodulation and parameters extractions (e.g. signal power, EVM, BER, residual I/Q and CFO mismatch calculations)

Some parameters of the received signals are shown in Fig. 8 and Fig. 9. They correspond to the IF signal shown in Fig. 4. The RX signal power values indicated in the figures are, in average, 4 dB weaker at the RX antenna aperture, after subtracting the RX gain shown in Fig. 6b. The signal power for the different channels varies due to the frequency response of the TX module (see Fig. 6a) and to the frequency-dependent

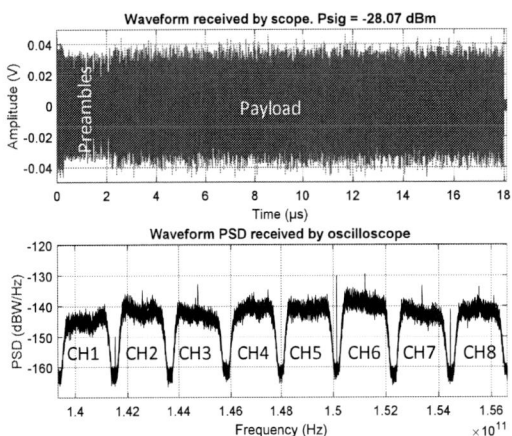

Fig. 8. Over-the-air received signal in time and frequency domain.

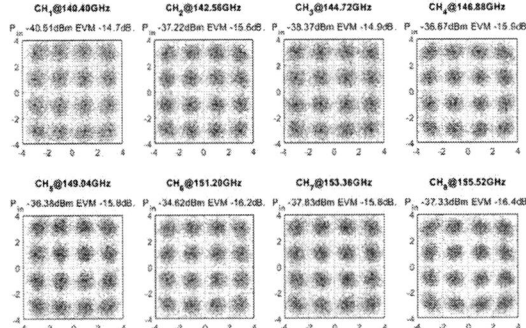

Fig. 9. Received 16-QAM constellation for 1.8 Gbauds.

Table 2. Performance comparison of recent D-band transmitters integrated with antenna systems.

Ref.	[3] TMTT'22	[4] ISSCC'22	[5] JSSC'22	**This Work**
Package Technology	LTCC Interposer	Radio on PCB	Radio on PCB	**Radio on PCB**
ICs Technology	22nm FDSOI + 0.25 InP HBT	130nm SiGe BiCMOS	45nm RFSOI	**45nm RFSOI**
TX architecture	Single channel BB to D-band direct conversion External LO@15 GHz	2×2 Channels RF beamformer	2x4 RF Front-end + IF Beamformer External LO@22 GHz	**BB-to-IF multi-channel up-converter + channel bonding RF Front-end with integrated LO generation**
Antenna system and gain (if available)	Antenna on LTCC (8 linear patch arrays) 12 dBi	Antenna on Glass (2×2 patch array)	Antenna on with on-chip feed (2×4 patch array)	**Antenna on PCB (2x2 patch)+ Flat lens on PCB 25 dBi**
Frequency (GHz)	131-137	130-150,150-165	136-147	**140-158**
EIRP@1dBCP*/Psat** (dBm)	27.5**	18**	17.5*	**26.4***
#Channels × Ch. BW (GHz)	1 × 6	1 × 6	1 × 3	**8 × 2.6**
Tx Pdc (mW)	760	1060	1848	**1580**
Peak DR (Gb/s)/Modulation	30 (64-QAM)	30 (64-QAM)	10.5 (64-QAM)	**57.6 (16-QAM)**
EIRP @ Peak DR (dBm)	21	8	10.5	**16.4**
Link distance (cm)	15	NA	65	**42**
TX energy efficiency (pJ/bit)	25.4	35.3	176	**27.4**

gain of the RX. The BB-to-IF controllable gain is set differently for each channel in the first TX IC in order to obtain a uniform channel power at IF. However, there is no gain control per channel on the D-band TX IC that could allow one to compensate the frequency-dependent losses of the D-band passive components. Nevertheless the RX DBB measured channel power varies less than ±2 dB among channels, with the exception of CH1 that is affected more severely by the drop of the RX downconverter gain. The EVM in all channels is better than -14 dB. For the signals in Fig. 9, the overall BER measured by the DBB is 6×10^{-3}, in line with the link budget analysis in Table 1. Some LO peaks can be observed at the center of the channels (as in Fig. 4). They are due to residual DC-offset on the I and Q signals of the BB-to-IF up-converter lanes that lead to LO leakage. The large spur between CH5 and CH6 is an artifact of the commercial down-converter RX mixer.

IV. DISCUSSION AND CONCLUSION

The presented TX chipset and module is compared with other recent D-band transmitters and links in Table 2. In [3], an interesting combination of CMOS TX with external InP PA is proposed to increase the output power. Its energy efficiency is comparable to that achieved in this work but it operates on a narrower band (6 GHz). The phased array modules reported in [4] and [5] also achieve a relatively narrowband performance and their energy consumption is significantly larger. The TX proposed in this work attains the largest data-rate among state-of-the-art D-band modules, as well as the lowest TX energy efficiency of 27.4 pJ/b including the LO generation, conversely to [3], thanks to the high-gain and low-cost flat lens antenna system and to the wideband, energy-efficient channel bonding architectures. The TX is demonstrated on a link over 42 cm with 25 dBi antenna gain at TX and 20 dBi horn at the RX. A data rate of 57.6 Gb/s is achieved with multi-channel16-QAM modulation resulting in BER better than 10^{-2}. Furthermore, the proposed channel-bonding architecture enables adaptable transceivers, because some lanes can be switched off, including

the digital signal processing and interfaces, if the peak data-rate is not required. This solution allows one to tune the power consumption with the data-rate and achieve a nearly constant energy efficiency, conversely to the other full-band, single-channel architectures [2]-[5]. The main drawback of the proposed approach is that the fixed-beam TA does not allows for self-beam-alignment. This limitation can be overcome by using electronically reconfigurable flat lenses such as [10].

REFERENCES

[1] A. Dascurcu et al., "A 60GHz phased array transceiver chipset in 45nm RF SOI featuring channel aggregation using HRM-based frequency interleaving," in *Proc. IEEE Radio Freq. Integr. Circuits Symp. (RFIC)*, Jun. 2022, pp. 67-70.

[2] S. Callender et al., "A fully integrated 160-Gb/s D-Band transmitter achieving 1.1-pJ/b efficiency in 22-nm FinFET," *IEEE J. Solid-State Circuits*, vol. 57, no. 12, pp. 3582-3598, Dec. 2022.

[3] A. A. Farid et al., "A fully packaged 135-GHz multiuser MIMO transmitter array tile for wireless communications," *IEEE Trans. Microw. Theory Techn.*, vol. 70, no. 7, pp. 3396-3405, July 2022.

[4] M. Elkhouly et al., "Fully integrated 2D scalable TX/RX chipset for D-band phased-array-on-glass modules," in *IEEE Int. Solid-State Circuits Conf. (ISSCC) Dig. Tech. Papers*, 2022, pp. 76-78.

[5] S. Li et al., "An eight-element 136–147 GHz wafer-scale phased-array transmitter with 32 dBm peak EIRP and >16 Gbps 16QAM and 64QAM operation," *IEEE J. Solid-State Circuits*, vol. 57, no. 6, pp. 1635-1648, June 2022.

[6] A. Hamani et al., "A 56.32 Gb/s 16-QAM D-band wireless link using RX-TX systems-in-package with integrated multi-LO generators in 45nm RFSOI," in *Proc. IEEE Radio Freq. Integr. Circuits Symp. (RFIC)*, Jun. 2022, pp. 75-78.

[7] A. Siligaris et al., "A multichannel programmable high order frequency multiplier for channel bonding and full duplex transceivers at 60 GHz band," in *Proc. IEEE Radio Freq. Integr. Circuits Symp. (RFIC)*, Aug. 2020, pp. 259-262.

[8] J. L. Gonzalez-Jimenez, et al., "A D-band high-gain antenna module combining an in-package active feed and a flat discrete lens," in *Proc. 52nd Eur. Microw. Conf. (EuMC)*, Sept. 2022, pp. 784-787.

[9] T. T. Nguyen-Ly, et al., "Analysis and design of cost-effective, high-throughput LDPC decoders," *IEEE Trans. Very Large Scale Integr. (VLSI) Syst.*, vol. 26, no. 3, pp. 508-521, Mar. 2018.

[10] S. Venkatesh et al., "A high-speed programmable and scalable terahertz holographic metasurface based on tiled CMOS chips," *Nat. Electron.*, vol. 3, pp. 785–793, 2020.

979-8-3503-2123-4/23 $31.00 © 2023 IEEE

RMo2C-4

A mm-sized implantable glucose sensor using a fluorescent hydrogel

Hyeonkeon Lee[$1], Honghyeon Park[*2], Taein Kim[#3], Mi Song Nam[^4], Yun Jung Heo[^5], Sanghoek Kim[#6*]

[$]LIG Nex1 Co., Ltd, South Korea
[*]Silicon Mitus Inc, South Korea
[#]Electronics and Information Convergence Engineering, Kyung Hee University, South Korea
[^]Mechanical Engineering, Kyung Hee University, South Korea
{[1]hkeon2, [3]kti0105, [4]misong333, [5]yunjheo, [6]sanghoek}@khu.ac.kr, [2]honghyeon47@gmail.com

Abstract — **This work proposes a millimeter-sized implantable glucose sensor using a fluorescent hydrogel based on boronic acids. The readout system is configured to provide an efficient wireless power transfer to the implantable sensor at 920-MHz frequency. The 675×959-μm^2 die fabricated in 180-nm CMOS consists of a rectifier, a bandgap reference, two regulators, a trans-impedance amplifier, two voltage-controlled ring oscillators, and a digital control with the total power consumption of 150 μW. The die is assembled with a 2.2×2.5-mm^2 antenna on PCB, an off-chip decoupling capacitor, an LED, and a photodiode. When the implantable sensor is surrounded by the fluorescent hydrogel and powered by the reader, the frequency of backscattered signals is encoded with the light intensity of the fluorescent glucose sensor, informing the glucose concentration. The miniaturization of the sensor minimizes the invasiveness, the potential discomfort, and immunity reaction against sensor, making it suitable for a long-term use after the implantation.**

Keywords — **implantable glucose sensor, fluorescence hydrogel, backscattered modulation, load-shift keying**

I. INTRODUCTION

Blood glucose monitoring is essential for patients with diseases such as diabetes. The blood glucose measurement by drawing blood, which is the conventional standard measurement, is not appropriate for continuous monitoring due to the induced pain by finger pricking and the cost of expensive test strips. Much research has been done to enable a non-invasive glucose measurement that overcomes such drawbacks of blood collection. Some of the previously reported works are a smart lens that checks the glucose concentration in tears [1], [2] and a method that exploits the correlation between the tissue permittivity and the glucose level [3]. However, those non-invasive methods suffer the issue of inaccuracy or insufficient selectivity yet [4].

In contrast, blood glucose measurement using a fluorescent hydrogel sensor exploits the reversible chemical reaction of boronic acids with glucose [5]. The fluorescence intensity scattered by the hydrogel inserted in the body changes selectively to the glucose level, and hence, by measuring the scattered light intensity, one can deduce the glucose concentration with high precision. It also shows high stability and bio-compatibility within the body, demonstrating its feasibility for the long-term, accurate continuous glucose monitoring. The opacity of skin for a light, however, significantly attenuates the fluorescence intensity after the penetration, making the system susceptible to the ambient light noise. To circumvent the problem, implantable sensors that also include the electronic components such as the light emitter and the light detector were proposed [4], [6]. As the consequence of system integration, however, those systems used to be bulky for implantation. For example, while a recent research presented a fully-implantable, wireless glucose sensor [6], the system necessitates several power-hungry components such as an ADC and a nonvolatile memory, ending up with the entire device size as large as 18.3 mm in length.

This work proposes a miniaturized wireless glucose sensor to minimize the invasiveness, the potential discomfort, and the immunity reaction against the sensor. This could be achieved by the following design choices; (i) we removed the bulky battery and instead delivered the power by wireless power transfer (WPT); (ii) the analog backscattered modulation scheme was adopted as in [4], which could eliminate the aforementioned auxiliary components; (iii) the circuit system including a rectifier, regulators, a trans-impedance amplifier (TIA), a voltage-controlled ring oscillator, and a digital control was custom-designed in integrated circuits. Especially, the digital control is used to manage the duty cycle of the power-hungry LED component, which cuts down the total power budget to 150 μW and reduces the entire implantable sensor size to 2.2×2.5-mm^2.

II. SYSTEM DESCRIPTION

In this section, we present the structure of the overall system and discussing the design choices made in the system implementation.

A. Overall System Architecture

The continuous glucose monitoring (CGM) system we present is to be implanted into the body. A schematic diagram of the system is shown in Fig. 1. To minimize the device size, the battery is removed from the device and power to operate the sensor system is wirelessly transferred from the external device, and blood glucose information is read through backscatter communication. Since the implantable device is miniaturized as a millimeter scale, power delivery through the weak wireless link under the safety regulation is a challenging task. For such a small implantable device, it is known that the optimal frequency of the operation lies at the near-GHz range. Based on the circulating pattern of the current distribution

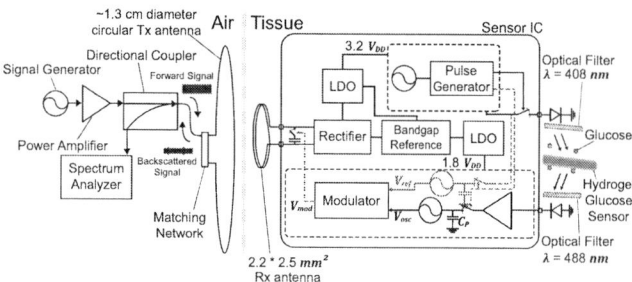

Fig. 1. Overall schematic diagram of the implantable glucose sensor system including the external reader

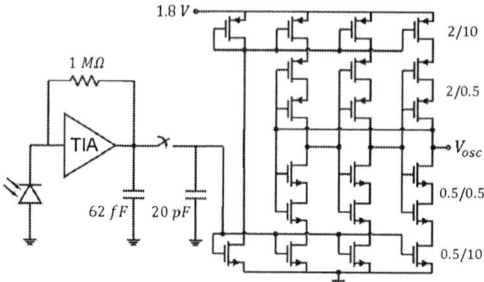

Fig. 2. Light-to-frequency converting circuit consisting of the PD, the TIA, and the current-starved ring-oscillator

that maximizes the power delivery for a given power loss in tissue [7], the transmit source was implemented with a loop antenna (8.54 nH) with a diameter of 1.3 cm. The AC power received by a small 2.2×2.5-mm^2 receiver loop antenna (3.40 nH) is converted into a stable DC power supply through a rectifier and a Low-Drop Out (LDO) regulator to power the entire circuit. The internal circuit of the integrated system is roughly composed of the module that controls the LED (SM0603UV-405) to stimulate the fluorescence sensor and the module that senses the intensity of the light scattered by the fluorescence sensor with a photodiode (PD; MicroFC-10035), amplifies, converts, and modulates it into a frequency. The former module requires 3.2 V for V_{DD} to operate the external LED while the latter does 1.8 V of V_{DD}. The fluorescent hydrogel that responds to glucose concentration is surrounding the LED and the PD. To prevent a direct coupling from the LED to the PD, optical filters are externally assembled with the system. Accounting for the wavelength difference between the incident and scattered light, the bandpass filter (ET405/40x) and the long-pass filter (450-nm OD2 ultra-thin long-pass filter) were applied in front of the LED and the PD, respectively.

Since fluorescence-type sensors generally suffer the degradation of function by bleaching if they are exposed to light for a long period of time [5], the proposed system turns on the LED and stimulates the fluorescence sensor for a short period of pulses (1% duty cycle). As a result, one can not only extend the life of the fluorescence sensor, but also save the power consumption of the system greatly. Since the LED is pulsed, the PD sensing module also uses a switch synchronized with the LED excitation to store glucose information in the capacitor C_p. In turn, the voltage stored in the capacitor controls the frequency of the current-starved ring oscillator (CSRO).

The modulator shapes the oscillating signal of the CSRO into a pulse signal for the backscattered modulation. The modulator switches the load attached to the receiver antenna with the period corresponding to the oscillating frequency of the CSRO. As the load impedance of the receiver antenna is distorted from the matching state periodically, the magnitude of the reflected signal at the transmit antenna also changes with the same period. When the reflected signal is analyzed in the frequency domain by an external spectrum analyzer, one

can observe the Load Shift Keying (LSK) signal on both sides of the carrier frequency. The intensity of the PD reception can be informed by the external reader from the separation of the sideband from the carrier frequency. Since the intensity of PD reception is directly related to the glucose concentration for a given LED luminance intensity, one can extract the glucose concentration wirelessly from the implantable sensor. In the WPT for implantable sensor, there can be an uncertainty in the coupling between the transmit and the receiver coil, which can be inherited to the uncertainty in the V_{DD} of the LED module and the LED luminance intensity. To calibrate out this uncertainty, an auxiliary CSRO (gray-colored CSRO in Fig. 1) is included to account for the luminance intensity. The modulator takes frequencies of both CSROs as inputs and creates the pulse signal of which the frequency corresponds to the intensity of the PD reception and the duty cycle does to the LED luminance intensity. By inspecting the V_{mod}, one can potentially deduce the glucose concentration more precisely.

B. Rectifier and Regulator

As an RF-DC converter, we employed the cross-coupled rectifier for high efficiency [8]. The number of stages for the cross-coupled configuration was chosen to be four to yield the highest efficiency of 78% at the output voltage of 3.3 V in simulation. The rectified power is stored in a large external capacitor (220 nF). To further stabilize the DC voltage and utilize multiple power supplies (1.8 V and 3.2 V), regulators are required. The LDO regulator is chosen because it is suitable to create a stable DC voltage for a low-voltage and low-power environment. The regulator is designed to have the phase margin of $60°$ at the frequency (8.2 MHz for 1.8 V_{DD} and 19.3 kHz for 3.2 V_{DD}) that corresponds to the maximum rate of the change of the current consumption in the circuit. Additionally, a limiter diode is placed between the rectifier and the regulator to protect the internal circuitry from the overvoltage.

C. Trans-impedance Amplifier and Current-starved Ring-oscillator

The induced current at the PD by the scattered light from the fluorescence hydrogel is encoded into the frequency through the TIA and CSRO shown in Fig. 2. The TIA first converts the current induced by the PD into the voltage across a

979-8-3503-2123-4/23 $31.00 © 2023 IEEE

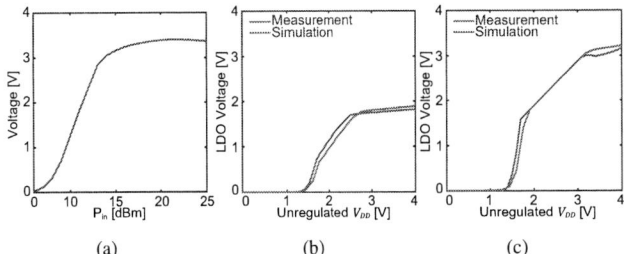

(a)

(b)

Fig. 3. Digital block diagrams for **(a)** a pulse generator to control the LED and the sampling switch, and **(b)** the frequency modulator.

capacitor, which is sampled by the switch that is synchronized with the LED-turn-on signal. The sampled voltage at the capacitor controls the amount of the current mirror source I_d for the CSRO. Depending on I_d, the oscillating frequency of the ring oscillator changes as

$$f_{osc} = \frac{I_d}{N \cdot V_{DD} \cdot C_{total}}, \tag{1}$$

where N is the number of stages, V_{DD} is the supply voltage and C_{total} is capacitance at the output of each stage [9]. Since I_d depends on the fluorescence intensity, the oscillating frequency f_{osc} infers the glucose concentration.

D. Pulse Generator and Modulator

Fig. 3(a) shows the pulse generator which generates the pulse signal (W_PULSE) to excite the LED and the signal (C_PULSE) to control the sampling switch between the TIA and CSRO. The pulse signal to turn on the LED has the duty cycle of 1% to reduce the system power consumption and the bleaching of the fluorescence sensor. The pulse signal for the sampling switch is synchronized with the LED excitation with a small offset. That is, to avoid the noise caused by overshoot at the moment when the LED turns on/off and instead captures the settled signal, the sampling signal goes high in the middle of the LED signal with a shorter duty cycle (0.5%).

Lastly, the modulator in Fig. 3(b) is necessary to convert the oscillating voltage out of the CSRO to a pulse signal with the same period, but a shorter duty cycle. This pulse signal is used to create LSK for the backscattered modulation. Because the turn-on of the pulse perturbs the power reception of the sensor, the duty cycle is nominally set to be short as 10%. Our design allowed the duty cycle to slightly deviate from its nominal value depending on the period of V_{ref} that represents the LED luminance information. In other words, the signal for the backscattered modulation contains the information of the PD reception in its period as well as the information of the LED luminance in its duty cycle, from which the glucose concentration can be more precisely deduced.

Fig. 4. The measurement results of **(a)** wireless RF power to DC rectifier with limiter and **(b)** LDO for 1.8 V sensing section and **(c)** for LED Vdd.

III. MEASUREMENT RESULTS

In this section, we show the performance of the fabricated chip, demonstrating the feasibility of the system as the wirelessly-powered implantable glucose sensor.

A. Rectifier and Regulator

Fig. 4 shows the measurement results for the rectifier and regulator. Fig. 4(a) plots the measured rectifier output voltage depending on the input power into the transmit antenna placed about 1 mm above the receiver loop. We can observe that the unregulated voltage saturates at ~3.3 V by the operation of the rectifier and the limiter when it receives power wirelessly. Fig. 4(b) is the plot comparing measured and simulated LDO operation with the unregulated voltage as input. The LDO for the sensing part regulates to the voltage of 1.8 V, while the LDO for the LED excitation regulates to 3.2 V. One can observe that a stable voltage is generated for the unregulated voltage of 3.3 V. This confirms that the system should be supplied with a stable V_{DD} and be tolerant to the variations of the distance and the orientation in WPT to some extent.

B. Light Responsiveness with Wired Powering

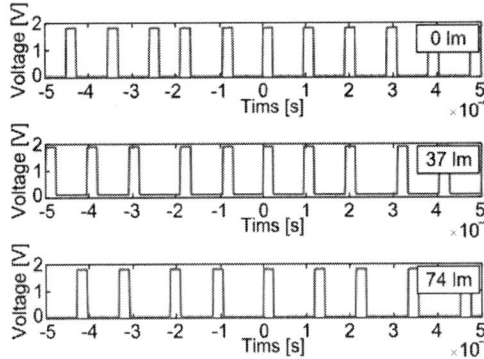

Fig. 5. The measurement results of pulses V_{mod} generated by the modulator depending on the incident light intensity.

Fig. 5 shows the oscilloscope measurement of the modulator output V_{mod} as the function of light intensity. For the experiment, the PD and the die chip were soldered on the PCB, while the power was directly supplied from the external DC power supply. The light intensity was adjusted in three different levels at the distance of 10 cm

979-8-3503-2123-4/23 $31.00 © 2023 IEEE

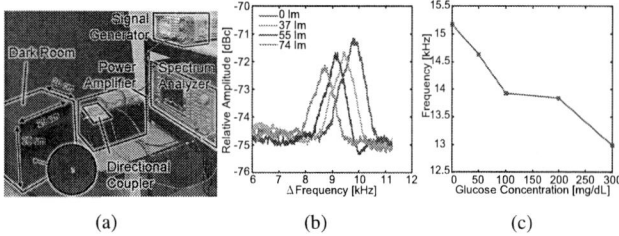

| (a) | (b) | (c) |

Fig. 6. (**a**) Electronic system test setup (**b**) Electronic system test result with light intensity change; (**c**) Measured results of backscattered frequency versus glucose concentration

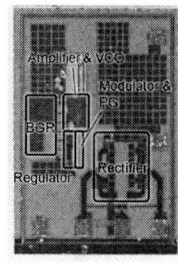 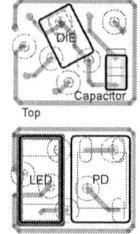 

| (a) | (b) | (c) |

Fig. 7. (**a**) Photograph of a die and (**b**) layout of PCB for external capacitor, PD, LED and antenna and (**c**) fabricated wireless glucose monitoring sensor.

above the PD. We can visually verify that the brighter the light is, the farther the interval between pulses are placed. Since the fluorescence intensity of the fluorescence sensor increases as the glucose concentration increases, the frequency of backscattering decreases as the glucose concentration increases.

C. Sensor Responsiveness with Wireless Powering

The test setup to check the functionality of the electronic system of the sensor is shown in Fig. 6(a). The -6-dBm RF signal generated by the signal generator passes through a power amplifier with 25-dB gain and goes into the input port of the directional coupler. The thru power of the coupler is fed into the transmit antenna placed in the dark room. Above the transmit antenna, there is a PDMS-coated sensor with the PD assembled, but not the LED, plunged in a glucose solution. The backscattered signal from the sensor is recaptured by the transmit loop and routed toward the coupled port of the coupler and analyzed by the spectrum analyzer. The light source is placed above the dark room, of which the light can penetrate into the room through a small hole made on the ceiling of the room. Without a change in the concentration of the glucose solution, the light intensity is controlled to four different levels. Fig. 6(b) shows the appearance of the backscattered signal in the spectrum analyzer according to the light intensity. The offset frequency Δf from the carrier frequency (920 MHz) clearly shifts depending on the light intensity, which shows the responsiveness of the system to the light intensity.

The experimental setup to test the functionality of the entire sensor system is more or less the same as Fig. 6(a). The only change is that now the sensor is fully assembled including the LED and the optical filters, surrounded by the fluorescence hydrogel. When the glucose concentration changes from 0 to 300 mg/dL, which is the range of interests for medical purpose, the offset of the frequency shifts from 15 kHz to 13 kHz as shown in Fig. 6(c). Therefore, the glucose concentration can be informed from the frequency offset of the backscattered signal which can be wirelessly observed.

IV. CONCLUSIONS

The circuit system is fabricated using TSMC 1P6M 180-nm CMOS process shown in Fig. 7(a). Including the external capacitor, LED, PD, and the loop antenna on PCB shown in Fig. 7(b), the entire system size is merely 2.2×2.5

mm^2. Such a small size of the entire implantable sensor is crucial to make the implantable sensor more stable and bio-compatible in a long term. The glucose concentration can be wirelessly extracted through backscattered modulation while the sensor is wirelessly powered.

ACKNOWLEDGMENT

H. Lee and H. Park contributed equally to this work. Heo and S. Kim are the corresponding authors. The chip fabrication and EDA tool were supported by the IC Design Education Center (IDEC), Korea. This work was supported in part by the Agency for Defense Development, the National Research Foundation of Korea under Grant NRF- 2017R1C1B2009892, and in part by the Ministry of Science and ICT (MSIT), Korea, through the Information Technology Research Center (ITRC) support program under Grant IITP-2021-0-02046 supervised by the Institute for Information & Communications Technology Planning & Evaluation (IITP).

REFERENCES

[1] Y.-T. Liao, H. Yao, A. Lingley, B. Parviz, and B. P. Otis, "A 3-uw cmos glucose sensor for wireless contact-lens tear glucose monitoring," *IEEE Journal of Solid-State Circuits*, vol. 47, no. 1, pp. 335–344, 2011.

[2] D. H. Keum, S.-K. Kim, J. Koo, G.-H. Lee, C. Jeon, J. W. Mok, B. H. Mun, K. J. Lee, E. Kamrani, C.-K. Joo *et al.*, "Wireless smart contact lens for diabetic diagnosis and therapy," *Science Advances*, vol. 6, no. 17, p. eaba3252, 2020.

[3] R. S. Hassan, J. Lee, and S. Kim, "A minimally invasive implantable sensor for continuous wireless glucose monitoring based on a passive resonator," *IEEE Antennas and Wireless Propagation Letters*, vol. 19, no. 1, pp. 124–128, 2019.

[4] H. Lee, J. Lee, H. Park, M. S. Nam, Y. J. Heo, and S. Kim, "Batteryless, miniaturized implantable glucose sensor using a fluorescent hydrogel," *Sensors*, vol. 21, no. 24, p. 8464, 2021.

[5] Y. J. Heo, H. Shibata, T. Okitsu, T. Kawanishi, and S. Takeuchi, "Long-term in vivo glucose monitoring using fluorescent hydrogel fibers," *Proceedings of the National Academy of Sciences*, vol. 108, no. 33, pp. 13 399–13 403, 2011.

[6] A. DeHennis, S. Getzlaff, D. Grice, and M. Mailand, "An nfc-enabled cmos ic for a wireless fully implantable glucose sensor," *IEEE journal of biomedical and health informatics*, vol. 20, no. 1, pp. 18–28, 2015.

[7] S. Kim, J. S. Ho, and A. S. Y. Poon, "Midfield wireless powering of subwavelength autonomous devices," *Physical review letters*, vol. 110, no. 20, p. 203905, 2013.

[8] K. Kotani, A. Sasaki, and T. Ito, "High-efficiency differential-drive cmos rectifier for uhf rfids," *IEEE Journal of Solid-State Circuits*, vol. 44, no. 11, pp. 3011–3018, 2009.

[9] S. Suman, K. G. Sharma, and P. K. Ghosh, "Analysis and design of current starved ring vco," in *2016 International Conference on Electrical, Electronics, and Optimization Techniques (ICEEOT)*, 2016, pp. 3222–3227.

RMo3A-1

A 14.2 mW 29–39.3-GHz Two-Stage PLL with a Current-Reuse Coupled Mixer Phase Detector

Yuan Liang [#1], Chirn Chye Boon [$2], Qian Chen [$3]

[#] Guangzhou University, China

[$] Nanyang Technological University, Singapore

[1]yliang017@e.ntu.edu.sg, [2]eccboon@ntu.edu.sg, [3]e170029@e.ntu.edu.sg

Abstract—A *Ka*-band millimeter wave (mmW) integer-*N* phase-locked loop (PLL) exploiting a novel current-reuse coupled mixer (CRCM) phase detector (PD) is proposed. Aiming to attenuate the reference spurs in the PLL, the CRCM PD is realized by a pair of coupled mixers folded to each other, achieving mutual spur compensation without consuming extra power or narrowing the PLL loop bandwidth. A mmW signal source is constructed by a two-stage PLL followed by a frequency tripler. Realized in a 28 nm CMOS process, the signal source attains a locking range of 29–39.3-GHz, maximum reference spur of −73.7 dBc, and 160.6 fs$_{rms}$ integrated jitter (integrated from 1 k to 100 MHz). It consumes 14.2 mW power and occupies an active area of only 0.1 mm², achieving a figure of merit (FoM) of −244.4 dB when using a 150-MHz reference.

Keywords—CMOS, millimeter wave, phase-locked loop (PLL), reference spur, signal source, spur reduction, phase detector.

I. INTRODUCTION

CMOS phase-locked loops (PLLs) continue to play a critical role in existing or future 5G/6G wireless transceivers. Standard metrics to evaluate PLL performance involves 1) integrated jitter, 2) reference spur, 3) frequency tuning range (FTR), 4) power dissipation, and 5) area. Owing to the poor-quality-factor of inductor coils on chip, attenuating the voltage-controlled oscillator (VCO) phase noise is commonly realized by extending the PLL loop bandwidth (f_{BW}). However, doing so degrades the level of spur rejection [1]. Exploiting a high-frequency (e.g., > 1 GHz) reference and lowering the PLL division ratio N is attractive to attain better noise and spur performance [2], [3], but commercial crystal frequencies are normally < 200 MHz.

As one of the most critical parts in a PLL, phase detector (PD) determines the PLL in-band noise and spur performance. Conventional mmW PLLs employ PFD/CP [3]–[5], or mixer PD [6], [7]. PFD/CP has unlimited acquisition range, but its PD gain (i.e., $I_{cp}/2\pi$) is small, leading to large input-referred phase noise. Mixer PD has very low intrinsic noise, and is therefore suitable for low-jitter mmW PLLs. Yet, it generates large reference spur at the PLL output, with limited acquisition range that commonly necessitates a frequency-tracking loop (FTL) to assist the locking. Using harmonic traps attenuates the reference spurs [8], [9], but the rejection level is limited to 14 dB [8], and it introduces extra poles that degrades the PLL loop stability [9]. Moreover, external calibration is mandatory. Sub-sampling PLLs [10]–[16] and injection-locked PLL [17] are attractive for low-jitter performance, but their reference spur are ∼−50 dBc.

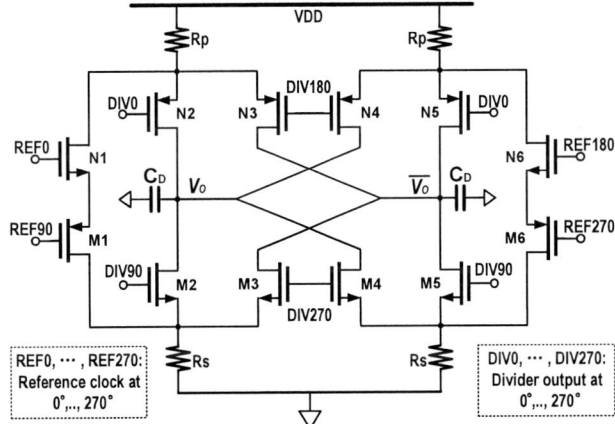

Fig. 1. Proposed CRCM PD for spur compensation.

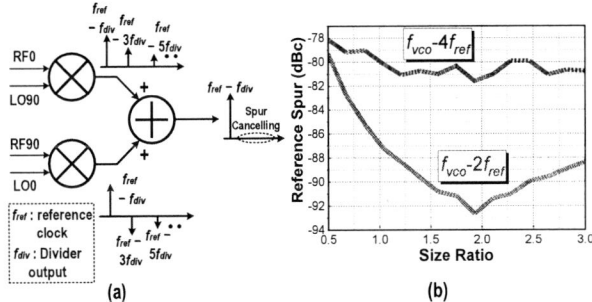

Fig. 2. (a) Equivalent circuit and phasor diagram of CRCM PD, and (b) simulated signal source reference spur against the size ratio.

This paper introduces a novel current-reuse coupled mixer (CRCM) PD that is dedicated to overcoming the trade-off between reference spur suppression and phase noise. With such a PD, a low-reference-spur and low-noise signal source is proposed. Section II details the operating principle of the PD, and a *Ka*-band mmW signal source is briefed in Section III. Section IV provides experiment results. This paper is summarized in Section V.

II. CRCM PD AND MMW SIGNAL SOURCE

Fig.1 sketches the proposed CRCM PD. PMOS N1–N6 and R_p forms one mixer, whereas NMOS M1–M6 and R_s forms the other one. Therefore, the topology is essentially featured by two mixers folded and coupled to each other. If only one branch exists, either NMOS or PMOS branch, the PD would degenerate to a conventional mixer PD with a *RC* load. The folded topology ensures the current conducted by one

979-8-3503-2123-4/23 $31.00 © 2023 IEEE   105   2023 IEEE Radio Frequency Integrated Circuits Symposium

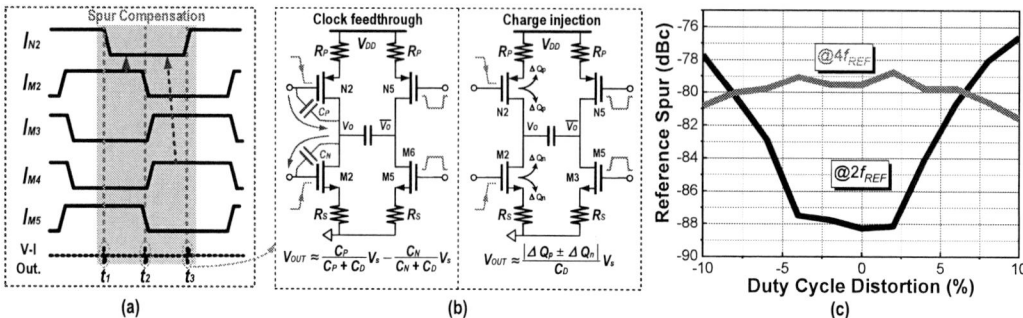

Fig. 3. Spur compensation, (b) compensation of clock feedthrough and charge injection, and (c) reference spur against duty cycle distortion.

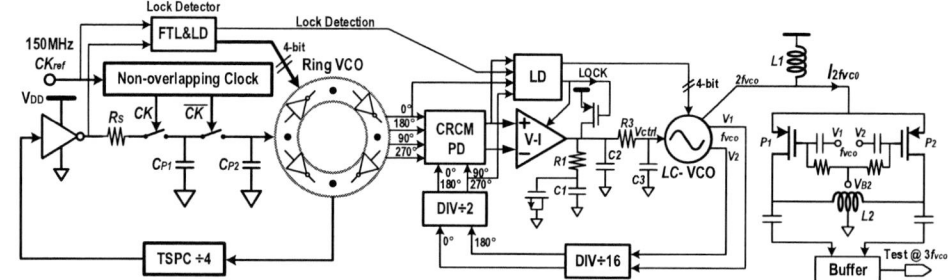

Fig. 4. Complete architecture of the proposed mmW signal source composed of a two-stage PLL and an ×3 up-conversion mixer.

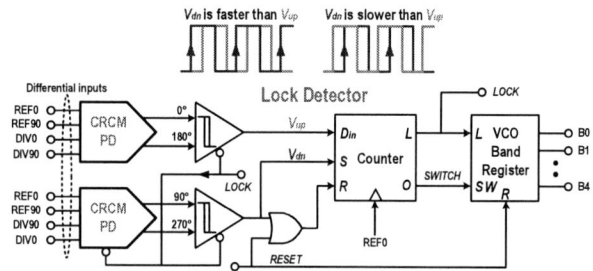

Fig. 5. Lock detector (LD) realized by dual CRCM PDs.

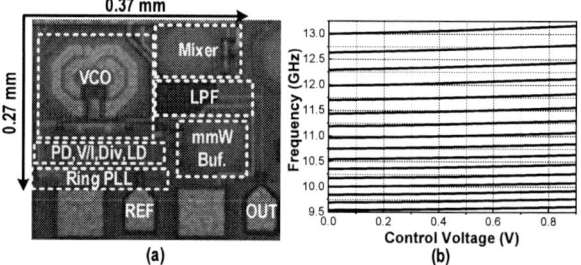

Fig. 6. (a) Chip micrograph, and (b) measured LC-VCO FTR.

branch is reused by the others, consuming the power by half. Unlike conventional mixer PD where the reference signal is small-signal sinusoidal wave whose amplitude affects the PD gain [6], all inputs here are periodic square pulses. The result is a rich of harmonics located at $nf_{ref} \pm mf_{div}$ generated at the PD output, where n and m are integers, perturbing the control line of VCO (V_{ctrl}) and incurring reference spurs. The equivalent circuit and the phasor diagram are provided in Fig. 2(a). With quadrature inputs to the CRCM PD, the spurious tones generated by one mixer is cancelled out by the other, operating akin to the image rejection realized by the quadrature mixer in a homodyne receiver.

The time domain of spur compensation is illustrated in Fig. 3(a). In the equilibrium state, the current out of transistor N_2, which is I_{N2}, is compensated by those from M2 and M4. In detail, I_{N2} is compensated by I_{M2} from t_1 to t_2, whereas it is compensated by I_{M4} from t_2 to t_3. Consequently, the current injected to the load capacitor C_D is nulled in one reference cycle. Moreover, the PD compares the phase two times in half of the reference cycle, meaning that the disturbance rate at the

PD output is four times of f_{ref}, boosting the spur frequency to $4f_{ref}$ when mismatches exist. Herein, it is essential to maintain a similar response to the clock edge by the two branches. Fig. 2(b) provides the simulated reference spur against the size ratio from PMOS branch to NMOS branch. The optimal size ratio is ~1.9, for sufficiently suppressing the spur at $2f_{ref}$, leaving the $4f_{ref}$ term to dominate, which is attenuated by the loop filter (LF).

The proposed PD can also mutually neutralize the charge injection and clock feedthrough, as illustrated in Fig. 3(b), when both branches are matched. Monte Carlo statistics implies a standard variation of 8.8 dB and 1.4 dB, respectively, for reference spur at $2f_{ref}$ and $4f_{ref}$. However, the spur compensation is degraded by the reference clock duty cycle distortion (DCD), as shown in Fig. 3(c), due to the generation of pulses running at $2f_{ref}$ that cannot be cancelled. By keeping DCD < 7.5%, the spur at $2f_{ref}$ is smaller than that at $4f_{ref}$.

III. K_A-BAND MMW SIGNAL SOURCE

The signal source is sketched in Fig. 4. It is composed of a two-stages cascaded PLL and a frequency tripler featured by

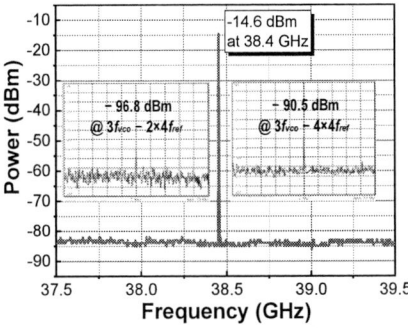

Fig. 7. Reference spur measured at 38.4 GHz (f_{ref} = 150 MHz).

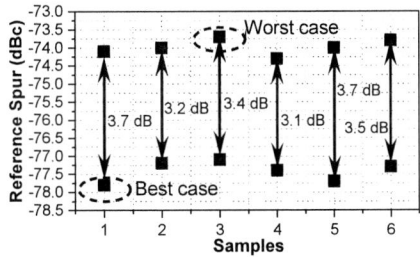

Fig. 8. Collected measured reference spurs for 6 samples.

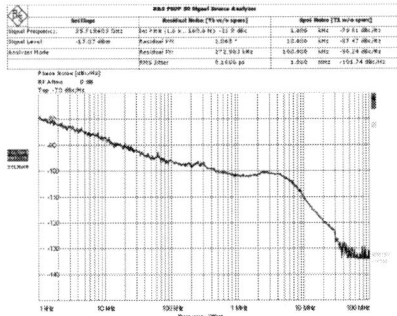

Fig. 9. Signal source phase noise measured at 35.7 GHz

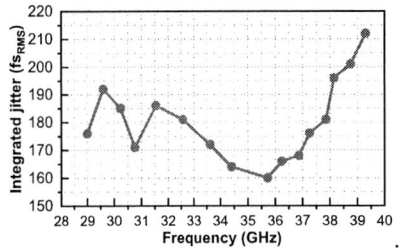

Fig. 10. Collected phase noise measured over the entire FTR.

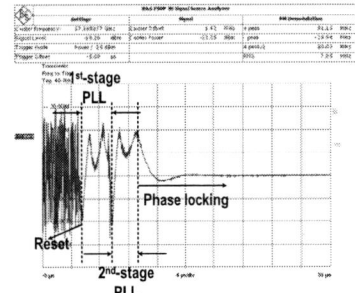

Fig. 11. Measured PLL locking behavior.

an up-conversion mixer. The first stage PLL is a ring-based PLL which generates quadrature clock signal for the subsequent CRCM PD. The sample-and-hold switch has a steep $V/\Delta\Phi$ transfer curve, implying a large PD gain that sufficiently attenuates its input-referred noise. FTL and lock detector (LD) are used for both PLLs for frequency tracking and VCO automatic band searching.

As the reference spur suppression relies on mutual spur compensation rather than heavy filtering like [6] and [7], the PLL needs not use a low corner RC-type LF, which sacrifices the acquisition range [9]. Simulation reveals that the PD output can sweep all the possible value of V_{ctrl} and without the need for an FTL. A digital LD based on dual CRCM PD is introduced in the 2nd-stage PLL, as shown in Fig. 5, operating like a Pottbacker frequency detector by using one PD output (V_{dn}) to sample the other (V_{up}). The counter measures the length of the beating signal (i.e., $|f_{ref}-f_{div}|$) and checks if $N\times f_{ref}$ is close enough to f_{vco}, then hands over the loop to CRCM PD for phase alignment with spur compensation.

The frequency tripler is formed by PMOS $P1$ and $P2$, where their inputs are from LC-VCO large-signal output (at f_{vco}) and the center tap of the VCO tank coil (at $2f_{vco}$). Coils $L1$ and $L2$

help to boost the amplitude of the final output at $3f_{vco}$. With such a configuration, the effective division ratio of the two-stage cascaded PLLs is reduced, leading to better noise performance and smaller LF size.

IV. EXPERIMENTAL RESULTS

The proposed Ka-band signal source was implemented in a 28 nm bulk CMOS process. The chip micrograph is shown in Fig. 6(a), and the complete design occupies only ~0.1 mm² active area. The ring-based PLL, 2nd PLL and the frequency tripler consumes 3.1 mW, 7.4 mW, and 3.7 mW, respectively, with a 1-V power supply. The mmW buffer burns 16 mW for testing. Fig. 6(b) shows the measured VCO tuning curve, with a measured FTR from 28.9 to 39.3 GHz. Fig. 7 provides the signal source spectrum measured at 38.4 GHz. As the output power of the mmW signal source was weak, the reference spurs can be observed by zooming in the spectrum, using relatively small resolution bandwidths (i.e., 100 Hz). The largest spur of −75.9 dBc was observed at $3f_{vco}-4\times4f_{ref}$, which was ~6 dB larger than that at $3f_{vco}-2\times4f_{ref}$, verifying the spur compensation and that the DCD of ring VCO outputs was within the tolerable margin. Fig. 8 summarizes collected reference spur (at $3f_{vco}-4\times4f_{ref}$) performance measured for the entire working range over 6 samples. The magnitudes of other spurious tones were at least 5.7 dB smaller for all samples.

The phase noise measured at 35.7 GHz is shown in Fig. 9, demonstrating the minimum integrated jitter of 160.6-fs$_{rms}$ (1 kHz to 100 MHz). Integrated jitters lower than 213 fs$_{rms}$ were attained for the entire FTR, as summarized in Fig. 10. The signal source locking behaviour is shown in Fig. 11, confirming the operation of CRCM PD and LD that leads the PLL to lock. Recent mmW signal source performance are summarized in Table I. The proposed signal source attains the lowest reference spur and Spur$_{Norm}$ by occupying the smallest area, and its $\mathcal{L}_{norm}$, FoM, FoM$_{JRP}$, and FoM$_{JIT,N}$ compare favourably with recent arts.

Table 1. Performance comparison of state-of-the-art mmW CMOS synchronized signal sources

	ISSCC'19 [10]	TMTT'21 [11]	JSSC'22 [3]	ISSCC'22 [12]	ISSCC'19 [13]	ISSCC'19 [14]	JSSC'20 [17]	ISSCC'20 [18]	**This work**
PD Type	iSSPD	PFD/CP +SSPD	PFD/CP	SSPD	SSPD	SSPD +ILFM	RSPD + ILFM	CSL	**CRCM +V/I**
Technology	65nm	65nm	22nm SOI	28nm	65nm	65nm	45nm	65nm	**28nm**
f_{ref} (GHz)	0.103	0.1	2.5	0.06	0.1	0.1	0.08	0.25	**0.15**
f_{out} (GHz)	25.4−29.5	40.5	18.1−21.2	13.2−17.3	30.6−34.2	28−31	33.6−36	21.7−26.5	**29−39.3**
PN@1MHz	−109.58	−96.6	−109	−106	−97.4	−106	−94.9	−107	**−101.1**
$\mathcal{L}_{norm}$ (dBc/Hz²)@1MHz	−238	−229	−221	−231	−228	−236	−227	−231	**−231**
Spur (dBc)	−63.2	−42	−43	−57.2	−67.5	−58	−60	−45	**<−73.7**
f_{BW} (MHz)	3	3	1	3	0.3	2	2	8	**6**
Spur$_{Norm}$ (dBc)	−80.7	−51.5	−15	−71.2	−57	−64	−70	−55	**−93.9**
Jitter (fs$_{rms}$)	71	228	82.7	116.5	197.6	76	251	75.9	**160.6**
Integration Range	1kHz−100MHz	10kHz−100MHz	10kHz−100MHz	1kHz−100MHz	30kHz−10MHz	1kHz−30MHz	10kHz−10MHz	10kHz−30MHz	**1kHz −100MHz**
Power (mW)	15.2	8.8	36.5	6.6	35	41.8	20.6	16.5	**14.2**
Area (mm²)	0.24	0.6	0.3	0.24	0.79	0.32	0.41	0.5	**0.1**
FoM (dB)	−251.1	−243.4	−246	−250.5	−238.6	−246.1	−238.9	−250.2	**−244.4**
FoM$_{JRP}$ (dB)	−251	−243.4	−232	−252.7	−238.6	−246.2	−239.8	−246.2	**−242.6**
FoM$_{JIT,N}$ (dB)	−275	−269	−255	−274.3	−264	−270.7	−265	−270	**−268**

$\mathcal{L}_{norm} = \mathcal{L}_{in\text{-}band} - 20\log(N) - 10\log(f_{ref})$ [15]	Spur$_{Norm}$ = Reference Spur $- 20\log(f_{BW}/f_{ref}) - 20\log(N)$
FoM $= 10\log[(\sigma_{rms}/1s)^2 \times (Power/1mW)]$ [1]	FoM$_{JRP} = 10\log[(\sigma_{rms}/1s)^2 \times (Power/1mW) \times (f_{ref}/100MHz)]$ [16]
FoM$_{JIT,N} = 10\log[(\sigma_{rms}/1s)^2 \times (Power/1mW)] + 10\log(f_{ref}/f_{out})$ [14]	

V. CONCLUSION

A novel CRCM PD is introduced for spur compensation in a mmW signal source. It is featured by a pair of mixers folded to each other, achieving spur suppression without extra power or extending the PLL loop bandwidth, and thus disengages the trade-off between VCO phase noise and spur suppression. Implemented in a 28 nm bulk CMOS process, the mmW signal source shows a 29–39.3 GHz locking range, <−73.7 dBc reference spur collected for 6 chips, an integrated jitter of 160.6-fs$_{rms}$ (from 1 kHz to 100 MHz), −244.4 dB FoM, and −268 dB FoM$_{JIT,N}$, while occupying only 0.1 mm² active area.

ACKNOWLEDGMENT

This work is supported by the National Research Foundation, Singapore and Infocomm Media Development Authority under its Future Communications Research & Development Programme.

REFERENCES

[1] C. Ko et al, "A 7-nm FinFET CMOS PLL with 388-fs Jitter and −80-dBc Reference Spur Featuring a Track-and Hold Charge Pump and Automatic Loop Gain Control," IEEE JSSC, vol. 55, no. 4, pp. 1043-1050, Apr. 2020.

[2] S. Ikeda et al, "A -244-dB FOM high-frequency piezoelectric resonator-based cascaded fractional N PLL with sub-ppb-order channel-adjusting technique," IEEE JSSC, vol. 52, no. 4, pp. 1123–1133, Apr. 2017.

[3] S. Kalia et al, "A Sub-100 Fs RMS$_{jitter}$ 20 GHz Fractional-N Analog PLL With a BAW Resonator Based On-Chip 2.5 GHz Reference," IEEE JSSC, vol. 57, no. 5, pp. 1372–1384, May 2022.

[4] S. Shahramian et al, "Design of a dual W- and D-band PLL," IEEE JSSC, vol. 46, no. 5, pp. 1011–1022, May 2011.

[5] K.-H. Tsai and S.-I. Liu, "A 43.7 mW 96 GHz PLL in 65 nm CMOS," ISSCC, Feb. 2009, pp. 276–277.

[6] G. Liu et al "A 64–84 GHz PLL with low phase noise in an 80 GHz SiGe HBT technology," IEEE TMTT, vol. 60, no. 12, pp. 3739–3748, Dec. 2012.

[7] S. Kang et al, "A W-band low-noise PLL with a fundamental VCO in SiGe for millimeter-wave applications," IEEE TMTT, vol. 62, no. 10, pp. 2390–2404, Oct. 2014.

[8] C.-C. Lin, H. Hu, S. Gupta, "Spur Minimization Techniques for Ultra-Low-Power Injection-Locked Transmitters," IEEE TCAS-I, vol. 67, no. 11, pp. 3643–3655, Jul. 2020.

[9] L. Kong and B. Razavi, "A 2.4 GHz 4 mW integer-N inductorless RF synthesizer," IEEE JSSC, vol. 51, no. 3, pp. 626–635, Mar. 2016.

[10] Z. Yang et al, "16.8 A 25.4-to- 29.5GHz 10.2mW Isolated Sub-Sampling PLL Achieving -252.9dB Jitter-Power FoM and −63dBc Reference Spur," ISSCC, Feb. 2019, pp. 270-272.

[11] H. Wang et al, "Low-Power and Low-Noise Millimeter-Wave SSPLL With Subsampling Lock Detector for Automatic Dividerless Frequency Acquisition," IEEE TMTT, vol. 69, no. 1, pp. 469–481, Jan. 2021.

[12] L. Zhang et al., "A 480-Multiplication-Factor 13.2-to-17.3GHz Sub-Sampling PLL Achieving 6.6mW Power and −248.1dB FoM Using a Proportionally Divided Charge Pump," ISSCC, Feb. 2022, pp. 1–3.

[13] L. Grimaldi et al., "A 30GHz digital sub-sampling fractional-N PLL with 198fsrms jitter in 65nm LP CMOS," ISSCC, Feb. 2019, pp. 268–269.

[14] J. Kim et al., "A 76fs$_{rms}$ Jitter and −40dBc Integrated-Phase-Noise 28-to-31GHz Frequency Synthesizer Based on Digital Sub-Sampling PLL Using Optimally Spaced Voltage Comparators and Background Loop-Gain Optimization", ISSCC, Feb. 2019, pp. 258-260.

[15] X. Gao et al., "A low-noise subsampling PLL in which divider noise is eliminated and PD/CP noise is not multiplied by N²," IEEE JSSC, vol. 44, no. 12, pp. 3253–3263, Dec. 2009.

[16] H. Zhang et al., "0.2mW 70fsrms-Jitter Injection-Locked PLL Using De-Sensitized SSPD-Based Injecting-Time Self-Alignment achieving −270dB FoM and -66dBc Reference Spur," VLSIC, Jun. 2019, pp. 38–39.

[17] D. Liao et al, "An mm-Wave Synthesizer With Robust Locking Reference-Sampling PLL and Wide-Range Injection-Locked VCO," IEEE JSSC, vol. 55, no. 3, pp. 536–546, Mar. 2020.

[18] Y. Hu et al., "17.6 A 21.7-to-26.5GHz Charge-Sharing Locking Quadrature PLL with Implicit Digital Frequency-Tracking Loop Achieving 75fs Jitter and −250dB FoM," ISSCC, Feb. 2020, pp. 276-278.

RMo3A-2

A Radiation-Hardened by Design 15-22GHz LC-VCO Charge-Pump PLL Achieving -240dB FoM in 22nm FinFET

David Dolt, Samuel Palermo

Texas A&M University, Analog and Mixed-Signal Center, USA

davidj972@tamu.edu

Abstract— This works presents a 15.0-22.0 GHz radiation hardened PLL for space applications designed in a 22nm FinFET process with radiation hardening techniques for single event upset (SEU) mitigation implemented in all key blocks off the PLL. The performance tradeoffs in the VCO, PFD, and CP with regards to the proposed radiation hardening techniques are analyzed and experimentally verified with the PLL characterization yielding a jitter FoM of -240.89dB at 15GHz with -64dB spur levels and heavy ion testing performed at the Texas A&M Cyclotron validating our robust radiation hardened performance across an LET range from 10 to 70 MeV.cm^2/mg.

Keywords— CMOS, jitter, PLL, radiation effects, single-event upset (SEU), VCO.

I. INTRODUCTION

Current space applications are experiencing an increased demand for high performance radiation hardened phase-locked loop (PLL) systems for both serial link and radio frequency applications. These PLLs must be able to support both low jitter <1ps, good spur performance <-50dBm, and a low radiation cross section to offer a complete solution. Currently there is a gap in radiation hardened PLL capabilities covering the Ku band as there are commercial designs with reasonable hardening and great electrical performance [1], meanwhile, academic work offers exceptional radiation hardening but has poor electrical performance [2]. In this work we present the design, electrical characterization, and single event radiation characterization of radiation hardened by design PLL covering 15-22GHz of tuning range with a jitter FoM of -240.89dB with -64dB spurs and a SEU cross section of 5.94E-7 cm^2 at an LET of 50 MeV.cm^2/mg. In order to achieve these performance metrics, we leverage radiation hardened techniques in all key PLL building blocks with a focus on the VCO due to its sensitivity to single event effects [3]. In the VCO we employ a radiation hardened varactor [4] configuration along with LC tail filtering for both noise and SEU suppression [5,6] for the first time in a radiation-hardened VCO design at our target frequency range of 15-22GHz. Furthermore, we explore radiation hardening techniques for the input buffer [7], divider, and analyze the trade-offs for PFD and charge pump radiation hardening [8] when it comes to its impact on system level noise performance. In carefully analyzing the performance trade-offs of our radiation-hardening techniques, to the author's best knowledge, we achieve a combination of the highest operating

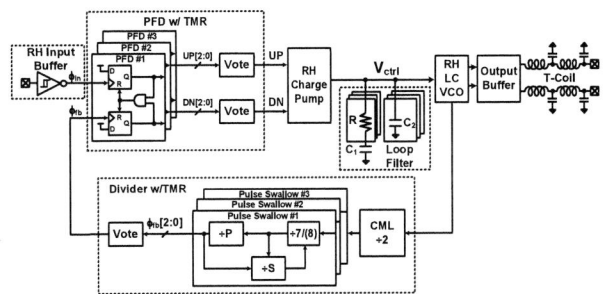

Fig. 1. Radiation hardened PLL system diagram.

frequency and state-of-the-art jitter FoM when compared to all other published radiation-hardened PLL designs.

II. PLL ARCHITECTURE

The proposed PLL architecture in Fig. 1 uses a classic LC-VCO and charge pump architecture with optimized loop parameters to achieve low jitter performance. Moreover, we drive the PLL via a 500MHz (RFPRO33-500.00) low noise crystal oscillator in order to maintain a low reference clock phase noise and to keep the loop divider value N low to reduce in band phase noise multiplication. Our minimum value of N is limited to 30 due to the CML %2 pre-scalar and (7/8) pulse swallow divider architecture. Furthermore, our sinusoidal reference clock is conditioned onto the chip with a large Schmitt trigger buffer which drives a tri-state phase frequency detector (PFD). The PFD drives a programmable charge pump whose current is set between 1 and 2 mA to keep the charge pump contributed phase noise low. The loop bandwidth is set to an optimum point based on the LC-VCO phase noise within a given band which is controlled via a programmable capacitor bank and varactor fine tuning which allows us to synthesize a wide range of output frequencies with continuous tuning from 15 to 22 GHz. To support a wide range of stable operating points and loop bandwidths, we rely on programmable values of R1 which is nominally ~1.6kohm and C2 which is nominally ~1pF to help maintain 60 degrees of phase margin across the PLL operating bands.

III. RADIATION-HARDENED CIRCUIT TECHNIQUES

The PLL reference clock is buffered onto the chip via a Schmitt trigger which operates with an upper and lower hysteresis window set by the device thresholds and the supply voltage. This type of input buffer provides immunity to noise

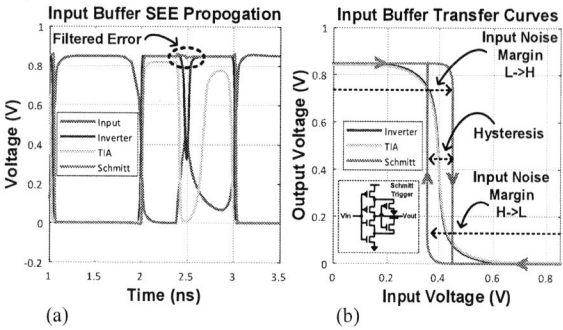

Fig. 2. Input buffer: (a) propagated error; (b) transfer characteristics.

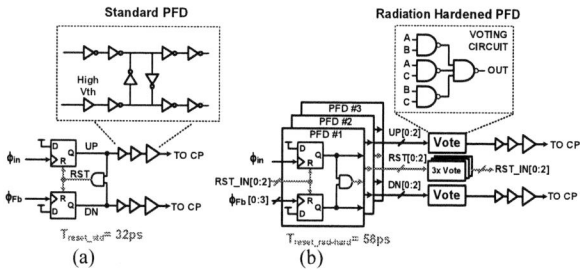

Fig. 3. PFD design: (a) standard; (b) radiation hardened.

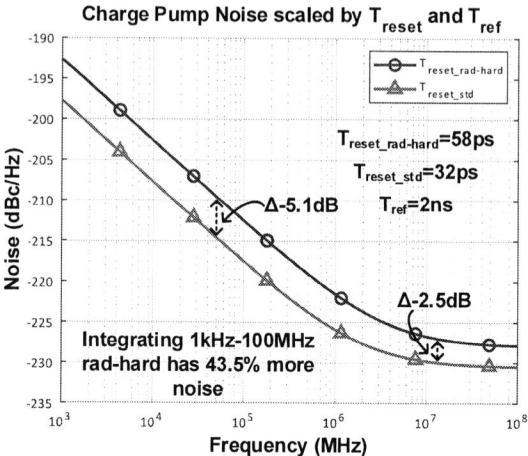

Fig. 4. Charge pump noise scaling from RH PFD.

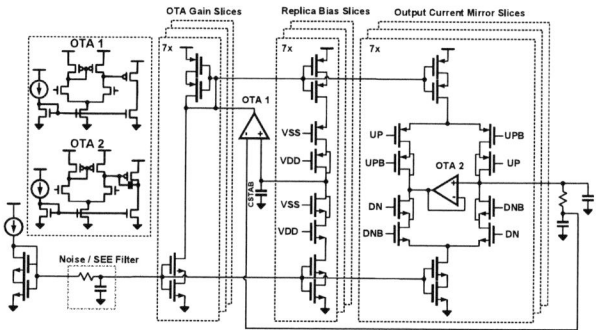

Fig. 5. Charge pump schematic.

fluctuations at the input and which gives it the ability to filter out (SEUs) [8]. Moreover, Fig. 2(a) shows a transient simulation for three common input buffer types when subjected to an SEU at the input. Both the inverter and TIA have their output corrupted by the erroneous input while the Schmitt trigger is able to filter this out. The Schmitt trigger's transfer characteristic is shown in Fig. 2(b) alongside the other two input buffer styles. Overall, the Schmitt trigger input buffer has to be sized large in order to both increase Qcrit which is the amount of charge needed to induce a SEU, and also to maintain low jitter performance when rectifying a single tone clock input into a CMOS signal. When comparing this input buffer to other single ended input buffers in terms of electrical performance, the jitter/mW of this input buffer is equivalent to the two aforementioned input buffer designs implying that there is no performance penalty for using the Schmitt trigger input buffer. A drawback of both the Schmitt trigger and inverter input buffer is that the output duty cycle is sensitive to the input swing and common mode level whereas the ac coupled TIA biases the input buffer at the optimum trip point.

The phase detector used is a NAND phase frequency detector that is triplicated in order to provide triple modular redundancy (TMR). We compare a standard PFD design Fig. 3(a) with a radiation hardened design Fig. 3(b) in terms of the circuit level differences between the two designs. Moreover, the reset path as well as the UP and DN signals are voted on to provide logical masking of bit flips from an induced single event upset. For the reset path, three separate reset voters are used to further increase the masking effort on this critical path. The overhead with radiation hardening in the PFD is an increase in power/area as well as an increase in the T_{reset} time from an unhardened value of 32ps to 58ps due to the voter

logic in the reset path. Furthermore, the implications of this is increased charge pump noise due to the charge pump noise being scaled by $(T_{reset}/T_{ref})^2$ in the flicker noise region and (T_{reset}/T_{ref}) in the thermal noise region. Moreover, Fig. 4 shows the output noise of a 1.5mA charge pump noise power spectral density after it has been scaled by the two different reset times given a Tref of 2ns. Integrating the charge pump noise across frequency from 1kHz to 100MHz we find that the increase in the reset path delay leads to an integrated noise that is 43.5% higher than would be the integrated noise of a standard PFD design. Moreover, this implies a much higher charge pump noise within our loop bandwidth necessitating either extra charge pump current (increasing power) to decrease our in-band noise, or lowering the loop bandwidth and placing a more stringent jitter requirement on the VCO.

The charge pump used in this PLL is a slice-based design where current mirror/switching pairs can be programmed into the output to provide a wide range of output currents Fig. 5. At the input of the charge pump current bias there is a 1.06kohm resistor and a 22pF capacitor which filter noise from the diode connected input pair as well as providing SEU filtering on the input bias current node. In this design, the charge pump output current range is 300uA to 2.8mA with seven programmable slices. With this high charge pump current, we are able to recover from charge accumulation on the loop filter from

979-8-3503-2123-4/23 $31.00 © 2023 IEEE

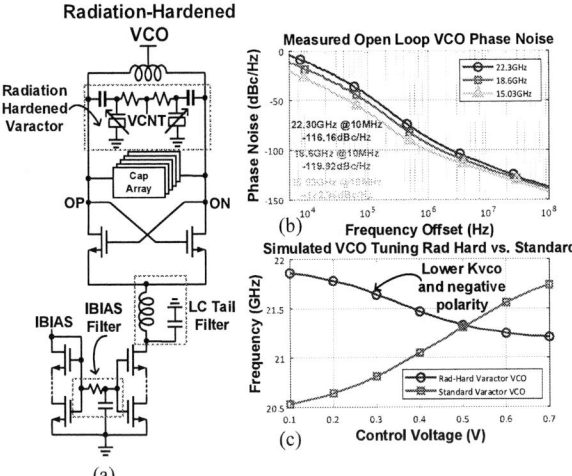

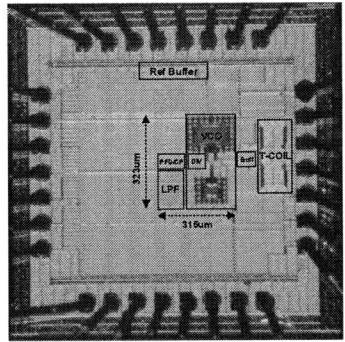

Fig. 6. RH LC-VCO: (a) schematic; (b) phase noise; (c) tuning polarity.

Fig. 7. PLL chip micrograph.

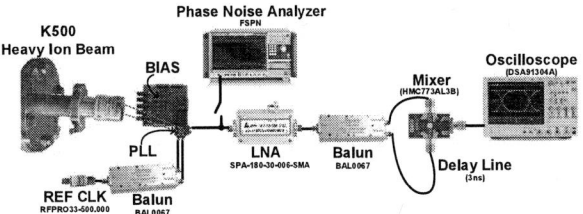

Fig. 8. PLL chip electrical and radiation testing setup.

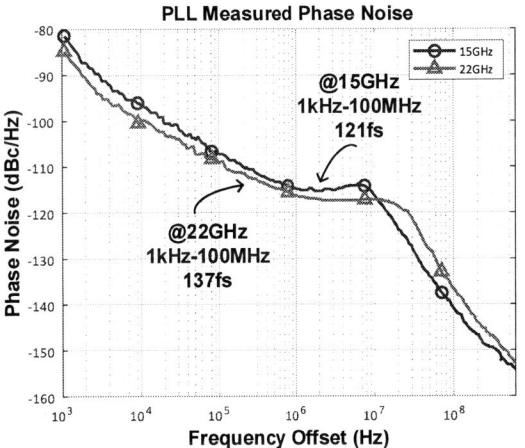

Fig. 9. PLL measured phase noise.

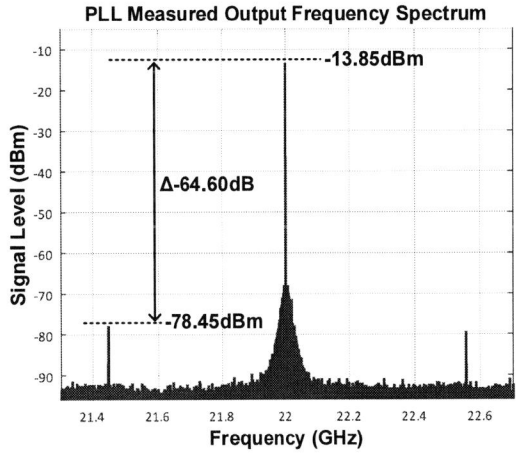

Fig. 10. PLL measured spurs.

SEUs quicker since the rate at which accumulated stray charge is flushed away is directly proportional to the charge pump current [8]. Moreover, previous work has developed radiation hardened voltage mode charge pumps that reduce the cross section by ~100x however, this voltage mode topology was also found to increase the phase noise of the charge pump of by ~10x.

The VCO as shown in Fig. 6(a) employs a radiation hardened topology for the varactor [4] and includes an LC tail filter for second harmonic/SEE filtering [5,6] as well as a large RC filter on the current bias node for filtering noise/SEUs. Overall, there is a significant area overhead for using the LC tail filter and the RC bias filter to assist with both SEU and noise performance. The radiation hardened varactor scheme grounds the NWELL of the varactor reducing the SEU sensitivity of the reverse biased junction [4]. However, in order to bias the varactors, we introduce resistance to the tank which degrades the overall tank Q increasing the overall measured VCO phase noise is shown in Fig. 6(b) at three spot frequencies within the VCO capacitor bank. Additionally, the ac coupling capacitors reduce the overall capacitive tuning of the varactor hence reducing the overall Kvco for an equivalently sized varactor tuned standard VCO as can be seen in the tuning curve in Fig. 6(c). In terms of functionality, the radiation hardened varactor flips the Kvco

polarity from a +Kvco to a -Kvco when compared to the standard varactor configuration meaning that the UP and DN signals of the PFD need to be swapped to account for this. Overall, the overhead for radiation hardened techniques in the VCO to alleviate SEEs result in primarily increased area due to the LC tail filter and RC bias filter as well as reduced tuning and phase noise performance due to the introduction of extra capacitance and resistance in the tank. The 6-bit capacitor bank uses a complimentary switch which reduces the

979-8-3503-2123-4/23 $31.00 © 2023 IEEE   111

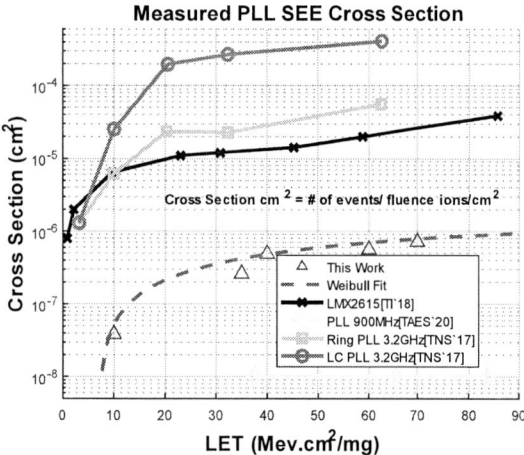

Fig. 11. PLL cross section comparison.

Table 1. Performance Summary

Work	LMX2615 TI [1]	Z. Chen TAES'20 [2]	J. Prinzie TNS'17 [3]	Z. Zhang TNS'20 [5]	[This Work]
Architecture	CPPLL	CPPLL	CPPLL LC/Ring	CPPLL	CPPLL
Technology	N/A	130nm	65nm	28nm	22nm
Frequency (GHz)	0.04-15	0.45-0.90	2.2-3.2	9.95-12.5	15-22
Power (mW)	1296	14.3	6/6	N/A	55.7
Jitter	45fs	3.2ps	325fs/5.6ps	292fs	121fs/137fs
Spurs(dB)	N/A	N/A	N/A	N/A	-64.10
FoM (dB)	-235.81	-218.34	-241.98/-217.25	N/A	-240.89/-239.56
Cross Section cm² @50 MeVcm²/mg	1.6E-5	1.0E-8	3.4E-4/3.8E-5	N/A	5.94E-7
Radiation Hardening	Yes	Yes	Yes	Yes	Yes

$$FoM = 10 * log10((\sigma_{rms}/1\,s)^2 * (Power/1\,mW))$$

V. CONCLUSION

We have presented a high-performance 15-22 GHz radiation hardened PLL in a 22nm FinFET process. By carefully considering the performance trade-offs in key radiation hardened design techniques, we were able to achieve a great jitter FoM and SEU robustness which was verified through testing at the Texas A&M Cyclotron.

ACKNOWLEDGMENT

We would like to thank our sponsor AFRL for funding this project, and the Intel University Shuttle Program for chip fabrication.

REFERENCES

[1] S. Damphousse, "Single Event Effect Report LMX2615-SP 40MHz to 15 GHz Wideband Synthesizer," 03-May-2018. [Online]. Available: https://www.ti.com/jp/lit/pdf/snak006. [Accessed: 22-Nov-2022].

[2] Z. Chen, D. Ding, Y. Dong, Y. Shan, Y. Zeng and J. Gao, "Design of a High-Performance Low-Cost Radiation-Hardened Phase-Locked Loop for Space Application," in *IEEE Transactions on Aerospace and Electronic Systems*, vol. 56, no. 5, pp. 3588-3598, Oct. 2020, doi: 10.1109/TAES.2020.2975448.

[3] J. Prinzie, J. Christiansen, P. Moreira, M. Steyaert and P. Leroux, "Comparison of a 65 nm CMOS Ring- and LC-Oscillator Based PLL in Terms of TID and SEU Sensitivity," in *IEEE Transactions on Nuclear Science*, vol. 64, no. 1, pp. 245-252, Jan. 2017, doi: 10.1109/TNS.2016.2616919.

[4] J. Prinzie, J. Christiansen, P. Moreira, M. Steyaert and P. Leroux, "A 2.56-GHz SEU Radiation Hard LC -Tank VCO for High-Speed Communication Links in 65-nm CMOS Technology," in *IEEE Transactions on Nuclear Science*, vol. 65, no. 1, pp. 407-412, Jan. 2018, doi: 10.1109/TNS.2017.2764501.

[5] Z. Zhang, H. Djanshahi, C. Gu, M. Patel and L. Chen, "Single-Event Effects Characterization of LC-VCO PLLs in a 28-nm CMOS Technology," in *IEEE Transactions on Nuclear Science*, vol. 67, no. 9, pp. 2042-2050, Sept. 2020, doi: 10.1109/TNS.2020.3008142.

[6] E. Hegazi, et al., "A filtering technique to lower LC oscillator phase noise," in *IEEE Journal of Solid-State Circuits*, vol. 36, no. 12, pp. 1921-1930, Dec. 2001, doi: 10.1109/4.972142.

[7] M. J. Gadlage, J. R. Ahlbin, P. Gadfort, A. H. Roach and S. Stansberry, "Characterization of Single-Event Transients in Schmitt Trigger Inverter Chains Operating at Subthreshold Voltages," in *IEEE Transactions on Nuclear Science*, vol. 64, no. 1, pp. 637-642, Jan. 2017, doi: 10.1109/TNS.2016.2629448.

[8] T. D. Loveless *et al.*, "A Single-Event-Hardened Phase-Locked Loop Fabricated in 130 nm CMOS," in *IEEE Transactions on Nuclear Science*, vol. 54, no. 6, pp. 2012-2020, Dec. 2007, doi: 10.1109/TNS.2007.908166.

overall off state leakage which improves tank Q across the 64 programmable coarse tuning states.

In the divider, the pulse swallow divider structure is synthesized and triplicated with the three outputs being voted on to provide feedback to the PFD. The overhead of the hardening in the divider is primarily an increase in the total power consumption and an area increase due to the triplicated cells. To save power we do not triplicate the CML divider at the cost of an increase in the radiation cross section.

IV. EXPERIMENTAL RESULTS

A micrograph of the PLL chip that was designed and fabricated in a 22nm FinFET process is shown in Fig. 7. Furthermore, electrical and radiation testing was performed with the setup as is shown in Fig. 8. The measured phase noise spectrum and jitter for two output frequencies is shown in Fig. 9 with measured jitters of 121fs at 15GHz and 137fs of jitter at 22GHz both integrated from 1k-100MHz. Moreover, in Fig. 10 we plot the PLL output spurs which were found to be -64.60dBs. Compared to the other radiation hardened PLLs in Table 1, the only design with a better FoM is [3] however that design operates at a much lower frequency of ~3GHz.

Heavy ion testing was performed at the Texas A&M Cyclotron Institute, and a delay line discriminator test setup as shown in Fig. 8 was used to capture frequency jumps at the output of the PLL. Overall, the capture time and resolution of the setup is limited by the 250ps hold time resolution of the oscilloscope and the +/-35mV (+/-2.75 deg) hysteresis window used in the scope for capturing the delay line output. To verify the setups repeatability, we tested the same ions at two different fluence rates (same flux) and verified that the calculated cross sections were similar. An error was counted if the common mode of the delay line output left the scope hysteresis window of +/-2.75 degrees for longer than 250ps. Overall, from the measured cross section in Fig. 11, we achieved excellent radiation hardening when compared to the work in [1], [5], and [3]. While the work in [2] has a better cross section, the electrical figure of merit is significantly worse.

RMo3A-3

A Fast-Startup 80MHz Crystal Oscillator with 96x/368x Startup-Time Reductions for 3.0V/1.2V Swings Based on Un-interrupted Phase-Aligned Injection

Chien-Wei Chen, Chao-Ching Hung, Yu-Li Hsueh

Mediatek Inc., Taiwan

pierce.chen@mediatek.com

Abstract—**This paper presents an 80MHz crystal oscillator with fast-startup capability based on phase-aligned clock injection. The instantaneous phase misalignment due to frequency error is detected in real time without interrupting the injection process. The injection clock's frequency is corrected accordingly and gradually approaches the crystal oscillator's intrinsic oscillation frequency. Very effective start-up acceleration is demonstrated as benchmarked against ideal injections with zero frequency error. A prototype circuit in 55nm CMOS process achieves 96x and 368x startup time reductions for 3.0V and 1.2V target swings, respectively.**

Keywords—**WiFi 7, IEEE 802.11be, crystal oscillator, fast startup.**

I. INTRODUCTION

Typical crystal oscillators exhibit long start-up time of a few ms due to the resonator's high-quality factor. Modern applications demand short start-up times, eg. for reducing overhead when circuit wakes up, for power-saving purposes in duty-cycled operations, etc. An exemplary duty-cycled operation is illustrated in Figure 1 which is a popular power-saving strategy in modern RF systems. In listening mode, the RF transceiver and the XO are only awakened for short durations separated by pre-determined time intervals and turned off otherwise to save power. The XO needs to be awakened before the RF transceiver to provide the clock source. Due to the crystal's high quality factor, a conventional XO's startup time is typically quite long, eg. 3~10ms, which can dominate the length of the awakened duration. A XO with shorter startup time allows fast response and helps reduce power. When proceeding to data mode, both fast XO settling and larger XO swing are desire from functionality and phase noise point of view. To meet the phase noise spec, target swings of 1.2Vpp and 3.0Vpp are required in listen and data mode respectively. To support these two scenarios, XO operates at Vdd 3.0V. Besides, data mode time can be long or short depending on user's requirement.

The chirped injection technique [1-2] accelerates XO start-up by injecting a clock with time-varying frequencies near the XO's intrinsic frequency which is not very efficient to reliably accelerate XO's start-up [3-4]; an ideal injection should exhibit an identical frequency and constructively aligned phases for optimal acceleration. In [3], a free-running ring oscillator (RO) is injected into the XO for a short duration and disconnected, a PLL locks the RO's frequency, and the RO is re-connected to the XO for start-up. In [4], the injection process is periodically stopped for phase alignment between

the injected clock and the XO's intrinsic oscillation. However, both [3] and [4] need to stop the injection process for frequency detection or phase alignment, inevitably slowing down the XO's start-up process. This paper proposes a XO start-up technique which successively adjusts the injection frequency and maintains constructive phase alignment without interrupting the injection process.

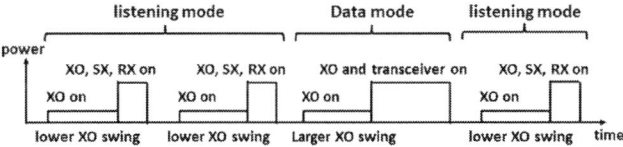

Fig. 1. Duty-cycled power-saving operation in a typical RF system.

II. BEATING BEHAVIOR AT THE OSCILLATION NODE

Figure 2 illustrates injection of an external clock to the XO's oscillation node. As shown in the top left plot, the injection signal I_{INJ} is injected at X0 node and X0 distorted waveform is monitored for fast start-up. As shown in the top right plot, if the injection frequency deviates from the XO's intrinsic oscillation frequency, beating occurs between them, and the crystal resonator's motional current I_{MO} (which resonates within the crystal) magnitude goes up and down depending on their phase relationship. A constructive phase relationship between the injection current I_{INJ} and the motional current I_{MO} accelerates the XO's start-up, while a deconstructive phase relationship drags the amplitude down and decelerates the XO's start-up. Typical on-chip injection clock sources, eg. RO, RC-osc, low Q LC oscillator etc., suffer from large frequency variations of a few thousand ppm or more. It is desirable to detect and align the injection clock's frequency and phase with the XO's intrinsic oscillation signal.

The injection signal can be modelled as a rail-to-rail square wave, and the XO's intrinsic oscillation signal is a sinusoidal wave. During the start-up process, the strong injection signal is modulated by the crystal resonator's motional current, resulting in waveform distortion as shown in the timing diagram of Fig. 2. When I_{INJ} and I_{MO} have aligned phases, ie. $\emptyset \approx 0°$, the modulated waveform V_{XO} shows a concave distortion. When their phases are in quadrature, ie. $\emptyset \approx 90°$, the distortion exhibits a sinusoidal shape. When they are completely out-of-phase, ie. $\emptyset \approx 180°$, the distortion is convex. Since different phase relationships lead to different pulse shapes, the pulse's enclosed area $S(\emptyset)$ can serve as an indicator of the phase alignment. As illustrated in Fig. 3, assuming a unity peak-to-peak swing of the injected signal, it can be shown that $S(\emptyset) = S_0 - A_0 \cdot T/\pi \cdot cos(\emptyset)$ in steady state, where $S(\emptyset)$ is the pulse area as a function of $\emptyset$, $S_0 = T/4$ is the pulse area without distortion, A_0 is the amplitude of the sinusoidal distortion, and T is the clock period. The minimum of $S(\emptyset)$ occurs at $\emptyset = 0°$, allowing a phase detection mechanism to be developed by searching the minimum of the pulse area function. Furthermore, during the XO's start-up, its amplitude actually grows up with time. The XO's amplitude can be modelled to grow linearly during a relatively short period of time, ie. $A(t) = A_0 + \alpha \cdot t$, and the phase error $\emptyset$ accumulates linearly with a given frequency error Δf, $\emptyset = 2\pi \cdot \Delta f \cdot t$. It can be derived that $S(\emptyset) = S_0 - (A_0 + k \cdot \emptyset) \cdot T/\pi \cdot cos(\emptyset)$ and its minimum satisfies $tan(\emptyset_{min}) = k/(A_0 + k \cdot \emptyset_{min})$, where $k \equiv \alpha/2\pi/\Delta f$. It can be shown that $-90° < \emptyset_{min} < 90°$ for positive A_0 and α. Therefore, a phase alignment technique can be developed by detecting the minimum of $S(\emptyset)$ to ensure $\emptyset$ is confined within $\pm 90°$. In practical XO design cases, $\emptyset_{min}$ are typically around $\pm 30°$ or so, leading to efficient acceleration if $|\emptyset| \leq |\emptyset_{min}|$ can be maintained during the injection process.

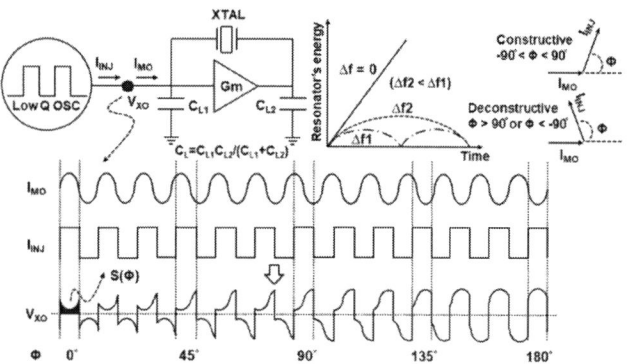

Fig. 2. XO startup by external injection. A frequency error Δf and phase errors $\emptyset$ lead to beating behavior and waveform distortions.

III. WORKING PRINCIPLE OF THE PROPOSED TECHNIQUE

Fig. 4 shows the circuit block diagram of the proposed frequency and phase alignment technique. The oscillation node is connected to an envelope detection circuit consisting of a resistor R1 in series with a diode, and a capacitor C1 with a resetting switch. The voltage on C1 is periodically reset to GND to initiate each envelope detection process. The diode is turned ON when V_{XO} is high, during which R1 and C1 attempt to take an average value of the distorted V_{XO}. The diode remains OFF when V_{XO} is low and does not contribute to the result. After a few clock cycles, the settled voltage on C1 is a representation of $S(\emptyset)$. The final voltage on the C1 is amplified and fed to a comparison circuit functioning in an auto-zero fashion with two operating phases. The amplifier gain does not need to be accurate; it provides additional gain to help discriminate the small voltage changes, and its gain should not be too large to saturate the output. In phase 1, the comparator is configured as a unity-gain buffer; the capacitor C2 stores the input voltage as well as the comparator's input

offset voltage. In phase 2, the switch is disconnected, and the circuit compares the new input voltage with the previously stored voltage on the C2. The comparison circuit's output reveals whether $S(\emptyset)$ is increasing or decreasing and is provided to the digital engine for processing. Note that the load capacitance C_L in the XO is required to resonate with the crystal. The effective C_L of a typical Pierce topology as in Fig. 2 is the capacitors C_{L1} and C_{L2} in series. A larger C_L leads to longer start-up time [3].

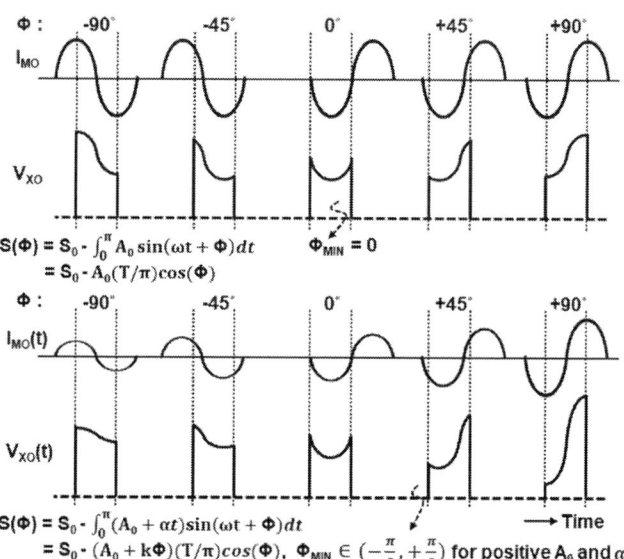

Fig. 3. Waveform distortions in steady-state and during XO startup.

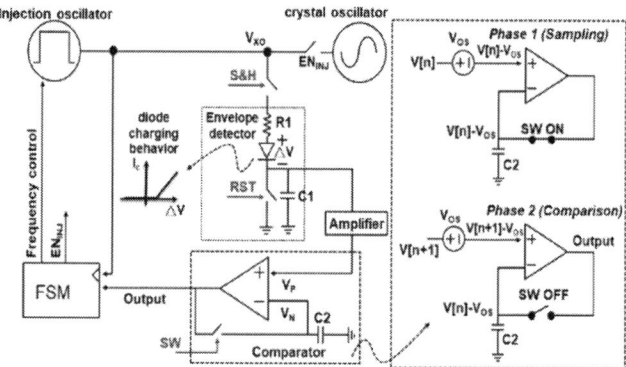

Fig. 4. Proposed circuit block diagram.

Fig. 5 shows the timing diagram to illustrate the detailed behavior of the proposed technique. In the beginning of each processing cycle, the envelope detector's output voltage is reset to GND. Afterwards, the envelope detector processes the distorted clock waveform and outputs a detected voltage. The voltage is amplified and fed to the comparator which compares VP value with the previously stored voltage VN from the previous processing cycle. If comparison circuit's output is 0, it means $S(\emptyset)$ is decreasing, and $|\emptyset| \leq |\emptyset_{min}|$.

On the other hand, if comparison circuit's output is 1, it means $S(\emptyset)$ starts to increase, and $|\emptyset|$ has drifted out of $|\emptyset_{min}|$. Upon detection of this event, the digital engine takes

979-8-3503-2123-4/23 $31.00 © 2023 IEEE

action to make phase error accumulates in the opposite direction. Besides, proposed technique can be compatible with low-steady-state-swing XOs, eg, 0.2~0.3V by using low Vth diode device or resetting to negative voltage. In this design, conventional diode device is used.

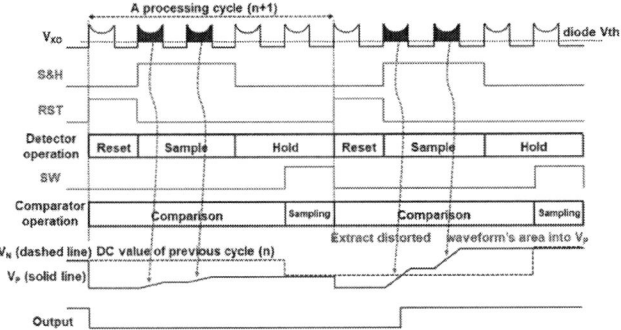

Fig. 5. Timing diagram of the proposed technique, in which the phase error is detected by waveform distortions.

A frequency-searching algorithm is developed which successively adjusts the injection frequency toward the XO's oscillation frequency, while maintaining the phase relationship $|\emptyset| \leq |\emptyset_{min}|$. The timing diagram in Figure 6 illustrates its working principle. In the beginning, the injection frequency can be purposely set as lower (or higher) than the XO's intrinsic frequency. The frequency discrepancy leads to phase accumulation, and the comparison circuit's output exhibits a minimum when the phase error reaches $\emptyset_{min}$. Upon detection of the minimum, the digital engine shifts the injection frequency upwards by a certain step, eg. +10000ppm in the figure, such that the injection frequency is higher than the XO's intrinsic frequency. The phase accumulates in the opposite direction, crosses zero, and becomes negative. A minimum of $S(\emptyset)$ is detected again, and the injection frequency is shifted downwards. The process is repeated and the step size is successively decreased to gradually approach the target frequency. Choice of the initial frequency step, ie. 10000ppm, depends on the frequency accuracy of the injection clock source. A smaller frequency step can be used if a more accurate injection clock source is adopted. It is worth noting that if the target frequency falls outside of the successively decreased searching range, as shown in Fig. 7, $S(\emptyset)$ will keep increasing instead of decreasing when the injection frequency is shifted. Upon detection of this event, the digital engine immediately alters the searching direction and sets new searching ranges in the following process, such that the target frequency always falls within the searching range. It is possible to shift the injection clock's phase, instead of adjusting its frequency, to maintain phase alignment for start-up acceleration. However, shifting phase requires additional phase-shifting or multi-phase generation mechanisms, eg. higher-frequency clock source. This work adjusts the injection clock frequency rather than shifting phase to fundamentally remove the root cause of phase misalignment, the realization is more straightforward.

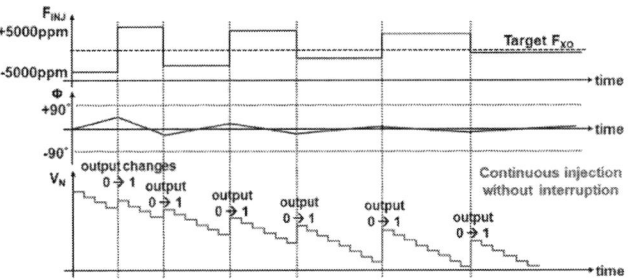

Fig. 6. Timing diagram of the developed frequency-searching algorithm.

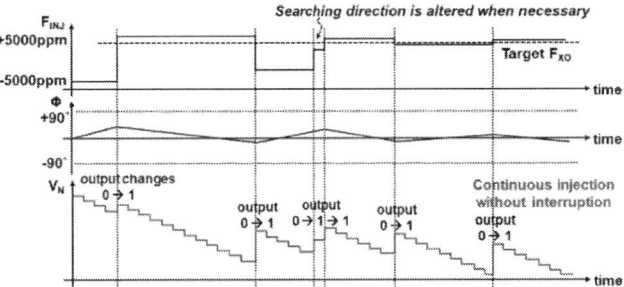

Fig. 7. Timing diagram of the developed frequency-searching algorithm (target frequency falls outside of searching range).

IV. MEASUREMENT RESULTS

A prototype circuit is manufactured in a 55nm CMOS process. The XO's oscillation frequency is 80MHz and phase noise is -166dBc/Hz at 100kHz offset, and the power consumption of 5.4mW with Vdd=3.0V. The peak-to-peak swing is 3.0V, targeting at low-noise applications such as WiFi-7. Fig. 8(a) shows the measurement results. The ideal injection shows a smoothly growing envelope, while the proposed technique exhibits obvious saw-tooth modulation on the XO's amplitude due to the beating behavior and proposed frequency-searching process. The start-up time is defined as the duration between powering-up and when the XO amplitude settles within 95% of its final value and the frequency error is within 20ppm. When the XO is started up alone without the injection clock, it requires 3ms to reach the target swing of 3.0Vpp. When the proposed technique is activated with an injection swing of 3.0Vpp and an initial injection frequency error of ±6000ppm, the start-up time is significantly shrunk to 31us. We also check an ideal injection's case by manually setting the injection frequency to match the XO's intrinsic frequency. The ideal injection's case has a start-up time of 28us, which is very close to the proposed technique. The bottom right plot shows the measured XO frequency error of the proposed technique. Figure 8(b) shows measured start-up times are ≤31us across various initial frequency errors in the ±6000ppm range. We also measured the start-up time across -40~85 ℃. The conventional XO takes 2.7~3.5ms to start up, while the proposed technique reliably shows start-up times of ≤33us. Other than frequency accuracy and phase alignment, a larger injection swing compared to the XO's target swing is another factor that also helps reduce start-up time. In this work, if the injection swing is kept at 3.0Vpp to accelerate a smaller target

XO swing of 1.2Vpp, the measured start-up time is 12us, while a conventional counterpart takes 4.4ms. Start-up energies are 29.8nJ and 11.5nJ for target XO swings of 3.0Vpp and 1.2Vpp, respectively, for a CL of 10pF. Fig. 9 shows the micrograph of the fabricated chip. Active areas of the XO and the fast start-up circuits (including frequency-searching algorithm) are 0.079mm2 and 0.023mm2, respectively.

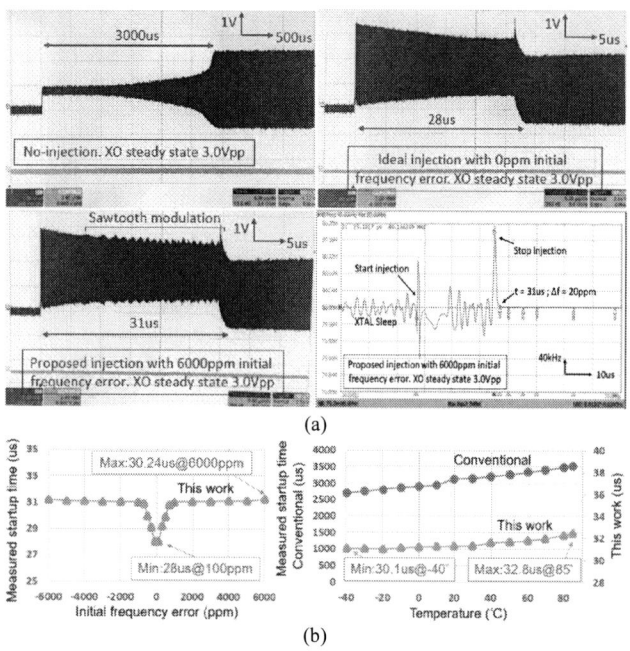

(a)

(b)

Fig. 8. (a) Measured XO startup time results of the conventional XO, the ideal injection, and the proposed technique; (b) measured XO startup time over temperature at ΔF=6000ppm and sensitivity to initial frequency.

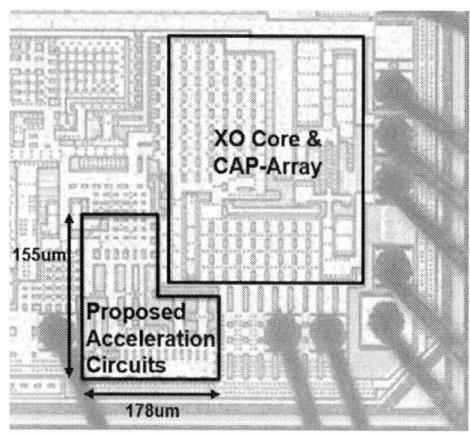

Fig. 9. Micrograph of the circuits.

V. CONCLUSION

The prototype circuit allows a high swing of 3.0Vpp with a phase noise -166dBc@100kHz for low-noise applications, eg. 4096-QAM in WiFi-7, and a lower swing of 1.2Vpp for lower-power applications, eg. Bluetooth, ZigBee, etc. This work shows competitive start-up energy values with a relatively large load capacitance, C_L, of 10pF. In addition, a

smaller C_L also helps reduce the start-up energy, which is the integrated power during the XO's start-up process, due to smaller startup time. However, smaller C_L suffers worse tuning sensitivity degrading XO frequency stability which is an important spec for WiFi-7.

Table 1 is a comparison table with prior arts. In the table, the start-up time reduction ratio (STRR) is defined as the ratio of a conventional XO's start-up time against the reported start-up time. The start-up time expansion ratio (STER) is defined as the ratio of the reported start-up time against an ideal injection's start-up time. An efficient start-up approach corresponds to a high STRR and a low STER. This work shows excellent results in both metrics.

Table 1. Performance summary.

Technique	This Work		Griffith ISSCC 16 [1]	Iguchi JSCC'16 [2]	Megawer ISSCC 19 [3]	Verhoef ISSCC'19 [4]		Jung ISSCC'22 [5]
	Un-interrupted Phase-aligned Injection		Dithered Injection	Chirped Injection	2-Step Injection	SSI		2-Step Injection
Technology (nm)	55		65	180	65	55		28
Injection CLK swing (Vpp)	3.00		1.68	1.50	1.00	2.40		1.20
Frequency (MHz)	80		24	39	54	32		75.8
Load capacitance (pF)	10*		6	6	6	6	12	7
XO's steady-state swing (Vpp)	3.00*	1.20	NA	NA	0.70	0.75	0.75	1.20
Startup time of conventional XO (ms)	3.00*	4.40**	0.43	2.10	0.60	0.40	2.90	0.72
Startup time of proposed technique (us)	31	12	64	158	19	23	32	39.6
STRR ***	96.8	368.6	6.7	13.3	31.5	17.4	90.6	18.2
STER ****	1.10	1.07	NA	NA	1.5	NA	NA	NA
Phase noise @100kHz (dBc/Hz)	-166.0	-156.5	NA	-172.5	-157.8	NA		-156.2
Startup energy (nJ) *****	29.8	11.5	NA	349	34.9	20.2	44.2	92.8
Startup time variation over temperature	+/-0.91% -40 to 85C		+/-35% -40 to 90°C		+/-7% -30 to 125°C	+/-1.25% -40 to 85C		+/-10% -40 to 140C	NA
Area (mm²)	0.023		0.08	0.12	0.069	0.049		NA

* To support WiFi-7, XO operates at Vdd 3.0V for larger swing and C_{LOAD}=10pF for better frequency stability
** The longer startup time is due to the weaker -gm for a smaller steady-state swing.
*** STRR ∈ [1, ∞]; the larger the better.
**** STER ∈ [1, ∞]; the smaller the better.
***** Defines as the total dissipated energy during XO's start-up process.

ACKNOWLEDGMENT

The authors would like to thank Yao-Chi Wang and Keng-Meng Chang for helps and discussions.

REFERENCES

[1] D. Griffith et al., "A 24MHz Crystal Oscillator with Robust Fast Start-Up Using Dithered Injection," ISSCC, pp. 104-105, Feb. 2016.

[2] S. Iguchi et al., "Variation-Tolerant Quick-Start-Up CMOS Crystal Oscillator with Chirp Injection and Negative Resistance Booster," IEEE JSSC, vol. 51, no. 2, pp. 496-508, Feb. 2016.

[3] K. M. Megawer, et al., "A 54MHz Crystal Oscillator with 30× Start-Up Time Reduction Using 2-Step Injection in 65nm CMOS," ISSCC 2019, pp. 302-303.

[4] B. Verhoef, et al., "A 32MHz Crystal Oscillator with Fast Start-up Using Synchronized Signal Injection," ISSCC 2019, pp. 304-305.

[5] J. Jung, et al., "A Single-Crystal-Oscillator-Based Clock-Management IC with 18x Start-Up Time Reduction and 0.68ppm/°C Duty-Cycled Machine-Learning-Based RCO Calibration," ISSCC 2022, pp.58-59.

RMo3A-4

Transformer-Coupled 2.5GHz BAW oscillator with 12.5fs RMS-Jitter and 1-kHz Figure-of-Merit (FOM) of 210dB

Bichoy Bahr, Sachin Kalia, Baher Haroun, Swaminathan Sankaran

Kilby Labs, Texas Instruments, USA

bichoy@ti.com

Abstract — **This paper reports two high-performance parallel resonance 2.5GHz BAW oscillators in 130nm SiGe BiCMOS technology. The proposed architecture concurrently improves 1kHz-FOM by more than 3.8dB together with jitter, and close-in phase-noise performance over prior works and further extends the state of the art in MEMS oscillator design. The design optimizes multiple transformer windings to step-down the high-Q resonator impedance and gain multiple benefits: i) cross-coupled pair size increase without performance tradeoff, ii) reduction in base swing, non-linearity and, flicker noise up-conversion and iii) low-voltage compatible operation. Resistive degeneration and inductive coupling at the BJT emitter are considered. Both topologies achieves <12.5fs RMS-Jitter. Resistive degeneration achieves 1kHz-FOM of 210dB, while the inductively coupled emitter allows for more robust operation.**

Keywords — **BAW, MEMS, Oscillator, resonators**

I. INTRODUCTION

Frequency reference oscillators are essential components in communication and sensing systems. Broadband 100+Gbps multi-lane wireline, multi-channel clocking and network synchronizer applications require ultra-stable, ultra-low reference jitter ($< 50fs$) to maintain acceptable Bit-Error-Rates (BER). It is of equal critical importance to maintain clock-reference fidelity for both close-in offset ($< 100kHz$) and over wider integration bandwidths ($\geq 100MHz$) in nQAM ($n > 64$) wireless communication, E/W-band, emerging xG and guided-wave systems [1]. Automotive radar and emerging long-range sensing applications also mandate stringent constraints on far-out phase noise (>1MHz) [2] It is widely observed in prior works that achieving both a good close-in phase noise and jitter performance is an unsurmountable challenge for reference oscillators. Most existing technologies rely on multiple oscillators with phase or frequency locking at the expense of higher complexity and power consumption. Bulk-Acoustic-Wave (BAW) resonators are very promising in breaking this trade-off [3]. With quality factors Q well over 1000 at multi-GHz frequencies, they offer a great solution for ultra-low-jitter clocks with low power consumption. Such high Q also helps with better close-in phase noise, that can be further enhanced by flicker-noise suppression techniques, making it possible to construct a BAW-based oscillator that is simultaneously optimized for ultra-low jitter and close-in phase noise while improving FOM.

The BAW resonator used in this work is an AlN Dual-Bragg BAW (DBBAW) similar to [3] as shown in Fig. 1.a. The Bragg mirrors confine the acoustic vibration energy

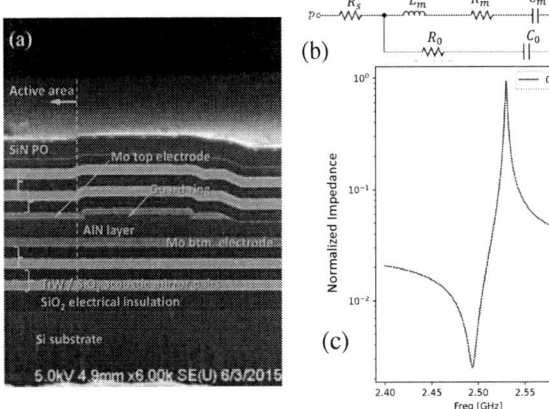

Fig. 1. (a) Cross-section of Dual-Bragg BAW resonator [3] (b) Modified Butterworth-Van-Dyke (MBVD) model (c) Unloaded BAW resonator impedance.

to the piezoelectric material, which preserves the resonator performance and eliminates the need for vacuum packaging, enhancing the manufacturability of the device.

The electrical response of BAW resonators can be modeled by the Modified Butterworth-Van-Dyke (MBVD) shown in Fig. 1.b. The resonator exhibits a series resonance at a frequency f_s followed by a parallel resonance at f_p (Fig. 1.c). The latter is characterized by a higher quality factor Q_p as it is not impacted by the electrodes and routing resistance R_s, making it ideal for low-noise oscillator realization. However, the associated higher impedance Z_p restricts its use to low-power designs due to swing limitations in most IC technologies. This was expanded further by the work in [4], where a tuned transformer is used to step-up the impedance Z_p for further reduction in power consumption, at the expense of phase noise performance.

In this work, a tuned transformer is used to step-down the impedance Z_p of a BAW resonator, leveraging the parallel resonance mode with high-Q_p simultaneously with increased core power consumption for significantly reduced phase noise and improved FOM. An additional coil is used to couple the tank to the BJT bases with even smaller swing, hence reducing the BJT non-linearity and flicker-noise up-conversion. Furthermore, emitter coupling through a fourth coil is explored for enhanced impulse-sensitivity-function (ISF) shaping and flicker-noise suppression.

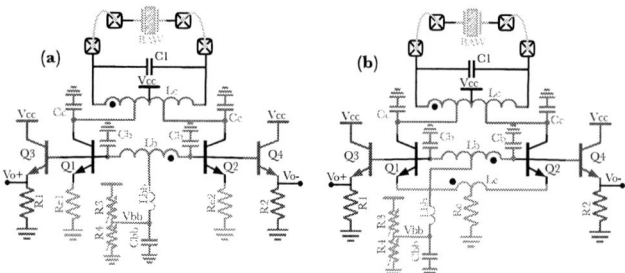

Fig. 2. Proposed transformer coupled BAW oscillator showing (A) architecture A with resistive degeneration, and (B) architecture B with inductive degeneration.

II. PROPOSED OSCILLATOR ARCHITECTURE

The proposed parallel-BAW oscillator architecture is shown in Fig. 2.a. A doubly-tuned auto-transformer couples the BAW resonator to the collectors of Q_1 and Q_2, with stepped-down impedance. The auto-transformer is implemented as a figure-8 to reduce EMI as shown in Fig. 3.a. The BJT bases are coupled to the collector through a third coil embedded in the auto-transformer. The architecture of Fig. 2.b introduces a fourth coil to couple the emitters as shown in Fig. 3.b. The inductance and mutual coupling coefficients of resulting four-way figure-8 transformer are optimized to achieve multiple goals on the collector, base, and emitter circuits simultaneously.

The base-coupling coil is designed to step-down the swing on the base for reduced non-linearity, and less flicker up-conversion. It also provides the required DC bias through a programmable resistor DAC (R_3 and R_4). The capacitance C_b is tuned for differential parallel resonance at the base of Q_1 and Q_2 at the fundamental frequency. The best flicker noise suppression occurs when C_{bb} is tuned for parallel resonance with the common mode L_{bb} at the second harmonic. Finally, R_{e1} and R_{e2} provide emitter degeneration for enhanced linearity. The degeneration also allows for finer current control in conjunction with the base bias resistor DAC. Fig. 4 shows the differential impedance of the tuned BAW resonator tank at different ports, normalized to the unloaded BAW resonator parallel resonance impedance Z_{BAW0}. The step-down in impedance (and hence voltage swing) from BAW, to collector, to base at the fundamental frequency is clearly demonstrated.

Fig. 2.b shows another implementation, where Q_1 and Q_2 emitters are inductively coupled to their respective collectors to create a three-tank transformer feedback oscillator. This significantly enhances the loop gain and provides flicker noise suppression by ISF shaping [5], [6]. The associated figure-8 transformer is shown in Fig. 3.b. For both proposed architectures, a full EM model for the transformer tank has been constructed, including ground return path and distributed capacitance for accurate tuning and optimization of the tank response.

The BiCMOS oscillator chip is packaged side-by-side with the BAW resonator and over-molded in a standard QFN package as shown in Fig. 5. Wirebonds are used to connect the BAW resonator to the SiGe chip.

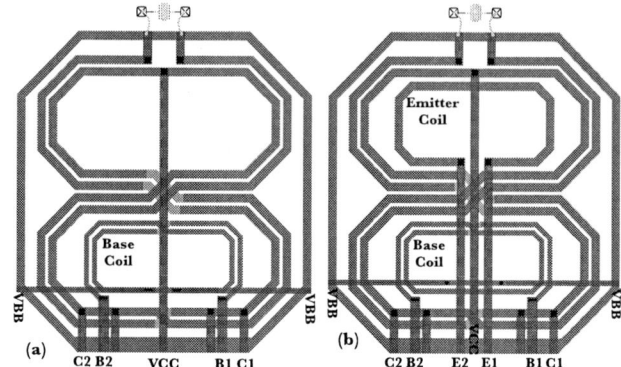

Fig. 3. Layout of figure-8 coupling coupling transformer, showing BAW, Collector, Base, and Emitter coils for architecture A and B.

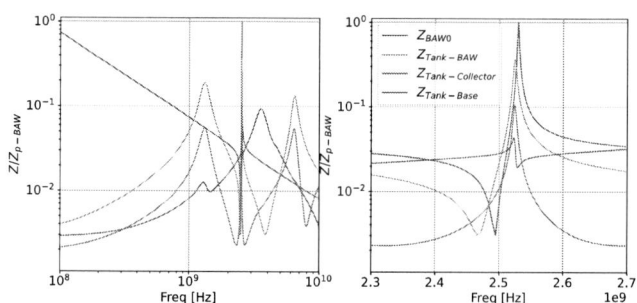

Fig. 4. Normalized differential tank impedance response at the BAW resonator port, collector port, and base port.

III. MEASUREMENT RESULTS AND COMPARISON

The testing setup block diagram is shown in Fig. 6. An external LDO with noise power spectral density (PSD) of $4nV/\sqrt{Hz}$ at 1kHz offset is used to supply the core oscillator, emitter follower buffer, and 50Ω driver. All the circuit blocks are operating from 1.2V supply. A mini-circuit ZRL-3500+ amplifier is used to gain the 2.52GHz output signal by 18dB for accurate jitter measurements. A Keysight E5052B signal source analyzer is used for measuring the phase noise and estimating the jitter of both oscillators.

Fig. 7 compares the best phase noise measured for both architectures, while Fig. 8 shows an E5052B screen capture for architecture B, highlighting the 12.4fs jitter performance. Fig. 9 shows the repeatability of phase noise performance from architecture B measured across 4 oscillator samples after bias current trimming, as well as its correlation to simulation results for similar current consumption. The observed discrepancy at low frequencies (< 1kHz) can be attributed to the lack of flicker noise model for the BAW resonator itself [7], or inadequate PDK flicker noise model, non-linearity, and junction capacitance models for the BJTs under large signal swings. The far-out phase noise discrepancy can be attributed to the ZRL-3500+ noise floor, as well as the 50Ω driver, and PCB trace losses that is not captured in the simulation. Fig. 10 compares the measured close-in phase noise and jitter of both architectures. Architecture A performance degrades quickly as bias current is changed, whereas architecture B shows good performance over a wide range of bias currents.

979-8-3503-2123-4/23 $31.00 © 2023 IEEE

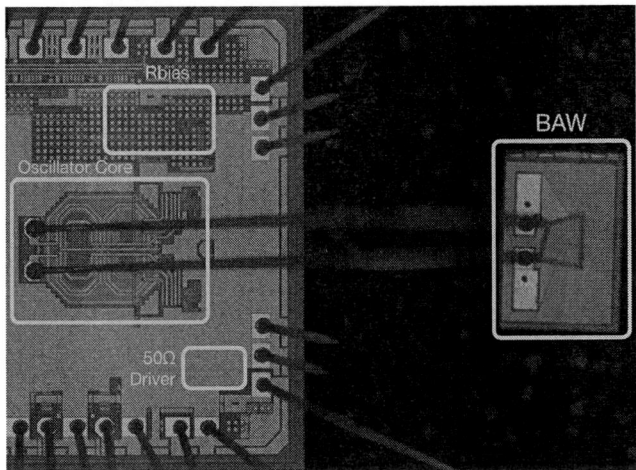

Fig. 5. Die micrograph showing the BAW resonator side-by-side with the SiGe BiCMOS oscillator chip.

Also, architecture B can operate reliably at lower current levels. This can be explained owing to the large loop-gain nature of architecture B, as well as the ISF shaping and flicker noise suppression introduced by the tuned transformer feedback architecture.

The measured oscillators show an RMS phase jitter of 12.4fs, and 12.3fs, with a close-in 1kHz offset phase noise of -92.0dBc/Hz and -91.5dBc/Hz, for architectures A and B, respectively. They consume 9.5mW and 20mW, from 1.2V supply, putting their close-in 1kHz FOM at 210.3dB and 206.5dB for A and B, respectively. Table 1 compares the presented oscillator architecture to state-of-the-art MEMS-based low-phase noise oscillator [1], [4], [6], [8]–[12]. The proposed oscillator architectures, clearly demonstrate both superior jitter performance as well as close-in phase noise and FOM, breaking the observed trade-off among state-of-the-art high-performance oscillators.

IV. CONCLUSION

Two architectures of transformer-coupled 2.52GHz BAW oscillators have been implemented and tested in 130nm SiGe BiCMOS, comparing resistive degeneration and inductively coupled emitters. Both oscillator topologies achieve an integrated RMS phase jitter of 12.4fs, far exceeding state-of-the-art BAW oscillators. Moreover, close-in phase noise performance is better than -91.5dBc/Hz

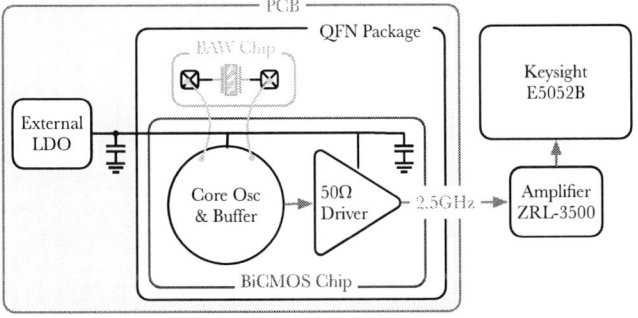

Fig. 6. Block Diagram of the complete BAW oscillator characterization setup.

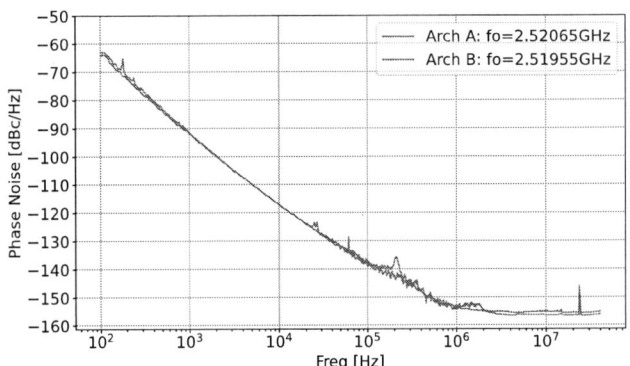

Fig. 7. Measured phase noise for architectures A and B.

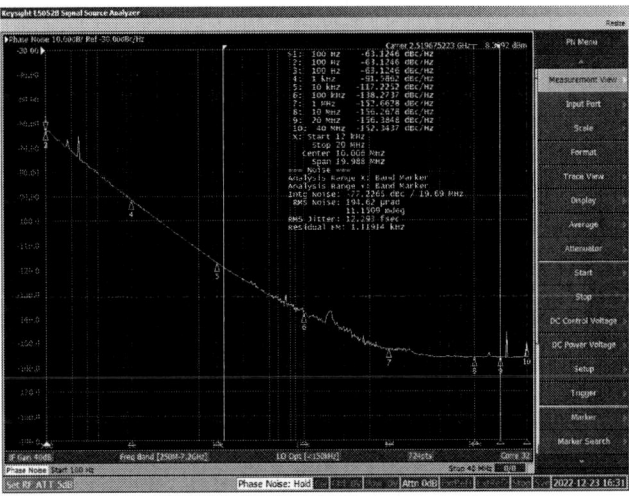

Fig. 8. Measured phase noise from architecture B.

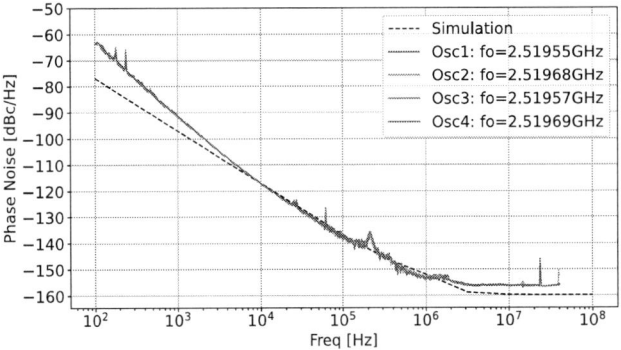

Fig. 9. Measurement results showing phase noise for 4 different architecture B oscillators as tested.

Table 1. Performance comparison to state of the art.

	This Work Arch-A	This Work Arch-B	ASSCC 2021	RFIC 2022	UFFC 2014	RFIC 2021	ISSCC 2020	TCAS-II 2019	CICC 2013	ISSCC 2010
Process	130nm	130nm	130nm	65nm	250nm	130nm	65nm	65nm	130nm	0.18μm
	BiCMOS	BiCMOS	BiCMOS	CMOS	BiCMOS	CMOS	CMOS	CMOS	CMOS	CMOS
Resonator	AlN	AlN	AlN	AlN	AlN	AlN	AlN	AlN	AlN	AlN
	BAW	BAW	BAW	BAW	BAW	BAW	BAW	FBAR	CMR	LBAR
Resonator Q	1000	1000	375	1000	1176	1000	1000	-	3991	7100
Oscillator Mdoe	Parallel	Parallel	Series	Parallel	Series	Parallel	Parallel	Parallel	Series	Series
Core Freq [GHz]	2.52	2.52	2.5	2.52	2.1	2.53	2.52	2	1.16	1.006
Output Freq [GHz]	2.52	2.52	0.626	2.44	2.1	0.632	0.048	2	1.16	1.006
Core Supply [V]	1.2	1.2	2.1	1.3	-	1.65	1.7	0.65-1.0	1	1.5
Core Power [mW]	9.5	20.7	67.2	0.675	21.6	3.95	1.1	0.35	4.2	10.7
PN [dBc/Hz] @ 1kHz	-92	-91.5	-104.8	-77	-87.5	-91.6	-110	-68.6	-82.3	-94
PN fo= 2.52GHz [dBc/Hz]	-92.0	-91.6	-92.7	-76.7	-85.9	-79.6	-75.6	-66.6	-75.6	-86.0
FOM [dB] @ 1kHz	210.3	206.5	202.5	206.5	200.6	201.6	203.2	199.2	197.4	203.8
FOM-Q [dB] @ 1kHz	150.3	146.5	151.0	146.5	139.2	141.6	143.2	-	125.3	126.7
Jitter [fs]	12.4	12.3	28	-	-	46	-	104	24	32
Area [mm²]	0.38	0.38	0.32	0.1	1.6	0.27	0.05	0.36	0.6*	0.33

$$FOM = \left(\frac{f_o}{\Delta f}\right)^2 \frac{1}{\mathcal{L}(\Delta f) \times P_{DC}} \quad , \quad FOM_Q = \frac{FOM}{Q^2}$$

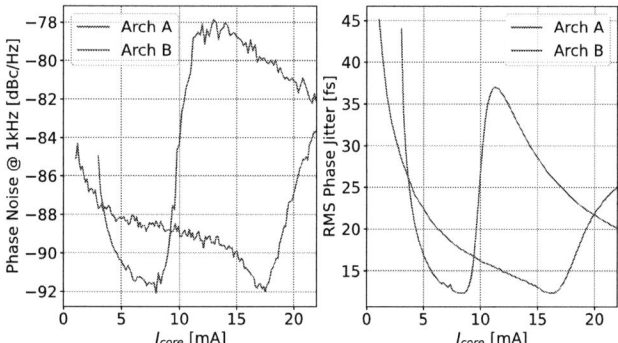

Fig. 10. Measured phase noise for architecture A and B at 1kHz offset, as well as RMS phase jitter.

at 1kHz, with a FOM of 210dB and 207.5dB for resistive degeneration and inductively-coupled emitters, respectively. The inductively-coupled emitters architecture offers higher loop gain and robust performance across large range of bias current. Simultaneously achieving such low jitter and close-in phase noise performance, breaks the trade-off observed in previous state-of-the-art, especially for oscillators with a BAW resonator.

REFERENCES

[1] S. Kalia, B. Bahr, T. Dinc, B. Haroun, and S. Sankaran, "An ultra-low close-in phase noise series-resonance baw oscillator in a 130-nm bicmos process," in *2021 IEEE Asian Solid-State Circuits Conference (A-SSCC)*, 2021, pp. 1–3.

[2] K. Dandu, S. Samala, K. Bhatia, M. Moallem, K. Subburaj, Z. Ahmad, D. Breen, S. Jang, T. Davis, M. Singh, S. Ram, V. Dudhia, M. DeWilde, D. Shetty, J. Samuel, Z. Parkar, C. Chi, P. Loya, Z. Crawford, J. Herrington, R. Kulak, A. Daga, R. Raavi, R. Teja, R. Veettil, D. Khemraj, I. Prathapan, P. Narayanan, N. Narayanan, S. Anandwade, J. Singh, V. Srinivasan, N. Nayak, K. Ramasubramanian, B. Ginsburg, and V. Rentala, "2.2 high-performance and small form-factor mm-wave cmos radars for automotive and industrial sensing in 76-to-81ghz and 57-to-64ghz bands," in *2021 IEEE International Solid- State Circuits Conference (ISSCC)*, vol. 64, 2021, pp. 39–41.

[3] E. T.-T. Yen, K. Martin, M. Chowdhury, J. Segovia-Fernandez, D. Griffith, B. Zhang, N. Dellas, T. Tarsi, D. Trombley, B. Goodlin, R. Jackson, A. Poddar, B. Haroun, and A. Bahai, "Integrated high-frequency reference clock systems utilizing mirror-encapsulated baw resonators," in *2019 IEEE International Ultrasonics Symposium (IUS)*, 2019, pp. 2174–2177.

[4] B. Bahr, A. Kiaei, M. Chowdhury, B. Cook, S. Sankaran, and B. Haroun, "Near-field-coupled bondless baw oscillators in wcsp package with 46fs jitter," in *2021 IEEE Radio Frequency Integrated Circuits Symposium (RFIC)*, 2021, pp. 155–158.

[5] S. Guo, P. Gui, T. Liu, T. Zhang, T. Xi, G. Wu, Y. Fan, and M. Morgan, "A low-voltage low-phase-noise 25-ghz two-tank transformer-feedback vco," *IEEE Transactions on Circuits and Systems I: Regular Papers*, vol. 65, no. 10, pp. 3162–3173, 2018.

[6] J. Koo, K. Wang, R. Ruby, and B. P. Otis, "A 2-GHz FBAR-Based Transformer Coupled Oscillator Design With Phase Noise Reduction," *IEEE TCAS-II: Express Briefs*, vol. 66, no. 4, pp. 542–546, 2019.

[7] H. J. Kim, S. I. Jung, J. Segovia-Fernandez, and G. Piazza, "The impact of electrode materials on 1/f noise in piezoelectric aln contour mode resonators," *AIP Advances*, vol. 8, no. 5, 2018.

[8] B. Bahr, D. Griffith, A. Kiaei, T. Tsai, R. Smith, and B. Haroun, "Class-c baw oscillator achieving a close-in fom of 206.5db at 1khz with optimal tuning for narrowband wireless systems," in *2022 IEEE Radio Frequency Integrated Circuits Symposium (RFIC)*, 2022, pp. 311–314.

[9] M. Li, S. Seok, N. Rolland, P. A. Rolland, H. E. Aabbaoui, E. D. Foucauld, P. Vincent, and V. Giordano, "Ultralow-phase-noise oscillators based on baw resonators," *IEEE Transactions on Ultrasonics, Ferroelectrics, and Frequency Control*, vol. 61, no. 6, pp. 903–912, 2014.

[10] D. Griffith, E. T. Yen, K. Tsai, H. U. R. Mohammed, B. Haroun, A. Kiaei, and A. Bahai, "An Integrated BAW Oscillator with < ±30 ppm Frequency Stability Over Temperature, Package Stress, and Aging Suitable for High-Volume Production," in *2020 IEEE International Solid-State Circuits Conference (ISSCC)*, 2020, pp. 58–60.

[11] J. Koo, A. Tazzoli, J. Segovai-Fernandez, G. Piazza, and B. Otis, "A −173 dBc/Hz at 1 MHz offset Colpitts oscillator using AlN contour-mode MEMS resonator," in *Proceedings of the IEEE 2013 Custom Integrated Circuits Conference*, 2013, pp. 1–4.

[12] H. M. Lavasani, W. Pan, B. Harrington, R. Abdolvand, and F. Ayazi, "A 76 dBΩ 1.7GHz 0.18 μm CMOS Tunable Transimpedance Amplifier Using Broadband Current Pre-amplifier for High Frequency Lateral Micromechanical Oscillators," in *2010 IEEE International Solid-State Circuits Conference (ISSCC)*, 2010, pp. 318–319.

979-8-3503-2123-4/23 $31.00 © 2023 IEEE

RMo3B-1

A mm-Wave Wideband/Reconfigurable LNA Using a 3-Winding Transformer Load in 22-nm CMOS FDSOI

Mohammad Ghaedi Bardeh, Jierui Fu, Navid Naseh, Jeyanandh Paramesh, Kamran Entesari

Electrical and Computer Engineering, Texas A&M University, USA

{m.ghaedi95, jierui.fu, nnaseh, kentesar}@tamu.edu, jeyanandh.paramesh@ieee.org

Abstract— A mm-Wave wideband/reconfigurable LNA using a 3-winding transformer load has been implemented in 22-nm CMOS FDSOI technology for 5G applications. The proposed LNA has three stages with the first two stages utilizing a novel 3-winding transformer as a load to provide three parallel paths, one main and two auxiliary paths. The load expands the frequency band of interest in the wideband mode when the main path is active and reconfigures the LNA bandwidth in the reconfigurable mode when either of the auxiliary paths are active. The third stage acts as a buffer. The LNA shows a measured S21 by the peak gain of 32 at 30 GHz and 3-dB bandwidth of 12.6 GHz for the wideband mode and peak gains of 32.4 dB and 33.4 dB at 22 and 29 GHz for 3-dB bandwidths of 5 and 5.3 GHz for two auxiliary paths in the reconfigurable mode, respectively. The measured NF is lower than 4.5 dB for the entire frequency band and measured OP1dB and OIP3 are better than -6 dBm and 2 dBm for the entire frequency band of interest, respectively. The total power consumption of the proposed LNA is 35 mW in each mode of working, and the chip area is 1155 μm $\times$ 642 μm excluding pads.

Keywords— mm-wave LNA, reconfigurable, wide-band, 3-winding transformer, double-tuned transformer.

I. INTRODUCTION

The demand for having higher data rates communication systems has derived the motivation of moving forward from RF frequency operation to mm-Wave frequencies. In mm-Wave frequency bands, the free space path loss is much higher than lower frequencies [1], [2]. To mitigate the high path loss problem in mm-wave systems, multiple-input multiple-output (MIMO) techniques are extensively used.

One of the key elements of a MIMO system front-end (FE) channel is the low-noise amplifier (LNA). Having wide bandwidth, low noise performance, low power consumption, and high linearity are the main challenges of designing an LNA that covers 5G frequency band of operation. To cover the frequency band requirement for 5G applications, several approaches are introduced that are categorized as wideband and reconfigurable. Using source-gate feedback and inductor peaking in [3], [4], gate-drain mutually induced feedback in [5], double-tuned transformers in [6], and one-port coupled resonator as load in [7], [8] are the techniques that are reported in recent state-of-the-art wideband LNAs. Another approach to realize mm-Wave LNAs is using reconfigurable techniques. There are several different approaches that can be used to design mm-Wave frequency reconfigurable LNAs. In [9], a reconfigurable LNA for 28/60 GHz is presented using an extra inductor controlled by a switch. By turning the switch OFF

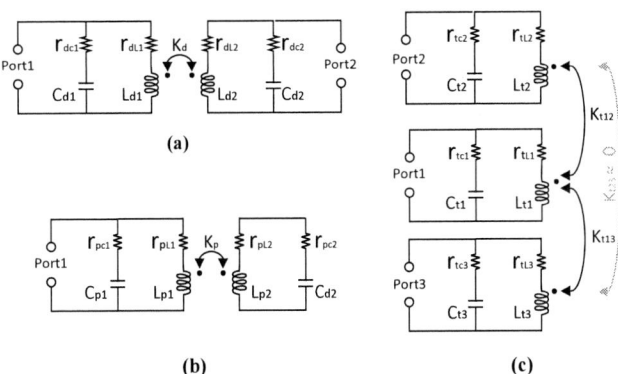

Fig. 1. Different coupled transformers as resonance load. (a) Double-tuned. (b) One-port double-tuned. (c) The proposed 3-winding transformer.

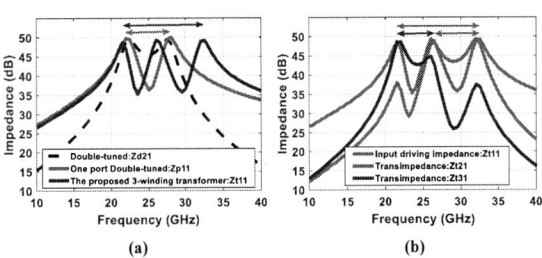

Fig. 2. (a) Frequency response comparison of three transformers. (b) Frequency response of the proposed 3-winding transformer in different modes.

and ON, the frequency changes from 28 to 60 GHz. The drawback here is limiting the inductor Q while switching its value directly at mm-Wave range. In [10], by using a variable inductor based on open and ground shielding area around the inductor another mm-Wave reconfigurable LNA is presented where the drawback is the complexity of providing effective ground in the LNA layout. One main challenge in designing reconfigurable mm-Wave LNAs is having equal specifications for LNA in different bands in terms of S21, NF, power consumption, and linearity. Another major challenge is that switching is done at mm-Wave frequencies where on-resistance and off-capacitance of the switch limit the S21 performance.

To address the aforementioned issues, a mm-Wave wide-band (21.6-34.2GHz)/ reconfigurable LNA is presented in this paper that uses a novel 3-winding transformer as a resonance load to enable three different frequency modes:

979-8-3503-2123-4/23 $31.00 © 2023 IEEE   121   2023 IEEE Radio Frequency Integrated Circuits Symposium

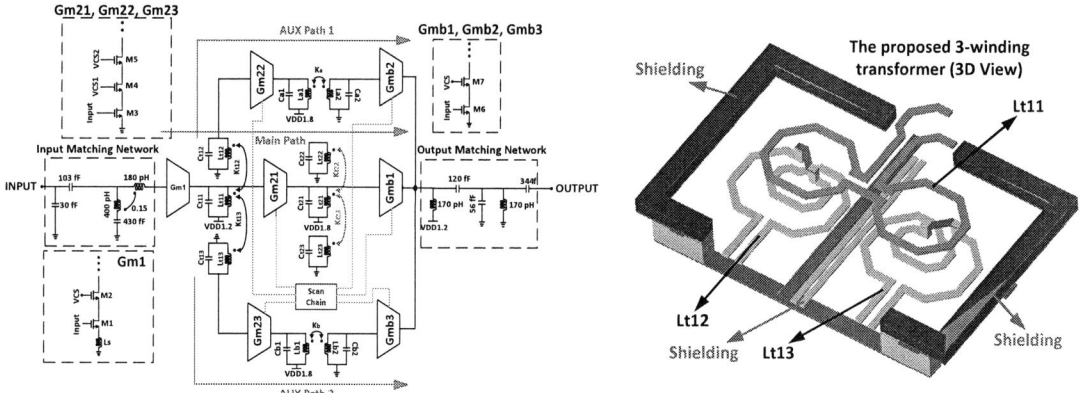

Fig. 3. Block diagram of the proposed LNA(left) and 3D layout of the 3-winding transformer (right).

1) Wide-band (21.6-34.2 GHz) frequency operation, 2) Low frequency (21.5-26.5 GHz) operation, and 3) High frequency (27.5-33 GHz) operation. Therefore, the proposed mm-Wave LNA with the frequency band of operation of 21.6-34.2 GHz, peak S21 of 32 at 30 GHz, NF of < 4.5 dB for 21.6-34.2 GHz, and S11 and S22 < -10 dB for the entire frequency band of interest have advantageous of both (wideband and reconfigurable) LNAs. One reason to move from wide-band operation to the reconfigurable one is to increase out-of-band IIP3 to handle interferers which are common in MIMO phased array systems.

II. 3-WINDING TRANSFORMER TOPOLOGY

Double-tuned transformers are a popular choice as wideband resonance loads for mm-wave LNAs to extend the operating frequency of the LNA. Traditionally, to achieve a wide-band load, the transimpedance parameter (Z_{21}) is used in a double-tuned transformer (Fig. 1(a)). But, as shown in Fig. 1(b), the input driving point (Z_{11}) can potentially have higher gain-bandwidth (GBW) compared to transimpedance (Z_{21}) as Z_{11} has an additional complex-conjugate zero pair that helps to increase the bandwidth of Z_{11} compared to Z_{21} [7]. Having inspired by this fact and the need to have both wide-band and reconfigurable LNAs to take advantage of both modes, a novel 3-winding transformer is presented that can simultaneously achieve reconfigurability, and wide bandwidth as shown in Fig. 1(c) which three ports (1, 2, and 3) can be used to achieve different operating frequency band. The proposed structure in Fig. 1(c) has one more inductor compared to two structures in Fig. 1(a), (b), which means having one pair of poles, therefore; it will have wider bandwidth compared to the conventional double-tuned transformer. The transformer is designed such that the coupling coefficient from the primary to each secondary is low, yet well controlled (e.g, $k_{12} = k_{13} = 0.21$), but the secondaries have negligible mutual coupling (i.e., $k_{23} \cong 0$). The resonators can be designed to have equal un-coupled resonant frequencies, but magnetic coupling causes pole-splitting action resulting in a wideband response. Fig. 2(a) shows the frequency response comparison between three cases: 1) Transimpedance (Z_{d21})

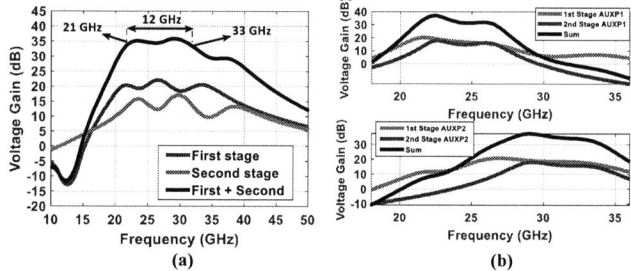

Fig. 4. (a) Simulated small signal voltage gain of the first and second stage in wide-band mode (ripple compensation). (b) low- frequency and high-frequency reconfigurable modes.

of a conventional double-tuned transformer, 2) Input driving impedance (Z_{p11}) of a conventional double-tuned transformer, and 3) Input driving impedance (Z_{t11}) of the proposed 3-winding transformer. As can be seen, for roughly equal values of inductors and coupling factors in three scenarios, Z_{t11} has higher frequency of operation (≈ 4 GHz) compared to other approaches. Also, Fig. 2(b) shows the frequency response for Z_{t11}, Z_{t21}, and Z_{t31}, where Z_{t11} mode covers entire frequency band, Z_{t21} supports lower frequency band, and Z_{t31} supports higher frequency band of operation. Z_{t11} response can be tailored to achieve wide bandwidth with low ripple, but with gradual stop-band skirts. Z_{t21} and Z_{t31} responses have lower ripple, but with lower bandwidth and sharper stop-band skirts. This load is used in the proposed LNA to reconfigure between these responses and trade bandwidth for out-of-band linearity.

III. LNA DESIGN

A. System-Level Architecture

The block diagram of the proposed mm-Wave wideband/reconfigurable LNA is shown in Fig. 3. The circuit employs three gain stages. The first stage starts with an input matching network to provide proper power transmission in the frequency band of interest. After the input matching network, there is the first G_m cell that amplifies the signal through the first stage of the proposed 3-winding transformer. At this stage, there are three possible

979-8-3503-2123-4/23 $31.00 © 2023 IEEE

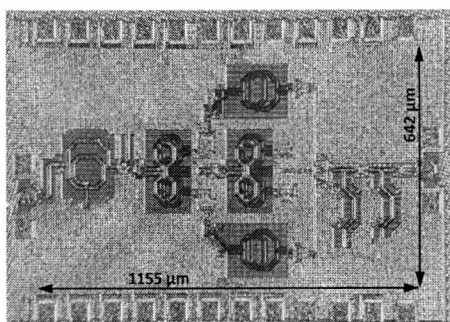

Fig. 5. Chip photograph of the proposed LNA.

paths for the signal, the main path in the middle marked as 'Main Path', and the two auxiliary paths marked as 'Aux Path1' and 'Aux Path 2' in Fig. 3. Then, the auxiliary paths (reconfigurable modes) each employ a G_m cell and a conventional double-tuned transformer to adjust the amplified signal to have a flat response in low and high frequency bands. The middle path is the main path where the signal is amplified and adjusted through the second stage of the proposed 3-winding transformer to have acceptable gain and flatness in the whole frequency band of operation. The selection of the function of the LNA between the three modes is done through the scan chain by turning on proper G_{m2i} and $G_{mbi}(i = 1, 2, 3)$ transconductance cells in Fig. 3. Finally, the signals are fed into a combiner that is consisted of three parallel G_m cells ($G_{mbi}(i = 1, 2, 3)$) where each one is ON depending on the selected mode. The buffer stage G_m cells ($G_{mb1}, G_{mb2}, G_{mb3}$) are designed in a cascode manner and all three parallel output paths of G_m cells are combined together and connected to the output matching network. Also, the capacitors C_{t1i} and C_{t2i} ($i = 1, 2, 3$) are tunable MOM capacitors which are optimized to provide maximum quality factor.

B. Circuit Design

The first stage consists of a cascode common-source (CS) with source degeneration including M1 and M2 which are biased and the layout of two transistors are customized to provide minimum NF over the entire frequency band of operation [11]. To compensate for the ripples out of the first 3-winding transformer, the second stage is designed to ensure gain flatness of all three parallel paths by skewing the pole positions of Z_{t11} the second 3-winding transformer in the main path compared to the Z_{t11} of the first 3-winding transformer and by properly locating the poles of Z_{d21} of the double-tuned transformers in both upper and lower paths. To ensure better isolation between the first and second stages, a triple-stacked structure is used in the G_m cells of the second stage ($G_{m21}, G_{m22}, G_{m23}$). Fig. 4(a) shows the simulated voltage gain of each stage and also the total voltage gain for the wideband mode when the main path is active. Also, Fig. 4(b) top and bottom show the simulated voltage gain for auxiliary paths 1 and 2 (low- and high-frequency modes), respectively.

Table 1. Performance comparison with the state of the art mm-wave LNAs

Ref. & Tech.	Freq. (GHz)	Peak Gain (dB)	Min NF (dB)	Type	Min OP1dB (dBm)	Power (mW)
This	**21.6-34.2**	**32**	**2.3**	**WB[1]**	**-6**	**35@1.2/1.8**
work	**21.5-26.5**	**32.4**	**2.7**	**RC[2]**	**-7**	**35@1.2/1.8**
22 FDSOI	**27.5-33**	**33.4**	**2.3**	**RC[2]**	**-6.2**	**35@1.2/1.8**
[11] 22 FDSOI	23.7-30.3	22	2.55	CC[3]	-10.5[4]	18
	38-42.7	16			-7	@0.8/1
[7] 65 CMOS	24.4-32.3	24.4	4	WB[1]	0.4	22 @1.1
[9] 130 SiGe	28/60	16.2	2.8	RC[2]	3.2	8.2
		15	3.35		7	21
[4] 22 FDSOI	24-43	23	3.1	WB[1]	-4	20.5 @1/1.6
[3] 22 FDSOI	19-36	21.5	1.7	WB[1]	-	17.03 @1.05
[10] 45 SOI	24/38 /39	9.5/12 /15.5	4.5	RC[2]	-3.5 -4.1	20.28 @1.5

[1] WB = Wideband. [2] RC = Reconfigurable. [3] CC = Concurrent. [4] Approximated as $(Gain_{dB} - 1) + iP_{1dB}$ when measurement result not available.

The supply voltage for the first, second, and buffer stages are 1.2 v and 1.8 v, and 1.2, and the current consumptions are 12 mA, 8 mA (each of the three parallel paths in the second stage consumes 8 mA but only one of which is ON at a time), and 10 mA respectively.

C. Input/Output Matching Circuits

The input matching circuit is designed to provide one zero in low frequency (around 12 GHz) to ensure filtering unwanted low frequency out-of-band interference and has two poles at 24 and 28 GHz where their location is determined by the coupling factor of the two inductors (180 pH and 400 pH) set ot be ~ 0.15. Also, to ensure maximum power transfer at the output, a second order output matching network with two poles at 23 and 34 GHz is designed to ensure S22 < -10 dB for the entire frequency band of interest.

IV. MEASUREMENT RESULTS

The proposed mm-wave wideband/reconfigurable LNA is fabricated in 22-nm CMOS FDSOI technology from Global Foundries with the area of 1155 μm × 642 μm (excluding pads). The die photograph is shown in Fig. 5. According to Fig. 5, the chip is measured using dc-probes on top and bottom of the chip for dc-biases and RF-probes for input (left side of the chip) and output (right side of the chip). The effect of probes are de-embedded the using through-open-short-match (TOSM) calibration substrate. The S-parameter, P1dB, and IIP3 are measured using Rohde & Schwarz ZVA67 network analyzer. Also, NF is measured using the Rohde & Schwarz FSV40 spectrum analyzer and Keysight 346CK01 noise source. The proposed LNA consumes 35 mW for both wideband and reconfigurable modes, respectively.

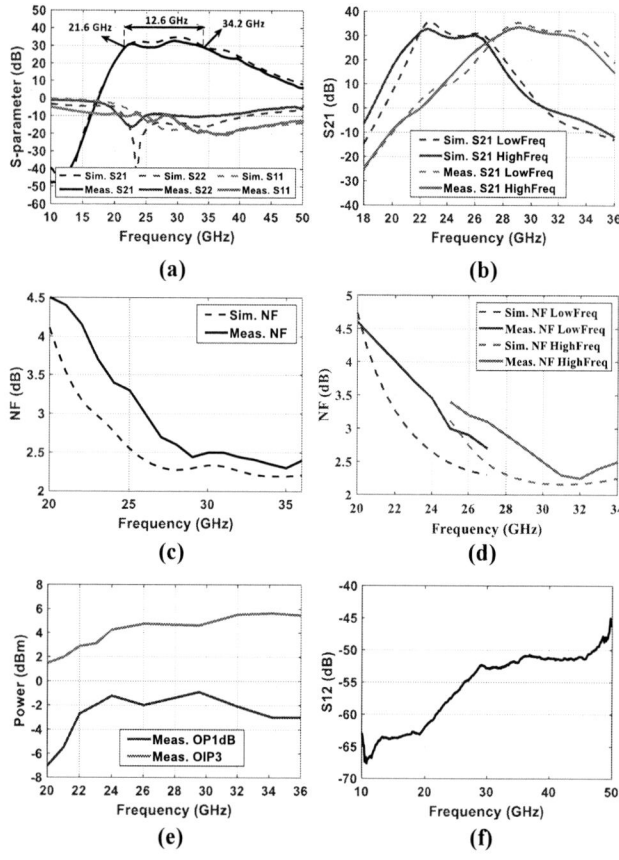

Fig. 6. Measured and simulated results of the proposed LNA. (a) The S-parameter for wideband mode. (b) S21 for reconfigurable mode. (c) Wide-band mode NF. (d) Reconfigurable mode NF. (e) OIP3 and OP1dB for wideband mode. (f) S12 for wideband mode.

Fig. 6(a), (b) show the measured results for S21, S11, and S22 versus frequency for three modes of operation. As shown, the measured 3-dB gain bandwidth of S21 for wide-band mode is from 21.6 to 34.2 GHz with a peak gain of 32 dB at 30 GHz which matches the simulation results of S21. The measured low-frequency mode of S21 is shown in Fig. 6(b) with 3-dB gain bandwidth from 21.5 to 26.5 GHz by the peak gain of 32.4 at 22 GHz. Also, the 3-dB gain bandwidth of the measured high-frequency mode of S21 is from 27.5 to 33 GHz by the peak gain of 34.4 dB at 29 GHz. Both measured S21 values for low-and high-frequency modes are also very close to the simulation results. The measured S11 and S22 are < -10 dB from 20 to 36 GHz which covers the entire frequency band of interest. Also, S11 and S22 for reconfigurable modes are identical to wide-band mode since the first stage is the same. Fig. 6(c), (d) show the measured and simulated NF for both wide-band and reconfigure modes. The minimum measured NF is 2.3 at 35 GHz. Fig. 6(e) shows the OIP3 and OP1dB measurement results of the proposed LNA which the maximum measured OIP3 and OP1dB of -0.9 and 5.8, are achieved. Finally, Fig. 6(f) shows measured S12 < -45 dB for the entire frequency.

The chip performance and comparison with the state-of-the-art is summarized in Table 1. The proposed LNA in wideband mode has comparable and even better results compared to other wideband LNAs. To our knowledge, this is the first LNA that can work in both wideband and reconfigurable modes.

V. CONCLUSION

A mm-Wave wideband/reconfigurable LNA using a 3-winding transformer load has been implemented in 22-nm CMOS FDSOI technology for 5G applications. The LNA shows measured S21 by the peak gain of 32 at 30 GHz and 3-dB bandwidth of 12.6 GHz for wide-band mode and peak gain of 32.4 dB and 34.4 dB for 3-dB bandwidth of 5 and 5.5 GHz using two auxiliary paths for low- and high-frequency modes, respectively. The measured NF is lower than 4.5 dB for the entire band for both modes. The total power consumption of the proposed LNA is 35 mW for both modes of operation.

ACKNOWLEDGMENT

This project is founded through NSF award No. 2116498. The authors would like to thank Global Foundries for the chip fabrication.

REFERENCES

[1] M. G. Bardeh, J. Fu, N. Naseh, J. Paramesh, and K. Entesari, "A wideband low RMS phase/gain error mm-wave phase shifter in 22-nm CMOS FDSOI," *IEEE Microwave and Wireless Technology Letters*, 2023.

[2] M. G. Bardeh, N. Naseh, J. Fu, J. Paramesh, and K. Entesari, "A mm-wave RC PPF quadrature network with gain boosting in 22nm CMOS FDSOI," in *2023 IEEE Radio and Wireless Symposium (RWS)*. IEEE, 2023, pp. 108–110.

[3] B. Cui and J. R. Long, "A 1.7-dB minimum NF, 22–32-GHz low-noise feedback amplifier with multistage noise matching in 22-nm FD-SOI CMOS," *IEEE Journal of Solid-State Circuits*, vol. 55, no. 5, pp. 1239–1248, 2020.

[4] L. Gao and G. M. Rebeiz, "A 24-43 GHz LNA with 3.1-3.7 dB noise figure and embedded 3-pole elliptic high-pass response for 5G applications in 22 nm FDSOI," in *2019 IEEE Radio Frequency Integrated Circuits Symposium (RFIC)*. IEEE, 2019, pp. 239–242.

[5] A. Ershadi, S. Palermo, and K. Entesari, "A 22.2-43 GHz gate-drain mutually induced feedback low noise amplifier in 28-nm CMOS," in *2021 IEEE Radio Frequency Integrated Circuits Symposium (RFIC)*. IEEE, 2021, pp. 27–30.

[6] M. Elkholy, S. Shakib, J. Dunworth, V. Aparin, and K. Entesari, "A wideband variable gain LNA with high OIP3 for 5G using 40-nm bulk CMOS," *IEEE Microwave and Wireless Components Letters*, vol. 28, no. 1, pp. 64–66, 2017.

[7] R. Singh, S. Mondal, and J. Paramesh, "A millimeter-wave receiver using a wideband low-noise amplifier with one-port coupled resonator loads," *IEEE Transactions on Microwave Theory and Techniques*, vol. 68, no. 9, pp. 3794–3803, 2020.

[8] M. El-Nozahi, E. Sánchez-Sinencio, and K. Entesari, "A millimeter-wave (23–32 GHz) wideband BiCMOS low-noise amplifier," *IEEE Journal of Solid-State Circuits*, vol. 45, no. 2, pp. 289–299, 2010.

[9] A. A. Nawaz, J. D. Albrecht, and A. Ç. Ulusoy, "A Ka/V band-switchable LNA with 2.8/3.4 dB noise figure," *IEEE Microwave and Wireless Components Letters*, vol. 29, no. 10, pp. 662–664, 2019.

[10] R. A. Shaheen, T. Rahkonen, and A. Pärssinen, "Millimeter-wave frequency reconfigurable low noise amplifiers for 5G," *IEEE Transactions on Circuits and Systems II: Express Briefs*, vol. 68, no. 2, pp. 642–646, 2020.

[11] J. Fu, M. G. Bardeh, J. Paramesh, and K. Entesari, "A Millimeter-Wave Concurrent LNA in 22-nm CMOS FDSOI for 5G applications," *IEEE Transactions on Microwave Theory and Techniques*, 2022.

979-8-3503-2123-4/23 $31.00 © 2023 IEEE

RMo3B-2

High-Performance Broadband CMOS Low-Noise Amplifier with a Three-Winding Transformer for Broadband Matching

Joon-Hyung Kim[#1], Jeong-Taek Son[#], Jung-Taek Lim[#], Jae-Eun Lee[#], Jae-Hyeok Song[#], Min-Seok Baek[#], Han-Woong Choi[#], Eun-Gyu Lee[#], Sunkyu Choi[#], Chong-Min Lee[*], Sung-Ku Yeo[*], Choul-Young Kim[#]

[#]SCD Lab, Chungnam National University, Republic of Korea
[*]Smart Device Team, Samsung Research in Samsung Electronics, Republic of Korea

Abstract—**This study presents a 32-to-46-GHz two-stage low noise amplifier (LNA) with a three-winding transformer-based input matching network. The proposed matching network provides a broadband noise and input matching for the entire operating range without performance degradation. To demonstrate the feasibility of the proposed circuit configuration, the LNA is implemented using a 65-nm bulk complementary metal-oxide-semiconductor (CMOS) process. The measured LNA achieved a gain of > 19.2 dB at 32–46 GHz with a peak of 21.5 dB at 32 and 45 GHz, simultaneously. The minimum noise figure of the fabricated LNA was 2.2 dB at 36.5 GHz, and remains below 3.2 dB across 14 GHz. The input and output return losses were < -10dB from 32 to 45.5 GHz (effective bandwidth). The third-order input intercept (IIP₃) point was -7.6 dBm at 38 GHz at the lowest gain when dissipating 22 mA with a 1-V supply voltage.**

Keywords—**fifth-generation (5G), broadband, complementary metal-oxide-semiconductor (CMOS) Q-band, transformer, low noise amplifier (LNA).**

I. INTRODUCTION

A low noise amplifier (LNA) is a key block in the receiver circuit. The increase in bandwidth and decrease in the noise figure at the receiver front-ends contribute to the development of wireless communication networks in millimeter waves (mm-waves). Systems on a Chip (SoCs) integrated with array antennas are expected to use millimeter-wave frequency bands to provide gigabit/s data throughput at low costs. Therefore, it is essential to obtain an improved chip performance over a small area. The bulk - complementary metal-oxide-semiconductor (CMOS) process, which can be mass-produced at a low cost, has a disadvantage as the performance degradation due to passive components is severe because the Q-factor is low. Therefore, high-performance CMOS LNAs represent a bottle-neck to realizing the effects SoCs. The conventional input-matching network of LNAs has shown a limited low-noise performance owing to the presence of low-quality passive networks on the lossy silicon substrate. To reduce the degradation resulting from the input matching network of the CMOS process, an LNA with a large transistor was proposed in [1]. The large transistor enabled an ultra-low noise performance by eliminating or reducing the lossy series gate inductor. However, this approach is limited in terms of obtaining broadband noise characteristics. To achieve a broadband low-noise performance, a transformer-based LNA is selected was identified as a feasible solution, and research using these systems has been conducted for more than 20 years since the 2000s [6]–[12].

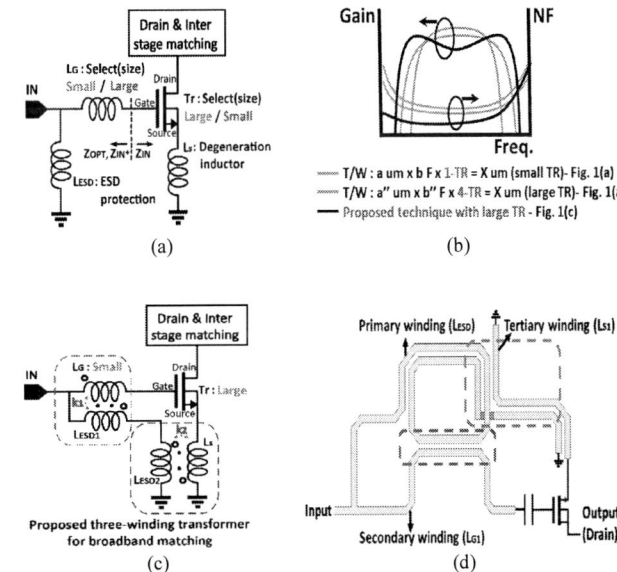

Fig. 1. (a) Simplified architecture of the large transistor used in [2]. (b) Characteristic of the gain and noise figure (NF) according to the transistor width and proposed technique. (c) Proposed three-winding transformer for the broadband matching technique. (d) Structure of Fig. 1(c).

In this study, we propose a three-winding transformer structure with a large transistor for broadband matching. This broadband LNA has 2.7 ± 0.5 dB NF and 20.3 dB average gain in frequency band (32 to 46 GHz). We defined a case where the |S₂₁| gain is within -3 dB of the peak and |S₁₁| and |S₂₂| are less than -10dB as BW_eff; It is 13.5 GHz (32–45.5 GHz).

The remainder of this paper is organized as follows. Section II introduces the proposed LNA broadband-matching technique. Section III provides the measurement results for the S-parameter, third-order input intercept (IIP₃), and noise figure. A comparison with state-of -the-art broadband LNAs using the transformer technique for low-noise and bandwidth was also carried out. Finally, the conclusions are presented in Section IV.

II. CIRCUIT DESIGN

Prior to the LNA design, we set a gain target of > 20 dB, sub-3 dB NF, effective bandwidth (BW_eff) > 10 GHz, input and output return losses < -10 dB, power consumption < 25 mW, and IIP3 > -10 dBm. This is the minimum target to achieve a performance comparable to those of state-of-the-art broadband LNAs using the CMOS process (bulk, silicon on insulator

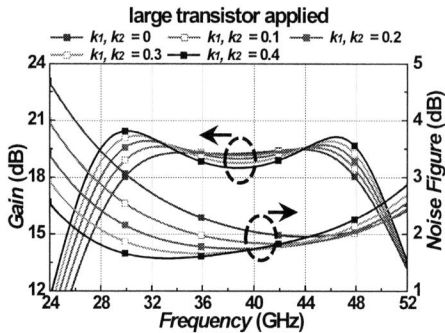

Fig. 2. Comparison of the gain and noise figures according to changes in k_1 and k_2 in a circuit with applied a large transistor (k = coupling factor).

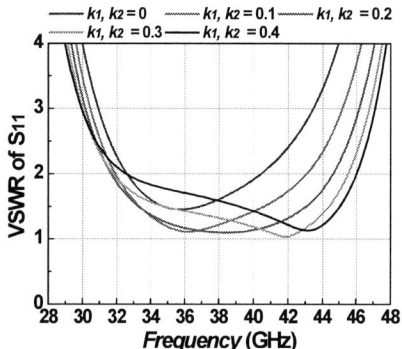

Fig. 3. Simulated VSWR of S_{11}

(SOI)). In the first stage of the proposed LNA, the system was constructed as a single structure for NF performance. Z_{IN}^* and Z_{OPT} (optimum noise impedance) were relocated to the admittance circle ($j0.02$ S) using the L_{G1} (series gate inductor), and Z_{IN}^* and Z_{OPT} were relocated to the 50-Ω point with L_{ESD} (ESD protection inductor). L_{S1} (degeneration inductor) has been widely used in LNA optimization techniques [3]–[5]. This method can simultaneously perform noise and input matching. In the second stage, a differential structure is constructed for linearity. To obtain a good LNA performance, several techniques, including the large-transistor technique, proposed three-winding transformer technique for broadband matching, and peak gain distributions using L_{G1}, L_{G2}, L_{D1}, L_{D2}, and L_{S2}, were applied.

A simplified architecture representing a large transistor is shown in Fig. 1(a). Using this system, a wide range of effective bandwidths can be achieved. Fig. 1(a) shows the improved noise performance achieved using a large transistor [2]. The size of L_{G1} was determined according to the size of the transistor (M1). If L_G is large and M_1 is small, the power consumption is low; however, NF increases. Conversely, if L_G is small and M_1 is large, the power consumption is high, but NF is reduced. In addition, as shown in the measurement results in [1], it is difficult to obtain broadband characteristics when L_G is eliminated. We used L_G for its low-power consumption and broadband characteristics. Fig. 1(b) shows the gain and NF characteristics according to the transistor width and the proposed three-winding transformer technique. Fig. 1(c) shows

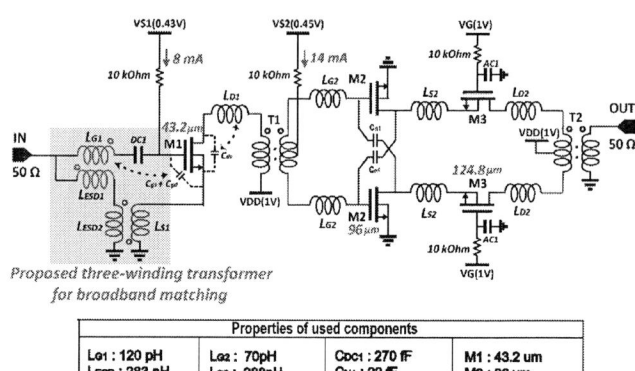

Fig. 4. Full schematic of the broadband LNA

the proposed three-winding transformer structure with a large transistor for broadband matching. Here, the primary winding system (L_{ESD}) was coupled with the secondary winding (L_{G1}) and tertiary winding (L_{S1}) systems. Fig. 1(d) shows the structure of the proposed three-winding transformer for broadband matching. L_{ESD}, L_{G1}, and L_{S1} comprise only the top metal layer to minimize high ohmic losses. According to the theory presented in [3], the degeneration inductor L_{S1} theoretically enables simultaneous noise and input matching without increasing NF_{min}. However, NF_{min} is degraded owing to the physical limitations of L_{S1}. This problem is overcome via the primary winding (L_{ESD}) and tertiary winding (L_{S1}) systems. The L_{ESD} and L_{S1} coupling method is widely used to reduce the loss of passive networks [6]–[7]. This method has been proven to enable bandwidth expansion, leading to a lower NF. However, it has limitations in achieving improved broadband characteristics.

The proposed three-winding transformer matching technique with a large transistor provides a wider BW. This matching technique consists of two feedback transformers: the coupling between L_{ESD} and L_{S1}, and the coupling between L_{ESD} and L_{G1}. As shown in Fig. 1(d), the primary and secondary windings generate mutual inductors (L_M) arranged in series with C_P (parasitic capacitor present in L_{ESD} owing to coupling in L_{ESD} and L_{G1}) arranged in the shunt and C_S (parasitic capacitor present in L_{G1} owing to coupling in L_{ESD} and L_{G1}) arranged in the shunt. It is composed of a band pass filter (series L_M and shunt capacitor C_P or C_S) and resonates. As a result, two feedback transformers, L_{ESD}-L_{G1} and L_{ESD}-L_{S1}, acquired two band-pass filter characteristics. These two filters generated a total of four resonant frequencies, which were distributed at low and high frequencies. These transformers achieved a wide bandwidth when optimized while adjusting the coupled area. Therefore, the proposed winding technique achieved a wide broadband NF performance.

Fig. 2 shows the simulated gain and NF achieved by sweeping k_1 and k_2 in Fig. 1(c) from 0 to 0.4. An increase in k_1 and k_2 disperse the gain and improves NF by alleviating the lossy silicon substrate. Fig. 3 shows the VSWR of S11 according to the k_1 and k_2 sweeps of the proposed broadband LNA. As shown in Figs. 2 and 3, an increase in the magnetic

Fig. 5. Chip microphotograph of the LNA.

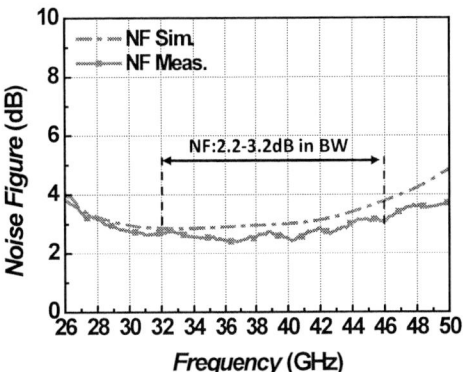

Fig. 7. Simulated and measured NF

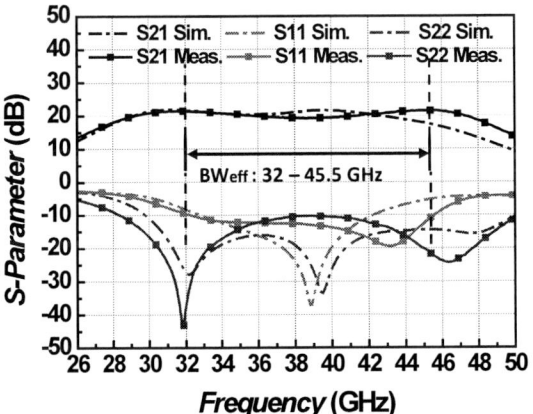

Fig. 6. Simulated and measured S-parameter values

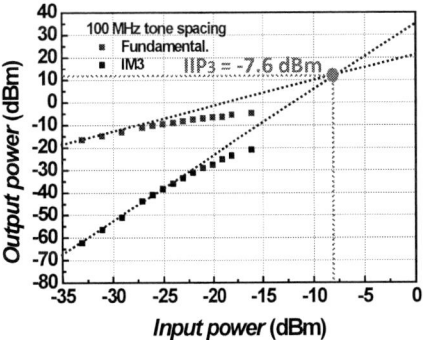

Fig. 8. Measured IIP₃ with100-MHz tone spacing at 38GHz.

coupling of the proposed three-winding transformer structure assists in the gain, NF and the input return loss ($|S11|$) bandwidth expansion. The fractional bandwidth of BW_{eff} is 22% when k_1 and k_2 are 0, and 36% when k_1 and k_2 are 0.4. Parasitic capacitors (C_{gs}, C_{gd}, and C_{ds}) are presented in the transistors, and they have a significant impact on the circuit performance as the frequency increases. Each parasitic capacitance generates a pole that determines the frequency response of the LNA [13]. These components, together with the three-winding transformers, generate band-pass filters. These can be used to adjust the gain distribution. The results show that the poles along the left-hand y-axis (gain) in Fig. 2 are distributed over a wide frequency range according to changes in k_1 and k_2. Fig. 4 shows a full schematic of the broadband LNA. M1 comprises a 43.2-μm total width (unit-transistor size of 1.8 μm x 6 fingers) and consumes 8 mA. The frequency response of the second stage was expanded using inductive peaking (L_{D1}, L_{G2}, L_{S2}, and L_{D2}). In the second stage, linearity was obtained using a differential structure. C_{n1}(neutralization cross capacitor) was used to compensate for the parasitic capacitance between the gate and drain, thereby increasing gain and stability.

III. SIMULATION AND MEASUREMENT

A micrograph of the LNA prototype fabricated using the 65-nm bulk CMOS process is shown in Fig. 5; the core chip size was 0.56× 0.28 mm. All measurements, including the NF and

linearity, were performed using GSG on-wafer probing conditions and a Keysight N5247B network analyzer (PNA-X). For a supply voltage of 1 V and gate bias voltages $V_{S1} = 0.43$ V and $V_{S2} = 0.45$ V, the proposed LNA consumed a power of 22 mW. Fig. 6 shows the measured and simulated input ($|S_{11}|$), output ($|S_{22}|$) return losses, and $|S_{21}|$ gain. The peak gain was 21.5 dB at 32 and 45 GHz. -3-dB bandwidth was 20 GHz (28.1 to 48.1 GHz), and measured BW_{eff} was 13.5 GHz (32–45.5 GHz). Fig. 7 shows the simulated and measured NF values obtained using the PNA-X with the cable loss calibrated. Within BW_{eff}, the measured NF was 2.2–3.2 dB. Fig. 8 shows the measured IIP₃ at 38 GHz. Two input tones near 38 GHz (i.e., at the lowest gain) with a frequency separation of 100 MHz were applied at the input to determine IP₃. Equal power levels of -33 dBm (i.e., near small-signal) were used for the fundamental tones. The measured IIP₃ was -7.6 dBm. Table I summarizes the state-of-the-art mm-wave broadband LNAs using the transformer technique for performance comparison. Our LNA shows the lowest NF within BW_{eff} among all compared topologies. Moreover, a comparable gain, -3 dB BW, power consumption, IIP₃, and chip size were obtained.

$$FOM_1 = \frac{Gain[lin.]}{(NF[lin.]-1)\times P_{DC}[mW]} \quad (1)$$

$$FOM_2 = \frac{BW[GHz]\times Gain[lin.]}{(NF[lin.]-1)\times P_{DC}[mW]} \quad (2)$$

IV. CONCLUSION

A novel input matching technique featuring broadband low-NF, input matching, and gain characteristics was presented. The proposed matching technique generated feedback transformers

Table 1. Performance comparison with state-of-the-art broadband LNAs.

	This work	**MWCL'22[8]**	**TMTT'21[9]**	**MWCL'22[10]**	**RFIC'19[11]**	**RFIC'21[12]**
Topology	65-nm CMOS	65-nm CMOS	65-nm CMOS	28-nm CMOS FDSOI	22-nm CMOS FDSOI	28-nm CMOS
Number# of Stages	2	2	2	2	3	3
RF frequency [GHz]	32-46	37-40	19.2-38.9	21.3-41.1	24-43	22.2-43
Gain [dB]	19.2-21.5	13.04	13.5	20.3	17-23	21.1
-10 dB |S11|, |S22| BW$_{eff}$ [GHz]	32-45.5	37-40	NR*	22.5-31**	20-35**	32-45**
-3 dB BW [GHz]	28.1-48.1	NR*	19.2-42.6	21.3-41.1	20-44	21.8-43
NF [dB]	2.2-3.2 (avg. 2.7)	4.47	3.1-4.5	2.92-3.61	3.1-3.7	3.5-5.3***
Power consumption [mW]	22	11.6	6.36	27.5	20.5	22.3
IIP$_3$ [dBm]	-7.6 @38	-6.85	-3@28	-15.7@28	-13 ~ -19	-3
Core area [mm^2]	0.16	0.17	0.13	0.13	0.22	0.22
FOM1	9.7	1	3.4	4.1	9.3	4.7
FOM2	194.7	-	79.1	80.5	224.2	98.9

* NR: not reported **graphically estimated ***measured up to 40GHz

using three passive components (L_G, L_{ESD}, and L_{S1}) to obtain broadband characteristics. In addition, the parasitic capacitors of the transistor generated a band-pass filter with each inductor and achieved pole distribution. These techniques increased the bandwidth without performance degradation. The proposed LNAs demonstrated superior noise and BW$_{eff}$ performances over comparable CMOS (bulk or SOI) LNAs using the transformer technique in the 5G mm-wave frequency band.

ACKNOWLEDGMENT

This work was supported by the National Research Foundation of Korea(NRF) grant funded by the Korea government(MSIT) (No. NRF-2019R1A2C1004805 and NRF-2021R1A4A1032580).

REFERENCES

[1] H. -W. Choi, S. Choi and C. -Y. Kim, "Ultralow-Noise Figure and High Gain Ku-Band Bulk CMOS Low-Noise Amplifier With Large-Size Transistor," in IEEE Microwave and Wireless Components Letters, vol. 31, no. 1, pp. 60-63, Jan. 2021, doi: 10.1109/LMWC.2020.3037296.

[2] H. -W. Choi, C. -Y. Kim and S. Choi, "6.7–15.3 GHz, High-Performance Broadband Low-Noise Amplifier With Large Transistor and Two-Stage Broadband Noise Matching," in IEEE Microwave and Wireless Components Letters, vol. 31, no. 8, pp. 949-952, Aug. 2021, doi: 10.1109/LMWC.2021.3092742.

[3] Trung-Kien Nguyen, Chung-Hwan Kim, Gook-Ju Ihm, Moon-Su Yang and Sang-Gug Lee, "CMOS low-noise amplifier design optimization techniques," in IEEE Transactions on Microwave Theory and Techniques, vol. 52, no. 5, pp. 1433-1442, May 2004, doi: 10.1109/TMTT.2004.827014.

[4] Hung-Wei Chiu, Shey-Shi Lu and Yo-Sheng Lin, "A 2.17-dB NF 5-GHz-band monolithic CMOS LNA with 10-mW DC power consumption," in IEEE Transactions on Microwave Theory and Techniques, vol. 53, no. 3, pp. 813-824, March 2005, doi: 10.1109/TMTT.2004.842510.

[5] C. -Y. Wu, Y. -K. Lo and M. -C. Chen, "A 3–10 GHz CMOS UWB Low-Noise Amplifier With ESD Protection Circuits," in IEEE Microwave and Wireless Components Letters, vol. 19, no. 11, pp. 737-739, Nov. 2009, doi: 10.1109/LMWC.2009.2032022.

[6] J. R. Long, "Monolithic transformers for silicon RF IC design," in IEEE Journal of Solid-State Circuits, vol. 35, no. 9, pp. 1368-1382, Sept. 2000, doi: 10.1109/4.868049.

[7] M. T. Reiha and J. R. Long, "A 1.2 V Reactive-Feedback 3.1–10.6 GHz Low-Noise Amplifier in 0.13 µm CMOS," in IEEE Journal of Solid-State Circuits, vol. 42, no. 5, pp. 1023-1033, May 2007, doi: 10.1109/JSSC.2007.894329.

[8] B. Bae, E. Kim, S. Kim and J. Han, "Dual-Band CMOS Low-Noise Amplifier Employing Transformer-Based Band-Switchable Load for 5G NR FR2 Applications," in IEEE Microwave and Wireless Components Letters, 2022, doi: 10.1109/LMWC.2022.3218001.

[9] H. Chen, H. Zhu, L. Wu, W. Che and Q. Xue, "A Wideband CMOS LNA Using Transformer-Based Input Matching and Pole-Tuning Technique," in IEEE Transactions on Microwave Theory and Techniques, vol. 69, no. 7, pp. 3335-3347, July 2021, doi: 10.1109/TMTT.2021.3074160.

[10] J. Lee and S. Hong, "A 21‐41-GHz Common-Gate LNA With TLT Matching Networks in 28-nm FDSOI CMOS," in IEEE Microwave and Wireless Components Letters, vol. 32, no. 9, pp. 1051-1054, Sept. 2022, doi: 10.1109/LMWC.2022.3169066.

[11] L. Gao and G. M. Rebeiz, "A 24-43 GHz LNA with 3.1-3.7 dB Noise Figure and Embedded 3-Pole Elliptic High-Pass Response for 5G Applications in 22 nm FDSOI," 2019 IEEE Radio Frequency Integrated Circuits Symposium (RFIC), 2019, pp. 239-242, doi: 10.1109/RFIC.2019.8701782.

[12] A. Ershadi, S. Palermo and K. Entesari, "A 22.2-43 GHz Gate-Drain Mutually Induced Feedback Low Noise Amplifier in 28-nm CMOS," 2021 IEEE Radio Frequency Integrated Circuits Symposium (RFIC), 2021, pp. 27-30, doi: 10.1109/RFIC51843.2021.9490469.

[13] H. -W. Choi, S. Choi and C. -Y. Kim, "A CMOS Band-Pass Low Noise Amplifier with Excellent Gain Flatness for mm-Wave 5G Communications," 2020 IEEE/MTT-S International Microwave Symposium (IMS), 2020, pp. 329-332, doi: 10.1109/IMS30576.2020.9224097.

RMo3B-3

A 28-GHz 12-dBm IIP3 Low-Noise Amplifier Using Source-Sensed Derivative Superposition of Cascode for Full-Duplex Receivers

Jonghoon Myeong, Byung-Wook Min

School of Electrical & Electronic Engineering, Yonsei University, Republic of Korea

aprilsky@yonsei.ac.kr, bmin@yonsei.ac.kr

Abstract—This paper presents a highly linear CMOS low-noise amplifier (LNA) to handle self-interference in a full-duplex array. This LNA is designed using the derivative-superposition technique and source-sensed cascode topology. The third-order intermodulation (IM3) suppression is performed by applying different biases to each of the two cascode amplifiers. By using a thick-oxide transistor for the auxiliary amplifier, it operates reliably even at high VDD of cascode. The LNA achieves noise figure (NF) of 3.3-4.3 dB, 8.6-dB peak gain, −8-dBm input 1-dB gain compression point (IP1dB) and 12-dBm third-order input intercept point (IIP3) while consuming 14-mA current from 1.8-V supply voltage. The measured results show IM3 suppression of 8 dB depending on whether the auxiliary amplifier is turned on/off for linearization.

Keywords — derivative superposition, full-duplex, IIP3, IM3, linearity, low noise amplifier, self-interference, suppression.

I. INTRODUCTION

In recent 5G and 6G next-generation communication, full-duplex wireless has been frequently mentioned as one of the candidates. In a full-duplex system, transmission and reception are performed simultaneously, which can greatly improve spectrum efficiency. In order to enable full-duplex operation, a high power self-interference (SI) signal leaking from the transmitter (TX) to the receiver (RX) must be suppressed, and RX needs to have sufficiently high linearity. Various techniques for achieving isolation of TX-RX in the antenna interface and SI cancellation (SIC) in the RF/analog [1], [2] and digital [3] domains are being studied. Although several isolation improvement techniques have been studied, SI that is not sufficiently reduced in the antenna interface may saturate the LNA in the RX or degrade the overall sensitivity of RX.

In a phased-array system, the linearity required for a transceiver is significantly increased due to power combine, division loss and cross-talk interference as shown in Fig. 1. For this reason, LNA requires a linearization technique to alleviate the deterioration of sensitivity due to the increasing interference signals of array. To improve the linearity, several LNA linearization techniques have been studied extensively in low-frequency silicon process. As one of the techniques, the derivative superposition (DS) technique called multi-gate transistor (MGTR) [4], which cancels the third-order transconductance by adjusting the bias condition of two parallel transistors, is widely used. However, since this method shares the gate node, the minimum NF deteriorates due to the effect of parasitic capacitance at high frequencies. To solve this problem, a source-sensed derivative superposition

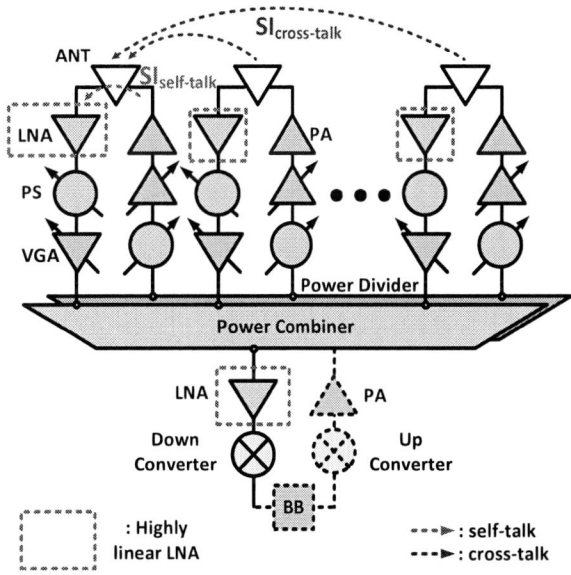

Fig. 1. Shared antenna full-duplex phased array system.

(SSDS) [5] technique that can minimize the deterioration of NF and input matching has been studied.

In this paper, we present a SSDS cascode LNA with high IIP3 and IP1dB which does not saturate even at high SI and does not deteriorate the sensitivity of RX. In Section II, input matching and noise of the conventional DS LNA and SSDS LNA are compared and the proposed SSDS cascode LNA for high linearity is introduced. A linearization mechanism using the third-order nonlinear component cancellation of SSDS cascode LNA and the role of thick-oxide transistor is explained. The measurement results are presented in Section III. Finally, Section IV concludes this article.

II. SSDS CASCODE LNA DESIGN

A. Multi-gate versus Source-Sensed Derivative Superposition

Since other previously studied LNAs perform IM3 suppression at the last stage among multiple stages, the DS structure, which shares the gate has been widely used as shown in Fig. 2(a). However, this is not suitable in a phased-array full-duplex radio. Because it can worsen input matching and NF in a system where large power comes from the front-end. Therefore, using a topology that receives a signal from the source without sharing the signal from the gate in the auxiliary amplifier, as shown in Fig. 2(b), has

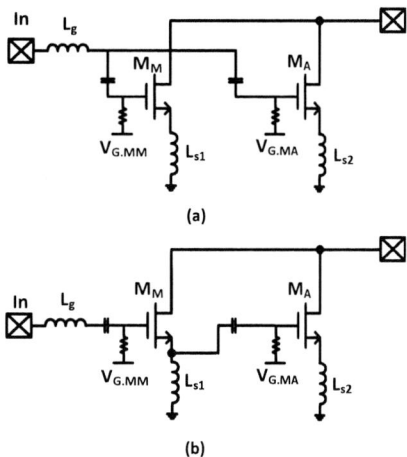

Fig. 2. Schematic of (a) multi-gate transistor DS LNA and (b) source-sensed DS LNA.

advantages in several respects. In the case of a conventional DS structure called multi-gate transistor (MGTR), the input impedance is changed and the operating frequency is reduced by adding C_{gs} capacitance compared to when there is only main transistor. Also, since the input capacitance is increased, a larger L_s is required for 50-Ω matching which has a great adverse effect on NF. However, if source-sensed topology is used, input matching can be performed with the same L_s and L_g as conventional LNAs [5]. At the same time, since the output capacitance increases, there is an advantage in that the size of L_d can be reduced because a smaller inductor is used to resonate. L_{s2} of the auxiliary amplifier is made of an appropriate size inductor according to the IM3 magnitude and phase value. At high frequencies, g_{m2} of the auxiliary amplifier affects IM3 due to feedback through L_s and C_{gs}. However, it can be treated as a constant regardless of the circuit components, and it does not affect because auxiliary amplifier operates with weak inversion [5]. Therefore, high IIP3 can be obtained by properly selecting L_{s2} of the auxiliary amplifier and tuning g_{m3}.

B. SSDS Cascode LNA Design

In the mm-Wave band, LNAs of two topologies, common source and cascode, are widely used in design. The cascode topology is preferred because it has many advantages over single transistor. It is one of the good candidates because the available gain is higher than single transistor. There have been many attempts to trade off other performance to improve the linearity of the cascode cell. As shown in Fig. 3(a), the proposed LNA consists of two separate cascodes. Among the RF transistors provided by Samsung 28nm LPP process, thin-oxide transistor with a nominal voltage of 1 V and thick-oxide transistor with a nominal voltage of 1.8 V were used. The drain voltage is 1.8 V, and a higher VDD than common-source LNA is selected to prevent saturation due to high power level of the SI. Since it is not a amplifier that requires a very high P1dB like a power amplifier, it was designed with a transistor that is not too large, and the gate

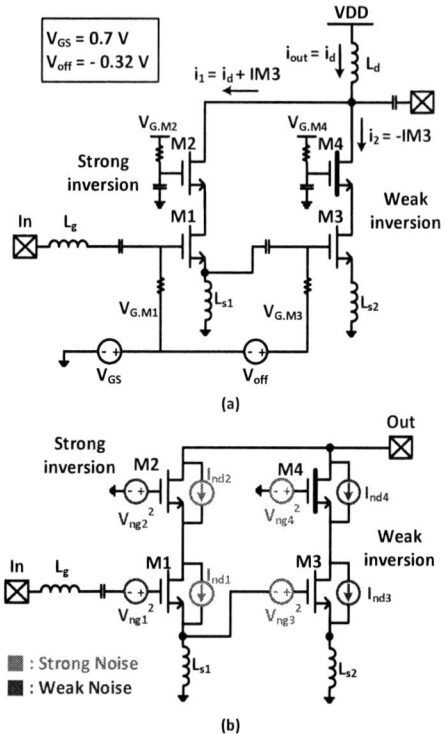

Fig. 3. Schematic of (a) the proposed SSDS cascode LNA and (b) noise source of SSDS cascode LNA.

width of 60 μm is used both thin- and thick-oxide transistor, respectively, considering power and NF. A DC block capacitor was used to separate the gate bias of the common-source amplifier. The cascode cell linearity can be adjusted by DC bias control. The main cascode composed of M1 and M2 operates with strong inversion, and the auxiliary cascode composed of M3 and M4 operates with weak inversion. The bias of M1 and M2 is set to operate with high gain like conventional LNA. We set $V_{G.M4}$ low to make M3 operate as a triode region. Since the average capacitance of the CMOS input varies according to the bias of the common-source amplifier, the auxiliary amplifier operating with weak inversion produces a third-order transconductance of opposite polarity to that of the main amplifier. If the bias of auxiliary amplifier is set well, positive third order transconductance can be obtained, which is cancelled by combining with the negative third-order transconductance created by the main amplifier. Phase tuning with a source degeneration inductor obtains better cancellation of IM3 for higher linearity. By using this characteristic, when the signals of the two paths are combined, the effect of cancelling each other in the third-order components is achieved, and higher linearity can be obtained than single cascode.

For higher linear operation than common-source amplifier, cascode is adopted and VDD is set to 1.8 V. At this time, if $V_{G.M4}$ is lowered, $V_{DS.M4}$ exceeds 1 V, the nominal voltage of thin-oxide transistor. In order to prevent breakdown, thick-oxide transistor is used. Even if $V_{G.M4}$ is sufficiently

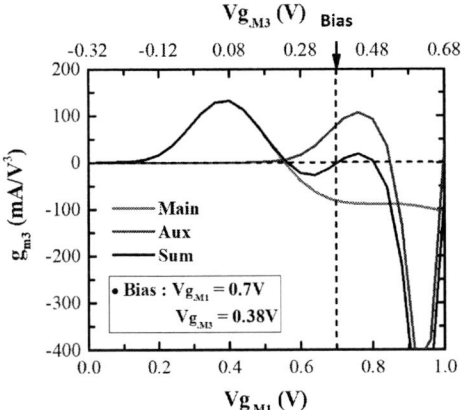

Fig. 4. Third-order transconductance coefficients versus Vgs of main amplifier M1,M2 and auxiliary amplifier M3,M4 in Fig.3(a).

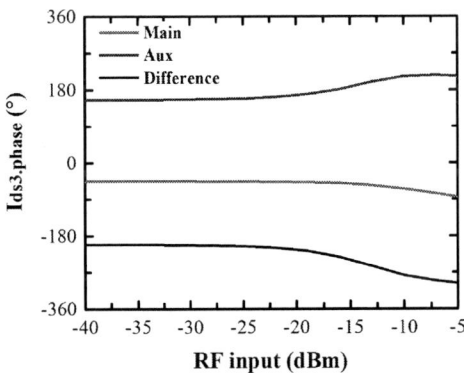

Fig. 5. Simulated phase differences of Ids3, third-order intermodulation current of main and auxiliary amplifier.

lowered to lower $V_{DS.M3}$ to operate as a triode, $V_{DS.M4}$ of thick-oxide transistor satisfies the nominal voltage of 1.8 V, so it can operate reliably. In addition, if the gate bias of the common-gate amplifier of the auxiliary cascode is set low, the drain voltage of the common-source amplifier of the auxiliary cascode can be lowered to boost the degree of weak inversion and nonlinearity components.

The DC bias of each device determines the magnitude and polarity of g_{m3}, and device size affects the scaling of the transconductance. The overall bias was adjusted appropriately by observing at the transconductance plot of both the main and the auxiliary amplifiers, as shown in Fig. 4. The size of M3 is determined by considering the magnitude of the third-order term of M1 and the sizing of L_{s2} for its phase tuning [10]. At low frequencies, it was not necessary to greatly consider the phase of the third-order term due to parasitic. This phase mismatch gets worse as the frequency increases. To tune this, two-tone simulation was performed and the IM3 current phase is checked while tuning the circuit parameters, as shown in Fig. 5. Since L_{s1} of the main amplifier was determined for input impedance and noise matching, the proposed LNA is designed using L_{s2} and $V_{G.M3,4}$ of auxiliary amplifier. Tuning L_{s2} can adjust the IM3 phase of the auxiliary amplifier, which

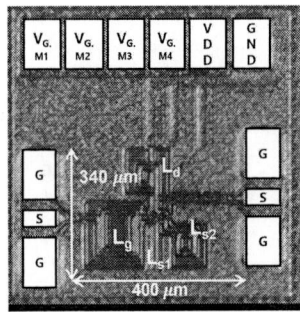

Fig. 6. Die micrograph of the proposed CMOS SSDS cascode LNA

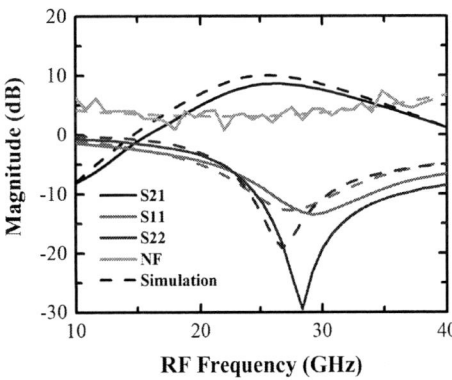

Fig. 7. Measured and simulated S-parameters of proposed LNA

does not greatly deteriorate input matching and NF due to the advantage of the source-sensed topology, as shown in Fig. 3(b). The two-tones are set to 28 GHz and 28.2 GHz, respectively, and cancellation can be seen through the phase difference of IM3 current at 27.8 GHz. When $V_{G.M1,2,3,4}$ are 0.7, 1.6, 0.38, 0.85 V and $L_{g,d,s1,s2}$ are 780, 190, 61, 184 pH, the IM3 phase difference between the two amplifiers is about 180 °, and cancellation is improved by making out of phase.

C. Noise Analysis of SSDS Cascode LNA

The noise contribution of MGTR DS and SSDS topology can be analyzed as follows. The drain current noise of transistor operating in weak inversion is negligible because the drain current is very small. On the other hand, induced gate noise cannot be ignored because it increases in inverse proportion to the drain current [6]. In the case of the DS method that shares the input at gate, the NF is degraded because the induced gate noise from the weak inversion transistor directly adds to the overall NF of LNA. However, as shown in fig. 3(b), in the SSDS topology, induced gate noise of the auxiliary transistor is added to drain noise of the main transistor, which is divided by gain of the LNA when input-reffered, so it does not significantly affect the overall NF. And the noise contribution from the cascode is small due to the inductor degeneration. Therefore, the SSDS cascode topology can achieve high linearity while consuming minimal additional power without significant degradation of gain and NF even if the auxiliary cascode operates in weak inversion.

979-8-3503-2123-4/23 $31.00 © 2023 IEEE

Table 1. Performance comparision of LNAs for mm-Waves.

	Process	Topology	Freq.* (GHz)	Gain (dB)	NF (dB)	IIP3 (dBm)	OIP3 (dBm)	IP1dB (dBm)	Power(mW)	VDD (V)
This work	28nm CMOS	SSDS cascode (Aux on)	21-34	8.6	3.3-4.3	12	21	-8	25.2	1.8
		SSDS cascode (Aux off)	21-34	8.0	3.2-4.2	8	17	-7	23.4	1.8
[7]	22nm FDSOI	Cascode Forward body bias	21-32	10.2	2.2	7.5	17.7	-3	15	1.6
[8]	45nm FDSOI	Cascode	14-31	12.8	1.4	5	17.8	-5	15	1.6
[9]	65nm CMOS	MGTR	35-40	18.2	4.2-5.6	7	25	-10.6	30.5	1.2
[10]	90nm CMOS	SSDS Common Source	58-65	11.3	4.8	0	11.3	-18	10.8	2.4

*3-dB bandwidth.

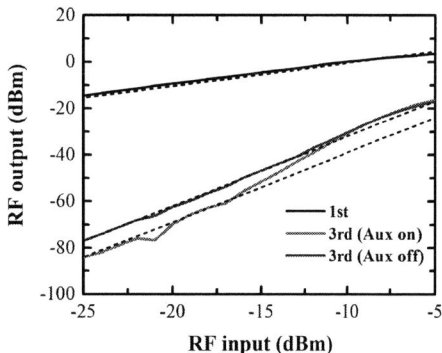

Fig. 8. Measured IM3 versus input power with/without the linearizer.

III. MEASUREMENTS

The SSDS cascode LNA is designed with chip size 0.136 mm² in Samsung 28nm CMOS process and the die micrograph is shown in Fig. 6. All LNA measurements were performed with GSG on-chip probing. A vector network analyzer was used for small-signal measurement, and 2-tone test was measured using two signal generators and a spectrum analyzer. The measured peak gain is 8.6 dB at 26.5 GHz, and the 3-dB bandwidth is 13 GHz, as shown in Fig. 7. The input and output return losses are better than 10 dB from 26 GHz to 33 GHz. The measured NF of the proposed LNA is 3.3-4.3 dB within 3-dB bandwidth. The proposed LNA consumes 14 mA from 1.8-V supply voltage. The two tones are combined using a 3-dB coupler and fed into the LNA input. Using two-tones of 28 GHz and 28.2 GHz with 200 MHz interval, the Pout measurement result according to Pin is shown in Fig. 8. IP1dB is -8 dBm and IIP3 is 12 dBm. Table 1 summarizes the performance of turning on/off the auxiliary amplifier of this work and the performance of other LNAs. When comparing proposed LNA to LNA without DS technique [7], [8], NF is worses but IP3 is higher. And when compared to LNAs where the DS thechnique [9], [10] is used, it is better in NF.

IV. CONCLUSION

We demonstrated a SSDS cascode LNA suppressing IM3 in mm-Wave on the 28nm CMOS technology. The LNA employs source-sensed derivative superposition topology to cancellate third-order nonlinear component and improves linearity using cascode topology. The SSDS cascode LNA achieves high IIP3 and 8-dB IM3 suppression at 28 GHz. This SSDS cascode LNA with high linearity can be one of the solutions that does not deteriorate the sensitivity of Rx with high power level of SI in a full-duplex system using a phased array.

ACKNOWLEDGMENT

This work was supported by Samsung Electronics Co., LTD (IO201209-07875-01), and IITP grant funded by the Korea government (MSIT) (No.2020000218). The EDA tool was supported by the IC Design Education Center(IDEC), Korea.

REFERENCES

[1] D. Lee and B. -W. Min, "Demonstration of Self-Interference Antenna Suppression and RF Cancellation for Full Duplex MIMO Communications," *2020 IEEE Wireless Communications and Networking Conference Workshops (WCNCW)*, 2020, pp. 1-4.

[2] K. Park, J. Myeong, G. M. Rebeiz and B. -W. Min, "A 28-GHz Full-Duplex Phased Array Front-End Using Two Cross-Polarized Arrays and a Canceller," in *IEEE Transactions on Microwave Theory and Techniques*, vol. 69, no. 1, pp. 1127-1135, Jan. 2021.

[3] D. Bharadia, E. McMilin, and S. Katti, "Full duplex radios," in *Proc. ACM SIGCOMM*, 2013, pp. 375–386.

[4] V. Lammert, P. Sakalas, A. Werthof, R. Weigel and V. Issakov, "Design and measurements of a 28 GHz High-Linearity LNA in 45nm SOI-CMOS," *2021 IEEE International Conference on Microwaves, Antennas, Communications and Electronic Systems (COMCAS)*, Tel Aviv, Israel, 2021, pp. 275-279.

[5] S. Ganesan, E. Sanchez-Sinencio and J. Silva-Martinez, "A Highly Linear Low-Noise Amplifier," in *IEEE Transactions on Microwave Theory and Techniques*, vol. 54, no. 12, pp. 4079-4085, Dec. 2006.

[6] V. Aparin and L. E. Larson, "Modified derivative superposition method for linearizing FET low-noise amplifiers," in IEEE Transactions on Microwave Theory and Techniques, vol. 53, no. 2, pp. 571-581, Feb. 2005.

[7] O. El-Aassar and G. M. Rebeiz, "Design of Low-Power Sub-2.4 dB Mean NF 5G LNAs Using Forward Body Bias in 22 nm FDSOI," in *IEEE Transactions on Microwave Theory and Techniques*, vol. 68, no. 10, pp. 4445-4454, Oct. 2020.

[8] C. Li, O. El-Aassar, A. Kumar, M. Boenke and G. M. Rebeiz, "LNA Design with CMOS SOI Process-1.4dB NF K/Ka band LNA," *IEEE/MTT-S International Microwave Symposium - IMS, 2018*, pp. 1484-1486.

[9] C. -N. Chen, Y. Chen, Y. Wang, T. -Y. Kuo and H. Wang, "38-GHz CMOS Linearized Receiver With IM3 Suppression, P1 dB/IP3/RR3 Enhancements, and Mitigation of QAM Constellation Diagram Distortion in 5G MMW Systems," in *IEEE Transactions on Microwave Theory and Techniques*, vol. 68, no. 7, pp. 2779-2795, July 2020.

[10] W. -T. Li et al., "Parasitic-Insensitive Linearization Methods for 60-GHz 90-nm CMOS LNAs," in *IEEE Transactions on Microwave Theory and Techniques*, vol. 60, no. 8, pp. 2512-2523, Aug. 2012.

979-8-3503-2123-4/23 $31.00 © 2023 IEEE

RMo3B-4

A SiGe BiCMOS D-Band LNA with Gain Boosted by Local Feedback in Common-Emitter Transistors

Guglielmo De Filippi, Lorenzo Piotto, Andrea Bilato, Andrea Mazzanti

Department of Electrical, Computer and Biomedical Engineering, University of Pavia, Italy

Abstract— The performance of silicon amplifiers in D-band is limited by the low gain of transistors operated close to f_{max}. Cascode stages, yielding higher gain than a single device, are commonly preferred, but the issue is only partially alleviated. Recognizing that the common-emitter (CE) transistor limits the gain of the cascode, this work investigates the use of a local reactive feedback to trade gain for stability. Feedback shifts the CE into a conditionally-stable operating region, and enables a gain beyond its maximum available gain (MAG). Then, when the CE is combined with a common-base, by properly selecting impedance terminations, the resulting cascode displays unconditional stability with superior gain. The concept is exploited in the design of a D-band LNA in BiCMOS 55 nm technology which shows 22.8 dB gain, 130-165 GHz operating frequency and NF down to 5 dB with 40 mW power consumption. Measured results compare favorably against state of the art.

Keywords— BiCMOS, LNA, D-Band, feedback, stability, millimeter wave integrated circuits

I. INTRODUCTION

The never ending demand for faster transfer rate in mobile applications and ubiquitous connectivity are posing severe requirements to the network infrastructure. To address this challenge, the wide bandwidth available beyond 100 GHz opens to wireless links with unprecedented capacity, competitive with fiber solutions. Significant spectrum portions have been already reserved in D-Band (110 - 170 GHz) for point-to-point links in 5G and beyond [1]. For this kind of applications, BiCMOS technology represents the optimal compromise between cost and performance. Several D-Band low noise amplifiers (LNAs) in Bipolar or BiCMOS have been demonstrated [2][3][4], proving the technology viability. Nonetheless, in current technologies, transistors are pushed to operate close to their frequency limit (f_{max}). The low available gain mandates careful selection of the active stage configurations and, possibly, investigation of unconventional design solutions to get the most out of the technology. The common-emitter (CE) gives the best noise performance and simplifies impedance matching, but its maximum available gain (MAG) is limited to few dBs only in D-band. Approaches aimed at improving power gain have been investigated, but end up with a cumbersome design approach and lead to a narrow-band response [5][6]. Most commonly, D-band amplifiers leverage straightforward cascode structures to rise the gain, stacking a common-base (CB) onto the CE [7][8]. In [9][10], interstage matching between CE and CB is also introduced to improve both gain and noise performance of the stack. Solutions to boost the gain for CB transistors, such as inductive base degeneration, have been deeply studied, with the drawback of further increasing the already high

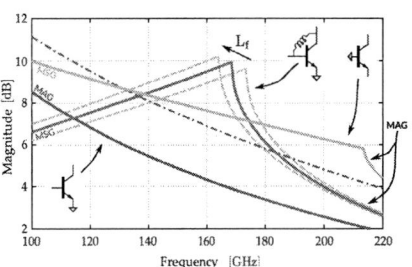

Fig. 1. Maximum power gain of CE, CB, CEf with L_f around 50 pH.

output impedance [11] and complicating matching networks implementation. Recognizing that the gain of a conventional cascode is limited by the CE, this paper investigates the use of local feedback to enhance the CE gain by shifting the device into a conditionally-stable operating region. Then, provided the impedance levels and matching networks are carefully selected, when the CE is combined with a CB, the resulting cascode structure displays superior gain with unconditional stability. The concept is exploited in the design of a two-stage D-band LNA which shows 22.8 dB gain with 33.5 GHz bandwidth around 148 GHz, NF down to 5 dB and 0 dBm output compression point with 40 mW power consumption. Measurements compare favorably against previously reported silicon amplifiers in D-band.

II. INDUCTIVE FEEDBACK ON THE COMMON EMITTER

As the operating frequency increases, the available power gain of a transistor drops, being unity at f_{max}. Moreover, beyond a certain corner frequency, which depends on the transistor configuration (e.g. CE or CB), the losses by parasitic elements make the device unconditionally stable. In this case, a useful figure of merit to evaluate gain capability is the maximum available gain (MAG):

$$\text{MAG} = \frac{S_{21}}{S_{12}} \left(K - \sqrt{K^2 - 1} \right) \quad (1)$$

with S_{21}, S_{12} the forward and reverse scattering parameters of the device and K > 1 the stability factor. In the unconditionally stable region, MAG defines an upper bound to power gain, achieved with simultaneous input/output conjugate match. On the other hand, under a certain frequency, K falls below unity and stability is constrained by a careful selection of input/output terminations. Within this regime, an alternative figure of merit is the maximum stable gain (MSG), which is (1) with K = 1:

$$\text{MSG} = \frac{S_{21}}{S_{12}} \quad (2)$$

979-8-3503-2123-4/23 $31.00 © 2023 IEEE   133   2023 IEEE Radio Frequency Integrated Circuits Symposium

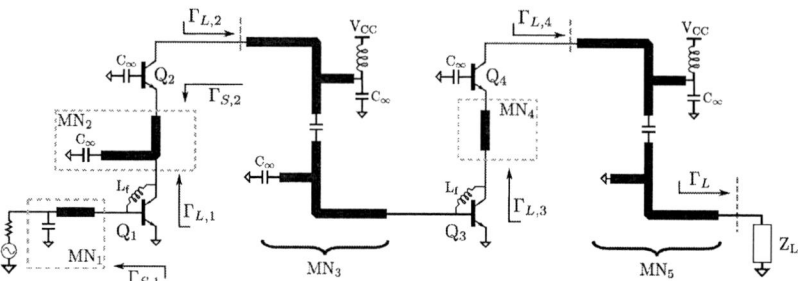

Fig. 2. Schematic of the proposed LNA.

It is worth noticing that the maximum available power gain for a conditionally stable device is unbounded, and MSG only defines the gain that could be achieved once the stage is resistively loaded and kept at the boundary of unconditional stability (K = 1) [12]. Fig. 1 shows the MAG/MSG curves for different transistor configurations in D-Band. The CB stage is conditionally stable with the corner frequency (K = 1) beyond 200 GHz. On the other hand, the CE is more sensitive to device parasitic losses which make it unconditionally stable above 80 GHz and limit the power gain to only ∼ 4 dB at 160 GHz. When the CE and CB are combined in a cascode structure, no more than MAG can be extracted from the CE, thus the CB provides most of the gain. A possibility to boost the CE gain consists in pushing its stability corner to higher frequency, such that the device is operated into a conditionally stable region. The solution proposed in this work is based on the introduction of a feedback inductor, L_f, from the collector to the base. The MAG/MSG curves for the CE with L_f (referred to as CEf) are plotted in Fig. 1. The positive feedback introduced by L_f and the parasitic capacitors of the device (base-to-emitter and collector-to-ground, C_{BE}, C_C) shifts the stability corner upwards, to a frequency set by L_f. Looking at Fig. 1, this solution provides an evident benefit to the achievable power gain, but mandates a careful design flow not to compromise the stability of the overall amplifier. Finally, it is important to highlight the conceptual difference of this technique from device unilateralization (i.e. neutralization of the feedforward base-to-collector capacitance, C_{BC}). Unilateralization, easily implemented in differential stages with explicit cross-coupled capacitors, brings the advantage of unconditional stability but with limited gain enhancement. To gain insight, the MAG for a neutralized CE at 160 GHz, shown with a dash-dot line in Fig. 1, is remarkably lower than the MSG of the CE with L_f.

III. CIRCUIT DESIGN

The schematic of the two-stage cascode LNA is shown in Fig. 2. All the multi-emitter HBTs have the same drawn emitter area, $A_e = 2 \times 5 \times 0.2 \, \mu m^2$, and are biased at $5 \, mA/\mu m^2$ current density. Shielded coplanar Tlines, optimized in the 9 metal layers ($Z_0 = 60 \, \Omega$ and quality factor $Q_{TL} = 30$ at 160 GHz), are used in matching networks. AC shorts for bias and supply decoupling (labeled C_∞) are realized with a stack of MIM ad MOM capacitors to ensure the best compromise between low impedance over the band of interest and self-resonance. The -3 dB bandwidth is primarily limited

by matching the high impedance at the collectors of CB devices, which show the highest nodal Q. Stagger tuning, as in [2][4], extends the bandwidth but sacrifices noise and linearity. In this design, matching networks MN_3 and MN_5, at the collectors of the CB transistors, implement capacitively coupled resonators with a 4th order response. The two complex-conjugate pole pairs match the impedance at two resonance frequencies, set nominally to 140 GHz and 160 GHz, yielding a flat wideband response. The remaining matching networks (MN_1, at the input, and MN_2, MN_4, between CE and CB) are characterized by a low nodal Q and do not limit the bandwidth; hence, simpler single resonance networks are implemented. The feedback inductance in CE transistors

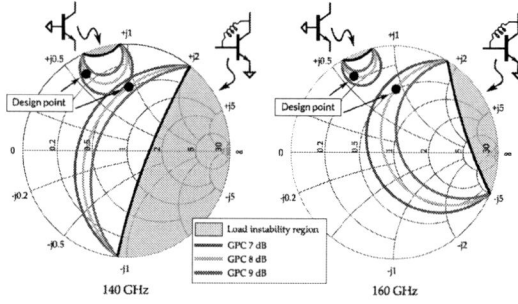

Fig. 3. G_P and load stability circles of the CB and CEf at 140 GHz, 160 GHz.

is set to 50 pH and shifts the stability corner frequency to 170 GHz. Fig. 3 shows the operating power gain (G_P) and load stability circles for the CEf and CB devices at 140 GHz and 160 GHz. The load impedances are carefully selected such that, once accounting for the matching networks losses, the amplifier is unconditionally stable. The black dots in Fig. 3 represent the selected load reflection coefficients, $\Gamma_{L,3}$, $\Gamma_{L,4}$ (or, equivalently, the load impedances $Z_{L,3} = 26 + j57$, $Z_{L,4} = 6 + j31$) for the second stage transistors, Q_3, Q_4. The points lie on the $G_P = 8$ dB circles at both 140 GHz and 160 GHz. The loss (~ 2 dB) introduced by MN_5 from the output pad (nominally terminated to $Z_L = 50 \, \Omega$) to the collector of Q_4 compresses the variation of $\Gamma_{L,4}$ against variations of Z_L and ensures stability of the CB device. The network MN_4, implemented with a simple series Tline, transforms the impedance at the emitter of Q_4 to the target load for the CEf, $Z_{L,3}$. The latter is selected sufficiently far from the CEf potentially-unstable region to still preserve the stability against variation of the off-chip load impedance Z_L. To gain insight, Fig. 4 shows the load reflection coefficient for Q_3

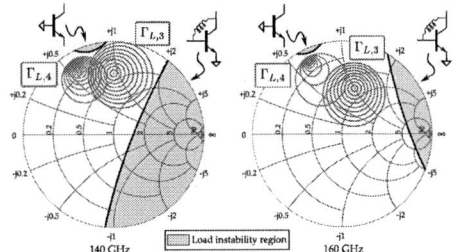

Fig. 4. CB and CE load stability circles and possible variation of $\Gamma_{L,3}$ and $\Gamma_{L,4}$ at 140 GHz and 160 GHz.

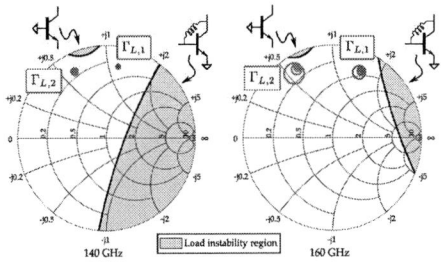

Fig. 6. CB and CE load stability circles and possible variation of $\Gamma_{L,1}$ and $\Gamma_{L,2}$ at 140 GHz and 160 GHz.

and Q_4, obtained by sweeping Z_L across the whole Smith chart (thus considering any possible passive termination at the GSG output pad). $\Gamma_{L,3}$ and $\Gamma_{L,4}$ remain always out of the CB and CEf load potentially-unstable regions (the gray areas in Fig. 4), confirming the unconditional stability of the implemented cascode structure. The nominal load impedance of Q_2 (the CB device in the first stage) is the same as for Q_4. Thus, the dual-resonance network MN_3 is designed such that the impedance at the base of Q_3 is matched to the load of Q_2, $Z_{L,2} = Z_{L,4} = 6 + j31$ at 140 GHz and 160 GHz. Networks MN_1 and MN_2, instead, are designed to optimize the noise performance. The amplifier noise factor, dominated by Q_1 and Q_2, is given by:

$$F \approx F_1 + \frac{F_2 - 1}{G_{A,1}} \quad (3)$$

where F_1, F_2 are the noise factors of Q_1, Q_2, and $G_{A,1}$ is the available power gain of Q_1. The noise factor of Q_1 and Q_2 is minimized by well-defined source terminations, while a trade-off may exist between F_1 and $G_{A,1}$. Fig. 5 shows the noise circles and constant G_A curves for Q_1, in CEf configuration, at 150 GHz. The center of the circles defines the source impedance $Z_{OPT,F1}$ that gives $F_1 = F_{1,\min}$. With $Z_{OPT,F1}$, $G_{A,1} \approx 10$ dB. Thus, the input matching network MN_1 transforms the 50 Ω off-chip nominal source impedance into $Z_{OPT,F1}$. Once MN_1 is set, the impedance at the collector of Q_1 is defined and matching network MN_2 is designed such that the input termination of Q_2, the CB device, is $Z_{OPT,F2}$ leading to $F_2 = F_{2,\min}$. Since Q_1 is a conditionally stable

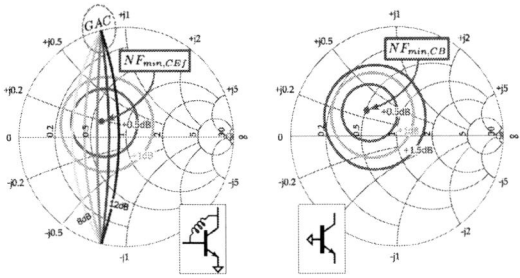

Fig. 5. G_A and noise circles of CEf, noise circles of CB. $f = 150$ GHz.

device, once MN_1 and MN_2 are fixed, its stability conditions must be verified. To this aim, Fig. 6 reports load stability circles of Q_1 and the values of $\Gamma_{L,1}$ for any passive load impedance Z_L. It is clearly seen that, with respect to Fig. 4, the increased isolation from the output provided by the second stage makes stabilization easier, and the matching networks

designed to minimize the noise factor do not compromise the amplifier unconditional stability.

IV. MEASUREMENTS

The amplifier (chip photo in Fig. 7) is realized in STMicroelectronics' SiGe BiCMOS 55 nm technology. The occupied area is $742 \times 256 \, \mu m^2$, including GSG pads. RF signals are conveyed through Infinity Waveguide GSG probes to VDI WR6.5-VNAX extension modules for the Agilent E8361C VNA. Fig. 8 compares small signal measurements and simulations, showing good agreement. Farran WGNS-06 noise source, FBC-06 frequency extender and Keysight N8975A analyser constitute the noise figure (NF) measurement equipment. The minimum measured NF is 5 dB and rises by 1.5 dB within the bandwidth. Notably, the measured NF is slightly lower (~ 1 dB) than simulations. The gain at 148 GHz is 22.8 dB and the -3 dB bandwidth extends from 131 GHz to 165 GHz. The K factor, reported in Fig. 9,

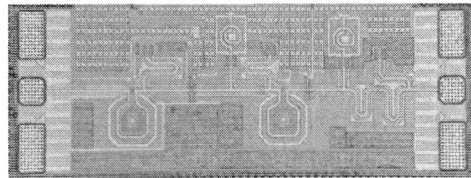

Fig. 7. Chip microphotograph.

confirms unconditional stability. The output power at 1 dB gain compression, OP_{1dB}, measured with ELVA-1 DPM-06 power meter is 0 dBm. Experimental results are summarized in Table I together with state-of-the-art D-Band amplifiers in bipolar or BiCMOS technology. The performance of the presented amplifier is compared against previous works with two figures of merit;

$$FoM = \sqrt[n]{G_T} \left(\frac{f_0}{f_{max}}\right)^2 \quad (4)$$

normalizes gain, G_T, at the center frequency f_0 to the number of devices n and the cut-off frequency f_{max} [6]. This FoM is a useful benchmark for the amount of gain extracted from devices operated close to their frequency limit. The overall LNA performance, like noise figure (NF), linearity (IIP_3[1]), and power consumption, are included into the FoM_{ITRS}, given by

$$FoM_{ITRS} = \frac{G_T \cdot IIP_3 \cdot f_0}{(NF - 1) \cdot P_{DC}} \quad (5)$$

[1]Where unreported, IIP₃ was estimated as $P_{IP1dB} + 9.6$ dB

Table 1. Comparison with the state-of-the-art.

	This work	[7]	[13]	[9]	[4]	[14]	[3]	[11]	[10]	[2]	[8]
Technology	55 nm[1]	55 nm[1]	120 nm[1]	130 nm[1]	55 nm[1]	130 nm[1]	90 nm[1]	130 nm[1]	130 nm[1]	130 nm[1]	130 nm[1]
f_t/f_{max}	320/370	320/370	200/265	250/370	320/370	250/370	300/350	300/500	300/500	300/500	300/500
# of stages	2	2	4	3	2	3	4	2	2	4	4
# of actives	4	4	8	6	4	6	7	4	6	8	6
G_T [dB]	22.8	20.1	26	32.8	20.6	23	30	27.5	26	32.6	25.3
f_0 [GHz]	148	150	130	140	144	157	136	126	150	140	134
-3 dB BW [GHz]	33.5	24.5	13	23.2	65.8	55	28	16	60	52	44
P_{DC} [mW]	40	27	57	39.6	24	25/60	45	12	70	28	30
NF [dB]	5-6.5	-	-	7.8-9.5*	8.1-10.2*	6-7.9	6.2-8	5.5-6.5	-	4.8-6.1	5.9-7
P_{OP1dB} [dBm]	0	-	-	3.2	-1.9*	-	-	-6.5	2	-6	-
FoM_{ITRS}	**19.65**	-	-	16.88	8.15	-	-	10.59	-	7.14	-
FoM	**0.59**	0.52	0.51	0.5	0.5	0.44	0.41	0.31	0.24	0.2	0.19

[1] SiGe BiCMOS   * simulated value

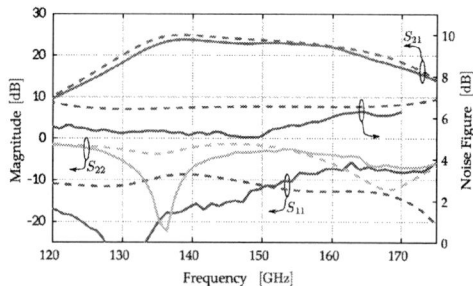

Fig. 8. Small signal measurements (solid) vs. simulations (dashed).

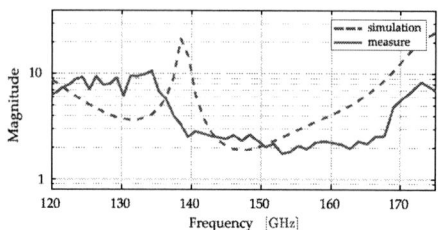

Fig. 9. Measured stability factor K.

All the quantities in the figures of merit are expressed in linear scale. The good achieved FoM and FoM_{ITRS}, beyond what demonstrated by previous works, support the proposed local inductive feedback in CE transistors as an effective solution to improve D-Band amplifiers performance.

V. CONCLUSIONS

Inductive feedback applied to a common-emitter device has been investigated as a solution to trade gain for stability. The technique forces the transistor to operate in a conditionally stable region with gain above the device maximum available gain. The concept has been applied to a D-band LNA and a design procedure has been described such that the overall amplifier features unconditional stability with high gain and minimum noise figure. The measured results, combined in figures of merit typically used as benchmark for high-frequency low-noise amplifiers, confirm the effectiveness of the proposed solution to push the state-of-the-art performance.

ACKNOWLEDGMENT

This work received funding from the Commission of the European Union within the H2020 DRAGON project (Grant Agreement No. 955699) and KDT SHIFT project (Grant Agreement No. 1010962).

REFERENCES

[1] (2018) Point-to-Point Radio Links in the Frequency Ranges 92-114.25 GHz and 130-174.8 GHz – ECC Report 282.

[2] E. Turkmen *et al.*, "A SiGe HBT D-Band LNA with Butterworth Response and Noise Reduction Technique," *IEEE Microwave and Wireless Components Letters*, vol. 28, no. 6, pp. 524–526, 2018.

[3] R. B. Yishay, E. Shumaker, and D. Elad, "A 122-150 GHz LNA with 30 dB gain and 6.2 dB noise figure in SiGe BiCMOS technology," in *2015 IEEE 15th Topical Meeting on Silicon Monolithic Integrated Circuits in RF Systems*, 2015, pp. 15–17.

[4] I. Petricli, H. Lotfi, and A. Mazzanti, "Analysis and Design of D-Band Cascode SiGe BiCMOS Amplifiers With Gain-Bandwidth Product Enhanced by Load Reflection," *IEEE Transactions on Microwave Theory and Techniques*, vol. 69, no. 9, pp. 4059–4068, 2021.

[5] H. Bameri and O. Momeni, "A High-Gain mm-Wave Amplifier Design: An Analytical Approach to Power Gain Boosting," *IEEE Journal of Solid-State Circuits*, vol. 52, no. 2, pp. 357–370, 2017.

[6] D.-W. Park *et al.*, "A 230–260-Ghz Wideband and High-Gain Amplifier in 65-nm CMOS Based on Dual-Peak Gmax-core," *IEEE Journal of Solid-State Circuits*, vol. 54, no. 6, pp. 1613–1623, 2019.

[7] I. Petricli, H. Lotfi, and A. Mazzanti, "Design of Compact D-Band Amplifiers With Accurate Modeling of Inductors and Current Return Paths in 55-nm SiGe BiCMOS," *IEEE Solid-State Circuits Letters*, vol. 3, pp. 250–253, 2020.

[8] B. Ustundag, E. Turkmen, B. Cetindogan, A. Guner, M. Kaynak, and Y. Gurbuz, "Low-Noise Amplifiers for W-Band and D-Band Passive Imaging Systems in SiGe BiCMOS Technology," in *2018 Asia-Pacific Microwave Conference (APMC)*, 2018, pp. 651–653.

[9] E. Aguilar, A. Hagelauer, D. Kissinger, and R. Weigel, "A low-power wideband D-band LNA in a 130 nm BiCMOS technology for imaging applications," in *2018 IEEE 18th Topical Meeting on Silicon Monolithic Integrated Circuits in RF Systems (SiRF)*, 2018, pp. 27–29.

[10] P. Stärke, V. Rieß, C. Carta, and F. Ellinger, "Wideband Amplifier with Integrated Power Detector for 100 GHz to 200 GHz mm-Wave Applications," in *2018 IEEE BiCMOS and Compound Semiconductor Integrated Circuits and Technology Symposium (BCICTS)*, 2018, pp. 160–163.

[11] A. C. Ulusoy, P. Song, W. T. Khan, M. Kaynak, B. Tillack, J. Papapolymerou, and J. D. Cressler, "A SiGe D-Band Low-Noise Amplifier Utilizing Gain-Boosting Technique," *IEEE Microwave and Wireless Components Letters*, vol. 25, no. 1, pp. 61–63, 2015.

[12] M. Gupta, "Power gain in feedback amplifiers, a classic revisited," *IEEE Transactions on Microwave Theory and Techniques*, vol. 40, no. 5, pp. 864–879, 1992.

[13] E. Shumakher, J. Elkind, and D. Elad, "Key components of a 130 Ghz dicke-radiometer SiGe RFIC," in *2013 IEEE 13th Topical Meeting on Silicon Monolithic Integrated Circuits in RF Systems*, 2013, pp. 255–257.

[14] T. Maiwald, J. Potschka, K. Kolb, M. Dietz, K. Aufinger, A. Visweswaran, and R. Weigel, "A Full D-Band Low Noise Amplifier in 130 nm SiGe BiCMOS using Zero-Ohm Transmission Lines," in *2020 15th European Microwave Integrated Circuits Conference (EuMIC)*, 2021, pp. 13–16.

RMo3B-5

A D-Band to J-Band Low-Noise Amplifier with High Gain-Bandwidth Product in an Advanced 130 nm SiGe BiCMOS Technology

Marcel Andree[#], Janusz Grzyb[#], Bernd Heinemann[*], Ullrich Pfeiffer[#]

[#]IHCT, University of Wuppertal, Germany

[*] IHP - Leibniz-Institut für innovative Mikroelektronik, Germany

andree@uni-wuppertal.de

Abstract—In this work, a broadband low noise amplifier fabricated in an advanced 130 nm SiGe BiCMOS technology with f_t/f_{max} of 470/650 GHz is presented. The amplifier comprises 5 pseudo-differential cascode stages. Double-peak transformer-based matching stages in the first 3 stages are used to achieve high gain and low noise across a large bandwidth. A suitable set of baluns was developed to facilitate on-wafer measurements from 100 GHz to 325 GHz. The amplifier provides a peak small-signal gain of 34.6 dB at 160 GHz and 235 GHz. The 3 dB bandwidth is 131 GHz - 277 GHz. The measured noise figure is between 8.4 dB to 12 dB, including baluns. With the balun losses de-embedded, the amplifier's peak small-signal gain increases to 37 dB and stays above 30 dB from 117 GHz to 285 GHz, while the lowest noise figure is 7.1 dB at 155 GHz. The achieved OP_{1dB} stays in between 0 dBm - 1 dBm from 150 GHz to 265 GHz. The amplifier consumes 152 mW with a V_{cc} of 3 V.

Keywords—Low noise amplifier (LNA), millimeter-wave (mmW), SiGe BiCMOS integrated circuits (ICs)

I. INTRODUCTION

High gain and large bandwidth (BW) are key features for many applications in the millimeter-wave (mmW) and (sub-) terahertz (THz) frequency range, including the upcoming 6G communications [1], radar systems [2], or radiometry [3]. Offering high fabrication yield, low cost, and a good integration level in combination with f_t/f_{max} of up to 505/720 GHz [4], silicon technologies are well suitable for RF front-end circuits. Low-noise amplifiers (LNAs) are the most critical receiver part that impacts the systems' Signal-to-Noise ratio (SNR) and bandwidth. Especially for radiometry, LNAs have to provide the challenging combination of high gain and low noise figure (NF) across a large bandwidth to allow minimum noise-equivalent temperature differences (NETD) for real-time passive imaging [3].

Most common LNA designs in the literature are based on classical transmission line matching networks. Either a high gain of around 32.8 dB in the D-Band [5], 28.5 dB in the J-Band [6], or a large bandwidth of 80 GHz centered at 180 GHz with a gain of 15.5 dB [7] is achieved by staggering of multiple frequency tuned stages. With such design techniques, a flat high-gain over more than 100 GHz bandwidth is difficult to achieve. It typically requires a large number of densely tuned stages which compromises the LNA NF across its bandwidth where the minimum Noise Figure (NF_{min}) is only achieved in a small section defined by the gain/bandwidth of the first LNA stages.

In this work, a design approach based on broadband transformer-based double-peak input/interstage/output matching is introduced to achieve high gain and low noise across a large bandwidth for real-time radiometric applications in the D-Band to J-Band (110 GHz to 325 GHz). Thereby, the implemented LNA achieves a gain of more than 30 dB in a close to 150 GHz bandwidth.

II. CIRCUIT FABRICATION AND DESIGN

The LNA is implemented in a development lot of IHP's next generation 130 nm SiGe HBT BiCMOS technology SG13G3 with target f_t/f_{max} values of 470/650 GHz. It comprises 5 pseudo-differential cascode stages with a transistor emitter size of $2\times(0.96\times0.1)$ μm² as shown in Fig. 1 a) and b). The cascode topology increases available gain and isolation at the cost of potential instability, while the relatively small device size aims for low power consumption. The first 3 stages are biased with a current density (J_c) of 22 mA/μm² (I_c = 4.2 mA), ensuring a good noise figure and a broadband high gain. The last two stages are biased for peak f_{max} with J_c of 34 mA/μm² (I_c = 6.6 mA). A 3 V V_{cc} bias enables a high gain towards 300 GHz and a linearity increase.

A broadband and high-gain LNA design requires careful frequency planning supported by modeling and co-simulating of all device interconnect parasitics with custom matching networks. This ensures the first stages achieve the required gain for a broadband NF_{min} across the whole band. Therefore, the cascode core layout shown in Fig. 1 c) has been optimized by full-wave EM simulations considering all important core layout parasitics, i.e., the common base (CB) stage's inductive base reactance, common emitter (CE) to CB stage interconnections and ground paths. Furthermore, broadband input and interstage matching are facilitated through six custom-designed, distributed, asymmetric coupled-line transformers with a typical implementation loss of 1 to 1.5 dB, allowing complex, frequency-dependent impedance transformation ratios for minimum noise figure in the targeted band (110 to 300 GHz) and maximum gain towards 300 GHz.

The input transformer provides a broadband match to the input balun. The transformer's output impedance trajectory approximates the complex conjugate of the first stage's input impedance, which ranges from 40-j80 to 44-j26 (140 GHz/280 GHz), within its 1 dB noise circles. To achieve a good input match from 150 GHz to 325 GHz, an additional set of 2-shunt transmission lines at the transformer output is applied, resulting in 3 return loss minima (Fig. 4). These lines are further used for device biasing and connected

979-8-3503-2123-4/23 $31.00 © 2023 IEEE   137   2023 IEEE Radio Frequency Integrated Circuits Symposium

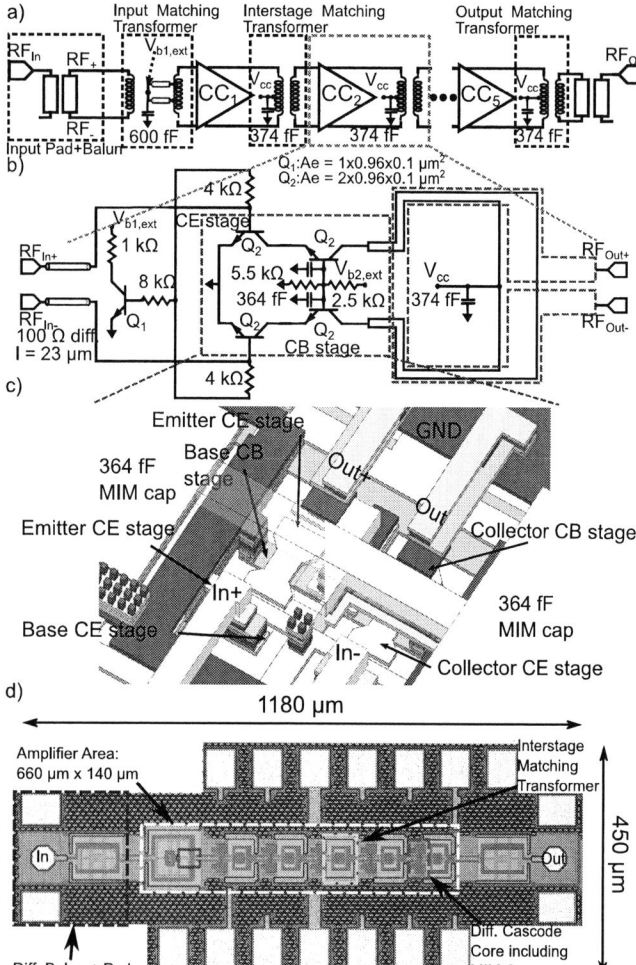

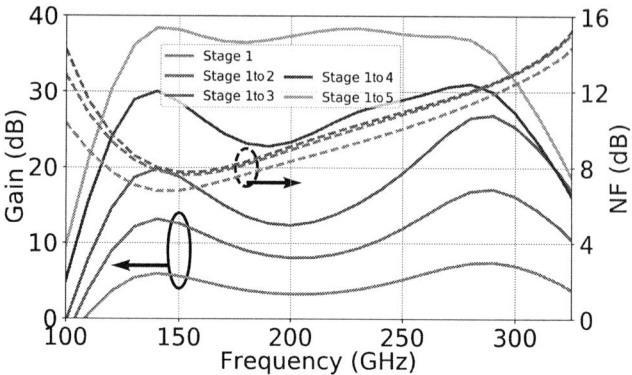

Fig. 2. Simulated small-signal gain (solid) and NF (dashed) for an increasing number of LNA stages (1 to 5) with implemented matching transformers, emphasizing the double-peak design process. Please note, in the simulations, the output of the considered stage combination is loaded with the input impedance of the following stage.

Fig. 1. a) Block diagram of the 5-stage pseudo-differential LNA. b) simplified schematic of a single transformer-coupled cascode (CC) stage, including bias network. c) Full-wave EM simulated (diff.+comm. mode) optimized core layout, including GND paths and 364 fF MIM bypass capacitors at the base of the CB stage. MIM caps on the left side are partly removed to show connections on lower metal layers. d) Chip micrograph of the LNA.

to 600 fF MIM capacitors in the transformer center for common-mode suppression. The first 3 stages utilize the same interstage matching transformer. The biasing voltage V_{cc} is provided through the center tap of the interstage transformer connected to a 374 fF MIM capacitor in the transformer layout center to avoid any common mode gain. A more detailed theoretical analysis of coupled transmission line based matching transformers can be found in the literature [8].

To achieve a flat gain and a low NF across a large BW with a minimum number of gain stages, the first 3 stages are designed to be broadband. This is facilitated by the interstage transformer's impedance trajectory between the in- and output, resulting in a double-peak gain curve for each stage. As the gain per stage is limited by the gain at the highest operating frequency (G_{max}), the transformer provides an impedance at the stage output that steadily derivates to that corresponding to G_{max} with frequency decrease. That way, the large possible

gain at frequencies below 200 GHz (>20 dB) is attenuated, ensuring device stability and sufficient gain to minimize the noise contribution of the consecutive stages. The simulated aggregate small-signal gain after 3 stages is shown in Fig. 2. Here, the aggregate gain of the first 2 stages is critical for a broadband NF and was traded against the LNA BW. The gain dip in the band center is compensated by the last 2 stages.

The presented LNA is intended to be implemented with a power detector for radiometry applications (passive imaging) which requires a complex impedance match for broadband operation of the entire system. Therefore, in the concept breakout, the LNA output match to the balun is not optimized but w.o. noticeable change in the LNA gain curve (Fig. 4).

III. MEASUREMENTS

The LNA micrograph is shown in Fig. 1 d). A Marchand balun compensates the parasitic on-chip pad capacitance for broadband on-wafer measurements. The balun is based on a coupled-line section with a non-uniform characteristic impedance profile and was placed at the input and output of the amplifier. The balun was measured in a separate set of back-to-back configurations, and a maximum insertion loss better than 1.5 dB (per balun) was de-embedded in the 136-325 GHz band. The core amplifier area, excluding the baluns and signal pads, is 0.09 mm², while the total chip area for this circuit is 0.53 mm².

The measurement setup is shown in Fig. 3. Two WR-6 (110-170 GHz) and WR-3 (220-325 GHz) vector network analyzer (VNA) extender modules from OML with a maximum output power (P_{max}) and conversion gain (CG) of -6.5 dBm/-20 dB (D-Band) and -20.5 dBm/-10 dB (J-Band) have been used. The modules were connected to a Keysight P8361A PNA network analyzer for small-signal measurements. Due to the lack of equipment covering the whole G-Band (140 GHz - 220 GHz), the D-Band and J-Band modules were calibrated out of band between $110 - 185$ GHz and $200 - 325$ GHz as the modules provide sufficient SNR. This results in a gap of 15 GHz present in the measured

979-8-3503-2123-4/23 $31.00 © 2023 IEEE

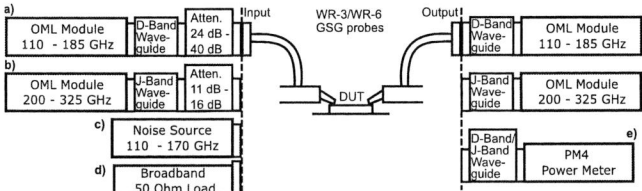

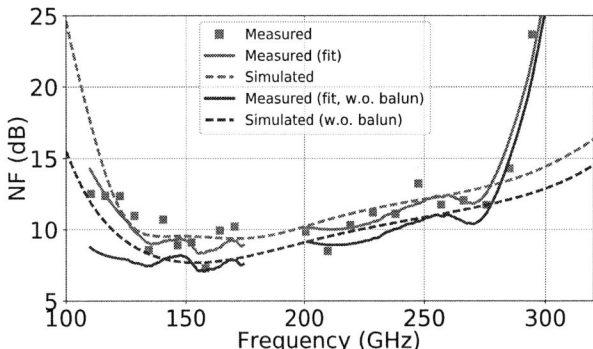

Fig. 3. Measurement setup for small-signal (a b), noise (c - d), and large signal (e) measurements. Two different OML extension module pairs cover the frequency bands from 110 GHz to 185 GHz and 200 GHz to 325 GHz.

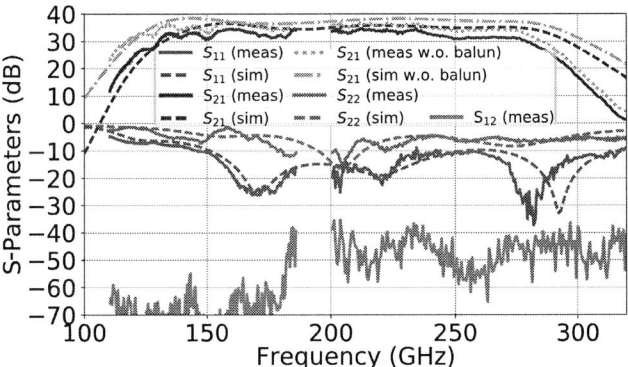

Fig. 4. Measured and simulated small-signal s-parameters of the presented LNA. S_{21} is shown with and without balun and pad. Note: Measured S_{12} is limited by the VNA noise floor of -70 dB (D-Band) and -50 dB (J-Band), which is 60 dB to 80 dB higher than the simulated LNA S_{12} of around -130 dB.

Fig. 5. Measured and simulated NF across 110 GHz to 325 GHz with and without balun de-embedded. The 30 GHz gap in the band center is due to the ENR calibration of the D-Band noise source not provided out-of-band.

curves. The small-signal measurements were performed with an additional set of waveguide attenuators at the LNA input. That way, linear amplifier operation is ensured, and a limited dynamic range of the modules due to the high LNA gain is prevented. All measurements were performed after an initial through-reflect-line (TRL) calibration on a standard calibration substrate which was also used to de-embed the insertion loss of each interconnect component in the measurement setup. To facilitate the TRL calibration, the attenuator values were set to the reliable minimum, whereas the LNA measurement required high attenuation values. The load impedance changes present at the attenuator outputs on a calibration substrate and the on-chip balun resulted in a small inaccuracy of their insertion loss calibration manifested as parasitic waviness (∼1 dB) in the de-embedded small-signal curves.

Simulated and measured s-parameters are shown in Fig. 4. S_{22} is mainly below -5 dB from 130 GHz to 325 GHz. Due to the novel transformer matching approach, the input is well matched with S_{11} smaller than -10 dB from 150 GHz to 325 GHz. Only at higher frequencies, around 290 GHz, a slight frequency shift can be observed between measurement and simulation. S_{12} is limited by the VNA noise floor of -50 dB (J-Band) and -70 dB (D-Band), which is much higher than the simulated S_{12} of around -130 dB. For this purpose, the stability has additionally been verified by breakout measurements of the first and last LNA stage showing a minimum k-factor of 5 and that the amplifier stages remain stable from 110 to 325 GHz.

A maximum S_{21} of 34.6 dB is achieved at 160 GHz and 235 GHz with a small-signal 3-dB bandwidth of 146 GHz (131-277 GHz). With the baluns de-embedded, these values increase to a peak gain of 37 dB and a BW of 155 GHz (124-279 GHz). For radiometry, it is important to note that the equivalent noise bandwidth (ENB) was also calculated from the measured gain curve and is 151 GHz which is close to the LNA BW. In total, an excellent model-to-hardware correlation across the large operation bandwidth validates the complex modeling approach.

The NF has been measured by two approaches depending on the observed frequency range. In both, it was verified that the VNA extender modules' parasitic conversion gain of the other harmonics is sufficiently low to minimize the NF measurement error. In the D-Band, the Y-factor method was used to measure the NF due to the availability of an ELVA-1 ISSN-06 noise source with a typical ENR of 12 dB connected to the input probe (Fig. 3 c)). In the J-Band, where no noise source was available, the input probe was terminated with a broadband 50 Ω match. Due to the high LNA gain, the noise floor of the OML modules could be overcome without using an additional external LNA. With the module CG and the LNA gain from previous small-signal measurements, the NF was estimated from the noise measurements at room temperature (293 K) by the Friis noise equation. Measured and simulated NF are shown in Fig. 5 and show a good correlation with minimum values of 8.4 dB at 155 GHz and 7.1 dB with the input balun de-embedded. The large slope of the measured NF above 280 GHz is based on a faster gain roll-off shown in the s-parameter measurements than in the simulations (Fig. 4).

For large signal measurements, the output OML module was exchanged with a pre-calibrated PM4 power meter from VDI (Fig. 3 e)). The measured compression curves are shown in Fig. 6. From 150 GHz to 265 GHz, the LNA exhibits an OP_{1dB} from 0 dBm to 1 dBm. With the output balun de-embedded OP_{1dB} is in between 1 to 2.1 dBm. Too high probe/waveguide losses of around 9 dB prevent the LNA from full saturation in the J-Band.

979-8-3503-2123-4/23 $31.00 © 2023 IEEE

Table 1. State-of-the-art of silicon LNAs operating in D-Band/J-Band

Technology	f_t/f_{max} [GHz]	Center frequency [GHz]	Peak Gain [dB]	GBP [THz]	BW [GHz]	Minimum NF [dB]	OP_{1dB} [dBm]	P_{DC} [mW]	Size [mm²]	Ref.
130 nm SiGe	250/370	140	32.8	44.2	23.2	7.8	-	39.6	0.07 (w.o. pads)	[5]
130 nm SiGe	300/500	144.5	32.6	94.6	52	4.8	-	28	1.0	[9]
130 nm SiGe	250/370	167	23	11	55	6	-	25-60	0.1 (w.o. pads)	[10]
130 nm SiGe	300/500	180	15.5	2.8	80	6.1	-2.2	46	0.48	[7]
130 nm SiGe	300/500	240	28.5	9.9	14	13.7	-	97.2	0.45	[6]
130 nm SiGe	300/450	290	12.9	0.45	23	16	-	136	0.25	[11]
130 nm SiGe	470/700	291	10.8	0.82	68	11	-6.5	119	0.26	[12]
130 nm SiGe	**470/650**	**204/201.5†**	**34.6/37†**	**421/777†**	**146/155†**	**8.4/7.1†**	**1/2.1†**	**152**	**0.53**	**This work**

† (w./w.o. baluns)

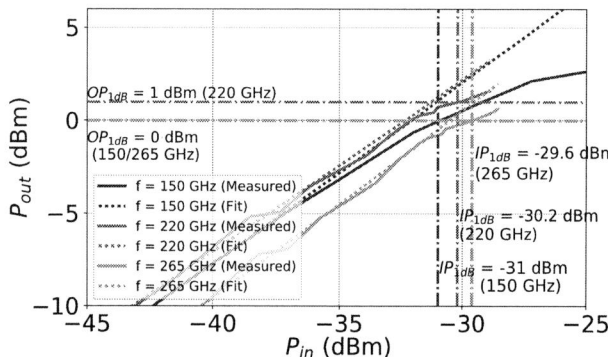

Fig. 6. Measured and linearly fitted compression curves of the LNA from 150 GHz to 265 GHz showing OP_{1db} of 0 dBm to 1 dBm.

IV. Conclusion

In this paper, a 5-stage pseudo-differential, cascode-based LNA implemented in an advanced 130 nm SiGe BICMOS technology operating across the full D-Band to J-Band is presented. Key parameters for a broadband, high-gain design with a low NF are novel distributed coupled-line matching transformers combined with precise modeling of the transistor core layout, including all parasitics. The interstage transformer matching realizes a close to G_{max} performance at the upper corner frequency (280 GHz) and an appropriate gain ensuring device stability at lower frequencies. This results in a broadband double-peaking gain behavior of the first three stages. Subsequently, the last two stages realize a broad gain peak in the band center, compensating for the previous stages' gain dip.

With a peak gain of 34.6 dB at 160 GHz and 235 GHz and a measured 3-dB BW from 131 GHz to 277 GHz, the LNA shows a record performance with a 421 THz gain-bandwidth product (GBP) compared to the current state-of-the-art of silicon-integrated LNAs operating in the D-Band and J-Band (Table 1). The Noise figure was characterized based on the Y-factor method in the D-Band and by applying the Friis Noise equation in the J-Band (due to the high LNA gain > 30 dB), where it lacks commercially available noise sources. A broadband minimum NF reaching from 8.4 dB to 12 dB was measured from 155 GHz to 270 GHz (7.1 dB/10.5 dB after input balun de-embedding). With an OP_{1dB} from 0 dBm to 1 dBm, it shows a good linearity which is uniform across 150 GHz to 265 GHz due to the broadband stages.

Acknowledgment

The research work presented in this paper was funded by the German Research Foundation ("Deutsche Forschungsgemeinschaft") (DFG) under Project-ID 287022738 TRR 196 for Project C08 and within the project DotSeven2IC (DFG PF 661/15-1).

References

[1] T. S. Rappaport, Y. Xing, O. Kanhere et al., "Wireless Communications and Applications Above 100 GHz: Opportunities and Challenges for 6G and Beyond," *IEEE Access*, vol. 7, pp. 78 729–78 757, 2019.

[2] C. Mangiavillano, A. Kaineder, K. Aufinger et al., "A 1.42-mm2 0.45–0.49 THz Monostatic FMCW Radar Transceiver in 90-nm SiGe BiCMOS," *IEEE Transactions on Terahertz Science and Technology*, vol. 12, no. 6, pp. 592–602, 2022.

[3] E. Dacquay, A. Tomkins, K. H. K. Yau et al., "D -Band Total Power Radiometer Performance Optimization in an SiGe HBT Technology," *IEEE Transactions on Microwave Theory and Techniques*, vol. 60, no. 3, pp. 813–826, 2012.

[4] B. Heinemann, H. Rücker, R. Barth et al., "SiGe HBT with ft/fmax of 505 GHz/720 GHz," in *2016 IEEE International Electron Devices Meeting (IEDM)*, 2016, pp. 3.1.1–3.1.4.

[5] E. Aguilar, A. Hagelauer, D. Kissinger et al., "A Low-Power Wideband D-band LNA in a 130 nm BiCMOS Technology for Imaging Applications," in *2018 IEEE 18th Topical Meeting on Silicon Monolithic Integrated Circuits in RF Systems (SiRF)*, 2018, pp. 27–29.

[6] M. Najmussadat, R. Ahamed, M. Varonen et al., "Design of a 240-GHz LNA in 0.13 μm SiGe BiCMOS Technology," in *2020 15th European Microwave Integrated Circuits Conference (EuMIC)*, 2021, pp. 17–20.

[7] Y. Mehta, S. Thomas, and A. Babakhani, "A 140–220-GHz Low-Noise Amplifier With 6-dB Minimum Noise Figure and 80-GHz Bandwidth in 130-nm SiGe BiCMOS," *IEEE Microwave and Wireless Technology Letters*, vol. 33, no. 2, pp. 200–203, 2023.

[8] T. Bücher, J. Grzyb, P. Hillger et al., "A Broadband 300 GHz Power Amplifier in a 130 nm SiGe BiCMOS Technology for Communication Applications," *IEEE Journal of Solid-State Circuits*, vol. 57, no. 7, pp. 2024–2034, 2022.

[9] E. Turkmen, A. Burak, A. Guner et al., "A SiGe HBT D -Band LNA With Butterworth Response and Noise Reduction Technique," *IEEE Microwave and Wireless Components Letters*, vol. 28, no. 6, pp. 524–526, 2018.

[10] T. Maiwald, J. Potschka, K. Kolb et al., "A Full D-Band Low Noise Amplifier in 130 nm SiGe BiCMOS using Zero-Ohm Transmission Lines," in *2020 15th European Microwave Integrated Circuits Conference (EuMIC)*, 2021, pp. 13–16.

[11] S. P. Singh, T. Rahkonen, M. E. Leinonen et al., "A 290 GHz Low Noise Amplifier Operating above $f_{max}/2$ in 130 nm SiGe Technology for Sub-THz/THz Receivers," in *2021 IEEE Radio Frequency Integrated Circuits Symposium (RFIC)*, 2021, pp. 223–226.

[12] A. Gadallah, M. H. Eissa, T. Mausolf et al., "A 300-GHz Low-Noise Amplifier in 130-nm SiGe SG13G3 Technology," *IEEE Microwave and Wireless Components Letters*, vol. 32, no. 4, pp. 331–334, 2022.

979-8-3503-2123-4/23 $31.00 © 2023 IEEE

RMo3C-1

A 0.32-THz 6.6-dBm Single-Chain CW Transmitter Using On-Chip Antenna With 2.65% DC-to-THz Efficiency

Georg Zachl, Christoph Mangiavillano, Rohish Kumar Reddy Mitta, Tim Schumacher, Harald Pretl, Andreas Stelzer

LIT/SAL mmW-Lab, Johannes Kepler University Linz, Austria

{georg.zachl, christoph.mangiavillano, rohish.mitta, tim.schumacher, harald.pretl, andreas.stelzer}@jku.at

Abstract — A 0.324-THz multiplier-based (x16) transmitter with an on-chip patch antenna has been implemented in a 130-nm SiGe:C BiCMOS technology with an f_T/f_{max} of 350/450 GHz. Power management and biasing are integrated on-chip for post-silicon optimizations of output power and dc-to-THz efficiency. Each circuit block can be programmed over a serial peripheral interface for bias current. Three individually programmable low-dropout voltage regulators supply the x8 20-to-160-GHz multiplier chain, the two-stage 160-GHz power amplifier, and the 0.32-THz frequency doubler, respectively. Measurements after optimization reveal a dc-to-THz efficiency of 2.65% with an output power of 6.6 dBm and a dc power consumption of 170 mW operating at 0.324 THz.

Keywords — Frequency multiplier, on-chip antenna, power amplifier, SiGe:C bipolar CMOS (BiCMOS), terahertz (THz).

I. INTRODUCTION

Terahertz frequencies have recently gained further relevance with the introduction of the IEEE 802.15.3d standard from 252 GHz to 325 GHz [1]. Silicon-based technologies such as CMOS and SiGe:C bipolar CMOS (BiCMOS) are optimal candidates for mass-market applications at these frequencies due to their low-cost and high integration levels. A major challenge presented by these technologies is the efficient generation of high output power (P_{out}), as state-of-the-art production silicon-integrated technologies typically only reach f_{max} of 500 GHz.

J-band power amplifiers [2], [3] have been proposed in an advanced SiGe BiCMOS technology with f_T/f_{max} of 470/650 GHz. Nevertheless, measured P_{out} are less than 7.5 dBm above 0.32 THz, with efficiencies below 2% and a dc power consumption (P_{dc}) of up to 710 mW. In comparison, high output power phase-locked signal generation at around 0.3 THz using frequency multiplication [4]–[6] achieves up to 6.6 dBm with dc-to-THz efficiencies up to 0.7%.

We present a multiplier-based THz transmitter (TX) with an on-chip patch antenna, for continuous-wave (CW) operation, which has been implemented using a 130-nm SiGe:C BiCMOS technology with an f_T/f_{max} of 350/450 GHz. A 20-GHz radio frequency (RF) input signal is multiplied by ×8 to 160 GHz, where a differential two-stage power amplifier (PA) provides the required drive for a push-push frequency doubler to the final output at 0.32 THz. Integrated low-dropout voltage regulators (LDO) and current sources for the individual circuit blocks allow for post-silicon optimization of P_{out} and dc-to-THz efficiency of the frequency multiplier chain. After optimization, a state-of-the-art dc-to-THz efficiency of 2.65% has been

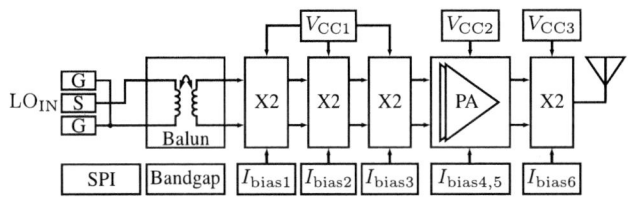

Fig. 1. Block diagram of the 0.32-THz single-chain CW transmitter, showing the 20-GHz input balun, the ×8 bootstrapped frequency doublers, the 160-GHz two-stage PA, the class-B push-push doubler, and the on-chip antenna. Auxiliary circuits like bandgap, SPI, programmable bias current generation, and LDOs are integrated as well.

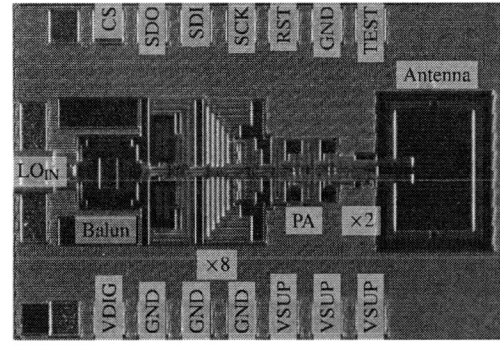

Fig. 2. Micrograph of the 0.324-THz transmitter chip ($1.5 \times 1.0\,mm^2$).

measured at 0.324 THz with a P_{out} of 6.6 dBm and a P_{dc} of 170 mW.

II. TRANSMITTER DESIGN

The RF path of the transmitter, shown in Fig. 1, consists of five building blocks, with the first one being the 20 GHz ground-signal-ground (GSG) RF input pad and a passive single-ended-to-differential conversion using a transformer. Then, a differential ×8 frequency multiplier feeds a two-stage 160-GHz PA. The high-power output signal drives a final push-push frequency doubler, resulting in an effective ×16 frequency multiplication from the input. The single-ended 0.32 THz output from the final doubler is radiated by an on-chip patch antenna.

Additionally, the RF-chain circuitry is supported by three programmable on-chip low-dropout voltage regulators (LDO) and six programmable bias-current generators, both using a bandgap as a reference. The micrograph is shown in Fig. 2.

979-8-3503-2123-4/23 $31.00 © 2023 IEEE   141   2023 IEEE Radio Frequency Integrated Circuits Symposium

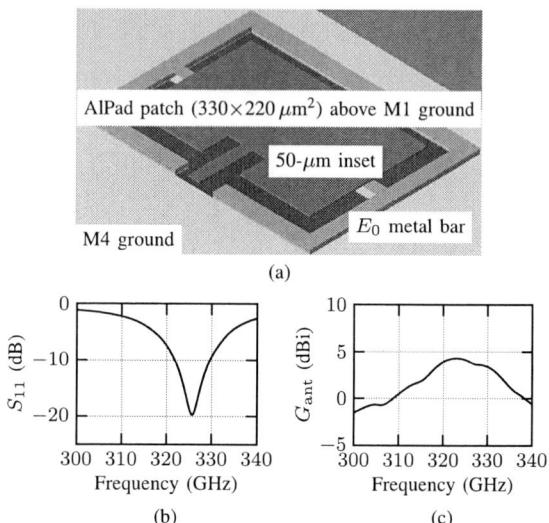

Fig. 4. On-chip antenna (a) simulation layout, (b) simulated S_{11}, and (c) simulated antenna gain.

Fig. 3. Simplified schematics of single stages of the RF-chain circuits: (a) the bootstrapped doubler (TL$_1$ and TL$_2$ are replaced with an inductor in the first doubler stage), (b) one stage of the 160-GHz power amplifier, (c) the push-push doubler. (d) Measured harmonics output power related to the carrier.

A. Bootstrapped Frequency Doubler

For the frequency multiplication from the 20 GHz input to the intermediate frequency of 160 GHz, three frequency doublers in cascade, each implementing the bootstrapped-doubler topology [7], are used, detailed in Fig. 3(a). The doublers use an identical core layout with adapted passive components. The first bootstrapped doubler uses 530 pH inductors instead of TL$_1$ and TL$_2$ for the required phase-shift between the top of the differential pair and the emitters of the switching quad. This results in a significantly reduced silicon area compared to using transmission lines at frequencies below 40 GHz. All transmission lines are single-ended microstrip lines with $Z_0 = 67\,\Omega$. For the stages 2 and 3, TL$_{1,2}$ are 560 µm and 98 µm long, respectively. For the stages 1 to 3, the length of TL$_{3,4}$ is 570 µm, 230 µm and 98 µm, respectively. The transistors are single NPN devices with an emitter length of 2 µm, the capacitors $C_{1,2}$ are 80 fF.

The differential pair operates in a class-AB mode and is biased through a current mirror, connected via two $\lambda/4$ transmission lines at IN$_P$ and IN$_N$. The switching quad is biased with a constant voltage derived from the supply voltage V_{CC1} via a resistive divider.

B. Two-Stage 160-GHz Power Amplifier

To achieve sufficient drive levels for the final 0.32-THz push-push frequency doubler, a differential cascoded high-power two-stage amplifier, shown in Fig. 3(b), was designed and placed after the ×8 frequency multiplier output.

A typical small-signal gain of 11 dB was simulated with a 3.3-V supply and a current consumption of 34 mA for a single stage. A two-stage design was chosen for a saturated P_{out} of more than 15 dBm, considering a typical input power (P_{in}) of around 0 dBm from the preceding ×8 frequency multiplier.

The emitter length of each transistor is 2.5 µm, and a total of four NPN devices are used in parallel for each transistor. TL$_1$ is an edge-coupled microstrip line with $Z_0 = 50\,\Omega$ / $l = 17$ µm, TL$_{2,3}$ are microstrip lines with $Z_0 = 67\,\Omega$ / $l = 75$ µm.

C. Push-Push Frequency Doubler

A common-emitter push-push frequency doubler, shown in Fig. 3(c), has been designed with class-B biasing to maximize the second-harmonic output current at 0.32 THz. Four parallel devices are used for each transistor, with a total emitter length of 4×0.9 µm. The transmission lines TL$_1$, TL$_2$ and TL$_3$ have Z_0 / length of 57 Ω / 10 µm, 57 Ω / 45 µm, and 48 Ω / 25 µm, respectively.

D. On-Chip Patch Antenna

For this 0.32-THz transmitter, an on-chip patch antenna was designed using the highest metal layer as the radiating element, and the lowest metal as the ground plane to maximize the antenna gain and efficiency. As some metal layers require a higher local density, a metal bar was placed at the E_0 of the patch antenna, to fulfill all required design rules without waivers and to prevent the unwanted impact of automatically generated filling structures. The antenna design was optimized and verified with the finite element method simulation using Keysight EMPro, on the diced chip, to account for any effects due to asymmetries. In Fig. 4, a three-dimensional (3D) model of the antenna is shown, as well as the simulated reflection coefficient S_{11} and the realized gain (G_{ant}).

979-8-3503-2123-4/23 $31.00 © 2023 IEEE   142

E. Supporting Circuits and SPI

An on-chip bandgap generates the analog reference voltage for all LDOs and bias current sources. The two high-voltage LDOs ($LDO_{HV1,2}$ for $V_{CC1,2}$) can generate supply voltages in the range of 2.16 V to 4.5 V, and the low-voltage LDO (LDO_{LV} for V_{CC3}) between 1.4 V and 2.13 V with 32 steps. The LDOs can be put in a bypass mode and feed the external supply voltage directly to the RF circuits. All the bias currents are programmable to 25/50/75/100 µA with a multiplier between 0 and 31, resulting in a maximum bias current of 3.1 mA.

For debugging and verification purposes, an 8-to-1 test multiplexer connected to a pad was implemented and allows measuring internal voltages (like LDO and bandgap voltages). With the included SPI, all those functions can be programmed while only using five pads on the chip, which significantly reduces the required external components for testing and application of the chip.

III. Measurement

After initial testing, an automated optimization was conducted. All measurements were performed on the wafer prober except for the beam pattern measurement. The RF input signal was generated using the Agilent Technologies E8257D. For RF over-the-air measurements, an R&S FSQ40 signal and spectrum analyzer with an external RPG ZRX330 up/down-conversion mixer, operating as a receiver, is used. The receiver was calibrated and compensated using an R&S ZC330 frequency extender as a transmitter and a VDI Erickson PM4 power meter. Both the extender and the mixer were used beyond (up to 340 GHz) their specified frequency range (220 GHz to 330 GHz) (marked by a gray area and the note "out of meas. range"). All measurements have been performed in the far field with a minimum distance of 17.5 cm between the on-chip antenna and the extenders horn antenna.

A. Automated Optimization Algorithm and Results

Due to the high integration of the chip, only two external power supplies, a signal source, and a digital controller are required to operate the chip, which allows for a compact setup for optimization. The search for the optimum configuration has been conducted in two steps: First, a set of 50 000 randomly generated configurations was tested on the setup, which provided a rough mapping of the parameter space. In the second step, a limited amount of the previously measured points was selected based on the optimization criterion. Each parameter was systematically altered, effectively condensing the parameter map at those points. Using the latest configurations that fulfilled the optimization criteria best, this step was repeated multiple times to fine-tune the result.

The algorithm was applied to optimize the chip configuration for two criteria, the highest effective isotropic radiated power (EIRP) and the highest dc-to-THz efficiency. Both optimizations yielded configuration parameter sets that significantly outperform the simulated *default* configuration in both output power and efficiency.

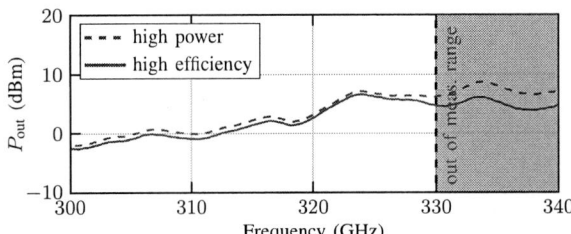

Fig. 5. Measured output power for two automatically optimized operating modes (high output power and high efficiency).

For the high-efficiency parameter set, the optimization yielded the following settings: $V_{CC1,2} = 2.16$ V, $V_{CC3} = 2.03$ V, $I_{bias1} = 500$ µA, $I_{bias2} = 600$ µA, $I_{bias3} = 1.5$ mA, $I_{bias4} = 600$ µA, $I_{bias5} = 1.5$ mA, and $I_{bias6} = 0$ µA. Technically, it is not possible to measure the individual blocks' supply current, and therefore, P_{dc} has been determined using the total supply current (with the RF turned on) and the respective LDO output voltages (with V_{CC3} increased to 2.16 V in the calculation) to $P_{dc} = 2.16$ V $\cdot$ 78.7 mA $= 170$ mW.

The resulting P_{out} versus frequency, calculated from the measured EIRP and the simulated realized antenna gain from Fig. 4(c), can be seen in Fig. 5 for both optimized parameter sets. For the high-power setting, this resulted in a peak EIRP of 11.3 dBm, P_{out} of 7.1 dBm and a dc-to-THz efficiency of 1.89% at 324 GHz. Using the receiver out-of-measurement-range at 333.9 GHz reveals a P_{out} of 8.9 dBm, resulting in an efficiency of 2.79%. The high-efficiency parameter set yielded a peak efficiency of 2.65% with an EIRP of 10.8 dBm and a P_{out} of 6.55 dBm at 324 GHz. In Fig. 3(d), the measured harmonics from the frequency multiplication are shown. The conversion gain (CG) versus P_{in} is shown in Fig. 6(a), with a peak CG of 23.8 dB at $P_{in} = -18.0$ dBm.

These parameter sets have been verified to be well-behaved, stable, and reproducible across different dies by overriding the internal bandgap reference voltage with an external supply and varying the nominal value of 1.025 V by $\pm 5\%$ from the nominal value, effectively shifting all internally generated voltages and currents linearly. P_{out} varied by -0.3 dB and -4.3 dB for the high-efficiency parameter set and 0.0 dB and -1.9 dB for the high-power setting, respectively.

B. Phase Noise

The automated phase noise (PN) measurement of an R&S FSW-85 has been used, resulting in a mostly flat phase noise difference $\Delta\mathcal{L}$ of 24 dB, as can be seen in Fig. 6(b). This is on par with the phase noise behavior of ideal frequency multiplication, which scales the PN by $20\log_{10}(16) = 24.1$ dB.

C. Beam Pattern

For the beam pattern measurement setup as depicted in Fig. 7, a robust pivoting device using stepper motors and a custom 3D-printed mounting adapter was used. The chip was wire-bonded onto a printed circuit board (PCB) and mounted

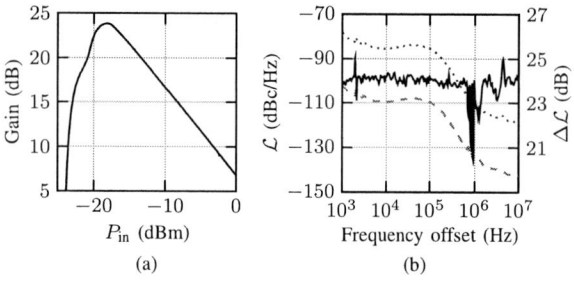

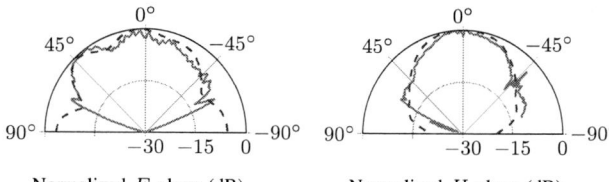

Fig. 8. Measurement (——) and simulation (– –) of normalized antenna pattern at 0.323 THz of E-plane and H-plane.

Fig. 6. (a) Measured CG versus P_{in} at 0.324 THz for the high efficiency parameter set. (b) Phase-noise measurement at 0.32 THz (···), the 20 GHz source (– –) and the respective phase scaling $\Delta\mathcal{L}$ (——).

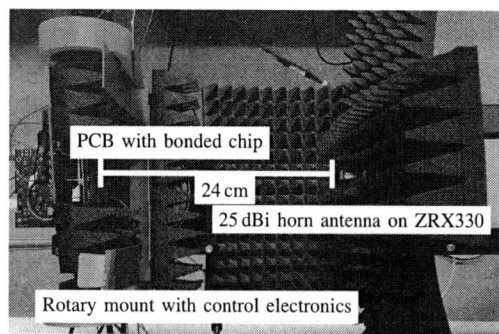

Fig. 7. Photograph of the measurement setup for the beam pattern.

Table 1. Phase-locked silicon-integrated J-band transmitters above 0.3 THz

f_0 (THz)	P_{out} (dBm)	P_{dc} (mW)	dc-to-THz efficiency (%)	Topology/ Ref.
0.313	−3.3	372	0.1	PLL+×4, [10]
0.32	2.9	1000	0.2	PLL+×16, [11]
0.312	2.5	535	0.3	×2, [12]
0.31	2.5	542	0.3	×8, [13]
0.32	3.0	433	0.5	PLL+×2, [14]
0.317	5.2	660	0.5	PLL+×2, [5]
0.304	5.2	505	0.7	×2, [4]
0.324	**7.1** / 6.6	270 / **170**	1.9 / **2.7**	×16, This work

in the adapter with the chip in the pivot point. Everything, including the R&S ZRX330, was set up on an optical bench for reproducible measurements. The results shown in Fig. 8 match with the simulation very well, but clearly depict a cutoff resulting from obstruction from the mechanical setup. The E-plane was measured first, and the H-plane was measured by rotating the PCB in the mounting mechanism by 90°.

IV. Conclusion

A 0.32 THz CW transmitter with an on-chip antenna has been implemented in a 130-nm SiGe:C BiCMOS technology, with the integrated SPI, bias, and power management circuits enabling a high level of configurability and simple application. Due to careful optimization, the achieved P_{out} compares favorably with other silicon-integrated transmitters at frequencies above 0.3 THz while also achieving a dc-to-THz efficiency of up to 2.65% (summarized in Table 1) and a close-to-ideal phase-noise performance due to frequency multiplication. A comparison to similar works with integrated phase-locked loop (PLL) sources is justifiable by the relatively small power addition on the order of tens of mW [8]. State-of-the-art free-running oscillators achieve comparatively lower P_{out} of 2.8 dBm with a dc-to-THz efficiency of 2.9% at 0.301 THz [9].

Acknowledgment

This work is supported by Silicon Austria Labs (SAL) and Johannes Kepler University (JKU). The authors would like to thank Richard Hüttner for PCB design and assembly.

References

[1] "IEEE standard for high data rate wireless multi-media networks–amendment 2: 100 Gb/s wireless switched point-to-point physical layer," *IEEE Std 802.15.3d-2017*, pp. 1–55, 2017.

[2] T. Bücher et al., "A broadband 300 GHz power amplifier in a 130 nm SiGe BiCMOS technology for communication applications," *IEEE JSSC*, vol. 57, no. 7, pp. 2024–2034, 2022.

[3] E. Mohamed et al., "220–320-GHz J-band 4-way power amplifier in advanced 130-nm BiCMOS technology," *IEEE MWC-L*, vol. 32, no. 11, pp. 1335–1338, 2022.

[4] S. Breun et al., "A 268-325 GHz 5.2 dBm Psat frequency doubler using transformer-based mode separation in SiGe BiCMOS technology," in *Proc. IEEE BCICTS*, 2021.

[5] R. Han et al., "A SiGe terahertz heterodyne imaging transmitter with 3.3 mW radiated power and fully-integrated phase-locked loop," *IEEE JSSC*, vol. 50, no. 12, pp. 2935–2947, 2015.

[6] C. Jiang et al., "A fully on-chip frequency-stabilization mechanism for terahertz sources eliminating frequency reference and dividers," *IEEE T-MTT*, vol. 67, no. 7, pp. 2523–2536, 2019.

[7] S. Yuan and H. Schumacher, "90–140 GHz frequency octupler in Si/SiGe BiCMOS using a novel bootstrapped doubler topology," in *Proc. Eur. Microw. Integr. Circuits Conf.*, Oct 2014, pp. 158–161.

[8] Y. Zhao et al., "A 20-GHz PLL with 20.9-fs random jitter," *IEEE JSSC*, pp. 1–13, 2022.

[9] Z. Peng et al., "A 300 GHz push-push coupling VCO employing T-embedded network in CMOS technology," *IEEE T-THz*, vol. 12, no. 4, pp. 426–429, 2022.

[10] Y. Liang et al., "A low-jitter and low-reference-spur 320 GHz signal source with an 80 GHz integer-N phase-locked loop using a quadrature XOR technique," *IEEE T-MTT*, vol. 70, no. 5, pp. 2642–2657, 2022.

[11] X.-D. Deng et al., "A 320-GHz 1×4 fully integrated phased array transmitter using 0.13-μm SiGe BiCMOS technology," *IEEE T-THz*, vol. 5, no. 6, pp. 930–940, 2015.

[12] S. Breun et al., "A 295-337 GHz 2.5 dBm Psat cascode-based frequency doubler in SiGe BiCMOS technology," in *Proc. IEEE SiRF*, 2022, pp. 66–69.

[13] A. Ali et al., "220–360-GHz broadband frequency multiplier chains (x8) in 130-nm BiCMOS technology," *IEEE T-MTT*, vol. 68, no. 7, pp. 2701–2715, 2020.

[14] C. Jiang et al., "A fully integrated 320 GHz coherent imaging transceiver in 130 nm SiGe BiCMOS," *IEEE JSSC*, vol. 51, no. 11, pp. 2596–2609, 2016.

979-8-3503-2123-4/23 $31.00 © 2023 IEEE

RMo3C-2

A 26-Gb/s 140-GHz OOK CMOS Transmitter and Receiver Chipset for High-Speed Proximity Wireless Communication

Qiuyu Peng[#], Haikun Jia[#], Ran Fang[*], Pingda Guan[#], Mingxing Deng[#], Jiamin Xue[#], Wei Deng[#], Xin Liang[*], Baoyong Chi[#]

[#] School of Integrated Circuits, BNRist, Tsinghua University, China
[*]BriRadio Technology Inc., China
jiahaikun@tsinghua.edu.cn

Abstract—This paper presents a high data rate 140-GHz on-off keying (OOK) CMOS transmitter and receiver chipset implemented in 28nm CMOS process for short-range wireless communications. The transmitter comprises a power amplifier (PA), modulator, CML-to-CMOS converter, and a fundamental voltage-controlled oscillator (VCO). The receiver consists of a low-noise amplifier (LNA), demodulator, and a single-ended to differential converting amplifier. Continuous-time linear equalizers (CTLE) are implemented on transmitter and receiver baseband to extend the baseband bandwidth. The proposed chipset is wire-bonded to an on-board Vivaldi antenna and a FR4 PCB carrier board to reduce the system cost. Despite the low-cost packaging, it demonstrates measured 26-Gb/s and 17-Gb/s error-free (BER<10^{-12}) wireless OOK links at 1 cm and 5 cm distance, respectively. The transmitter and receiver consume 99 mW and 63 mW power consumption, respectively.

Keywords—CMOS, D-band, on-off keying (OOK), wireless, proximity communication, high data rate, on-board antenna

I. INTRODUCTION

The proximity wireless high-speed communication is extremely appealing these days due to the enormous demand for continuously increasing data traffic. For this short-range wireless distance, the millimeter-wave (mm-wave, 30 GHz to 300 GHz) on-off keying (OOK) modulation is one of the most attractive schemes. On one hand, the OOK modulation has a simpler architecture, leading to lower power consumption and more compact chip size, compared with high-order modulation systems. On the other hand, the mm-wave frequency band offers a wide frequency spectrum to support tens of Gbps high-speed wireless communication systems.

The reported OOK transceivers in CMOS process mostly span from 60 GHz [1]-[2] to frequency above 100GHz [5]-[9]. In the lower mm-wave band, such as V-band and E-band, the reported OOK wireless transmission data rate is mostly restricted to about 10-Gbps owing to the limitation of the available bandwidth. Thus, to improve the data rate at those bands, much more complex modulation schemes are needed, which would compromise the simplicity and low-power characteristics [10]. However, mm-wave frequency range above 100-GHz, such as D-band, is more promising to achieve more than 10-Gbps and even multi-decade-Gbps wireless communications for OOK modulation thanks to its sufficient spectrum resource.

This paper presents a D-band high-speed proximity wireless communication transmitter and receiver chipset based

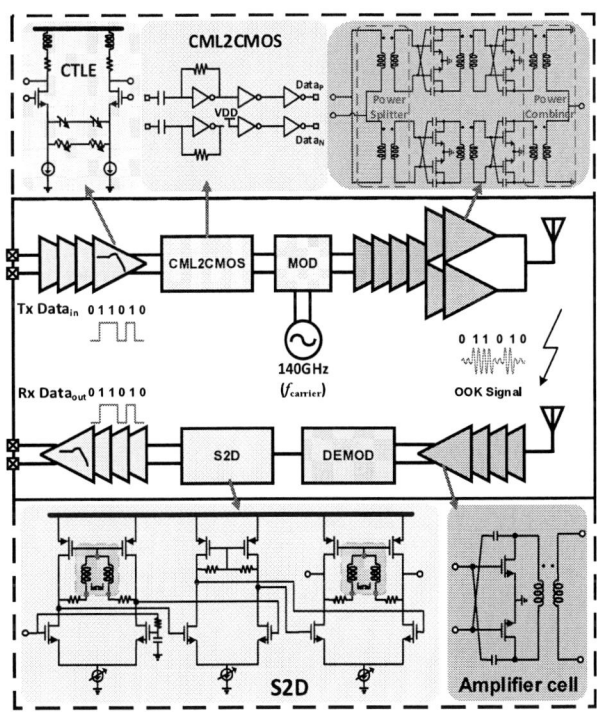

Fig. 1. Block diagram of the 140-GHz OOK chipset: (top) TX; (bottom) RX.

on OOK modulation scheme. To achieve a practical distance for short-range communication with high data rate, increasing the RF power while maintaining wideband is necessary. A D-band six-stage power amplifier (PA) including power splitter and power combiner is adopted in the transmitter (TX). An on-chip voltage-controlled oscillator (VCO) operating at 140-GHz directly modulates the baseband signal, which simplifies the system design. It is noteworthy that a cancellation-based OOK modulator is proposed to improve the data rate and ON-OFF ratio in this work. To improve the sensitivity of the receiver, a four-stage low-noise amplifier (LNA) is adopted in the receiver (RX). And the demodulator based on the envelop detector (ED) maintains the simplicity of the system. As a result, the proposed chipset demonstrates a 26-Gb/s and 17-Gb/s error-free (BER<10^{-12}) wireless OOK link at 1 cm and 5 cm distance, respectively.

979-8-3503-2123-4/23 $31.00 © 2023 IEEE

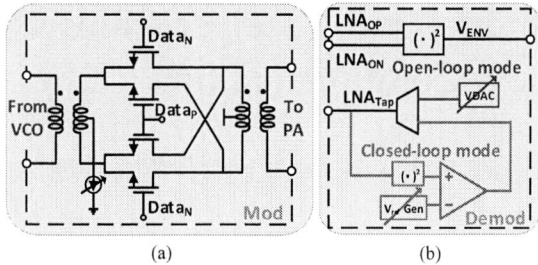

Fig. 2. Simplified schematic of the cancellation-based OOK modulator: (a) and demodulator: (b).

II. SYSTEM ARCHITECTURE

Fig. 1 shows the block diagram of the proposed 140-GHz OOK proximity communication system, which comprises of TX (top), RX (bottom), and on-board Vivaldi antenna. The TX incorporates a continuous-time linear equalizer (CTLE), CML-to-CMOS converter (CML2CMOS), cancellation-based OOK modulator, 140-GHz VCO, and a PA. The RX incorporates a LNA, OOK demodulator, single-ended-to-differential converting amplifier (S2D), and a CTLE same as the TX. The basic amplifier cell of multi-stage PA and LNA is shown in the bottom right of Fig. 1. And the detailed schematic of the CML2CMOS and S2D is presented in Fig. 1. The top left of Fig. 1 depicts the basic cell of the four-stage CTLE. And the simplified schematic of the proposed modulator and demodulator is shown in Fig. 2. In addition, The Tx/Rx chip and on-board antenna are connected by the wire bonding assembly, which benefits a simple and low-cost implementation of the system. The proposed transceiver system is validated using the over-the-air (OTA) measurement.

III. CIRCUIT IMPLEMENTATION

A. Transmitter

The transmitter chip architecture is depicted in Fig. 1 (top). To achieve wireless transmission at D-band, an on-chip 140-GHz VCO that generates the D-band carrier signal is applied in the TX. The schematic and the measured frequency tuning range of the VCO are shown in Fig. 3. Multiplier-based architecture is adopted in many works to obtain sub-terahertz

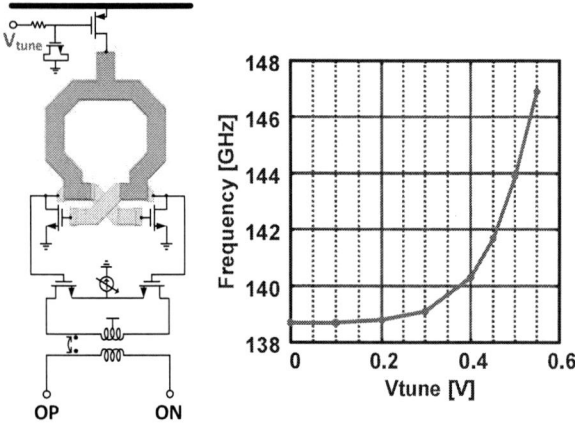

Fig. 3. 140-GHz VCO schematic and the its measured frequency tuning range.

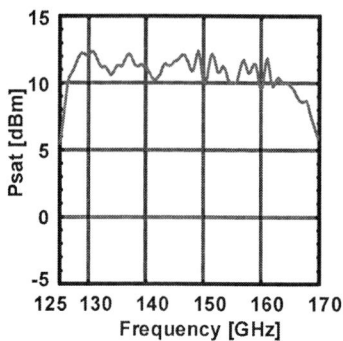

Fig. 4. Measured saturated output power of the TX.

carrier frequency output, which consumes large power and large chip area due to considerable insertion loss in frequency-multiplying operation and the necessary buffer stages. Therefore, a fundamental cross-coupled oscillator working at D-band is used in this design. Inductors are inserted into the drain-to-gate connection to boost the VCO's output power and loop-gain. The frequency is tuned through the bias voltage of the current tail transistor. A measured frequency tuning range from 138.7 GHz to 146.9 GHz is obtained as shown in Fig. 3. And a variable gain amplifier (VGA) stage is utilized to adjust the output power of the VCO not to saturate the following modulation and amplification stages.

The design of the OOK modulator is one of the keys that guarantee high data rate operation, which compromises such as ON-OFF isolation, conversion loss, and data rate capacity. Fig. 2 (a) shows the simplified schematic of the OOK modulator. The proposed OOK modulator is implemented as a cancellation-based topology. In this cancellation-based OOK modulator, the baseband signal modulates the transistor gate. The $Data_N$ signal level is set to VDD, which always renders the modulated transistors on-state, while $Data_P$ turns on and off the corresponding transistors according to the data, leading to in-phase and out-of-phase combination of the carrier signal, and generating the OOK modulation. The cancellation-based topology has three advantages compared with the switch-controlled amplifier or oscillator scheme. 1) The dc path is always on, reducing the bias current fluctuation, and therefore improving the switching speed. 2) The fluctuation of modulator output impedance is also reduced, which is good for the wideband characteristic of the transformer matching network. 3) The cancellation operation of the RF current eliminates the feedforward effect of the parasitic capacitor, leading to the improving of the ON-OFF ratio. Thanks to this cancellation-based OOK modulation scheme, an error-free (BER<10-12) data rate as high as 26-Gb/s is achieved.

In the front-end of the TX, a six-stage PA is adopted to boost the output power. The output power of the PA is measured and shown in Fig. 4. The saturated output power is about 12.7 dBm. Four-stage RC source-degeneration CTLEs with inductive peaking are used in both the TX and RX baseband to compensate for the channel insertion loss. Additionally, a resistance feedback inverter-based CML-to-CMOS converter is presented to convert the CML signal from the CTLE to CMOS signal to the modulator in the TX.

979-8-3503-2123-4/23 $31.00 © 2023 IEEE

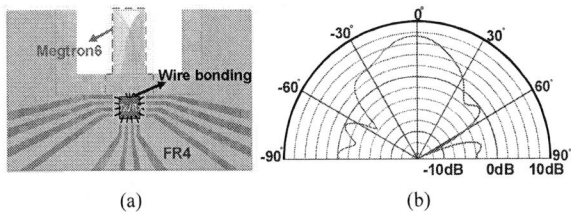

(a)   (b)

Fig. 5. Package design and on-board antenna simulated performance.

B. Receiver

As demonstrated in Fig. 1 (bottom), the receiver chip comprises a four-stage LNA, demodulator, S2D, and CTLE. As for 140-GHz wideband LNA design, the active gm-stage is implemented as neutralized differential common-source amplifiers, and the passive inter-stage matching network adopts the transformer-based wideband matching network, which leads to more than 23-GHz 3 dB bandwidth. A demodulator based on an envelope detector (ED) subsequently translates the LNA output signal down to the baseband. Fig. 2 (b) shows the design of the demodulator. A push-push differential pair using a resistor load is adopted to realize square-law detection. It is noteworthy that the performance of the demodulator, such as gain and bandwidth, is highly sensitive to the bias voltage of the ED. Therefore, two optional bias schemes, namely the closed-loop scheme and the open-loop scheme, are implemented in this work for robustness. The closed-loop operation mode is based on a replica loop that compensates for process, voltage, and temperature (PVT) variations. The open-loop operation mode provides more flexible bias generation using an on-chip voltage digital-to-analog converter (VDAC). A three-stage single-ended-to-differential circuit is designed to converter the single-ended demodulator output to the differential in the RX path. PMOS-based active inductors and passive inductors are combined to extend the bandwidth of the demodulator. The inter-winding coupling of the passive inductors in the S2D amplifier improves the common-mode signal suppression.

C. Systems-in-package and antenna integration

Fig. 5 (a) shows the package of the TX/RX chip and on-board antenna integration. As depicted in Fig. 5, the Vivaldi antenna is utilized to radiate electromagnetic wave. The Vivaldi antenna is appropriate for D-band application because of its high gain and wideband performance. And Fig. 5 (b) shows the simulated antenna radiation pattern, which exhibits high directivity and a 7.5 dBi antenna gain. To balance the antenna performance and the system cost, the antenna is constructed on a small Megtron6 board which is pasted to a

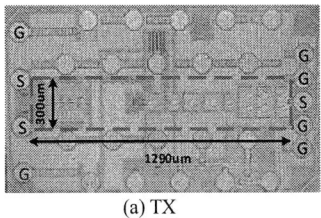   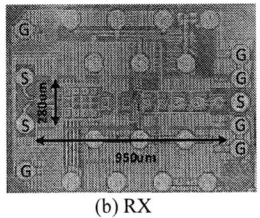

(a) TX   (b) RX

Fig. 6. Micrographs of the fabricated chipset: (a) transmitter, (b) receiver.

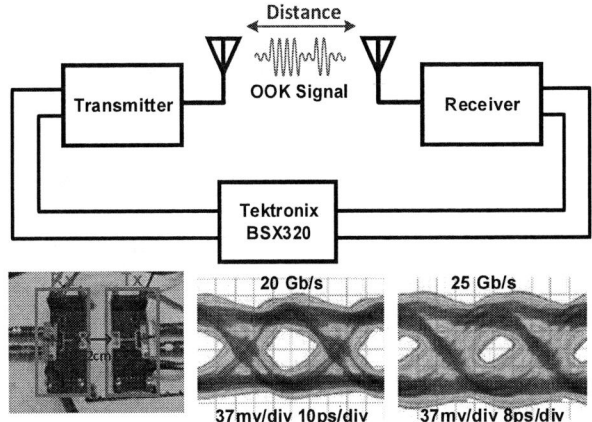

Fig. 7. Measurement setup for proximity communication and measured eye diagrams.

low-cost FR4 carrier board. Both TX and RX chips are wire-bonded to both the antenna board and the carrier board.

IV. EXPERIMENTAL RESULTS

The proposed chipset is fabricated in 28nm CMOS process. Fig. 6 shows the chipset including TX and RX die photos. And the TX and RX are implemented on core areas of 1.29×0.3 and 0.95×0.28 mm^2, respectively. For 26-Gb/s wireless data transmission, the TX and RX consume a dc power of 99 mW and 63 mW at supply voltage of 0.9 V.

Fig. 7 presents the measurement setup and the measured eye diagrams. The baseband differential input signal is generated by a Bit-Error-Rate-Tester (Tektronix BSX320). A pseudorandom binary sequence (PRBS) of length 2^7-1 is applied to the TX. This is modulated by the TX, and the OOK signal is sent to the RX module over-the-air. The differential output of the RX is connected to the BSX320 to observe the eye diagram and the bit error rate (BER). The results in Fig. 7 show the measured 20-Gb/s and 25-Gb/s eye diagrams at the RX's output. The measured BER performance at different data rates and wireless transmission distances is exhibited in Fig. 8. At a carrier frequency of 140-GHz, the measured peak error-free (BER<10^{-12}) data rate at 1 cm, 2 cm, 3 cm, 4 cm, and 5 cm distance is 26-Gb/s, 25-Gb/s, 20-Gb/s, 18-Gb/s, and 17-Gb/s, respectively. The comparison of the measured results

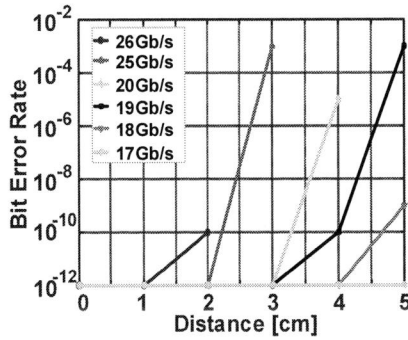

Fig. 8. Measured BER performance for different data-rate and link length.

Table 1. Performance comparison of the OOK wireless transceivers.

Reference	Technology	Frequency (GHz)	Modulation	Distance (cm)	Data rate (Gb/s)	BER	Power consumption (mW)	Integration	Package solution
[1]	65 nm CMOS	60	OOK	2	12.5	< 1e^{-12}	TX: 12.1 * RX: 21	TX, RX, on-board Yagi-Uda antenna	Wire-bond
[3]	40 nm CMOS	80	OOK	0.5	20	< 1e^{-12}	TX: 52 RX: 85	TX, RX, on-board Dipole antenna	Wire-bond
[4]	65 nm CMOS	100	OOK	1	7.6	< 1e^{-11}	TX: 32.8 RX: 75.6	TX, RX, on-board Yagi-Uda antenna	Flip-chip
[6]	45 nm CMOS	135	OOK	10	10	< 1e^{-11}	TX: 18 * RX: 80	TX, RX, Horn antenna	n/a
[8]	32 nm SOI CMOS	210	OOK	3.5	10	< 1e^{-5}	TX: 240 RX: 68	TX, RX, on-chip Dipole antenna	Wire-bond
[9]	65 nm CMOS	260	OOK	4	6	n/a	TX: 688 RX: 485	TX, RX, on-chip Half-wave LW antenna	n/a
This work	28 nm CMOS	140	OOK	5	17	< 1e^{-12}	TX: 99** RX: 63	TX, RX, on-board Vivaldi antenna	Wire-bond
				3	20	< 1e^{-12}			
				1	26	< 1e^{-12}			

*Does not include PA. **Measured results @26-Gb/s

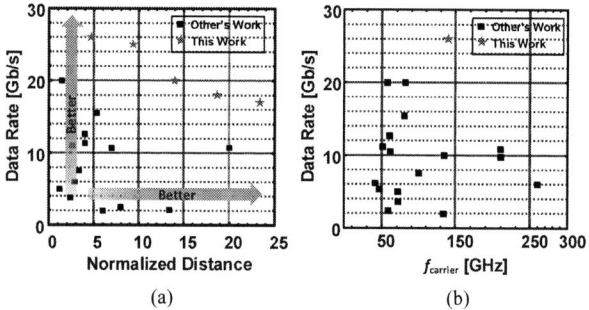

(a)   (b)

Fig. 9. The comparison of the reported CMOS wireless OOK transceivers with BER<10^{-12}.

with the prior reported CMOS wireless short-range OOK transceivers with BER<10^{-12} is shown in Fig. 9. Note that in Fig. 9 (a) the communication distance is normalized to the carrier wavelength for a fair comparison. Among those works, our proposed OOK chipset achieves the highest error-free (BER<10^{-12}) data rate for OOK wireless communication and a decent OTA communication distance. Table 1 presents a summary of the performance of the proposed transceiver chipset and compares it with other similar works.

V. CONCLUSION

In this paper, a D-band transceiver chipset is demonstrated with a maximum data rate of 26-Gb/s utilizing OOK modulation for wireless proximity communication. The package and on-board antenna integration adopt the efficient and low-cost wire-bonding scheme, which is in consistent with simplicity of the OOK system. The chipset was verified with an OTA test with different distances. As a result, the proposed transceiver realized an error-free (BER<10-12) data rate wireless link as high as 26-Gb/s at 1 cm distance.

ACKNOWLEDGEMENT

This work was supported by the National Key R&D Program of China under Grant No.2019YFB2204700 and the Beijing Innovation Center for Future Chips (ICFC).

REFERENCES

[1] C. W. Byeon et al., "A 2.65-pJ/Bit 12.5-Gb/s 60-GHz OOK CMOS Transmitter and Receiver for Proximity Communications," in IEEE Transactions on Microwave Theory and Techniques, vol. 68, no. 7, pp. 2902-2910, July 2020.

[2] C. W. Byeon et al., "A 67-mW 10.7-Gb/s 60-GHz OOK CMOS Transceiver for Short-Range Wireless Communications," in IEEE Transactions on Microwave Theory and Techniques, vol. 61, no. 9, pp. 3391-3401, Sept. 2013.

[3] Y. Tanaka et al., "A versatile multi-modality serial link," 2012 IEEE International Solid-State Circuits Conference, 2012, pp. 332-334.

[4] K. Nakajima et al., "23Gbps 9.4pJ/bit 80/100GHz band CMOS transceiver with on-board antenna for short-range communication," A-SSCC, 2014, pp. 173-176.

[5] N. Ono et al., "A 113 GHz 176 mW transmitter and receiver chipset using 65 nm CMOS technology," 2012 Asia Pacific Microwave Conference Proceedings, 2012, pp. 439-441.

[6] M. Fujishima et al., "98 mW 10 Gbps Wireless Transceiver Chipset With D-Band CMOS Circuits," in IEEE Journal of Solid-State Circuits, vol. 48, no. 10, pp. 2273-2284, Oct. 2013.

[7] B. Suh et al., "A D-Band Multiplier-Based OOK Transceiver With Supplementary Transistor Modeling in 65-nm Bulk CMOS Technology," in IEEE Access, vol. 7, pp. 7783-7793, 2019.

[8] Z. Wang et al., "A CMOS 210-GHz Fundamental Transceiver With OOK Modulation," in IEEE Journal of Solid-State Circuits, vol. 49, no. 3, pp. 564-580, March 2014.

[9] J. -D. Park et al., "A 260 GHz fully integrated CMOS transceiver for wireless chip-to-chip communication," VLSIC, 2012, pp. 48-49.

[10] A. Hamani et al., "A 56.32 Gb/s 16-QAM D-band Wireless Link using RX-TX Systems- in-Package with Integrated Multi-LO Generators in 45nm RFSOI," 2022 IEEE Radio Frequency Integrated Circuits Symposium (RFIC), 2022, pp. 75-78.

RMo3C-3

A 189GHz three-stage super-gain-boosted amplifier with power gain of 10.7 dB/stage at near-f_{max} frequencies in 65nm CMOS

Fei He, Menghu Ni, Qian Xie, Zheng Wang

University of Electronic Science and Technology of China, China

wangzheng@uestc.edu.cn

Abstract—In this paper, a 189GHz three-stage super-gain boosted amplifier with power gain of 10.7dB/stage in 65nm CMOS is presented. Based on the *U*-boosted core and Y/Z-embedding network, a super-gain-boosting technique is proposed to improve the power gain of amplifier at near-f_{max} frequency. The cross-conductance technique is analyzed and exhibits the potential to improve the mason's *U* of an active two port network leading to the further improvement of the theoretical upper limit of the maximum available gain (G_{ma}). Furthermore, by employing the sensitized inductor, the sensitivity to improve mason's *U* due to the process variations and modeling errors can be decreased. On top of the *U*-boosted core, additional Y/Z-embedding networks are employed to make G_{ma} reach the boosted $G_{ma\_upper\_limit}$. Based on the proposed super-gain-boosting technique, a three-stage amplifier is implemented in 65nm CMOS process, exhibits a peak small-signal gain of 32.1dB and a saturated power of -1.96dBm at 189GHz.

Keywords—terahertz, gain-boosting, CMOS, near-f_{max}, embedding network

I. INTRODUCTION

THz and sub-THz technologies attracts extensive attention in applications like radar, communication and imaging systems [1]. CMOS technology, due to its superiority in size, cost, and integration level, has been proven to be a promising candidate in sub-THz systems. However, as the operating frequency approaches the maximum oscillation frequency (f_{max}) of a transistor, the insufficient power gain of front-end amplifier becomes the most stringent challenge in sub-THz systems.

To boost the power gain, great efforts have been made to introduce all types of "embedding network" to the core device [2-8]. The representative gain-boosting techniques are based on linear, lossless and reciprocal embedding network (LLREN), such as Y/Z-embedding and Y/PreZ-embedding networks. The analysis of the LLREN has been well established and it is found that, even with lossless embedding networks, there exists a theoretical upper limit of the maximum available gain (G_{ma}), which is $2U-1+2\times\sqrt{U(U-1)}$, where U denotes the Mason's U. Recent work attempts to introduce lossy components to break the limitation of $2U-1+2\times\sqrt{U(U-1)}$ by boosting U instead of G_{ma}. Based on the over-neutralization capacitor technique, [4] introduces an extra resistor in series with the over-neutralization capacitor which exhibits the potential to improve U. However, this technique cannot guarantee that G_{ma} reaches the improved $G_{ma\_upper\_limit}$.

In this work, to improve mason's U and achieving the upper limit of the boosted $G_{ma\_upper\_limit}$, we present a super-gain-

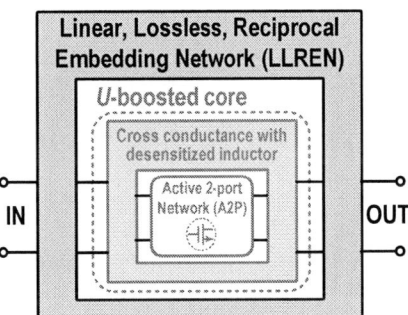

Fig. 1. The block of the proposed super-gain-boosted structure.

boosting technique, the structure of the embedded core is shown in Fig. 1. Based on the super-gain-boosting technique, a 189GHz three-stage super-gain-boosted amplifier in 65nm CMOS is implemented, achieving measured power gain of 32.1dB (10.7dB/stage) and FoM of 3.41 at near-f_{max} frequencies. The proposed amplifier features a cross-conductance U-boosting technique with desensitized inductor to effectively improve the U of G_{ma}-core to $U_{boosted}$ by 3.8 dB.

II. THREE-STAGE SUPER-GAIN-BOOSTED AMPLIFIER

A. U-boosted technique based on cross-conductance network

The basic idea to boost the mason's U is to introduce lossy component in a proper way which break the limitation of LLREN.

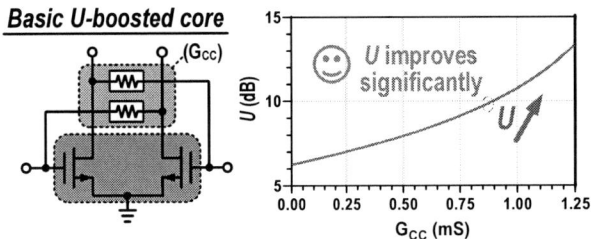

Fig. 2. The structure of the basic U-boosted core and simulation results of U versus G_{CC}.

For the basic U-boosted core as shown in Fig. 2, by adding cross conductance G_{CC} to the differential pair, G_{CC} will be added to each term of the Y-parameters matrix of the differential pair:

$$\begin{bmatrix} Y_{11\_DIFF} & Y_{12\_DIFF} \\ Y_{21\_DIFF} & Y_{22\_DIFF} \end{bmatrix} + \begin{bmatrix} G_{CC} & G_{CC} \\ G_{CC} & G_{CC} \end{bmatrix}, \quad (1)$$

where Y_{ij_DIFF} (i, j=1 or 2) denotes Y-parameters of the differential pair.

979-8-3503-2123-4/23 $31.00 © 2023 IEEE   149   2023 IEEE Radio Frequency Integrated Circuits Symposium

Since the expression of mason's U is equal to:

$$U = \frac{|Y_{21} - Y_{12}|^2}{4(\mathrm{Re}[Y_{11}]\mathrm{Re}[Y_{22}] - \mathrm{Re}[Y_{12}]\mathrm{Re}[Y_{21}])}, \qquad (2)$$

the expression of $U_{boosted}$ can be obtained by substituting the total Y-parameters of basic U-boosted core into (2):

$$\begin{cases} U_{boosted} = \dfrac{|Y_{21\_DIFF} - Y_{12\_DIFF}|}{4(M-N)} \\ M = \mathrm{Re}[Y_{11\_DIFF} + G_{CC}]\mathrm{Re}[Y_{22\_DIFF} + G_{CC}] \\ N = \mathrm{Re}[Y_{12\_DIFF} + G_{CC}]\mathrm{Re}[Y_{21\_DIFF} + G_{CC}] \end{cases} . \quad (3)$$

From the expression of $U_{boosted}$ (3), the added G_{CC} will not change the numerator of $U_{boosted}$. The denominator will decrease when the condition of (4) is satisfied:

$$\mathrm{Re}[Y_{12\_DIFF}] + \mathrm{Re}[Y_{21\_DIFF}] - \mathrm{Re}[Y_{11\_DIFF}] - \mathrm{Re}[Y_{22\_DIFF}] > 0 . \quad (4)$$

Eventually, the added G_{CC} leading to the rise of $U_{boosted}$. This is quite counterintuitive while it is proven to be effective from the simulation results of U versus G_{CC} in Fig. 2.

In the basic U-boosted structure, it can be also observed that as the increase of G_{CC}, the derivative of U with respect to G_{CC} also increases rapidly. Due to the process variations and modeling errors, it may be too sensitive if only G_{CC} is employed to the circuit to result in a huge discrepancy between simulation and measurement results.

In order to solve this aforementioned problem, a U-boosted core with desensitized inductors is proposed. As shown in Fig. 3, the desensitized inductors are placed at the gate and drain terminals of transistors. The total Y-parameters of the differential pair with desensitized inductors becomes:

$$\frac{1}{\Delta}\begin{bmatrix} Z_{22\_DIFF} + jX_{P2} & -Z_{12\_DIFF} \\ -Z_{21\_DIFF} & Z_{11\_DIFF} + jX_{P1} \end{bmatrix}, \quad (5)$$

with

$$\Delta = (Z_{11\_DIFF} + jX_{P1})(Z_{22\_DIFF} + jX_{P2}) - Z_{12\_DIFF} Z_{21\_DIFF}, \quad (6)$$

where Z_{ij_DIFF} (i, j=1 or 2) denotes Z-parameters of the differential pair. The added desensitized inductors will affect the impact when adding G_{CC} into the differential pair, eventually decrease the sensitivity of U over G_{CC}. From the simulation results of U versus G_{CC} at different X_{P1} and X_{P2} in Fig. 3, the curve becomes flatter and the sensitivity of U over G_{CC} is decreased. Eventually, combined with the technique of cross conductance and desensitized inductors, eventually a controllable U-boosted structure is obtained.

B. Proposed super-gain-boosted structure

On top of the U-boosted core, additional embedding networks are also required because the improvement of U only implies the upper limit of G_{ma} is boosted and the embedding networks would help to push the gain to the boosted $G_{ma\_upper\_limit}$. As is shown in Fig. 4(a), this work proposed a super gain-boosting structure with desensitized U-boosted core and Y/Z-embedding network which can not only boost U but also improve G_{ma} to the boosted $G_{ma\_upper\_limit}$. The condition of achieving the boosted $G_{ma\_upper\_limit}$ is:

$$\begin{cases} \mathrm{Re}[U_{boosted}/A_{tot}] = -U_{boosted}/G_{ma\_boosted\_upper\_limit} \\ \mathrm{Im}[U_{boosted}/A_{tot}] = 0 \end{cases}, \quad (7)$$

where $A_{tot} = Y_{21\_tot}/Y_{12\_tot}$ is the measure of reciprocity with Y_{12_tot} and Y_{21_tot} are the Y-parameters of the total super-gain-boosting structure.

By employing the cross conductance G_{CC} and desensitized inductor X_{P1} and X_{P2}, U is increased from 6.2dB to 10dB. Then, after adding the capacitive Y-embedding network jB_F and capacitive Z-embedding network jX_F, the G_{ma} can reach boosted $G_{ma\_upper\_limit}$. From the gain versus frequency simulation results in Fig. 4(b), for the proposed super-gain-boosting structure, compared to the core device without any embedding networks, G_{ma} is increased from 5.9 dB to 16 dB at 190GHz and the improvement of the power gain is nearly 10.1 dB.

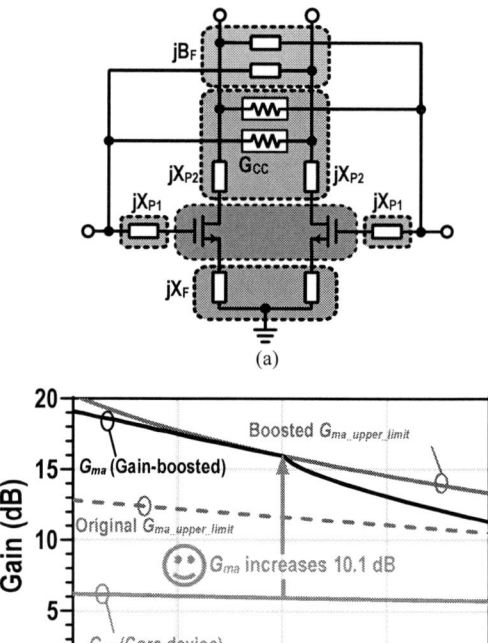

(a)

(b)

Fig. 4. Proposed super-gain-boosting (a) structure and (b) gain versus frequency.

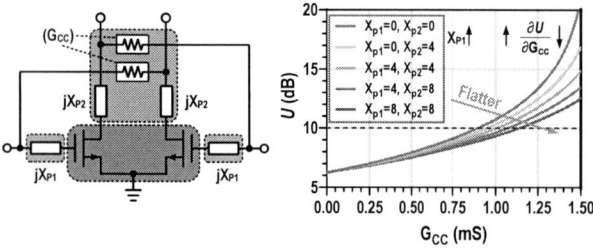

Fig. 3. The structure of the proposed U-boosted with desensitized inductors and U versus G_{CC} at different desensitized inductors conditions.

979-8-3503-2123-4/23 $31.00 © 2023 IEEE   150

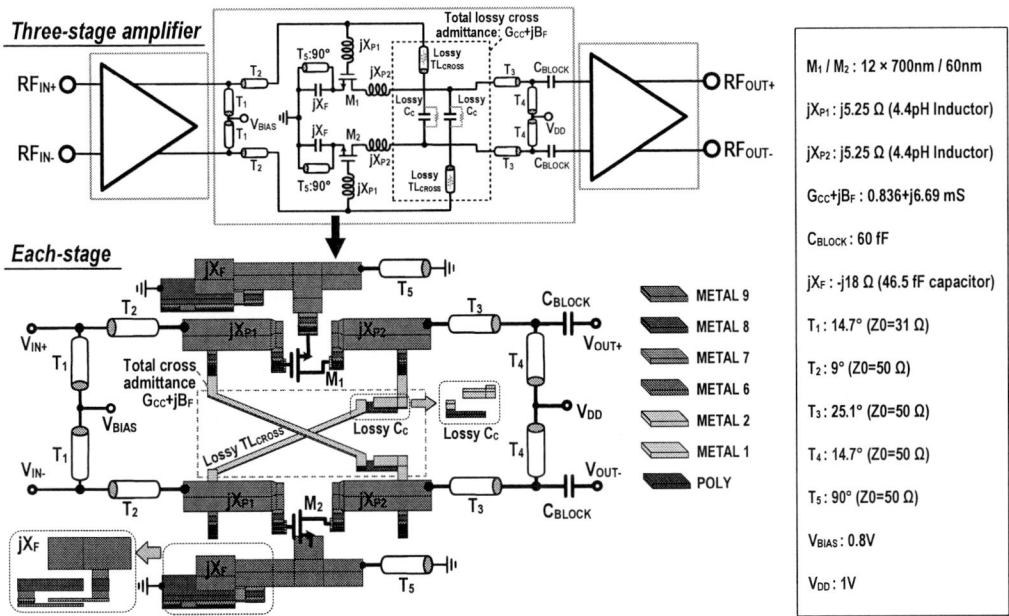

Fig. 5 The schematic of the three stage amplifier

C. Circuit implementation

To verify the analytic assessment of the proposed super-gain-boosted technique, a three-stage amplifier is implemented in 65nm CMOS. The circuit diagram of the three-stage amplifier is shown in Fig. 5. Each-stage of the amplifier exhibits the same configuration. The transistor size is optimized to W/L= 8.4μm/60nm with V_{BIAS}=0.8V and V_{DD}=1V to obtain optimal U of 6.2dB at 190GHz. Desensitized inductors of jX_{P1} and jX_{P2} are achieved by two transmission lines (TL). For the cross conductance G_{CC}, since it is difficult to realize a pure resistor in sub-THz band, G_{CC} is absorbed into the loss of the low quality factor (Q) capacitor in series with a lossy TL. This low-Q capacitor C_{CC} is a plate capacitor by adopting Metal2 and Poly layer. Eventually, the total cross admittance network is realized by the low quality-factor capacitor C_{CC} in series with the lossy TL_{CROSS} which exhibits an equivalent admittance of 0.836+j6.69 mS. Since the required Z-embedding element jX_F is also capacitive, it is implemented by a multi-layer plate capacitor, in which Metal 7 and Metal 9 are the signal layers while Metal 8 and Metal 6 are the ground layers. A 90° transmission line T_5, providing DC path to ground, is realized by a folded ground-shielded co-planar waveguide (GCPW) line to save area. Meanwhile, the GCPW is also utilized for all the input, output and inter-stage matching networks. To make the design more robust to process-dependent uncertainty at sub-THz frequency, all stages of the amplifier are matched to 50 Ω.

III. MEASUREMENT RESULTS

Fig. 6 shows the small-signal S-parameters. The S-parameters measurement has been performed in 140-220 GHz (G-band) using WR5 extender and vector network analyzer. The whole measurement setup is calibrated up to the GSG probe heads by standard Through-Reflect-Line (TRL) calibration procedure. For the three-stage differential amplifier,

the loss of the Marchand balun and GSG pad are de-embedded through a back-to-back structure. The three-stage amplifier exhibits a peak S21 of 32.1dB at 189GHz, both the S11 and S22 are nearly -10dB.

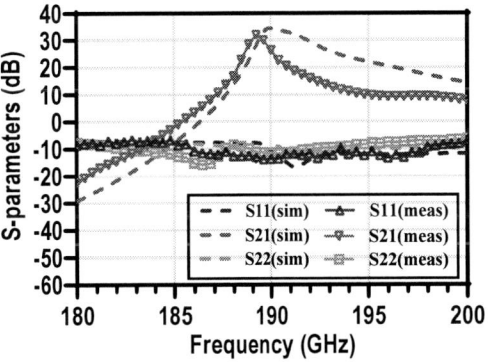

Fig. 6. S-parameters results.

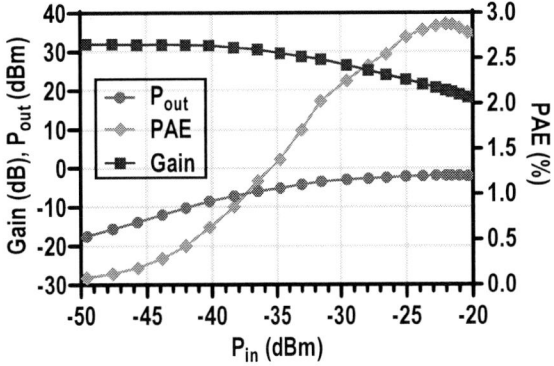

Fig. 7. Large-signal CW measurement results

979-8-3503-2123-4/23 $31.00 © 2023 IEEE

Large-signal CW measurements are performed using a WR5 extender connected to a signal source and G-band variable attenuator to provide variable input power. A power meter (Ceyear 2438PB) with sensor head and WR-5 to WR-10 tapered waveguide is used to measure the output power. As is shown in Fig. 7, at 189 GHz the amplifier exhibits a measured Psat of -1.95dBm, P1dB of -6dBm and maximum PAE of 2.87%.

Table I represent the performance summary and comparison with the state-of-the-art CMOS amplifiers with single-stage gain boosting technique around 190GHz. The three-stage amplifier based on the desensitized U-boosted core and Y/Z-embedding technique, proposed in this work, exhibits 10.7 dB/stage gain at 189GHz. To compare the gain-boosting performance between the proposed work with prior works with different silicon-based processed and working frequencies, a figure of merit (FoM) is adopted which is defined as [3]:

$$\text{FoM} = \sqrt[n]{Gain} \times (f/f_{max})^2, \qquad (8)$$

where $Gain$ is the measured gain, n is the number of stages and f is the working frequency. As is shown in Table I, compared to prior CMOS amplifiers with gain boosting techniques, the proposed amplifier exhibits better FoM, which is 3.41. Fig. 8 represents the chip micrograph of the proposed three-stage amplifier, which occupies a core area of 260×790 μm^2 and consumes a DC power of 22mW. The proposed amplifier with narrowband and high gain property has great potential for narrow-band application, such as high-sensitivity detection radar and terahertz source with clean spectrum.

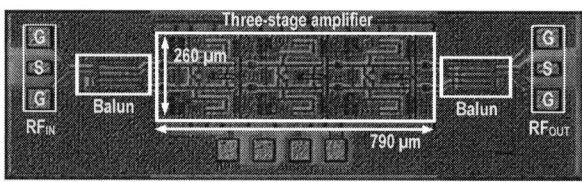

Fig. 8. Chip micrograph.

Table 1. Performance summary and comparison with the state-of-the-art CMOS amplifiers with one-stage gain boosting technique around 190GHz.

Ref	This work	[2]	[3]	[4]	[5]	[7]
Technology	**65nm CMOS**	65nm CMOS	130nm SiGe	28nm CMOS	28nm FD-SOI	32nm SOI
Frequency [GHz]	**189**	257	173	190	184	210
f_{max} [GHz]	**350**	350	280	420	390	320
f/f_{max}	**0.54**	0.73	0.62	0.45	0.47	0.656
Gain (dB)	**32.1**	9.2	18.5	14.3	7.6	15
Gain/stage (dB)	**10.7**	2.3	6.167	2.86	7.6	5
P_{sat} (dBm)	**-1.95**	-3.9	-1.8	1.5	-3.7	4.6
Max PAE (%)	**2.87**	0.8	-	2.6	4.2	6
Area (mm²)	**0.195**	0.08#	0.15#	0.09	0.09#	0.06
DC Power (mW)	**22**	27.6	63	45	5.1	40
FoM%	**3.41**	0.9156	1.58	0.395	1.28	1.36

%FoM= $\sqrt[n]{Gain} \times (f/f_{max})^2$

#Estimated from the layout

IV. CONCLUSION

We presented a super-gain-boosting technique based on cross-conductance, desensitized inductors and Y/Z-embedding network. The proposed gain-boosting technique features the controllable improvement of U and the achievement of the boosted $G_{ma\_upper\_limit}$. Based on the proposed gain-boosting technique, a three-stage amplifier is implemented based on 65nm CMOS process, offering a power gain of 32.1dB, a saturated power of -1.95dBm and a FoM of 3.41 at 189GHz. The narrowband and high gain property makes it quite suitable for narrow-band THz application which requires high frequency selectivity.

ACKNOWLEDGMENT

This work was supported by National Natural Science Fund of China under Grant 62034002.

REFERENCES

[1] M. Kim, C. Wang, L. Yi, H. -S. Lee and R. Han, "A Sub- THz CMOS Molecular Clock with 20 ppt Stability at 10,000 s Based on Dual-Loop Spectroscopic Detection and Digital Frequency Error Integration," 2022 IEEE Radio Frequency Integrated Circuits Symposium (RFIC), Denver, CO, USA, 2022, pp. 115-118, doi: 10.1109/RFIC54546.2022.9863147.

[2] H. Bameri and O. Momeni, "A High-Gain mm-Wave Amplifier Design: An Analytical Approach to Power Gain Boosting," in IEEE Journal of Solid-State Circuits, vol. 52, no. 2, pp. 357-370, Feb. 2017, doi: 10.1109/JSSC.2016.2626340.

[3] H. Khatibi, S. Khiyabani and E. Afshari, "A 173 GHz Amplifier With a 18.5 dB Power Gain in a 130 nm SiGe Process: A Systematic Design of High-Gain Amplifiers Above $f_{max}/2$," in IEEE Transactions on Microwave Theory and Techniques, vol. 66, no. 1, pp. 201-214, Jan. 2018, doi: 10.1109/TMTT.2017.2727038.

[4] D. Simic and P. Reynaert, "Analysis and Design of Lossy Capacitive Over-Neutralization Technique for Amplifiers Operating Near f_{max}," in IEEE Transactions on Circuits and Systems I: Regular Papers, vol. 68, no. 5, pp. 1945-1955, May 2021, doi: 10.1109/TCSI.2021.3060662.

[5] S. Sadlo, M. De Matos, A. Cathelin and N. Deltimple, "One stage gain boosted power driver at 184 GHz in 28 nm FD-SOI CMOS," 2021 IEEE Radio Frequency Integrated Circuits Symposium (RFIC), Atlanta, GA, USA, 2021, pp. 119-122, doi: 10.1109/RFIC51843.2021.9490441.

[6] J. Kim, C. -G. Choi, K. Lee, K. Kim, S. -U. Choi and H. -J. Song, "Analysis and Design of Dual-Peak Gmax-Core CMOS Amplifier in D-Band Embedding a T-Shaped Network," 2022 IEEE Radio Frequency Integrated Circuits Symposium (RFIC), Denver, CO, USA, 2022, pp. 87-90, doi: 10.1109/RFIC54546.2022.9863211.

[7] Z. Wang, P. -Y. Chiang, P. Nazari, C. -C. Wang, Z. Chen and P. Heydari, "A CMOS 210-GHz Fundamental Transceiver with OOK Modulation," in IEEE Journal of Solid-State Circuits, vol. 49, no. 3, pp. 564-580, March 2014, doi: 10.1109/JSSC.2013.2297415.

[8] Z. Wang and P. Heydari, "A Study of Operating Condition and Design Methods to Achieve the Upper Limit of Power Gain in Amplifiers at Near-f_{max} Frequencies," in IEEE Transactions on Circuits and Systems I: Regular Papers, vol. 64, no. 2, pp. 261-271, Feb. 2017, doi: 10.1109/TCSI.2016.2607231.

RMo3C-4

A Fully Integrated 400 GHz OOK Transceiver with On-Chip Antenna in 90 nm SiGe BiCMOS for Multi Gbps Wireless Communication

Sidharth Thomas, Sam Razavian, Aydin Babakhani

University of California, Los Angeles, USA

{sidhthomas, samrazavian24}@g.ucla.edu, aydinbabakhani@ucla.edu

Abstract — This paper demonstrates a fully integrated 400 GHz OOK transceiver in 90nm SiGe BiCMOS. The transmitter employs a PIN diode quadrupler driven by an on-chip oscillator to generate a 0.4 THz signal, which is modulated with OOK data. The receiver employs a fundamentally driven passive mixer-first architecture. The LO for this mixer is generated from two mutually locked PIN diode quadruplers, which generate sufficient power at 0.4 THz to demonstrate fundamental mixing. The transmitter has an EIRP of 17 dBm and consumes 80 mW DC power. The receiver has 25 dB noise figure, 17.3 dB conversion gain, and consumes 184 mW DC power. The transmitter achieves a data rate of 8 Gbps at 20 cm distance and 1 Gbps at 85 cm distance. The transmitter-receiver system achieves 2 Gbps at 20 cm distance. This is the first demonstration of a fully integrated multi-Gbps wireless transceiver above 300 GHz in silicon.

Keywords — mm-Wave, THz, transmitter, receiver, OOK, SiGe, multi-Gbps

I. INTRODUCTION

There has been increasing interest in opening up the THz band (0.1 - 10 THz) for wireless communication. In 2019, the US Federal Communications Commission recognized using EM spectrum between 95 GHz and 3 THz for new technologies and services [1]. The critical requirement for commercializing wireless communication using these frequency bands is designing energy-efficient silicon-based transmitters and receivers.

THz frequencies lie beyond the maximum oscillation frequency (f_{max}) of silicon technologies. This makes it challenging to generate THz power efficiently and achieve a low noise figure (NF). In a wireless transceiver, these factors, along with the large bandwidth requirements of multi-Gbps data, significantly reduce the SNR. This affects the data transmission range, making transceiver design challenging beyond f_{max}. Among fully integrated silicon-based works, only [2] has shown over-the-air (OTA) multi-Gbps data transmission above 300 GHz. There has been no demonstration of receivers with multi-Gbps data communication above 300 GHz.

This work presents a wireless transceiver at 400 GHz, with OOK modulation capability, tested over the air, for short-range communications. The reverse recovery in PIN diodes is used here to generate power at THz frequencies. The highly non-linear nature of the PIN diode allows high-power THz generation, enabling the design of THz transmitters and receivers with high effective isotropic radiated power (EIRP) and low NF. The energy-efficient transceiver can open up several applications, such as low-latency distributed computing, fixed wireless, and wireless backhaul.

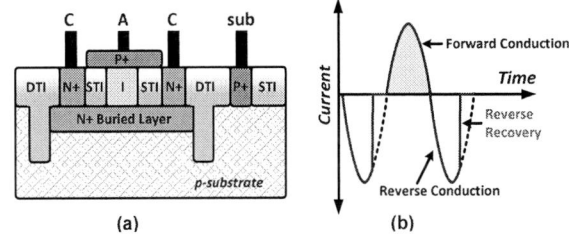

Fig. 1. (a) Structure of PIN diode in GlobalFoundries 90nm process (b) PIN diode reverse recovery

II. CIRCUIT DESIGN

Due to the limited f_{max} of silicon technologies, direct frequency generation is not possible at 400 GHz. Conventionally, harmonic generation using non-linear devices such as FETs, BJTs, or varactors is employed to generate power beyond f_{max} [3-5]. However, these devices have weak non-linearity and generate less THz power. The reverse recovery in PIN diodes is highly non-linear and has been used for THz generation [6]. This work uses PIN diode multipliers to generate high THz power at 400 GHz to design an efficient THz transceiver.

A. PIN Diode and Reverse Recovery

This section briefly explains the PIN diode and the reverse recovery mechanism. Fig. 1 (a) shows a PIN diode in the GlobalFoundries 90 nm 9HP process. It consists of an intrinsic 'I' region, separated by 'P+' and 'N+' regions. The diode goes through three regions of operation when a large-signal voltage is applied across it (Fig. 1 (b)). Initially, it enters forward conduction mode, where the 'I' region is flooded with carriers from the P+ and N+ regions. Because of these carriers, the diode continues to conduct in reverse mode, during which the excess carriers get depleted. Once all the excess carriers are depleted, the diode can no longer sustain the current and turns OFF. This is called reverse recovery. As seen in Fig. 1 (b), reverse recovery is a highly non-linear process where the diode switches a large amount of current in a short interval. This abrupt switching enables efficient THz generation. This work uses the PIN diode models directly from the 9HP process.

B. THz Transmitter Design

Fig. 2 (a) shows the structure of the THz OOK transmitter. A mm-wave Colpitts oscillator drives a PIN diode quadrupler, and the generated fourth harmonic at 0.4 THz radiates through

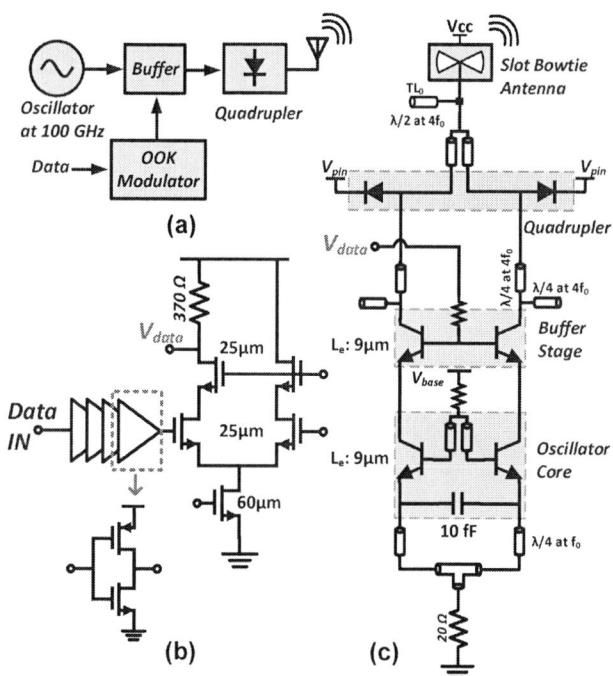

Fig. 2. (a) Block diagram of the THz transmitter (b) High-speed OOK modulator circuit (c) mm-wave Colpitts oscillator, buffer and PIN diode quadrupler

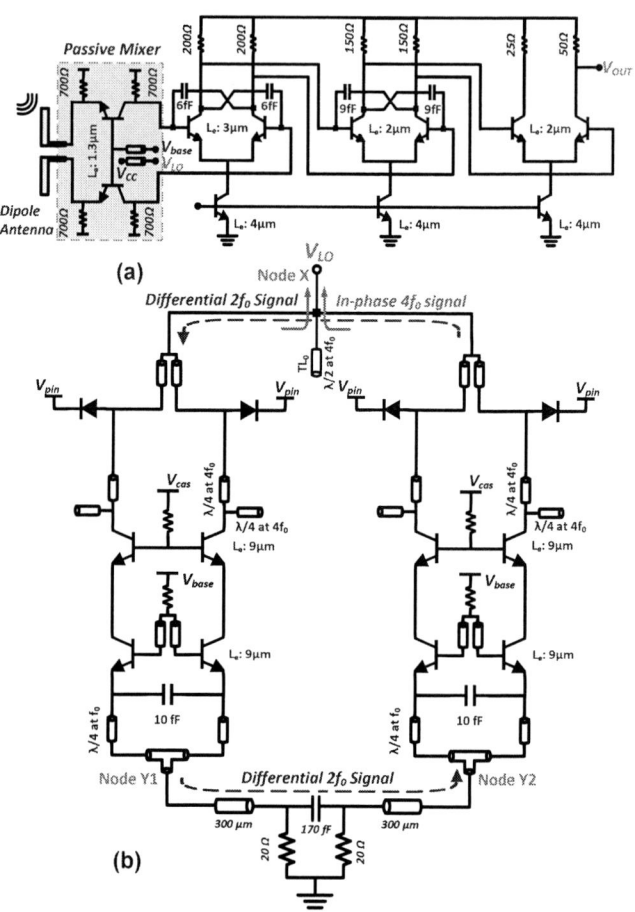

Fig. 3. (a) Schematic of the 400 GHz mixer-first receiver (b) LO generation and power-combining for driving the receiver mixer

an on-chip antenna. An OOK modulator is used to modulate the oscillator with data.

Fig. 2 (b) shows the schematic of the OOK modulator. It consists of a cascade of inverter buffers that amplify and sharpen the baseband binary data. This signal is then sent to a current mode logic (CML) buffer. The CML buffer performs voltage level shifting and modulates the oscillator with data.

Fig. 2 (c) shows the schematic of the mm-wave differential Colpitts oscillator. The oscillator core sets the oscillation frequency to 100 GHz, and a cascode stage is used to buffer and amplify this signal. The OOK modulator in Fig. 2 (b) is connected to the base of the cascode buffer stage and modulates the oscillator-quadrupler with data. Transmission line (TL) based matching networks connect the cascode transistors to the PIN diode quadruplers. These matching networks simultaneously ensure that the oscillator is large-signal impedance matched to the PIN diodes at the fundamental frequency f_0 and that the generated fourth harmonic can be appropriately extracted and radiated out from the antenna. A slot bowtie antenna is used in this work. It is matched to the PIN diode quadrupler at 400 GHz. The antenna is connected to TL_0, which ensures there is no second harmonic leakage without affecting the fourth harmonic.

C. THz Receiver Design

A mixer-first architecture is employed in this work, as front-end amplification is not possible at 400 GHz. Although a mixer cannot undergo hard-switching at 400 GHz, it can still perform the mixing operation due to resistance modulation by the LO. Sub-harmonic mixing is a popular approach where

a sub-harmonic of the LO drives a mixer. This relaxes the LO generation requirements. However, the noise figure and conversion loss (CL) depend on how non-linearly the mixer responds to the sub-harmonic LO. This work proposes using fundamental mixing, i.e., driving the mixer with a 400 GHz LO. This approach is typically not used due to the challenges in generating a high-power LO. However, in this work, the highly non-linear nature of the PIN diode enables generating sufficient LO power to perform fundamental mixing.

Fig. 3 (a) shows the structure of the receiver. A folded dipole antenna receives a differential RF signal which is downconverted using a passive mixer. The LO is provided through a coupled transmission line which performs impedance transformation and DC bias isolation. The conventional double-balanced architecture is avoided here since generating a differential LO requires a balun, which adds additional loss. The downconverted signal is then amplified using capacitive neutralized differential amplifiers. Neutralization is used to enhance the bandwidth. The final amplifier stage performs an active differential to single-ended conversion and drives a 50 load. With -2 dBm LO power, the mixer achieves a single-sideband (SSB) NF of 21 dB in

979-8-3503-2123-4/23 $31.00 © 2023 IEEE

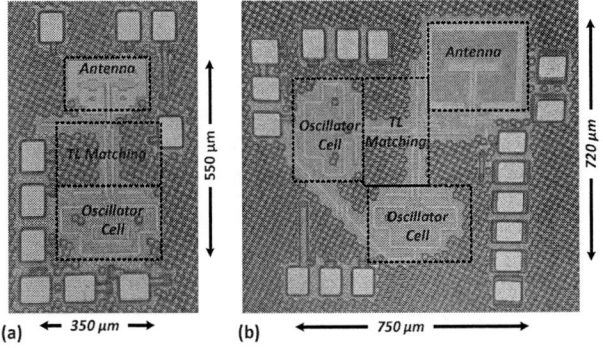

Fig. 4. Die micrograph of the 400 GHz OOK (a) transmitter (b) receiver

simulations.

The PIN diode-based quadrupler discussed in the THz transmitter section is used here to drive the mixer. The outputs of two such quadrupler cells are combined to increase the THz power. This is beneficial as the mixer provides a better NF when driven by higher LO power. Fig. 3 (b) shows the LO with power combining. Two ocillator-quadrupler cells are interconnected at node X. Since node X lies on a common-mode for individual quadruplers, they can interact only through even harmonics. The oscillators can lock in either odd-mode or even-mode of the second harmonic. But TL_0 connected to node X provides a short at the second harmonic, preventing the second harmonic from even-mode coupling. The oscillators thus couple in the odd-mode of the second harmonic, where node X is a virtual ground and is unaffected by TL_0. The quadrupler cells interlock and the 0.4 THz signal adds up in-phase at node X. This is fed to the mixer through a coupled line. Node Y1 and Y2 of the oscillators are connected through TLs, due to layout constraints.

III. MEASUREMENT RESULTS

The design is fabricated in GlobalFoundries 90nm SiGe BiCMOS 9HP process. The die micrographs of the chips are shown in Fig. 4. In nominal operation, the transmitter and receiver consume 80 and 184 mW DC power and occupy 0.2 and 0.54 mm^2 of active area, respectively. The chips are mounted onto a PCB and a hyper-hemispherical silicon lens is attached. This eliminates substrate modes and increases the directivity.

Fig. 5 (a) shows the measurement setup for characterizing the transmitter. A pre-calibrated VDI WR 2.2 mixer (SAX) is placed at a far-field distance of 30 cm from the chip. The EIRP is measured across different frequencies and is plotted in Fig. 5 (b) after de-embedding the losses. A peak EIRP of 17 dBm is measured at 397.5 GHz. The phase noise spectrum is plotted in Fig. 5 (c). A phase noise of -96.3 dBc/Hz is obtained at a 10 MHz offset. The chip is mounted on a rotary stage, and the radiation pattern is measured (Fig. 5 (c)). From this, a directivity of +26.4 dB is calculated, and the chip radiates -9.4 dBm power.

Fig. 6 (a) shows the measurement setup for characterizing the receiver. A pre-calibrated VDI WR2.2 source is placed

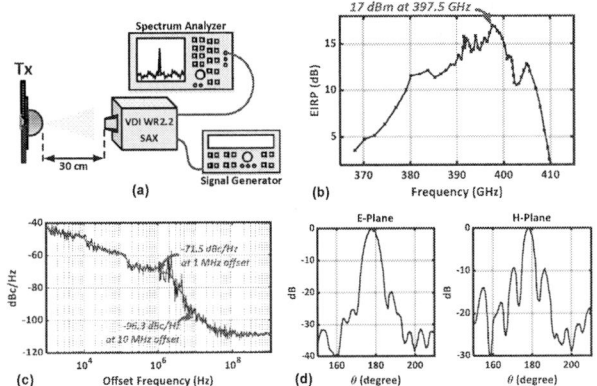

Fig. 5. (a) Transmitter characterization setup (b) Measured EIRP spectrum (c) Measured phase noise spectrum (d) Measured radiation pattern

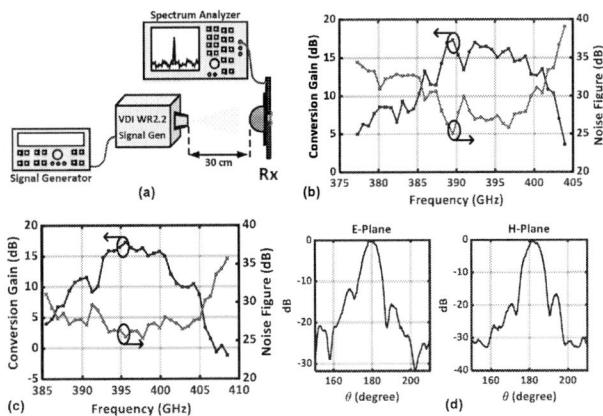

Fig. 6. (a) Receiver characterization setup (b) Measured CG and NF at a constant IF of 1 GHz (c) Measured CG and NF keeping the on-chip LO constant at 397 GHz (d) Measured radiation pattern

at a 30 cm far-field distance from the chip. The SSB NF and conversion gain (CG) are measured using the methods provided in [8] and plotted. Fig. 6 (b) shows the SSB NF and CG for a constant IF frequency of 1 GHz. A minimum NF of 25 dB and a conversion gain of +17.3 dB is measured at 389 GHz. Fig. 6 (c) shows the SSB NF and CG when the on-chip LO frequency is fixed to 397 GHz. Note that antenna efficiency is not de-embedded here. The radiation pattern is plotted in Fig. 6 (d), and a directivity of 27.2 dB is measured.

Fig. 7 (a) shows the measurement setup for characterizing the stand-alone transmitter. An M8190A AWG is used for data modulation. On the receiver side, a VDI mixer is used to downconvert the OOK signal to a low IF which is sent to an oscilloscope. The raw data is then demodulated in MATLAB using envelope detection. No equalization is used. The measured eye diagrams are plotted in Fig. 7 (c), (d), (e). At 20 cm distance, the eye is open at 5 Gbps but starts to close at 8 Gps (10 pJ/bit). At 85 cm distance, the eye remains fairly open at 1 Gbps.

Due to the lack of availability of a THz source with sufficient power and modulation speed, stand-alone receiver

979-8-3503-2123-4/23 $31.00 © 2023 IEEE 155

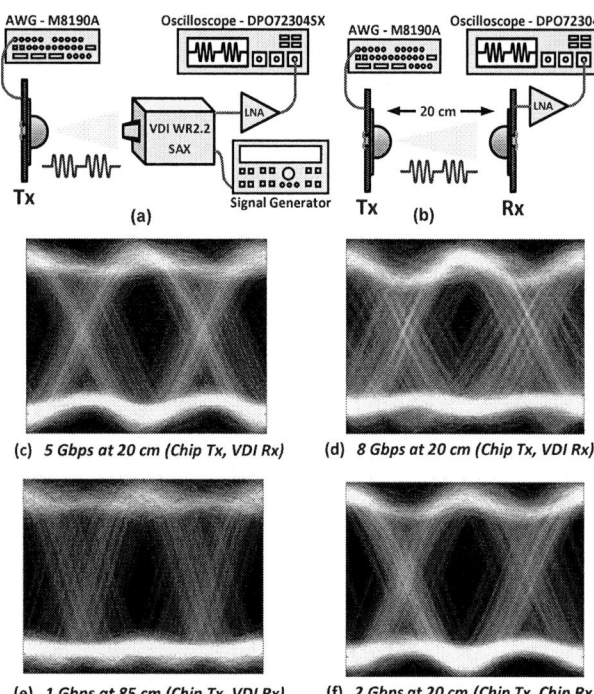

(a) Transmitter OTA measurement setup

(b) Transceiver OTA measurement setup

(c) *5 Gbps at 20 cm (Chip Tx, VDI Rx)*

(d) *8 Gbps at 20 cm (Chip Tx, VDI Rx)*

(e) *1 Gbps at 85 cm (Chip Tx, VDI Rx)*

(f) *2 Gbps at 20 cm (Chip Tx, Chip Rx)*

Fig. 7. (a) Transmitter OTA measurement setup (b) Transceiver OTA measurement setup (c)-(f) Measured eye diagrams

Table 2. Receiver Performance Comparison Table

References	This Work	[7] JSSC'21	[8] JSSC'19	[9] VLSI'15
Frequency (GHz)	398	420	380	410
SSB NF (dB)	25	30	29.2 (Avg.)	37.1
Conv. Gain (dB)	+17.3	+21	-	-16.8
Data Rate (Gbps) and OTA distance	2 at 20 cm OOK	-	-	-
DC Power (mW)	184	601	163	-
Efficiency (pJ/bit)	92	-	-	-
Antenna	On-chip + Silicon Lens	On-chip + Silicon Lens	On-chip	On-chip
Technology	90nm SiGe	40nm CMOS	65nm CMOS	65nm CMOS

Though [4] achieves better data rate and efficiency, it uses wafer probing and an external high-power 130 GHz source, which is not included in efficiency calculations.

Among silicon-based receivers, this is the only work that demonstrates multi-Gbps communication above 300 GHz. Table 2 compares this work with other receivers around 400 GHz which are designed for sensing applications. This work achieves the lowest NF with a very competitive DC power comsumption. Other works [7-9] use external high-power LO which consumes additional power.

IV. CONCLUSION

A fully integrated wireless transceiver is demonstrated at 400 GHz, using PIN diode reverse recovery for THz generation. The transmitter achieves a +17 dBm EIRP and a data rate of 8 Gbps with 10 pJ/bit efficiency. The receiver achieves 25 dB NF and a data rate of 2 Gbps with 92 pJ/bit efficiency. This is the first fully integrated wireless transceiver above 300 GHz and can pave the way for future 6G communications.

REFERENCES

[1] US Federal Communications Commission, "FCC Opens Spectrum Horizons for New Services & Technologies," 2019.

[2] S. Razavian et al., "A 0.4 THz Efficient OOK/FSK Wireless Transmitter Enabling 3 Gbps at 20 meters," in *2022 IEEE BCICTS*, 2022, pp. 1–4.

[3] A. Standaert and P. Reynaert, "A 390-GHz Outphasing Transmitter in 28-nm CMOS," *IEEE JSSC*, vol. 55, no. 10, pp. 2703–2713, 2020.

[4] C. D'heer and P. Reynaert, "A High-Speed 390GHz BPOOK Transmitter in 28nm CMOS," in *RFIC 2020*, pp. 223–226.

[5] A. Standaert and P. Reynaert, "A 410 GHz OOK Transmitter in 28 nm CMOS for Short Distance Chip-to-Chip Communications," in *RFIC 2018*, pp. 240–243.

[6] S. Razavian and A. Babakhani, "Silicon Integrated THz Comb Radiator and Receiver for Broadband Sensing and Imaging Applications," *IEEE TMTT*, vol. 69, no. 11, pp. 4937–4950, 2021.

[7] D. Simic et al., "A 420-GHz sub-5-μm Range Resolution TX-RX Phase Imaging System in 40-nm CMOS Technology," *IEEE JSSC*, vol. 56, no. 12, pp. 3827–3839, 2021.

[8] H. Saeidi et al., "THz prism: One-shot Simultaneous Localization of Multiple Wireless Nodes with Leaky-Wave THz Antennas and Transceivers in CMOS," *IEEE JSSC*, vol. 56, no. 12, 2021.

[9] W. Choi et al., "410-GHz CMOS Imager using a 4th Sub-Harmonic Mixer with Effective NEP of 0.3 fW/Hz$^{0.5}$ at 1-kHz Noise bandwidth," in *2015 Symposium on VLSI Circuits*, 2015, pp. C302–C303.

Table 1. Transmitter Performance Comparison Table

References	This Work	[2] BCICTS'22	[3] JSSC'20	[4] RFIC'20	[5] RFIC'18
Frequency (GHz)	398	390	390	390	410
EIRP (dBm)	+17	+13	-	-	-4
Pout (dBm)	-9.4	-	-16	-5.4	-14.6
Data Rate (Gbps) and distance	8 at 20 cm 1 at 80 cm	3 at 20 meters	6 (Waveguide)	28 (Probe)	5 (Waveguide)
OTA Measurements	Yes	Yes	No	No	No
Modulation	OOK	OOK, FSK	OOK, 8PSK, PAM8, STAR-16 QAM	OOK, BPSK, BPOOK	OOK
LO Generation	On-chip	On-chip	LO at 115 GHz	LO at 130 GHz	On-chip
DC Power (mW)	80	70	1100	114	120
Efficiency (pJ/bit)	10	23.3	183.33	4	41.66
RF Output	On-chip antenna + Silicon Lens	On-chip antenna + Silicon Lens+ Collimation	Waveguide Flange	Wafer Probing	Waveguide Antenna + Dielectric Waveguide
Technology	90nm SiGe	90nm SiGe	28nm CMOS	28nm CMOS	28nm CMOS

characterization is not performed. The setup shown in Fig. 7 (b) is used to characterize the wireless transceiver. A maximum data rate of 2 Gbps (92 pJ/bit) is measured at a 20 cm distance and the eye diagram is plotted in Fig. 7 (f). The data rate is suspected to be limited by the receiver IF bandwidth, due to the bondwire.

A comparison with other state-of-the-art THz transmitters operating above 300 GHz is given in Table 1. Only this work and [2] demonstrates OTA multi-Gbps transmission above 300 GHz. Compared to [2] this work achieves a higher data rate and better efficiency and does not require beam collimation.

RMo4A-1

A Double Balanced Frequency Doubler Achieving 70% Drain Efficiency and 25% Total Efficiency

Jesse Moody

Sandia National Laboratories, USA

JMoody@Sandia.gov

Abstract — This work presents a compact double-balanced frequency doubler achieving better than 70% drain and 25% total power efficiency. Complementary NMOS and PMOS devices enable a truly double-balanced frequency doubler. This work's complementary current reuse structure implements voltage scaling in the device. Voltage scaling enables each device to operate at half the effective supply voltage improving efficiency. The stacked design with inverted NMOS and PMOS positions allows deep class C biasing for an effective VGS of negative 0.5V without on-chip negative voltage generation. These techniques enable a high-efficiency frequency doubler, showing nearly 3x higher drain efficiency and 25% higher total efficiency than previously published frequency doublers. This device offers almost 60% higher efficiency than devices without active second harmonic gain. This work also shows wide-band operation with over 23GHz RF BW and excellent output power of 9.8dBm. Implemented in a commercial 45nm SOI technology, this device presents one of the smallest area consumptions in the literature thanks to the complementary current reuse implementation.

Keywords — Frequency doubler, high efficiency, millimeter-wave

I. INTRODUCTION

Millimeter-wave systems achieve high throughput and reliable link margins through beamforming. These systems often rely on a single master RF LO to ensure phase alignment between many transceivers. The distribution of LO signals is complicated due to the inherently high-frequency operation of these transceivers and the loss associated with signal distribution at very high frequencies. An alternative approach is the distribution of signals at sub-harmonics of the desired LO frequency followed by multiplying the signal on-chip to the desired frequency.

This work describes a small and efficient frequency doubler that can be integrated into phased array transceivers. Although millimeter-wave frequency multipliers have improved, designing high-efficiency frequency doublers still has challenges. Harmonic suppression is especially difficult, as it requires suppressing unwanted harmonics and preventing energy loss in these harmonics while also keeping desired harmonics from interfering with sensitive components. Ensuring wide bandwidth is also challenging, often due to high-impedance interfaces with high Q-factor impedances. Finally, selecting the right bias point is crucial for efficiency. This device presents a solution that suppresses unwanted harmonics, maintains good impedance across all ports, and has the highest overall efficiency reported so far, thanks to optimized bias point selection.

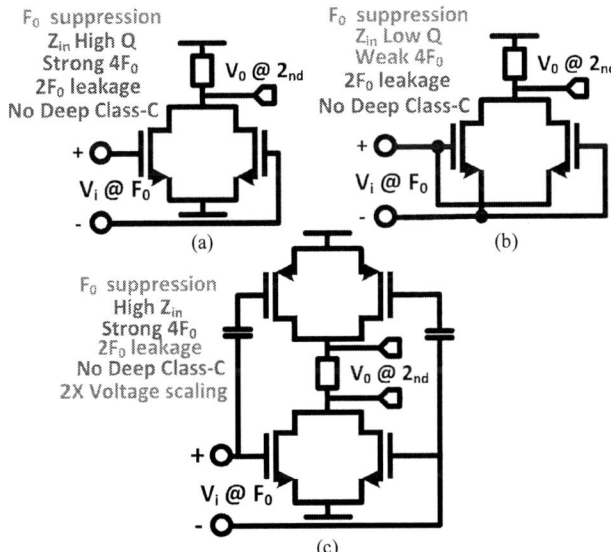

Fig. 1. Various doubler topologies presented in the literature.

II. FREQUENCY MULTIPLIER DESIGN

Recently several improvements have been made to the standard push-push frequency doubler shown in Fig 1 (a) [1], [2]. While this topology does provide inherent fundamental rejection, it presents a high-Q factor input impedance making broadband matching difficult. It also requires second harmonic trapping on the input, generates strong fourth-order currents, and cannot operate in a deep class-C biasing mode without negative voltage generation. Recently, this structure with embedded active gain achieved high efficiency (20%) [1] but required external negative bias voltage generation and presented limited RF bandwidth. The dual-source and gate injection-based structure in Fig. 1 (b) ([3]) solves several issues present in (a). This topology suppresses undesired fourth harmonic currents and presents a low Q-factor input impedance while maintaining the inherent fundamental cancellation present in (a). Unfortunately, this structure still presents issues with required input second harmonic termination and is not amendable towards a deep class-C bias point without negative bias voltage generation. Both designs in (a) and (b) utilize the full supply voltage across a single device; typically, RF devices use voltages around 1V or higher.

Reduced supply voltages can enable higher efficiency, so topologies which inherently scale supply voltages closer

979-8-3503-2123-4/23 $31.00 © 2023 IEEE   157   2023 IEEE Radio Frequency Integrated Circuits Symposium

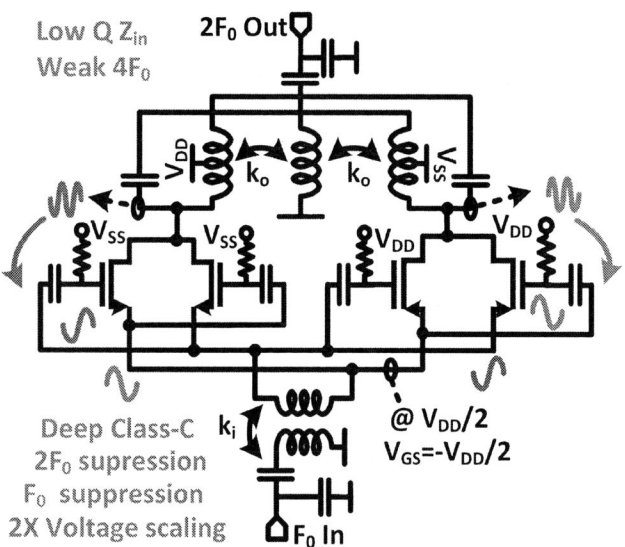

Fig. 2. Schematic diagram of frequency doubler presented in this work.

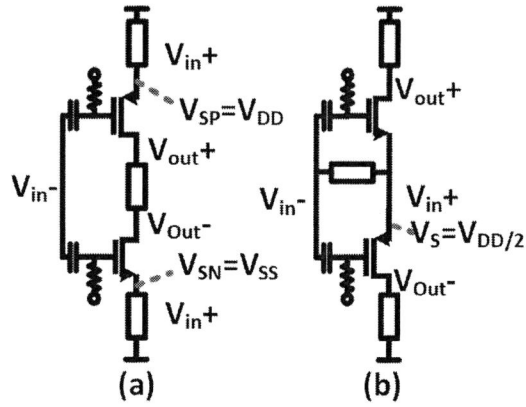

Fig. 3. (a) Inverter like stacking for complementary dual gate-source injection frequency doubler, (b) reverse inverter configuration enabling negative gate source voltage generation.

to optimum efficiency levels are desirable. Fig. 1 (c) [4], [5] shows a doubler topology that enables complementary voltage scaling. This topology also provides an inherent second harmonic trap on the input removing the need for a second harmonic trapping network [4]. While the device in Fig. 1 (c) has many desirable properties, it lacks the advantages of the dual source and injection gate structure, namely excellent input impedance and inherent fourth-order current suppression.

The doubler presented here achieves the properties shown in Fig. 1 (b) and (c) and is depicted in Fig. 2. It uses complementary NMOS and PMOS devices to suppress second-order leakage and scale the supply voltage by a factor of two for high efficiency. This design saves area by omitting second-order trapping networks commonly used in traditional doublers.

Fig. 3 presents an alternative half circuit for the device in Fig. 2. Fig. 3 (a) uses a standard CMOS inverter-like configuration, while Fig. 3 (b) has a reversed structure. Both circuits have the same RF voltages and relative impedances, but their DC voltages differ. Fig. 3 (b) allows negative gate-source bias voltages without needing on-chip negative voltage generation, as the source nodes are close to $VDD/2$.

Frequency doubler bias points correspond to traditional amplifier classes based on conduction angle. Unlike typical RF power amplifiers, current is optimized at the desired harmonic, not the fundamental. When RF devices are biased for shallow conduction angles, the relative strength of higher order harmonics increases compared to DC, as the current pulse generated by the transistor approximates an impulse function, as seen in [6].

Fig. 4 (a) compares drain current between differing bias points, a class-C point using $V_{GS} = 0V$ and a deep class-C with $V_{GS} = -.5V$; with both devices for 10mA DC

consumption. The deep class-C bias point provides a stronger second harmonic current, roughly 2dB higher than the class-C bias point. Fig 4 (b) compares the relative harmonic current amplitudes across varying bias points. The deep class-C bias point increases the relative strength of the second harmonic compared to the fundamental and DC terms. The increased magnitude of the second harmonic current indicates that the saturated output power should increase as the device is biased deeper into class C. Fig. 4 shows the measured P_{Sat} of the fabricated doubler shown in Fig. 2 with varying gate-source voltage. In measurement, the saturated output power increases from 7.4dB with a 0V bias on V_{GS} to 9.8dB with $V_{GS} = -0.5V$, while peak efficiency increases from <20% to 25%.h

Short-channel effects decouple the dependence of the drain current on the applied drain voltage as driving signal power increases due to velocity and channel mobility saturation. Eqn. (1) shows this effect. Fig. 5 (a) demonstrates that scaling the supply voltage for the NMOS half circuit shown in Fig. 1 (b) only increases peak current by 20% for a 9x increase in drain voltage. Fig. 5 (b) shows that decreasing the supply voltage decreases the second harmonic current by roughly 1dB, reducing output power at the second harmonic while decreasing DC power consumption by a factor of 4. This decrease in DC power consumption for a slight decrease in harmonic power enables improvements in the drain and total efficiency, improving the circuit's total efficiency by roughly 60% and drain efficiency by 3x.

$$I_{SAT} = WC_{ox}[V_{GS} - V_T - V_{DSAT}]v_{sat} \quad (1)$$

Where I_{SAT} is the saturated device drain current, and V_{DSAT} is defined as $L/\mu_n v_{sat}$ and v_{sat} is the saturation velocity.

The device uses transformer-based fourth-order impedance matching networks to extend bandwidth on the input and output. However, the output network layout is complex due to the need for different DC voltage levels for the NMOS and PMOS devices, which is required because of the configuration shown in Fig. 3 (b). A triple-stacked transformer is used

979-8-3503-2123-4/23 $31.00 © 2023 IEEE

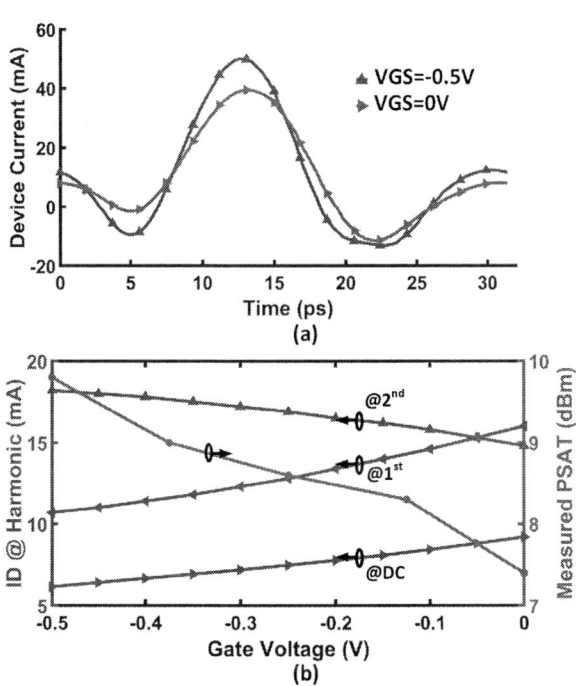

Fig. 4. (a) Drain current waveforms comparison for varying bias points, (b) relative strength of DC fundamental and second harmonic components vs bias, along with measurement of saturated RF output power level across applied bias.

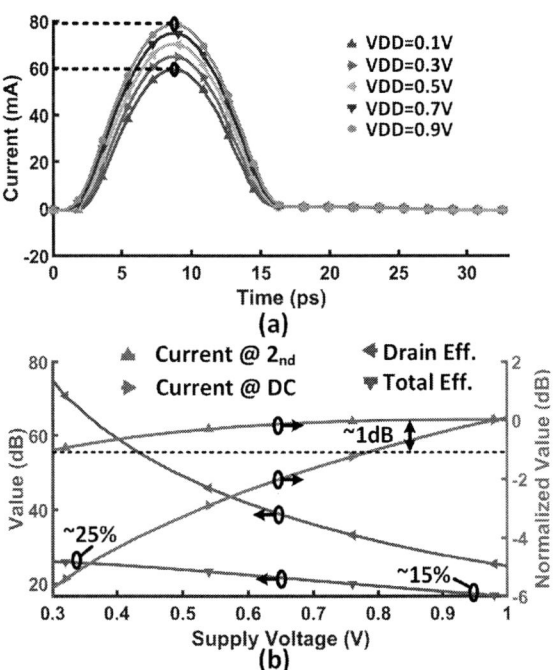

Fig. 5. (a) Drain current generation for class-C doubler across varying supply voltage levels indicating strong channel velocity saturation, (b) device currents and doubler efficiency across varied supply voltages.

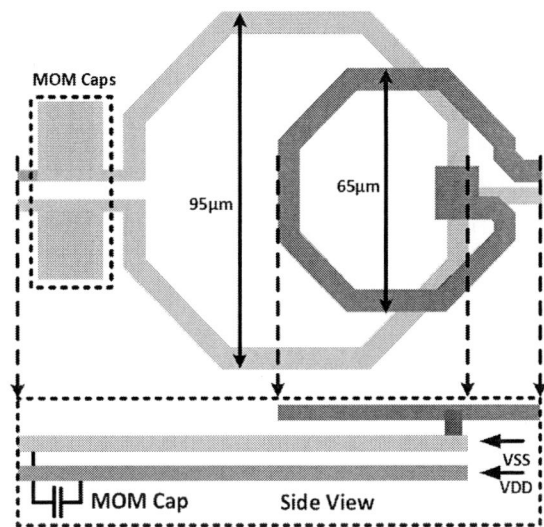

Fig. 6. Layout of output impedance matching network, enabling complmentary current reuse structure

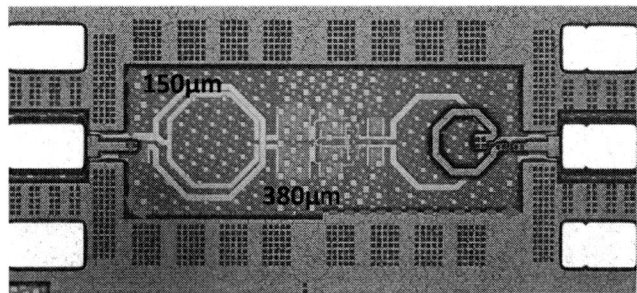

Fig. 7. Die photo of implemented frequency doubler

(Fig. 6), with twin primaries stacked directly over each other and capacitively coupled at their terminals, enabling a single transformer to provide different DC voltages to each terminal.

III. MEASUREMENT RESULTS

The presented frequency doubler is fabricated in a commercial 45nm RF-SOI CMOS process where the die photograph is shown in Fig. 7. The device consumes a small active area of $0.058m^2$. The device is supplied with a standard 1V supply and consumes between 7 and 8mA of bias current.

The device is characterized through on-wafer probing. The RF input is supplied through an E8257C Agilent signal generator, while the RF output has been measured with a WR15SAX VDI Spectrum analyzer extender and a V8486A power meter, while fundamental rejection is measured through an E4446A spectrum analyzer.

The device performance across input power is shown at 60GHz in Fig. 8, where RF output power, conversion gain, and efficiencies are characterized. The saturated output power is measured as 9.8dBm, and the peak conversion gain is above -4dB. The device is shown to achieve maximum efficiency near its 1dB compression point. It maintains an overall efficiency of over 10% for more than 10dB of output power backoff while

979-8-3503-2123-4/23 $31.00 © 2023 IEEE

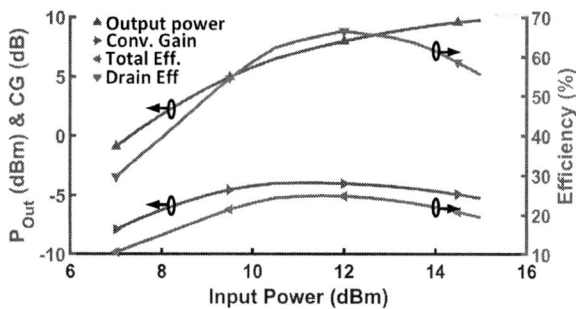

Fig. 8. Measurement of doubler performance across RF input power at 60GHz.

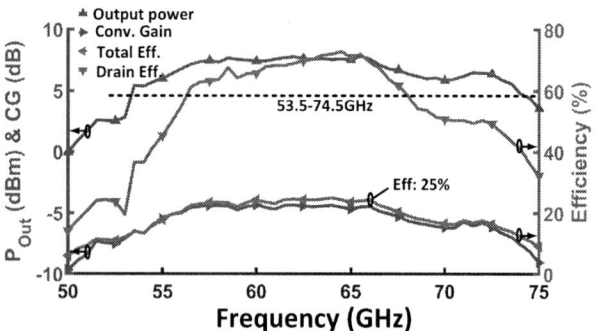

Fig. 9. Measurement of doubler performance across RF frequency.

maintaining greater than 30% drain efficiency across the same output power range.

Performance across frequency is shown at the 1dB compression point in Fig. 9. The device shows excellent broadband, operating with over 23GHz of RF bandwidth. The efficiency across the 3dB bandwidth varies from 14 to 25%, the drain efficiency ranges from 34 to 74%. Peak conversion gain is measured as -4.3dB. P_{SAT} is measured across applied bias in Fig. 4 (b), where the total device efficiency varies from 17% to 25% as V_{GS} varies from 0 to -0.5V.

Table 1 compares this work and state-of-the-art frequency doublers in the V-band. This device outperforms other

recently published frequency doublers regarding total and drain efficiency. The total efficiency of this device is roughly 25% higher than previously demonstrated works, and the drain efficiency is nearly three times as high as other reported frequency doublers. This work shows an excellent saturated output power of almost 10dBm while maintaining competitive RF bandwidth. The measured fundamental rejection is better than -32dB at all frequencies, and the device has a significantly smaller area than works with an efficiency of over 10%. By adopting the complementary current reuse dual gate-source injection doubler architecture and voltage scaling and deep class-C bias points, this work substantially improves frequency doubler efficiency while maintaining a tiny active area.

IV. CONCLUSIONS

This work presents a compact and power-efficient frequency doubler operating in the V-band. Small area and broad bandwidth are achieved through the adoption of a complementary current reuse structure combined with the dual gate and source injection doubler architecture. This device achieves exceptional efficiency thanks to the implicit voltage scaling from the architecture combined with a deep class C bias point. The combination of the small area, high output power, and small area make this device an excellent candidate for frequency generation in large-scale phased array systems.

ACKNOWLEDGMENT

The Author acknowledges Stefan Lepkowski and Travis Forbes for fruitful technical discussions.This work was supported by the Laboratory Directed Research and Development program at Sandia National Laboratories, a multimission laboratory managed and operated by National Technology and Engineering Solutions of Sandia LLC, a wholly owned subsidiary of Honeywell International Inc. for the U.S. Department of Energy's National Nuclear Security Administration under contract DE-NA0003525.

REFERENCES

[1] M. Eladwy, J. Xia, A. B. Ayed and S. Boumaiza, "A 60 GHz CMOS-SOI Stacked Push-Push Frequency Doubler with 12 dBm Output Power and 20% Efficiency,"*2021 IEEE MTT-S International Microwave Symposium (IMS)*, Atlanta, GA, USA, 2021, pp. 297-300.

[2] Z. Chen, Y. Yu, Y. Wu, H. Liu, C. Zhao and K. Kang, "A 19.5% Efficiency 51–73-GHz High-Output Power Frequency Doubler in 65-nm CMOS,," *IEEE Microwave and Wireless Components Letters*, vol. 29, no. 12, pp. 818-821, Dec. 2019.

[3] S. Li, T. Chi, T. -Y. Huang, M. -Y. Huang, D. Jung and H. Wang, "A Buffer-Less Wideband Frequency Doubler in 45-nm CMOS-SOI With Transistor Multiport Waveform Shaping Achieving 25% Drain Efficiency and 46–89 GHz Instantaneous Bandwidth,"*IEEE Solid-State Circuits Letters*, vol. 2, no. 4, pp. 25-28, April 2019

[4] J. Moody, S. Lepkowski and T. Forbes, "A 67 GHz 23 mW Receiver Utilizing Complementary Current Reuse Techniques,"*2022 17th European Microwave Integrated Circuits Conference (EuMIC)*, Milan, Italy, 2022, pp. 292-295

[5] X. Wu, Z. Kang, Y. Wang and L. Wu, "A 53–78 GHz Complementary Push–Push Frequency Doubler With Implicit Dual Resonance for Output Power Combining,"*in IEEE Transactions on Circuits and Systems I: Regular Papers*

[6] F. E. Terman, "Analysis and design of harmonic generators,"*Transactions of the American Institute of Electrical Engineers*, vol. 57, no. 11, pp. 640-645, Nov. 1938

Table 1. Comparison to the State of the Art

	This work	SSCL 19	IMS 21	MWCL 19	TCAS 1 23
Proccess	45-SOI	45-SOI	45-SOI	65	65
Gain @2F0?	No	No	Yes	Yes	Yes
Drain Eff. (%)	74 (>33%)*	25	23.2	26	8.4
Total Eff. (%)	25 (>14%)*	15.5	20.3	19.5	7.9
P_{SAT} (dBm)	9.8 (>7dBm)*	7.4	12.1	2.5	7.1
DC Power (mW)	7-8	17-20	79.8	12	21.5
Peak CG (dB)	-4.1	-4.3	2.1	-2.5	2.5
3dB BW (GHz)	21	43	16	22	5
F_C (GHz)	64	67	65	62	61
Fund. Rej. (dB)	>32	>38	>24	>30	>36
Core Area [mm²]	0.058	0.1	0.3	0.33	0.04

* Performance range over 3dB BW

979-8-3503-2123-4/23 $31.00 © 2023 IEEE

RMo4A-2

A 47 GHz to 70 GHz Frequency Doubler Exploiting 2nd-Harmonic Feedback with 10.1 dBm P_{sat} and η_{total} of 22% in 65 nm CMOS

Amin Aghighi, Mostafa Essawy, Arun Natarajan

High-Speed Integrated Circuits Lab, Oregon State University, USA

{aghighia, essawym, nataraja}@oregonstate.edu

Abstract — A wideband millimeter wave (mm-wave) frequency doubler (FDB) architecture is proposed, where a feedback network is employed to increase the 2nd-harmonic signal at the output. Unlike traditional approaches where the 2nd-harmonic is nulled at the input, common-mode/differential-mode signals are exploited to create frequency-dependent networks at f_0 and $2f_0$. The stand-alone FDB in 65-nm CMOS achieves a saturated output power (P_{sat}) of > 10.1 dBm. A wideband power amplifier (PA) follows the FDB to achieve a total P_{sat} of 15 dBm with a maximum DC-to-RF efficiency (η_{total}) of 24.5%, demonstrating the applicability of the FDB for mm-wave radar.

Keywords — frequency doubler, frequency multiplier, power amplifier, PA, efficiency, FMCW radar.

I. INTRODUCTION

Communication and sensing applications that leverage wide available bandwidths have led to mm-wave transceiver and large-element array implementations. While RF-in and RF-out arrays have been considered at lower frequencies, typical arrays at frequencies > 40 GHz include frequency translation [1], [2]. Given the challenges with wide tuning range VCOs, frequency multipliers are commonly included in the LO path of wideband mm-wave ICs. For example, such wideband multipliers are utilized in frequency-modulated radars where chirp generation can occur at lower frequencies and chirp bandwidth is doubled with multiplication [1].

Critical frequency multiplier design challenges include (i) operating bandwidth, (ii) output power, (iii) efficiency and (iv) conversion gain. In [3], a push-push pair is followed by a transformer-based Gm-boosted buffer to achieve a peak η_{total} of 19.5% and a P_{sat} of 5.7 dBm at 66 GHz. Similarly, [4] utilizes a push-push pair with a stacked common-gate buffer to achieve a peak η_{total} of 20.3%. The stacked buffer reaches a high P_{sat} of 12 dBm but requires a 2 V power supply in 45-nm CMOS at 60 GHz, impacting reliability. The doubler also requires a negative voltage bias (generated off-chip) for peak η_{total}. The challenge of wideband input matching is addressed in [5] where a gate-source driven push-push pair with multi-resonance input matching is used to achieve a wide fractional bandwidth of 64% with a P_{sat} and η_{total} of 7.4 dBm and 15.5% at 60 GHz, respectively.

In this paper, we present a 2nd-harmonic feedback topology for a frequency doubler (FDB) that creates different networks at the fundamental frequency, f_0, and second harmonic, $2f_0$, to enable wideband input matching at f_0 and high output power at $2f_0$. The proposed doubler achieves $P_{sat} > 10.1$ dBm, drain efficiency $> 20\%$ across 21.5 GHz

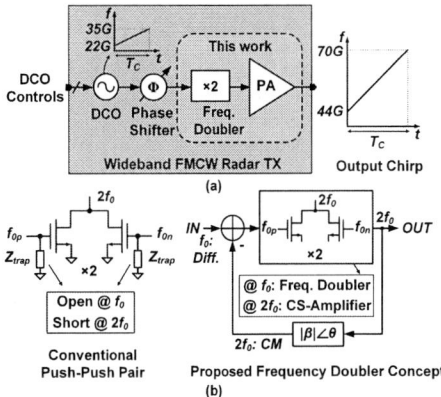

Fig. 1. (a) Block diagram of an FMCW radar transmitter and (b) Conventional push-push pair and conceptual model of the proposed frequency doubler

from 48.5 GHz to 70 GHz, demonstrating state-of-the-art bandwidth/efficiency/output power in CMOS. The doubler drives a two-stage wideband power amplifier (PA) to deliver a 15 dBm P_{sat} demonstrating the feasibility of this architecture for a wideband mm-wave frequency-modulated continuous wave (FMCW) radar TX shown in Fig. 1(a). The architecture and implementation of the proposed doubler and PA are presented in Sec. II. Measurements are presented in Sec. III and conclusions and future work are presented in Sec. IV.

II. ARCHITECTURE OF THE PROPOSED WIDEBAND MILLIMETER-WAVE FREQUENCY DOUBLER

A. Proposed Buffer-Less Frequency Doubler

Achieving efficient, wideband frequency doubling operation requires a wideband input match as well as efficient harmonic generation. The conceptual model of the proposed FDB is shown in Fig. 1(b). The push-push pair is differentially driven at f_0 and generates a $2f_0$ at the output. The balanced topology and differential input lead to the rejection of odd harmonics at the output. The signal at $2f_0$ is fed back to the input and generates an input common-mode (CM) signal at $2f_0$, which is typically nulled. In this work, the push-push pair operates as a common-source amplifier at the 2nd-harmonic and amplifies the $2f_0$ feedback signal. Therefore, with the proposed feedback network optimized at $2f_0$, better performance is expected as compared with the conventional push-push FDB, where nulling the 2nd harmonic is a design objective (Fig. 1(b)).

Fig. 2(a) shows the implementation of the proposed doubler. Also, the equivalent circuits at f_0 and $2f_0$ frequencies

979-8-3503-2123-4/23 $31.00 © 2023 IEEE   161   2023 IEEE Radio Frequency Integrated Circuits Symposium

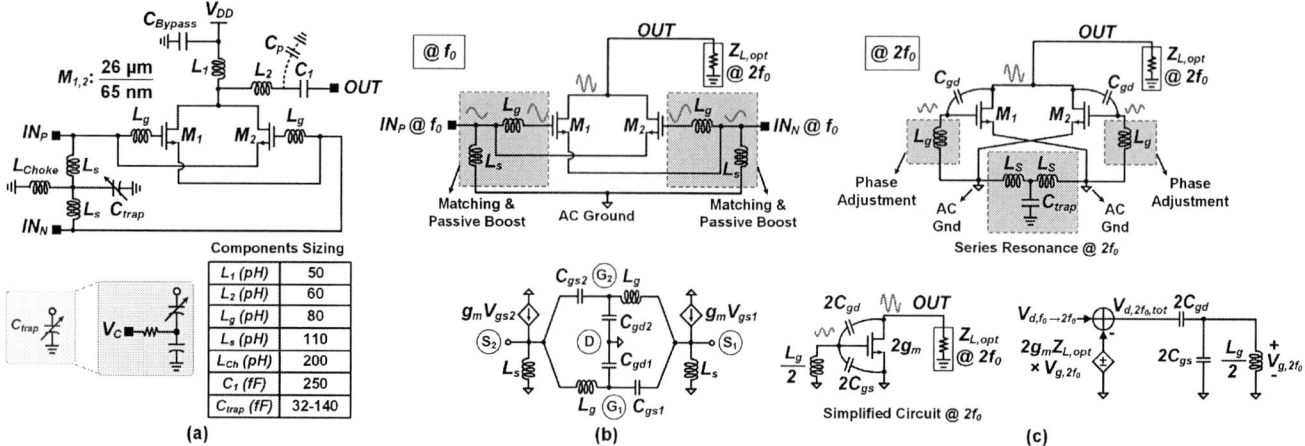

Fig. 2. Circuit implementation of the proposed FDB (a), its equivalent circuit and small-signal model at fundamental frequency (b), and at second harmonic (c)

are shown in Fig. 2(b) and Fig. 2(c), respectively to motivate the topology selection.

As discussed in [5], wideband input match at f_0 is achieved through a combination of common-gate and common-source input differential drive. While the input pair is biased at $0\,V$ through L_{Choke} for increased harmonic generation, the gate and source terminals are driven differentially. Since each of the differential inputs sees a capacitive gate and a resistive source impedance, wideband matching is feasible. At f_0, the output signals for the differential push-pull drive (odd harmonics) cancel each other. Hence, the drain node acts as an AC ground at f_0 as shown in Fig. 2(b). Similarly, the mid-point of the L_s inductors is AC grounded because of the differential input signal at f_0. Therefore, L_{Choke} and C_{trap} do not impact the input impedance at f_0. It is important to ensure gate and source signals at f_0 are $180°$ out of phase to achieve a higher V_{gs} swing. From small-signal model in Fig. 2(b):

$$V_{g,f_0} = -V_{s,f_0} \times \frac{1 + L_g C_{gs}\omega^2}{1 - L_g(C_{gs} + C_{gd})\omega^2} \qquad (1)$$

$$F(\omega) = \frac{V_{gs,f_0}}{V_{s,f_0}} = -\frac{2 - L_g C_{gd}\omega^2}{1 - L_g(C_{gs} + C_{gd})\omega^2} \qquad (2)$$

where, V_{g,f_0} and V_{s,f_0} are the fundamental tone voltage swing at the gate and source, respectively. Given that $\omega < 1/\sqrt{L_g(C_{gs} + C_{gd})}$, not only is the $180°$ phase difference between source and gate signals guaranteed but there is also a passive voltage boost from input (V_{s,f_0}) to V_{g,f_0} that increases the generated second harmonic at the output. L_g and device sizes in the proposed FDB are chosen such that for $20\,\text{GHz} < f_0 < 40\,\text{GHz}$, $\angle V_{g,f_0} - \angle V_{s,f_0}$ and $|F(\omega)|$ are kept within $180° \pm 0.5°$ and 2.2 ± 0.1, respectively.

At $2f_0$, as shown in Fig. 2(c), the output signal couples back to the input through C_{gd} of the push-push pair devices and generates a $2f_0$ CM signal at each gate terminal. From the small-signal model at $2f_0$, the total output signal at $2f_0$ ($V_{d,2f_0,tot}$) depends on two networks:

- the non-linearity of the input pair and the differential input signal at f_0 that leads to a signal at $2f_0$: $V_{d,f_0 \to 2f_0}$

- The feedback network operating at $2f_0$: $2g_m Z_{L,opt} \times V_{gs,2f_0}$

Since the impact of the $Z_{L,opt}$ in the small-signal model of Fig. 2(c) is already considered in $V_{d,f_0 \to 2f_0}$ and $2g_m Z_{L,opt} \times V_{gs,2f_0}$, it is not shown as a separate load at the drain node. A varactor (C_{trap}) is employed at the mid-point of the two L_S inductors to form a series resonance network at each side and force an AC-ground at the source terminal at $2f_0$. Hence, the CM $2f_0$ components at both gates are amplified through common-source amplifiers and added together with the $V_{d,f_0 \to 2f_0}$. In order to ensure that $V_{d,2f_0,tot}$ is higher than the non-linear component alone, $V_{g,2f_0}$ and $V_{d,2f_0,tot}$ should be $180°$ out of phase. Therefore:

$$H(\omega) = \frac{V_{g,2f_0}}{V_{d,2f_0,tot}} = -\frac{L_g C_{gd}\omega^2}{1 - L_g(C_{gs} + C_{gd})\omega^2} \qquad (3)$$

$$V_{d,2f_0,tot} = V_{d,f_0 \to 2f_0} - 2g_m Z_{L,opt} H(\omega) \times V_{d,2f_0,tot} \qquad (4)$$

$$V_{d,2f_0,tot} = \frac{V_{d,f_0 \to 2f_0}}{1 + 2g_m Z_{L,opt} H(\omega)} \qquad (5)$$

where, $V_{g,2f_0}$ is the 2nd harmonic voltage swing at the gate due to the feedback. Hence, $\angle H(\omega) = 180°$ up to $\omega < 1/\sqrt{L_g(C_{gs} + C_{gd})}$ which is sufficiently high ($> 95\,\text{GHz}$) in the proposed FD. In order to ensure stability, $0 < 2g_m Z_{L,opt}|H(\omega)| < 1$. Under this constraint, the proposed frequency doubler can have a higher output swing as $V_{d,2f_0,tot} > V_{d,f_0 \to 2f_0}$. With the same device sizes and optimized load for the proposed and conventional FDB without feedback, simulation results with $P_{IN} = 6$, 8, and $10\,\text{dBm}$ at $30\,\text{GHz}$ show a higher P_{OUT} for the proposed FDB by 2.6, 1.8 and 1 dB, respectively. The C_{trap} can be implemented with a tunable varactor to enhance bandwidth.

B. Power Amplifier

Fig. 3 shows the proposed PA that loads the FDB. The 2-stage PA employs transformers for input, output, and inter-stage matching. While input and output matching networks only utilize magnetic coupling due to unbalanced

979-8-3503-2123-4/23 $31.00 © 2023 IEEE

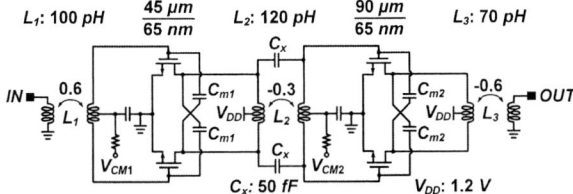

Fig. 3. Wideband 60 GHz transformer-coupled 2-stage PA

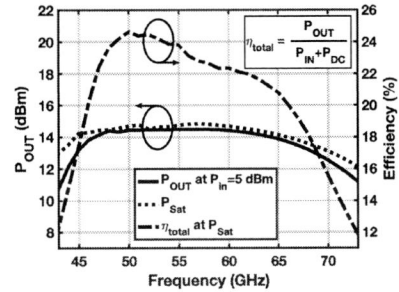

Fig. 4. Simulated performance of the wideband 60 GHz PA in 65nm CMOS

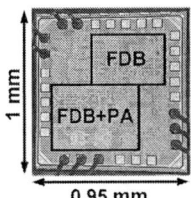

Fig. 5. Die photo of the 65nm CMOS IC with FDB and FDB/PA structures

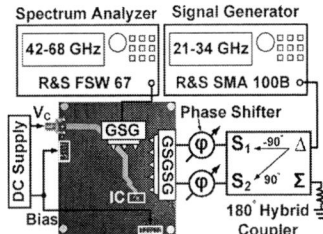

Fig. 6. Probe-based measurement setup of the FDB and FDB/PA

configuration, the inter-stage transformer which is inherently balanced employs both magnetic and electric coupling to achieve a wideband matching [6]. Additionally, $C_{m1} = 21$ fF and $C_{m2} = 42$ fF realize cross-coupled C_{gd} neutralization to improve stability and increase the gain-bandwidth product. Simulated stand-alone PA performance is shown in Fig. 4, with $P_{Sat} \approx 15$ dBm across $BW_{1dB} = 22.5$ GHz and $\eta_{total} > 20\%$ at P_{sat} for 20 GHz BW from 46 GHz to 66 GHz. PA simulations also show a $BW_{3dB} = 24$ GHz for $P_{IN} = 5$ dBm.

III. MEASUREMENT RESULTS

The stand-alone FDB (which occupies 0.18mm^2) and the FDB/PA (occupying 0.28mm^2) are implemented on the same 65-nm CMOS IC (Fig. 5). The probe-based setup used to characterize the stand-alone FDB and the FDB/PA structures is shown in Fig. 6. Fig. 7 summarizes measured and simulated results of the stand-alone FDB. Measured P_{OUT} and drain efficiency (DE)/DC-to-RF efficiency (η_{total}) of the proposed FDB at a fixed $P_{IN} = 12$ dBm, is plotted in Fig. 7(a) and Fig. 7(b) respectively. FDB and PA operation extend beyond 70 GHz, as shown by the simulations in Fig. 7 that match well with measured data. However, measurements are limited to 68 GHz by existing setup. Measurements beyond this frequency will be performed in an updated setup.

As shown in Fig. 7(a), the FDB achieves a measured P_{OUT} $BW_{3dB} > 21$ GHz from 47 GHz to 68 GHz. Extrapolation from measurement and simulations suggests an overall BW_{3dB} of 23.5 GHz from 47 GHz to 70.5 GHz, representing a fractional bandwidth (FBW) of 40%. The FDB achieves 10.1 dBm $P_{OUT,max}$ at 60 GHz. At 1.2V supply, the FDB reaches a DE_{max} and peak η_{total} of 32.7% and 21.8%, respectively, with $DE > 20\%$ for 48.5-70 GHz.

The FDB was also characterized at lower supplies for a low-power operating mode. At $V_{DD} = 0.6$ V, the FDB achieves a high DE_{max} of 44.4% at 60 GHz, while providing 8 dBm of P_{OUT} which is still higher than P_{sat} of state-of-the-art FDBs.

Fig. 7(c) shows P_{OUT}, η_{total}, and DE of the proposed FDB across input power for 60 GHz output at 1.2 V supply. FDB at $P_{IN} = 12.2$ dBm achieves a $P_{OUT} = 10.1$ dBm with an η_{total} and DE of 21.8% and 32.7%, respectively. The measured maximum conversion gain (CG) is -1.7 dB at $P_{IN} = 9$ dBm, which is higher than prior CMOS FDBs (excluding output buffers).

Fig. 8(a) shows the measured and simulated performance of the FDB/PA test structure, with $V_{DD} = 1.2$ V, at a fixed P_{IN} of 7 dBm. The FDB/PA achieves a flat response with $BW_{1dB} = 18.5$ GHz and $P_{OUT,max} = 14.7$ dBm. Measurements show a peak η_{total} of 24.5% for the FDB/PA at 48 GHz, while $\eta_{total} > 19\%$ for more than 19 GHz from 46.5 to 66 GHz. As shown in Fig. 8(b), the FDB/PA achieves a $P_{sat} = 15$ dBm at 60 GHz and a peak $\eta_{total} \approx 20\%$ when driven by $P_{IN} = 8$ dBm at 30 GHz.

The FDB includes a varactor at the source of the differential pair (as shown in Fig. 2(c)). Fig. 8(c) compares FDB performance with a fixed and variable C_{trap} to evaluate the sensitivity to C_{trap}. With a fixed C_{trap}, the maximum degradation in P_{OUT} for FDB is < 1.1 dB compared to a scenario where C_{trap} is tuned, showing limited sensitivity to C_{trap} variations. Finally, the fundamental rejection ratio (FRR) of the proposed FDB is shown in Fig. 9 across frequency. Measurements show a good rejection of the f_0 at the output, with FRR ranging from 29.5 dB to 47 dB with the minimum and maximum at 44 GHz and 60 GHz, respectively and FRR > 33 dB within the BW_{3dB} of the FDB.

The performance of the FDB and FDB with PA are summarized and compared to the prior art in Table 1. The proposed feedback network ensures wide bandwidth and higher output power in the FDB, resulting in higher efficiency. The 2nd harmonic feedback also achieves higher output power in a lower f_T, f_{MAX} CMOS node.

IV. CONCLUSION

This paper presents a mm-wave frequency doubler that exploits a differential-mode/common-mode feedback

979-8-3503-2123-4/23 $31.00 © 2023 IEEE

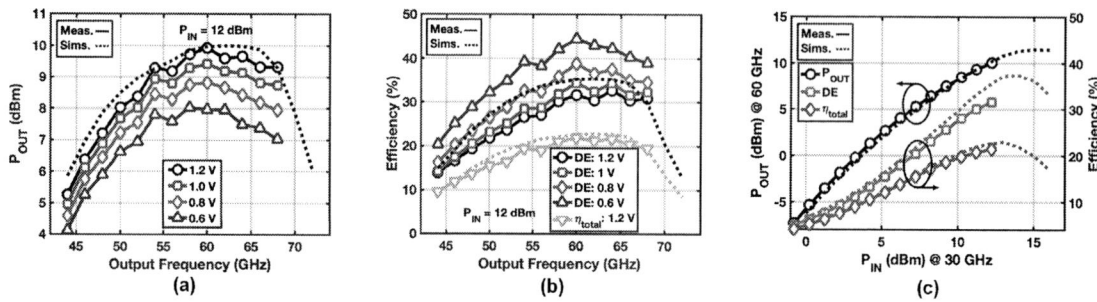

Fig. 7. Measured FDB Performance: (a) FDB output power across frequency, (b) FDB efficiency across frequency, (c) FDB output power and efficiency across P_{IN} at 60 GHz output

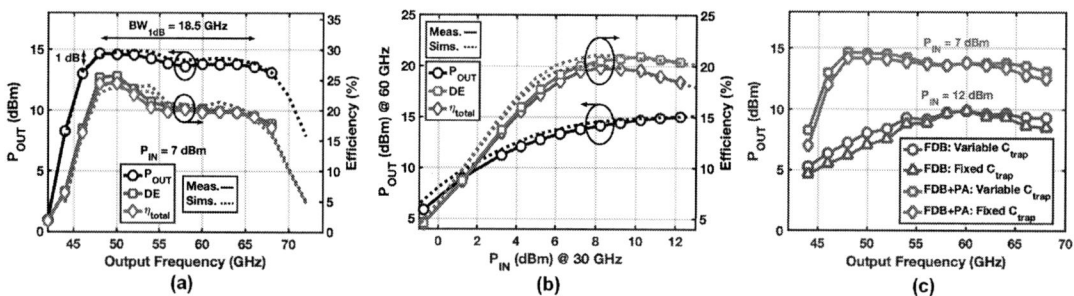

Fig. 8. Measured FDB/PA Performance: (a) FDB/PA output power and efficiency across frequency, (b) FDB/PA output power and efficiency across P_{IN} at 60 GHz output, (c) Impact of the tunable C_{trap} on FDB and FDB/PA output power

Table 1. Performance Summary and Comparison with Prior Art

Ref.	[3]	[5]	[7]	[9]	This Work: FDB		This Work: FDB+PA
Tech.	65 nm CMOS	45 nm CMOS SOI	65 nm CMOS	130 nm SiGe	65 nm CMOS		65 nm CMOS
Output Buffer?	Yes	No	No	No	No		Yes
V_{DD} (V)	1	0.9-1.1	1	1.5	0.6	1.2	1.2
BW_{3dB} (GHz)	51-73	46-89	62-90	54.4-62.8	45.5-71.5#	47.5-70.5#	45.5-69.5#
FBW	35.4%	64%	37%	14%	45.3%	40%	41.7%
P_{sat} (dBm)	5.7	7.4	2.5	2	>8 (8.9#)	>10.1 (11.5#)	15
DE_{Peak} =P_{OUT}/P_{DC}	N/A	25% (>15%, 49-87GHz)	11.5%**	16%**	44.4%# (>20%, 44-73GHz)	32.7%# (>20%, 48.5-70GHz)	25.6% (>20%, 46-65GHz)
$\eta_{total,Peak}$ =$P_{OUT}/(P_{IN}+P_{DC})$	19.5%* (>15%, 54-70GHz)	15.5% (>10%, 50-85GHz)	9.7% (>6%, 62-78GHz)	14%* (>10%, 55-60GHz)	22.2% (>19%, 54-68GHz)	21.88% (>19%, 54-68GHz)	24.5% (>19%, 46.5-66GHz)
FRR (dBc)	>30	>38	>20	>40*	>26	>33	N/A
CG_{Peak} (dB)	0.8	-4.3	-2.5	0.6	-3.2	-1.8	8-12
Area (mm²)	0.327	0.1	0.27	0.16	0.18	0.18	0.28
P_{DC} (mW)	14	17-20	9-14	7.2	12-15	26-32	103-115

*Graphically estimated. **Calculated based on reported values. #Due to measurement limitation, the higher edge is estimated based on simulation results.

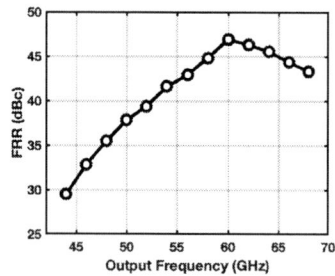

Fig. 9. Measured fundamental rejection ratio of proposed FDB

topology that enhances the generated second harmonic of the conventional push-push pair while achieving a wideband input matching. An integrated CMOS implementation of the proposed doubler demonstrates state-of-the-art output power and efficiency across a wide mm-wave bandwidth. Future work includes integrating the proposed doubler in a low-power wideband mm-wave radar sensor array.

ACKNOWLEDGMENT

The authors thank CDADIC for project funding and Rohde & Schwarz (Radu Fetche) for providing measurement equipment and measurement support.

REFERENCES

[1] A. Visweswaran et al., "A 28-nm-CMOS Based 145-GHz FMCW Radar: System, Circuits, and Characterization," in *IEEE JSSC*, July 2021.

[2] S. Shahramian, et al., "A Fully Integrated 384-Element, 16-Tile, W-Band Phased Array ...," in *IEEE JSSC*, Sept. 2019.

[3] Z. Chen, et al., "A 19.5% Efficiency 51–73-GHz High-Output Power Frequency Doubler in 65-nm CMOS," in *IEEE MWCL*, Dec. 2019.

[4] M. Eladwy, et al., "A 60 GHz CMOS-SOI Stacked Push-Push Frequency Doubler with 12 dBm Output Power ...," in *IEEE MTT-S IMS*, June 2021.

[5] S. Li, et al., "A Buffer-Less Wideband Frequency Doubler in 45-nm CMOS-SOI ...," in *IEEE SSLC*, April 2019.

[6] V. Bhagavatula, et al., "An Ultra-Wideband IF Millimeter-Wave Receiver With a 20 GHz Channel Bandwidth ...," in *IEEE JSSC*, Feb. 2016.

[7] Y. Ye, et al. "A High Efficiency E-Band CMOS Frequency Doubler With a Compensated Transformer-Based Balun ...," in *IEEE MWCL*, Jan. 2016.

[8] O. Momeni and E. Afshari, "A Broadband mm-Wave and Terahertz Traveling-Wave Frequency Multiplier ...," in *IEEE JSSC*, Dec. 2011.

[9] B. Sutbas and G. Kahmen, "A 7.2-mW V-Band Frequency Doubler With 14% Total Efficiency ...," in *IEEE MWCL*, June 2022.

979-8-3503-2123-4/23 $31.00 © 2023 IEEE

RMo4A-3

A 91.9-113.2 GHz Compact Frequency Tripler with 44.6 dBc Peak Fundamental Harmonic-Rejection-Ratio Using Embedded Notch-filters and Area-Efficient Matching Network in 65 nm CMOS

Xiangrong Huang, Haikun Jia, Wei Deng, Zhihua Wang, Baoyong Chi

School of Integrated Circuits, BNRist, Tsinghua University, China

jiahaikun@tsinghua.edu.cn

Abstract—This article presents a W-band mixer-based frequency tripler. The folded four-coil transformer is proposed to achieve the input matching and input power distribution to the push-push frequency doubler and the mixer simultaneously. The mixer mixes the second-harmonic current with the input fundamental harmonic to obtain the third harmonic. A single-stage class-AB amplifier follows to drive the output load. The fundamental harmonic notch-filters composed of parallel inductor and capacitance tanks are embedded into the interstage and output matching networks to save the chip area and improve the harmonic-rejection-ratio (HRR). The proposed frequency tripler has been fabricated in 65nm CMOS process with a 160 μm × 420 μm core chip area. Measurement results show a conversion gain of -2.35 dB, a 44.6 dBc peak fundamental HRR and a 3.88% DC-RF efficiency for an input power of 6 dBm at 102 GHz. The measured output 3 dB bandwidth is 91.9-113.2 GHz.

Keywords—CMOS process, harmonic rejection ratio, notch filter, transformer, frequency tripler.

I. INTRODUCTION

The wireless communication at millimeter wave (mm-wave) has drawn a lot of attention in recent years because of the wide available bandwidth to support high data-rate transmission. Among the mm-wave frequency bands, W-band is considered to have great potential for applications like radar, imaging, satellite communications and auto-driving [1] due its wide bandwidth, small form factor, and the achievability in silicon process. Local-oscillator (LO) signal generator is the key component in W-band transceiver. Due to the parasitic effect and passive loss in W-band, it is difficult to design a high-performance voltage-controlled oscillator (VCO) directly [2]. A popular alternative approach is to cascade a lower frequency VCO with a frequency multiplier.

The odd-order frequency multipliers, such as frequency triplers, are typically realized by a transistor biased at class-C mode which can generate rich harmonic current. The desired third harmonic is selected by a band-pass filter. However, the fundamental harmonic-rejection-ratio (HRR) is around only 20 dB, dominated by the leakage of input signal. The higher HRR of the input signal needs more filter stages which means larger power consumptions, larger chip area, and lower conversion efficiency [3]. To save power consumption and improve the conversion efficiency, the mixer-based frequency tripler, which employs the stronger second-order nonlinearity instead of direct third harmonic nonlinearity, can be chosen. However, the power loss of second-harmonic tone from doubler to mixer increases due to the increasing parasitic effect in W-band.

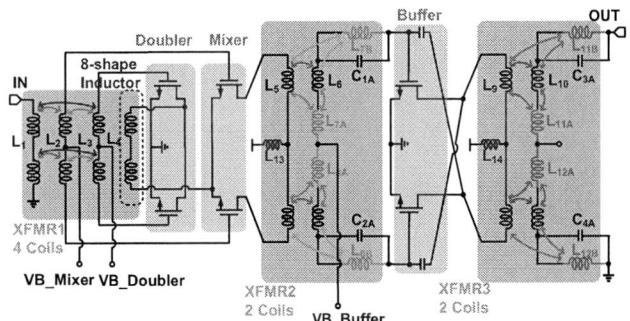

Fig. 1. Schematic of the proposed compact frequency tripler.

In this work, a novel frequency tripler circuit topology is proposed which achieves a high HRR by embedded notch-filters and a compact chip size by folded four-coil transformer for input match. As shown in Fig. 1, the proposed frequency tripler is composed of a frequency doubler followed by a mixer and an amplifier buffer. The desired third-harmonic tone is obtained by mixing the input fundamental tone and the second-harmonic tone generated from the frequency doubler. Meanwhile, the extra area of notch-filters is minimized by embedding them into the transformers. The frequency tripler bas been fabricated and verified in 65 nm CMOS process.

II. CIRCUIT DESIGN

A. Folded four-coil transformer

Fig.2(a) shows the conventional architecture of mixer-based frequency tripler. As described in [4], orthogonal signals are required for the input frequency doubler and mixer to generate a constructive second-harmonic voltage. However, orthogonal signal generators are usually either power lossy or area-consuming in mm-wave for CMOS process. Furthermore, as the frequency becomes higher, the power loss and impedance mismatch at intermediate node N_1 deteriorate the second-harmonic power. Instead of providing the orthogonal signals at the input ports, a series inductor L_4 is inserted into the node N_1 in this work, forming a C-L-C tank together with the parasitic capacitance as shown in Fig. 2(b), introducing a certain amount of second-harmonic current phase shifting. Meanwhile, the inductor L_{13} is placed on the center tap of the interstage transformer. A proper choice of L_4 and L_{13} delivers the second harmonic power efficiently from the frequency doubler to the mixer at $2*f_0$, and thus improving the conversion gain. Fig. 3(a)

979-8-3503-2123-4/23 $31.00 © 2023 IEEE

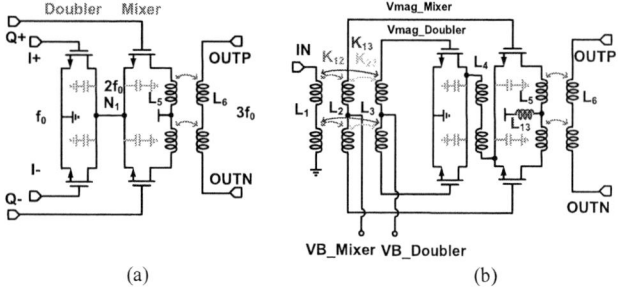

(a)   (b)

Fig. 2. The basic circuit topology of (a) conventional tripler and (b) the proposed tripler.

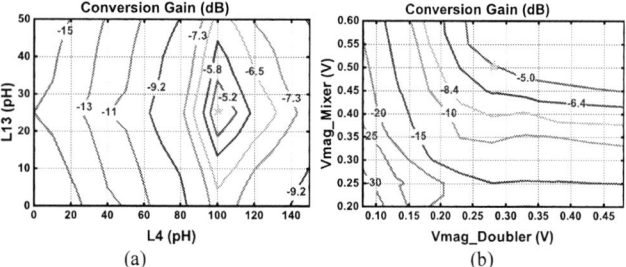

(a)   (b)

Fig. 3. Simulated results of conversion gain versus (a) different L_4 and L_{13}. (b) different signal magnitudes of frequency doubler and mixer.

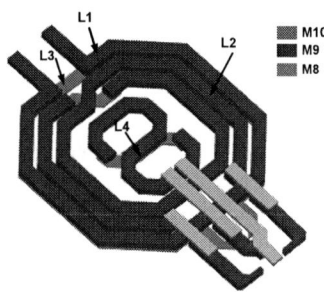

Fig. 4. 3D view of the XFMR1 layout.

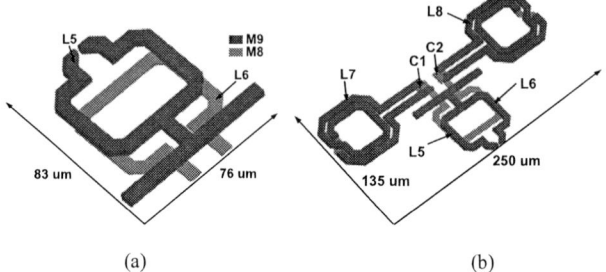

(a)   (b)

Fig. 5. 3D view of (a) XFMR2 without notch-filters and (b) XFMR2 with separated notch-filters.

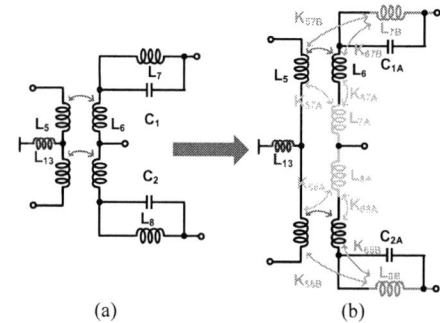

(a)   (b)

Fig. 6. Matching network with (a) separated notch filters and (b) embedded notch filters.

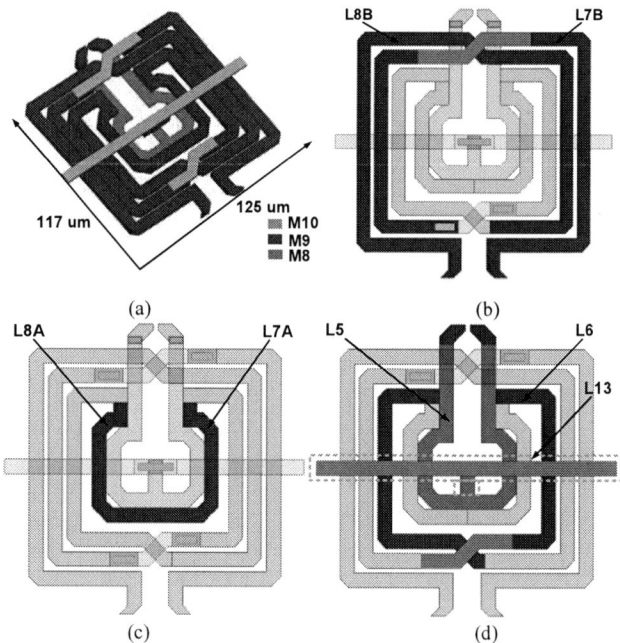

Fig. 7. (a) 3D view of XFMR2 with embedded notch-filters. (b) Embedded inductors L_{7B} and L_{8B}. (c) Embedded inductors L_{7A} and L_{8A}. (d) Embedded match network composed of L_5 and L_6.

shows the simulated conversion gain with different inductance L_4 and L_{13}. There is a global peak corresponding to $L_4 = 100$ pH and $L_{13} = 25$ pH. The conversion gain is improved by about 10 dB after inserting L_4 and L_{13}.

Besides the series inductor L_4 and central-tap inductor L_{13}, the input power is delivered to the frequency doubler and the mixer through a three-coil transformer in this work. As a comparison, dc block capacitors are used in [5] to deliver the input voltage to the frequency doubler and the mixer and separate the dc bias voltage. The three-coil transformer approach has several advantages over the dc block capacitors approach. First, the three-coil transformer naturally provides the dc separation, therefore eliminate the needs for the dc block capacitors, as well as the parasitic capacitance introduced by the dc block capacitors. Second, using the three-coil transformer, it is possible to flexibly allocate the voltage signal to the frequency doubler and the mixer to optimize the conversion gain by manipulating the number of turns and coupling coefficient of each coil. Third, it saves the chip area consumption of the dc blocking capacitors.

To improve the conversion gain, both the frequency doubler and the mixer require a high input signal magnitude. However,

at a given input power, trade-off exists between the magnitude of the signal delivered to the frequency doubler and the mixer. Fig. 3(b) shows the simulated conversion gain of the frequency tripler versus different signal magnitude at those two nodes with 6 dBm input power. The signal magnitude at the frequency doubler and the mixer is chosen to be 0.28V and 0.50 V, respectively, which is achieve by setting the turn ratio of L_1, L_3,

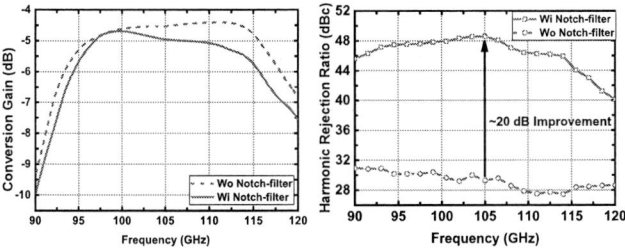

Fig. 8. Simulated results with and without the notch-filters. (a) Frequency response. (b) First-order HRR.

Fig. 9. Micrograph of the chip.

and L_2 to be 1:1:2. Note that larger signal magnitude is required by the mixer not to significant degrade its conversion gain.

To further reduce the chip area consumption, the series inductor L_4 at intermediate note is winded as an 8-shape topology and folded into the three-coil transformer composed of L_1, L_2 and L_3, constituting a four-coil transformer. The detailed 3D view of transformer layout is shown in Fig. 4.

B. Embedded notch-filter

To achieve high conversion gain of the frequency tripler, the input power of fundamental harmonic is relatively high (6 dBm in this work). The desired third-harmonic tone is obtained by mixing the input fundamental tone and the second-harmonic tone generated from the frequency doubler. Meanwhile, the dc current from the frequency doubler will also be mixed up to first-harmonic tone which results in significant first-harmonic power leakage. Thus, it is necessary to strengthen first-harmonic HRR. Additional notch-filters composed of L-C tanks can be added in the signal feedforward path to filter out the first harmonic component. As shown in Fig. 6(a), L_7 and C_1 resonant at fundamental frequency f_0, forming a high impedance at the fundamental frequency while ac-shorted at the third harmonic frequency. Therefore, the third-order harmonic signal can be transmitted while the fundamental signal is choked.Fig. 5(a) and (b) shows the layout of XFMR2 with and without additional notch-filters. The two notch filters also consume additional chip area.

In this work, instead of using separated notch filters, the notch filters are embedded into the matching network to reduce the chip area. Since L_7/L_8 and C_1/C_2 resonant at f_0 while L_5 and

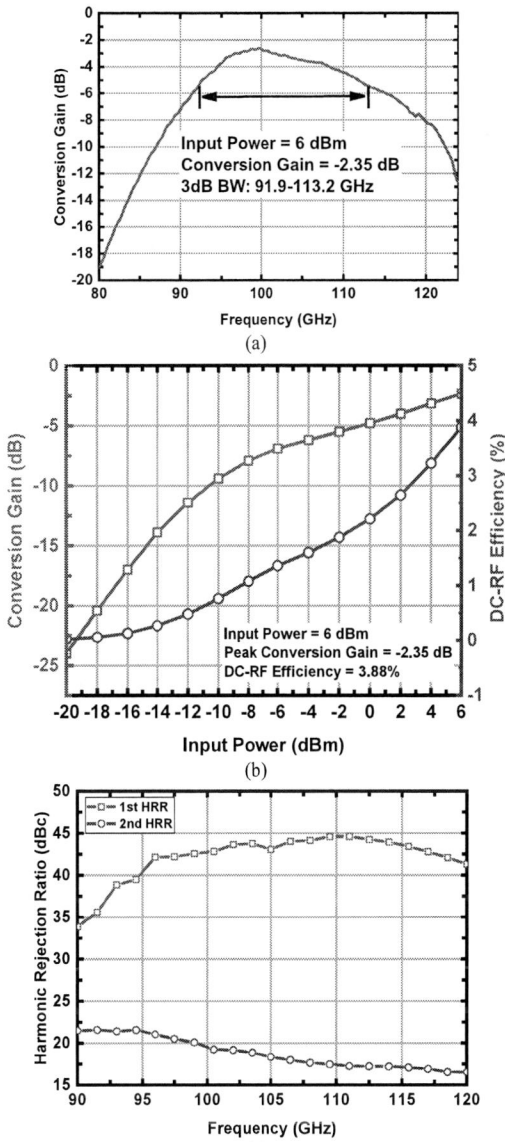

Fig. 10. Measurement results of the proposed tripler, (a) Conversion gain and dB BW, (b) Maximum conversion gain and efficiency versus input power , (c) HRR of first and second-order harmonic tones

L_6 match the impedance at $3*f_0$, the value of L_7/L_8 is about twice the value of L_5/L_6. L_7/L_8 can be separated into two smaller inductors L_{7A}/L_{8A} and L_{7B}/L_{8B} as shown in Fig. 7. By proper arrange the inductor position, the current flow in L_{7A}/L_{8A} and L_{7B}/L_{8B} has a opposite direction. Therefore, their magnetic coupling with the match network L_5/L_6 can self-cancelled when the following conditions are satisfied:

$$K_{57A}\sqrt{L_5 L_{57A}} + K_{57B}\sqrt{L_5 L_{57B}} = 0 \qquad (1)$$

$$K_{67A}\sqrt{L_6 L_{67A}} + K_{67B}\sqrt{L_6 L_{67B}} = 0 \qquad (2)$$

In the final XFMR2, the notch filter capacitors C_1/C_2 is implemented using the distributed side-wall parasitic capacitance between the thick metals. Fig. 8 shows the simulated results of frequency response at $3*f_0$ and fundamental HRR with 6 dBm input power. It can be seen that there is little

Table 1. Performance summary and comparison with the state of the art

	MWCL 2020 [1]	JSSC 2022 [5]	MWCL 2021 [6]	MWCL 2020 [7]	IMS 2021 [8]	This Work
Technology	130 nm SiGe	28 nm CMOS	65 nm CMOS	22 nm SOI	130 nm SiGe	65nm CMOS
Out Frequency (GHz)	75-96	135-155	55-65.4	93-123	69-86	91.9-113.2
Bandwidth (GHz)	21	20	10.4	30	17	21.3
Conversion Gain (dB)	-0.1	-6.5^	-0.6	-4^	7.9	-2.35
Pin (dBm)	3	6	5	20	2	6
DC-RF Efficiency (%)	2.21	4.8	3.92	N/A	6.12	3.88
Fundamental HRR (dBc)	31	17#	51.1^	N/A	39.6	44.6
Power (mW)	88	6^	74.4*	N/A	158*	18.3*
Chip Area (mm²)	0.24	0.10^	0.48	0.96	0.53	0.06

^ Estimated from graphics. # Simulation results. * Including output buffer.

influence on the frequency response at $3*f_0$ while the first-harmonic HRR has a peak improvement of about 20 dB after adding the notch-filters into XFMR2 and XFMR3.

III. MEASUREMENT

The proposed frequency tripler has been fabricated in 65 nm CMOS process. The frequency tripler has a 160 μm × 420 μm core chip area, as shown in Fig. 9. Under 1.2V supply voltage, the frequency tripler consumes 15.3 mA current, including 4.8 mA by the frequency doubler and mixer and 10.5mA by the output buffer.

The frequency tripler is measured through an on-chip probing system. The measurement results are shown in Fig. 10. With the increase of input power, the conversion gain and efficiency also increase. When the input power is 6 dBm, the measured peak conversion gain is -2.35 dB. The output 3 dB bandwidth is 21.3 GHz from 91.9 GHz to 113.2 GHz. The maximum DC-RF efficiency is about 3.88% around 102 GHz.

Fig. 10(c) shows the measured first-harmonic and second-harmonic HRR. The peak fundamental HRR is 44.6 dBc at 109 GHz. And in the frequency range from 90 GHz to 120 GHz, the fundamental HRR is greater than 33.8 dBc.

The measured peak HRR of the second harmonic is 21.5 dBc. The second harmonic HRR is not as good as the fundamental HRR. The main reason for the less second-harmonic HRR is the mismatch of the single-ended-to-differential and differential-to-single-ended balun at the input and output. When the frequency tripler is used in the LO chain of the whole transceiver, where those two single-ended-to-differential conversion and verse vice are not necessary, the HRR of second harmonic can be improved.

The performance comparison with recent reported works is shown in Table 1. Among those works, our proposed frequency tripler prototypes have the highest fundamental HRR in W-band and the smallest chip area at the same time. It also has the smallest power consumption including the output buffer. Note that the 6 mW power consumption in [5] doesn't include the output buffer. The comparison demonstrates the advantage of our proposed techniques in high-performance and low-cost mm-wave transceiver.

IV. CONCLUSION

In this article, we proposed a novel W-band frequency tripler using folded four-coil transformer and embedded notch-filters. The folded four-coil transformer achieves the input matching and input power distribution to the push-push frequency doubler and the mixer simultaneously with compact chip size. The embedded notch-filters improves the fundamental HRR with minimal extra area consumption. The frequency tripler is designed and fabricated in 65nm bulk CMOS process. Measurement results show that the conversion gain is -2.35 dB with 44.6 dBc peak and greater than 33.8 dBc fundamental HRR within the whole 90-120 GHz range and 3.88% DC-RF efficiency for an input power of 6 dBm at 102 GHz. The output 3 dB bandwidth is 91.9-113.2 GHz.

ACKNOWLEDGEMENT

This work was supported by the National Natural Science Foundation of China (No. 62074090) and the Beijing Innovation Center for Future Chips (ICFC).

REFERENCES

[1] J. A. Qayyum, J. D. Albrecht, J. Papapolymerou and A. C. Ulusoy, "A Compact W-Band Frequency Tripler Using Single-Balanced Topology," in *IEEE Microwave and Wireless Components Letters*, vol. 30, no. 8, pp. 806-809, Aug. 2020.

[2] B. Razavi, *RF Microelectronics*, 2nd ed. New York, NY, USA: Prentice-Hall, 2011.

[3] N. Mazor et al., "A high suppression frequency tripler for 60-GHz transceivers," in *IEEE MTT-S International Microwave Symposium*, 2015.

[4] A. Kankuppe, S. Park, P. T. Renukaswamy, P. Wambacq, and J. Craninckx, "A wideband 62-mW 60-GHz FMCW radar in 28-nm CMOS," IEEE Trans. Microw. Theory Techn., vol. 69, no. 6, pp. 2921–2935, Jun. 2021.

[5] S. Park et al., "A D-Band Low-Power and High-Efficiency Frequency Multiply-by-9 FMCW Radar Transmitter in 28-nm CMOS," in *IEEE Journal of Solid-State Circuits*, vol. 57, no. 7, pp. 2114-2129, July 2022.

[6] D. -J. Shin, U. -G. Choi and J. -R. Yang, "High-Power V-Band CMOS Frequency Tripler With Efficient Matching Networks," in *IEEE Microwave and Wireless Components Letters*, vol. 31, no. 8, pp. 1020-1022, Aug. 2021.

[7] N. Zhang, L. Belostotski and J. W. Haslett, "D-Band Broadband Passive Frequency Tripler Using Antiparallel Diode-Connected nMOS Transistor Pair in 22-nm CMOS SOI," in *IEEE Microwave and Wireless Components Letters*, vol. 30, no. 7, pp. 689-692, July 2020.

[8] P. Zhou et al., "A high-efficiency E-band SiGe HBT frequency tripler with broadband performance," in *IEEE MTT-S Int. Microw. Symp. Dig.*, Philadelphia, PA, USA, Jun. 2018, pp. 690–693.

RMo4A-4

A Compact 70-86 GHz Bandwidth Frequency Quadrupler with Transformer-Based Harmonic Reflectors in 28nm CMOS

Paolo Ricco [1], Gianfranco Avitabile [2], Danilo Manstretta [1]

[1] Microlab, University of Pavia, Italy
[2] ETLC Lab, Polytechnic of Bari, Italy
paolo.ricco01@universitadipavia.it

Abstract—**A frequency quadrupler based on cascaded push-push frequency doublers is presented in this work. Push-push frequency doublers suffer from limited power efficiency and conversion gain, mainly due to second-harmonic feedback. Conventional harmonic reflectors minimize this undesired feedback introducing a common-mode second-harmonic resonance, at the price of increased area and reduced bandwidth. In this design the harmonic reflector is embedded into the input matching network, resulting in a more compact design. The coupling coefficient between the multiple windings of the transformer secondary is used to decouple the differential-mode inductance from the common-mode inductance, that acts as a wideband harmonic reflector. A common-gate transistor is stacked with the push-push pair to further boost the output power while reusing the same current. Two push-push frequency doublers are cascaded without additional power amplification stages. The quadrupler, implemented in 28nm CMOS, achieves a peak output power of 0 dBm and peak power efficiency of 5% at 77 GHz and the 3-dB bandwidth is from 70 to 86 GHz.**

Keywords—**frequency doubler, frequency quadrupler, wideband, millimeter wave, CMOS.**

I. INTRODUCTION

Millimeter-wave transceivers for radars and mobile communications require on-chip oscillators with high spectral purity. A common solution is the use of a sub-harmonic oscillator followed by a frequency multiplier. This allows to maximize the oscillator figure-of-merit, resulting in a better trade-off between power dissipation and phase noise. On the other hand, frequency multipliers suffer from high power dissipation, low power efficiency, and limited bandwidth. As a result, the frequency multiplier can consume more power than the oscillator itself. This has motivated extensive research on frequency multiplier solutions [1]-[2]. In this work an E-band frequency quadrupler based on cascaded push-push frequency doublers is proposed.

II. PUSH-PUSH FREQUENCY DOUBLERS DESIGN

A. Conventional push-push doubler design issues

In a push-push doubler the second harmonic current generated by the input transistors flows back to the gates through the voltage divider formed by the gate-drain capacitance and the common-mode gate impedance. This feedback results in a nonzero second-harmonic voltage swing at the gates, which generates a second harmonic current that is 180° out of phase compared to the original one, lowering the output current [1]-[2]. Biasing the doubler in deep class-C [3]

can improve drain power efficiency, but it requires large input powers. As a result, deep class-C doublers cannot be directly cascaded without the use of power-hungry interstage power amplification stages.

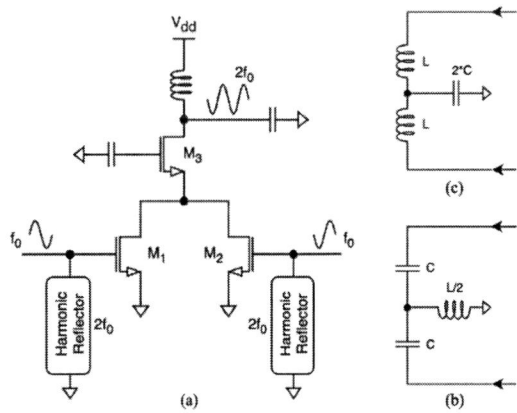

Fig. 1. Conventional implementation of (a) cascode push-push frequency doubler with lumped-elements harmonic reflectors: (b) differential capacitance and common-mode inductance and (c) differential inductance and common-mode capacitance.

A cascode transistor can be added to boost the available output power and improve power efficiency thanks to current reuse. The resulting power conversion gain, however, is still typically lower than one. To improve gain and power efficiency, harmonic reflectors can also be used. Harmonic reflectors form a short circuit for common-mode signals at the gates around the second harmonic of the input signal. Depending on the operating frequency range, the short is achieved adding a transmission line stub in parallel to the input nodes (either a l/2 short stub or a l/4 open stub) or a series-resonant LC network. In both cases, harmonic reflectors also affect the bandwidth. This is easily verified with the idealized push-push frequency doubler reported in Fig. 1, where the harmonic reflector is implemented using two differential capacitors and a common-mode inductance, resonating at the second harmonic (around 80 GHz), and is loaded by the optimum output impedance found by a load-pull analysis. The output power as a function of frequency is reported with no reflector and with different values of the reflector capacitance, for a fixed 600 mV differential input voltage swing and 100 W differential driving impedance. Using harmonic reflectors, the output signal at 80 GHz is improved by 2.2 dB, corresponding to a 66% improvement in power efficiency, but the bandwidth is also

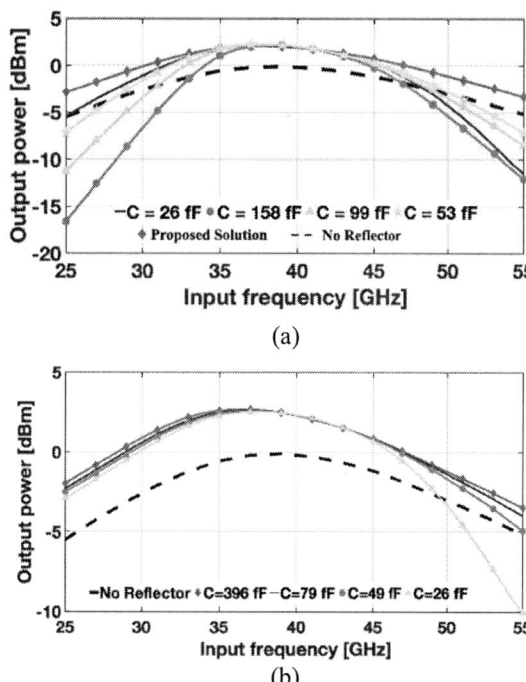

(a)

(b)

Fig. 2. Push-push frequency doubler second harmonic output power as a function of input frequency with no harmonic reflector (dashed lines), with the reflector in Fig. 1(b), for different values reflector capacitance C, and with the proposed reflector solution. Simulations with a driving impedance of (a) 100 Ohm and (b) 1 Ohm.

reduced. The main reason is the reduction of the input bandwidth due to the additional differential capacitance introduced by the reflector. As the reflector capacitance is reduced to increase the input bandwidth, however, another effect becomes prominent, i.e. the parasitic resonance induced in the push-push core output impedance [2], causing a notch in the output power. For large reflector capacitance this parasitic resonance is at very high frequency, but as the reflector capacitance is reduced, the resonance comes closer to the doubler center frequency, eventually limiting the maximum operating frequency of the doubler. This can be better appreciated using a very low differential driving impedance (Fig. 2.b), so that input bandwidth limitations are eliminated. In this case, a large reflector capacitance is actually advantageous, while for small reflector capacitance values a notch of the output power in the upper part of the band is clearly observed, as already reported in [2] for both open and short stub reflector implementations. A similar trade-off exists also for the alternative solution (Fig. 1.c), i.e. using a differential inductor with a center-tap capacitance. A large common-mode capacitor and a small differential shunt inductor are desirable to push the parasitic resonance at a sufficiently high frequency, but this lowers the input bandwidth since the network quality factor increases. These issues are exacerbated when two push-push frequency doublers are cascaded in order to build a frequency quadrupler. In fact, due to the typical class-C behavior, a 3-dB drop in the output power of the first doubler results in an even greater power reduction in the second doubler. Pushing the

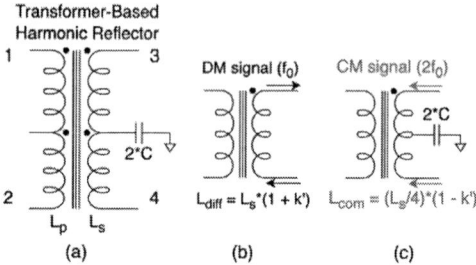

Fig. 3. Proposed transformer-based harmonic reflector (a) and equivalent circuit for (b) differential-mode and (c) common-mode signals.

doubler into deep class-C (e.g. using a 0 V voltage bias as in [3]) to improve power efficiency lowers the conversion gain, making things even worse.

B. Proposed push-push frequency doubler design

The solution proposed here is to embed the harmonic reflector into the doubler input matching network, as shown in Fig. 3. A transformer-based matching network is used, where the secondary has multiple windings. Thanks to the mutual couplings between the secondary windings, a large differential mode and a small common-mode inductance are obtained. In this way, a larger common-mode capacitance can be used for the harmonic reflector, pushing the parasitic resonance of the push-push doubler output impedance at very high frequencies, while the differential mode inductance can be set to the desired value. This can be seen from the simulation results in Fig. 2(a), showing an improved bandwidth. In this example, which is close to our second multiplier design, with a coupling factor k=0.52, the differential inductance of 360 pH resonates with a differential capacitance of 44 fF at 40 GHz, while the common-mode inductance of 28 pH, resonates with a reflector capacitance of 140 fF at 80 GHz.

III. FREQUENCY QUADRUPLER DESIGN

The complete schematic of the proposed frequency quadrupler is reported in Fig. 4. The design is implemented in a bulk 28 nm CMOS technology and has a single 1 V supply voltage. The quadrupler input connects through a 100 Ω differential transmission line to the input pads. Offset between the primary and secondary transformer windings is used to control their coupling factor and extend the bandwidth. Transformer-based matching networks are implemented for the two doublers, as reported in Fig. 5. The first transformer has a differential secondary inductance of 663 pH and a common-mode inductance of 60 pH. The second transformer has a differential secondary inductance of 360 pH and a common-mode inductance of 28 pH. In both cases a step-up transformer is used in order to boost the effective voltage swing driving the push-push doublers gates. The first doubler is designed to reach the saturation level of the following doubler. This results in a reduced peak power efficiency, but it allows to keep the quadrupler output power constant across a wider bandwidth. A transformer with single-turn inductors is used for the output matching network. No power amplification is added to reach the desired 0 dBm output power.

979-8-3503-2123-4/23 $31.00 © 2023 IEEE   170

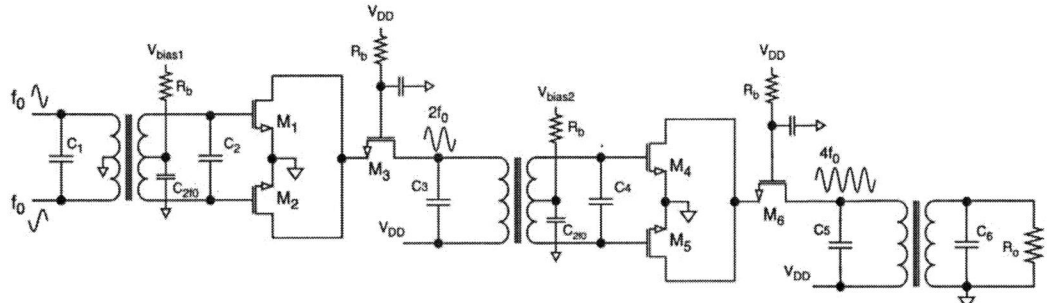

Fig. 4. Complete schematic of the frequency quadrupler.

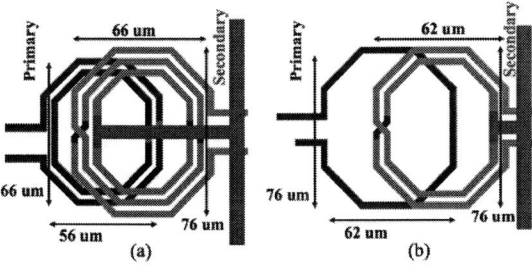

Fig. 5. Physical layout of the transformers used for the input matching network of (a) the first frequency doubler and (b) the second frequency doubler. The T-shaped lines in blue connect the secondary center-tap to the reflector capacitors.

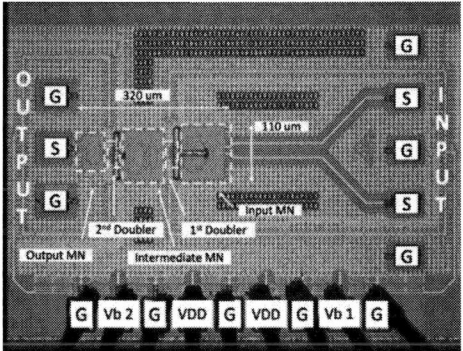

Fig. 6. Chip microphotograph.

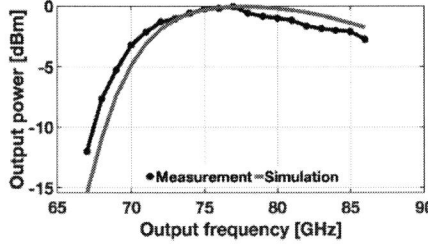

Fig. 7. Measured and simulated quadrupler fourth harmonic output power as a function of frequency for Pin=0 dBm.

IV. MEASUREMENT RESULTS

The chip microphotograph is reported in Fig. 6. The core area only occupies 320 mm by 110 mm. The measured quadrupler output power (Pout) as a function of output

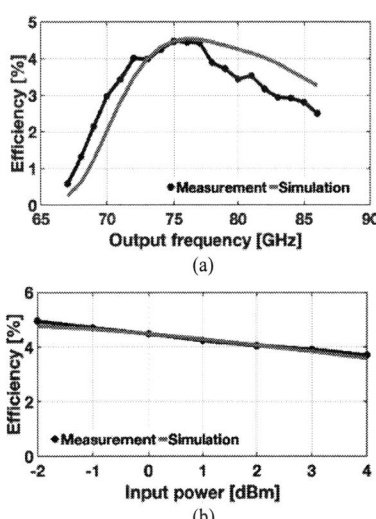

Fig. 8. Measured power efficiency as a function of (a) frequency and (b) Pin.

frequency for a fixed input power (P_{in}) of 0 dBm is reported in Fig. 7. P_{out} reaches a peak of 0 dBm at 77 GHz and is kept within 3-dB variation from 70 GHz up to 86 GHz. Measurements agree well with simulations, showing a down-shift of the characteristic by approximately 1 GHz. Measured power efficiency as a function of frequency for P_{in} = 0 dBm and at 77 GHz output frequency as a function of P_{in} are reported in Fig. 8. A peak efficiency of 5% is achieved for P_{in} = -2 dBm. The measured Pout at 77 GHz as a function of P_{in} is reported in Fig. 9. The measurements agree very well with simulations showing less than 1 dB variation for P_{in} going from -2 dBm up to 4 dBm. Fundamental and second-harmonic output power levels are compared to the fourth harmonic power in Fig. 10. Better than 40 dB second harmonic rejection (SHR) and 60 dB fundamental rejection (FRR) was measured across the whole bandwidth, in good agreement with simulations. A summary of the measured performance is reported in Table 1 and compared with state-of-the-art quadrupler implementations covering a similar frequency range. The proposed solution achieves a peak power efficiency of 5%, similar to [8], which has the highest reported efficiency in CMOS when considering the output buffer. Notice, however, that the power efficiency in [8] drops to 2.5% without buffer. Moreover, a much lower input power is required in our design, which is desirable since it would save

979-8-3503-2123-4/23 $31.00 © 2023 IEEE

Table 1. Performance summary and comparison with other works.

	This Work	[4]	[8]	[7]	[5]	[6]
Type	**x4 cascaded push-push**	x4	Single-stage push-push	Cascaded push-push	x4	x4 cascaded Gilbert mult.
PA stage	**NO**	NO	YES	YES	YES	NO
Technology	**28 nm CMOS**	22nm FDSOI	40nm CMOS	40nm CMOS	120nm SiGe	130nm SiGe
V_{dd} [V]	**1**	0.8		1.5	2	3.3
Fout [GHz]	**70-86**	71-81	65.6-75.2	74-82	70-110	72-80
BW-3dB [%]	**20.8**	14.8	13.6	10.2	44.4	10.5
P_{OUT} @ P_{IN} [dBm]	**0@ 3**	3.1 @ 0	-0.2 @ 8.9	1.7 @3	3 @0	-4 @0
P_{DC} [mW]	**20**	70	11.4	34	170-240	57
Peak η_{DC} *[%]	**5**	2.9	5 (2.5†)	4.4	1.2	0.7
FRR /SHR [dB]	**>60 / >40**	35	45 / 30	45 / 20	>30	N/A
Core Area [mm^2]	**0.03**	0.38	0.1	0.92	1.9	0.2 ^

* η_{DC}= P_{OUT}/(P_{DC}+P_{IN}); † without buffer; ^ estimate from chip photo

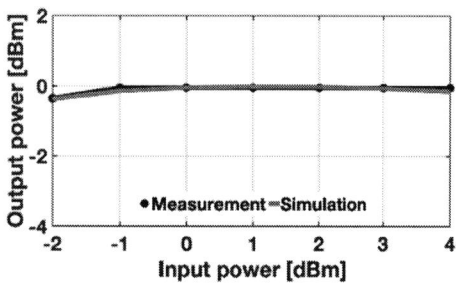

Fig. 9. Measured and simulated Pout at 77 GHz as a function of Pin.

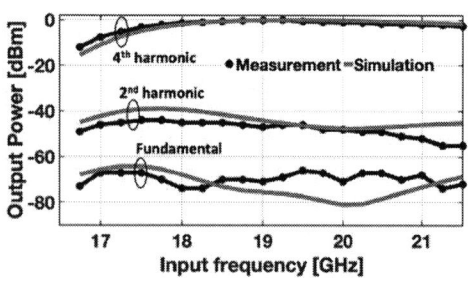

Fig. 10. Measured and simulated Pout at the fundamental, second and fourth harmonics as a function of output frequency for Pin=0dBm.

power from the driving buffer. The proposed solution achieves a fractional bandwidth of 20.8%, which is better than prior CMOS implementations. In [5] a bipolar implementation reports a much wider bandwidth, but with poor power efficiency and very high power dissipation. Finally, the proposed design occupies the smallest die area.

V. CONCLUSION

The design of a compact E-band frequency quadrupler implemented by cascading two frequency doublers without additional power amplification stages is proposed in this work. Both frequency doublers adopt a cascoded push-push structure with a second-harmonic reflector for power efficiency enhancement. The proposed reflector re-uses the secondary windings of the transformer used for the input matching network, resulting in 2.2 dB higher output power (66% higher power efficiency for a single push-push doubler) without area or bandwidth penalties. The quadrupler achieves an output power of 0 dBm at 76 GHz with 5% peak efficiency and has a 3-dB bandwidth from 70 to 86 GHz (20.8% fractional bandwidth), improving over prior CMOS implementations.

ACKNOWLEDGMENT

This work was supported by Huawei Technology Co.

REFERENCES

[1] Juo-Jung Hung, T. M. Hancock and G. M. Rebeiz, "High-power high-efficiency SiGe Ku- and Ka-band balanced frequency doublers," in IEEE Trans. on MTT, vol. 53, no. 2, pp. 754-761, Feb. 2005.

[2] K. Wu, S. Muralidharan and M. M. Hella, "A Wideband SiGe BiCMOS Frequency Doubler With 6.5-dBm Peak Output Power for Millimeter-Wave Signal Sources," in IEEE Transactions on Microwave Theory and Techniques, vol. 66, no. 1, pp. 187-200, Jan. 2018.

[3] S. Li, et al., "A Buffer-Less Wideband Frequency Doubler in 45-nm CMOS-SOI With Transistor Multiport Waveform Shaping Achieving 25% Drain Efficiency and 46–89 GHz Instantaneous Bandwidth," in IEEE Solid-State Circuits Letters, vol. 2, no. 4, pp. 25-28, April 2019.

[4] S. Vehring, Y. Ding, P. Scholz, and F. Gerfers, "A 3.1-dBm E -Band Truly Balanced Frequency Quadrupler in 22-nm FDSOI CMOS," IEEE MWCL, vol. 30, no. 12, pp. 1165–1168, 2020.

[5] B. -H. Ku, H. Chung and G. M. Rebeiz, "A Milliwatt-Level 70–110 GHz Frequency Quadrupler With >30 dBc Harmonic Rejection," in IEEE Trans. on MTT, vol. 68, no. 5, pp. 1697-1705, May 2020.

[6] H. P. Forstner, et al., "Frequency quadruplers for a 77 GHz sub-harmonically pumped automotive radar transceiver in SiGe," in Proc. EuMIC, Rome, Italy, Sep. 2009, pp. 188–191.

[7] X. Liao, D. Zhao and X. You, "An E-Band CMOS Frequency Quadrupler with 1.7-dBm Output Power and 45-dB Fundamental Suppression," 2021 IEEE International Conference on Integrated Circuits, Technologies and Applications (ICTA), 2021, pp. 119-120.

[8] K. Lee, K. Kim, G. Shin and H. -J. Song, "65.6–75.2-GHz Phase-Controlled Push–Push Frequency Quadrupler With 8.3% DC-to-RF Efficiency in 40-nm CMOS," in IEEE Microwave and Wireless Components Letters, vol. 31, no. 6, pp. 579-582, June 2021

RMo4B-1

A 0.75 mW Receiver Front-end For NB-IoT

Hossein Rahmanian Kooshkaki, Patrick P. Mercier

University of California, San Diego, USA

{hosseinr, pmercier}@ucsd.edu

Abstract — This paper presents a sub-mW receiver front-end for Narrowband IoT (NB-IoT) applications. A low-power low noise amplifier (LNA) provides a sub-3dB minimum noise figure (NF) using a transformer with a turns ratio of less than 1 and further improves the linearity and NF using a local feedback and derivative superposition techniques, without any power overhead. Mathematical expressions are presented to enable an optimum choice of a small head resistor in a low-power class-D voltage-controlled oscillator to improve the phase noise and reduce power consumption. A feed-forward technique is proposed in a baseband amplifier to increase gain and reduce the input-referred noise while maintaining linearity. Measurement results of a prototype fabricated in a 65nm CMOS process achieved a sensitivity of -110dBm and an IIP_3 of -5dBm, which meet NB-IoT requirements, all while dissipating 0.75mW.

Keywords — NB-IoT, low-power, low-noise, RX front-end.

I. INTRODUCTION

Internet of Things (IoT) devices have been largely built on the backs of low-power wireless communication standards such as BLE and Zigbee. However, such standards only operate over limited ranges, and further expansion of smart connections to wider ranges - such as in smart cities - is the next trend in the IoT world. This requires a low-power wide-area network (LPWAN). However, current WAN infrastructure is built upon cellular standards such as GSM and LTE, and these do not offer low power communication solutions required for many emerging IoT applications. This has led to the introduction of the Narrowband IoT (NB-IoT) standard - also known as LTE Cat NB1, emphasizing that NB-IoT is in fact a subset of LTE [1]. NB-IoT covers the same range as cellular networks (100m - 1km) while its assigned bandwidth and data rate ($\leq$200kb/s) are similar to low-power standards such as Zigbee or BLE.

Based on [1], various sub and above GHz frequency bands (460MHz - 2690MHz) of LTE and GSM were defined to be usable by NB-IoT when available. This work targets the 729-960MHz frequency range. The channel bandwidth is 200kHz, which is further divided into 12 sub channels with 15kHz bandwidth each, in an OFDMA multiple access scheme. The duplexing method is frequency division duplex (FDD), and half-duplex (HD). A SAW filter should not be used at the front end for cost and size reasons. Two types of modulation can be used for NB-IoT: binary phase shift keying (BPSK) and quadrature phase shift keying (QPSK), with QPSK utilized in this work. A summary of NB-IoT RX specifications is provided in Fig. 1.

It can be seen from the specifications that in contrast to the conventional short-range IoT standards such as BLE, NB-IoT

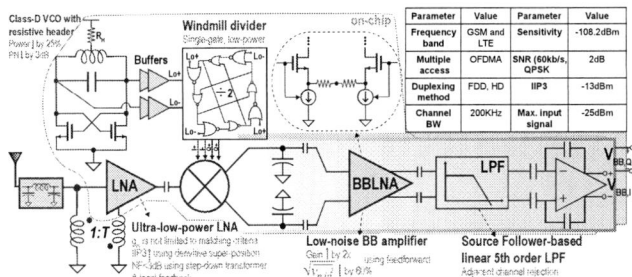

Fig. 1. Block diagram of the proposed RX front-end and the required specs.

aims to achieve a low-power yet rather high performance, as it can be concluded by comparing the -108.2dBm required NB-IoT sensitivity with -70dBm in BLE, for instance.

There have been a few prior works focusing on NB-IoT RX design in the past few years. However, most have not focused on low power consumption. In [2], for instance, an NB-IoT TRX was combined with a GNSS SoC, including a power management unit, a digital phase-locked-loop, a wake-up RX, and a clock system. A sensitivity of -125dBm was achieved by consuming 50mW power to minimize NF in a direct power-noise trade-off. One of the earliest designs for the 750 - 960MHz frequency bands of NB-IoT TRX was proposed in [3], which achieved 5.1dB NF in a 180nm process yet with a power consumption of 25mW. In [4], two identical analog Received Signal Strength Indicator (RSSI) modules with an 8-bit ADC for blocker detection were proposed, enabling a smart automatic gain control mechanism that led to a SAW-less RX with -112.5dBm sensitivity and 53mW power.

Achieving better than -100dBm sensitivity with sub-mW power consumption for NB-IoT RX front-ends remains an ongoing challenge. The proposed structure, shown in Fig. 1, presents a solution to overcome this. It meets all the required specifications of the NB-IoT standard with only 0.75mW of power consumption. The key factors to achieve this are explained in the next section.

II. PROPOSED RX ARCHITECTURE

Fig. 1 shows the proposed block diagram of the NB-IoT RX front-end. An off-chip passive matching network along with a step-down transformer (turns ratio less than 1) boosts the input impedance and reduces the noise contribution of the main transistor, relaxing the practical limitation between the minimum required g_m for matching and minimum NF. Furthermore, a local feedback is utilized to re-use the LNA current source transconductance, minimizing its noise

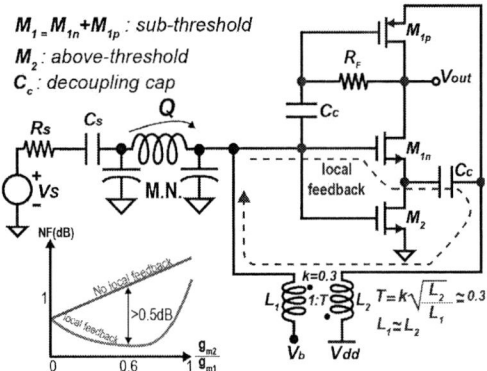

Fig. 2. Proposed LNA schematic utilizing a transformer with turns ratio less than 1 and a local feedback to improve NF and linearity.

contribution. The LNA circuit design is based on [5], [6]. However, when the LNA is used in an NB-IoT RX, its linearity should also be considered which is addressed in this work. Specifically, a current reuse derivative superposition technique has been utilized that improves IIP_3 of the LNA without any power overhead, which is described in the following section.

The output of the LNA drives a passive mixer with a 4 phase non-overlap clock generated by a low-power single-gate windmill divider [7], enabling channel selection at RF using voltage-based N-path filtering properties of the passive mixer. This filtering is important since a SAW-less front-end is of great interest for NB-IoT applications.

An ultra-low-power class-D oscillator using an optimized-size head resistor provides the required clock for the divider [8]. In this work, mathematical expressions have been provided to ensure an optimum value for the small resistive header, resulting in more than 25% power reduction and up to 3dB phase noise improvement, all without any significant power and area overhead.

Although around 20dB of RF gain is provided by the LNA, the strict sensitivity requirements of NB-IoT enforce a careful design of the first Baseband (BB) stage in terms of the input-referred noise. In addition, the zero-IF nature of the RX requires a decent amount of gain (about 30 dB) for the first BB block to enable power-efficient detection of the BB signal without any need for a high-performance, high-power ADC. Moreover, the SAW-less requirement of NB-IoT imposes a rather linear amplification for this stage. All the above-mentioned challenges make the design of the BB amplifier, called BBLNA in this work, difficult. To overcome these challenges, a feed-forward technique is proposed that lowers the input-referred noise by 60% and increases the gain by 2× all in a linear source degenerated structure with overall transconductance improvement. This will also be described in detail in the next section. The output of the BBLNA goes through a source follower-based linear 5th-order LPF [9] to meet the adjacent channel rejection requirements of NB-IoT. The last stage of the front-end is a two-stage OPAMP to provide up to 15dB of additional linear gain using a capacitive feedback structure.

III. CIRCUIT IMPLEMENTATION

A. Low-power sub-3dB minimum NF LNA

Utilizing a transformer with a turns ratio of more than 1 in a gate-boosted CG structure to provide passive gain before LNA is a well-known design method, lowering the NF and relaxing the matching criteria [10]. However, achieving a reasonable quality factor with a turns ratio of more than 1 at RF is challenging for both on-chip and off-chip designs since the secondary side of the transformer should be physically larger than the primary side, which creates more parasitic capacitance and a reduced coupling coefficient [11]. As a consequence, the self-resonant frequency and the effective turns ratio is typically lower than expected.

Nevertheless, [5] proposed a source-boosted LNA with a feed-forward transformer with a turns ratio of 2, resulting in 3.3dB NF with only 30μW of power consumption at 2.4 GHz in simulations. To make the NF less than 3dB without any power overhead, [6] proposed the notion of using a transformer with a turns ratio of less than 1 (step-down) along with a passive matching network, resulting in a minimum NF of 2.8dB with 36μW power consumption in measurements. The design, shown in Fig. 2, is used in this work - but with improved linearity to satisfy the NB-IoT IIP_3 requirements - which achieves a minimum noise factor of less than 3dB:

$$F = 1 + \frac{\gamma T}{1 + T} \ (T < 1), \qquad (1)$$

when the passives and the transformer are ideal. To further save power, a current-reuse CMOS structure (M_{1n} and M_{1p}) is utilized. Transistor M_2 is initially used as a current source to provide a DC current and more importantly to decouple the source of M_{1n} and M_{1p} from GND to be able to connect to the secondary side of the transformer. However, if M_2 does not contribute to the overall transconductance, it will degrade the NF performance. To alleviate this, the gate of M_2 is also connected to the gate of M_1, forming a local feedback with the transformer. In order to keep all the transistors in saturation and also to save more power, M_1 is in sub-threshold while M_2 is above-threshold. The noise factor, shown in (2), suggests that as long as gm_2 is less than gm_1, the noise contribution of M_2 is attenuated - which can improve NF up to 0.5dB as shown in Fig. 2. Since M_1 and M_2 have the same DC current but they work in different regions, this criterion is always satisfied. It can be shown that $gm_2 < gm_1$ is the stability requirement as well. The noise factor of the LNA of Fig. 2 is :

$$F = 1 + T\frac{(\gamma(1+T)^2\frac{g_{m2}}{g_{m1}} + \frac{n}{2}((1+T) - 2\frac{g_{m2}}{g_{m1}})^2)}{(1+T)^2(1+T-\frac{g_{m2}}{g_{m1}})}, \qquad (2)$$

where n is the sub-threshold coefficient of M_1 and γ is the excess noise coefficient of M_2. Since a SAW-less front-end is the ideal goal for NB-IoT, the linearity of the LNA should be considered carefully, especially because the passive gain at the front-end ($= Q$) reduces the IIP_3 of the LNA and the RX front-end. To overcome this, a derivative superposition technique is utilized [12]: if the LNA output AC current is a

979-8-3503-2123-4/23 $31.00 © 2023 IEEE   174

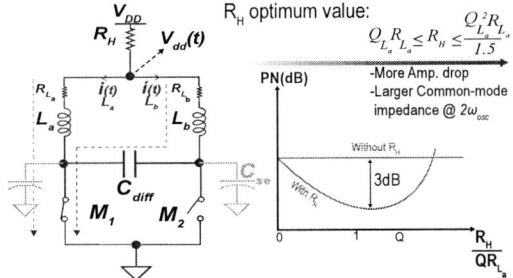

Fig. 3. Class-D OSC with resistive header and its optimum range.

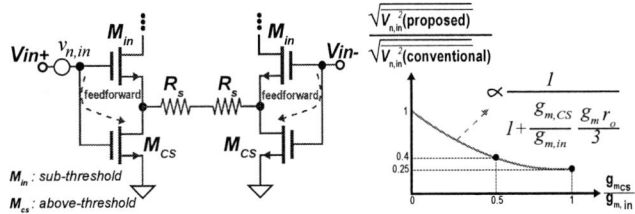

Fig. 4. Proposed BBLNA, improving gain by $2\times$ and decreasing input referred noise by 60%.

combination of a sub-threshold and an above-threshold current, there is a region at which the third-order transconductance non-linearity of the drain current (g_3) can be canceled out, with appropriate sizing of M_1 and M_2. This has been performed in this work, re-using the same structure shown in Fig. 2, without adding any additional circuit that would otherwise add to the LNA NF. It can be shown for the proposed LNA that :

$$IIP_{3,LNA} \propto \frac{IIP_{3,M_2}}{Q\sqrt{1 - \frac{g_{3,M1}}{g_{3,M2}}(1+T)^3}}, \qquad (3)$$

suggesting that as long as $g_{3,M1}$ is close to but less than $g_{3,M2}$, IIP_3 improvement happens since $T < 1$.

B. Low-power class-D oscillator with resistive header

It is shown by simulation and measurement results in [8] that utilizing a small head (or tail) resistor (R_H) in a class-D OSC can save power consumption and improve the phase noise. The goal of this work is to prove this through math and derive an optimum value for the head resistor. To do so, it is worth reviewing the expressions of phase noise, oscillation amplitude, and DC current for the original class-D OSC, introduced in [13]. The phase noise mainly depends on the oscillator supply voltage (V_{DD}) which also determines the maximum oscillation amplitude. All noise from the transistor is modeled by parameter n, including the current noise and the effect of other parameters such as the limited common mode impedance of the tank. The DC power consumption depends on V_{DD} and the loss of the inductor is modeled by R_L.

In this work, the expressions are re-derived to consider the effect of a small head resistor in order to find an optimum range. Adding the small head resistor makes two major changes: first, it changes the effective supply voltage, which is shown by $V_{dd}(t)$ in Fig. 3. Second, it changes the common mode input impedance of the circuit which will effectively change the parameter n in the phase noise expression. If the head resistor is small enough ($R_H I_{DC} \ll V_{DD}$), $V_{dd}(t)$ can be derived by calculating the IR drop across the head resistor considering a similar process proposed in [13] for deriving the inductor currents, $i_{L_a}(t)$ and $i_{L_b}(t)$ in Fig. 3. This leads to (4) and (5) for the maximum oscillation amplitude and DC current of class-D with resistive header, respectively:

$$V_{pick} = V_{DD}(1 + \sqrt{\frac{\alpha^2\pi^2}{4} + 1} - \frac{5R_H}{Q_{L_a}{}^2 R_L}) \approx \qquad (4)$$

$$V_{pick,classD}(1 - \frac{1.5R_H}{Q_{L_a}{}^2 R_L}).$$

$$I_{DC} = I_{DC\,classD}(1 - \frac{5R_H}{Q_{L_a}{}^2 R_L}), \qquad (5)$$

where α is a constant defined in [13]. The more important effect, however, can be seen in the phase noise expression when the common-mode input impedance of the circuit is derived at $2\omega_{osc}$:

$$Z_{CM} \approx \frac{R_{L_a}}{(1-4X)^2}\sqrt{1 + Q_{L_a}{}^2(1-4X)^2} + R_H, \quad (6)$$

$$X = \frac{C_{se}}{C_{se} + C_{diff}}.$$

This impedance depends on the tank quality factor and also the ratio of the single-ended (C_{se}) to differential-ended (C_{diff}) tank capacitor. Having this, one can find the optimum value of R_H to increase the common mode input impedance without dropping the oscillation amplitude too much in such a way that an overall phase noise improvement is achieved. The optimum value is depicted in Fig. 3, showing that as long as $R_H/Q_{L_a}R_{L_a}$ is chosen to be between 1 and Q_{L_a} (for a typical $Q_{L_a} \geq 10$), a phase noise improvement up to 3 dB can be achieved theoretically. In this work, a 15 - 30Ω resistor satisfies this condition.

C. Linear BBLNA with feedforward

As mentioned before, the BBLNA needs to provide a relatively high but linear gain with a low input referred noise. These requirements cannot be satisfied by using a simple conventional CMOS differential amplifier in sub-threshold region with a source degeneration resistor, for instance, as this would degrade the amplifier gain and add extra noise directly at the input - both from the source degeneration resistor and the current source.

A feed-forward technique, shown in Fig. 4, is proposed to tackle these issues. In a conventional differential amplifier, the current source (M_{CS}) only participates in the output noise without improving the overall transconductance. When its gate is connected to the input, however, its transconductance is reused in the overall transconductance through the proposed feed-forward path, decreasing the overall input-referred noise. Interestingly, if the transconductance of the current source is chosen to be exactly half of the main transistor (M_{in}) and the value of R_S is chosen to be equal to the value of the r_o of the current source (M_{CS}), the overall gain of this structure is

979-8-3503-2123-4/23 $31.00 © 2023 IEEE

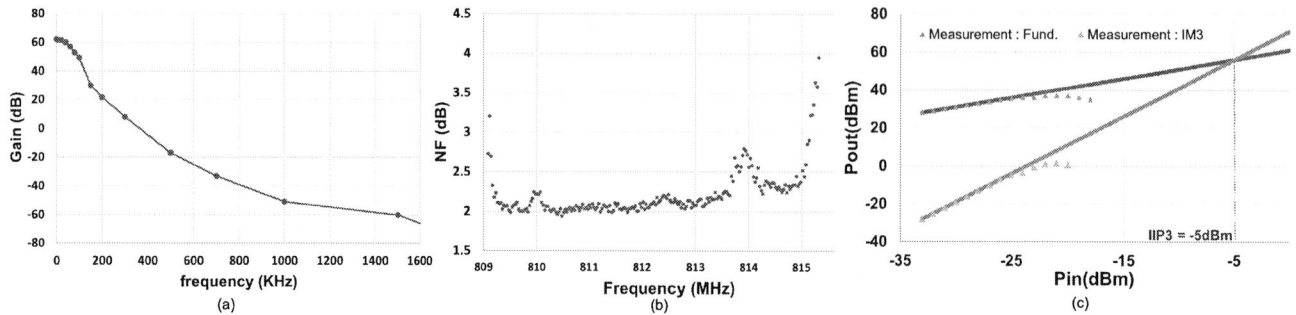

Fig. 5. Measurement results : (a) RX RF to BB transfer function (b) LNA NF at f_{RF} = 810.3MHz (c) RX IIP_3.

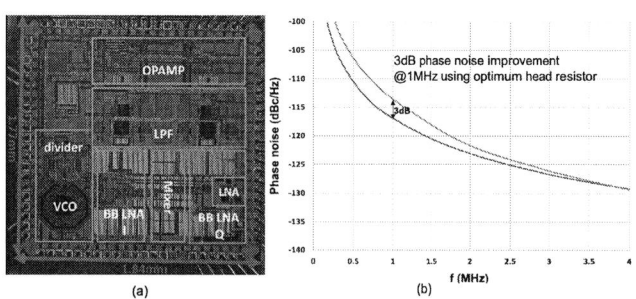

Fig. 6. (a) Chip micrograph (b) VCO phase noise at $2f_{RF}$.

Table 1. Performance comparison with state-of-the-art NB-IoT RXs.

NB-IoT RX Standard	TCASI'17 [3]	RFIC'19 [14]	ISSCC'20 [4]	ISSCC'20 [2]	RFIC'21 [15]	This work
Freq. Range (MHz)	460-2690	750-960	450-2200	450-960, 1561-2220	450-2200	750-960	729-960
Sensitivity (dBm)	-108.2	-110	-140 w/ repetition	-112.5 (LB) -111.9 (HB)	-125	-109 (wake-up RX)	-110*
IIP3 (dBm)	-13	——	——	——	——	——	-5
PN @ 1MHz (dBc/Hz)	-105	-120	——	——	——	-111	-117
V_{DD} (V)	——	1.7	0.9/1.1	0.9	2.5-5	0.9	0.55/0.8
P_{DC} (mW)	——	25	11.8	53-56	50	2.1	0.75
Active Area (mm²)	——	< 8.75	2.23	7.1 (TRX)	24.6 (SoC)	1.08	2.8
Technology (nm)	——	180	55	40	28	28	65

*calculated based on measured NF

proved to be $3g_{m,in}r_o$. In the actual implementation, a CMOS structure has been used, but it is not shown here for simplicity. In conclusion, the proposed structure shows $2\times$ improvement in gain, 60% decrease in input referred noise while maintaining linearity with source degeneration at the same time, meeting the design goals. Post-layout simulations show $12nV/\sqrt{(Hz)}$ input referred noise while 30dB of gain with $14\mu W$ power consumption was achieved in measurement.

IV. MEASUREMENT RESULTS AND CONCLUSION

The chip was fabricated in a 65nm CMOS process and achieved an RF to BB maximum gain of 61dB with more than 100dB rejection at a 1MHz offset, enabling a SAW-less solution (Fig. 5(a)). The achieved LNA NF at f_{RF} = 810.3MHz is around 2dB (Fig. 5(b)) while the measured IIP_3 is -5dBm (Fig. 5(c)). The chip is shown in Fig. 6(a) occupying an area of 2.8mm². The optimum 15Ω head resistor improved phase noise by 3dB (Fig. 6(b)) while reducing the power consumption of the VCO from 0.488mW to 0.353mW (more than 25%). In conclusion, the proposed front-end is the only sub-mW design that meets the requirements of NB-IoT among state-of-the-art RXs, as shown in Table 1.

ACKNOWLEDGMENT

This design was made possible by grant NPRP11S-0104-180192, Qatar National Research Fund.

REFERENCES

[1] LTE; Evolved Universal Terrestrial Radio Access (E-UTRA); User Equipment (UE) radio transmission and reception (3GPP TS 36.101 version 15.3.0 Release 15), 2018.

[2] J. Lee et al.,"30.2 NB-IoT and GNSS All-in-One System-on-Chip Integrating RF Transceiver, 23dBm CMOS Power Amplifier, Power Management Unit and Clock Management System for Low-Cost Solution," in *ISSCC*, 2020, pp. 462-464.

[3] Z. Song et al., "A Low-Power NB-IoT Transceiver With Digital-Polar Transmitter in 180-nm CMOS," in *IEEE TCAS I*, vol. 64, no. 9, pp. 2569-2581, Sept. 2017.

[4] H. Guo et al., "30.3 A SAW-Less NB-IoT RF Transceiver with Hybrid Polar and On-Chip Switching PA Supporting Power Class 3 Multi-Tone Transmission," in *ISSCC*, 2020, pp. 464-466.

[5] E. Kargaran et al., "Design and Analysis of 2.4 GHz 30 μW CMOS LNAs for Wearable WSN Applications," in *IEEE TCAS I*, vol. 65, no. 3, pp. 891-903, March 2018.

[6] H. R. Kooshkaki and P. P. Mercier, "A 36 μW 2.8–3.4 dB Noise Figure Impedance Boosted and Noise Attenuated LNA for NB-IoT," in *IEEE TCAS I*, vol. 70, no. 1, pp. 101-113, Jan. 2023.

[7] B. J. Thijssen et al., "2.4-GHz Highly Selective IoT Receiver Front End With Power Optimized LNTA, Frequency Divider, and Baseband Analog FIR Filter," in *IEEE JSSC*, vol. 56, no. 7, pp. 2007-2017, July 2021.

[8] H. R. Kooshkaki and P. P. Mercier, "A 0.55mW Fractional-N PLL with a DC-DC Powered Class-D VCO Achieving Better than -66dBc Fractional and Reference Spurs for NB-IoT," in *CICC*, 2020, pp. 1-4.

[9] Y. Xu et al., "A 0.65mW 20MHz 5th-order low-pass filter with +28.8dBm IIP3 using source follower coupling," in *CICC*, 2017, pp. 1-4.

[10] Xiaoyong Li et al., "G$_m$-boosted common-gate LNA and differential colpitts VCO/QVCO in 0.18μm CMOS," in *IEEE JSSC*, vol. 40, no. 12, pp. 2609-2619, Dec. 2005.

[11] A. Zolfaghari et al., "Stacked inductors and transformers in CMOS technology," in *IEEE JSSC*, vol. 36, no. 4, pp. 620-628, April 2001.

[12] H. Zhang et al., "Linearization Techniques for CMOS LNAs: A Tutorial," in *IEEE TCASI*, vol. 58, no. 1, pp. 22-36, Jan. 2011.

[13] L. Fanori and P. Andreani, "Class-D CMOS Oscillators," in *IEEE JSSC*, vol. 48, no. 12, pp. 3105-3119, Dec. 2013.

[14] P. S. Tseng et al., "A 55nm SAW-Less NB-IoT CMOS Transceiver in an RF-SoC with Phase Coherent RX and Polar Modulation TX," in *RFIC*, 2019, pp. 267-270.

[15] T. J. Odelberg et al., "A 2.1mW 109dBm NB-IoT Wake-Up Receiver," in *RFIC*, 2021, pp. 235-238.

RMo4B-2

A C-Band Compact High-Linearity Multibeam Phased-Array Receiver With Merged Gain-Programmable Phase Shifter Technique

Jingying Zhou[#1], Nayu Li[#*], Yuexiaozhou Yuan[#], Huiyan Gao[#], Shaogang Wang[#], Hang Lu[#], Chunyi Song[#],
Yen-Cheng Kuan[^], Qun Jane Gu[$], Zhiwei Xu[#2]

[#]Zhejiang University, China
[*]Donghai Lab, Zhoushan, China
[^]National Yang Ming Chiao Tung University, Taiwan
[$]University of California, Davis, USA
[1]22034224@zju.edu.cn, [2]xuzw@zju.edu.cn

Abstract—This paper presents a C-band four-element eight-beam phased-array receiver. By utilizing the proposed merged gain-programmable phase shifter (GPS) technique, the chip achieves a 360° phase-shifting range with <2.4° rms phase error and a 20.5-dB gain range with <0.18 dB rms amplitude error at 4.5–7 GHz. The chip demonstrates a 23.5-dB gain and a 6.5-dB noise figure (NF) at 5 GHz. By utilizing the multigated transistor (MGTR) technique, the receiver realizes a −10.8-dBm input 1-dB gain compression point (IP$_{1dB}$) and a −4.7-dBm input third-order intercept point (IIP$_3$) at 5.5 GHz. The proposed receiver occupies 6.4 × 3.1 mm^2 area and consumes 1265 mW, which achieves a state-of-the-art number of concurrent reconfigurable beams and an excellent linearity among silicon-based beamformers.

Keywords—gain-programmable phase shifter, multibeam, multigated transistor, phased array receiver.

I. INTRODUCTION

The increasing demand for real-time high-definition video data transfer drives the need of ad-hoc high-speed wireless data link for portable devices, unmanned aerial vehicles (UAVs), etc. To enable high throughput and low latency wireless communication, the emerging Wi-Fi 7 adopts a new 6 GHz band cooperating with 2.4/5 GHz and expands the single channel bandwidth from 160 MHz in Wi-Fi 6 to 320 MHz. The employment of 4096 quadrature amplitude modulation (QAM) with high peak-to-average power ratio (PAPR) imposes stringent requirements to in-band distortion. Silicon-based analog phased-array receiver can be utilized to overcome the path loss and improve the signal quality through beamforming technique. Compared with digital beamformers, RF implementations offer low complexity and high energy efficiency by saving power hungry analog-to-digital converters and wideband digital signal processors [1]-[3].

Multibeam capability is highly desirable to receive data from multiple users simultaneously, forming a space division duplexing (SDD) communication. Fully-connected multibeam former can support multiple concurrent users with high combining gain and communication flexibility [4]. However, with the number of elements N_E and the number of beams N_B in a chip increase, its power and area consumption become overwhelming to integrate $N_E \times N_B$ complex weighting paths on a single beamformer. The receiver in [3] demonstrates the feasibility of integrated RF multibeam forming while shows limited amplitude-control dynamic range (DR), which could

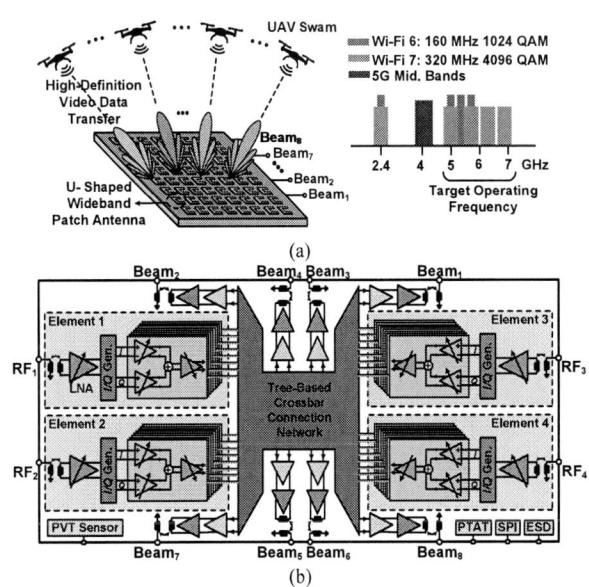

Fig. 1. (a) Wideband multibeam phased array with multiple wireless data links for UAV applications. (b) Block diagram of the proposed multibeam receiver.

not provide tapering capability to reduce sidelobes for a large-scale array.

Fig. 1(a) shows the proposed co-aperture 64-element multibeam phased array and its applications. To further reduce the receiver noise figure (NF) for a high system gain-to-noise temperature (G/T), GaAs LNAs can be used. However, this approach imposes a high linearity requirement to the silicon beamformer. In this paper, a merged gain-programmable phase shifting and combining architecture utilizing multigated transistor (MGTR) technique is proposed to realize a compact high-linearity multibeam receiver in 65-nm CMOS.

II. HIGH-LINEARITY MULTIBEAM RECEIVER DESIGN

A. Receiver Architecture

Fig. 1(b) shows the block diagram of the four-element eight-beam receiver. In each element, the low-noise amplifier (LNA) amplifies the receiving signal and drives a reconfigurable I/Q generator. The receiver integrates a crossbar beamforming matrix with 32 sets of gain-programmable phase shifters (GPSs). The tree-based crossbar connection network located at

979-8-3503-2123-4/23 $31.00 © 2023 IEEE   177   2023 IEEE Radio Frequency Integrated Circuits Symposium

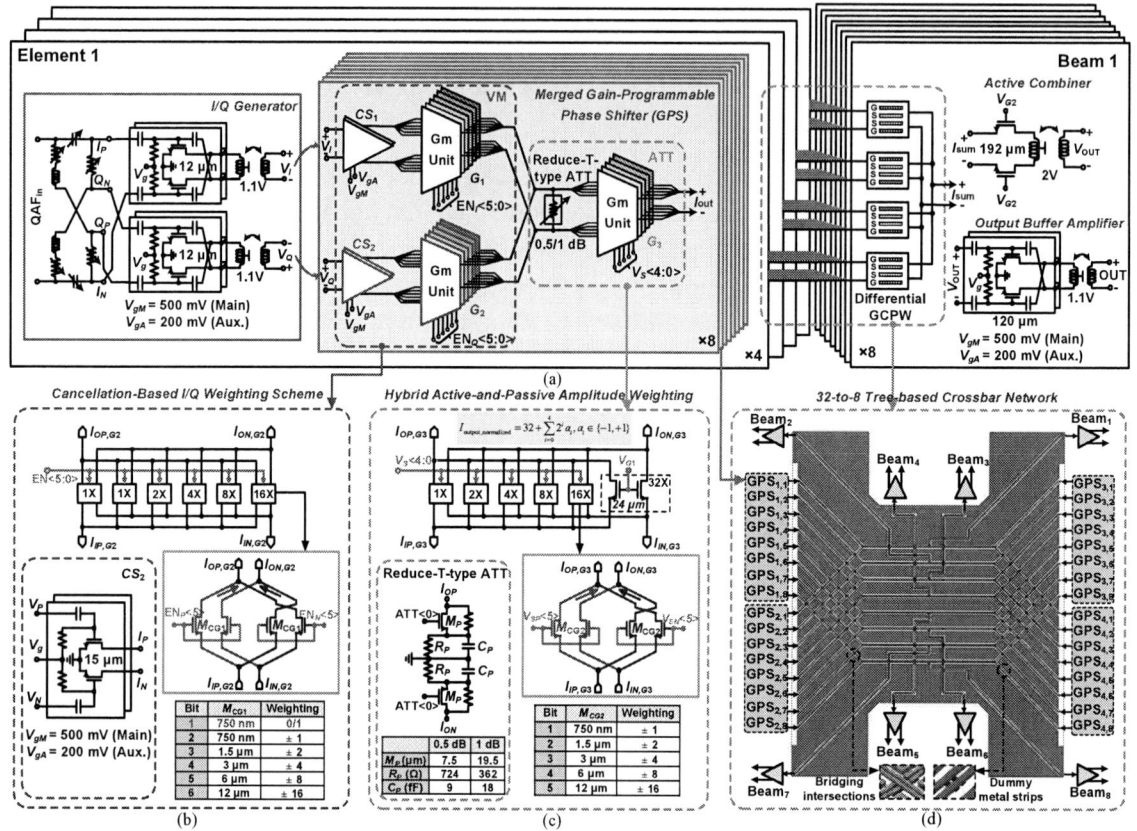

Fig. 2. (a) Architecture of the crossbar beamforming matrix. (b) Cancellation-based *I/Q* weighting scheme. (c) Hybrid active-and-passive amplitude weighting. (d) Floorplan of the 32-to-8 tree-based crossbar connection network.

the center of the chip groups and combines the 32 processed signals into 8 streams in the current domain, and then feds them into active combiners. Correspondingly, eight buffer amplifiers with output matching baluns drive off-chip 50-Ω loads. In this implementation, each beam is synthesized by the complex-weighted input signals from all the available elements, which fully utilizes the antenna aperture. Besides, the receiver integrates a sensor monitoring the process corner, voltage, and temperature (PVT) under different working environments and adapts itself to achieve its best performance. In addition, high isolation among different elements and beams is critical to reduce array calibration complexity [3]. Hence, four elements are placed at the left/right sides with large space, and eight output beams are spread at the top/bottom sides.

B. Merged Gain-Programmable Phase Shifter

Fig. 2(a) shows the architecture of the proposed crossbar beamforming matrix. Eight GPSs in each element share an identical *I/Q* generator. The GPS stacks the phase shifting and amplitude tuning with a current-sharing scheme and effectively saves the chip area. Each GPS utilizes the vector-interpolation scheme to synthesize the phase and the hybrid active-and-passive structure to program the amplitude.

1) Multibeam Phase Shifting

The *I/Q* generator is implemented by the quadrature all-pass filter (QAF) utilizing programmable resistor and capacitor

arrays to provide tunability against PVT variations and a pair of buffer amplifiers to isolate the QAF from the heavy loading of the input capacitance of the eight parallel GPSs. To preserve a constant loading to the *I/Q* generator and minimize cross-beam interference, the vector modulator (VM) adopts cancellation-based transconductance (g_m) arrays (G_1, G_2) to program the *I/Q*-path amplitude precisely while keeping a steady current flow for the common-source (CS) input transistors. Each g_m array consists of 5-bit binary-weighted g_m units with normalized output current of ± 1, ± 2, ± 4, ± 8, and ± 16, and an additional least significant bit (LSB) with a weighting of 0/1, as shown in Fig. 2(b). As a result, each g_m array can achieve normalized output current with a range of -31 to $+31$ and a resolution of 1. Such cancellation-based implementation guarantees nearly-invariant input and output impedance, which is key to decouple the phase and amplitude control.

2) Hybrid Active-and-Passive Amplitude Weighting

Conventional passive attenuators provide high resolution and large DR at the cost of large area. In contrast, active implementations are difficult to meet the DR requirement in a single stage due to the linear-in-dB gain control [5]. To render a compact design, the amplitude weighting can be arranged in a hybrid active-and-passive scheme, as shown in Fig. 2(c). Given a 5-bit g_m array (G_3), the polarity control signals (V_{SP}, V_{SN}) in each g_m unit turn on the left or the right signal amplification path and assign polarity to the output differential

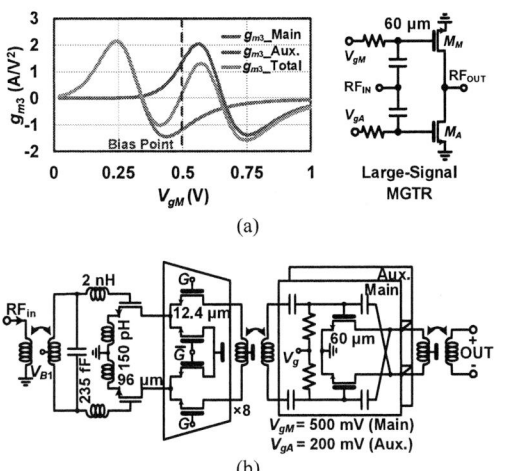

(a)

(b)

Fig. 3. (a) Simulated g_{m3} of the proposed MGTR. (b) Schematic of the LNA.

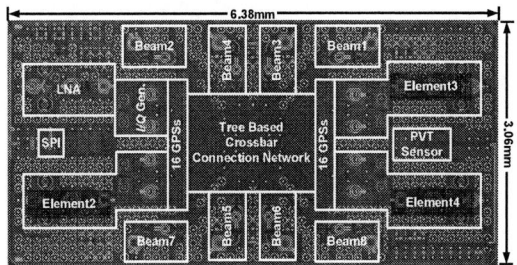

Fig. 4. Die photograph of the multibeam phased-array receiver in 65-nm CMOS.

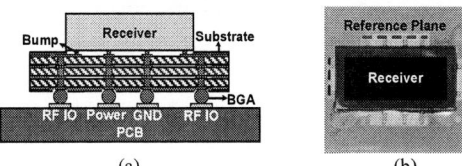

(a)   (b)

Fig. 5. (a) Cross-section structure of the flip-chip package. (b) Top view of the package for probe measurements.

current. By introducing an offset bit with a normalized ×32 output current weight, all the current values are positive and G_3 can achieve normalized output currents ranging from 1 to 63 with a resolution of 2. In addition, a 2-bit reduce-T-type attenuator with 0.5/1-dB attenuation is introduced to provide fine gain steps between the coarse gain states governed by the g_m array. Theoretically, if define the maximum acceptable gain error is ±0.25 dB and the attenuation step is 0.5 dB, the G_3 alone can only achieve an 8-dB DR, while with the 2-bit passive attenuator, the DR can be expanded to 23 dB sufficient for phased-array applications.

3) Current-Sharing Active Combining

Fig. 2(d) shows the floorplan of the tree-based crossbar connection network, where 32 phase-shifted and amplitude weighted currents are fed into the network and combined at the end. The utilization of the differential grounded coplanar waveguide (GCPW) structure with a perpendicularly crossing scheme mitigates electromagnetic interference and guarantees high isolation between different beams, which is over 50 dB in simulation. To equalize the phase delay of four signal transmission paths of each beam, dummy metal strips are placed under the paths with bridging intersections. The output beams adopt a nonsequential arrangement where signals from adjacent GPSs are fed into non-adjacent active combiners, thus further improving the isolation. Besides, the supply of the active combiner is raised up to 2 V to provide enough voltage headroom and ensure the proper operation of the stacked transistors. It also mandates delicate design of the bias and control voltages to guarantee circuit reliability.

C. MGTR-Based High-Linearity Drivers for the GPS

To accommodate the high-PAPR concurrent multibeam receiving signals, the receiver adopts a MGTR-based nonlinearity cancellation technique, which utilizes a pair of transistors with the same dimension but different bias conditions to offset their third-order transconductances (g_{m3}). The simulated g_{m3} is shown in Fig. 3(a). Compared with the g_{m3} of the conventional MGTR to achieve its minimum in a small region catering to small-signal operation, the proposed scheme

targets to optimize the g_{m3} in a large region. Although with more evident fluctuations, it accomplishes a large-signal third-order intermodulation (IM$_3$) suppression [6] and provides a better third-order linearity when with large signals. To guarantee a high overall linearity, the proposed nonlinearity cancellation technique is applied to all the driver amplifiers except for the LNA first stage. Fig. 3(b) shows the schematic of the LNA, which delivers flat gain response and excellent input matching in nearly a double frequency range. Post-layout simulation shows that the LNA achieves an 18-dB gain and a 2.3-dBm input third-order intercept point (IIP$_3$) at 5 GHz with a 7.3-dB improvement compared with the LNA without the auxiliary transistor pair. Note that common-gate (CG) transistors usually would not constrain the linearity, hence the implementation of signal weighting and combining through CG-based g_m arrays and active combiners facilitates a high-linearity design.

III. MEASUREMENT RESULTS

The proposed four-element eight-beam phased-array receiver is fabricated in 65-nm CMOS and occupies 6.4 × 3.1 mm², as shown in Fig. 4. The entire chip consumes 1265 mW with 750 mA from 1.1 V and 220 mA from 2.0 V. As shown in Fig. 5, the receiver adopts a flip-chip package where the input/output ports are fan out to the four sides of the interposer to improve the off-chip isolation. On-package probing measurements are conducted to characterize the chip performance.

Fig. 6(a) shows the simulated and measured S parameters. The measured power gain from the four inputs to the BM$_1$ is 20.3–23.6 dB at 4.5–7 GHz. The measured S_{11} and S_{22} are <−10 dB due to wideband input and output matchings. As shown in Fig. 6(b), each element achieves a minimum NF of 6.5 dB at 5 GHz. The measured IIP$_3$ and input 1-dB gain compression point (IP$_{1dB}$) are −5.6 to −3.6 dBm and −13 to −10 dBm at 4.5–7 GHz, Respectively, Fig. 7(a) shows the measured large-signal intermodulation performance. Two fundamental tones are produced at 6/6.1 GHz and the measured IIP$_3$ is −4.61 dBm. Fig. 7(b) shows the measured error vector magnitude (EVM) at 6 GHz. With 100-Msps 256-QAM signals, the chip demonstrates

979-8-3503-2123-4/23 $31.00 © 2023 IEEE   179

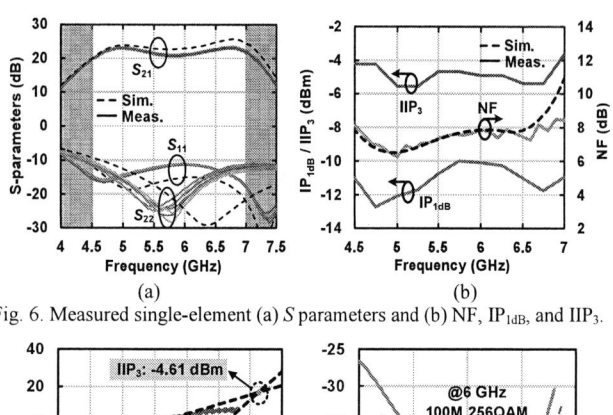

Fig. 6. Measured single-element (a) S parameters and (b) NF, IP$_{1dB}$, and IIP$_3$.

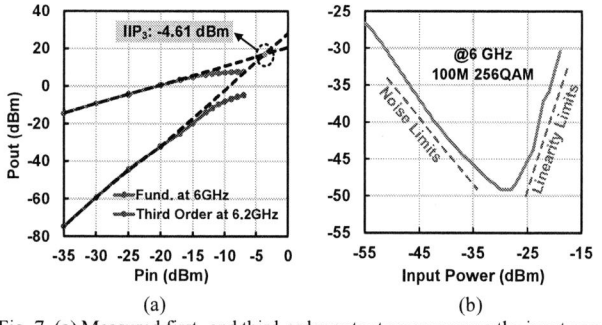

Fig. 7. (a) Measured first- and third-order output power versus the input power. (b) Measured EVM versus the input average power.

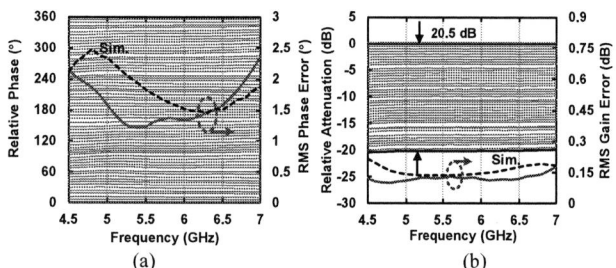

Fig. 8. (a) Measured 360° phase shifting and rms phase error. (b) Measured relative attenuation and rms gain error.

a wide input average power range from −47 to −21 dBm to achieve a <−35 dB EVM. The results verify the superior linearity of the proposed receiver.

Fig. 8(a) shows the 360° phase shifting and the measured rms phase error is <2.4° at 4.5–7 GHz without calibration. Fig. 8(b) shows the 20.5-dB gain tuning ability with a 0.5-dB step and the measured rms gain error is <0.18 dB. Beam-to-beam isolation is quantified by enabling the BM$_3$ and BM$_4$ and varying the phase settings of the VMs connected to the BM$_4$, the measured rms phase and amplitude errors of the BM$_3$ are smaller than 0.2° and 0.03 dB, respectively, which verifies a high isolation is achieved. Table 1 compares the performance of the chip with other receivers and it demonstrates the state-of-the-art multibeam forming capability and an excellent linearity.

IV. CONCLUSION

This paper presents a C-band high-linearity four-element eight-beam receiver in 65-nm CMOS for broadband wireless data communications. The merged gain-programmable phase shifting and combining architecture with the MGTR-based nonlinearity cancellation technique is proposed to address the multibeam integration challenge and fulfil the high-linearity

Table 1. Performance Comparison

	[1]	[2]	[3]	[7]	This Work
Technology	0.13 um BiCMOS	0.18 um BiCMOS	65 nm CMOS	55 nm CMOS	65 nm CMOS
Architecture	RX	TRX	RX	RX	RX
Frequency (GHz)	2–16	8–16	7.5–9	5–6	4.5–7
Phase Control	360°, 5 bit	360°, 7 bit	360°, 6 bit	360°, 4 bit	360°, 6 bit
RMS Phase Err. (°)	<8.5	<2.8	<2	<10	<2.4
Gain Control (dB)	8/1	31/0.5	8/0.25	15.5/0.5	20.5/0.5
RMS Gain Err. (dB)	<1	0.3	0.3–0.62	<0.48	<0.18
Gain (dB)	6–11	10–16	20	18	20.3–23.6
NF (dB)	11.5–12.3[a]	<10	3.6–6	<4.3	6.5–8.5
IP$_{1dB}$ (dBm)	−17 to −14	−16	−23 to −19	−23.2 to −15.8	−13 to −10
IIP$_3$ (dBm)	N/A	N/A	N/A	N/A	−5.6 to −3.6
EVM (dB)	N/A	N/A	N/A	−38.3 dB @5.8G OFDM 80M 256-QAM −40 dB	<−35 dB @6G 100M 256-QAM −47 to −21 dBm
$N_F \times N_B$	8	4	32	1	32
Power (mW)	2000	858[b]	860	32	1265
Die Area (mm²)	5 ×2.5	4 × 4[c]	5.4 × 6[d]	2.8	6.4 ×3.1

[a] NF of the two-input four-beam mode.
[b] RX only.
[c] Including the area of TX.
[d] Excluding pads.

requirement. The chip consumes 1265 mW and each element achieves a 23.6-dB peak power gain, a 6.5-dB minimum NF, and a best-case −10 dBm IP1dB.

ACKNOWLEDGMENT

This research was supported by Zhejiang Provincial Natural Science Foundation of China under Grant No. LQ23F040010, the National Key Research and Development Program of China under Grant No. 2020YFB1806304, and the Leading Innovative and Entrepreneur Team Introduction Program of Zhejiang under Grant 2018R01001.

REFERENCES

[1] M. Sayginer and G. M. Rebeiz, "An Eight-Element 2–16-GHz Programmable Phased Array Receiver With One, Two, or Four Simultaneous Beams in SiGe BiCMOS," *IEEE Transactions on Microwave Theory and Techniques*, vol. 64, pp. 4585-4597, Dec. 2016.

[2] P. Saha, S. Muralidharan, J. Cao, O. Gurbuz and C. Hay, "X/Ku-Band Four-Channel Transmit/Receive SiGe Phased-Array IC," in *IEEE Radio Frequency Integrated Circuits Symposium*, 2019, pp. 51-54.

[3] N. Li et al., "A Four-Element 7.5–9-GHz Phased-Array Receiver With 1–8 Simultaneously Reconfigurable Beams in 65-nm CMOS," *IEEE Transactions on Microwave Theory and Techniques*, vol. 69, no. 1, pp. 1114-1126, Jan. 2021.

[4] S. Mondal, R. Singh, A. I. Hussein and J. Paramesh, "A 25–30 GHz Fully-Connected Hybrid Beamforming Receiver for MIMO Communication," *IEEE Journal of Solid-State Circuits*, 2018, vol. 53, pp. 1275-1287.

[5] W. Zhu et al., "A 24–28-GHz Four-Element Phased-Array Transceiver Front End With 21.1%/16.6% Transmitter Peak/OP1dB PAE and Subdegree Phase Resolution Supporting 2.4 Gb/s in 256-QAM for 5-G Communications," *IEEE Transactions on Microwave Theory and Techniques*, vol. 69, pp. 2854-2869, June 2021.

[6] C. -N. Chen, Y. Chen, Y. Wang, T. -Y. Kuo and H. Wang, "38-GHz CMOS Linearized Receiver With IM3 Suppression, P1 dB/IP3/RR3 Enhancements, and Mitigation of QAM Constellation Diagram Distortion in 5G MMW Systems," *IEEE Transactions on Microwave Theory and Techniques*, vol. 68, pp. 2779-2795, July 2020.

[7] X. Lei et al., "A 5–6-GHz CMOS Beamforming Transceiver Front-End for Fiber-to-the-Room All-Optical Wi-Fi Solution," *IEEE Microwave and Wireless Components Letters*, doi: 10.1109/LMWT.2022.3220868.

RMo4B-3

A Wi-Fi Tri-band switchable Transceiver with 57.9fs-RMS-jitter Frequency Synthesizer, Achieving -42.6dB EVM floor for EHT320 4096-QAM MCS13 signal

Tsung-Ming Chen[1], Ming-Chung Liu[1], Pi-An Wu[1], Wei-Kai Hong[1], Ting-Wei Liang[1], Wei-Pang Chao[1], Po-Yu Chang[1], Yu-Ting Chou[1], Chien-Wei Chen[1], Sen-You Liu[1], Chang-Cheng Huang[1], Hsiu-Hsien Ting[1], Min-Shun Hsu[1], Yao-Chi Wang[1], Chao-Ching Hung[1], Yu-Li Hsueh[1], Eric Lu[2], Yuan-Hung Chung[1], Jing-Hong Conan Zhan[1]

[1]MediaTek Inc., Taiwan
[2]MediaTek Inc., USA
tm.chen@mediatek.com

Abstract—**This paper presents a Wi-Fi RF transceiver with a 2.4GHz/5GHz/6GHz tri-band switchable design. To support the wide 320MHz channel BW for Wi-Fi 7, the RF LC-tank response and TXLPF drooping are compensated via a proposed TX flatness calibration scheme that flattens the amplitude difference over the 320MHz signal bandwidth and improves the EVM over each sub-carrier. This work also proposes a reset-pulse XO design to significantly reduce the XO phase noise. A VCO pushing compensation and calibration technique is developed to suppress the sensitivity to LDO noise and DC-DC spurs. The integrated PLL RMS jitter is 57.9fs at 7.115GHz. The measured TX EVM floor achieves -42.6dB at 0dBm output power with EHT320 4096-QAM signals. This RF Transceiver occupies 3.74mm² in 55nm CMOS technology.**

Keywords—**Wi-Fi 7, IEEE 802.11be, RF transceiver.**

I. INTRODUCTION

In recent years, the explosive growth of handheld smart devices has demanded increasing network capacity and higher data rate. Compared to Wi-Fi 6, Wi-Fi 7 (802.11be) offers great flexibility by utilizing a wider signal bandwidth and more complex modulation scheme (4096QAM) to achieve the PHY rate up to 3.466Gbps with EHT320 2x2 MIMO with the combined around 2GHz bandwidth in 5GHz and 6GHz UNII frequency bands. However, the increased signal bandwidth from 160MHz (Wi-Fi 6) to 320MHz (Wi-Fi 7) poses stringent design challenges for radio transceiver, such as tighter frequency synthesizer phase noise requirement for better EVM floor in fading channels, and overcoming the TX in-band spectrum flatness ripple for 320MHz signal bandwidth induced by the effect of TX LPF drooping and TX RF front-end L-C tank frequency response. The RF transceiver adopts a fractional-N PLL with a low RMS jitter of 57.9fs at 7115MHz, a 2.4GHz/5GHz/6GHz tri-band switchable RF front-end circuits achieving -42.6dB EVM at 0dBm Pout for EHT320 4096-QAM MCS13 signals, and a TX spectrum flatness calibration achieving TX spectrum flatness < +/- 0.5dB over the 320MHz signal bandwidth.

The conventional dual-band selectable (DBS) RF transceiver is shown in Fig.1. In the conventional NxN DBS RF transceiver, there are 2N front-end radios on-chip but only N-radios are on at the same time, resulting in a 50% utilization

rate of the radio area. To improve the radio utilization rate, a 2.4/5/6GHz tri-band switchable transceiver is proposed in Fig.2. Most of the dual-band front-end circuits are highly combined together, achieving a smaller area with 0.65X of the conventional DBS RF front-end design. The radio transceiver consists of a 2.4/5/6GHz tri-band switchable transceiver and a fractional-N frequency synthesizer. The direct-conversion architecture is chosen for both TX and RX. Furthermore, to meet the targeting EVM floor requirement of EHT320, each circuit block impairment contributing to EVM should be 15dB better than -42.6dB, such as the LO phase noise, circuits noise, linearity, image rejection ratio (IRR) and TX in-band spectrum flatness.

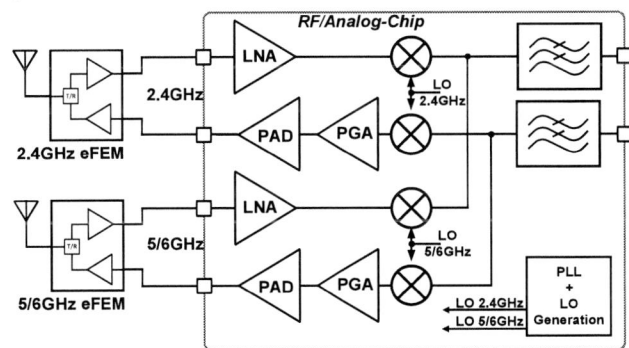

Fig. 1. The conventional DBS transceiver system diagram

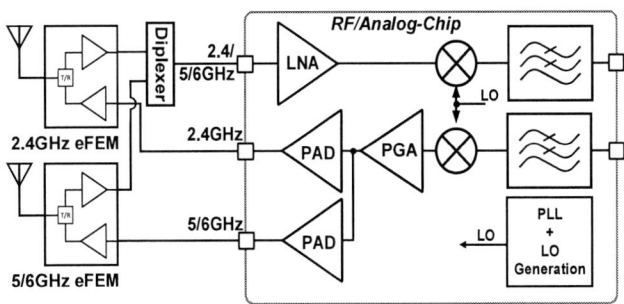

Fig. 2. The proposed 2.4/5/6GHz tri-band switchable transceiver system diagram

979-8-3503-2123-4/23 $31.00 © 2023 IEEE

II. Tri-Band Switchable TX/RX RF Design

As illustrated in Fig.3, the TX RF front-end consists of an up-converter (IQM), followed by a PGA, and a PA Driver (PAD). To support the 2.4/5/6GHz tri-band operation in a single TX RF front-end, the switch inductor design is adopted both in IQM's and PGA's load. The IQM is designed with gilbert-cell active mixer whose open drain output is connected to the switchable LC network. As shown in FIG.3, the switch inductor technique is adopted in the secondary winding of the switch-L transformer to support the 2.4/5/6GHz operation. In 2.4GHz operation mode, S1 is open, and S2 is short, resulting in the secondary winding $L_{total\_TX}=L2+L3$ with larger inductance which is required for 2.4GHz L-C tank optimization. In 5/6GHz operation mode, S1 is short, and S2 is open, resulting in the secondary winding $L_{total\_TX}=L2$ with smaller inductance which is required for 5/6GHz L-C tank optimization. Both S1 & S2 are implemented with Deep-N-Well devices and large-resistor biased technique to prevent the device drain-to-bulk leakage during the switch is open. In PGA, a common-source cascode amplifier design is adopted, and the open drain output is connected to another switchable LC network for the tri-band operation. In the PA Driver (PAD) design, the gm stage is shared in 2.4/5/6GHz modes, and the cascode stages are separated. The inductance of XF1 is designed at 2.4GHz, and the inductance of XF2 is designed for both 5 and 6GHz, to meet the 50ohm RF matching requirement at different frequency band.

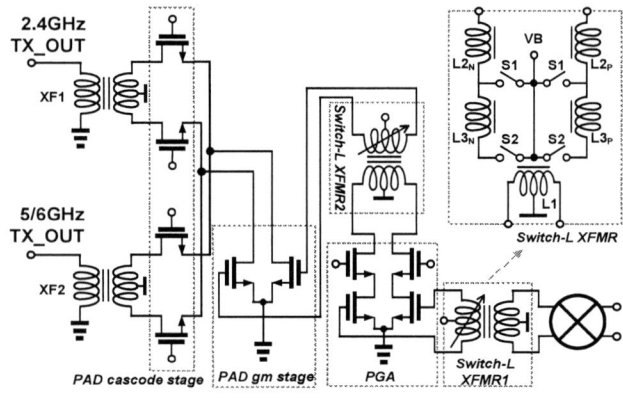

Fig. 3. Tri-band switchable TX front-end block diagram

The RX chain block diagram is shown in Fig.4. A single-ended LNA with resistor shunt-shunt feedback and inductor-degeneration technique is adopted. This LNA is meant to interface with an external LNA and act as a second amplifier stage, so the NF requirement is relaxed, and the broadband input matching is achieved with a low feedback resistor. A transformer load of LNA converts single-ended signal to differential signals for the passive mixer, followed by the receiver LPF. To support the 2.4/5/6GHz tri-band operation, the switch inductor design is adopted in the primary winding as shown in Fig.4. In 2.4GHz operation mode, S1 is open, and S2 is short, resulting in the $L_{total\_RX}=L1_{RX}+L2_{RX}$ for larger inductance which is required for 2.4GHz L-C tank design. In 5/6GHz operation mode, S1 is short, and S2 is open, resulting

in the $L_{total\_RX}=L1_{RX}$ for smaller inductance which is required for 5/6GHz L-C tank design. The measured RX noise figures for 2.4GHz and 5/6GHz bands are 3.5dB and 4.5/5.5dB respectively

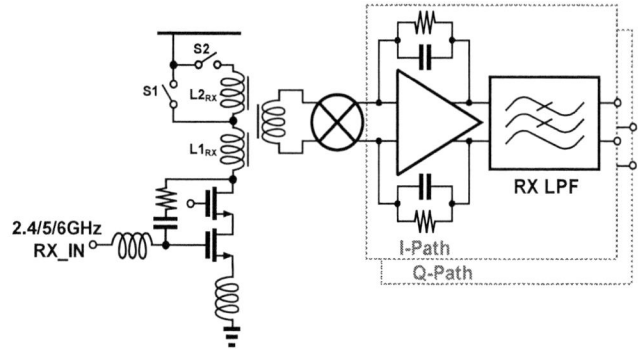

Fig. 4. Tri-band switchable RX chain block diagram

III. TX Flatness Calibration For Improved BW320 TX In-Band Spectrum Flatness

The TX in-band spectrum ripple over 320MHz wide signal bandwidth is induced by the TX LPF drooping effect in baseband frequency, and the TX RF front-end L-C tank frequency response in RF frequency. The in-band spectrum ripple will degrade the EVM, especially the resource-unit (RU, minimum 2MHz) EVM with limited SNR located at edges of the signal bandwidth. To further mitigate the in-band spectrum ripple contributing to 4096QAM EVM, a TX in-band spectrum flatness calibration is proposed in Fig.5. The calibration engine generates TX baseband tones and sweeps the frequency from -160MHz to +160MHz in digital domain, records the in-band ripple by calculating the loopback gain of each baseband frequency tones, then compensate it in the digital PHY. In the detection path, a Test-Tone-Generator (TTG) is implemented as the LO of the loopback mixer. In order to get rid of the RXLPF drooping caused calibration error, the LO_TTG frequency is adjusted accordingly while the TX tone frequency sweeping for keeping the loopback RX baseband tone frequency to be constant and much lower than the RLXPF corner frequency. After the calibration, the TX in-band spectrum flatness ripple could be improved from +/- 1.5dB to +/-0.5dB, and minimize the EVM contribution impacted by TX in-band spectrum flatness to a negligible level.

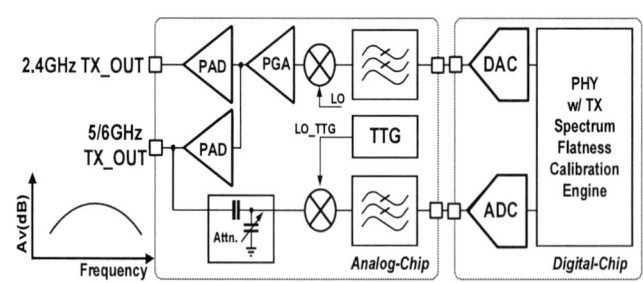

Fig. 5. The TX spectrum flatness calibration block diagram

IV. HIGH PERFORMANCE XO DESIGN

A low-phase-noise crystal oscillator (XO) is mandatory to support advanced modulation schemes. As reported in [1], the noise of Rbias is a major dominant phase noise contributor of XO. A dynamic impedance-switching biasing technique is introduced in [1] to decouple the trade-off between noise contribution and loss. However, an accurate delay timing control calibration state machine is required to align with the XO signal's transition edges. In this work, a periodic short-pulse resetting technique is developed to significantly reduces the noise of R_{bias} with negligible loss to the resonator and get rid of the delay timing control calibration requirement in [1].

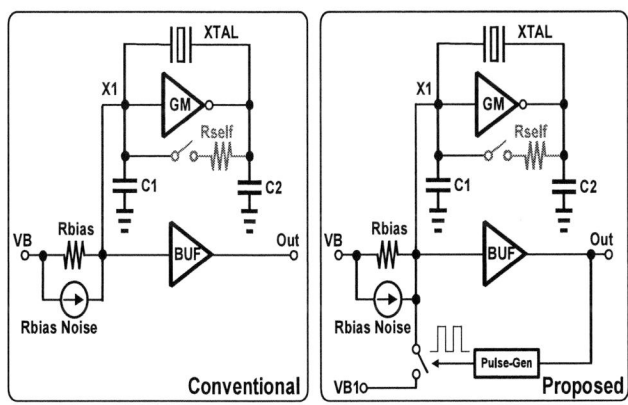

Fig.6. Low-Phase-Noise XO with periodic short-pulse resetting technique.

The XO block diagram is shown in Fig.6. At the XO start-up, R_{self} is connected initially, then R_{self} is disconnected and R_{bias} is connected instead to maintain the biasing voltage at node X1 during normal operation. The noise of R_{bias} can be modelled as a noise current injecting into the XO's oscillation node X1. The time-varying noise gives rise to voltage fluctuations at node X1, resulting in jitters after the XO buffer's sampling action. Suppressing the noise's of R_{bias} could improve XO output's phase noise, as well as the overall PLL's phase noise. However, the R_{bias} and the load capacitor at node X1 correspond to a large time-constant, leading to accumulation of time-varying noise charges at node X1 which cannot be quickly discharged in a conventional design. In this work, very short pulses, eg. 100ps, is used to periodically reset the node X1 to VB1, which is chosen to be close to XO waveform's voltages at the resetting event, and remove the noise's low-frequency spectral components. Timing of the resetting event can be anywhere except for the transition thresholds of the XO buffer to prevent the XO out transient waveform distortion. Compared with [1], a similar phase noise improvement of >5dB is achieved without the need of delay timing control calibration.

V. FREQUENCY SYNTHESIZER DESIGN

An analog fractional-N PLL frequency synthesizer is adopted in this work. To suppress the dominant noise contributors from supply noise and DC-DC spurs via VCO pushing mechanism, a pushing compensation technique is illustrated in Fig.7, in which an array of pushing-compensation-varactors (PCV) with controllable polarities is inserted in the VCO tank. Each PCV can be configured in different ways to connect to the two biasing voltages, Vb1 and Vb2, leading to either positive or negative contributions to the VCO's pushing coefficient. Vb1 is generated by taking voltage division of the supply voltage, which exhibits a similar spectral response of the supply voltage. On the other hand, Vb2 is a constant bias voltage source which exhibits independent spectral response of the supply voltage. The VCO pushing coefficient can be extracted from its frequency variation at different VCO supply voltages. The number of VCO cycles within a pre-defined duration, which is an integer number of reference clock cycles, can be counted at slightly different supply voltages. The difference between the counting results reveals the VCO's pushing coefficient. A searching algorithm finds the optimal PCV configuration to minimize the VCO's pushing coefficient. In this work, the VCO's pushing coefficient can be improved by 10~15dB after calibration.

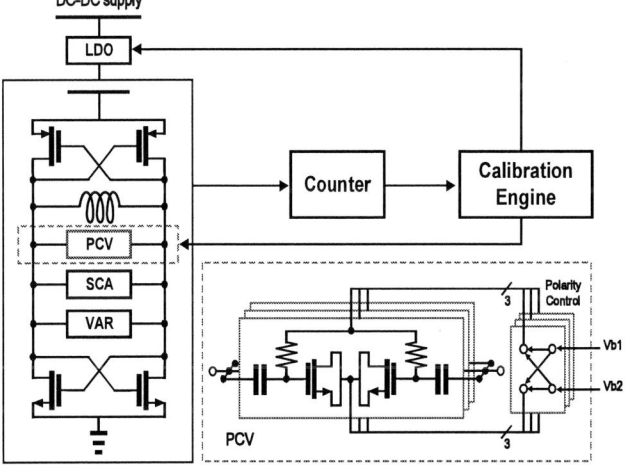

Fig.7. The VCO pushing compensation technique

VI. MEASUREMENT RESULTS

With the proposed tri-band switchable RF architecture, a chip-output EVM floor < -42.6dB at 0dBm output power for Wi-Fi 7 EHT320 4096QAM signal in 6GHz frequency band can be achieved, as shown in Fig.8. The transmit EVM vs. output power curve is shown in Fig. 9. The TX EVM floor is -48/-43.4/-42.6 dB at 0dBm output power in 2.4/5/6GHz respectively. The power consumption of the RF transceiver in EHT320 MCS13 is 380mW in RX mode, and 899mW at 0dBm TX output power, respectively.

Thanks to the switching impedance technique of XO and VCO pushing compensation technique, the measured PN of PLL is improved to 57.9fs, corresponding to 0.149° at 7115MHz. The Transceiver performance comparison is summarized in Table 1. Compared with recently published sub-100fs PLLs, this work achieves low RMS jitter with a low reference clock frequency is summarized in Table 2. The die micrograph is shown in Fig.10. The die size of this transceiver is 3.74 mm².

979-8-3503-2123-4/23 $31.00 © 2023 IEEE   183

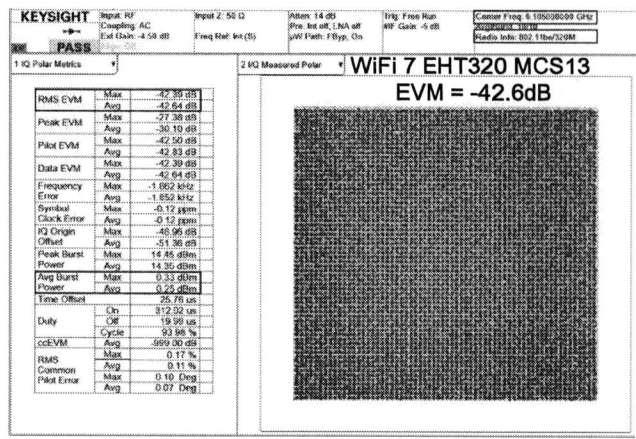

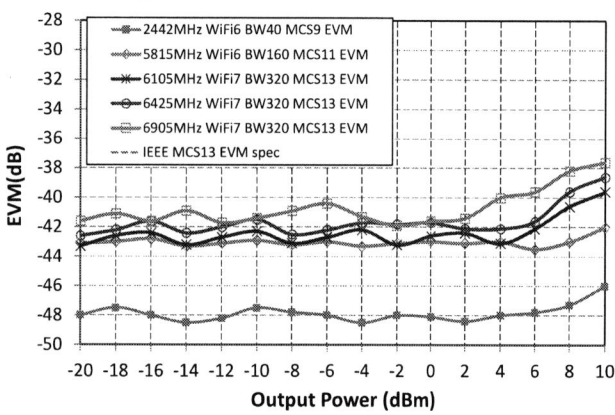

Fig. 8. The measured EHT320 MCS13 TX EVM

Fig.9. The measured tri-band TX EVM curves from HE40 MCS9~EHT320 MCS13

Table 2. PLL Performance comparison Summary

	This Work	ISSCC 2022 [3]	ISSCC 2022 [4]	ISSCC 2021 [5]	ISSCC 2020 [6]	ISSCC 2020 [7]
Process	55nm	28nm	7nm	14nm	28nm	28nm
PLL Type	Frac-N Analog	Frac-N BB-PLL	Frac-N HM-PLL	Frac-N Analog	Frac-N BB-PLL	Frac-N Sampling
Ref.Frequency	40MHz	250MHz	74MHz	76.8MHz	500MHz	500MHz
RMS Jitter	57.9fs	68.8fs	88fs	80fs	66.2fs	58.2fs
Reference Spur	-95.8dBc	-70.2dBc	-83dBc	-66dBc	-80.1dBc	-73.5dBc
Power	38.24mW	20mW	12.9mW	14.2mW	19.8mW	18mW
FoM	-248.9dB	-250.3dB	-250dB	-250.4dB	-250.6dB	-252.15dB
FoM$_N$	-272.3dB	-266.3dB	-275.8dB	-266.5dB	-265.4dB	-266.7dB
Active Area	0.33mm² *	0.23mm²	0.24mm²	0.31mm²	0.17mm²	0.16mm²

$$FoM = 10 \cdot \log_{10}[(Jitter/1s)^2 \cdot (Power/1mW)] \qquad FoM_N = FoM + 10 \cdot \log_{10}(F_{REF}/F_{OUT})$$

Fig. 10. Die photo

VII. CONCLUSION

Comparing with state-of-the-art prior arts, this work achieves the best TX EVM floor with EHT320 MCS13 signal and the lowest RMS jitter. The competitive area of 3.74mm²is achieved with the proposed tri-band switchable front-end topology.

ACKNOWLEDGMENT

The authors would like to thank RF/DE, VE, CTD/CD, CSD, SA, SE and CT teams for the great support.

REFERENCES

[1] Eric Lu, et al., "A 4×4 Dual-Band Dual-Concurrent WiFi 802.11ax Transceiver with Integrated LNA, PA and T/R Switch Achieving +20dBm 1024-QAM MCS11 Pout and -43dB EVM Floor in 55nm CMOS" *ISSCC Dig. Tech. Papers*, pp. 178-179, Feb. 2020

[2] S. Kawai, et al., "An 802.11ax 4×4 spectrum-efficient WLAN AP transceiver SoC supporting 1024QAM with frequency-dependent IQ calibration and integrated interference analyzer "*ISSCC Dig. Tech. Papers*, pp. 442-444, Feb. 2018

[3] Simone Mattia Dartizio, et al., "A 68.6fs rms -Total-Integrated-Jitter and 1.56 μ s-Locking-Time Fractional-N Bang-Bang PLL Based on Type-II Gear Shifting and Adaptive Frequency Switching" *ISSCC Dig. Tech. Papers*, pp. 386-387, Feb. 2022

[4] Dihang Yang, et al., "A Sub-100MHz Reference-Driven 25-to-28GHz Fractional-N PLL with -250dB FoM" *ISSCC Dig. Tech. Papers*, pp. 384-385, Feb. 2022

[5] Wanghua Wu, et al., "A 14nm Analog Sampling Fractional-N PLL with a Digital-to-Time Converter Range-Reduction Technique Achieving 80fs Integrated Jitter and 93fs at Near-Integer Channels" *ISSCC Dig. Tech. Papers*, pp. 444-445, Feb. 2021

[6] Alessio Santiccioli, et al., "A 66fs rms Jitter 12.8-to-15.2GHz Fractional-N Bang-Bang PLL with Digital Frequency-Error Recovery for Fast Locking" *ISSCC Dig. Tech. Papers*, pp. 268-269, Feb. 2020

[7] Mario Mercandelli, et al., "A 12.5GHz Fractional-N Type-I Sampling PLL Achieving 58fs Integrated Jitter" *ISSCC Dig. Tech. Papers*, pp. 274-275, Feb. 2020

Table 1. Transceiver Performance Summary

		This work	ISSCC 2020 [1]	ISSCC 2018 [2]
Support WLAN standards		Wi-Fi 7	Wi-Fi 6	Wi-Fi 6
RF Bandwidth		320MHz	160MHz	80MHz
Integrated PA & LNA		No	Yes	No
Chip-in RX NF (dB)	2.4GHz	3.5	4.2	N/A
	5GHz	4.5	4.8	N/A
	6GHz	5.5	Not Supported	Not Supported
2.4GHz TX EVM (dB)	MCS7,40MHz	-48@0dBm	-44@0dBm	-42.5@-5dBm
	MCS11,40MHz	-48@0dBm	-44@0dBm	-
5GHz TX EVM (dB)	MCS9,80MHz	-45@0dBm	-42@0dBm	-38.4@-5dBm
	MCS11,80MHz	-43.8@0dBm	-41.5@0dBm	-38.1@-5dBm
	MCS11,160MHz	-43.4@0dBm	-41@0dBm	N/A
6GHz TX EVM (dB)	MCS13,320MHz	-42.6@0dBm	Not Supported	Not Supported
RF Power Consumption (mW)	RX 2.4GHz	254 (1SS,40M)	322 (4SS+1LO)	354 (4SS+1LO)
	TX 2.4GHz	548@0dBm (1SS,40M)	4324@20dBm (4SS+1LO)	844@-5dBm (4SS+1LO)
	RX 5GHz	359 (1SS,160M)	420 (4SS+1LO)	447 (4SS+1LO)
	TX 5GHz	831@0dBm (1SS,160M)	4850@20dBm* (4SS+1LO)	832@-5dBm (4SS+1LO)
	RX 6GHz	380 (1SS,320M)	Not Supported	Not Supported
	TX 6GHz	899@0dBm (1SS,320M)	Not Supported	Not Supported
PLL/LO IPN (degree)		0.149@7.1GHz	0.105@2.4GHz 0.15@5.925GHz	- 0.32@5.6GHz
Technology		55 nm	55nm	28nm

RMo4C-1

A 26-40 GHz 4-Way Hybrid Parallel-Series Role-Exchange Doherty PA with Broadband Deep Power Back-Off Efficiency Enhancement

Edward Liu[#$1], Hua Wang[#2]

[#]ETH Zürich, Switzerland

[$]Georgia Institute of Technology, USA

[1]edwliu@iis.ee.ethz.ch

Abstract—This paper presents a hybrid parallel-series Doherty power amplifier (PA) architecture that supports broadband operation and deep power back-off (PBO) efficiency enhancement to support next-generation wireless communication. The architecture builds on hybrid use of broadband parallel- and series-type coupled-line Doherty power combiners with four PA paths. The wide carrier bandwidth is achieved by exploiting different frequency dependent active load-modulations of the parallel/series Doherty combiners and PA path biasing reconfigurations, while the deep PBO is realized by the optimum turn-on sequences of the four PA paths. As a proof-of-concept design, a 26-40 GHz PA is implemented in GlobalFoundries 45nm CMOS SOI process. The PA measures 22.5-23.9 dBm OP_{1dB}, 26.2-38.2% PAE_{OP1dB}, 18.1-34.5% $PAE_{6dB\,PBO}$, and 10.7-22.1% $PAE_{12dB\,PBO}$. Note the PBO levels are referenced from OP_{1dB}. Compared to a normalized ideal class B PA, this design achieves 1.3-1.8x PAE boost at 6dB PBO, and 1.6-2.3x boost at 12dB PBO. With a 250MSym/s 64-QAM signal, this PA achieves 16.1 dBm P_{ave} and PAE_{ave} of 30.7% at an EVM_{rms} of -25.3 dB. With a 100Msym/s 256-QAM signal, this PA achieves P_{ave} of 13.5 dBm and PAE_{ave} of 26.2% at an EVM_{rms} of -31.3 dB.

Keywords— 5G, CMOS, Doherty, Mm-wave, Power amplifier.

I. INTRODUCTION

The exponential growth of wireless data traffic relies on spectrally efficient modulations to achieve high-speed, high-throughput links. However, these schemes have high peak-to-average power ratio (PAPR) often above 6dB. To process these modulated signals power amplifiers (PAs) must operate at large power back-off (PBO) levels. For most PAs, this results in a heavy penalty on their average energy efficiency in operation [1]-[3]. Moreover, advanced array operations often require aggressive tapering for sidelobe reduction. For cost saving, it is desired to use the same or limited frontend circuit variants in these tapered arrays, thus necessitating PA designs with high-efficiency operations at deep PBO. Additionally, PAs that cover multiple 5G mm-Wave bands (24-28 GHz and 37-40 GHz) can support international roaming for their flexibility and re-usability. Consequently, broadband PAs with deep PBO efficiency enhancement have become highly desired.

One technique for efficiency enhancement is the Doherty PA, where a Main and Auxiliary (Aux) PA path cooperate to perform active load modulation [4]. However, a standard symmetric 2-way Doherty PA is limited to one efficiency enhancement at 6dB PBO. A practical way to increase the number of efficiency peaks is to add more auxiliary

stages. Recently published higher-order Doherty PAs [5][6] have shown deep PBO efficiency enhancement but with relatively narrow carrier bandwidth. A 3-way Doherty radiator in [7] achieves large PAE enhancement at 9dB PBO, but similarly suffers from narrow carrier bandwidth. Also, the relatively large impedance transformation ratios required of the high-order Doherty output network further limit the PA bandwidth.

To address the challenges of broadband deep PBO efficiency enhancement, we propose a 4-way hybrid series-parallel Doherty PA that achieves deep PBO efficiency enhancement over almost the entire K_a-band (26-40 GHz).

Section II presents the proposed 4-way parallel-series Doherty output network. Section III described the overall PA design, and section IV demonstrates the measurement results.

II. PARALLEL-SERIES DOHERTY OUTPUT NETWORK

The LMBA and its variants achieve intrinsic wideband PBO efficiency enhancement [8][9]. However, they are limited mm-Wave LMBA designs with wideband PBO efficiency enhancement, and they do not guarantee multi-way deep PBO efficiency enhancement either. Alternatively, [10] shows that a coupler balun based parallel Doherty PA together with Main/Auxiliary PA role exchange can achieve PA PBO efficiency enhancement over an almost 3:1 carrier bandwidth. Similarly, a series coupler balun PA can nearly cover a 3:1 carrier bandwidth as well [11]. However, these two coupler balun PA architectures are intrinsically 2-way and symmetric Doherty PAs with efficiency enhancement only at 6dB PBO.

To achieve wideband deep PBO enhancement, we propose the concept of combining a series continuous coupler Doherty PA [11] with the parallel coupler Doherty PA [10] to realize a hybrid 4-way, parallel-series Doherty combiner with role exchange operations. Shown in Fig.1(a), the output 90° coupler with an open termination at the isolation port acts as a parallel Doherty combiner [12] for PA1 and PA2, while PA2/PA3 and PA1/PA4 are series Doherty combined pairs. The four coupled line baluns provide the optimal load-pull impedance for each PA and absorb their output device capacitance.

The proposed 4-way Doherty PA operates in two modes. Shown in Fig.1(b), the PA Mode-1 spans 27-36 GHz with PA1 as the main PA and PA2/3/4 as the Aux PAs. On the other hand, the PA Mode-2 covers 22-27 GHz and 36-42 GHz where PA2 is the main PA and PA1/3/4 are the Aux PAs. In deep PBO and

979-8-3503-2123-4/23 $31.00 © 2023 IEEE   185   2023 IEEE Radio Frequency Integrated Circuits Symposium

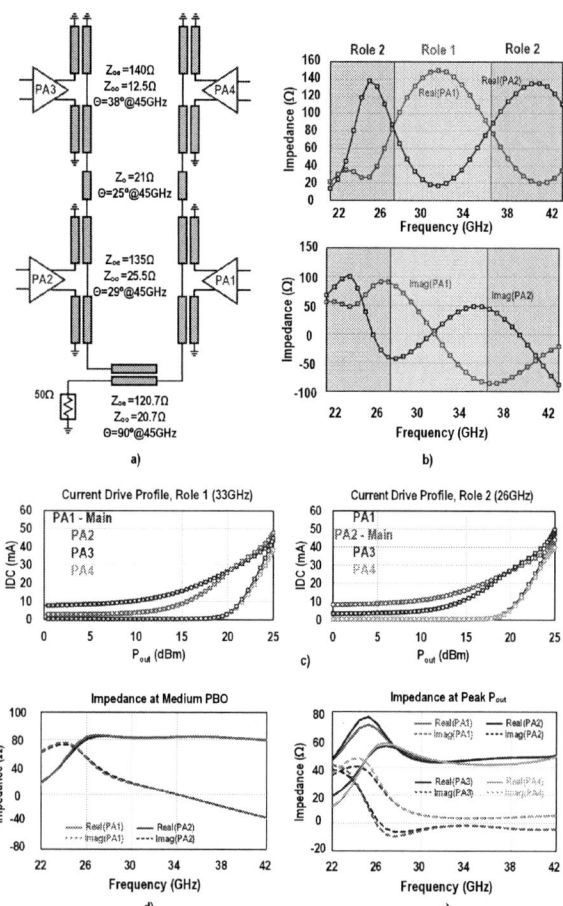

Fig. 1. (a) Schematic of 4-way Parallel-Series Doherty Combiner, (b) impedance seen by PA1/PA2 and role exchange behavior, (c) current drive profiles of the four paths in both roles, and (d)-(e) impedance seen by the PA paths at 6dB PBO and peak P_{OUT}.

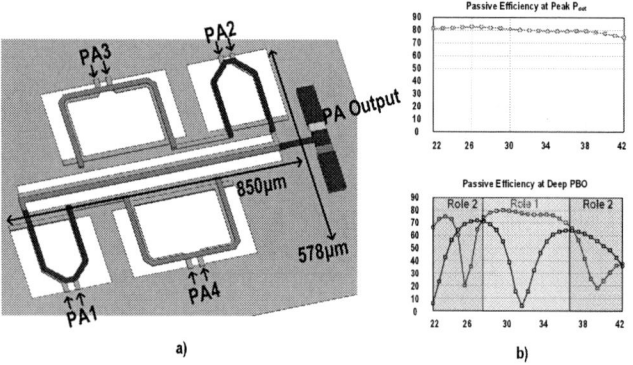

Fig. 2. (a) 3D EM model of the Output network and (b) passive efficiency at peak P_{OUT} and deep PBO.

low input drive, all the Aux PA paths are off, and the main PA will see its optimum load impedance of $Z_{OPT}=150\Omega$. As the PA input drive increases, the first Aux path (PA2 in Mode-1 or PA1 in Mode-2) will turn on, modulating the impedance such that the main PA continues to see the optimum load value and operate with high efficiency. Finally, PA3 and PA4 will

Table 1. Summary of different PA modes.

	Mode		PBO Turn On Point
	1	2	
Main PA	PA1	PA2	Always On
Aux1 PA	PA2	PA1	12dB
Aux2 PA	PA3/4	PA3/4	6dB

gradually turn on together for high PA input power. Fig.1(c) shows the simulated DC currents of the four PA branches at 33 GHz (Mode-1) and 26 GHz (Mode-2). Fig. 1 (d) and (e) show the impedance presented by the Doherty network to the PAs at medium PBO (only PA1 and PA2 are on, while PA3 and PA4 are off), and peak P_{POUT} (all PAs are on). Table 1 summarizes the PA operation in different modes.

3D electromagnetic (EM) models are constructed and simulated in ANSYS HFSS for all the passives. Fig.2(a) shows the 4-way coupler-based hybrid series-parallel Doherty combiner network at the PA output. Thick top metal layers are used to minimize the coupler loss. Each coupler balun uses its center tap to provide V_{DD} biasing, and is sufficiently bypassed to ensure near ideal AC ground. As shown in Fig.2(b), the passive efficiency at peak P_{OUT} is >74% across 22-42 GHz. At deep PBO with only the main PA on, the passive efficiency is maintained >50% from 26-40 GHz.

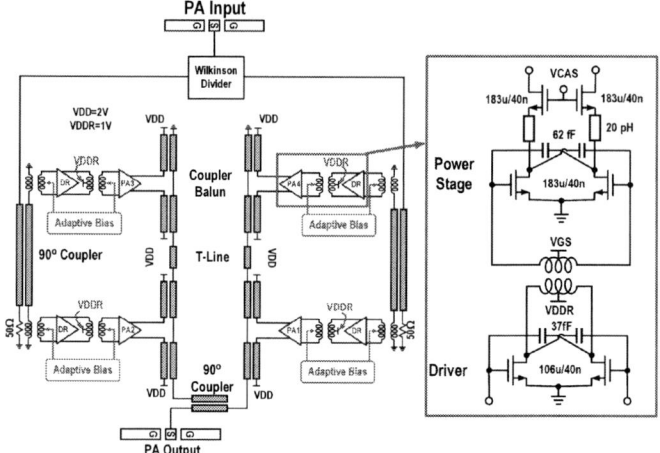

Fig. 3. Schematic of the proposed 26-40 GHz 4-way hybrid parallel-series role-exchange Doherty with broadband deep PBO efficiency enhancement.

III. PROPOSED BROADBAND PARALLEL-SERIES DOHERTY PA

The full schematic of the proposed PA is shown in Fig.3. The input signal is equally divided by a 2-way Wilkinson divider. Then, the input signal in each path is fed into a broadband 90° coupler that generates the required phase shift at the input of each PA path for Doherty operation. A balun then generates differential signals to drive each PA path. The driver and output stages are both differential, using

979-8-3503-2123-4/23 $31.00 © 2023 IEEE   186

neutralization capacitors for stability enhancement and gain boosting. The power stage uses a cascode topology to increase the output power. Additionally, a 20pH transmission line is used in the middle of the cascoded power stage to absorb the parasitic capacitance at the drain of the CS devices and source of the CG devices, which improves the PA core efficiency and stability. Each path is controlled by an adaptive biasing circuit, allowing for optimal coordination and turn on/off of different paths [11].

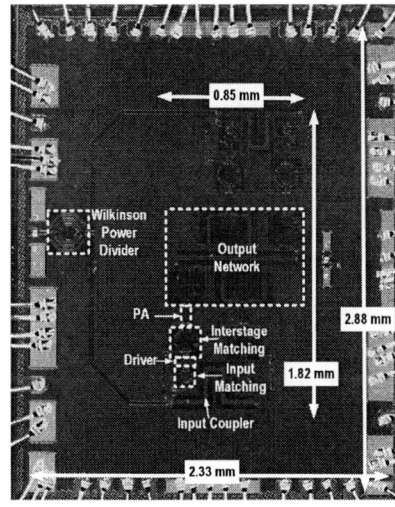

Fig. 4. Chip microphotograph in GlobalFoundries 45nm RF CMOS SOI.

IV. MEASUREMENT RESULTS

The proposed PA is fabricated in GlobalFoundries 45nm CMOS RFSOI process (Fig.4). The PA is measured by chip-on-board packaging and direct probing. The measured PA large signal continuous-wave (CW) results shown in Fig.5. At 33 GHz, the PA achieves 15dB gain, 23.9 dBm P_{1dB}, 24.1 dBm P_{SAT}, and PAE at OP_{1dB}/6dB-PBO/12dB-PBO of 38.2%/34.5%/22.1%, respectively. A summary of the results are presented in Fig.5(e). For 26-40 GHz, the PA achieves 22.5-23.9 P_{1dB}, 26.2-38.2% PAE_{1dB}, 18.1-34.5% PAE at 6dB PBO, and 10.7-22.1% PAE at 12dB PBO. The PA shows significant back-off efficiency enhancement at both 6dB and 12dB PBO, verifying the efficacy of the coupler based hybrid parallel-series active load modulation network. The proposed PA is further characterized using single carrier wideband 64-QAM and 256-QAM signals with no DPD (Fig.6). At a carrier frequency of 33 GHz, the PA achieves 16.1 dBm P_{avg} and 30.7% PAE_{avg} with -25.3dB EVM_{rms} for 64-QAM. With 256-QAM, it achieves 13.5 dBm P_{avg} and 26.2% PAE_{avg} at an EVM_{rms} of -31.3dB.

V. CONCLUSION

The proposed parallel-series Doherty PA demonstrates deep PBO efficiency enhancement over a wide frequency range, achieving 22.5-23.9 dBm P_{1dB}, 26.2-38.2% PAE_{1dB}, 18.1-34.5% PAE_{6dB}, and 10.7-22.1% PAE_{12dB} from 26-40 GHz. From the comparison table (Table 2), this PA achieves

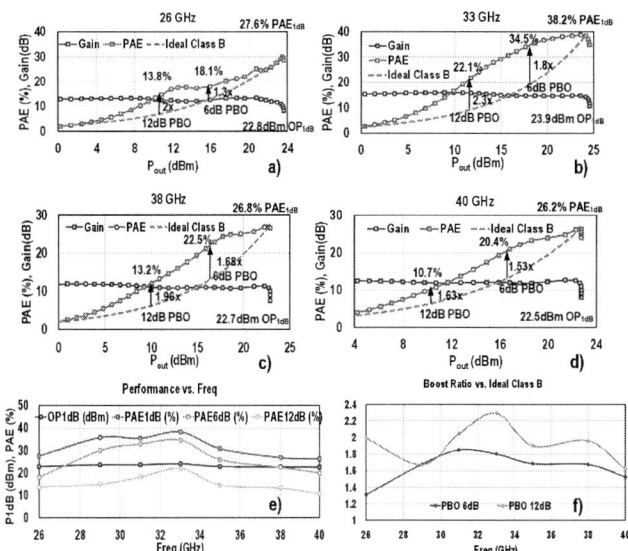

Fig. 5. Measured large signal CW results at (a) 26 GHz (b) 33 GHz (c) 38 GHz (d) 40 GHz. (e) Measured large-signal performance vs. frequency. (f) Comparison of PBO PAE efficiency enhancement ratio over an ideal Class-B PA.

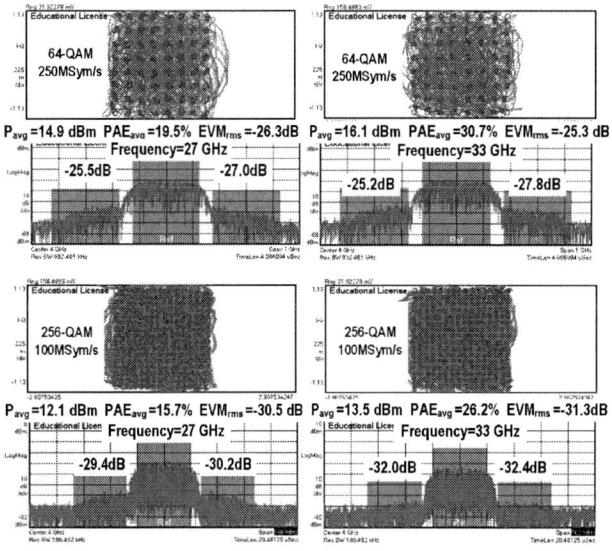

Fig. 6. Modulation measurement results with single-carrier 64 and 256-QAM at 27GHz and 33GHz.

the highest bandwidth of all higher order Doherty PAs, in addition to achieving the highest PAE at 12 dB PBO.

ACKNOWLEDGMENT

This work was in part supported by Qorvo and DARPA YFA program. Additionally, the authors thank GlobalFoundries for chip fabrication and Keysight for measurement equipment. The authors also thank Yuqi Liu for chip dicing.

REFERENCES

[1] F. Wang and H. Wang, "24.6 an instantaneously broadband ultra-compact highly linear pa with compensated distributed-balun output network

Table 2. Comparison Table with Recent Mm-Wave Doherty PAs and Linear PAs in CMOS/CMOS SOI Technologies.

	This Work				20-40 GHz Doherty PAs									Linear Mm-Wave PAs	
					ISSCC 2022 [5]	RFIC 2022 [6]		ISSCC 2021 [11]				JSSC 2022 [13]	JSSC 2021 [9]	ISSCC 2020 [1]	ISSCC 2021 [2]
Technology	45nm SOI CMOS				55nm CMOS	45nm SOI CMOS		45nm SOI CMOS				40nm CMOS	28nm CMOS	45nm SOI CMOS	45nm SOI CMOS
Architecture	Coupler Based Parallel-Series Doherty PA				Transformer based 3-way Parallel-Series Doherty	Coupled inductor based three-way Doherty		Continuous Coupler Doherty				Digital Polar	Doherty-like LMBA	Distributed Balun	Class W Dual-Drive
Supply (V)	2 (PA), 1 (Driver)				2.4	1.8 (PA), 1 (Driver)		2				1	1	2	1.9
Freq. (GHz)	26	33	38	40	28	38		26	32.5	37.5	42.5	29.5	36	28	30
Gain (dB)	12.6	15.3	12.4	11.8	16.1	15		11.5*	16*	13.5*	13*	N/A	18	20.5	20.4
P_{SAT} (dBm)	23.5	24.1	22.8	22.7	25.5	18.9		20.8	22	21.8	21.8	18.7	22.6	20.4	20.1
P_{1dB} (dBm)	22.8	23.9	22.7	22.5	24.3	23		18.3	21.5	20.7	20.7	24#	19.6	19.1	19
PAE_{1dB} (%)	27.6	38.2	26.8	26.2	24.4	39.1		20.4	39.9	29.5	26.3	N/A	30.5	42.5	47.1
PAE@6dB PBO (%)	18.1	34.5	21.9	20.4	20.4	17.1		13.4	32.8	23.9	19.5	15#	24.2	19.5*	27*
PAE@12dB PBO (%)	13.8	22.1	12.5	10.7	14.2	11.0[a]		N/A	N/A	N/A	N/A	10#a	N/A	9*	10*
Modulation	64-QAM		256-QAM		64-QAM	5G NR 1-CC 64-QAM OFDM	5G NR 2-CC 64-QAM OFDM	64-QAM		5G NR FR2		64-QAM OFDM	64-QAM	5G NR FR2 64-QAM 2-CC OFDM	5G NR FR2 64-QAM 1-CC
Data Rate (Gb/s)	1.5		0.8		1.5	0.6	0.6	3		1.2		1.8	18	4.8	1.2
Freq. (GHz)	33				28	38	38	32.5		32.5		29.5	36	28	30
EVMrms (dB)	-25.3		-31.3		-25.2	-25	-25.1	-25		-25.4		-27.58	-25.11	-25.1	-25.05
$Pout_{avg}$ (dBm)	16.1		13.5		17.7	11.3	10	13.4		9.5		7.9	15.5	11.3	11.4
PAE_{avg} (%)	30.7		26.2		17.5	14.7	13.4	24.8		15.5		8#	20	16.6	17

*Estimated from figures   **[a]At 11.5dB PBO**

#Including the power of DPAs, DPMs, and digital circuits excluding SRAM

achieving >17.8dbm p1db and >36.6% paep1db over 24 to 40ghz and continuously supporting 64-/256-qam 5g nr signals over 24 to 42ghz," in *2020 IEEE International Solid- State Circuits Conference - (ISSCC)*, 2020, pp. 372–374.

[2] E. F. Garay, D. J. Munzer, and H. Wang, "26.3 a mm-wave power amplifier for 5g communication using a dual-drive topology exhibiting a maximum pae of 50% and maximum de of 60% at 30ghz," in *2021 IEEE International Solid- State Circuits Conference (ISSCC)*, vol. 64, 2021, pp. 258–260.

[3] S. Shakib, H.-C. Park, J. Dunworth, V. Aparin, and K. Entesari, "20.6 a 28ghz efficient linear power amplifier for 5g phased arrays in 28nm bulk cmos," in *2016 IEEE International Solid-State Circuits Conference (ISSCC)*, 2016, pp. 352–353.

[4] W. Doherty, "A new high efficiency power amplifier for modulated waves," *Proceedings of the Institute of Radio Engineers*, vol. 24, no. 9, pp. 1163–1182, 1936.

[5] Z. Ma, K. Ma, K. Wang, and F. Meng, "A 28ghz compact 3-way transformer-based parallel-series doherty power amplifier with 20.4%/14.2% pae at 6-/12-db power back-off and 25.5dbm psat in 55nm bulk cmos," in *2022 IEEE International Solid- State Circuits Conference (ISSCC)*, vol. 65, 2022, pp. 320–322.

[6] X. Zhang, S. Li, D. Huang, and T. Chi, "A 38ghz deep back-off efficiency enhancement pa with three-way doherty network synthesis achieving 11.3dbm average output power and 14.7% average efficiency for 5g nr ofdm," in *2022 IEEE Radio Frequency Integrated Circuits Symposium (RFIC)*, 2022, pp. 239–242.

[7] H. T. Nguyen, S. Li, and H. Wang, "4.6 a mm-wave 3-way linear doherty radiator with multi antenna coupling and on-antenna current-scaling series combiner for deep power back-off efficiency enhancement," in *2019 IEEE International Solid- State Circuits Conference - (ISSCC)*, 2019, pp. 84–86.

[8] D. J. Shepphard, J. Powell, and S. C. Cripps, "An efficient broadband reconfigurable power amplifier using active load modulation," *IEEE Microwave and Wireless Components Letters*, vol. 26, no. 6, pp. 443–445, 2016.

[9] V. Qunaj and P. Reynaert, "A ka-band doherty-like lmba for high-speed wireless communication in 28-nm cmos," *IEEE Journal of Solid-State Circuits*, vol. 56, no. 12, pp. 3694–3703, 2021.

[10] N. S. Mannem, T.-Y. Huang, and H. Wang, "Broadband active load-modulation power amplification using coupled-line baluns: A multifrequency role-exchange coupler doherty amplifier architecture," *IEEE Journal of Solid-State Circuits*, vol. 56, no. 10, pp. 3109–3122, 2021.

[11] T.-Y. Huang, N. S. Mannem, S. Li, D. Jung, M.-Y. Huang, and H. Wang, "26.1 a 26-to-60ghz continuous coupler-doherty linear power amplifier for over-an-octave back-off efficiency enhancement," in *2021 IEEE International Solid- State Circuits Conference (ISSCC)*, vol. 64, 2021, pp. 354–356.

[12] N. S. Mannem, M.-Y. Huang, T.-Y. Huang, S. Li, and H. Wang, "24.2 a reconfigurable series/parallel quadrature-coupler-based doherty pa in cmos soi with vswr resilient linearity and back-off pae for 5g mimo arrays," in *2020 IEEE International Solid- State Circuits Conference - (ISSCC)*, 2020, pp. 364–366.

[13] M. Mortazavi, Y. Shen, D. Mul, L. C. N. de Vreede, M. Spirito, and M. Babaie, "A four-way series doherty digital polar transmitter at mm-wave frequencies," *IEEE Journal of Solid-State Circuits*, vol. 57, no. 3, pp. 803–817, 2022.

RMo4C-2

A 26 GHz Balun-First Three-Way Doherty PA in 40 nm CMOS with 20.7 dBm Psat and 20 dB Power Gain

Anil Kumar Kumaran[#], Masoud Pashaeifar[#], Hossein Mashad Nemati[*], Leo C.N. de Vreede[#], Morteza S. Alavi[#]

[#]Electronic Circuits and Architecture (ELCA) Research Group,
Delft University of Technology, The Netherlands
[*]Huawei Technologies, Sweden
a.k.kumaran@tudelft.nl

Abstract — This paper presents a 40 nm CMOS mm-wave 3-way Doherty power amplifier (PA) suitable for 5G mm-wave transmitters. It features a bandwidth-enhanced technique using a compact single-supply balun-first 3-way Doherty combiner. The realized front-end with a core area of $0.77\,mm^2$ delivers a peak power/gain of more than 20 dBm/16 dB and a drain efficiency (DE) of better than $15\%/22\%/33\%$ at 9.5 dB/6 dB/0 dB power back-off across a 24-to-30 GHz band. At 26 GHz, it achieves an EVM/ACLR of -23.5 dB/-29.5 dBc for an 800 MHz 64-OFDM signal with 9.8 dBm average output power and a 15% average DE.

Keywords — Doherty, 3-stage Power amplifier, Compact, Millimeter wave, Lumped components, Norton transformation.

I. INTRODUCTION

Millimeter-wave (mm-wave) 5G transmitters (TXs) are key enablers to streamline a multi-Gbit/s data throughput. In these mm-wave TXs, high integration, compact area, low cost, and high yield are prime reasons for exploiting nanoscale CMOS technologies. Moreover, they generally utilize spectrally efficient complex modulation schemes with high peak-to-average power ratios (PAPRs) that entail the TX operating in deep power back-off (PBO), thus degrading its average efficiency. Also, these mm-wave 5G front-ends must provide sufficient power gain and radiated power so as to relax the design constraints for the preceding up-converting stages and overcome free space path loss. Accordingly, techniques such as N-way Doherty configuration [1]–[8], comprising main and peak power amplifiers (PAs), are investigated to deliver the required TX power levels and power gain with decent efficiency at peak and PBOs.

The recently published 4-way Doherty combiner for sub-6 GHz operation [5] appears superior to conventional 2-way Doherty PAs because in deep PBO, when only the main PA is active, the 4-way Doherty has a direct (uncompromised) signal path between the main PA output and the overall output of the Doherty combiner, which minimizes power loss. Furthermore, in this configuration, only the off-state impedance of the peak-1 PA affects Doherty PA's efficiency. In contrast, at mm-wave frequencies, including practical constraints for the losses in an N-way combiner, analysis of the N-way Doherty PAs shows that 3-way Doherty PA is the best candidate since at 12 dB PBO its performance is comparable to 4-/5-way Doherty networks while offering a more compact combiner and lower design complexity [8]. In addition, a 2-way Doherty PA doesn't fully address the

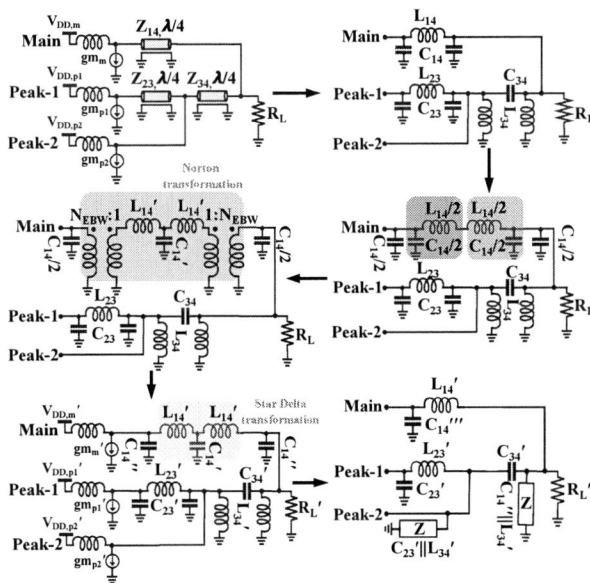

Fig. 1. The proposed compact EWB technique on a 3-way Doherty combiner.

issue of average efficiency improvement for signals with a large PAPR, while all output power must be generated using only two power devices. Furthermore, practical 3-way Doherty PAs have moderate bandwidth for drain efficiency/output power at deep PBO and full power. Besides, they typically exploit two supply voltages in their front-end stages to obtain sufficient power gain and radiated power. This paper proposes a compact single-supply three-stage balun-first 3-way Doherty PA operating in a 25-to-26 GHz band.

II. DESIGN OF THE BALUN-FIRST 3-WAY DOHERTY

Figure 1 unveils progressively how to implement the proposed 3-way Doherty network featuring an enhanced bandwidth technique (EBW). This technique is implemented on the main PA's path as it is the most dominant power loss segment. First, the three transmission lines (TLs) of the 3-way combiner are replaced by their low-pass (LP) and high-pass (HP) lumped element equivalents. Next, the inductor and the capacitors in the main path are split, and a Norton transformation [9] is applied. Subsequently, the transformer obtained by Norton transformation is supplanted by lowering the main supply source (V_{DD}) by N_{EBW}, increasing the transconductance of the power device (gm) by N_{EBW},

979-8-3503-2123-4/23 $31.00 © 2023 IEEE   189   2023 IEEE Radio Frequency Integrated Circuits Symposium

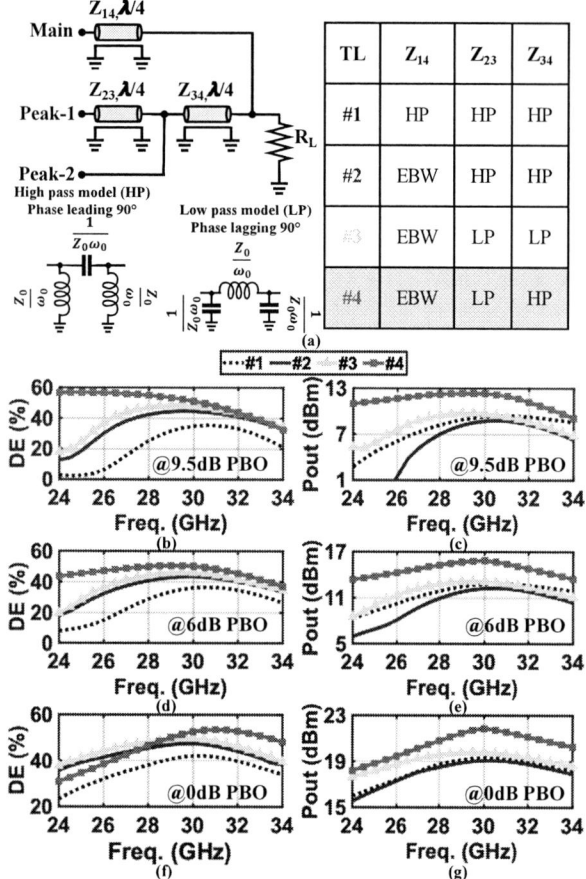

Fig. 2. (a) The 3-way Doherty structure and its HP and LP options for implementing the TLs, (b) DE, (c) output power, across frequency at 9.5 dB PBO, (d) DE, (e) output power, across frequency at 6 dB PBO, (f) DE, and (g) output power, across frequency at full power.

increasing capacitors by N_{EBW}^2, and reducing inductors and impedance seen at the drain node by N_{EBW}^2. Finally, a star-to-delta transformation makes the 3-way combiner even more compact.

Figure 2a exhibits different variations of 3-way Doherty, which is obtained by alternating between the LP and HP models for the TLs. All simulations are performed with an ideal PA model and a quality factor of 25/15 for the capacitors and inductors, showing that alternative #4 in Fig. 2a has the best performance across frequency among its peers at 9.5/6/0 dB PBO. Intuitively, its equivalent passive networks' magnitude and phase variation are lower than the others since it comprises a combination of LP and HP circuits in the peak PAs, similar to an inverted 2-way Doherty PA [10], thus enhancing its operational bandwidth. Furthermore, having an LP structure adjacent to the active power devices absorbs the parasitic drain-source capacitance. Also, push-pull PAs traditionally incorporate a balanced-to-unbalanced (balun) transformer required for a single-ended antenna. However, the push-pull PAs don't see identical impedances in such designs due to unwanted coupling between the various traces of the balun. Besides, the final circuit in Fig. 1 needs additional

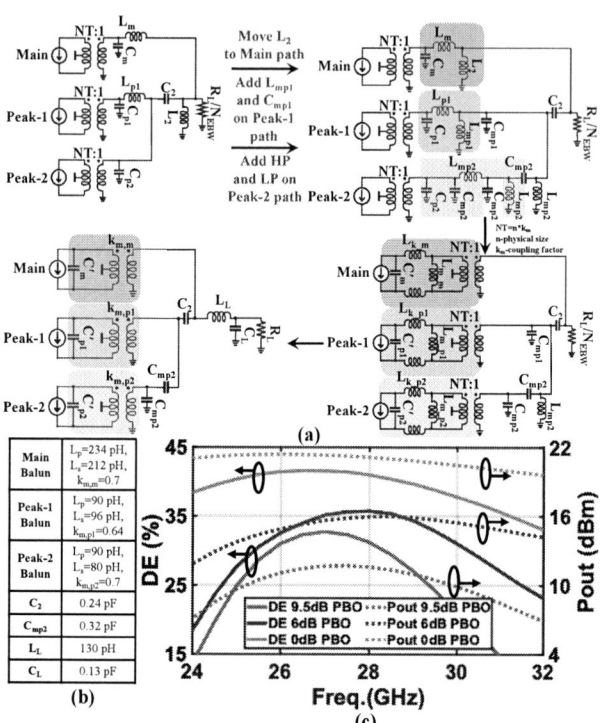

Fig. 3. (a) Procedure to design the proposed balun-first 3-way Doherty, (b) components' value, and (c) DE and output power across frequency.

RF chokes to supply DC voltages to the main and peak-1 PA, increasing the complexity of the circuit. To mitigate these issues, the output balun is split into three baluns and repositioned to the drain of the PAs [1].

Figure 3a illustrates the developing steps for designing a balun-first 3-way Doherty combiner. First, L_2 is moved to the main path. Likewise, L_{mp1}/C_{mp1} are added to the peak-1 path. Also, LP/HP models are added to the peak-2 path. Next, C_m, L_m, and L_2 are repositioned to the drain side of the main PA. Similarly, C_{p1}, L_{p1}, L_{mp1}, C_{p2}, C_{mp2}, L_{mp2}, and L_{mp2} are relocated to the drain side of peak-1/peak-2 PA. Afterward, the ideal transformer, magnetizing (L_m), and leakage (L_k) inductance are superseded with practical balun transformers with coupling factors of $k_{m,m}/k_{m,p1}/k_{m,p2}$. Eventually, an L-match (L_L and C_L) is utilized to transform R_L/N_{EBW} to R_L. The final circuit is compact, with only one inductor, four capacitors, and three baluns whose values are shown in the Fig. 3b. The baluns' turn ratio (NT), C_{mp1}, and L_{mp2} are design parameters that can be adjusted. C_{mp1} and L_{mp2} are chosen to resonate at the desired operational frequency. Additionally, single-turn baluns are exploited to achieve higher self-resonance. This arrangement demands that the NT becomes 0.67 to obtain a similar physical size (n) for the balun's primary (L_p) and secondary (L_s) inductors, making them easier to implement. The simulations with an ideal PA model, EM model of the proposed 3-way Doherty combiner, and capacitors with a quality factor (Q_C) of 25 provide drain efficiency (DE) greater than 15/20/37 % at 9.5/6/0 dB PBO across the 24-to-30 GHz band (Fig. 3c).

979-8-3503-2123-4/23 $31.00 © 2023 IEEE

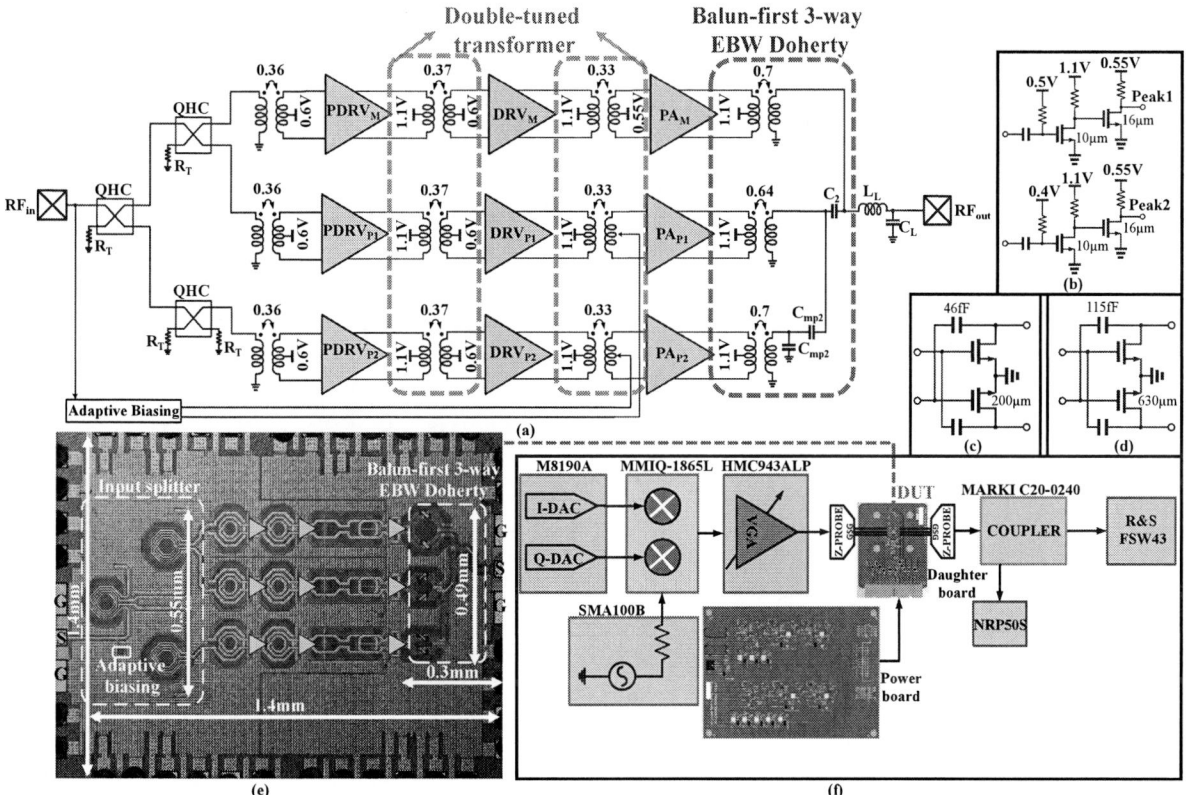

Fig. 4. (a) Top-level of the proposed PA, (b) a detailed schematic of adaptive biasing, (c) PDRV/DRV with a bias of 0.6 V, (d) PA with a bias of 0.55 V, (e) die micrograph of the proposed balun-first 3-way Doherty PA, and (f) measurement setup.

III. CIRCUIT IMPLEMENTATION

Figure 4a depicts a detailed schematic of the proposed PA. The phase advance of 90° and 180° for the 3-way Doherty PA is generated using three quadrature hybrid couplers (QHCs) which also provide wideband input matching. Each branch (main/peak-1/peak-2) consists of an input balun, pre-driver (PDRV), inter-stage matching, driver (DRV), inter-stage matching, and a PA. The required power gain is achieved with a single-supply, three-stage neutralized common source PA (PDRV/DRV/PA in Fig. 4c/d using a bias of 0.6 V and 0.55 V) but at the expense of lower power-added efficiency (PAE). The inter-stage matching utilizes a double-tuned transformer network to obtain wideband matching. Furthermore, the biases of the peak PAs are modulated by using turn-on voltage of 0.5 V and 0.4 V in the adaptive biasing circuits to perform Doherty load modulation (Fig. 4b).

IV. MEASUREMENT RESULTS

The proposed PA is fabricated in 40 nm bulk CMOS technology (Fig. 4e). The core area of the proposed balun-first 3-way Doherty is $1.4 \times 0.55 \text{ mm}^2$. Figure 4f shows the measurement setup. The small-signal s-parameters measurement results of the chip show that the proposed PA achieves more than 5 GHz 3 dB S21 bandwidth while S11/S12 are less than $-10/-50$ dB over a 23-to-28 GHz band (See Fig. 5a). Figure 5b indicates that the proposed PA achieves

more than 20 dBm peak output power and a DE of better than 10 %/15 %/22 %/33 % at 12 dB/9.5 dB/6 dB/0 dB across the 24-to-30 GHz band. Likewise, Fig. 5c exhibits the measured DE and gain versus output power across the 25-to-27 GHz band, indicating active load modulation over the 25-to-26 GHz band. Note that the operational frequency is limited due to the high impedance transformation ratio required by the interstage between DRV and PA. Figure 6a exhibits 9.8 dBm/15 % average output power/DE are measured for a 4.8 Gb/s OFDM 64-QAM signal with 9.6 dB PAPR. Its EVM/ACLR are -23.5 dB/-29.5 dBc, respectively. Furthermore, the spectral purity and constellation of a 400 MHz OFDM 64-QAM signal is measured at 26, 25, and 27 GHz with EVM/ACLR of -24.9 dB/-28.9 dBc, -24.1 dB/-30.7 dBc and -25.4 dB/-30 dBc which are illustrated in Fig. 6b/c/d. The performance of the proposed balun-first 3-way Doherty PA is summarized in Table 1 and compared to that of the prior art. It indicates that our compact single-supply three-stage mm-wave front-end, which operates over the 24-to-30 GHz band without using any digital pre-distortion can handle 4.8 Gb/s while delivering more than 20 dBm/16 dB/33 % peak output power/power gain/DE, respectively.

V. CONCLUSION

This work proposes a single-supply three-stage balun-first mm-wave 3-way Doherty PA with required bias voltages. Implemented in 40 nm CMOS with a core area of

979-8-3503-2123-4/23 $31.00 © 2023 IEEE   191

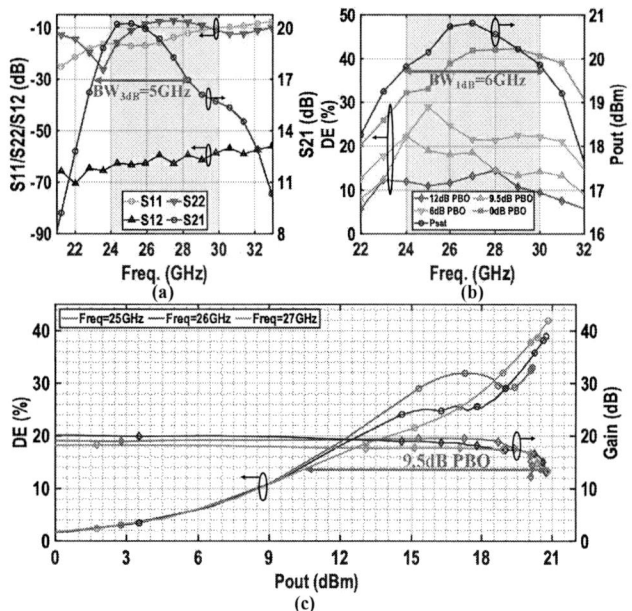

Fig. 5. (a) Small-signal S-parameters measurement results, (c) peak output power and DE across frequency at 12 dB PBO, 9.5 dB PBO, 6 dB PBO, full power, and (c) DE/gain versus output power.

Table 1. Performance summary and comparison to prior art

Specifications	This Work	Z.Ma ISSCC'22 [1]	X.Zhang RFIC'22 [2]	Pashaeifar ASSCC'21 [6]	Huang ISSCC'21 [3]
Architecture	Balun-first 3-way Parallel Doherty PA	3-way Parallel Series Doherty PA	Coupled-inductor based 3-way Doherty PA	2-step impedance inversion series Doherty PA	Continuous Coupler Doherty PA
PA structure	3-stage PAs	2-stage PAs	2-stage PAs	2-stage PAs	2-stage PAs
Technology	40 nm CMOS	55 nm CMOS	45 nm SOI	40 nm CMOS	45 nm SOI
Supply	1.1 V	2.4 V, 1.2 V+	1.8 V, 1 V	1.8 V, 0.9 V	2 V, 1 V
Freq. (GHz)	24-30	28	38	24 to 32	26 to 60
Core Area (mm²)	0.77	0.54	1.4	0.37	0.62
Gain (dB)	20 (26 GHz)	16.1	15	17.4	15.5*
P_{sat} (dBm)	20.7 (26 GHz)	25.5	18.9	20.4	22
DE_{0dB}/PAE_{0dB} (%)	39/22.3 (26 GHz)	32.5/25.2	31*/23.3	46*/38.2	NA/40.5
DE_{6dB}/PAE_{6dB} (%)	24.7/11.7 (26 GHz)	25.4/20.4	20*/17.1	39*/34	NA/32.5
$DE_{9.5dB}/PAE_{9.5dB}$ (%)	18.1/7 (26 GHz)	21*/17	17*/13.7	25*/20*	NA/15*
DE_{12dB}/PAE_{12dB} (%)	11.7/4.2 (26 GHz)	18.2/14.2	15*/10*	10*/10*	NA/10*
Modulation Scheme	64/64/256/1024 QAM OFDM (26GHz)	64 QAM	64 QAM OFDM	64 QAM OFDM	64 QAM
Data rate (Gb/s)	4.8/2.4/1.6/0.5	1.5*	0.6	0.6	3
Modulation BW (MHz)	800/400/200/50	250	100	100	500
EVM_{rms} (dB)	-23.5/-24.9/-26.8/-30	-25.2	-25*	-24	-25
ACLR (dBc)	-29.5/-28.9/29.9/-36	-27	-26.5*	-27.7	-28.8
P_{avg} (dBm)	9.8/9.4/9.3/8.6	17.7	11.3	9.35	13.4
PAE_{avg} (%)	15/14/13/12 (DE)	17.5	14.7	16.4	24.8
DPD	No	No	No	No	No

#limited by equipment. *Graphically estimated. +Nominal voltage of the technology.

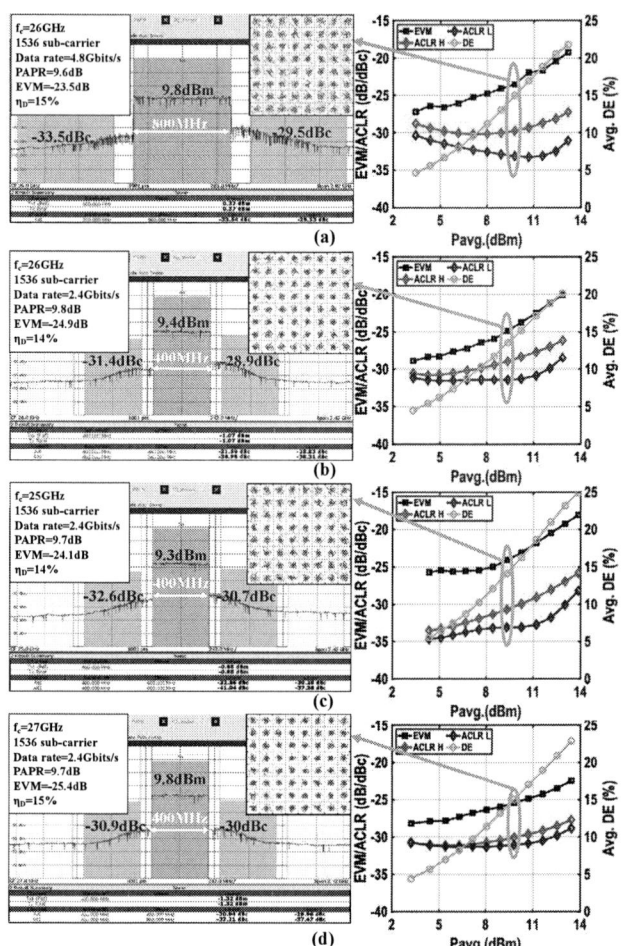

Fig. 6. (a) An OFDM 800 MHz 64-QAM at 26 GHz, (b) an OFDM 400 MHz 64-QAM at 26 GHz, (c) 25 GHz, and (d) 27 GHz measurement results.

$0.77\,mm^2$, the realized front-end at 26 GHz exhibits a power gain of 20 dB, a peak power of 20.7 dBm, and a DE of 39%/24.7%/18.1% at 0 dB/6 dB/9.5 dB PBO. It achieves EVM/ACLR of $-24.9\,dB/-28.9\,dBc$ for a 400 MHz 64-OFDM signal with 9.4 dBm average output power and a 14% average DE, making it an excellent candidate to exploit in 5G mm-wave TXs.

Acknowledgment

The authors would like to thank A. Akhnoukh, Z.Y. Chang, Dr. M. Spirito, C. De Martino, E. Shokrolahzade, Dr. Bueno, M.A. Montazerolghaem, G.D. Singh, and M.R. Beikmirza for their support and helpful discussions. They also thank IMEC Leuven for handling the tape-out.

References

[1] Z. Ma *et al.*, "A 28GHz Compact 3-Way Transformer-Based Parallel-Series Doherty Power Amplifier With 20.4%/14.2% PAE at 6-/12-dB Power Back-off and 25.5dBm PSAT in 55nm Bulk CMOS," in *IEEE ISSCC*, vol. 65, pp. 320–322, 2022.

[2] X. Zhang *et al.*, "A 38GHz Deep Back-Off Efficiency Enhancement PA with Three-Way Doherty Network Synthesis Achieving 11.3dBm Average Output Power and 14.7% Average Efficiency for 5G NR OFDM," in *IEEE RFIC*, pp. 239–242, 2022.

[3] T.-Y. Huang *et al.*, "A 26-to-60GHz Continuous Coupler-Doherty Linear Power Amplifier for Over-An-Octave Back-Off Efficiency Enhancement," in *IEEE ISSCC*, vol. 64, pp. 354–356, 2021.

[4] M. Pashaeifar *et al.*, "A Millimeter-Wave Mutual-Coupling-Resilient Double-Quadrature Transmitter for 5G Applications," *IEEE JSSC*, vol. 56, no. 12, pp. 3784–3798, 2021.

[5] M. Beikmirza *et al.*, "A Wideband Four-Way Doherty Bits-In RF-Out CMOS Transmitter," *IEEE JSSC*, vol. 56, no. 12, pp. 3768–3783, 2021.

[6] M. Pashaeifar *et al.*, "A 24-to-32GHz series-Doherty PA with two-step impedance inverting power combiner achieving 20.4dBm Psat and 38%/34% PAE at Psat/6dB PBO for 5G applications," in *IEEE A-SSCC*, pp. 1–3, 2021.

[7] M. Pashaeifar *et al.*, "A Millimeter-Wave CMOS Series-Doherty Power Amplifier With Post-Silicon Inter-Stage Passive Validation," *IEEE JSSC*, vol. 57, no. 10, pp. 2999–3013, 2022.

[8] A. K. Kumaran *et al.*, "Compact N-Way Doherty Power Combiners for mm-wave 5G Transmitters," in *IEEE ISCAS*, pp. 438–442, 2022.

[9] V. Bhagavatula *et al.*, "A Compact 77% Fractional Bandwidth CMOS Band-Pass Distributed Amplifier With Mirror-Symmetric Norton Transforms," *IEEE JSSC*, vol. 50, no. 5, pp. 1085–1093, 2015.

[10] S. Hu *et al.*, "A 28GHz/37GHz/39GHz multiband linear Doherty power amplifier for 5G massive MIMO applications," in *IEEE ISSCC*, pp. 32–33, 2017.

RMo4C-3

A 26-GHz Linear Power Amplifier with 20.8-dBm OP1dB Supporting 256-QAM Wideband 5G NR OFDM for 5G Base Station Equipment

Zhilin Chen[#*], Xiyu Wang[#$], Xiaoxiao Ma[#*], Min Lu[#*], Jie Hu[#*], Keqing Ouyang[#*], Zhijun Long[#*]

[#]State Key Laboratory of Mobile Network and Mobile Multimedia Technology, China

[*]Analog Design Dept., Sanechips Technology Co., Ltd., China

[$]ZTE Corparation, China

{lu.min1, hu.jie30}@sanechips.com.cn

Abstract—This paper presents a high output power and linearity power amplifier (PA) in 65nm CMOS SOI process for 5G communications. A two-way power combining topology with peaking inductive technique is utilized for high output power. Additionally, the PMOS compensate capacitor, second-order harmonic traps and low impedance network are proposed to improve linearity for wideband modulation. The PA achieves a gain of 19 dB at 26 GHz with 3-dB bandwidth from 22.3 GHz to 28.6 GHz. The PA also realizes 20.8 dBm OP1dB and 21.3 dBm Psat with a peak PAE of 26.15% at 26 GHz. In modulation signal test, using a 5G NR 400 MHz 1-CC 64-QAM and 256-QAM OFDM signal, this PA demonstrates 5% and 3% rms error vector magnitude (EVM) with average output power (Pavg) of 15.5 dBm and 14.4 dBm, respectivly.

Keywords—26GHz, fifth generation (5G), new radio (NR), power amplifier, 256-QAM, linearity.

I. INTRODUCTION

With the rapid growth of fifth-generation (5G) wireless communication, rigorous performance requirements for tranceiver frontend have been steadily increased. From 3GPP R17, the modulation scheme requires up to 256 QAM with EVM < 3.5%, and the bandwidth of single carrer component (CC) up to 400 MHz for 5G new radio (NR) FR2 [1]. Furthermore, the average EIRP of 5G base station (BS) equipment has exceeded 65 dBm nowadays [2]. Therefore, delivering sufficient average output power (Pavg) per channel is highly recommended to reduce the size and cost of the phase array. Typically, more than 12 dBm Pavg is required for each channel in a 256-element phase array system [2]. Such requirements leads to big challenges in power amplifier (PA) design.

Recently, many mm-wave PA design approaches have been reported [3]-[8]. In [3]-[4], class-AB PAs has good linearity, but the output power is limited. Power combining techniques are commonly used to improve the output power of PAs, but those PAs [5]-[6] are difficult to support the 5G NR 256-QAM OFDM without linearity enhancement. In addition, the combining network suffers from amplitude and phase imbalance to decrease the power combining effeciency. Doherty PA has also attracted a lot of attention due to its advantages in back-off efficiency, but its modulation bandwidth and linearity are extremely challenging [7]-[8].

In this paper, a high linearity and output power PA is presented to address above challenges. To the best of our knowledge, the proposed PA firstly supports 3% EVM with 256-QAM 5G NR 400MHz OFDM, achieving Pavg more than 13 dBm compared with other state of art silicon PAs.

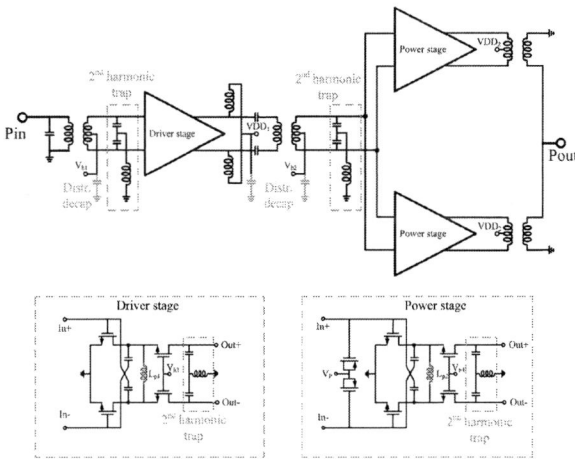

Fig. 1. Schematic of the proposed PA.

II. CIRCUIT DESIGN

Fig.1 shows the shematic of the proposed PA, which consists two-stage differential cascode amplifier with neutralized technique. To improve the output power and linearity of the PA, both the driver stage and output stage use 2nd harmonic traps and a parallel peaking inductor at the intermediate node between the cascode transistors. Additionally, a PMOS-based capacitor is added at the input of the power amplifier. A transformer-based parallel power combiner is utilized in 2nd stage. In the following sections, several techniques are introduced and analyzed.

A. Linearization Techniques

Based on [4]-[5], PA nonlinearity significantly limits modulation bandwidth and output power mainly due to amplitude-to-amplitude (AM-AM) distortion, amplitude-to-phase (AM-PM) distortion, and memory effects. The memory effects are mostly contributed by the 2nd-order harmonic distortion as well as the baseband impedance [9]. In this work, some linearization techniques are used to alleviate above nolinearity.

1)To compensate AM-PM distortion, the PMOS-based capacitor is employed in power stage amplifier. The PMOS-based capacitor shows the opposite variation characteristics

979-8-3503-2123-4/23 $31.00 © 2023 IEEE   193   2023 IEEE Radio Frequency Integrated Circuits Symposium

versus the input power with respect to the input capacitance of NMOS transistors [5]. Therefore, by adding the PMOS-based capacitor at input node, the change trend of the input capacitor remains constant with the input power. The control voltage V_p is used to tune the capacitance of the PMOS to actually compensate the capacitance variation of NMOS. It is worth that, this technology is only used in the power stage amplifier to reduce the gain loss, where the main AM-PM distortion comes from.

To mitigate the influence of the memory effects, the 2nd-order harmonic traps are introduced into the input and output matching network both in driving and power stages, as shown in Fig. 1. The unwanted common-mode 2nd-order harmonic component is resonanted by the LC trap to ground without additional sacrifice for differential foundamental signal [4]. Besides, the distributed capacitance arrays composed of capacitors with different sizes to form large decoupling capacitors are employed in the bias and VDD supply paths, offering a wideband low impedance to ground for the beat modulation signal.

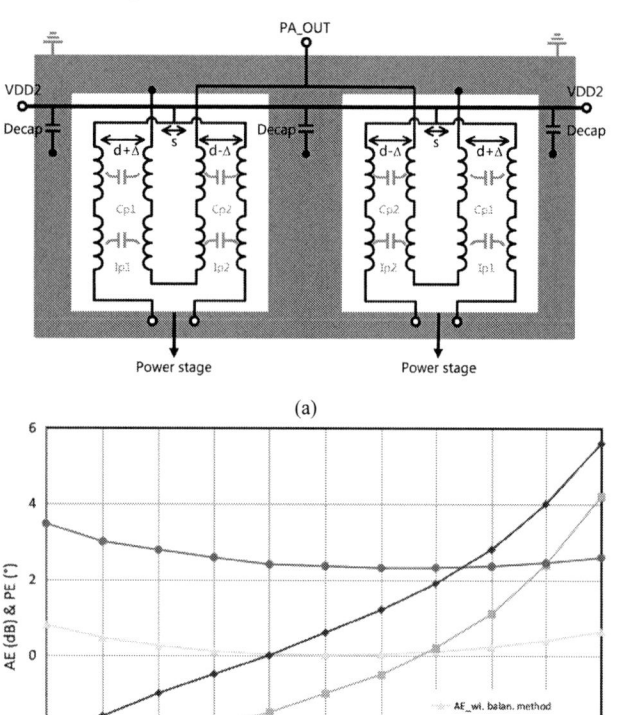

(a)

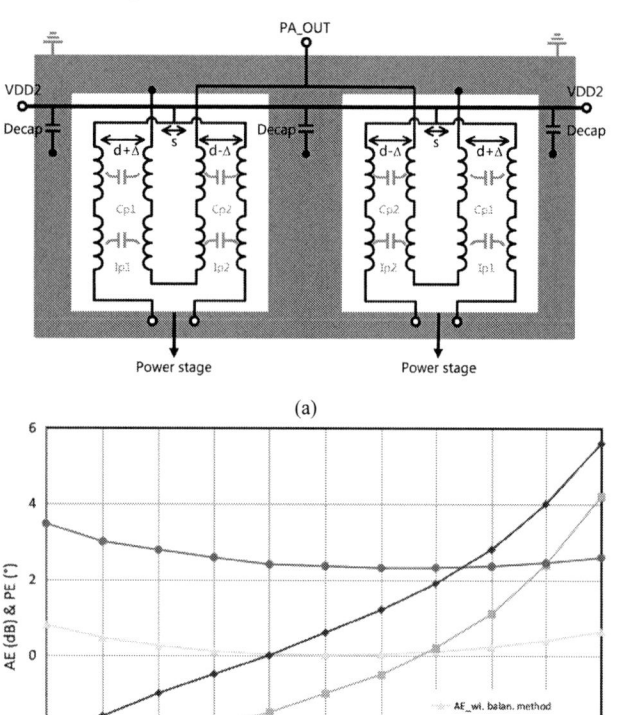

(b)

Fig. 2. (a) Structure of the proposed power combiner. (b) Simulated AE and PE comparision.

B. High Output Power Techniques

To improve the output power, the parallel peaking inductor is applied to resonant the parasitic capacitance at the node of the cascode structure. In addition, a two-way parallel

power combiner is proposed in the 2nd stage. Theoretically, it improves the power of 3 dB, however, the power combining performance heavily dependents on the balance performance of the power combiner. In this work, a balance compensation technique for the combiner is introduced, as shown in Fig.2a. A ground plane is added between two transformers to reduce the coupling and improve the balance. Another critical imbalance comes from the balun structure, where one port of the unbalanced coil is connected to the load and the other port is grounded. The parasitic capacitors (C_{p1} and C_{p2}) between primary and secondary coils lead to the unequal parasitic currents ($I_{p1} \neq I_{p2}$) due to the unequel loads, resulting in amplitude imbalance. As shown in Fig. 2a, a simplified method is used to change the C_{p1} and C_{p2} through shifting the overlapping region (d ± Δ shown in Fig. 2a) between primary and secondary coils, so that the parasitic currents can be equal ($I_{p1} = I_{p2}$). Therefore, the amplitude can achieve balanced. Meanwhile, the phase imbalance can be also optimized by moving the position of the center tap [10]. Eventually, the decaping in the VDD path should be as large as possible. Fig 2b shows the simulated amplitude error (AE) and phase error (PE) relative to 180° of the power combiner compared with and without above methods, where good balance performances are observed by using the above methods.

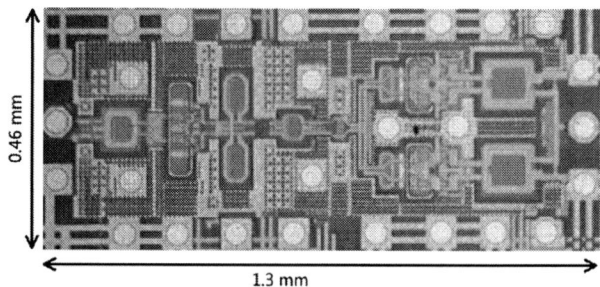

Fig. 3. Chip micrograph of the proposed PA.

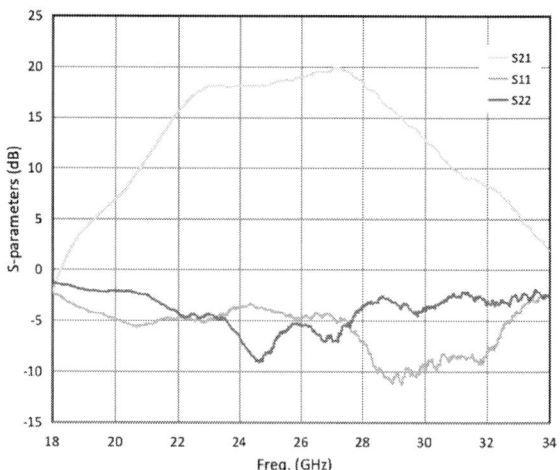

Fig. 4. Measured small-signal results.

III. MEASUREMENT RESULTS

The proposed PA is implemented in the 65 nm CMOS SOI process. The chip microphotograph is shown in Fig. 3, where

979-8-3503-2123-4/23 $31.00 © 2023 IEEE

the size is $1.3*0.46$ mm^2. Fig. 4 shows the measured small-signal results. with 2-V power supply, the PA achieves a gain of 19 dB at 26 GHz. The 3-dB bandwidth is from 22.3 GHz to 28.6 GHz. The S11 performance is around -5 dB within the operating frequency range, which is mainly due to the conjugate matching with the impedance of previous amplifier. The large-signal results are shown in Fig. 5 and Fig. 6. At 26 GHz, the PA realizes a OP1dB of 20.8 dBm and a Psat of 21.3 dBm with a peak PAE of 26.15%. From 24 GHz to 27.5 GHz, the PA also achieves over 20.2 dBm OP1dB, over 21 dBm Psat, and over 24% peak PAE.

In the modulation measurement, as shown in Fig. 7 to Fig. 9, the PA is featured using 5G NR FR2 400 MHz 1CC 64-QAM and 256-QAM OFDM. At 26 GHz, this PA achieves 5% EVM for 64 QAM with 15.5 dBm Pavg and 12.8% PAEavg, and 3% EVM for 256 QAM with 14.4 dBm Pavg and 9.1% PAEavg , respectivvely. From 24 GHz to 27.5 GHz, the Pavg and PAEavg are more than 14.5 dBm and 10.3% for 64 QAM, while the Pavg and PAEavg are more than 13 dBm and 7.2% for 256 QAM, respectively.

proposed to realize high output power and linearity. Compared with reported silicon PAs in Table 1, this work achieves the outstanding Psat and OP1dB, with highly competitive performances in terms of gain, PAEsat, Pavg, and EVM. More importantly, the PA firstly demonstrates the EVM of 3% and highest Pavg for 256 QAM using 5G NR FR2 400MHz OFDM, which satisfies the requirements of the lastest 3GPP standard for base station equipment.

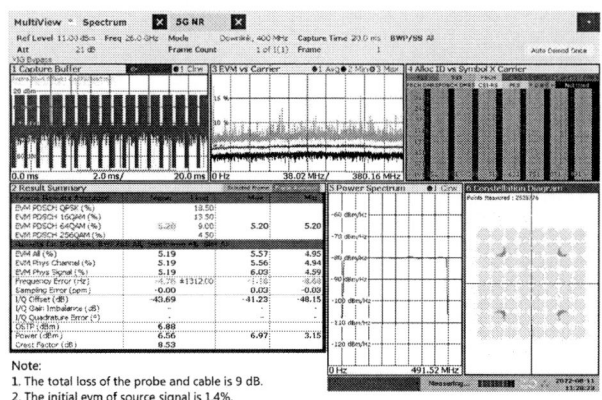

Fig. 7. Measured results for 5G-NR 400MHz 1-CC 64-QAM OFDM at 26 GHz.

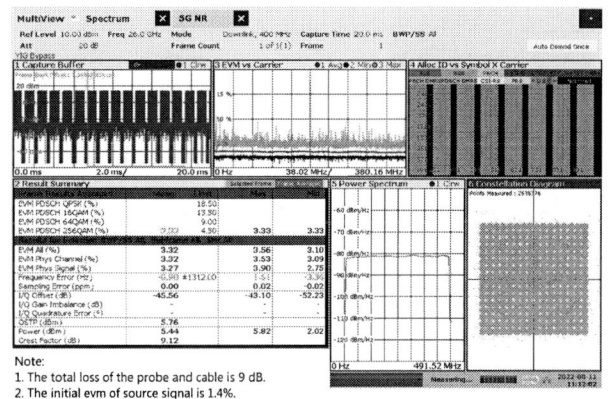

Fig. 8. Measured results for 5G-NR 400MHz 1-CC 256-QAM OFDM at 26 GHz.

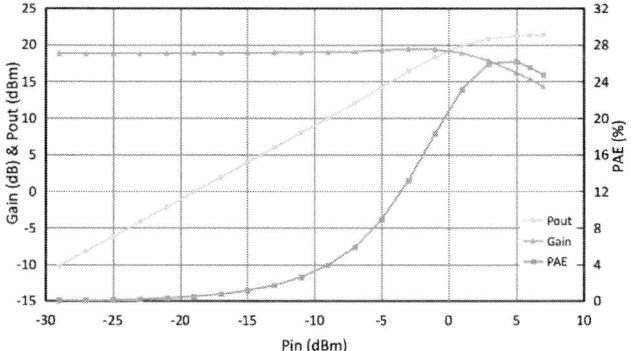

Fig. 5. Measured large-signal results at 26 GHz vs. input power (Pin).

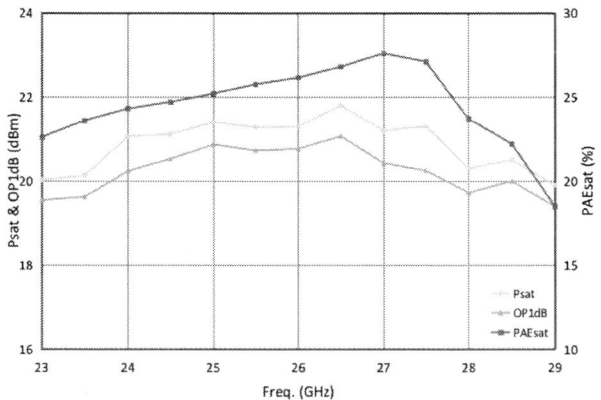

Fig. 6. Measured large-signal results vs. frequency.

IV. CONCLUSION

In this paper, a 26-GHz two-way power combined PA with peaking inductor, PMOS capacitor, high balance power combiner, and low impedance VDD/bias network are

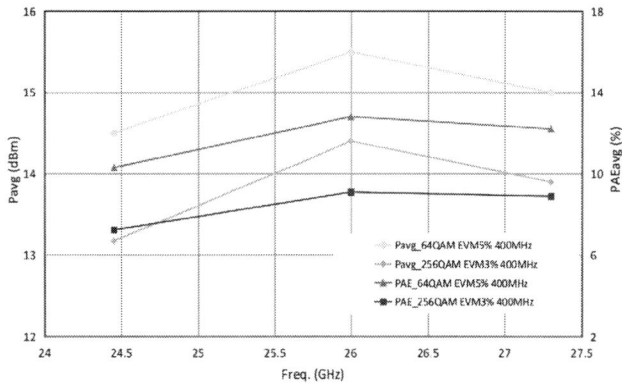

Fig. 9. Measued Pavg and PAEavg for 64QAM and 256QAM vs frequency.

Table 1. Performance Summary and Comparison.

References	This work		ISSCC 21 [3]	IMS 20 [4]	JSSC 18 [5]	SSCL 20 [6]	RFIC 21 [7]	RFIC 22 [8]
Technology	65-nm CMOS SOI		45-nm CMOS SOI	28-nm CMOS	28-nm CMOS	28-nm CMOS	28-nm CMOS	45-nm CMOS SOI
Freq. (GHz)	26		28	28	28	28	27	38
Gain (dB)	19		20.4	18.5	20.8	20.4	16.5	15
Psat (dBm)	21.3		20.1	18.9	16.6	21.5	18.8	18.9
OP1dB (dBm)	20.8		19.1	18.5	13.4	20.7	17.6	18.4
PAEsat (%)	26.15		48.3	39.7	24.2	26	30	23.3
Modulation Scheme	5G NR 1-CC 64-QAM OFDM	5G NR 1-CC 256-QAM OFDM	5G NR 64-QAM OFDM	256-QAM OFDM	64-QAM OFDM	64-QAM OFDM	64-QAM OFDM	5G NR 1-CC 64-QAM OFDM
Bandwidth (Hz)	400M	400M	200M	400M	675M	400M	100M	100M
Pavg (dBm)	15.5	14.4	10.7	6.6	6.8	11.4	12.4	11.3
EVM (%)	5	3	5.6	2.6	5.6	5.6	5.6	5.6
PAEavg (%)	12.8	9.1	15.5	5.6	2.9	6.2	20.2	14.7
Area (mm2)	0.6		0.21*	0.31*	0.16*	0.5	0.16*	1.4*

*Only active size

ACKNOWLEDGMENT

The authors would like to thank Yajuan Du, Yonghui Yang and Guojun Zhang in ZTE Corparation for technical discussions. They also thank Wei Zhao, Jiang Gao, Xiangmin Wang, Liheng He and all team members of RF Group II from Sanechips Technology Co., Ltd. for technical supports.

REFERENCES

[1] The 3GPP website. [Online]. Available :
https://www.3gpp.org/ftp/Specs/archive/38_series/38.104

[2] A. Khalil et al., "mm-Wave 5G Radios: Baseband to Waves," 2021 IEEE International Solid- State Circuits Conference (ISSCC), 2021, pp. 38-40.

[3] E. F. Garay, H. Wang, "A mm-Wave Power Amplifier for 5G Communication Using a Dual-Drive Topology Exhibiting a Maximum PAE of 50% and Maximum DE of 60% at 30GHz," 2021 IEEE International Solid- State Circuits Conference (ISSCC), 2021, pp. 258-260.

[4] Y. -W. Chang et al., "A 28 GHz Linear and Efficient Power Amplifier Supporting Wideband OFDM for 5G in 28nm CMOS," 2020 IEEE/MTT-S International Microwave Symposium (IMS), 2020, pp. 1093-1096.

[5] M. Vigilante and P. Reynaert, "A Wideband Class-AB Power Amplifier With 29 – 57-GHz AM – PM Compensation in 0.9-V 28-nm

Bulk CMOS," IEEE Journal of Solid-State Circuits, vol. 53, no. 5, pp. 1288-1301, May 2018.

[6] D. Manente et al., "A 28-GHz Stacked Power Amplifier with 20.7-dBm Output P1dB in 28-nm Bulk CMOS," IEEE Solid-State Circuits Letters, vol. 3, pp. 170-173, 2020.

[7] S. Kim et al., "A 24.5 – 29.5GHz Broadband Parallel-to-Series Combined Compact Doherty Power Amplifier in 28-nm Bulk CMOS for 5G Applications," 2021 IEEE Radio Frequency Integrated Circuits Symposium (RFIC), 2021, pp. 171-174.

[8] X. Zhang et al., "A 38GHz Deep Back-Off Efficiency Enhancement PA with Three-Way Doherty Network Synthesis Achieving 11.3dBm Average Output Power and 14.7% Average Efficiency for 5G NR OFDM," 2022 IEEE Radio Frequency Integrated Circuits Symposium (RFIC), 2022, pp. 239-242.

[9] N. Borges de Carvalho and J. C. Pedro, "A comprehensive explanation of distortion sideband asymmetries," IEEE Transactions on Microwave Theory and Techniques, vol. 50, no. 9, pp. 2090-2101, Sept. 2002.

[10] P. Huang et al., "Design and modeling of an on-chip asymmetric tap balun for CMOS millimeter-wave circuits," 2014 IEEE International Wireless Symposium (IWS 2014), 2014, pp. 1-4.

RMo4C-4

A 23-30 GHz 4-Path Series-Parallel-Combined Class-AB Power Amplifier with 23 dBm P$_{sat}$, 38.5% Peak PAE and 1.3° AM-PM Distortion in 40nm Bulk CMOS

Junjie Gu[#], Haoqi Qin[#], Hao Xu[#], Weitian Liu[#], Kefeng Han[$], Rui Yin[#], Lei Deng[&], Xiaoliang Shen[&], Zongming Duan[^], Hao Gao[%], Na Yan[#]

[#]Fudan University, China
[$]Jiashan Fudan Institute, China
[&]National Integrated Circuit Innovation Center, China
[^]East China Research Institute of Electronic Engineering, China
[%]Eindhoven University of Technology, the Netherlands
haoxu@fudan.edu.cn, yanna@fudan.edu.cn

Abstract—This paper presents a 4-path series-parallel combined highly-efficient class-AB power amplifier (PA) with broad bandwidth and low AM-PM distortion in CMOS process. Frequency staggered tuning scheme enables a wide passband of 23-30GHz. AM-PM distortion is minimized by utilizing PMOS varactors that mitigate the voltage dependence of transistor intrinsic capacitors and harmonic traps that minimize common-mode voltage swings at the second-harmonic frequency. Complete electromagnetic modeling ensures the proposed PA achieve its full potential. Fabricated in a 40nm CMOS process, the PA achieves 38.5% peak power added efficiency (PAE), 23.0dBm saturated output power (P$_{sat}$) and 20.4dBm output 1-dB compression point (P$_{1dB}$) with 29.5% PAE. The peak PAE is above 35% and P$_{sat}$/P$_{1dB}$ remains above 21.5dBm/19.5dBm across 23-30GHz respectively. The minimum normalized AM-PM distortion is less than 1.3° at 26 GHz and remains less than 4.4° across 26-30GHz. Measured EVM/ACLR is below -29dB/-29dBc with 64QAM 5G-NR modulated signal at 28GHz.

Keywords—CMOS power amplifier, Ka-band, harmonic traps, frequency staggered tuning, PMOS varactor.

I. INTRODUCTION

Communication systems utilizing mm-wave frequency bands have enabled higher transmission bandwidths for increased data rates, including 5G and satellite communication. In these applications, the increasingly higher-order complex modulation urges transceivers of higher bandwidth, efficiency and linearity, in which the power amplifier (PA) often dominates the power consumption with significant influence on signal quality. Limited by the voltage headroom in modern CMOS process, recent mm-Wave power amplifiers often utilize power combining structures to reach higher output power [1]. While class-AB operation ensures high efficiency, the linearity is degraded by non-linear transistor intrinsic capacitors and transconductance that lead to AM-PM distortion. Common approaches mitigating AM-PM distortion include applying PMOS capacitors [2] or voltage-controlled L-C network [3], which aim to cancel the undesired voltage dependence of transistor capacitors.

In this paper, we propose a compact 4-path series-parallel combined broadband class-AB PA (Fig.1) that simultaneously

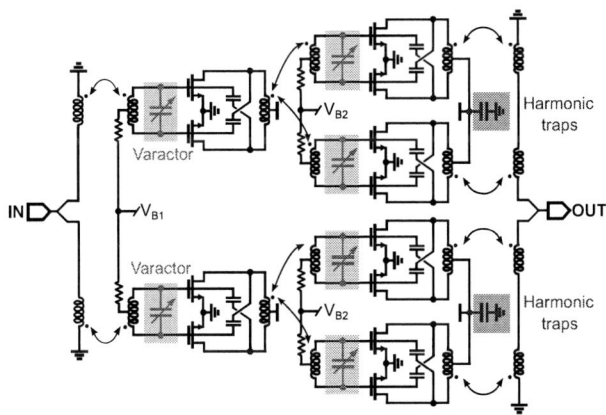

Fig. 1. Simplified overall schematic of the proposed PA.

achieves high output power and efficiency with low AM-PM distortion. Frequency staggering tuning scheme expands the bandwidth of the cascading two-stage amplifier. Beyond varactors at the transistor inputs mitigating the impact of nonlinear transistor intrinsic capacitors C_{gs} and C_{gd}, harmonic traps are introduced into the common-mode return path that minimize voltage swings at the transistor drain terminals. This further suppresses AM-PM distortion caused by the nonlinear transistor transconductance in a class-AB PA. Layout optimizations and full EM simulations ensure the fabricated PA achieve its full potential.

II. CIRCUIT IMPLEMENTATION

The simplified schematic of the proposed PA is shown in Fig.1. The series-parallel combined architecture allows high output power without stressing the power combiner that would otherwise degrade efficiency. The active amplifier adopts the common-source implementation with neutralization capacitors to achieve higher gain and bandwidth. Each common-source amplifier utilizes PMOS varactors for AM-PM suppression. Explicit capacitors are inserted between the center tap of each inductor and ground to form notches at the 2nd-order harmonic frequency in the common-mode return path.

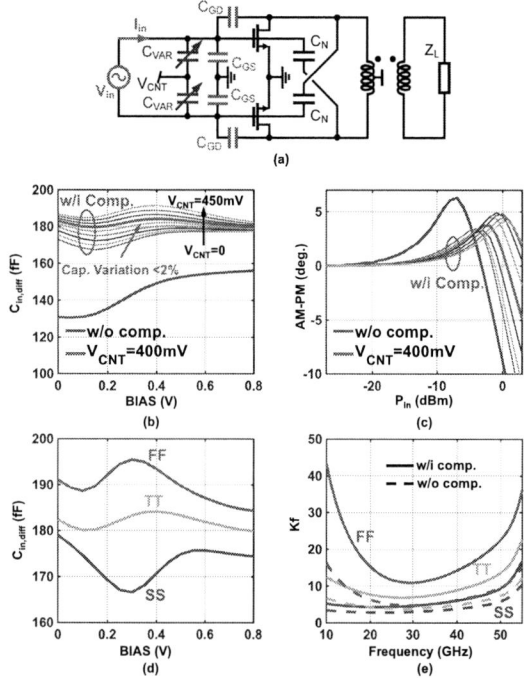

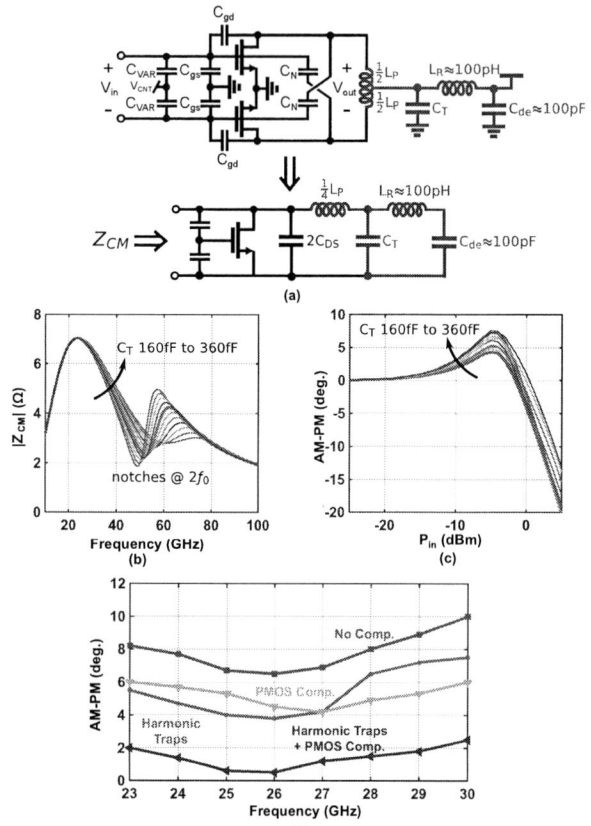

Fig. 2. (a) Simplified schematic of the common-source amplifier stage with neutralization capacitors and PMOS varactors; (b) Equivalent differential input capacitance versus the input voltage with and without the PMOS varactor, V_{cnt} swept linearly from 0 to 450mV; (c) Simulated AM-PM distortion with and without the PMOS varactor, V_{cnt} swept linearly from 0 to 450mV; (d) Impact of process variation on capacitance compensation; (e) Impact of the PMOS varactors on stability.

A. Suppression of AM-PM Distortion

1) PMOS Varactor Compensation

PMOS varactors at the input of the NMOS amplification transistors compensate AM-PM distortion caused by non-linear transistor intrinsic capacitors C_{gd} and C_{gs} as shown in Fig.2(a). The complementary capacitance-voltage dependence of the PMOS varactor compared to C_{gs}, C_{gd} from the amplification transistors leads to an input independent equivalent input capacitor $C_{in,diff}$ if the varactor bias V_{CNT} is properly chosen as shown in Fig. 5(b). AM-PM distortion is thus reduced from over $6°$ to less than $3.5°$ before P_{1dB} as shown in Fig.2(c). While the nonlinear capacitance compensation is process dependent, it still offers improvement with a fixed bias as shown in Fig.2(d). Practical application may apply voltage trimming to resolve this process dependence. Stability is slightly improved with PMOS varactors as their quality factors become lower at mm-Wave frequencies as shown in Fig.2(e).

2) Common-Mode Harmonic Traps

As revealed by [4], the amplitude dependent harmonic distortion induced by non-linear g_m leads to a phase shifted fundamental voltage at the drain terminal. Being modulated by the input amplitude, this phase shift results in AM-PM distortion. This mechanism is further prominent in the proposed PA that adopts a pseudo-differential implementation. To reduce the common-mode voltage

Fig. 3. (a) Schematic of the common-source amplifier and its common-mode equivalent circuit; (b) Simulated equivalent common-mode impedance with various C_T; (c) Simulated AM-PM distortion with various C_T; (d) AM-PM suppression in a class-AB PA with proposed techniques.

amplitudes at drain terminals [5] that would worsen AM-PM distortion through common-mode to differential-mode conversion, the common-mode capacitor C_T is introduced. Placed between the center tap and ground as shown in Fig.3(a), C_T is co-designed with the decoupling capacitors to create common-mode harmonic traps at the 2nd-order harmonic frequencies so that drain terminals experience less common-mode voltage swings. Fig.3(b) and (c) illustrate the impact of the common-mode harmonic traps, which suppress AM-PM distortion to below $4°$ before P_{1dB}.

With both PMOS varactors and common-mode harmonic traps, AM-PM distortion is reduced from $\sim 10°$ to below $2.2°$ at P_{1dB} across 23-30GHz as shown in Fig.3(d). The common-mode harmonic traps remain effective for the rather wide bandwidth with its quality factors set by the transistor as shown in Fig.3(a).

B. Bandwidth Extension with Frequency Staggering Tuning

Shown in Fig.4(a) is the simplified equivalent circuit of a two-stage cascading amplifier that consists of the input matching network, inter-stage matching network and the output matching network. The natural resonance frequencies of the primary and secondary tanks within each matching network, realized with transformers of different coupling strengths, are

979-8-3503-2123-4/23 $31.00 © 2023 IEEE   198

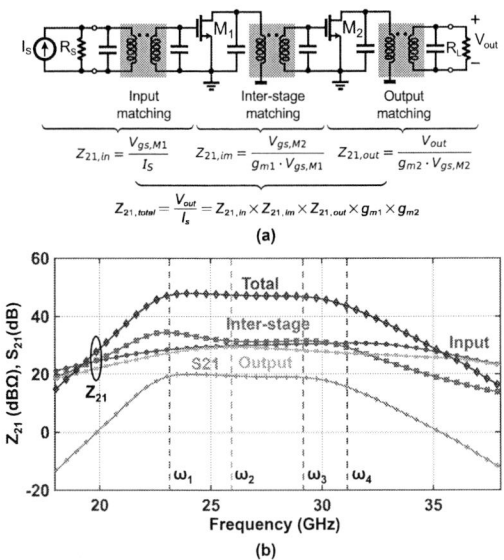

(a)

(b)

Fig. 4. (a) The simplified equivalent circuit of a two-stage cascading PA; (b) The frequency response of the proposed PA and its matching networks.

purposely skewed to achieve an overall flat passband as shown in Fig.4(b) [6]. The moderately-coupled transformer is adopted in the input balun to create high frequency poles at ω_4=32GHz, which set the higher boundary of the passband. The output power combiner utilizes the closely-coupled transformer for higher efficiency[7], which create complex frequency poles at ω_2=26GHz that dominate the center frequency of the passband. The inter-stage network is realized by weakly-coupled transformers for the required high impedance transformation ratio and its poles at ω_1=23GHz and ω_3=29GHz set the lower frequency boundary and compensate the gain droop within the passband.

Fig.5 presents the EM implementations in the overall PA. Edge-coupling structure is used in the input matching network for a moderate coupling strength $k = 0.4$. Traces in the inter-stage matching networks are separated to create low coupling strength $k = 0.25$. Electrical couplings in the input and inter-stage matching networks are minimized to improve signal uniformity. The output hybrid-mode power combiner utilizes the semi-overlapping structure for higher efficiency with $k = 0.75$. EM structures in Fig.5 are simulated as a whole set and co-optimized with the transistors.

III. Measurement Results

The proposed PA is fabricated in a 40nm CMOS process with its chip micrograph shown in Fig.6. The chip occupies a core area of 0.39mm^2 and a total area of 0.6mm^2 with ESD circuits. The input and output pad are connected to the high-frequency probe station with DC power and bias provided through wire-bonded pads during measurement.

Fig.7(a) and (b) shows the measured small signal performance. The small signal gain (S_{21}) is above 19.7dB from 23GHz to 30GHz, with a peak gain of 22.7dB at 26.8GHz. Fig.7(c) shows the measured continuous-wave large signal performance of the proposed PA at 26GHz. The PA

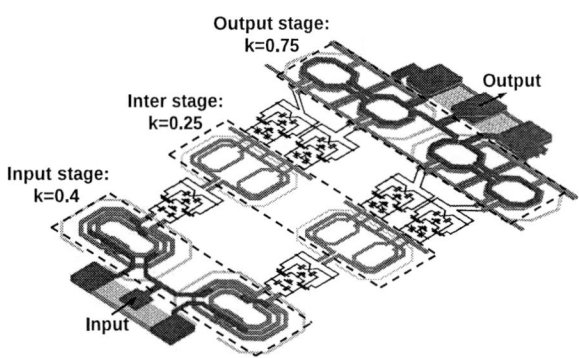

Fig. 5. The overall PA with corresponding EM implementations.

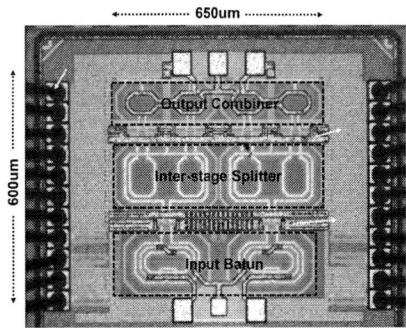

Fig. 6. Chip micrograph of the proposed PA.

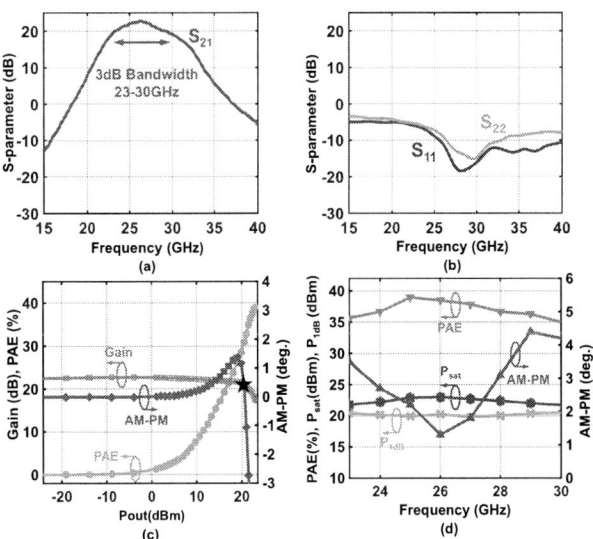

Fig. 7. Measurement results. (a) S_{21}; (b) S_{11} and S_{22}; (c) power gain, PAE and AM-PM distortion verses the output power at 26GHz; (d) P_{sat}, power gain, PAE, P_{1dB} and AM-PM distortion before P_{1dB} across 23-30GHz

achieves a P_{1dB} of 20.4dBm with 29.5% PAE and a P_{sat} of 23.0dBm with the peak PAE of 38.5%. The minimum AM-PM distortion is ~1.3° at 26GHz and remains less than 4.4° across 23-30GHz thanks to the common-mode harmonic traps and the compensating PMOS varactors. Enabled by the efficient power combiner and co-optimized matching networks, the peak PAE and P_{sat} remain above 35% and 21.5dBm from 23GHz to 30GHz as shown in Fig.7(d). With 5G-NR 64-QAM modulated signal with 200MHz bandwidth, the proposed PA achieves

979-8-3503-2123-4/23 $31.00 © 2023 IEEE

Table 1. CMOS Ka Band PA Performance Comparison.

Ref.	TMTT'16[8]	RFIC'18[9]	ISSCC'17[10]	RFIC'20[1]	JSSC'18[2]	ISSCC'22[11]	ISSCC'18[3]	**This work**
Process	28nm	90nm	40nm	65nm	28nm	55nm	65nm	**40nm**
Freq.[GHz]	25-35	24	27	28	29-57	28	28	**23-30**
VDD [V]	1.1	2.4	1.1	1.2	0.9	2.4	1.1	**1.2**
Gain [dB]	10	17.4	22.4	15.9	20.8	16.1	15.8	**22.7**
P_{sat} [dBm]	14.75	25.6	15.1	23.2	16.6	25.5	15.6	**23.0**
PAE_{peak} [%]	46.4	32.8	33.7	33.5	24.2	25.2	41	**38.5**
P_{1dB} [dBm]	13.2	23.6	13.7	22	13.4	24.3	14	**20.2**
PAE@P_{1dB} [%]	40*	N/A	31.1	26*	12.6	24.2	34.7	**29.8**
AM-PM [°]	N/A	N/A	N/A	N/A	0.7@30GHz 0.26@40GHz 0.45@50GHz	N/A	0.7@28GHz 3@24GHz 2.3@32GHz	**1.3@26GHz 3.8@23GHz 4.4@30GHz**
Modulation Scheme	64-QAM	256-QAM	64-QAM	256-QAM	64-QAM	64-QAM	64-QAM	**64-QAM**
Data Rate [Gb/s]	1.5	1	4.8	0.8	2	1.5	2	**1.2**
P_{avg} [dBm]	9.2	20	6.7	18.36	10.1	17.7	9.8	**14**
EVM [dB]	-25.75	-21	-25	-31.6	-25	-25.2	-26.4	**-29**
ACLR [dBc]	-32	-35	-29.41	-30	-32.1	-28.8	-30	**-29**
ITRS_FOM#	71	85.8	81.4	83.3	83.9	84.5	77	**90**

*Estimated from the Figure. #ITRS_FOM = Gain[dB] + P_{sat}[dBm] + $20\log(f_c[GHz]) + 10\log(PAE_{peak}[\%])$.

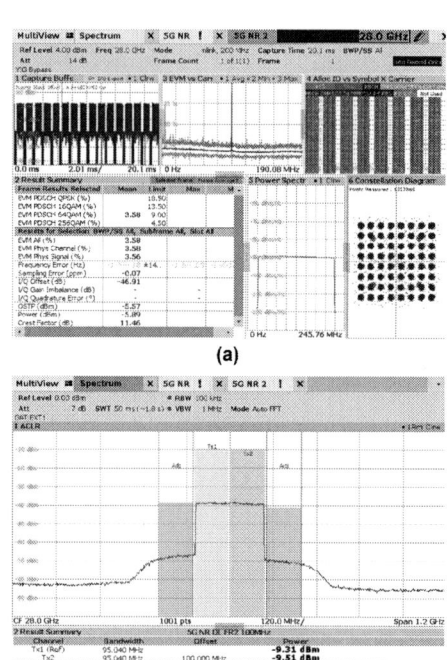

Fig. 8. Measured (a) EVM and (b) ACLR using a 64-QAM modulated 5G-NR signal with 200MHz Bandwidth.

an average output power of 14 dBm with -29dB EVM and -29/-31.5dBc ACLR at 28GHz as shown in Fig.8.

Table I compares the PA prototype with the state-of-the-art Ka band PAs in CMOS processes.

IV. CONCLUSION

In this paper, a Ka band broadband 4-path series-parallel combined class-AB PA with high efficiency and linearity has been implemented in a 40nm CMOS process. Frequency staggering tuning scheme enables a wide bandwidth of 23-30GHz. The PA operates in the hybrid power combining mode for high output power. PMOS compensation capacitors and common-mode harmonic traps suppress

AM-PM distortions that allows the class-AB PA to simultaneously achieve high efficient and high linearity. Measurement results validate the effectiveness of the design choices with state-of-art performances.

ACKNOWLEDGMENT

This work was supported by the National Key Research and Development Program of China (No. 2020YFB1806300).

REFERENCES

[1] H. Ahn, I. Nam, and O. Lee, "A 28-GHz Highly Efficient CMOS Power Amplifier Using a Compact Symmetrical 8-Way Parallel-Parallel Power Combiner with IMD3 Cancellation Method," in *IEEE RFIC*, 2020, pp. 187–190.

[2] M. Vigilante and P. Reynaert, "A Wideband Class-AB Power Amplifier With 29–57-GHz AM–PM Compensation in 0.9-V 28-nm Bulk CMOS," *IEEE JSSC*, vol. 53, no. 5, pp. 1288–1301, 2018.

[3] S. Ali, P. Agarwal, J. Baylon, S. Gopal, L. Renaud, and D. Heo, "A 28GHz 41%-PAE Linear CMOS Power Amplifier using a Transformer-Based AM-PM Distortion-Correction Technique for 5G Phased Arrays," in *IEEE ISSCC*, 2018, pp. 406–408.

[4] S. Golara, S. Moloudi, and A. Abidi, "Processes of AM-PM Distortion in Large-Signal Single-FET Amplifiers," *IEEE TCAS-I*, vol. 64, no. 2, pp. 245–260, 2017.

[5] B. Park, S. Jin, D. Jeong, J. Kim, Y. Cho, K. Moon, and B. Kim, "Highly Linear mm-Wave CMOS Power Amplifier," *IEEE TMTT*, vol. 64, no. 12, pp. 4535–4544, 2016.

[6] H. Jia, C. Prawoto, B. Chi, Z. Wang, and P. Yue, "A Full Ka-Band Power Amplifier With 32.9% PAE and 15.3-dBm Power in 65-nm CMOS," *IEEE TCAS-I*, vol. 65, no. 9, pp. 2657–2668, 2018.

[7] J. Gu, H. Xu, N. Yan, H. Qin, R. Yin, G. Jin, and X. Shen, "A 57-64 GHz Two-Way Parallel-Combined Power Amplifier with 16.6 dBm Psat and 23.6% Peak PAE in 40nm Bulk CMOS," in *IEEE RFIT*, 2022, pp. 177–180.

[8] S. Ali, P. Agarwal, S. Gopal, S. Mirabbasi, and D. Heo, "A 25–35 GHz Neutralized Continuous Class-F CMOS Power Amplifier for 5G Mobile Communications Achieving 26% Modulation PAE at 1.5 Gb/s and 46.4% Peak PAE," *IEEE TCAS-I*, vol. 66, no. 2, pp. 834–847, 2019.

[9] W. Huang, J. Lin, Y. Lin, and H. Wang, "A K-Band Power Amplifier with 26-dBm Output Power and 34% PAE with Novel Inductance-based Neutralization in 90-nm CMOS," in *IEEE RFIC*, 2018, pp. 228–231.

[10] S. Shakib, M. Elkholy, J. Dunworth, V. Aparin, and K. Entesari, "A Wideband 28GHz Power Amplifier Supporting 8×100MHz Carrier Aggregation for 5G in 40nm CMOS," in *IEEE ISSCC*, 2017, pp. 44–45.

[11] Z. Ma, K. Ma, K. Wang, and F. Meng, "A 28GHz Compact 3-Way Transformer-Based Parallel-Series Doherty Power Amplifier With 20.4%/14.2% PAE at 6-/12-dB Power Back-off and 25.5dBm PSAT in 55nm Bulk CMOS," in *IEEE ISSCC*, vol. 65, 2022, pp. 320–322.

RTu1A-1

A 15.6-GHz Quad-Core VCO with Extended Circular Coil Topology for Both Main and Tail Inductors in 8-nm FinFET Process

Suoping Hu, Zhiyu Chen, Wanghua Wu, Pei-Yuan Chiang, Zhanjun Bai, Chih-Wei Yao, Sangwon Son

Samsung Semiconductor Inc, USA

suoping.hu@samsung.com

Abstract— Although process scaling has brought numerous benefits in digital circuit design, it is still very challenging to design high-performance RF circuitry in advanced FinFET processes due to the increased flicker noise, reduced supply voltage, and worsened non-linearity. This paper explores the RF oscillator design in an 8-nm FinFET process and presents a quad-core oscillator with an extended circular inductor frame for both the main and tail inductors. Thanks to the extended circular topology, the tail inductance, and its quality factor are significantly increased to provide resonance efficiently at 2nd-order harmonic for better flicker noise suppression. Designed and fabricated in an 8-nm FinFET process, this oscillator achieves a record-low PN of −115.4 dBc/Hz at 1-MHz frequency offset and a competitive Figure-of-Merit (FoM) of −185 dBc/Hz at 15.6 GHz with a wide tuning range of 30%.

Keywords— Voltage-controlled oscillator (VCO), FinFET, LC-VCO, 2nd-order harmonic tail filtering, multi-core, and phase noise.

I. Introduction

The advanced FinFET process nodes (e.g., 8 nm) are developed to deliver superior performance for digital circuitry. However, high-performance analog and RF circuit design in the FinFET process remain a challenge. The shrinking power supply voltage, worsening flicker noise [1] and increasing non-linearity [2] along with the process scaling all attributed to it. To harness the benefits of the process scaling for digital circuit design, advanced RF transceiver system-on-chip (SoC) needs to migrate to the advanced FinFET process as digital baseband circuitry occupies a significant chip area (e.g., ≥50% of the total SoC die size [3], [4]). To close this gap, this work focuses on the low phase noise (PN), wide tuning range RF voltage-controlled oscillator (VCO) design for 5G cellular applications in an 8-nm FinFET process.

To meet the ever-growing demand on data rate, 5G cellular communication adopts high-order modulations [e.g., 64-quadrature amplitude modulation (QAM)] and multiple-input and multiple-output (MIMO) technology at mm-wave frequency bands (i.e., frequency range 2, from 24.25 to 52.6 GHz), to boost the data rate to a few gigabits per second. Thus, the corresponding error vector magnitude (EVM) specification becomes much more stringent. For example, to support the 64-QAM and MIMO under non-ideal channel conditions (i.e., fading and non-ideal orthogonality), a minimum link EVM of ≤ 2.2% is required [5]. At the 28-/39-GHz band, it translates to a 125-/90-fs integrated phase noise (IPN) of a local oscillator (LO) respectively [see Eq. (1)], where f_o represents for oscillation frequency]. To achieve an

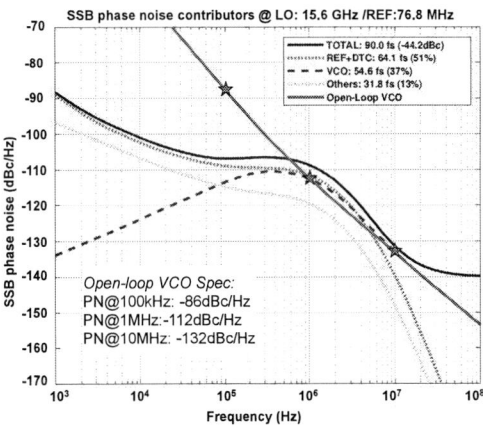

Fig. 1. PLL PN Spectrum based on a linear s-domain model in [5]

optimal integrated phase noise (IPN) at a PLL output, the PLL bandwidth is typically selected so that the VCO and the reference path circuitry each contribute ∼50% of the IPN. Figure 1 shows the PN spectrum of a PLL with total RMS jitter of 90 fs (integrated from 1 kHz to 100 MHz) base on a linear s-domain model in [5]. With the given noise of REF_{CLK} and other circuitry in the reference path, including digital-to-time converter (DTC), phase detector, and the multi-modulous divier, the open-loop VCO PN requirement can be derived, which is shown in Fig. 1. In this example, VCO needs to achieve a very low PN of −86, −112 and −132 dBc/Hz at 10-kHz, 100-kHz, and 1-MHz offset frequency, respectively.

$$\text{Jitter}_{RMS} \leq \frac{\text{EVM}}{2\pi f_o} \qquad (1)$$

Phase noise in an LC-VCO can be described by Lesson's equation [see (2)], where f_o is the resonance frequency of VCO, Δf is a frequency offset from the carrier resonance frequency, k is the Boltzmann's constant, T is the absolute temperature, F is equal to the noise coefficient γ of active devices, Q, R_T and L is the quality factor, in-parallel resistance, and inductance of the LC tank, respectively. A_0 represents the VCO swing.

$$\mathcal{L}(\Delta\omega) \propto \frac{4kTR_T}{A_0{}^2}\left(\frac{f_o}{2Q\Delta f}\right)^2 \propto \frac{f_o{}^3 L}{A_0{}^2 Q} \qquad (2)$$

Equation (2) suggests that to achieve a better VCO PN, we should maximize the A_0, however, there is a maximum allowed operating voltage for a given process due to the reliability

979-8-3503-2123-4/23 $31.00 © 2023 IEEE   201   2023 IEEE Radio Frequency Integrated Circuits Symposium

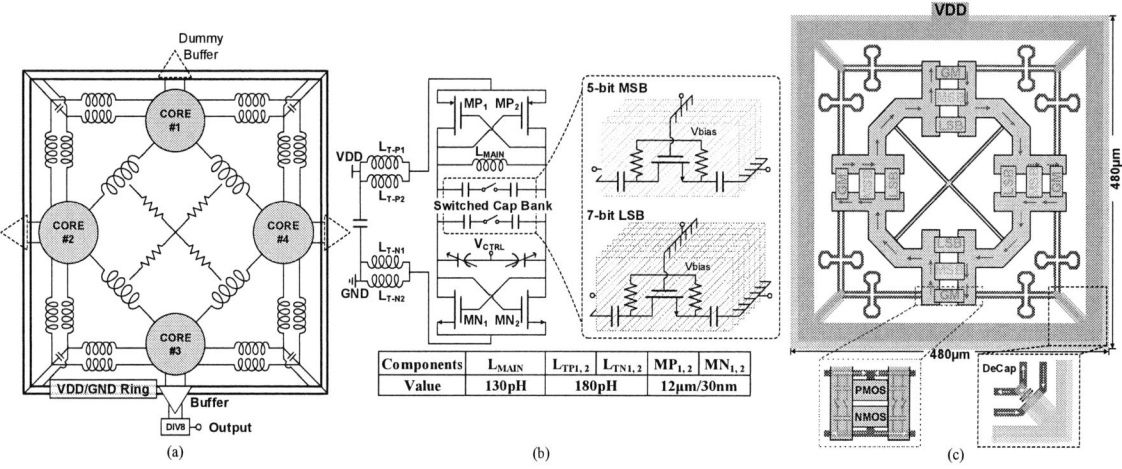

Fig. 2. (a) Circuit diagram of proposed circular-geometry quad-core VCO with both main and tail inductors, and (b) schematic of a single-core oscillator. (c) Layout and implementation of the proposed circular-geometry main and tail inductor frame

concern. Minimizing L while maintaining a great Q is another way. Unfortunately, at 16-GHz, the value of L is already quite small, generally at the level of 100~200 pH with a reasonable Q value of ~20. By further scaling down L, the mutual coupling from the inner edges of one single-turn coil dramatically degrades the Q. This trade-off between smaller L and Q degradation sets the lowest PN a single-core VCO can achieve in a given process.

To break this limit, multi-core VCOs are demonstrated in the literature [6], [7]. Ideally, merging N extract same single-core VCO boosts the PN by $10 \times \log_{10}(N)$ dB, with a cost of $N\times$ power dissipation and ambiguous start-up issues. A quad-core VCO operating at 27 GHz with circular inductor topology is proposed in [6] to boost the inductor Q up to 30, and a 4th-order harmonic tail filtering is implemented for flicker noise suppression. However, the tail inductor is not well defined, and as stated in [8], a tail LC tank resonating at $2 \times f_{LO}$ provides better flicker noise suppression.

In this paper, we extend the circular inductor topology not only for the main inductance but also for the tail inductance. There are three main advantages to implementing this circular tail inductor: 1) a larger Q value for the tail inductor, which leads to sharper filtering; 2) more area-efficient compared to the classic standalone tail inductor, hence, a larger inductance value to support 2nd-order harmonic resonance; 3) providing a well-defined common-mode return path. Furthermore, this work explores to design high-performance RF VCOs in an 8-nm FinFET process with a targeted PN much lower than the previously published RF VCOs in planar CMOS process nodes (e.g., 40 nm in [6] and [9], 28 nm in [10], and 65 nm in [11], [7]). The proposed VCO runs at 15.6 GHz with a record-low absolute PN of −115.4 dBc/Hz at 1-MHz offset and 30% tuning range in an 8-nm FinFET process. The VCO achieves an FoM of −184.8, which is competitive with the published low-PN VCOs at similar oscillation frequencies in 65-nm or 40-nm bulk planar CMOS. Section II discusses the detailed implementation of the proposed quad-core VCO, followed by the measurement results in section III. Section IV draws the conclusion.

II. VCO DESIGN

A. Oscillator Design

Figure 2 (a) shows the overall quad-core VCO circuit diagram. Fig. 2 (b) shows the schematic of a single core, and all four cores are identical in the quad-core VCO. A complementary transconductance (gm) cell with the thin-oxide device is chosen in this work. In this FinFET process, NMOS and PMOS transistor offers comparable gm and transit frequency so that a CMOS cross-coupled topology is in favor as it reduces the DC current consumption by half due to the current reuse between PMOS and NMOS gm stages. Since the targeted VCO frequency is at ~16 GHz, thin-oxide devices are used as it provides a faster transit frequency (f_T) compared to a thick-oxide device, resulting in lower parasitic capacitance desired for a wide tuning range. To achieve both a wide tuning range (i.e., 30%) and fine-tuning step (resolution of 10 MHz), the VCO core is segmented into five unit-weighted MSB capacitors and seven binary-weighted LSB capacitors. A varactor of 240 MHz/V is implemented, which provides sufficient tuning range to compensate for the VCO frequency shift over temperature variation from −25 to 125 °C. The size of the CMOS transistor is set to a total of 12 μm with a fin of 8 and 72 fingers to ensure startup over PVT. Tail inductors ($L_{T-P1,2}$ and $L_{T-N1,2}$) are used to filter out flicker noise both from PMOS and NMOS transistors. In addition, four differential tail tuning capacitors are used to make sure flicker noise is suppressed across the whole tuning range. Bypass capacitors are placed between the VDD and the ground rings. The output buffer is directly driven by one of the cores (core #3) and followed by a divide-by-8 (DIV8) frequency divider for measurement. Dummy buffers are added at all other cores to avoid asymmetry.

979-8-3503-2123-4/23 $31.00 © 2023 IEEE

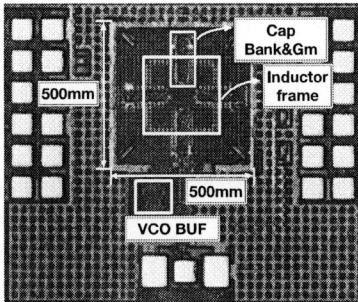

Fig. 3. Chip Micrograph

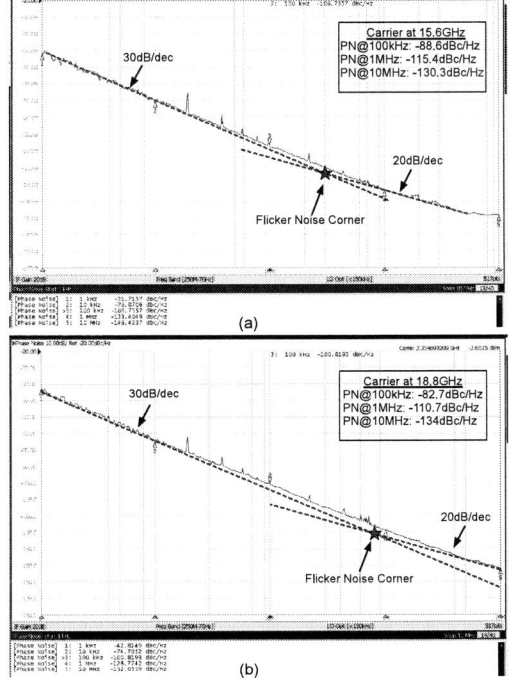

Fig. 4. Measured VCO Phase Noise (after dividing by 8) at (a) 15.6 and (b) 18.8 GHz.

B. Circular-Geometry Inductor Frame Design

Figure 2 (c) illustrates the layout and implementation of the quad-core VCO passive structure. A circular-geometry design is utilized for the primary inductor similar to [6], [7]. This design can boost the main inductor Q without introducing any mode ambiguity. By adding narrow metal traces across the circular geometry, Q is significantly degraded in all other unwanted modes while Q in the desired mode is preserved since the traces act as virtual grounds. The main inductor frame is formed using the top copper layer ($\sim 3 \mu$m thickness) and the simulated inductance for a single core is ~ 130 pH with Q of ~ 30. LSB capacitor bank is sitting inside the primary inductor to save layout area and reduce capacitor bank frame length. Gm cells are placed at the end of the frame instead of the middle, giving shorter ground access and tail inductor connection. Since transistors cannot rotate in FinFET technology, drain connections to the inductor frame

of the gm cell in vertical and horizontal cores are slightly different. Tail resonator around the 2nd-order harmonic in the common mode path is a well-known method to improve the phase noise of an oscillator [8]. However, conventional tail inductors occupy a large area, especially for all four cores. Resonance at the 4th-order harmonic is used in [6] which is less effective than that of the 2nd-order harmonic. Besides, the VCO PN performance can be degraded if tail inductors are not placed properly and the common mode current path is not well-defined. To mitigate this, the circular-geometry tail inductors are proposed in this design. As shown in Fig. 2 (c), two parallel tail inductors for both PMOS and NMOS gm cells in each core are placed outside the primary inductor and connected at corners. The CMOS transistor pairs are arranged in such a way that both sources are connected from both sides with metal traces formed by the second top copper layer ($\sim 1 \mu$m thickness) to act as tail inductors as well as the common-mode return path. The simulated inductance of one single tail inductor is ~ 180 pH with Q of ~ 15 at 30 GHz. Since the parallel tail inductor has a relatively high Q, four units of tail tuning capacitor are added to cover the whole tuning range. These capacitors are placed close to gm cells and beneath the main inductor frame. Compared to the tail inductor topology in [7], parallel tail inductors implemented in our work lead to a more symmetrical layout and a better simulated PN. Moreover, the parallel-inductor structure minimizes the VCO fundamental tone coupling from the main inductor to the tail inductor. The coupled fundamental tone will be canceled out at the sources of the CMOS gm cell. VDD and ground are connected from all four corners. By adding an AC decoupling capacitor at each corner, each VCO core can minimize its AC current return path, resulting in higher tolerance against supply and ground parasitic inductors.

III. MEASUREMENT RESULTS

The prototype is fabricated in an 8-nm FinFET process. Figure 3 shows the chip micrograph with a core area of 0.23 mm^2. It burns 28.8-mW dc power consumption with a supply voltage of 0.9 V at 15.6 GHz. An on-chip DIV8 frequency divider is implemented for test purposes. The measured phase noise after DIV8 at f_{norm} (i.e., 15.6 GHz) and f_{max} (i.e., 18.8 GHz) are shown in Figs. 4 (a) and (b), respectively. Fig. 4 (a) shows a VCO PN of -88.6 at 100-kHz and -115.4 dBc/Hz at 1-MHz offset with the operating frequency of 15.6 GHz. The noise floor observed ~ 5 MHz offset [see Fig. 4 (a)] is due to the divider and the test buffer path, which is driven by an LC buffer that could be slightly off-tuned compared to the VCO oscillation frequency. This could be resolved by adding a tunable capacitor bank in the LC buffer.

As shown in Fig. 5, the measured tuning range is as wide as 30%, which is best-in-class compared to other works in Table 1. This helps to combat the process-voltage-temperature (PVT) variation. Furthermore, the measured PN is extremely consistent across different samples and process corners, as shown in Fig. 5. Although the measured VCO PN at 1 MHz

979-8-3503-2123-4/23 $31.00 © 2023 IEEE

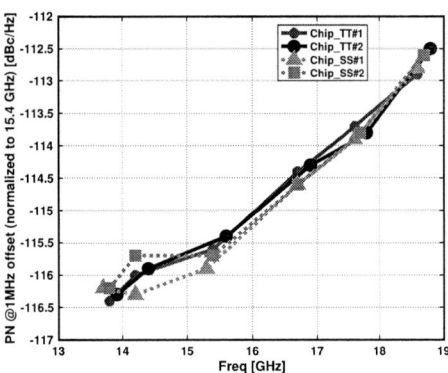

Fig. 5. Measured PN at 1 MHz offset vs. frequency with various silicon samples and corners (normalized to 15.6 GHz)

Table 1. Performance Summary and Comparison With State-of-he-Art High-Performance VCOs.

	This work	JSSC'18 [6]	RFIC'22 [9]	JSSC'20 [10]	CICC'20 [11]	ESSCIRC '21 [7]
Freq. [GHz]	15.6	27	12.11	19.5	27.5	19
Process [nm]	**8 (FinFet)**	40	40 (SOI)	28	65	65
Process Robustness	**Yes**	not shown	not shown	not shown	not shown	not shown
TR [%]	**30**	26	21	12	13	16
Area [mm²]	0.27	0.1	0.15	0.07	0.04	0.13
Supply [V]	0.9	0.95	0.3	0.9	0.9	0.75
Power [mW]	28.8	16.9	1.4	20.7	3.4	16.4
PN @1MHz [dBc/Hz]*	**-115.4**	-114.1	-105.2	-114	-110.6	-116.7
1/f3 Corner Freq. [kHz]	**300 to 800**	500	1000	500	450	900
FoM_1M [dB]	-184.8	-185.7	-187.6	-184.7	-189.1	-188.3
FoM_1M=PN-20log(f_{OSC}/1MHz)+10log(P_{DC}/1mW); *Normalized to 15.6GHz						

offset varies from -116.5 dBc/Hz to -112.5 dBc/Hz across the tuning range, it is still sufficient for the intended 5G FR2 cellular applications.

Table 1 compares the state-of-the-art high-performance VCOs. The PN noise and Figure-of-Merit reported in Table 1 are normalized to 15.6 GHz for a fair comparison. This work achieves a record-low PN of -115.4 dBc/Hz with a flicker noise corner of 300 to 800 kHz. Moreover, the reported tuning range is $2\times$ wider compared to [7], [10], [11] and 50% more compared to [9]. This VCO consumes 32 mA from 0.9 V, resulting in an FoM of -185-dBc/Hz at 1 MHz offset, which is among the best-in-class compared to its counterparts [6], [10]. Refs [11] and [9] have a 4- to 3-dB better FoM thanks to the low-power consumption. However, VCO phase noise power does not reduce proportionally with its power consumption, especially for very low-PN VCO design. Besides, as mentioned in Section I, RF or mmWave VCO design in a FinFET process is intrinsically more challenging than it does in a planar MOSFET process (e.g., 28, 40, 65 nm, etc.).

IV. CONCLUSION

A 15.6-GHz quad-core VCO with extended circular inductor topology for both the main and tail passive frame is proposed in this paper. The extended tail inductor provides an efficient 2nd-order harmonic resonance to further suppress the flicker noise. To be integrated with digital processors in an SoC, it is designed and fabricated in an advanced 8-nm FinFET process, which is barely implemented in other works. It achieves a record-low PN of -115.4 dBc/Hz at 1-MHz offset with a wide tuning range of 30% and robust performance across the process.

REFERENCES

[1] P. Kushwaha, H. Agarwal, Y.-K. Lin, A. Dasgupta, M.-Y. Kao, Y. Lu, Y. Yue, X. Chen, J. Wang, W. Sy, F. Yang, P. C. Chidambaram, S. Salahuddin, and C. Hu, "Characterization and Modeling of Flicker Noise in FinFETs at Advanced Technology Node," *IEEE Electron Device Letters*, vol. 40, no. 6, pp. 985–988, 2019.

[2] T.-H. Tsai, R.-B. Sheen, S.-Y. Hsu, C.-H. Chang, and R. B. Staszewski, "A 55.9-fs Integrated Jitter (100 kHz–100 MHz) Hybrid LC-Tank PLL in 5-nm FinFET Using Programmable Phase Realignment and Dynamic Coarse Tuning," *IEEE Solid-State Circuits Letters*, vol. 4, pp. 230–233, 2021.

[3] T.-M. Chen, Y. Lu, P.-N. Chen, Y.-H. Chang, M.-C. Liu, P.-Y. Chang, C.-J. Liang, Y.-C. Chen, H.-L. Lu, J.-Y. Ding, C.-C. Wang, Y.-L. Hsueh, J.-C. Tsai, M.-S. Hsu, Y.-H. Chung, and G. Chien, "7.1 An 802.11ac dual-band reconfigurable transceiver supporting up to four VHT80 spatial streams with 116fsrms-jitter frequency synthesizer and integrated LNA/PA delivering 256QAM 19dBm per stream achieving 1.733Gb/s PHY rate," in *2017 IEEE International Solid-State Circuits Conference (ISSCC)*, 2017, pp. 126–127.

[4] B. Khamaisi, D. Ben-Haim, A. Nazimov, A. Ben-Bassat, S. Gross, N. Shay, G. Asa, V. Spector, Y. Eilat, A. Azam, E. Borokhovich, I. Shternberg, P. Skliar, E. Solomon, A. Beidas, T. A. Hazira, A. Lane, E. Shaviv, G. Nudelman, E. Dahan, M. S. Shemer, N. Kimiagarov, A. Ravi, and O. Degani, "A 16nm, +28dBm Dual-Band All-Digital Polar Transmitter Based on 4-core Digital PA for Wi-Fi6E Applications," in *2022 IEEE International Solid- State Circuits Conference (ISSCC)*, vol. 65, 2022, pp. 324–326.

[5] W. Wu, C.-W. Yao, K. Godbole, R. Ni, P.-Y. Chiang, Y. Han, Y. Zuo, A. Verma, I. S.-C. Lu, S. W. Son, and T. B. Cho, "A 28-nm 75-fsrms Analog Fractional- N Sampling PLL With a Highly Linear DTC Incorporating Background DTC Gain Calibration and Reference Clock Duty Cycle Correction," *IEEE Journal of Solid-State Circuits*, vol. 54, no. 5, pp. 1254–1265, 2019.

[6] D. Murphy and H. Darabi, "A 27-GHz Quad-Core CMOS Oscillator With No Mode Ambiguity," *IEEE Journal of Solid-State Circuits*, vol. 53, no. 11, pp. 3208–3216, 2018.

[7] S. Ming and J. Zhou, "A 19 GHz Circular-Geometry Quad-Core Tail-Filtering Class-F VCO with -115 dBc/Hz Phase Noise at 1 MHz Offset in 65-nm CMOS," in *ESSCIRC 2021 - IEEE 47th European Solid State Circuits Conference (ESSCIRC)*, 2021, pp. 303–306.

[8] E. Hegazi, H. Sjoland, and A. Abidi, "A filtering technique to lower LC oscillator phase noise," *IEEE Journal of Solid-State Circuits*, vol. 36, no. 12, pp. 1921–1930, 2001.

[9] M. Fang and T. Yoshimasu, "A 14-GHz-Band Harmonic Tuned Low-Power Low-Phase-Noise VCO IC with a Novel Bias Feedback Circuit in 40-nm CMOS SOI," in *2022 IEEE Radio Frequency Integrated Circuits Symposium (RFIC)*, 2022, pp. 167–170.

[10] A. Franceschin, P. Andreani, F. Padovan, M. Bassi, and A. Bevilacqua, "A 19.5-GHz 28-nm Class-C CMOS VCO, With a Reasonably Rigorous Result on 1/f Noise Upconversion Caused by Short-Channel Effects," *IEEE Journal of Solid-State Circuits*, vol. 55, no. 7, pp. 1842–1853, 2020.

[11] A. Masnadi, M. Mahani, H. M. Lavasani, S. Mirabbasi, S. Shekhar, R. Zavari, and H. Djahanshahi, "A Compact Dual-Core 26.1-to-29.9GHz Coupled-CMOS LC-VCO with Implicit Common-Mode Resonance and FoM of -191 dBc/Hz at 10MHz," in *2020 IEEE Custom Integrated Circuits Conference (CICC)*, 2020, pp. 1–4.

979-8-3503-2123-4/23 $31.00 © 2023 IEEE

RTu1A-2

A 10.8-14.5GHz 8-Phase 12.5%-Duty-Cycle Non-Overlapping LO Generator with Automatic Phase-and-Duty-Cycle Calibration for 60-GHz 8-Path-Filtering Sub-Sampling Receivers

Khoi T. Phan, Yang Gao, Howard C. Luong

Hong Kong University of Science and Technology, China

tkphan@connect.ust.hk, ygaoay@connect.ust.hk, eeluong@ust.hk

Abstract — A 10.8-14.5GHz 8-phase 12.5%-duty-cycle non-overlapping LO generator is proposed for 60-GHz 8-path-filtering sub-sampling receivers. A 4-stage ring oscillator is followed by reconfigurable injection-locked-oscillator NOR gates to generate 8-phase 12.5%-duty-cycle signals featuring automatic successive phase calibration and automatic frequency-domain duty-cycle calibration. The generator prototype measures minimum phase errors of ~0.1° and maximum 4th-harmonic output power of -5dBm with >30dB improvement while consuming 77.8mW from a 1V supply.

Keywords — mm-wave, oscillator, receiver, injection locking, N-path, phase calibration, duty-cycle calibration.

I. INTRODUCTION

Multiple-phase non-overlapping non-50%-duty-cycle LO signals have been widely used in N-path-filtering wireless receivers to achieve RF filtering with high quality factor Q. However, the system performance is limited by the errors of the LO output phases and duty cycles (DTCs). Accurate detection and calibration of phase and DTC errors at 15GHz are challenging not only because both the pulse widths and the phase differences become as small as 8.33 ps, which is comparable to the rise/fall times, but also because the output amplitudes become sensitive not only to the narrow pulse width but also to the DTC and PVT variations. This work proposes generation and automatic calibration of the phase and DTC errors for 8-phase 12.5-%-DTC non-overlapping LO signals at a fundamental frequency of ~15GHz, whose 4th-harmonic components at ~60GHz are to be used in a 60-GHz 8-path-filtering direct-conversion sub-sampling receiver.

II. CIRCUIT IMPLEMENTATION

A. System Design

A Windmill frequency divider with feedback loop can generate non-overlapping multiple-phase 25%-DTC LO signals with small phase errors and low phase noise [1], but its operation frequency is limited to 2.4GHz because it requires input signals at higher frequencies for division. At higher frequencies, non-overlapping multi-phase 25%-DTC signals can be generated with either logic gates [2] or tunable dc bias voltages [3, 4]. As compared to logic gates, tunable bias voltages can operate at higher frequencies and with less power. However, its output swing is limited because the sinusoidal signals are chopped to create the non-overlapping signals. As a result, high-supply buffers are needed to provide enough LO power which consume large power. Furthermore, the DTC

calibration is still limited while the phase errors still cannot be controlled for good calibration due to non-50% DTC.

DTC errors can be detected in the time domain by comparing the average output voltage with an external reference voltage corresponding to the targeted DTC, and the comparison results can be fed back to control the DTC [5]. However, either rail-to-rail amplitude would be required for the calibrated signals resulting in sub-GHz operation or advanced processes for sufficiently high time resolution [5]. For phase-error detection at high frequencies, analog phase detector is preferred to its digital counterpart. On the other hand, as the dc output of analog phase detector depends on both the amplitude and the phase difference of the input signals, this approach is sensitive to the DTCs. Moreover, in the presence of strong harmonic components at the inputs due to their non-50% DTCs, the phase detection and calibration become more complicated and challenging.

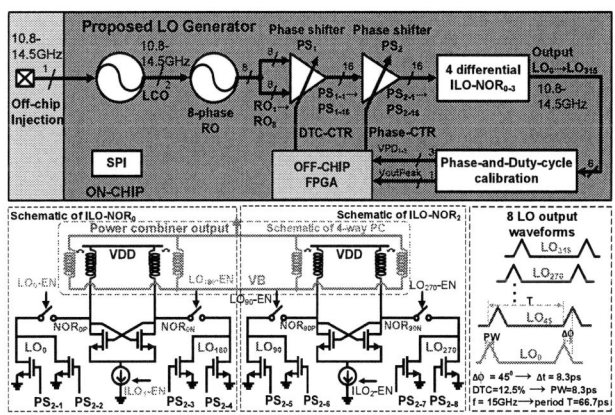

Fig. 1. Block diagram of the proposed 15-GHz 12.5%-DTC 8-phase clock generator, the schematic of ILO-NOR$_{0,2}$ together with 4-way power combiner, and 8 45° non-overlapping output waveforms LO$_0$ – LO$_{315}$.

Figure 1 shows the block diagram of the proposed LO generator together with the 8 45° output waveforms LO$_0$ – LO$_{315}$. As the core, a ring oscillator (RO) is designed with 4 differential active-load delay cells to generate 8 45° sinusoidal signals, and 4 differential injection-locked-oscillator NOR (ILO-NOR) gates are proposed to obtain the 8-phase 12.5%-DTC LO output signals with automatic phase and DTC calibration. An LC-based oscillator is included to provide low-phase-noise input signals to be injected to RO to improve the RO phase noise without large-area LC delay cells. A cascade of two groups of digitally-controlled phase shifters PS$_1$ and PS$_2$ is used to realize tunable phases to calibrate the output

979-8-3503-2123-4/23 $31.00 © 2023 IEEE   205   2023 IEEE Radio Frequency Integrated Circuits Symposium

phases and DTCs. Finally, the phase and DTC calibration blocks are implemented to detect the output phase and DTC errors and to generate appropriate control signals Phase-CTR and DTC-CTR for PS_2 and PS_1, respectively, via an off-chip FPGA. Notably, the 8 RO outputs are duplicated to drive PS_1, PS_2, and ILO-NORs to enable separate selection and control of the ILO-NOR inputs to tune and calibrate the phases and the DTC of the 8 LO outputs independently.

The detailed schematic of the key block ILO-NORs is shown in Fig. 1, which is uniquely designed to be reconfigured in two different modes. During the normal operation, ILO_1-EN = 0, LO_0-EN = LO_{180}-EN = 1, and ILO-NOR is configured as an NOR gate to create 12.5%-DTC outputs LO_0 and LO_{180}. During the phase calibration, ILO_1-EN = 1, ILO-NOR is reconfigured as injection-locked oscillator (ILO) while the LO outputs to be calibrated are selected as the injected inputs by appropriate LO-EN signals. Due to the injection locking, the selected 12.5-%-DTC LO signal is converted to differential 50%-DTC sinusoidal signals at the ILO-NOR outputs with the same phase information but without harmonic components.

Inductive peaking and pseudo loading techniques are employed for the ILO-NORs to boost up their frequencies while the embedded ILO is implemented with direct injection for wide locking range and with slab inductors for high quality factor and small unwanted mutual coupling. Also embedded in the ILO-NOR, each of the 2 4-way power combiners consists of 4 transformers using the top two metal layers with low coupling coefficient of ~0.4 to achieve small loading and high bandwidth as compared to the direct combining.

B. Automatic Phase Calibration

Figure 2 shows the detailed block diagram of the proposed automatic phase detection and calibration scheme. The phases of the converted 12.5-%-to-50-% sinusoidal signals from ILO-NOR outputs are properly compared and detected by the phase detectors PD_{1-3} whose outputs VPD_{1-3} are connected to an off-chip FPGA to generate control signals Phase-CTR to tune the phase shifters PS_2 for phase calibration.

The proposed phase calibration extends the successive-phase-calibration algorithm in [6, 7] to compare progressively the relative phases of two adjacent LO outputs, but it is uniquely modified to be applicable to 12.5%-DTC signals, first for (LO_0, LO_{90}, LO_{180}, LO_{270}) and then for (LO_{45}, LO_{135}, LO_{225}, LO_{315}). Illustrated in Fig. 2 are the steps and the corresponding control signals for the phase calibration algorithm. In Step 1, to calibrate LO_0 and LO_{90}, ILO_1-EN = ILO_3-EN = 1, LO_0-EN = LO_{90}-EN = 1, LO_{180}-EN = LO_{270}-EN = 0 while LO_0 and LO_{90} are selected as injected input signals to ILO-NOR_1 and ILO-NOR_3, and $\phi_1 = \angle LO_{90} - \angle LO_0$ is measured with PD_1 and stored by the off-chip FPGA. In Step 2, $\angle LO_{180}$ is tuned until $\phi_2 = \angle LO_{180} - \angle LO_{90} = \phi_1$. In Step 3, $\angle LO_{270}$ is tuned until $\phi_3 = \angle LO_{270} - \angle LO_{180} = \phi_2$. In Step 4, $\phi_4 = \angle LO_0 - \angle LO_{270}$ is measured and compared with ϕ_1. If $\phi_4 \neq \phi_1$, $\angle LO_{90}$, $\angle LO_{180}$ and $\angle LO_{270}$ are tuned in Step 5 until $\phi_4 = \phi_1$, or equivalently, $\phi_1 = \phi_2 = \phi_3 = \phi_4 = 90^0$, when the 4 signals LO_0, LO_{90}, LO_{180} and LO_{270} are completely calibrated. In Step 6, PI-

EN = 1, and the phase interpolator (PI) acts like a MUX providing an interpolating signal PI_{45} between LO_0 and LO_{90}. In Step 7, $\angle LO_{45}$ is tuned and calibrated when $\phi_5 = \angle LO_{45} - \angle PI_{45} = 0^0$. Similar procedures are then repeated to calibrate LO_{45}, LO_{135}, LO_{225}, and LO_{315} with control signals properly swapped.

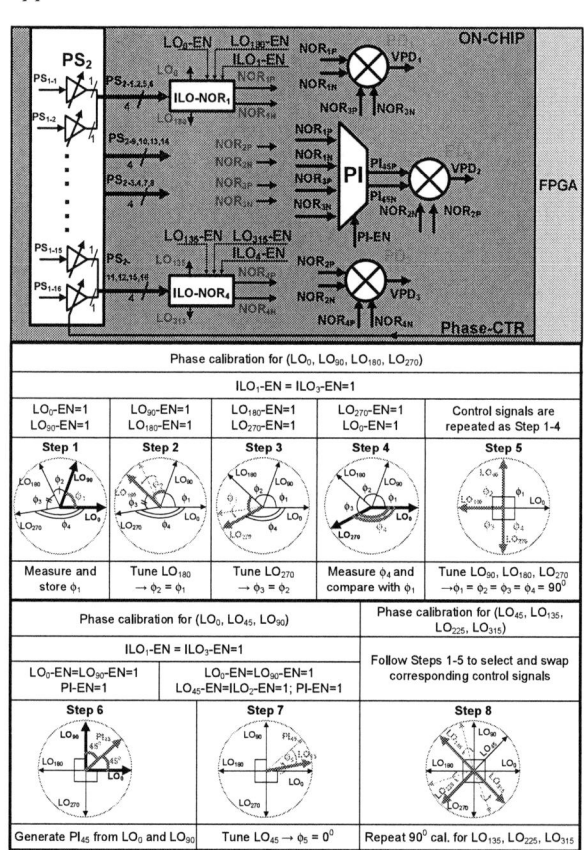

Fig. 2. Proposed automatic phase calibration: block diagram and successive phase calibration algorithm.

Thanks to the ILO scheme, only the fundamental mixing products are presented at the outputs VPD_{1-3}, which enables accurate detection of the phase difference of the two selected LO outputs independent of their 12.5% DTCs. In other words, the proposed ILO configuration breaks the dependency of the phase calibration on the DTC. Moreover, the proposed phase calibration scheme does not need accurate external reference signals while requiring only one phase detector for 45^0 phase calibration, which minimizes both the power consumption and the static phase offset. The phases (LO_0, LO_{90}, LO_{180}, LO_{270}) and (LO_{45}, LO_{135}, LO_{225}, LO_{315}) use phase detector PD_1 and phase detector PD_3, respectively, for the 90-degree phase calibration, hence, eliminating the mismatch. Furthermore, three phase detectors are laid out close to each other to minimize the effects of PVT variations.

C. Automatic Duty-Cycle Calibration

A novel DTC calibration in the frequency domain is also proposed based on the observation that the 4th-harmonic components of 4 90^0 12.5%-DTC output signals are in phase and thus their combined output power becomes maximum. Figure 3 shows the block diagram of the proposed DTC calibration scheme together with its operation principle. During the DTC detection, which is done after the phase calibration, the 8 LO outputs are divided into 2 separate groups: (LO_0, LO_{90}, LO_{180}, LO_{270}) and (LO_{45}, LO_{135}, LO_{225}, LO_{315}), and coupled to 4-way power combiners PC_1 and PC_2 to extract their combined 4th-harmonic components while rejecting and attenuating other unwanted harmonics. The output of each power combiner is selected by a 2-to-1 MUX and amplified by a shared 60-GHz amplifier before being applied to a single-stage cascode peak detector. The output of the peak detector is connected to the off-chip FPGA to generate control signals DTC-CTR to tune the phase shifters PS_1 until the output becomes maximum.

Conventionally, only the phase of one input of a NOR gate needs to be tuned to control the output DTC. Unfortunately, such a tuning mechanism would also move the peaks of the output waveforms and thus affect the output phase. Consequently, the calibrated 90^0-phase-difference information would be lost. In this work, to keep the peaks constant to preserve the calibrated phase information while tuning the DTCs, the phases of the two inputs of the ILO-NOR gate are independently tuned in 2 opposite directions by two different phase shifters, as illustrated in Fig. 3. The DTC control signals of the 4 LO outputs are first simultaneously and coarsely tuned, and each control signal is then separately and finely tuned to achieve local optimization.

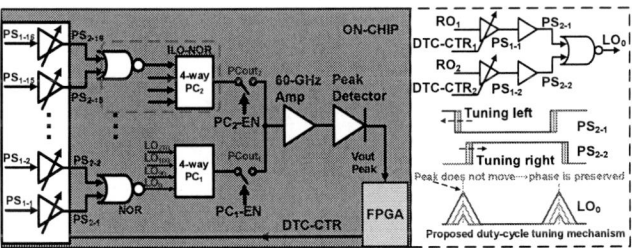

Fig. 3. Proposed DTC calibration: block diagram, tuning mechanism.

III. MEASUREMENT RESULTS

The proposed LO generator is fabricated in a 28-nm CMOS process with a core area of 0.81 mm^2, and the die photo is shown in Fig. 4. To minimize parasitics and loading effects in the 8 signal paths, the active components in the phase shifters and the NOR gates are placed underneath the signal paths, and the phase detectors are also laid out underneath ILO-NORs. The 8 LO outputs are calibrated by sharing the same buffer and peak detector to eliminate mismatches. The proposed LO generator measures a frequency range from 10.8GHz to 14.5GHz at 1-V supply while consuming 77.8 mW for the LO core and 17.7 mW for the calibration circuitry, which can be powered off after calibration. The effectiveness of the phase and DTC calibration schemes are illustrated in Figs. 5, 6 and 7, respectively, for 5 different samples. When the phase calibration is enabled, the minimum phase error among all the

outputs is reduced from 8.2^0 to 0.1^0 while the maximum phase error is reduced from 14.3^0 to 1.0^0. The phase error is calculated from the measured voltages VPD_{1-3} using FPGA with the successive-phase-calibration algorithm.

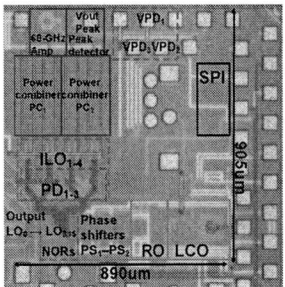

Fig. 4. Die micrograph.

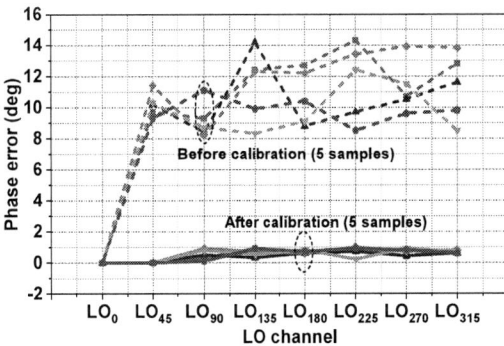

Fig. 5. Measurement results of the phase error for the 8 LO output signals before and after phase calibration for 5 different samples.

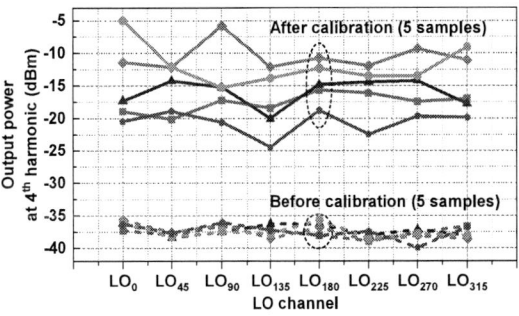

Fig. 6. Measured 4th-harmonic output power of the 8 LO output signals for 5 different samples at 58GHz before and after DTC calibration (for maximum output power).

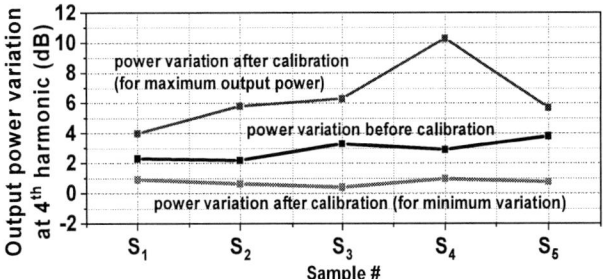

Fig. 7. Measured variations of the 4th-harmonic output power of the 8 LO output signals for 5 different samples at 58GHz before and after DTC calibration (for maximum output power and for minimum output power variations).

979-8-3503-2123-4/23 $31.00 © 2023 IEEE   207

Table 1. Performance Summary and Comparison with Existing State-of-the-Art LO Generators

	This work		[2] RFIC'21	[3] JSSC'20	[4] JSSC'21	[5] ISSCC'20	[1] JSSC'21
	f_{out}	$4*f_{out}$					
Technology	28nm		45nm SOI	28nm	65nm	7nm FinFET	22nm FDSOI
Topology	ILO+NOR gates		Divider + Transmission gates	HBC + bias voltage		Pulse generator	Windmill divider
No of phases	8		4	4	4	4	4
Duty cycle	12.5%		25%	25%	25%	25%	25%
LO calibration	Auto phase and DTC		No	Manual DTC only		Auto phase and DTC	No
f_{out} (GHz) (%)	10.8-14.5 (29.2%)	43.2-58.0 (29.2%)	6-31 (135.1%)	70-100 (35.3%)	21-29 (32%)	14	2.4
f_{in} (GHz)	10.8-14.5		12-62	70-100	21-29	14	4.8
Pnoise @ 1MHz (dBc/Hz)	-112.9	-100.9[a]	NA	NA	NA	NA	-154[b]
Power (mW)	Core: 58.3 - 77.8		35 - 228	8	13.2	NA	0.04
	LO Calibration: 14.7- 17.7[c]		0	NA	NA	NA	0
Supply Voltage (V)	1		1-1.3 (LO)	1	1.2	0.85	0.7
FOM	177.2	165.2[a]	NA	NA	NA	NA	235.5
Phase error (⁰) [Min; Max]							
Before phase calibration	[8.2; 14.3]	NA	NA	4 (HBC) [b]	NA	NA	0.54[b]
After phase calibration	[0.1; 1]	NA	NA	NA	NA	1.1[d]	NA
Output power (dBm) [Min; Max]							
Before DTC calibration	NA	[-39.9; -35.0]	NA	-3 (HBC) [b]	NA	NA	NA
After DTC calibration	NA	[-24.5; -5.0]	NA	NA	NA	NA	NA

[a]extrapolation; [b]estimated from simulation; [c]turn off after calibration; [d]extracted from measurement. FOM = |PN(Δf)| + 20log10(f_{osc}/Δf) − 10log10(P_{DC}/1mW)

The maximum achievable 4th-harmonic output power of the 8 individual LO outputs before and after DTC calibration are measured and plotted in Fig. 6 for 5 different samples. Without calibration, the 4th-harmonic output power is measured from -39.9dBm to -35dBm, and with calibration, it varies from -24.5dBm to -5.0dBm, corresponding to a maximum improvement of 30dB. In addition, the DTC calibration is also done independently to achieve maximum 4th-harmonic output power but with minimum variations among the 8 LO outputs for 5 different samples. The lowest maximum output power of single channel is chosen as reference output power, after that the remain LO channels are calibrated to achieve the minimum power variation. From the measurement results shown in Fig. 7, the 4th-harmonic power variation is measured from 2.2dB to 3.8dB without calibration and from 0.4dB to 1.0dB with calibration for minimum variation and from 4.0dB to 10.3dB with calibration for maximum output power. It demonstrates the ability of the proposed work to achieve either maximum output power or minimum variations for different system applications. Table 1 summarizes the measured performance of the LO generator and compared to existing state-of-the-art multiple-phase non-50%-DTC LO generators. The measurement setups and the power breakdown are shown in Fig. 8.

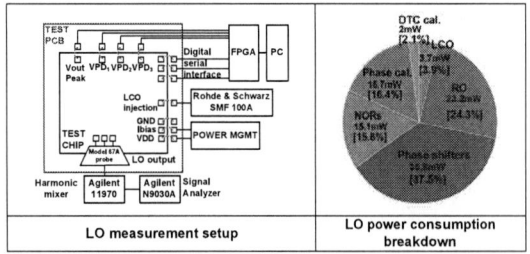

Fig. 8. Measurement setup and power breakdown.

IV. CONCLUSION

An 8-phase 12.5%-DTC non-overlapping LO signal generator intended for 60-GHz 8-path-filtering sub-sampling receivers employs automatic phase-and-duty-cycle calibration to achieve maximum 4th-harmonic output power with small phase errors and minimum power variation. Thanks to the calibrations, over a frequency range from 10.8GHz to 14.5GHz with 77.8mW from 1V supply, the prototype measures minimum phase error of < 0.1⁰ and maximum 4th-harmonic output power of -5dBm (improved by 30dB from -35dBm) for 5 different samples from 43 to 58GHz.

ACKNOWLEDGMENT

This project was partially supported by Hong Kong General Research Fund under Grant 16209220 and by ACCESS – AI Chip Center for Emerging Smart Systems, sponsored by InnoHK Funding, HKSAR.

REFERENCES

[1] Bart J. Thijssen, et al., "2.4-GHz Highly Selective IoT Receiver Front End with Power Optimized LNTA, Frequency Divider, and Baseband Analog FIR Filter," *IEEE JSSC*, pp.2007-17, Jul. 2021.

[2] S. Hari, C. J. Ellington, and B. A. Floyd, "A 6-31 GHz Tunable Reflection-Mode N-Path Filter," *IEEE RFIC*, pp. 143-146, June 2021.

[3] L. Iotti, G. LaCaille, and A. M. Niknejad, "A Low-Power 70-100-GHz Mixer-First RX Leveraging Frequency-Translational Feedback," *IEEE JSSC*, vol. 55, pp. 2043-2054, August 2020.

[4] P. Song, et al., "Mm-Wave Mixer-First Receiver with Selective Passive Wideband Low-Pass Filter," *IEEE JSSC*, vol.56, pp.1454-1463, May 2021.

[5] T. Ali, et al., "A 460mW 112Gb/s DSP-based Transceiver with 38dB Loss Compensation for Next Generation Data Centers in 7nm Finfet Technology," *IEEE ISSCC Dig. Tech. Papers*, pp. 118-120, Feb 2020.

[6] L. Wu, et al., "A 4-path 42.8-to-49.5GHz LO Generation with Automatic Phase Tuning for 60GHz Phased-Array Receiver," *ISSCC* Feb. 2012.

[7] S-M Lee, et al., "A 64Gb/s Downlink and 32Gb/S Uplink NRZ Wireline Transceiver with Supply Regulation, Background Clock Correction and EOM-Based Channel Adaptation for Mid-Reach Cellular Mobile Interface in 8nm FinFet," *IEEE ESSCIRC*, Sep. 2022.

RTu1A-3

A 4.4 mW Inductorless 2-20 GHz Single-Ended to Differential Frequency Doubler in 45 nm RFSOI CMOS Technology

A. Meyer[$][*], M. L. Leyrer[#][^][*], C. Ziegler[$], M. Maier[$], V. Lammert[#], V. Issakov[$]

[$] Institute for CMOS Design at Technical University Braunschweig, Germany
[#] Infineon Technologies AG, Germany
[^] ale.meyer@tu-braunschweig.de

Abstract — This work presents a miniature inductorless frequency doubler with high fundamental rejection, wide bandwidth, and single-ended to differential conversion. By utilizing an NMOS/PMOS pair with symmetrical loads attached to drain and source terminals, the second harmonic of an input signal is extracted. Fundamental rejection of up to 35 dB and a wide output range of 2 to 20 GHz is shown in measurement. The doubler is implemented in 45 nm CMOS RFSOI technology and draws 4.4 mW including biasing and buffer stages. The circuit consumes an active area excluding pads of only 50 x 70 μm^2.

Keywords — frequency doubler, radar, CMOS

I. INTRODUCTION

Precise frequency generation is crucial for high-frequency applications. Realizing Phased-Locked-Loops (PLLs) with a fundamental VCO operating at millimeter-wave (mm-wave) frequencies is highly challenging due to the trade-off between phase noise, frequency tuning range and power consumption. The quality factor of an LC-based VCO's resonant tank drastically decreases at higher frequencies, leading to a severe degradation of phase noise performance and output power [1], [2]. Thus, to achieve low phase noise at mm-wave frequency, a local oscillator (LO) signal can be generated in the low-GHz regime (< 10 GHz) and translated into the mm-wave domain by a frequency multiplier chain, while preserving the noise performance [1], [3]. Integrated frequency multipliers are also required in case the LO signal is generated externally by an off-the-shelf high-precision, low-noise commerical PLL operating at few GHz, typically providing a single-ended interface. Hence, frequency multipliers operating at low-GHz frequencies and enabling a single-ended to differential output conversion, e.g. to drive a high-ohmic mixer input, are required. Frequency multipliers, such as the differential-pair [4], push-push [5], [6], common-source [7], push-pull [8], [9], waveform shaping [10] or injection-locking [11] topology, do not perform a single-ended to differential output conversion and often require a large area due to bulky passive components. In this work, we propose a novel, inductorless, single-ended to differential frequency doubler topology implemented in a 45 nm RFSOI CMOS technology.

II. THEORY OF OPERATION

The proposed frequency doubler comprises an NMOS and PMOS devices connected in parallel, loaded symmetrically by resistive loads R_L to the positive and negative supply rails, as

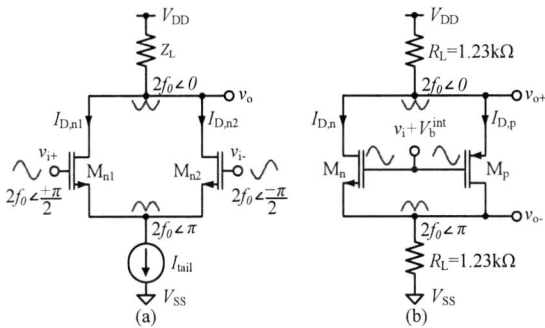

Fig. 1. (a) Classical differential to single-ended frequency doubler and (b) proposed single-ended to differential frequency doubler.

shown in Fig. 1. While the input signal v_i at the fundamental frequency with an additional DC bias V_b^{int} is applied to both MOS gates, the output voltage at twice the input frequency appears differentially across the output terminals v_{o+} and v_{o-}.

An accurate mathematical model of the MOSFETs' behavior is needed to derive an analytical expression for the differential output voltage v_o. For this purpose, the EKV-model is chosen, since it describes the MOSFETs' behavior across all operating regions [12]. $I_{D,n}$ of the NMOS is given by

$$I_{D,n} = \frac{W_n}{L_n} \frac{I_{spec,n,\square} \cdot 4\left(q_s^2 + q_s\right)}{2 + \lambda_{c,n} + \sqrt{4(1 + \lambda_{c,n}) + \lambda_{c,n}^2 \left(1 + 2q_s\right)^2}}, \quad (1)$$

where $I_{spec,n}$, is the technology-dependent specific current, q_s represents the normalized inversion charge at the source, and $\lambda_{c,n}$ is the velocity saturation parameter. Furthermore, for analytical calculations, the EKV model in (1) can be simplified by omitting the square term in the root resulting in

$$I_{D,n} \approx \frac{W_n}{L_n} \frac{I_{spec,n,\square} \cdot 4\left(q_s^2 + q_s\right)}{2 + \lambda_{c,n} + \sqrt{4(1 + \lambda_{c,n})}} = \alpha_n \cdot \left(q_s^2 + q_s\right), \quad (2)$$

with α_n comprising all technology and geometry parameters. This approximation is valid and sufficiently accurate in the weak and moderate inversion regions. The transistors in the proposed circuit are biased in the moderate inversion region.

By invoking the Wright omega function [13], the transcendental equation describing the relationship between the normalized inversion charge q_s and the overdrive voltage $V_{ov,n}$

$$V_{ov,n}/(n_n \cdot V_T) = \ln(q_s) + 2q_s, \quad (3)$$

with $V_{ov,n} = V_{gs} - V_{T0,n}$, slope factor n and $V_T = (k_B T)/q$ can be solved, leading to

---

[*] equal contribution

979-8-3503-2123-4/23 $31.00 © 2023 IEEE   209   2023 IEEE Radio Frequency Integrated Circuits Symposium

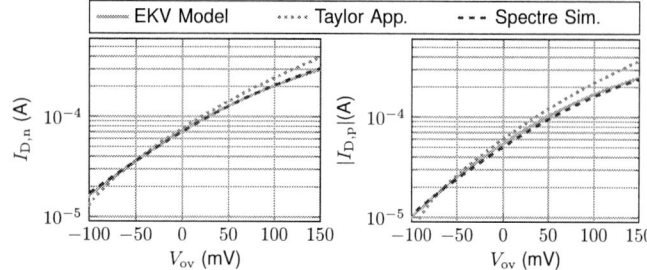

Fig. 2. Comparison of EKV model, Taylor approximation and Spectre simulation for the NMOS/PMOS drain current ($W_{n,p}/L_{n,p} = 2\,\mu\text{m}/40\,\text{nm}$).

Table 1. Coefficents α_i for NMOS and β_i for PMOS

Coefficient:	Expression:	Coefficient	Expression:
α_0	$\alpha_n \cdot (a^2 + a)$	β_0	$\alpha_p \cdot (a^2 + a)$
α_1	$\alpha_n \cdot \frac{2ab+b}{n_n \cdot V_T}$	β_1	$\alpha_p \cdot \frac{2ab+b}{n_p \cdot V_T}$
α_2	$\alpha_n \cdot \frac{b^2+2ac+c}{(n_n \cdot V_T)^2}$	β_2	$\alpha_p \cdot \frac{b^2+2ac+c}{(n_p \cdot V_T)^2}$
α_3	$\alpha_n \cdot \frac{2bc}{(n_n \cdot V_T)^3}$	β_3	$\alpha_p \cdot \frac{2bc}{(n_p \cdot V_T)^3}$
α_4	$\alpha_n \cdot \frac{c^2}{(n_n \cdot V_T)^4}$	β_4	$\alpha_p \cdot \frac{c^2}{(n_p \cdot V_T)^4}$

Table 2. Extracted parameters for NMOS and PMOS transistors

$I_{\text{spec,n}} = 3.75\,\mu\text{A}$	$n_n = 2$	$V_{T0,n} = 0.3247\,\text{V}$	$\lambda_{c,n} = 1$
$I_{\text{spec,p}} = 2.775\,\mu\text{A}$	$n_p = 1.88$	$V_{T0,p} = 0.3445\,\text{V}$	$\lambda_{c,p} = 1$

$$q_s = \frac{1}{2} \cdot \omega \left(\ln(2) + \frac{V_{\text{ov,n}}}{n_n \cdot V_T} \right). \qquad (4)$$

For small overdrive voltages $V_{\text{ov,n}}$ the Taylor approximation can be used to simplify (4). Neglecting terms with the order higher than three, results in

$$q_s|_{V_{\text{ov,n}}=0} \approx \frac{1}{2} \cdot \left[a + b \cdot \frac{V_{\text{ov,n}}}{n_n \cdot V_T} + c \cdot \left(\frac{V_{\text{ov,n}}}{n_n \cdot V_T} \right)^2 \right] \qquad (5)$$

with $a = w_a/2$, $b = \frac{w_a}{1+w_a}/2$, $c = 0.25 \cdot w_a/(1+w_a)^3$ and $w_a = \omega(\ln(2))$ as constants. With this expression derived, q_s in (2) can be substituted by the approximation in (5). This yields a fourth order polynomial expression for the drain current

$$I_{D,n} = \alpha_0 + \alpha_1 V_{\text{ov,n}} + \alpha_2 V_{\text{ov,n}}^2 + \alpha_3 V_{\text{ov,n}}^3 + \alpha_4 V_{\text{ov,n}}^4. \qquad (6)$$

A similar expression is valid for PMOS, yet with coefficients β_i, where α_i and β_i are summarized in Table 1. Consequently, Fig. 2 shows a comparison between Spectre simulation, EKV model and Taylor approximation (6) for small overdrive voltages V_{ov} and for parameters listed in Table 2. The Taylor approximation shows good agreement with the Spectre simulation results proving the validity of this approach.

Having arrived at a convenient expression for I_D (6), the output voltage v_o can be derived. For ease of calculation, albeit without changing the implications for the circuit's behavior, it is assumed that the frequency doubler operates around a symmetric supply $-V_S$ and $+V_S$, implying $V_S = V_{DD}/2$. The output voltage $v_o = v_{o+} - v_{o-}$ is expressed as

$$v_o = [V_S - (I_{D,n} + I_{D,p}) \cdot R_L] - [(I_{D,n} + I_{D,p}) \cdot R_L - V_S]$$
$$= 2 \cdot V_S - 2 \cdot (I_{D,n} + I_{D,p}) \cdot R_L = -2 \cdot v_{o-}, \qquad (8)$$

noting that due to the symmetry v_{o+} equals $-v_{o-}$. Hence, to determine the output voltage, it is sufficient to derive v_{o-} as a function of input voltage v_i. Next, referring to Fig. 1, V_{ov} for NMOS and PMOS can be written as

$$V_{\text{ov,n}} = [V_b^{\text{int}} + v_i - v_{o-}] - V_{T0,n} = V_{c,n} + v_i - v_{o-} \qquad (9)$$

$$V_{\text{ov,p}} = [v_{o+} - (V_b^{\text{int}} + v_i)] - |V_{T0,p}| = V_{c,p} - v_i - v_{o-} \qquad (10)$$

with $V_{c,n} = V_b^{\text{int}} - V_{T0,n}$ and $V_{c,p} = -V_b^{\text{int}} - |V_{T0,p}|$. Equations (9) and (10) are utilized to substitute the overdrive voltage in (6). Consequently, the resulting equations for the drain currents are inserted into (8) resulting in

$$v_{o-} = R_L(I_{D,n} + I_{D,p}) - V_S =$$
$$R_L \sum_{k=0}^4 \left[\alpha_k (V_{cn} + v_i - v_{o-})^k + \beta_k (V_{cp} - v_i - v_{o-})^k \right] - V_S \quad (11)$$

with the constant coefficients listed in Table 1. With α_4 and β_4 being much smaller than other coefficients, the fourth order term can be neglected. Coefficients α_i and β_i depend on circuit parameters such as W/L. Thus, they can be set equal, i.e. $\alpha_1 = \beta_1$. This results in cancellation of the linear term of v_i (for $k = 1$), which is the dominant contributor to the fundamental harmonic at the output voltage in (11)

$$v_{o-} = \ldots + R_L \left[\alpha_1 (V_{c,n} + \cancel{v_i} - v_{o-}) + \alpha_1 (V_{c,p} - \cancel{v_i} - v_{o-}) \right].$$

Hence, the fundamental (linear) input term is eliminated. Next, we assume v_i to be a sinusoidal signal with amplitude A_i ($v_i = A_i \cos(\omega t)$). Thus, v_{o-} contains multiple harmonics of the input signal. Since v_{o-} appears in (11) on both sides of the equation, we need to make some simplifying assumptions. First, we consider v_{o-} only up to the third harmonic

$$v_{o-} = V_{o,DC} + V_{o1}\cos(\omega t) + V_{o2}\cos(2\omega t) + V_{o3}\cos(3\omega t). \quad (12)$$

Second, given that V_{o1}, V_{o2}, V_{o3} are small compared to $V_{o,DC}$, the magnitude of v_{o-}^2 is dominated by the DC component $V_{o,DC}$ and is approximated as $v_{o-}^2 \approx V_{o,DC}^2$. This assumption about the form of input and output signals can be used to equate harmonic coefficients on both sides and simplify the expression. Demanding $\alpha_i = \beta_i$ for $i = \{0, 1, 2, 3\}$, (11) can be expanded, resulting in (7). Rearranging (7), introducing coefficients $C_{\text{out}}, C_0, C_1, C_2$ and plugging $v_i = A_i \cos(\omega t)$ into (7), yields an expression for the output voltage

$$v_{o-} = \left[\frac{C_0}{C_{\text{out}}} + \frac{1}{2}\frac{C_2}{C_{\text{out}}}A_i^2 \right] + \left[\frac{C_1}{C_{\text{out}}} \cdot A_i \right]\cos(\omega t)$$
$$+ \left[\frac{1}{2} \cdot \frac{C_2}{C_{\text{out}}} \cdot A_i^2 \right]\cos(\mathbf{2}\omega t). \qquad (13)$$

Apart from the second harmonic, also the unwanted fundamental component $\cos(\omega t)$ appears at the output due to the intermodulation products. However, by nullifying C_1, the fundamental harmonic can be suppressed. Additionally, the second harmonic can be enhanced by increasing C_2. It should be noted, that frequency dependency has been neglected in this derivation to not further complicate the analysis.

979-8-3503-2123-4/23 $31.00 © 2023 IEEE

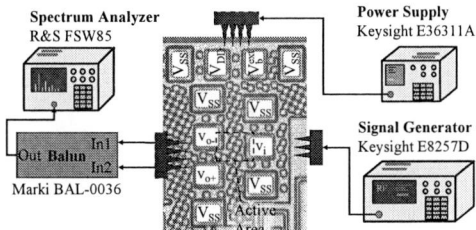

Fig. 3. Chip photograph of the frequency doubler (470 μm x 370 μm).

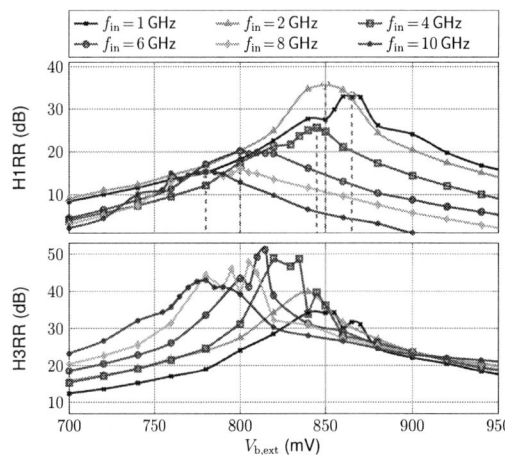

Fig. 4. Measured harmonic rejection ratios (HRR) at $P_{in} = 0$ dBm. Here the external bias V_b^{ext} instead of V_b^{int} is plotted on the x-axis.

III. CIRCUIT IMPLEMENTATION

The proposed frequency doubler as shown in Fig. 1 is followed by two self-biased, inverter based buffers that are used to amplify the output signal. An additional biasing block enables controlling of V_b^{int} by manipulating an external V_b^{ext} pin. No passive filters are incorporated in this design.

Three constraints necessary to achieve frequency doubling, while suppressing all other harmonics, can be deduced from the previous analytical derivations. First, the coefficients summarized in Table 1 have to match each other with respect to the NMOS and PMOS used. As only the width and length of the transistors are available as design variables, matching of all coefficients in an actual implementation is permitted. With the main goal of improving H1RR, matching $\alpha_1 = \beta_1$ is prioritized, leading to minor mismatch between other parameters and potentially degrading rejection ratio for higher harmonics. For $\alpha_1 = \beta_1$ this yields a width ratio constraint for NMOS and PMOS transistor, depending on the technology parameters

$$\frac{W_n}{W_p} \overset{!}{=} \frac{I_{spec,p,\square}}{I_{spec,n,\square}} \cdot \frac{L_n}{L_p} \cdot \frac{2 + \lambda_{c,p} + \sqrt{4(1 + \lambda_{c,p})}}{2 + \lambda_{c,n} + \sqrt{4(1 + \lambda_{c,n})}} \cdot \frac{n_p}{n_n} . \quad (14)$$

Second, C_1 should be nullified to suppress the fundamental harmonic, while C_2 should be maximized to amplify the second. Referring to (7), the term $(V_{c,n} - V_{c,p})$ has to be set to zero C_1 and thus the fundamental harmonic

$$C_2 = 2\alpha_2 - 3\alpha_3(V_{T0,n} + V_{T0,p}) - 6\alpha_3 V_{o,DC} ,$$
$$C_1 = C_2 \cdot (V_{c,n} - V_{c,p}) \overset{!}{=} 0 ,$$
$$\rightarrow (V_{c,n} - V_{c,p}) = 0 . \quad (15)$$

According to (9) and (10), $V_{c,n}$ and $V_{c,p}$ relate to the bias voltage V_b^{int}. Thus, a simple condition for V_b^{int} can be found

$$V_b^{int} \overset{!}{=} \frac{V_{T0,n} - |V_{T0,p}| + V_{DD}}{2} \quad (16)$$

To account for a single supply, $V_{DD}/2$ has to be added. With the MOSFET parameters listed in Table 2 and $V_{DD} = 1$ V, the ideal bias can be calculated, yielding $V_b^{int} = 490.1$ mV.

Therefore, by choosing the correct sizing and biasing as dictated by (15) and (16), the first order term of the PMOS, respectively NMOS, drain current $I_{D,n,p}$ in (6) cancel each other and hence the fundamental input frequency is suppressed.

It is important to note that V_b^{int} is generated by the bias circuit, which can be externally controlled by the biasing voltage V_b^{ext}. The conversion between V_b^{int} and V_b^{ext} can be easily obtained by simulation. Finally, as a third constraint C_{out} equaling the left side of the expression in (7) can be minimized by increasing the load resistance R_L. However, this is limited by headroom and will affect the second harmonic. Additionally, since the DC level varies with input amplitude, a self-biasing effect is seen according to (13).

The employed buffers achieve a simulated voltage gain of 20.6 dB and output amplitude of 800 mV, which drops under 100 Ω loading presented by measurement equipment to 0.4 dB and 128 mV respectively. As in the most likely use-cases in an integrated system, the doubler has to drive a high ohmic load, e.g. a mixer, the performance suffices.

IV. MEASUREMENT RESULTS

The proposed frequency multiplier including output buffer and biasing block is manufactured in Global Foundries 45 nm RFSOI CMOS technology. Fig. 3 depicts the die photograph and the measurement setup. The active area of the chip is covered by metal filling, but spans an area of approximately 70 μm x 50 μm (0.0035 mm²) excluding pads. All measurement results are reported for the whole frequency multiplier including the inverter based output buffers as well as the biasing block. In Fig. 4 the measured fundamental and

$$v_{o-} \cdot C_{out} = C_0 + v_i \cdot C_1 + v_i^2 \cdot C_2 \quad (7)$$
$$\rightarrow v_{o-}\{R_L^{-1} + 2\alpha_1 - 2\alpha_2(V_{T0,n} + V_{T0,p}) + \alpha_3(V_{T0,n} + V_{T0,p})^2 + 2\alpha_3(V_{c,n}^2 - V_{c,n}V_{c,p} + V_{c,p}^2 + V_{o,DC}^2)\} =$$
$$\{(\alpha_0 + \beta_0) - \alpha_1 \cdot (V_{T0,n} + V_{T0,p}) - \alpha_3 \cdot [(V_{c,n}^2 - V_{c,n}V_{c,p} + V_{c,p}^2 + V_{o,DC}^2) - 2V_{o,DC}^2] \cdot (V_{T0,n} + V_{T0,p}) - V_0 R_L^{-1} + \alpha_2[V_{c,n}^2 + V_{c,p}^2$$
$$+ 2V_{o,DC}^2]\} + v_i \cdot \{[2\alpha_2 - 3\alpha_3(V_{T0,n} + V_{T0,p}) - 6\alpha_3 V_{o,DC}] \cdot (V_{c,n} - V_{c,p})\} + v_i^2 \cdot \{2\alpha_2 - 3\alpha_3(V_{T0,n} + V_{T0,p}) - 6\alpha_3 V_{o,DC}\}$$

979-8-3503-2123-4/23 $31.00 © 2023 IEEE

Table 3. Performance summary and comparison with state of the art frequency doublers

Ref. :	Tech. (nm)	Topology	$2f_0$ (GHz)	P_{in,f_0} (dBm)	H1RR (dB,peak)	$P_{\text{out},2f_0}$ (dBm)	V_{DD} (V)	P_{DC} (mW)	η_{tot} (%)	Chip Area (mm^2)
This	CMOS RFSOI 45	PMOS+NMOS pair	$2-20$	0	$>15(35)$	-9.3	1.1	4.4	0.8	0.0035
[4]	SiGe 130	differential pair	$11.2-22.6$	0	$>40(64)$	5	3.3	376	0.84	1.91
[5]	CMOS SOI 22	push-push	$35-40$	0	>37	3.2	0.8	26.4	7.63	0.28
[6]	CMOS 180	push-push	$15-36$	5	>33	-5.2	-	11	2.1	0.32
[7]	CMOS SOI 90	common-source	$26.5-28.5$	-4.5	>11.2	-3	1.25	1.0	37.0	0.1
[8]	CMOS 110	push-pull	$21.8-25.8$	0	$>39(44)$	-4.2	1.2	6.0	5.43	0.45
[9]	CMOS 110	push-pull	$20.4-25.4$	0	$>28(36)$	-8.4	1.2	2.4	4.25	0.46
[10]	CMOS RFSOI 45	waveform shaping	$46-89$	11	$>38(45)$	6.7	1	17	15.5	0.10
[11]	CMOS 130	injection-locking	$11-15$	0	$>45(53)$	7	1.3	5.2	80.8	0.08

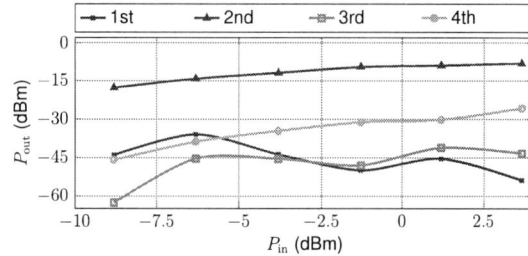

Fig. 5. Measured harmonic output power sweep across input power at $f_{\text{in}} = 2\,\text{GHz}$.

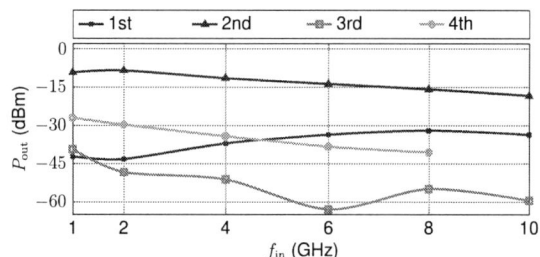

Fig. 6. Measured harmonic output power sweep over output frequency at $\text{P}_{\text{in}} = 0\,\text{dBm}$.

third order rejection ratio (HRR) over bias voltage for input frequencies between 1 to 10 GHz are shown for an input power of 0 dBm. By tuning $V_{\text{b}}^{\text{ext}}$ to 850 mV a peak rejection of 35 dB is achieved for an f_{in} of 2 GHz. Other input frequencies show a small deviation from the same $V_{\text{b}}^{\text{ext}}$ to achieve peak HRR. The HRR of the third harmonic reveals a peak rejection of up to 50 dB at almost the same biasing condition of 820 mV.

Fig. 5 presents the output power for different harmonics at an input frequency of 2 GHz swept across the input power. A strong suppression of the first, third and fourth order harmonics is observed. The output power at an input power of 0 dBm is measured to be $-9.3\,\text{dBm}$. By optimizing capability of the output buffer to drive an external $100\,\Omega$ load, this value can be further improved. It should be noted, that due to the strong self-biasing, $V_{\text{b}}^{\text{ext}}$ as indicated by (13) has been adjusted manually during measurement to maintain $V_{\text{b}}^{\text{int}}$ at optimum across P_{in}. By choosing, e.g. a replica bias or a feedback loop, this change in optimum biasing voltage due to self biasing or potentially across PVT variations could be sensed

and corrected without the need for manual adjustment of $V_{\text{b}}^{\text{ext}}$.

The output power up to the fourth harmonic at an input power of 0 dBm swept across the input frequency is visualized in Fig. 6. With increasing f_{in} the output power of the second harmonic slightly decreases.

Finally, Table 3 compares the implemented frequency doubler with other recent publications. Without large passive components the doubler consumes minimum power and occupies the smallest area compared to other publications.

V. CONCLUSION

We have presented a miniature frequency doubler in 45 nm CMOS RFSOI technology. High fundamental rejection up to 35 dB and a large output frequency range from 2 to 20 GHz are measured, while 4.4 mW is dissipated from a 1.1 V supply.

ACKNOWLEDGMENTS

The authors thank Global Foundries for chip area donation.

REFERENCES

[1] V. Issakov, "The State of the Art in CMOS VCOs: Mm-Wave VCOs in Advanced CMOS Technology Nodes," *IEEE Microwave Magazine*, 2019.
[2] V. Issakov et al., "A Dual-Core 60 GHz Push-Push VCO with Second Harmonic Extraction by Mode Separation," *RFIC*, 2018.
[3] A. Bilato et al., "Considerations on 120GHz LO Signal Generation and Distribution for Highly-Integrated Multi-Channel Radar Transceivers," *2019 COMCAS*, 2019.
[4] A. Thakkar et al., "Switched Capacitor Bank Design with Geometric Progression for Wideband Frequency Doubler," *SSC-L*, 2021.
[5] S. Vehring et al., "A 3.1-dBm E-Band Truly Balanced Frequency Quadrupler in 22-nm FDSOI CMOS," *MWC-L*, 2020.
[6] P.-H. Tsai et al., "Broadband Balanced Frequency Doublers With Fundamental Rejection Enhancement Using a Novel Compensated Marchand Balun," *TMTT*, 2013.
[7] F. Ellinger et al., "Ultracompact SOI CMOS frequency doubler for low power applications at 26.5-28.5 ghz," *MWC-L*, 2004.
[8] Y. Gong et al., "Ultralow Power K-Band Frequency Doubler with Differential Output," *MWC-L*, 2021.
[9] Y. Gong et al., "An Ultra-Low Power K band Balanced Frequency Doubler with a Novel Current-reused Structure," *ESSCIRC 2021*, 2021.
[10] S. Li et al., "A Buffer-Less Wideband Frequency Doubler in 45-nm CMOS-SOI With Transistor Multiport Waveform Shaping Achieving 25% Drain Efficiency and 46-89 GHz Instantaneous Bandwidth," *SSC-L*, 2019.
[11] E. Monaco et al., "Injection-Locked CMOS Frequency Doublers for u-Wave and mm-Wave Applications," *JSSC*, 2010.
[12] C. Enz et al., "Charge-Based MOS Transistor Modeling: The EKV Model for Low-Power and RF IC Design," Chichester, UK, *Wiley*, 2006.
[13] R. M. Corless et al., "The wright ω function," in *Artificial Intelligence, Automated Reasoning, and Symbolic Computation*, Springer, 2002.

979-8-3503-2123-4/23 $31.00 © 2023 IEEE

RTu1A-4

An Efficient 0.4 THz Radiator with 20.6 dBm EIRP and 0.2% DC-to-THz Efficiency in 90nm SiGe BiCMOS

Sidharth Thomas[*], Sam Razavian[*], Aydin Babakhani
University of California, Los Angeles, USA
{sidhthomas, samrazavian24}@g.ucla.edu, aydinbabakhani@ucla.edu

Abstract — This paper presents an efficient 0.4 THz single-element radiator implemented in 90nm SiGe BiCMOS. It consists of a PIN diode quadrupler, where a mm-wave Colpitts oscillator at 100 GHz drives a PIN diode switching-reactance-multiplier into reverse recovery. Because of this, the PIN diode abruptly switches between two impedance states and produces strong harmonics. Harmonic injection locking is also presented in this work, where two quadrupler cells are mutually interlocked, and their fourth harmonic power at 0.4 THz combines at the antenna. The radiator achieves a peak EIRP of +20.6 dBm and -5.8 dBm radiated power at 398 GHz, with a 10.7 % tuning range, and consumes 130 mW DC power. This work has a DC-to-THz generation efficiency of 0.2%, the highest reported efficiency above 320 GHz, and achieves the highest power and EIRP generated by a single-element radiator.

Keywords — PIN diode, multiplier, THz source, SiGe

I. INTRODUCTION

The THz band (0.1-10 THz) has several unique properties, making it a potential candidate for imaging, spectroscopy, and communication applications. These applications require efficient high-power THz generation over a wide frequency range. Due to the limited f_T/f_{max} of transistors, direct THz generation is not feasible, and harmonic generation using non-linear devices such as transistors or varactor diodes is popularly used [1-4]. However, these devices have weak non-linearity, limiting the amount of generated THz power. PIN diodes have shown strong non-linearity and have been used for THz generation [5], [6].

A large-scale array implementation, with free space power combining, can be employed to boost the radiated THz power. However, this approach faces major challenges: (1) Arrays consume high DC power. (2) Inter-element frequency and phase synchronization become challenging at THz frequencies. (3) When used with a silicon lens, arrays have less directivity than a single-element radiator due to lens off-axis misalignment. Because of this, a single-element radiator can provide superior EIRP than an array for the same DC power [7].

This work presents an efficient 0.4 THz single-element radiator based on PIN diode switching reactance frequency multipliers (SRM). Due to reverse recovery, PIN diodes are strongly non-linear and are hence used here to maximize DC-to-THz generation efficiency. To improve the power, on-chip power combining at the antenna node is adopted, where two quadruplers are mutually interlocked, and the 0.4

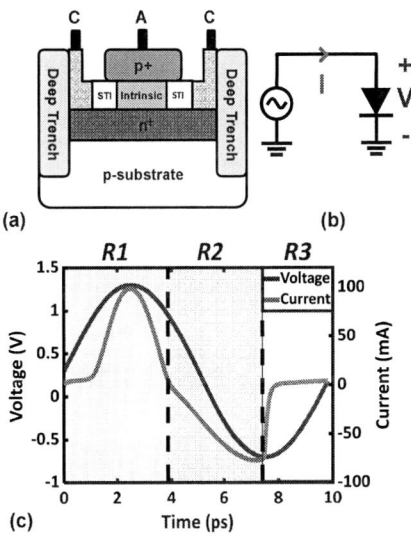

Fig. 1. (a) PIN diode illustration (b) PIN diode excited by a voltage source (c) Current through the PIN diode when excited by a 100 GHz voltage source

THz signal adds in-phase. Performing power combining at the antenna ensures low loss, and the THz signal is directly radiated.

II. PIN DIODE AND REVERSE RECOVERY

The reverse recovery of PIN diodes is used in this work for efficient THz generation. A PIN diode consists of p+ and n+ regions, separated by an intrinsic 'I' region. The presence of the 'I' region differentiates a PIN diode from conventional diodes. A standard PIN diode from GlobalFoundries 90nm process is used here and is illustrated in Fig. 1 (a). Fig. 1 (b), (c) demonstrate the operation of a PIN diode when a large-signal sinusoidal voltage is applied across it. In the forward mode of operation (R1), the diode conducts, and the I-region is filled with electrons and holes to facilitate forward conduction. Because of these excess carriers, the diode remains ON and continues current conduction in the reverse mode (R2) until all excess carriers are depleted. This is different from a conventional diode, where there is no carrier storage, and the diode stops conducting as soon as it leaves R1. Once the carriers in the PIN diode are depleted, the current snaps to zero (R3), exhibiting reverse recovery. This switching is abrupt, and the generated waveform is rich in harmonics. During a typical drive cycle, the diode operates across the conduction regions (R1 and R2) and the OFF state region (R3). In R1 and R2,

---

[*] equal contribution

979-8-3503-2123-4/23 $31.00 © 2023 IEEE   213   2023 IEEE Radio Frequency Integrated Circuits Symposium

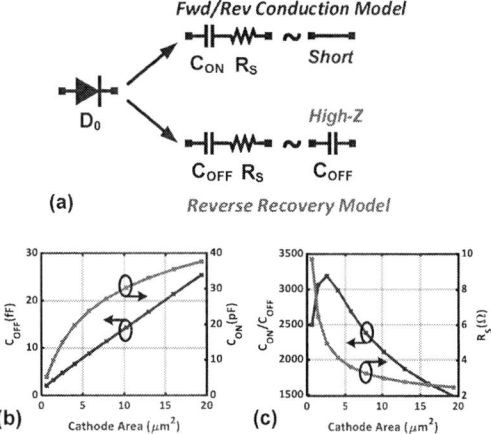

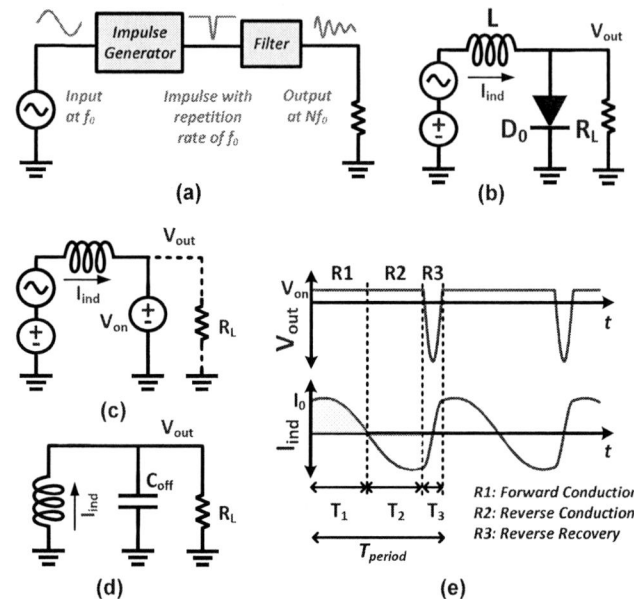

Fig. 2. (a) Large signal RF equivalent circuit model of a PIN diode (b) C_{OFF} and C_{ON} vs diode cathode area (c) C_{ON}/C_{OFF} and R_S vs diode cathode area

ideally, the diode can store any amount of charge at a constant voltage (turn-on voltage of the diode). In R3, the diode behaves like a small capacitance.

A large-signal equivalent for the PIN diode is presented to understand the circuit functionality better. Fig. 2 (a) shows a simplified large-signal RF model of the diode. During forward and reverse conduction (Regions R1 and R2), the diode presents low impedance and behaves as a large capacitance C_{ON} in series with the diode impedance R_S. Ideally, this can be approximated as a short. When turned OFF (Region R3), the diode presents a high impedance and behaves as a small capacitance C_{OFF} in series with R_S. Ideally, this can be approximated as just C_{OFF}. The parameters C_{ON}, C_{OFF}, and R_S for various diode sizes are plotted in Fig. 2 (b), (c).

III. CIRCUIT DESIGN

The PIN diode quadrupler design is explained in this section. It consists of a mm-wave oscillator at 100 GHz that drives a PIN diode. Two quadruplers are interlocked, and the power at 0.4 THz is combined and radiated through an on-chip antenna.

A. Switching Reactace Multiplier

PIN diode multipliers are Switched Reactance Multipliers (SRMs), where the diode switches abruptly from a short to a high impedance state. The fast switching behavior enables efficient THz generation using PIN didoes. This differs from conventional varactor or transistor-based multipliers where the reactance variation, although non-linear, is gradual (Variable Reactance Multiplier, VRM). SRMs were designed in the past using discrete microwave step-recovery diodes (SRD) [8]. However, they are not available in modern silicon processes. This work extends the SRM theory to PIN diodes since they behave similarly to SRDs under large-signal operation.

Fig. 3 (a) illustrates a general SRM circuit. It consists of an impulse generator circuit that is excited at a frequency f_0. This generates an impulse that can be filtered to extract

Fig. 3. (a) A switched reactance frequency multiplier (b) PIN diode impulse generator (c) Equivalent circuit during R1 and R2. (d) Equivalent circuit during R3. (e) V_{out} and inductor current

the harmonic of interest at 'Nf_0'. The impulse train generator circuit is shown in Fig. 3 (b). A PIN diode is connected to an RF source through an inductor and to a load R_L. The non-linear time-varying behavior of the PIN diode can be analyzed by replacing the diode with the linear switching model from Fig. 2 (a), and applying boundary conditions. The simplified circuit is shown in Fig. 3 (c), (d). This circuit is solved using linear analysis, and the resulting diode voltage (V_{out}) and inductor current (I_{ind}) are plotted in Fig. 3 (e). Regions R1, R2, and R3 correspond to forward conduction, reverse conduction, and reverse recovery mode. During forward and reverse conduction (R1, R2), the PIN diode is ON and clamped to the diode turn-on voltage V_{ON}. Charges stored in the 'I' region of the diode during R1 are depleted during R2, and the diode turns OFF at R3 when this charge becomes zero ($Q1 = Q2$). We bias the diode such that it carries peak reverse current at the onset of R3. Even though the diode turns OFF, current continuity through the inductor must be maintained. This results in a sharp negative impulse voltage at V_{out}. The height and width of this impulse depend on C_{OFF} and R_L. This impulse is then filtered to extract the harmonic of interest at Nf_0.

B. THz Transmitter Design

Fig. 4 (a) shows the schematic of the PIN diode-based THz SRM quadrupler. It consists of a differential mm-wave Colpitts oscillator, which oscillates at 100 GHz. A cascode stage is used to amplify the oscillator output and isolate the oscillator core. Transmission line (TL) matching networks connect the cascode to the diode and extract the 4th harmonic at 400 GHz. These matching networks ensure that (1) the 100 GHz signal is large-signal matched to the diode and (2) the antenna is matched to the diode at the 4th harmonic at 400 GHz. The

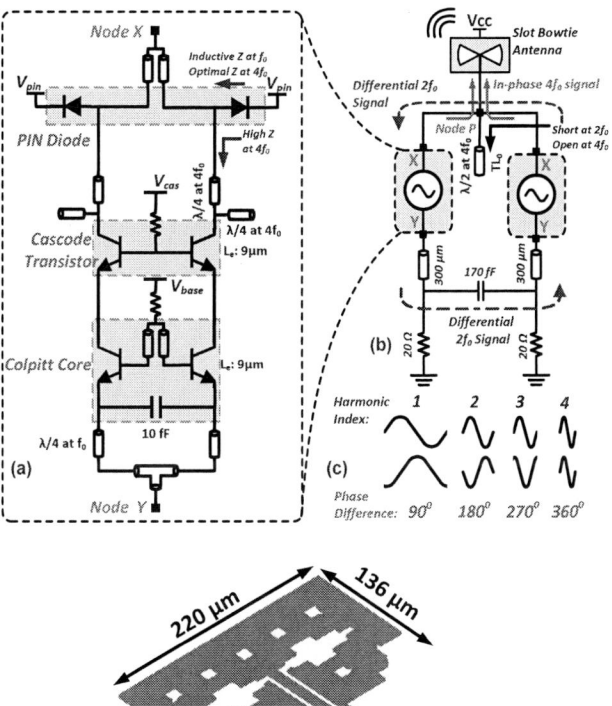

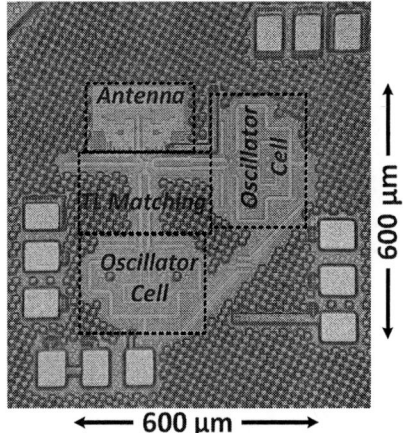

Fig. 5. Die Micrograph

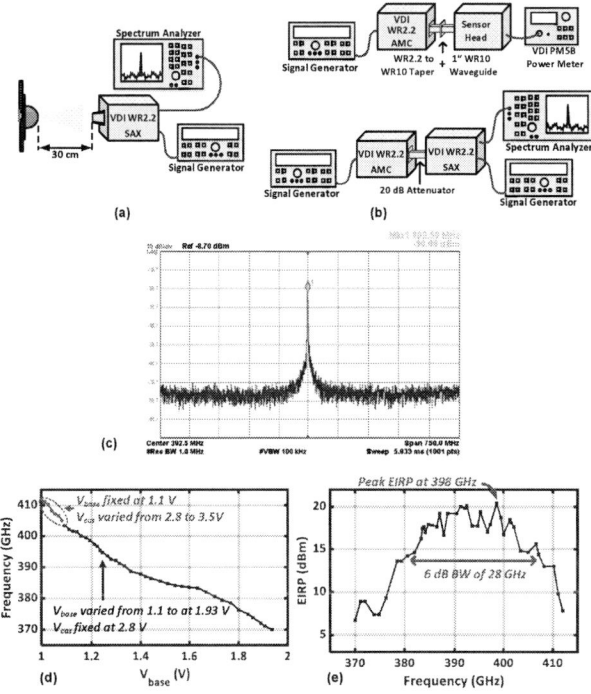

Fig. 6. (a) EIRP measurement setup (b) VDI WR2.2 SAX loss calibration (c) Measured spectrum at 398 GHz (d) Frequency tuning curve (e) EIRP variation vs frequency

Fig. 4. (a) PIN diode switched reactance multiplier at 100 GHz (b) Quadrature locking and power combining (c) Odd mode coupling at 2nd harmonic provides quadrature and in-phase signals at 1st and 4th harmonics respectively (d) Slot bowtie antenna

amount of generated THz power depends on the diode size. Choosing a diode size that has high (C_{ON}/C_{OFF}) ensures the diode switches between widely different impedance levels. Choosing a diode size with low R_S ensures high Q. Thus, the diode size, which maximizes the ratio (C_{ON}/C_{OFF})/R_S, is chosen and this consequently maximizes the generated THz power.

On-chip power combining at the antenna is used in this work. Two quadrupler cells are mutually locked at 100 GHz to boost the power at 400 GHz. Fig. 4 (b) shows the locking mechanism. Node X of both the quadrupler cells are directly connected to the antenna and to an open line TL_0, which is $\lambda/2$ length at 400 GHz. Only even harmonics of the quadrupler cells can interact with each other at node X since it lies on the common mode of the individual cells. The oscillators undergo mutual injection locking and can operate in the odd mode or even mode of the second harmonic. However, TL_0 creates a short at node P at the second harmonic. This prevents the even mode signals from coupling. For odd mode, node P is a virtual ground and remains unaffected by TL_0. Thus odd mode oscillation is sustained at the second harmonic. Consequently, due to harmonic injection locking, the oscillators become quadrature locked at the fundamental frequency and in-phase locked at the 4th harmonic (Fig. 4 (c)). At node Y, since the oscillators cannot be directly connected due to layout

constraints, they are connected through 300 μm TLs and coupling capacitors. Resistors are used for biasing.

A slot-bowtie antenna is used in this work. Fig. 4 (d) shows the structure of the antenna. Additional slots are added to the antenna to meet density requirements. The antenna is used with a hyper hemispherical silicon lens. Since only one antenna is used, it can be aligned to the axis of the lens.

IV. MEASUREMENT RESULTS

The design is fabricated in GlobalFoundries 90nm SiGe BiCMOS process. It consumes 130 mW DC power in nominal

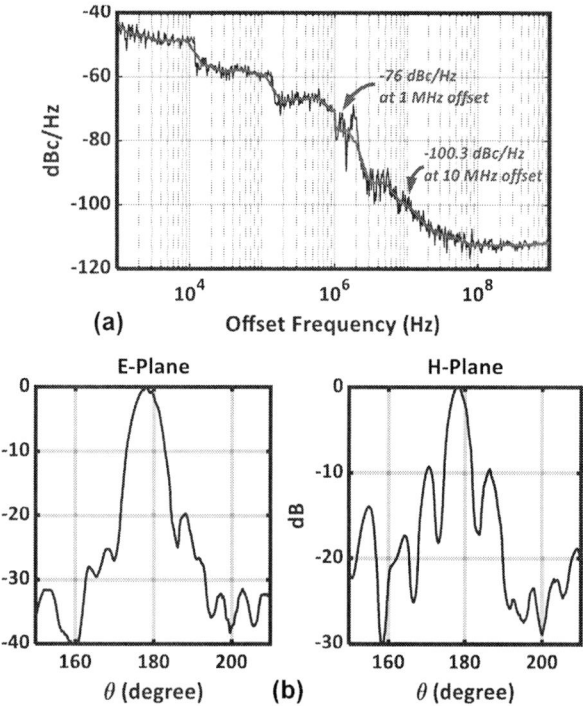

Table 1. Comparison Table

References	This Work	[1]H. Saeidi JSSC'22	[5]S. Razavian ISSCC'22	[2] H. Jalili JSSC'19	[3] L. Gao JSSC'22	[4] Y. Tousi JSSC'15
Frequency (GHz)	398	416	424	344	450	338
EIRP (dBm)	20.6	14	18.1	4.9	29.1	17.1
Radiated Power (dBm)	-5.8	-3	-5	-6.8	-3.2	-0.9
Antenna Array size	1	4 x 4	2 x 3	2 x 2	4 x 4	4 x 4
Radiated Power per Element (dBm)	-5.8	-15	-12.8	-12.8	-15.24	-12.9
DC Power (mW)	130	1450	400	450	347	1540
DC to Radiated Power Efficiency (%)	0.2	0.03	0.08	0.04	0.14	0.05
Tuning Range (%)	10.7	1.7	14.6	15.1	4.6	2.1
Phase Noise (dBc/Hz)	-76 (1 MHz)	-88 (1 MHz)	-104 (10 MHz)	-93.1 (10 MHz)	-76.4 (1 MHz)	-93 (1 MHz)
Area (mm²)	0.36	4.1*	0.98	1.2*	0.55	3.9*
Lens	Silicon Lens	No	Silicon Lens	No	Custom Teflon Lens	No
Technology	90 nm SiGe	65 nm CMOS	90 nm SiGe	130 nm SiGe	65 nm CMOS	65 nm CMOS

Fig. 7. (a) Measured phase noise spectrum (b) Measured radiation pattern

operation. Fig. 5 shows the die micrograph. The design occupies an active area of 0.36 mm^2.

The chip is wire-bonded to a PCB, and a hyper hemispherical silicon lens is attached. The lens helps improve radiation efficiency by removing substrate modes, and increases the directivity. No substrate thinning is employed. Fig. 6 (a) shows the free space measurement setup. The chip is kept at a 30 cm far-field distance from a VDI WR2.2 mixer (SAX). The conversion loss of the mixer is measured using a VDI WR2.2 signal generator extender (AMC), which is calibrated using a VDI PM5B power meter using the setup shown in Fig. 6 (b). The measured spectrum at 398 GHz is shown in Fig. 6 (c). An EIRP of 20.6 dBm is obtained after de-embedding the losses. The frequency is varied by changing the base and cascode bias. This is plotted in Fig. 6 (d). A tuning range of 42 GHz (10.7 %) is measured. The EIRP variation across frequency is plotted in Fig. 6 (e), and a 6 dB bandwidth of 28 GHz is measured. The measured phase noise spectrum is plotted in Fig. 7 (a). A phase noise of -76 dBc/Hz at 1 MHz offset and -100.3 dBc/Hz at 10 MHz offset is measured. The chip is mounted on a rotational stage to measure the radiation pattern. The radiation pattern in E and H planes are plotted in Fig. 7 (b), and a directivity of 26.4 dB is calculated. The total radiated power is -5.8 dBm.

A comparison with other state-of-the-art silicon-based THz radiators is given in Table 1. This work achieves the highest DC-to-THz efficiency (0.2 %), the highest radiated power per element (-5.8 dBm), and one of the highest EIRP (20.6 dBm). Though [3] has higher EIRP, it is achieved using

additional processing steps such as substrate thinning and by using a highly directive custom teflon lens. The radiated power achieved in this work is comparable to multi-element arrays and this high power can directly drive other on-chip circuits such as mixers and multipliers. Array radiators, with free-space combining, do not offer this.

V. CONCLUSION

This work presents a 0.4 THz single-element radiator in 90 nm SiGe BiCMOS. The reverse recovery phenomenon is used in this work to generate efficient THz power. Two multiplier cells are interlocked, and the THz power is combined at the antenna. This work achieves an EIRP of 20.6 dBm, radiated power of -5.8 dBm, and a wide tuning range of 10.7%. It achieves the DC-to-THz efficiency (0.2 %) and the highest radiated power per element (-5.8 dBm) above 320 GHz.

REFERENCES

[1] H. Saeidi et al., "A 4 × 4 Steerable 14-dBm EIRP Array on CMOS at 0.41 THz With a 2-D Distributed Oscillator Network," *IEEE Journal of Solid-State Circuits*, vol. 57, no. 10, pp. 3125–3138, 2022.

[2] H. Jalili and O. Momeni, "A 0.34-THz Wideband Wide-Angle 2-D Steering Phased Array in 0.13- μ m SiGe BiCMOS," *IEEE Journal of Solid-State Circuits*, vol. 54, no. 9, pp. 2449–2461, 2019.

[3] L. Gao and C. H. Chan, "A 0.45-THz 2-D Scalable Radiator Array With 28.2-dBm EIRP Using an Elliptical Teflon Lens," *IEEE Journal of Solid-State Circuits*, vol. 57, no. 2, pp. 400–412, 2022.

[4] Y. Tousi and E. Afshari, "A High-Power and Scalable 2-D Phased Array for Terahertz CMOS Integrated Systems," *IEEE Journal of Solid-State Circuits*, vol. 50, no. 2, pp. 597–609, 2015.

[5] S. Razavian and A. Babakhani, "A Highly Power Efficient 2×3 PIN-Diode-Based Intercoupled THz Radiating Array with 18.1dBm EIRP in 90nm SiGe BiCMOS," in *2022 IEEE International Solid- State Circuits Conference (ISSCC)*, vol. 65, 2022, pp. 1–3.

[6] S. Razavian, and A. Babakhani, "Silicon Integrated THz Comb Radiator and Receiver for Broadband Sensing and Imaging Applications," *IEEE TMTT*, vol. 69, no. 11, pp. 4937–4950, 2021.

[7] H. Jalili and O. Momeni, "A 0.46-THz 25-Element Scalable and Wideband Radiator Array With Optimized Lens Integration in 65-nm CMOS," *IEEE JSSC*, vol. 55, no. 9, pp. 2387–2400, 2020.

[8] S. Hamilton and R. Hall, "Shunt-mode harmonic generation using step recovery diodes," *Microwave Journal*, vol. 10, no. 5, pp. 69–78, 1967.

979-8-3503-2123-4/23 $31.00 © 2023 IEEE

RTu1B-1

A 28nm CMOS Dual-Band Concurrent WLAN and Narrow Band Transmitter with On-chip Feedforward TX-to-TX Interference Cancellation Path for Low Antenna-to-Antenna Isolation in IoT Devices

Sai-Wang Tam, Alireza Razzaghi, Alden Wong, Sridhar Narravula, Weiwei Xu, Timothy Loo,
Akash Kambale, Andrew Liu, Ovidiu Carnu, Yui Lin, Randy Tsang
NXP Semiconductors, USA
sai-wang.tam@nxp.com

Abstract—A dual-band, concurrent 2.4G WLAN and 2.4G narrow band (NB) transmitter (TX) with on-chip feedforward TX-to-TX interference cancellation path for low antenna-to-antenna isolation in IoT devices is proposed. An on-chip cancellation path generates a replica signal of the same magnitude but 180° out-of-phase with respect to the "aggressor" TX signal appeared at the "victim" TX output. With cancellation path properly calibrated, the measured IMD3 product is reduced by 25 dB. Additionally, the maximum output power during concurrent transmission, while meeting the FCC out-of-band emission specification, improves from 10 to 17 dBm across all WLAN channels. With this proposed architecture, the issue of TX-to-TX interference in multi-radio coexistence is finally addressed, opening the door to future high power concurrent multi-band transmitters in reconfigurable IoT devices.

Keywords— multi-radio coexistence, concurrent transmitter, power amplifier, class-AB, inverse class D, feedforward, IMD3, interference cancellation, IoT

I. INTRODUCTION

With an increasingly advanced IoT device wireless landscape, demands increase not only for seamless and secure connectivity for smart home and industrial use cases, but also for the development of future groundbreaking applications and connectivity protocols. Future IoT devices must support high data rates, be highly reconfigurable and energy efficient, conform to small form factors, and support multi-radio coexistence. Furthermore, the current trends in form factor reduction directly resulted in an increasing number of reconfigurable multi-radio systems coexisting in the same wireless environment with much lower isolation among antennas within the same platform. Two primary challenges arise when multiple radios coexist within low antenna-to-antenna isolation environments. First, the high-output-power transmitter (TX) signal directly couples to the receiver (RX) antenna, causing TX-to-RX self-interference. This leads to the desensitization of the received signal and; hence, degradation of the overall RX sensitivity and bit-error-rate (BER). Second, the high-output-power TX signal of one radio can also couple to the high-output-power TX of another radio during concurrent transmission, causing TX-to-TX interference. This results in two high-power TX signals mixing with each other and generating undesirable IMD3 products. Many TX-to-RX self-interference cancellation schemes have been proposed [1-5]. On the other hand, little has been done to resolve the issue of TX-to-TX interference. To fill this gap, this paper proposes a dual-band, concurrent 2.4G WLAN and 2.4G NB (Bluetooth/IEEE 802.15.4) transmitter with an on-chip feedforward TX-to-TX interference cancellation path for low antenna-to-antenna isolation application in IoT Devices.

II. TX-TO-TX INTERFERENCE

Concurrent transmission of two high-output-power TXs (~20 dBm), 2.4G WLAN and 2.4G NB, is illustrated in Fig. 1(a). With 20 dB TX-to-TX isolation, a 0 dBm aggressor signal appears at the output of each respective TX. This aggressor signal couples through the TX balun to the power amplifier (PA) output where the high-output-power signal (victim) is present. The non-linear parasitic drain-bulk diode at the PA output node subsequently mixes these two signals and generates the undesirable IMD3 product, violating the FCC out-of-band emission specification of -41.5 dBm/MHz. The NB signal only occupies a bandwidth of 1 MHz compared to the 40 MHz WLAN signal. Consequently, the NB signal has a higher power spectral density than WLAN signal, leading to a larger IMD3 product at WLAN TX output when the NB PA transmits at high output power. The conventional solution to reduce the IMD3 product is to simply back-off the output power of both transmitters. Therefore, the total output power would be limited by the back-off applied to meet the FCC out-of-band emission specification. Another solution is to directly cancel the IMD3 product. However, due to the high dynamic range between the victim signal and the IMD3 product (~60 dB), the cancellation and calibration circuitries cannot be implemented with reasonable complexity and power consumption. To overcome the above trade-off, this paper proposes a solution, Fig. 1(b), that generates an inverse replica of the aggressor NB TX signal injected directly at the victim WLAN TX output, subsequently reducing the undesirable IMD3 product. Because the dynamic range between the victim and aggressor signal is now much smaller (~20 dB), the requirements of the cancellation and calibration circuitries become significantly relaxed. Additionally, compared with TX-to-RX self-interference cancellation, the TX-to-TX interference cancellation operates in the large signal regime and; hence, becomes limited by the linearity, dynamic range, and large signal coupling rather than the electronic noise. Subsequent sections will discuss these design challenges with respect to the ultimate goal of reducing the IMD3 product.

979-8-3503-2123-4/23 $31.00 © 2023 IEEE

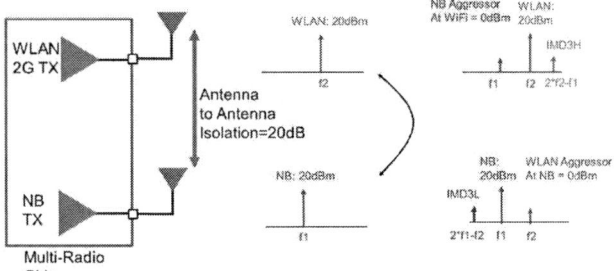

Fig. 1 (a): Transmitter IMD3 due to TX-to-TX interference

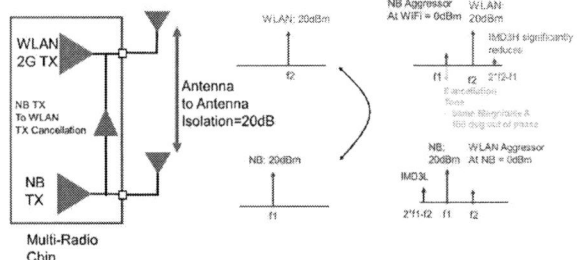

Fig. 1(b): Proposed on-chip TX-to-TX cancellation technique with the "aggressor" NB TX signal canceled at the "victim" WLAN TX output

III. PROPOSED TX-TO-TX CANCELLATION ARCHITECTURE

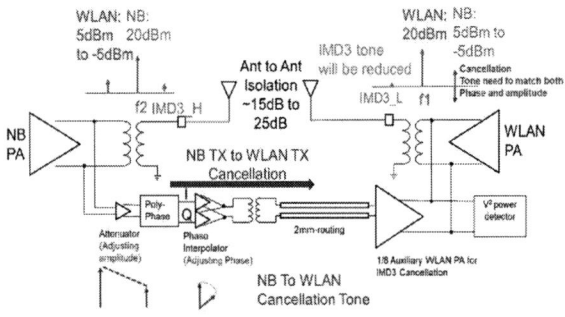

Fig. 2: Detailed architecture of proposed on-chip TX-to-TX cancellation with built-in calibration

Shown in Fig.2, the proposed TX-to-TX cancellation scheme consists of the NB PA, WLAN PA, NB TX signal attenuator, poly-phase filter, IQ-phase interpolator, 2 mm NB-TX-to-WLAN-TX routing trace, 1/8-sized auxiliary WLAN PA, and V^2 power detector.

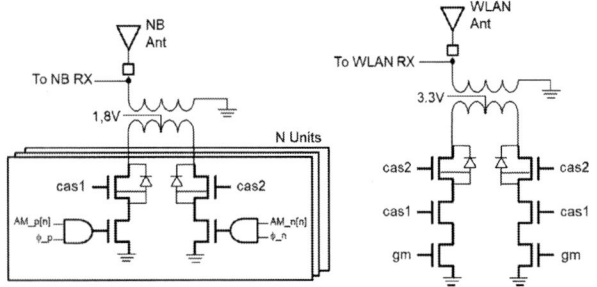

Fig.3: (a) Class D^{-1} Digital NB PA.   (b) Analog Class AB WLAN PA

Fig. 3(a) shows the schematic of the inverse class D digital NB PA with 1.8 V power supply and the integrated front-end

with the NB RX connected to the secondary of the NB TX balun. Since the NB transmitted signal is a constant envelope signal with only 1 MHz modulation bandwidth and 21 dBm output power, a single-cascode PA is employed with a static 10-b amplitude code, comprised of 3-b thermometer and 7-b binary segments. On the other hand, the WLAN TX signal is an OFDM signal with a 40 MHz modulation bandwidth, which has stricter linearity, spectrum mask, and saturated output power requirements than NB TX. Fig. 3(b) shows the schematic of the double-cascode analog class AB WLAN PA with a 3.3 V power supply and the integrated front-end with the WLAN RX connected to the secondary of the WLAN TX balun. During concurrent transmission at high output power (~20 dBm), strong coupling occurs between the NB TX and the WLAN TX due to the low antenna-to-antenna isolation. These signals mutually couple to each respective PA output and mix with each other through the non-linear junction capacitance of the drain-bulk diode of the cascode device, thus generating undesirable IMD3 products in both PAs.

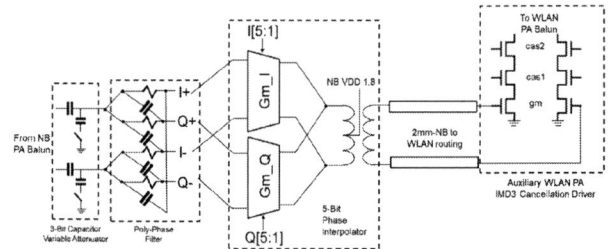

Fig.4: Schematic of the NB TX to WLAN TX Cancellation Path

Architecture selection of the cancellation path highly depends on the group delay of the over-the-air antenna-to-antenna channel and the modulation bandwidth of the aggressor signal. The antenna-to-antenna group delay (τ_{ant}) in a typical IoT device is on the order of several nano seconds, while the modulation bandwidth of the NB aggressor signal (BW_{NB}) is ~1 MHz. In this case, because the $\tau_{NB} = 1/BW_{NB} = ~1000\ ns$ time constant is much larger than the τ_{ant} ~1 ns group delay, the NB TX to WLAN TX aggressor signal appears as a sinusoidal, continuous wave (CW) tone with a time period equal to that of the NB TX carrier. This approximation greatly simplifies the delay requirement of the cancellation path. A simple 360° phase shifter to tune the phase of the NB TX carrier can now be implemented instead of a wide-band nano second true-time delay line [6-8], which is impractical in a CMOS process due to its large power and area. The schematic of the NB TX to WLAN TX 360° phase shifter is displayed in Fig.4. First, the 3-b coarse variable capacitive divider attenuates the large NB PA output signal. Second, the poly-phase filter decomposes the attenuated signal into complex I and Q components which are 90° out-of-phase. These IQ signals are then fed into a 5-b IQ phase interpolator to create phase rotations up to 360° along with additional fine adjustment of the magnitude through the programmable transconductors, Gm_I and Gm_Q. Finally, the phase-rotated output signal drives a 2 mm on-chip differential transmission line from the NB TX to the WLAN TX. The output transformer of the phase interpolator is located at the NB TX side rather than the WLAN

979-8-3503-2123-4/23 $31.00 © 2023 IEEE   218

TX side to avoid the long routing of the NB supply and hence its cross-talk with WLAN TX supply. On the WLAN TX side, an auxiliary PA driver with 1/8th the size of the WLAN PA delivers an NB cancellation signal with the same magnitude but 180° out-of-phase compared to NB aggressor signal at the WLAN PA drain. Moreover, this driver must be biased in a highly linear deep class A region instead of class AB to avoid any cross modulation between the NB cancellation signal, the victim WLAN signal, and the NB aggressor signal.

During calibration, the aggressor NB TX transmits a CW tone at saturated output power while the victim WLAN TX is turned on without transmitting any signal. The NB TX to WLAN TX cancellation path adjusts the phase and magnitude of the NB cancellation tone until the output of the WLAN TX V^2 power-detector reaches at minimum. This minimum point represents the maximum achievable cancellation of the aggressor signal at the output of the WLAN TX. Given the cancellation specification, the maximum WLAN and NB concurrent transmitting power is limited by the dynamic range of this V^2 power-detector and the antenna-to-antenna isolation.

IV. MEASUREMENT RESULTS

The multi-radio test-chip is fabricated in a 28nm CMOS process with the NB PA, WLAN PA, and NB TX to WLAN TX feedforward cancellation path as illustrated in Fig. 5. Fig. 6 shows the over-the-air channel measurement setup using two dipole antennas. The over-the-air antenna isolation is approximately 23 dB across the 2.4 to 2.5 GHz frequency range. A 16 dB directional coupler is connected to each TX output, coupling the respective output signals to two spectrum analyzers so that the IMD3 products can be monitored during the TX-to-TX interference cancellation. In the case of the conductive testing, an external digital RF attenuator replaces the pair of dipole antennas to allow for tunability on the NB TX to WLAN TX isolation.

Before measuring the IMD3 performance using the TX-to-TX interference cancellation path, calibration is performed on the cancellation path across all channel combinations between NB and WLAN. The output of the power detector during calibration, shown in Fig. 7, exhibiting a minimum point at the phase rotation of 280°. Once calibration is completed, the IMD3 performance is measured using the spectrum analyzer. The WLAN TX spectrum plot in Fig. 8 demonstrates a 25 dB reduction in the IMD3 product with the NB TX transmitting a 1MHz bandwidth BLE signal at 17 dBm and the WLAN TX concurrently transmitting an 802.11b signal at 17 dBm and 23 dB of antenna-to-antenna isolation between the two TXs. This represents the worst case IMD3 product for all combinations of NB and WLAN signal modulations. Fig. 9 shows the WLAN TX IMD3 product versus the WLAN TX output power at 2.472 GHz with constant NB TX output power of 20 dBm at 2.423 GHz. The cancellation significantly boosts the maximum concurrent output power from 10 dBm up to 20 dBm without violating the FCC out-of-band emission specification. Fig. 10 reveals the IMD3 performance at fixed WLAN frequencies of 2.472 GHz and 2.412 GHz for NB frequencies below and above 2.440 GHz, respectively. As seen in this figure, the proposed

TX-to-TX interference cancellation meets the FCC out-of-band emission specification across all channel combinations during concurrent NB and WLAN transmission at 17 dBm.

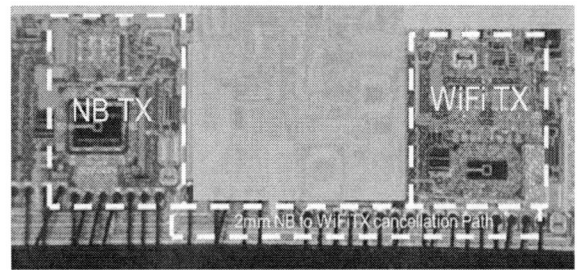

Fig. 5: Chip micrograph of the multi-radio test-chip in 28nm CMOS process

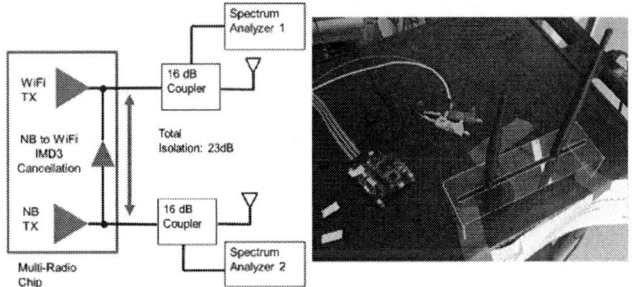

Fig. 6: Over-the-air IMD3 measurement setup diagram (left) and photograph (right)

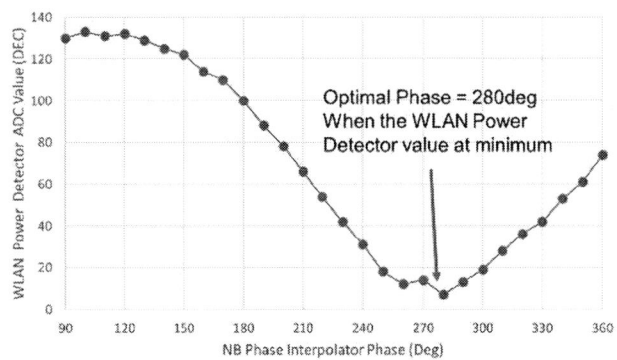

Fig. 7: WLAN power detector measured output during NB TX to WLAN TX calibration.

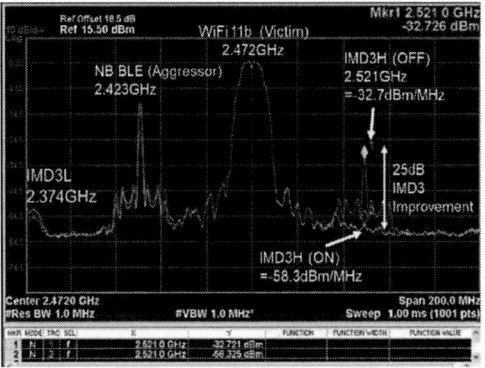

Fig. 8: Spectrum plot of the WLAN TX IMD3 measurement under the 23 dB antenna-to-antenna isolation with NB TX and WLAN TX concurrently transmitting 17 dBm of output power.

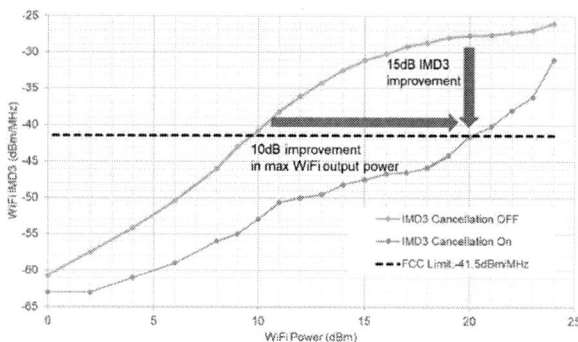

Fig. 9: WLAN TX IMD3 performance versus WLAN TX output power at 2.472 GHz with a 20 dBm NB TX aggressor at 2.423 GHz.

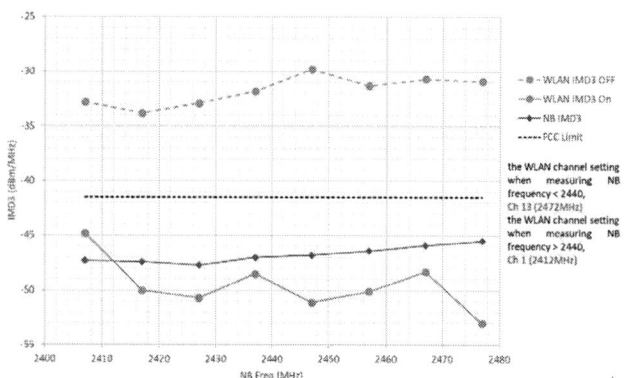

Fig. 10: IMD3 performance across all NB channels with an over-the-air antenna-to-antenna isolation of 23 dB and NB and WLAN concurrently transmitting at 17 dBm

Table 1: Performance Summary

Multi-Radio SoC Test-Chip in 28nm CMOS	
2.4G NB PA	
PA Architecture	Digital Class D^{-1}
Supporting Modulation	BLE, 802.15.4
Saturated Output Power	23.3 dBm
Drain Peak Efficiency	27% on 1.8 V supply
2.4G WLAN PA	
PA Architecture	Analog Class AB
Supporting Modulation	802.11a/b/g/n/ac/ax up to 40 MHz
Saturated Output Power	25.5 dBm
Drain Peak Efficiency	23.5% on 3.3 V supply
TX Output Power for IMD3 < -41.5 dBm/MHz	
Cancellation OFF (across all channel combinations)	NB=10 dBm (BLE) WLAN=10 dBm (11b)
Cancellation ON (across all channel combinations)	NB=17 dBm (BLE) WLAN=17 dBm (11b)
Power Consumption of the Cancellation Path	NB Phase Interpolator = 46 mW WLAN Auxiliary PA Driver = 38 mW

V. SUMMARY AND CONCLUSION

In summary, we demonstrated a robust 28nm dual-band, 2.4G WLAN and 2.4G NB, high output power concurrent transmitter with an on-chip feedforward TX-to-TX interference cancellation path attenuating TX IMD3 product by 25 dB with antenna-to-antenna isolation of 23 dB. Table 1 summarizes the overall performance of the proposed concurrent NB and WLAN

transmitter along with the power consumption breakdown of the cancellation path. An on-chip IMD3 calibration engine carries out the cancellation calibration across all channel combinations in NB and WLAN using an over-the-air antenna-to-antenna channel. With the cancellation off, both NB and WLAN can only transmit concurrently at 10 dBm before violating the FCC out-of-band emission specification of -41.5 dBm/MHz. With the cancellation on, both NB and WLAN can transmit concurrently at 17 dBm across all channels and achieve an IMD3 product of <-41.5 dBm/MHz. Comparing with the TX-to-RX self-interference cancellation [1-5], the TX-to-TX interference cancellation operates in the large signal regime rather than the small signal regime or digital domain. Moreover, the TX-to-TX interference cancellation is limited by the linearity, dynamic range, and large signal coupling rather than the electronic noise in the TX-to-RX self-interference cancellation. The proposed TX-to-TX interference cancellation architecture addresses and mitigates a significant challenge in the concurrent multi-radio coexisting environment, paving the way for the future high power concurrent multi-band transmitters in reconfigurable IoT devices.

ACKNOWLEDGMENT

The authors would like to thank the NXP wireless connectivity analog/RF layout team for the layout support.

REFERENCES

[1] J. Zhou, A. Chakrabarti, P. R. Kinget and H. Krishnaswamy, "Low-Noise Active Cancellation of Transmitter Leakage and Transmitter Noise in Broadband Wireless Receivers for FDD/Co-Existence," in *IEEE Journal of Solid-State Circuits*, vol. 49, no. 12, pp. 3046-3062, Dec 2014

[2] T. Zhang, C. Su, A. Najafi and J. C. Rudell, "Wideband Dual-Injection Path Self-Interference Cancellation Architecture for Full-Duplex Transceivers," in *IEEE Journal of Solid-State Circuits*, vol. 53, no. 6, pp. 1563-1576, June 2018

[3] K. E. Kolodziej, B. T. Perry and J. S. Herd, "In-Band Full-Duplex Technology: Techniques and Systems Survey," in *IEEE Transactions on Microwave Theory and Techniques*, vol. 67, no. 7, pp. 3025-3041, July 2019

[4] A. Nagulu *et al.*, "A Full-Duplex Receiver With True-Time-Delay Cancelers Based on Switched-Capacitor-Networks Operating Beyond the Delay–Bandwidth Limit," in *IEEE Journal of Solid-State Circuits*, vol. 56, no. 5, pp. 1398-1411, May 2021

[5] A. Hamza *et al.*, "A Code-Domain, In-Band, Full-Duplex Wireless Communication Link With Greater Than 100-dB Rejection," in *IEEE Transactions on Microwave Theory and Techniques*, vol. 69, no. 1, pp. 955-968, Jan. 2021

[6] F. Hu and K. Mouthaan, "A 1–20 GHz 400 ps true-time delay with small delay error in 0.13 μm CMOS for broadband phased array antennas," *2015 IEEE MTT-S International Microwave Symposium*, 2015

[7] E. Zolkov and E. Cohen, "A 0.2–3-GHz N-Path True Time Delay Circuit Achieving <1% Delay Variation Over Frequency," in *IEEE Transactions on Microwave Theory and Techniques*, vol. 70, no. 6, pp. 3224-3233, June 2022

[8] T. Forbes, B. Magstadt, J. Moody, A. Suchanek and S. Nelson, "A 0.2-2 GHz Time-Interleaved Multi-Stage Switched-Capacitor Delay Element Achieving 448.6 ns Delay and 330 ns/mm2 Area Efficiency," *2022 IEEE Radio Frequency Integrated Circuits Symposium (RFIC)*, 2022

979-8-3503-2123-4/23 $31.00 © 2023 IEEE

RTu1B-2

A Distributed Cascode Power Amplifier with an Integrated Analog SIC Filter for Full-Duplex Wireless Operation in 65 nm CMOS

Itamar Melamed, Nimrod Ginzberg, Omer Malka, Emanuel Cohen

Faculty of Electrical and Computer Engineering, Technion, Israel

Abstract— In this work, we propose a fully integrated transmitter front-end based on a balanced distributed cascode power amplifier and a passive second-order reconfigurable reflective self-interference cancellation (SIC) filter for full-duplex wireless applications. The balanced topology provides inherent passive transmit-receive (TX-RX) isolation complemented by the passive SIC filter, which accounts for the signal, noise, and nonlinearity components of the direct TX-RX leakages and the reflections from a commercial Wi-Fi antenna. A front-end chip prototype fabricated in TSMC's 65 nm CMOS process operating between 5-6 GHz and occupying the area of 1.2 mm² achieves 19.5 dBm P$_{sat}$ with 31% peak PAE, 17 dBm OP1dB, and 8-10 dB RX noise figure, along with 40 dB of TX-RX isolation and −30 dB TX EVM at 10 dB power backoff using a 20 MHz Wi-Fi OFDM signal without DPD.

Keywords — Full-duplex, self-interference cancellation (SIC), Power Amplifier, Electrical-Balanced Duplexer (EBD).

I. INTRODUCTION

Full-duplex wireless operation offers increased spectral utilization and reduced latency by allowing same-channel simultaneous transmission and reception (STAR), and has been the focus of intense research in recent years. The electrical-balanced duplexer (EBD) is a common STAR front-end topology, which usually involves a reflective load-balancing filter tuned to match the antenna impedance to create a virtual short circuit at the RX port. Various EBD front-ends complemented by load-balancing filters have been reported [1], [2], [3], [4], [5]. In these works, TX and RX insertion losses and the balancing filters' tuning capabilities remain significant challenges. An alternative balanced configuration of a quadrature balanced transmitter with digital self-interference cancellation (SIC) technique was proposed in [6], however, with limited noise figure (NF) performance as the SIC did not account for uncorrelated noise of the PAs.

In this work, we present a balanced distributed cascode power amplifier (BDC-PA) architecture incorporating the required characteristics of transmission, reception, and cancellation paths of a single-antenna front-end for full-duplex STAR operation. The BDC-PA comprises a common source (CS) driving stage followed by balanced low-noise and linear common-gate buffers, ensuring the leakage signal, noise, and nonlinearity of the CS driver get canceled at the RX port using a reflective passive bandpass SIC filter. This analog filter operates as a balancing network to offset a commercial Wi-Fi antenna and achieves comparable performance as the filters in [1] and [3], albeit with fewer degrees of freedom and controlled elements.

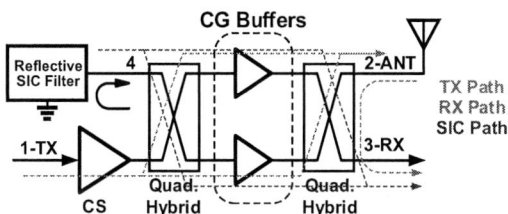

Fig. 1. The proposed distributed cascode PA RF front-end.

II. PRINCIPLE OF OPERATION

A. Balanced Distributed Cascode with a Reflective SIC Filter

The proposed BDC-PA is shown in Fig. 1. The TX path includes a CS driver that delivers the amplified TX signal to two low-noise CG buffers through a quad hybrid. Another quad hybrid at the CG buffers' output reconstructs the TX signal in-phase at the antenna and out-of-phase at the RX port, resulting in inherent passive TX-RX isolation. The RX path utilizes the buffers' output reflectivity to direct RX signals from the antenna to the RX port.

The TX interference at the RX port comprises antenna reflections and direct leakage due to the finite passive isolation. Interference cancellation is performed by limiting the input matching of the CG buffers such that sufficient energy reflects into the SIC filter, which scales it according to the leakage profile, and reflects it back into the circuit to arrive at the RX port to offset the TX leakage over frequency. Hence, the SIC filter accounts for the TX signal leakage and the noise and nonlinearity of the CS driver. Since the noise of the CG buffers cannot be considered by the SIC filter, the buffers were designed to have low gain, low noise, high linearity, and high output reflectivity to minimize their contribution to the RX noise floor and insertion loss.

The transfer function of the TX leakage at the RX port, denoted by $S_{31_{BDC}}$, can be approximated as:

$$S_{31_{BDC-PA}} = -S_{21_{CS}} S_{21_{CG}} (\Gamma_{ANT} S_{22_{CG}} + \Gamma_{SIC} S_{11_{CG}}) \quad (1)$$

where $S_{21_{CS}}$ is the forward transmission of the CS, $S_{21_{CG}}$, $S_{11_{CG}}$ are the forward transmission and input return loss of the CG buffers, respectively, Γ_{ANT} is the antenna reflection coefficient, and Γ_{SIC} is the reflection coefficient of the SIC filter. Cancellation condition, i.e., $S_{31_{BDC-PA}} = 0$, dictates:

$$\Gamma_{SIC} = -\Gamma_A \frac{S_{22_{CG}}}{S_{11_{CG}}} \quad (2)$$

For example, for a BDC-PA with $S_{11_{CG}} = 0.5$ (-6 dBr) and $S_{22_{CG}} = 0.8$ (-2 dBr), we find that $\Gamma_{SIC} = 1.6\Gamma_{ANT}$.

979-8-3503-2123-4/23 $31.00 © 2023 IEEE   221   2023 IEEE Radio Frequency Integrated Circuits Symposium

Table 1. Filter Comparison Table

Ref.	TMTT'16 [1]	ESSCIRC'14 [3]	This Work
# of components	7	7	5
Ind. Value (nH) (Peak Q)	2×1.8 (20)	0.77 (6.1), 0.18 (10.7)	0.4 (25), 0.35 (25)
Cap. Value (pF) (Peak Q)	4×1.25/0.3 (>19)	2×9.1/3.3 (>34)	2×2.4/0.8 (24)
Res. Value (Ω)	1×50	1×200/6.2	1×200/30
Variable Components	4	4	3
Frequency Range (GHz)	1.9-2.2	1.8-2	5-6
# of bits	4×8=32	NA	6+2×10=26
Technology	0.18um CMOS SOI	0.18um CMOS	65nm CMOS

To offset an antenna reflection of $\Gamma_A = 0.4$ (-8 dBr) and assuming the filter has 2 dB insertion loss (in each direction), the reflection required from the filter is $\Gamma_{SIC} = 4$ dBr.

Assuming sufficiently deep SIC in the RF domain, the RX NF in the BDC-PA font-end is determined according to the insertion loss of the RX path and the CG buffers' output-referred noise, as follows:

$$NF = NF_{CG} + S_{21_{CG}} + S_{22_{CG}} + 2IL_{hybrid} \quad (3)$$

where NF_{CG} is the CG buffers' NF and IL_{hybrid} is the insertion loss of the output hybrid. Therefore, the buffers should have low gain, low NF, and high output reflectivity to minimize NF. Also, this architecture provides IMD suppression since the cascode amplifier nonlinearity, dominated by the CS nonlinearity, is cancelled at the RX port, similar to the CS noise.

B. Electrical Balancing RLC Network

To maximize the magnitude of Γ_{SIC} and allow the cancellation of commercial Wi-Fi antennas, the reflective SIC filter insertion loss should be minimized. The return loss of a filter depends on the impedance transformation it performs relative to a 50 Ω resistor and is also highly dependent on the filter's quality factor [1]. Table 1 compares the RLC filter network in this paper with [1] and [3]. In [3], the authors employed a controlled resistor at the far end of the filter, adding an extra degree of freedom. However, despite achieving a quality factor Q>34 for the digital capacitor banks, the low Q of the transformers degraded the return loss. The controlled capacitors and the inductors in [1] show similar quality factors. Therefore, we decided to employ two inductors and decrease the number of capacitors from four to two by implementing a second-order bandpass filter with a controlled resistor, as shown in Fig. 2.

III. CIRCUIT IMPLEMENTATION

The BDC-PA circuit is shown in Fig. 2. The PA is based on a modified, lumped-element cascode structure, in which the output of the CS stage is split in quadrature using a hybrid coupler and fed into two identical CG buffers. The outputs of the CG buffers are then input to another hybrid that delivers them in-phase to the antenna port and out-of-phase to the RX port.

The first hybrid coupler, QC_1, which serves as the CS stage load, was designed with a characteristic impedance $Z_{01} = 16$ Ω for maximum power transfer into the CG buffers. Similar to regular cascode, the CS stage reuses the bias current of the

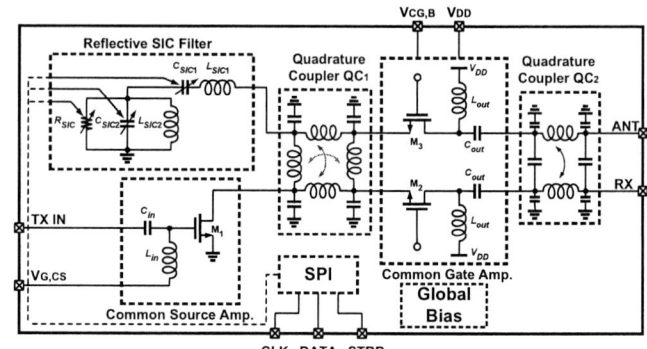

Fig. 2. BDC-PA Circuit Schematic

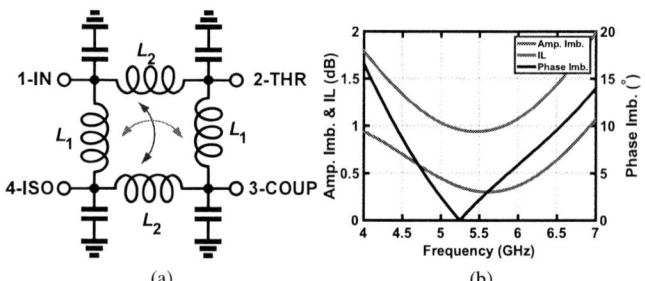

Fig. 3. (a) Schematic of QC_1, (b) QC_1 performance.

CG buffers for better power efficiency. Such a configuration requires a DC short between the hybrid ports; hence we have chosen an all-inductor hybrid implementation, as shown in Fig. 3(a). To minimize losses, we used two pairs of coupled inductors instead of four independent inductors. The EM simulated performance of this hybrid is shown in Fig. 3(b), having an insertion loss of 0.9 dB at 5.5 GHz.

The second hybrid coupler, QC_2, was designed with $Z_{02} = 50$ Ω characteristic impedance required for the antenna and RX matching. Since this coupler does not require a DC short, a more compact design with two coupled inductors and two capacitors was chosen, as shown in Fig. 4(a). EM simulations indicate that the insertion loss of this hybrid is 0.7 dB at 5.5 GHz (see Fig. 4(b)). Simple LC networks were used for input matching of the CS stage and output matching of the CG stages.

According to (2), the SIC filter needs to synthesize a reflection coefficient that coincidences with the antenna and is scaled by the input and output return losses of the CG buffers. Hence, we have implemented a second-order reflective bandpass SIC filter utilizing only three degrees of freedom instead of four, as in [1], while achieving comparable performance. Also, this filter configuration has fewer losses since the number of switched components is reduced.

Tunability is achieved by two switched variable capacitor banks (0.8-2.4pF) with 5-bit thermometric and 5-bit binary weighting and a single 6-bit variable resistor (30 − 200Ω). Two high-Q fixed inductors, optimized at the center frequency of the design, are employed to resonate with the capacitors. Note that the nonlinearity introduced by the SIC filter appears

979-8-3503-2123-4/23 $31.00 © 2023 IEEE

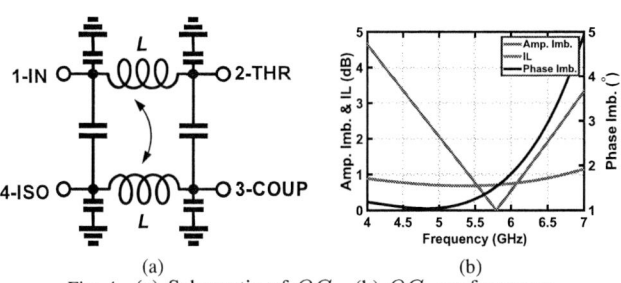

(a)   (b)

Fig. 4. (a) Schematic of QC_2, (b) QC_2 performance.

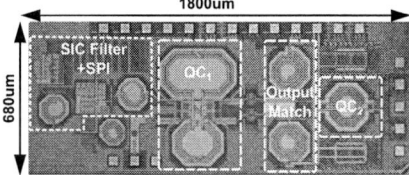

Fig. 5. Die photograph

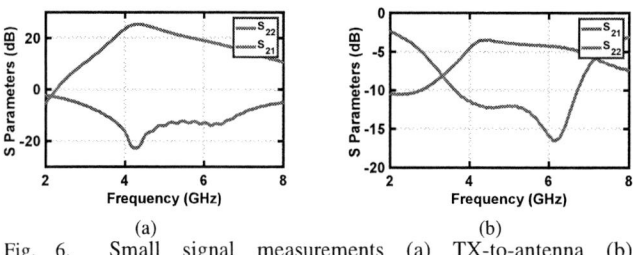

(a)   (b)

Fig. 6. Small signal measurements (a) TX-to-antenna (b) antenna-to-RX

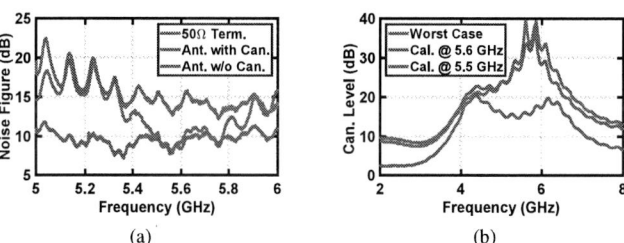

(a)   (b)

Fig. 7. Measurement of (a) Noise figure (b) Cancellation levels

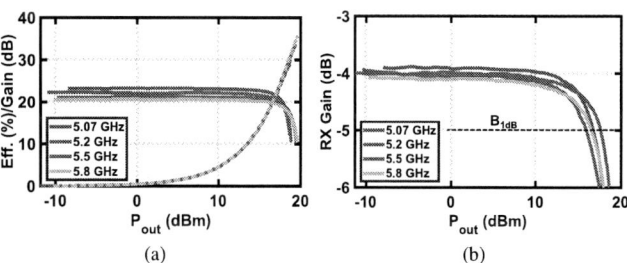

(a)   (b)

Fig. 8. Large signal measurements (a) PA efficiency and gain (b) antenna-to-RX loss

directly in the RX port; hence, we designed it to be below the TX-RX nonlinearity so it would not limit the linearity performance of the design.

IV. EXPERIMENTAL RESULTS

The die photograph is shown in Fig. 5. The chip was implemented in TSMC's standard 65 nm CMOS process. The PA with the SIC filter occupies an area of $1800\mu m \times 680\mu m$ and operates from a 3.3 V supply. The measurements were performed on-wafer with a commercial TAOGLAS FXP830.07.0100C Wi-Fi antenna. The SIC filter states were configured via an on-chip SPI.

A. Small-Signal Measurements

Small-signal characterization was done using a Keysight PNA N5227A network analyzer. Fig 6(a) presents the TX-to-antenna performance with a 50 Ω termination on the RX port. Between $5-6$ GHz, the BDC-PA achieves $19-22$ dB gain with -13 dB output return loss. Fig. 6(b) presents a measurement of the antenna-to-RX path, showing -4 dB RX path insertion loss, attributed to the output hybrid coupler QC_2 IL (back and forth) and the limited reflectivity, $S_{22_{CG}}$, of the CG buffers.

Fig. 7(a) presents the NF measurement in various modes. For a best-case scenario with a 50 Ω termination at the antenna port, the NF is 8 dB at 5.5 GHz. This result was achieved after tuning the SIC filter state for best cancellation at this frequency. With the commercial antenna in a non-optimized filter case, the NF degrades due to the antenna reflection. However, optimizing the SIC filter for maximum cancellation brings the NF back to 8 dB, similar to the 50 Ω case. Fig. 7(b) presents CW measurements of various cancellation states with a 50 Ω antenna termination. A maximum linear cancellation of 35 dB is achieved over an instantaneous bandwidth of 300 MHz.

B. Large-Signal Measurements

Large-signal TX-to-antenna characterization measurement is presented in Fig. 8(a). A CW power sweep was performed in different frequencies across the operating range. The BDC-PA achieves a power gain of 21-23 dB with 17 dBm OP_{1dB}. The peak efficiency is 31% at P_{sat} of 19.5 dBm. Antenna-to-RX loss versus TX output power, measuring power handling capability, is shown in Fig. 8(b), indicating power handling of 14 dBm, for which RX gain compresses by 1 dB.

Fig. 9 presents a two-tone power sweep centered at 5.5 GHz with 20 MHz spacing for an optimized SIC filter state. The graph shows the total antenna power and the first and third-order frequency tones at the RX port (two fundamentals and two IMD_3). At 10 dB power backoff, the linear cancellation is better than 40 dB, and the nonlinear terms are more than 60 dB below the transmitted signal.

C. Full System Measurements

The full system measurement setup is shown in Fig. 10. A Keysight M8190A arbitrary waveform generator was used to drive the PA input with a modulated 20 MHz OFDM Wi-Fi signal. The TX leakage was measured at the RX port using

979-8-3503-2123-4/23 $31.00 © 2023 IEEE   223

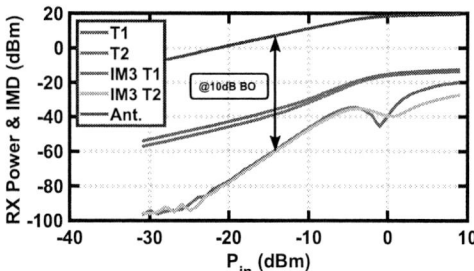

Fig. 9. TX-to-RX two-tone cancellation with 20 MHz spacing at 5.5 GHz

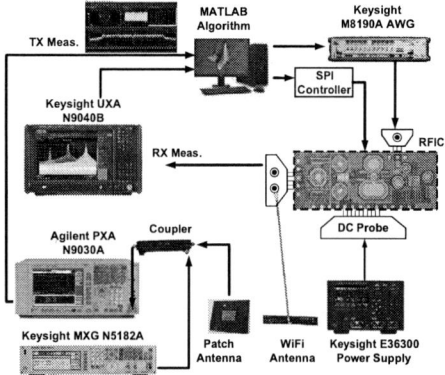

Fig. 10. Measurement setup

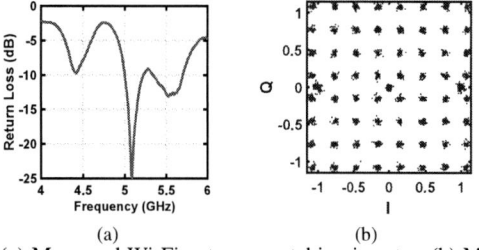

Fig. 11. (a) Measured Wi-Fi antenna matching in setup (b) Measured TX constellation with EVM of -30 dB

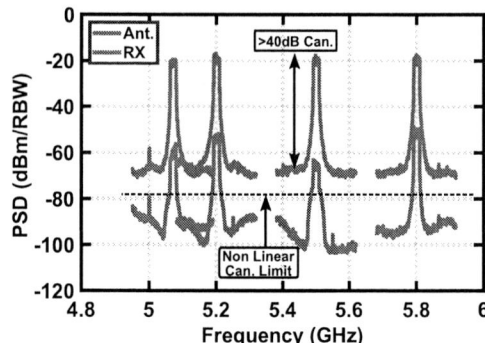

Fig. 12. Spectral measurement of TX and RX with Wi-Fi antenna and SIC

Table 2. Comparison Table

	TMTT'16 [1]	ESSIRC'14 [3]	MWCL'22 [6]	This Work
Technology	0.18um SOI	0.18 CMOS	0.18um CMOS	65nm CMOS
$Z_{ANT}(\Omega)$	1.4:1 VSWR	SkyCross Ant.	Patch antenna	Taoglas Ant.
SIC Architecture	EBD	EBD	QBPA	Reflective SIC+BDC-PA
SIC Test Method	CW	CW	OFDM	OFDM
Freq. Range (GHz)	1.9-2.2	1.78-2	5-6	5-6
Tunable Elements	4	4	Digital	3
Area (mm^2)	1.75	0.67	2.56	1.2 (+PA)
NF (dB)	NA	NA	21	8-10
TX P_{anl} (dBm)	NA	NA	19.3	19.5
TX Power @ OP$_{1dB}$ RX (dBm)	NA	NA	18	14
TX EVM	NA	NA	-33 @ 4 dBm	-30@ 9.5 dBm
RX IL (dB)	<3.9	11	-1.8	-4
TX-antenna IIP3 (dBm)	>70	>48	NA	31 (OIP3)
Antenna-RX IIP3 (dBm)	72	32	NA	33

V. CONCLUSIONS

In this work, we have proposed a fully integrated CMOS transmitter front-end based on a balanced distributed cascode power amplifier along with a reflective, passive, and reconfigurable SIC filter for STAR operation in full-duplex radios. A 65 nm CMOS front-end prototype measured with a commercial Wi-Fi antenna between 5−6 GHz achieves 40 dB of TX-RX isolation in the RF domain, along with competitive system efficiency, linearity, and noise figure performance.

REFERENCES

[1] B. van Liempd, B. Hershberg, S. Ariumi, K. Raczkowski, K.-F. Bink, U. Karthaus, E. Martens, P. Wambacq, and J. Craninckx, "A +70-dBm IIP3 electrical-balance duplexer for highly integrated tunable front-ends," *IEEE Transactions on Microwave Theory and Techniques*, vol. 64, no. 12, pp. 4274–4286, 2016.

[2] M. Mikhemar, H. Darabi, and A. A. Abidi, "A multiband RF antenna duplexer on CMOS: Design and performance," *IEEE Journal of Solid-State Circuits*, vol. 48, no. 9, pp. 2067–2077, 2013.

[3] B. van Liempd, J. Craninckx, R. Singh, P. Reynaert, S. Malotaux, and J. Long, "A dual-notch +27dBm TX-power electrical-balance duplexer," in *ESSCIRC 2014 - 40th European Solid State Circuits Conference (ESSCIRC)*, 2014, pp. 463–466.

[4] S. H. Abdelhalem, P. S. Gudem, and L. E. Larson, "Hybrid transformer-based tunable differential duplexer in a 90-nm CMOS process," *IEEE Transactions on Microwave Theory and Techniques*, vol. 61, no. 3, pp. 1316–1326, 2013.

[5] K. Shi, H. Darabi, and A. A. Abidi, "Design and analysis of an electrical balance duplexer with independent and concurrent dual-band TX-RX isolation," *IEEE Journal of Solid-State Circuits*, vol. 57, no. 5, pp. 1385–1396, 2022.

[6] N. Ginzberg, D. Regev, R. Keren, S. Shilo, D. Ezri, and E. Cohen, "A four-element 5–6-GHz CMOS quadrature balanced full-duplex MIMO transmitter with wideband digital interference cancellation," *IEEE Microwave and Wireless Components Letters*, vol. 32, no. 2, pp. 173–176, 2022.

a Keysight UXA N9040B signal analyzer. The commercial Wi-Fi antenna was directly connected to the die via a GSGSG probe, and the output TX signal was measured using Agilent PXA N9030A signal analyzer and a patch antenna. A Keysight MXG N5182A signal generator was used to measure the antenna-to-RX path loss.

Fig. 12 presents a measurement with the modulated 20 MHz QAM64 OFDM signal across the 5-6 GHz band. The transmitted spectrum at the antenna is located at the top, and the bottom is the RX spectrum. A maximum of 40 dB TX-RX isolation was measured around 5.5 GHz. The nonlinear limit marked in the RX spectrum shows 60 dB of cancellation, which agrees with the two-tone measurement. For the best case cancellation at 5.5 GHz, the measured TX EVM is -30 dB, shown in the constellation in Fig. 11(b) for an output power of 9.5 dBm.

Table 2 compares the performance of this proposal with the state-of-the-art.

RTu1B-3

Frequency-Domain-Equalization-based Full-Duplex Receiver with Passive-Frequency-Shifting N-Path Filters Achieving $>53\,$dB SI Suppression Across 160 MHz BW

Sastry Garimella[#], Sasank Garikapati[#], Aravind Nagulu[*], Igor Kadota[#], Alfred Davidson[#], Gil Zussman[#], Harish Krishnaswamy[#]

[#]Department of Electrical Engineering, Columbia University, USA
[*]Department of Electrical and Systems Engineering, Washington University in St. Louis, USA

Abstract — **Wideband full-duplex (FD) transceivers pose a significant challenge as they require >100dB of self-interference cancellation (SIC) over large bandwidths. This work utilizes (i) a near-zero-power rotary clock-path passive frequency shifting technique for N-path filters while requiring only a single common LO signal across all filters, and (ii) a closed-loop adaptation algorithm to find the optimal configurations of various RF canceler filters using analytical modelling of tap non-idealities caused by frequency shifting and quality factor variations. The FD receiver achieves (i) tunable operation from 200MHz to 1GHz, (ii) wideband SI suppression of up to 53dB across 160MHz BW when operating at 720MHz (4.44× more fractional bandwidth (FBW) compared to [1]), (iii) a power consumption of 1.8mW/DoF (degree of freedom for each tap, almost 2× better than [1]), while (iv) handling TX power of up to +15dBm across an initial circulator isolation of 26dB.**

I. INTRODUCTION

Time-domain equalization (TDE) and frequency-domain equalization (FDE) are the two common architectures utilized to achieve integrated cancellation in full-duplex (FD) and frequency-division duplex (FDD) applications. FIR-based TDE RF SI cancelers have been widely reported [1], [2], [3], [4], and require large (nanosecond-scale) delays at RF with fine resolution. To date, the state of the art is 65dB of SI suppression over 40MHz BW (5% FBW) (including circulator iso.), with 16 RF taps consuming 7.4mW power per RF tap and 8 BB taps [1]. One of the earliest FD works was based on FDE [5], where each canceller tap uses a programmable filter, as opposed to a true time delay. This necessitates multiple widely-tunable high-Q filters, and the cancellation reported in [5] was very limited (20dB across 26MHz) and the active gm-C-based frequency shifting of the N-path filters used was extremely power hungry (47mW per tap).

Fig. 1 compares the trade-offs associated with TDE and FDE cancelers. TDE cancelers typically use all-pass (RC-CR) filter-based delays [2], [3], [4] or switched-capacitor delays [1], with two DoFs per tap (*delay* and *gain*). FDE cancelers based on N-path filters [5] feature four DoFs per tap, namely filter *center frequency*, *Q*, *gain* and *phase*. System level simulations using ideal delay taps (TDE) with 2 DoFs and ideal 2nd order bandpass filter taps (FDE) with 4 DoFs, optimized to cancel a COTS circulator response, indicate that the cancellation levels are similar between the two approaches for the same number of DoFs across different bandwidths (Fig. 1). The added noise

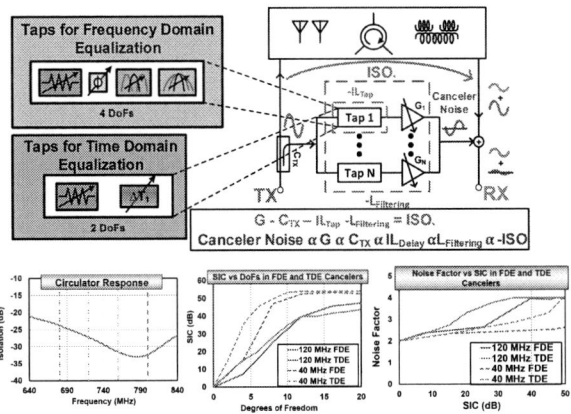

Fig. 1. Architectural differences between TDE- and FDE-based cancelers along with system level simulations showing SIC and NF degradation as a function of number of DoFs.

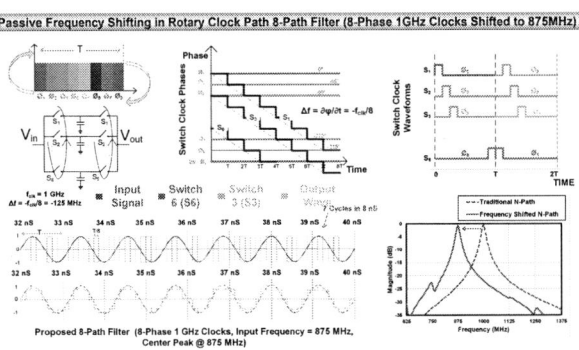

Fig. 2. Concept of passive frequency shifting of N-path filters using rotary clocking - case where $P = 1$; $Q = 1$; $f_c = 1$GHz shifted to 875MHz.

due to the canceler is proportional to the total sum of gains of individual taps which increases with the total number of taps due to higher loss associated with filtering. We plot the expected noise factor of the receiver obtained from the sum of gains in the system level simulations and interestingly, observe a lower expected noise figure penalty for FDE cancellers versus TDE.

In this paper, we revisit FDE-based RF cancellation, but address the power consumption challenge of active gm-C-based frequency shifting by using a near-zero-power

979-8-3503-2123-4/23 $31.00 © 2023 IEEE   225   2023 IEEE Radio Frequency Integrated Circuits Symposium

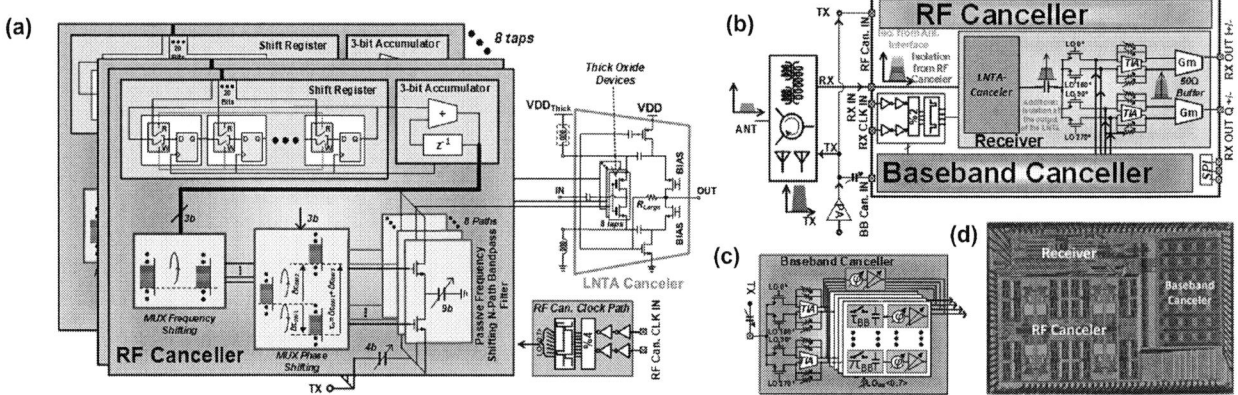

Fig. 3. (a) FDE based RF canceler architecture with rotary-clocked 8-path filters along with clock-path. (b) FD receiver architecture with RF and BB cancelers along with a broadband receiver. (C) TDE based BB canceler with 8 taps along with a down conversion mixer. (d) FD chip micro-photograph

rotary clock-path passive frequency shifting technique for N-path filters. While this technique was previously reported in [6] in another context, here we (i) exploit second-harmonic operation to lower the clock power consumption, and (ii) explore techniques to mitigate the additional spurs produced by rotary clocking. The implemented FD receiver achieves (i) tunable operation from 200MHz to 1GHz, (ii) wideband SI suppression of up to 53dB across 160MHz BW when operating at 720MHz ($2\times$ more bandwidth compared to [1] while achieving similar cancellation), (iii) a power consumption of 1.8mW/DoF (degree of freedom for each tap, almost $2\times$ better than [1]), while (iv) handling TX power of up to +15dBm across an initial circulator isolation of 26dB.

II. NEAR-ZERO-POWER ROTARY CLOCK-PATH PASSIVE FREQUENCY SHIFTING IN N-PATH FILTERS

A. Rotary Clocking in N-path filters

FDE without having separate LO synthesizers for each filter requires a frequency shifting approach for N-path filters. Using baseband gm cells, resulting in an active gm-C N-path filter, results in a high power consumption that scales linearly with the required SIC bandwidth [5]. The proposed rotary-clock-path passive frequency shifting technique (first reported in [6]) is shown in Fig. 2, where the clock phases driving the switches of the N-path filter are periodically rotated in clockwise or anti-clockwise manner, creating a staircase approximation of a phase ramp, or frequency shift. Specifically, rotating the clocks by P phases every Q clock cycles in an N-path filter in clockwise or anti-clockwise manner would result in a frequency shift of $\pm P f_c/QN$ (where f_c is the clocking frequency), as it would take QN/P clock cycles to rotate through a phase shift of 360°. Fig. 2 depicts the waveforms for an 8-path filter for the case of $P = Q = 1$ (frequency shift of $f_c/8$ (as $N = 8$)).

B. Power Reduction Through Second-Harmonic Operation

A traditional N-path filter has filter characteristics around multiples of the clock frequency ($m f_c$, $m \in \mathcal{N}$) except at

$m = kN$ ($k \in \mathcal{N}$)). However, the loss in the corresponding passband increases with m. For $N = 8$, this loss is only greater by $\approx$ 1dB between $m = 2$ and $m = 1$, which can be compensated for when the taps are used for FD RF SIC using gain control. Thus, using 8-path rotary clocking with clocks at half the operating frequency and exploiting the filtering response at $m = 2$ for SIC, the power consumption for each filter tap falls by a factor of 2.

C. Spur Mitigation

Rotary-clocked N-path filters generate additional spurs beyond regular N-path filters due to their LPTV nature (non-zero higher HTFs) at frequency offsets of $mP f_c/Q$ ($m \in \mathcal{Z}$). If the center frequency is $f = f_c - P f_c/QN$ and the input sinusoid is at the same frequency, the closest spur is at $f_{spur} = f_c + P f_c/QN$, and has the potential to be inside the SIC BW. This spur's power level depends on N and reduces as N gets larger, but for $N = 8$, this spur is $\approx$ 20 dB lower in magnitude. Having these spurs in the band of interest for SIC would therefore limit the SIC to be effectively lower than 20 dB. To mitigate this, in this work, we clock the filter taps with a frequency at one edge of the band of interest. That is, if the SIC band of interest is between $f_r - f_{bw}/2$ and $f_r + f_{bw}/2$, then to achieve power reduction and spur mitigation, we clock the filters with clocks having frequency $\frac{f_r + f_{bw}/2}{2}$ and only rotate the clocks in an anti-clockwise manner to shift the filters originally placed at $f_r + f_{bw}/2$ downwards to lie within the band of interest. In other words, *only lower frequency shifts are used*, causing all of the spurs to be pushed above the band of interest. *Therefore, a near-zero-power spur-mitigated fully-passive approach to frequency shifting is achieved.*

III. IMPLEMENTATION OF THE FULL DUPLEX RECEIVER WITH FDE BASED RF CANCELER WITH CLOSED LOOP MEASUREMENT ALGORITHM

A. FDE based RF Canceler

Fig. 3(a) shows the block diagram of the FDE-based RF canceler. The RF canceler has 8 filter taps, with each tap being

979-8-3503-2123-4/23 $31.00 © 2023 IEEE

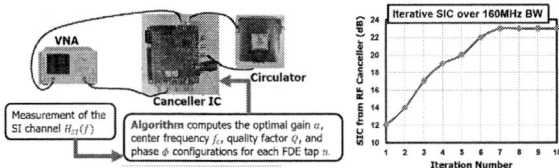

Fig. 4. Closed-loop algorithm to obtain optimal settings for both SIC and NF degradation.

a two-port 8-path rotary-clocked filter with 3-bit embedded phase shifting (45° resolution) realized by staggering the clocks of the switches at the two ports. Quality factor tunability is achieved through tunable 9-bit path capacitance providing quality factors up to 60. Each of the 8 RF taps has an on-chip 4-bit configurable highly-linear input coupling capacitor to couple a small amount of TX power into the tap while also isolating the N-Path filters from each other and giving us control over tap attenuation. The N-path filter taps drive 7-bit configurable thick-oxide inverter-based Gm cells for each tap that are co-designed into the LNTA and induce the cancellation current into the RX, thereby providing additional control of the individual tap gain.

A shift register stores the sequence for phase rotation, and is used to drive a MUX that selects the clock phase that drives each switch of the N-path filters. The minimum resolution of frequency shifting occurs for $P = 1$ (1 phase every Q cycles) and the maximum value of Q, which is equal to the shift register length (20 in our case). This provides frequency shifts with a min. step of $-f_c/160$ and a max. shift of $-f_c/8$.

B. FD Receiver

Fig. 3(b) shows the block and circuit diagram of the FD RX. The RX begins with a partial-noise-canceling LNTA canceler similar to [1] which incorporates the Gm cells mentioned earlier for injection of the cancellation current into the receiver with weighted summation of tap responses leading to partial and full SIC at the input and output of the LNTA, respectively. Additionally, we have a BB canceler (Fig. 3(c)) implemented similar to [1] providing a delay spread up to 45ns while being relatively flat in the bandwidth of interest when clocked using 80MHz clocks. The BB canceler cancellation current is injected after the downconversion mixer in the receiver. Fig. 3(d) shows the chip micrograph of the 65nm CMOS FD wideband receiver occupying 4.05mm×2.7mm chip area.

C. Closed-Loop Measurement Algorithm

A closed-loop algorithm (Fig. 4) which co-optimizes the SIC and NF degradation leveraging the measured tap responses for different configurations is used. Tap measurements for different frequency shifts, quality factors and phases are extensively measured and saved. This captures the non-idealities in the filter response as opposed to assuming ideal 2nd order filter responses. For instance, from Fig. 5, it can be noticed that the filter center frequency is dependent on

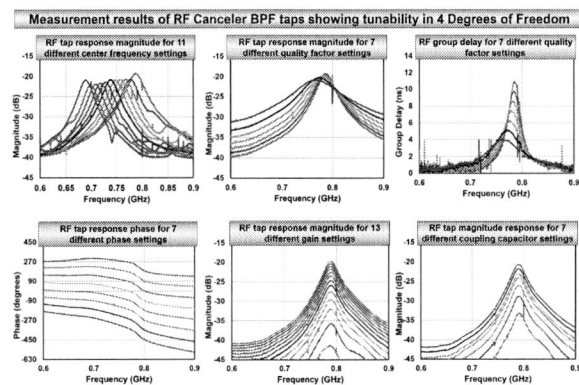

Fig. 5. Measurement results for the RF taps in the FD receiver having tunability in 4 degrees of freedom, namely *center frequency*, *quality factor*, *gain* and *phase*.

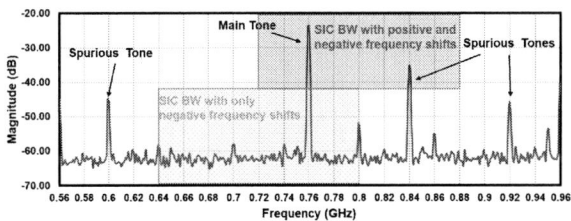

Fig. 6. Measured spurs from the RF canceler when configured for SIC and excited with a single tone at 760MHz.

the quality factor as well, which is due to the fact that the filters are driven directly using a coupling capacitor which leads to charge sharing between the N-path filter path capacitance and the coupling capacitor leading to a small frequency drift. A one-shot initial configuration and gain selection is achieved assuming the FDE response as the weighted sum of the tap responses and minimizing the measured SI. This is performed by employing a greedy mechanism that iteratively selects, for each tap, the configuration which yields the largest SI reduction. This is done with a constraint on the sum of all tap gains, which in turn limits the NF degradation. An iterative gain adaptation is then undertaken to reduce the residual SI by iterating over gain using gradient descent while keeping other parameters the same and measuring the residual SI. A similar approach is performed for the BB canceler as well but with much smaller configuration parameters (only gain and phase).

IV. MEASUREMENTS

Fig. 5. shows the *center frequency*, *quality factor*, *gain* and *phase* reconfigurability of the 2nd-order bandpass filters, demonstrating the 4 degrees of freedom of the RF canceler measured from the TX port to the RX input port and clocked using 400MHz clocks. The maximum frequency shift achieved is $2f_c/N$ (the factor 2 comes from the fact that we are using the second harmonic filtering response) which in this case is 100MHz (Fig. 5) which enables us to get cancellations over large bandwidths. Fig. 6 plots the measured spurs when a single tone at 760MHz is excited at the input of the RF

979-8-3503-2123-4/23 $31.00 © 2023 IEEE

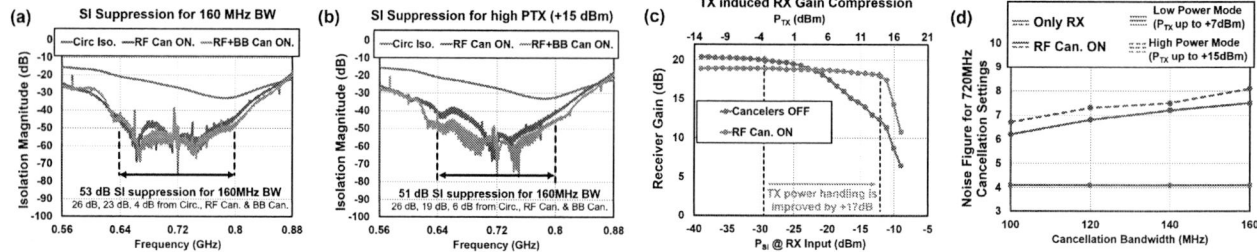

Fig. 7. (a) SIC over 160MHz bandwidth around 720MHz with 26dB initial isolation. (b) SIC over 160MHz bandwidth around 720MHz for high PTX (+15 dBm). (c) TX induced RX gain compression with and without SI cancellation. (d) NF degradation plot across BW around 720MHz with high and low power modes.

Fig. 8. Detailed comparison to state-of-the-art full-duplex receivers.

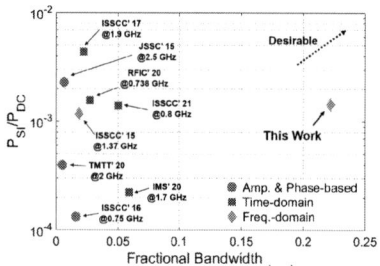

Fig. 9. Graphical comparison to a broader set of prior full-duplex receivers.

the RX input) (Fig. 7). Fig. 4 shows the SIC obtained across 160MHz BW as a function of iterations in the closed-loop algorithm. It can be seen that after a few iterations, the SIC converges. In the presence of SIC, a TX power of +14dBm (-12dBm at RX input) can be handled without compressing the receiver, which is an improvement of +17dB compared to when the cancelers are turned off.

The power consumption per DoF is 1.8mW for the RF canceler and NF degradation from the RF canceler is between 2.1dB and 3.4dB across cancellation BWs on top of a baseline RX NF of 4.1dB (Fig. 7). Fig. 8 shows the detailed performance comparison with prior art. Compared to prior state-of-the-art FDE [5] (TDE [1]) cancellers, we achieve similar SIC levels for $11.7\times$ ($4.44\times$) higher fractional BW while consuming 6-$12\times$ ($2\times$) lower power per DoF. Fig. 9 compares the performance with a broader cross-section of prior art. This work shows a significant improvement in fractional cancellation bandwidth product while achieving similar canceller efficiency (P_{SI}/P_{DC}).

V. CONCLUSION

This work demonstrates a new approach to FDE-based SIC, utilizing rotary-clocked N-path filters while lowering canceller power consumption and mitigating spurs. The implemented chip achieves record cancellation bandwidth, and moves us a step closer towards practical FD systems.

ACKNOWLEDGMENT

This work was supported by DARPA WARP.

REFERENCES

[1] A. Nagulu *et al.*, "6.6 Full-Duplex Receiver with Wideband Multi-Domain FIR Cancellation Based on Stacked-Capacitor, *N*-Path Switched-Capacitor Delay Lines Achieving >54dB SIC Across 80MHz BW and >15dBm TX Power-Handling," in *ISSCC 2021*.

[2] X. Li *et al.*, "A 2.4 GHz Full-Duplex Transceiver with Broadband (+120 MHz), Linearity-Calibrated and Long-Delayed Self-Interference Cancellation," in *ESSCIRC 2022*.

[3] T. Zhang *et al.*, "Wideband dual-injection path self-interference cancellation architecture for full-duplex transceivers," in *JSSC 2018*.

[4] K.-D. Chu *et al.*, "A broadband and deep-tx self-interference cancellation technique for full-duplex and frequency-domain-duplex transceiver applications," in *ISSCC 2018*.

[5] J. Zhou *et al.*, "Integrated wideband self-interference cancellation in the rf domain for fdd and full-duplex wireless," in *JSSC 2015*.

[6] M. Khorshidian *et al.*, "An Inductor-Less All-Passive Higher-Order *N*-Path Filter Based on Rotary Clocking in *N*-Path Filters," in *IMS 2022*.

canceler. It can be noticed that, as discussed earlier, only using negative frequency shifts and clocking at 400MHz (640 - 800MHz SIC BW) would place the spurs out of the band of interest. When both positive and negative frequency shifts are utilized (720 - 880MHz SIC BW), the main spur falls in the band of interest (840MHz tone in Fig. 6). The residual tone at 800MHz in Fig. 6 is the clock leakage.

The FD wideband receiver shows conversion gain of 15 to 40dB (24dB nominal), NF of 4.1dB, IIP3 of -14dBm, and IP1dB of -25dBm while consuming 34mW. For a BW of 120/160MHz, measured SIC for a circulator-based antenna interface operating at 740/720MHz is 60/53dB - 29/26dB, 20/23dB and 11/4dB from the circulator, RF canceler and BB canceler, respectively (Fig. 7). These SIC levels remain consistent across TX power levels up to +15dBm (-11dBm at

RTu1B-4

A Frequency-Tunable Dual-Path Frequency-Translated Noise-Cancelling Self-Interference Canceller RX with >16dBm SI Power-Handling in 65nm CMOS

Mostafa Essawy, Kareem Rashed, Amin Aghighi, Arun Natarajan

School of EECS, Oregon State University, USA

{essawym, rashedk, aghighia, nataraja}@oregonstate.edu

Abstract—A dual-path self-interference (SI) cancellation architecture is presented for high-power SI cancellation (SIC) for simultaneous transmit and receive. The proposed approach breaks SIC trade-offs between operating frequency, SI power and canceller noise/distortion by using a frequency-translated (FT) SI canceller and an auxiliary FT noise-cancelling path that cancels the noise/distortion from the primary canceller. The 65 nm CMOS implementation achieves ~45 dB SIC for 12 dBm RX SI peak power with only 4 dB NF degradation, blocker-1dB of >16 dBm with respect to SI (8x higher than prior work) while operating from 0.5 GHz to 1.3 GHz.

Keywords— self-interference, full duplex, TX leakage, wideband RX, frequency-translation, N-path mixers

I. Introduction

Simultaneous TX and RX schemes are attractive for communication and radar sensing. However, limited isolation between TX/RX antennas or duplexers leads to SNR degradation due to self-interference (SI) [1]. SI cancellation (SIC) has been implemented by leveraging TX/RX orthogonality in the frequency [2]–[7], code [8], [9] and polarization [10] domains. Noise and distortion in full-duplex (FD) and frequency-domain duplex (FDD) RX due to active/passive SIC degrades SNR, increasing effective noise figure. *Importantly, the SNR degradation increases with SI power incident on RX.* Therefore, SI power handling and noise degradation due to SIC emerge as key metrics. In prior work on freq. domain SIC, -7 dBm SI at RX input has been mitigated with NF degradation of 2.3 dB [6], and a distortion-cancellation scheme achieves -4 dBm SI cancelled with 0.6 dB NF degradation [11]. A dual-path approach was proposed in [12] where SIC is accomplished by breaking the high-linearity, low-noise (high dynamic range) trade-off using a dual-path SIC (Fig. 1(a)). In this scheme, the SI signal is cancelled using a high-linearity, high-noise path (SIC path) and the noise introduced by the SIC is cancelled using an auxiliary low-noise noise-cancelling (NC) path with high common-mode rejection. When the SIC and NC paths are enabled, the RX achieves +7 dBm Blocker-1dB for SI at RX input and 3.8 dB NF degradation when cancelling +1 dBm peak SI. However, as shown in Fig. 1(a), the SIC and the NC paths operate at RF in [12], limiting maximum operating frequency (~420 MHz) and bandwidth. Additionally, device parasitics and RF design constraints limit SI power handling and noise cancellation.

In this work, we demonstrate a N-path passive-mixer based dual-path cancellation scheme with both SIC and NC operating

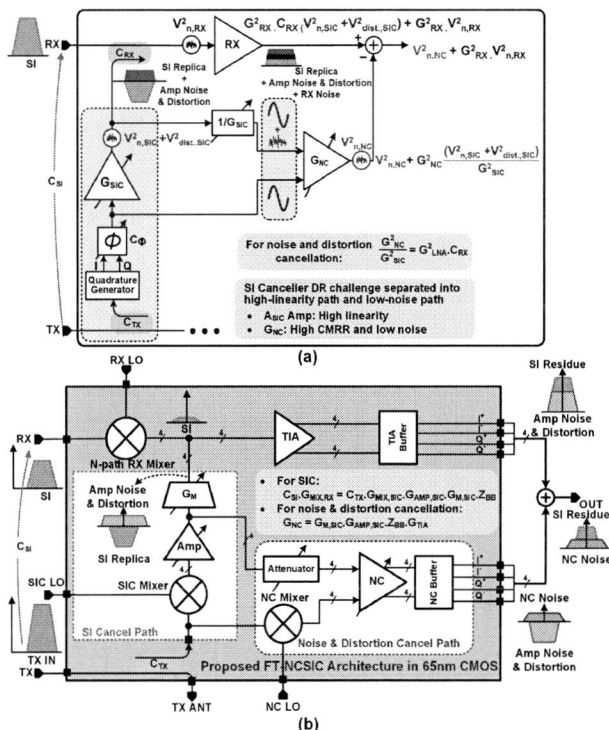

Fig. 1. (a) Dual-path SIC architecture where auxiliary path cancels noise and distortion introduced by the canceller, (b) Proposed frequency-translated dual-path approach implements SIC and noise-cancelling auxiliary path in baseband with N-path RX for higher operating frequency and power handling

at baseband (Fig. 1(b)), leading to the following benefits: (i) higher frequency of operation and wider operating range (500 MHz to 1.3 GHz), (ii) flexible frequency offset between TX SI and RX, and (iii) 8x increase in peak power handling to 16.3 dBm (since SIC operates at BB) with 3.55 dB NF degradation for 10 dBm peak SI. The proposed scheme in Fig. 1(b) is the first to achieve distortion/noise cancellation from SIC using a frequency-translated auxiliary path that senses the difference between canceller input and output signals to capture added distortion and noise. The architecture of the proposed N-path frequency translation noise-cancelling self-interference cancellation (FT-NCSIC) architecture and circuit implementation are detailed in Sec. II. Measurement results are presented in Sec. III and conclusions and future work are discussed in Sec. IV.

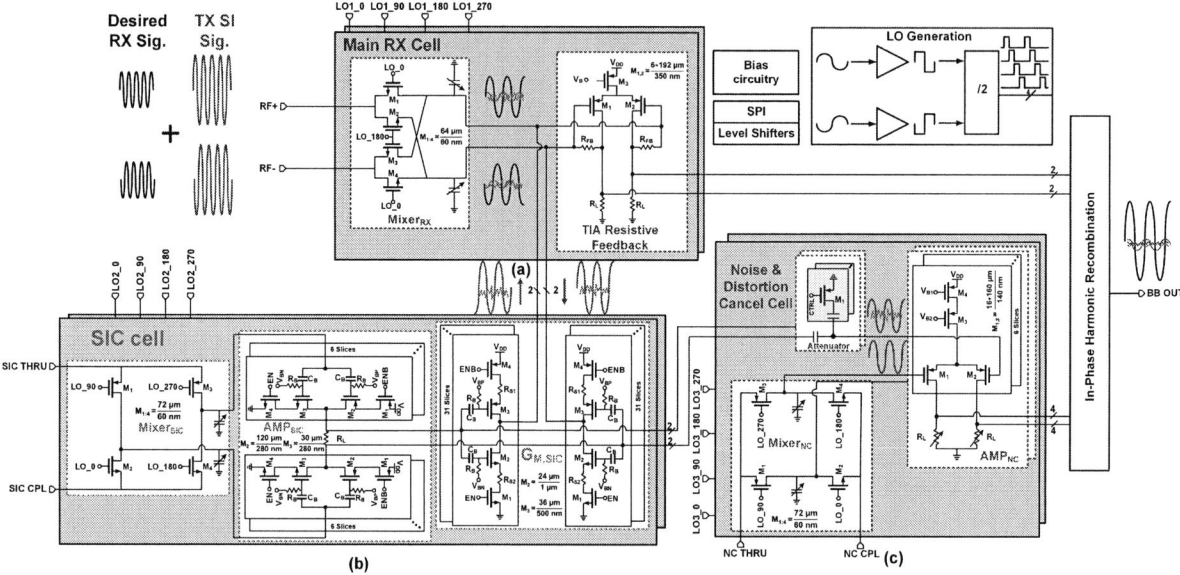

Fig. 2. Schematic of Frequency-Translated Noise-Cancelling Self-Interference Canceller:(a) Primary N-path RX (b) SI canceller with mixer, amplifier and baseband G_M, (c) noise-cancelling amplifiers which sense noise/distortion and cancel at RX output.

II. FREQUENCY-TRANSLATED NOISE-CANCELLING SIC (FT-NCSIC) ARCHITECTURE AND IMPLEMENTATION

The proposed FT-NCSIC architecure is presented in Fig. 1(b). Notably, the proposed approach targets flexible frequency separation between TX SI and RX in FDD schemes. Since the TX SI is mitigated by an active canceller rather than by filtering, a frequency-selective filter in the RX path is not needed and a wideband RX can be supported. Therefore, the primary N-path RX is designed to be relatively wideband ($\sim$ 250 MHz RF bandwidth). The primary N-path mixer-first RX in Fig. 1(b) provides linear translation from RF to IF. However, stand-alone in-band/out-of-band blocker-1dB compression are -10 dBm and 0 dBm respectively implying that only -10 dBm of in-band SI can be tolerated without SIC.

The FT-NCSIC approach in Fig. 1(b) applies the SIC at baseband (compared to the RF scheme in [12]) and therefore allows the use of longer channel-length devices, supporting high-power SIC. The cancelling SI signal is injected into the signal path with appropriate phase and amplitude to cancel SI at baseband input as shown in Fig. 1(b). In the FT-NCSIC scheme, the relative phase between the SI and the SIC signal can be adjusted using a phase shift in the LO driving the SIC mixers. Variable amplitude can be accomplished using a set of scalable, digitally-switched Gm unit cells. Notably, this scheme enables simpler SI phase/amplitude matching compared to the RF approach in [12] that requires weighted quadrature signal combining.

While the SIC path baseband linearity can be improved using longer channel length devices, SIC power handling is still limited by distortion and noise introduced by the SIC amplifier. This trade-off is broken using the auxiliary noise-cancelling (NC) path which senses the noise and distortion introduced by the SIC circuits. As shown in Fig. 1(b), the cancellation signal appears as a common mode

for the differential amp in the NC path while the distortion and noise introduced by the SI appears as a differential input. Since the NC path also operates at baseband, device capacitances such as input-pair Cgd and current source Cdb have a lower impact on common-mode rejection. In conjunction with the higher channel length, this enables high signal handling in the NCSIC path.

In order for the NCSIC path to cancel the noise and distortion, the SIC and NCSIC path gains must satisfy the constraint in Fig. 1(b). A programmable capacitive divider ensures that the SI signal is common-mode across the NCSIC amp input. VGAs in the NCSIC path further ensure that the gain constraint is satisfied in practice.

A. N-path Mixer-First Primary RX

A four-phase differential N-path mixer-first RX drives quadrature differential baseband TIAs as shown in Fig. 2(a). RX input matching is achieved using impedance translation of the resitive feedback TIA. As noted in Sec. I, the RX supports wide bandwidth given the objective of cancelling the SI using active cancellers as opposed to frequency-domain filtering. The desired N-path clocks are generated using a divide-by-two. The primary RX TIA consumes 25 mA from a 2 V supply, and the RX LO buffer consumes 13 mA from a 0.9 V supply in 0.6 GHz.

B. FT-NCSIC Building blocks

Complementary stages are used in SIC path to support high SI power. The N-path mixer in the SIC path, MIX_{SIC} creates quadrature SI replicas at base band that are used to cancel the RX SI. The baseband amplifier, AMP_{SIC} in Fig. 2(b) can provide >20 dB gain with an effective canceller $OIP3$ >34 dBm increasing the signal at the $G_{M,SIC}$ cell input. Since the noise of the $G_{M,SIC}$ cell is not cancelled, it is important to increase signal amplitude at

979-8-3503-2123-4/23 $31.00 © 2023 IEEE

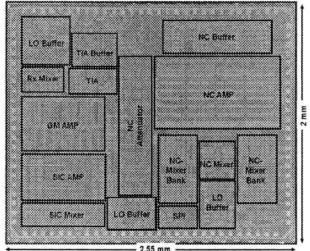

Fig. 3. Die photo of FT-NCSIC RX in 65 nm CMOS occupying 5.1 mm².

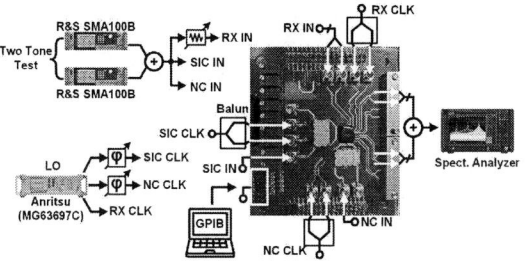

Fig. 4. Measurement setup including two-tone inputs for linearity testing. Leakage path losses are emulated in measurements through a variable attenuator.

its input to ensure low noise contribution (Fig. 1(b)). A variable capacitve divider couples the quadrature AMP_{SIC} outputs to the differential NCSIC amps, AMP_{NC}. The corresponding differential inputs of the AMP_{NC} are driven by the quadrature NCSIC mixer, MIX_{NC}. With appropriate scaling, the SI appears as common-mode to the NCSIC amps while the noise and distortion introduced by the AMP_{SIC} appears as differential-mode. The NCSIC amp, AMP_{NC}, is a PMOS differential amplifier with high common-mode rejection and programmable transconductances/resistances that provide variable gain (Fig. 2(c)). The power consumption and noise of the NCSIC amp can be programmed for targeted SI power and noise degradation.

III. MEASUREMENT

The FT-NCSIC IC is implemented in 65 nm CMOS with the IC occuppying 5.1 mm² of active area (Fig. 3). The TX input for the SI and NCSIC cancellation path are generated using a hybrid-coupler scheme similar to [12] to provide a wideband 3 dB passive voltage boost. However, the proposed approach is compatible with other TX coupling schemes. The IC is packaged using a chip-on-board scheme and measurement setup is shown in Fig. 4.

The LO-frequency defined input match, gain, and noise figure of N-path RX is shown in Fig. 5 (a,b). As shown in Fig. 5(b), the RX has an RF bandwidth of ∼250 MHz which can be tuned from 0.3 GHz to 1.5 GHz. The stand-alone RX has a NF of 3 to 4 dB. The measured small-signal RX SI cancellation is shown in Fig. 6 demonstrating *single-sided* SIC bandwidth > 50 MHz for 20 dB cancellation across 600 MHz to 1200 MHz LO. The objective of SIC is to improve linearity in the presence of SI at RX input. Fig. 7(a) shows the blocker 1dB compression where the blocker is offset from LO. Given the wide RX bandwidth, typical N-path out-of-band blocker 1dB of 0 dbm is achieved only >200 MHz offset

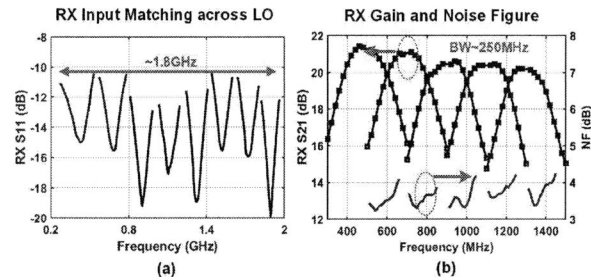

Fig. 5. (a) Measured LO-defined RX input match, (b) Measured stand-alone RX gain and NF; wider IF bandwidth supported to enable wideband RX with flexible TX SI

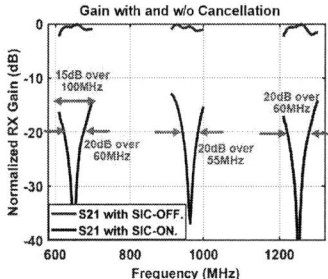

Fig. 6. Small-signal SIC demonstrated with >50 MHz bandwidth for 20 dB cancellation using active cancellers across operating range

with in-band blocker-1dB of ∼-8 to -10 dBm. Fig. 7(b) demonstrates the SIC capabilities with an improvement of 25 dB in blocker-1dB compression with SIC enabled. The RX 1dB gain compression occurs at 16.3 dBm in-band SI power (70 MHz offset). Measured SIC across input power is also shown in Fig 7(b) demonstrating > 35 dB cancellation up to 16 dBm SI power. The proposed canceller can handle a TX power >30 dBm assuming initial TX-RX isolation of ∼15 dB.

Distortion and noise cancellation improvement using the NCSIC path is measured using a two-tone SI input to the RX as shown in Fig. 4. Measured performance is shown in Fig. 8 for two-tones at 600 MHz and 900 MHz. As noted in Fig. 8(b), the 12 dBm SI at 600 MHz is cancelled by ∼45 dB. When the NCSIC is engaged, the NC path in Fig. 2(c) results in cancelling the distortion (as shown in the IM3 tones) by >30 dB. Similarly, the noise figure degradation is improved by ∼4.6 dB (RX NF before and after SIC of 12.1/7.5 dB). Measured cancellation for modulated input signal (40 MHz BW) with LO of 900 MHz is shown in Fig. 9 demonstrating

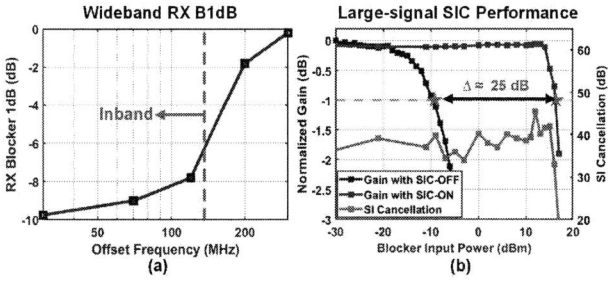

Fig. 7. (a) In-band and out-of-band blocker-1dB for RX (∼-10 dBm and ∼0 dBm respectively) without SIC, (b) SIC improves blocker-1dB for in-band SI to 16.3 dBm (30 MHz offset), improving blocker tolerance by 25 dB. FT-NCSIC supports >35 dB SIC up to 16.3 dBm input.

979-8-3503-2123-4/23 $31.00 © 2023 IEEE

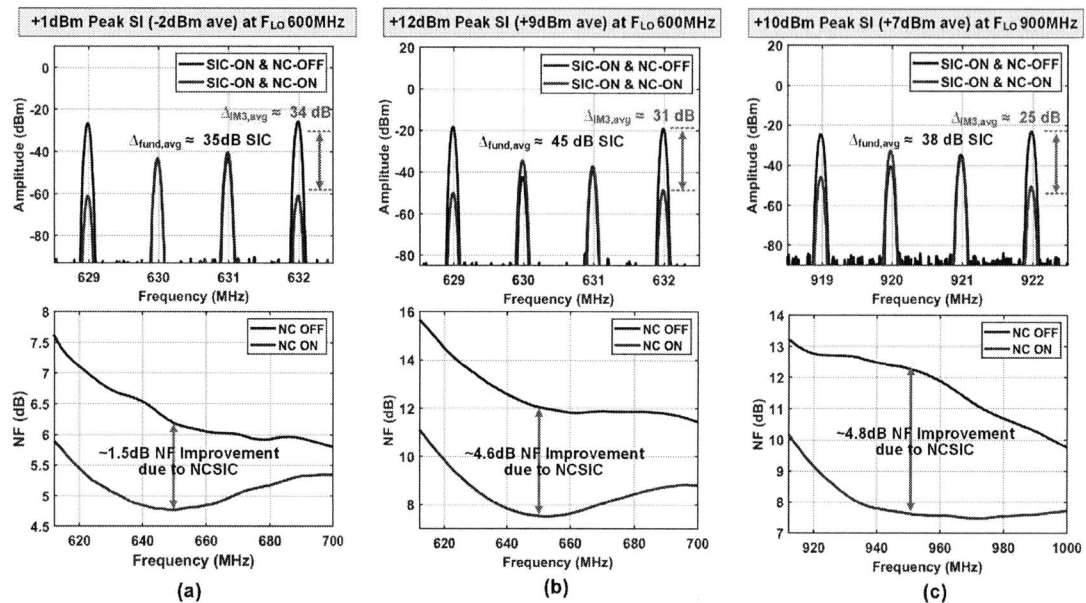

Fig. 8. Measured SIC for two-tone input and noise-canceller performance across power levels and input frequency demonstrates SIC for 12 dBm peak SI (9 dBm average) and distortion/noise cancellation when the NCSIC path is engaged (~20 dB IM3 tone cancellation, ~5 dB NF improvement)

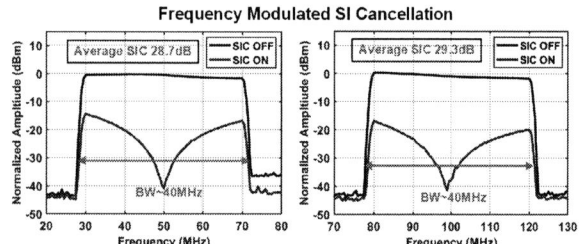

Fig. 9. Measured cancellation for 40 MHz-BW modulated SI at 50 MHz and 100 MHz offset frequencies (~29 dB ave. SIC), demonstrating wideband SIC.

Table 1. Comparison with state-of-the-art RX with SIC

	ISSCC'18 [4]	RFIC'17	JSSC'20 [8]	JSSC'15 [2]	JSSC'15 [5]	ISSCC'21 [6]	ESSIRC'22	RFIC'21 [12]	This Work
Architecture	Dual-path canceller	BB Hilbert SIC	Code-domain IAD N-path	Freq. Equalization SIC	Mixer-first SIC with Vector Mod.	RF and BB SIC using FIR filter and LNTA	Integrated EBD	NC-SIC with I/Q path	Frequency Translational NC-SIC
Frequency (MHz)	1600-1900	900	400-1100	800-1400	150-3500	100-1000	2100-2500	370-420	300-1500
Technology (nm)	CMOS 40	CMOS 130	GF 45 SOI	CMOS 65	CMOS 65	CMOS 65	CMOS 40	CMOS 180	CMOS 65
On-Chip SIC (dB) / BW(MHz)	34/40	20/80	49.5/13	20/20	24/16.25	47/80	62/120 68/40	30/35 20/50	15/100 20/60
RX NF (dB)	8.1	9.6	2.6 to 5.6	4.8	6.3	3.7	6.8	3.5	3 to 4
RX Blocker 1dB due to SIC (dBm)	N/A	N/A	12.1	-8	1.5	-14	-31	7.2	16.3
RX SI Power (dBm)	-29	-30	N/A	-16/-8	-18/-10	-12/-7	N/A	-9/+1	12
RX NF Degradation (dB)	1.6	1.4/2.1	N/A	1.1-1.3	4-6	0.3(-12dBm) 2.3(-7dBm)	1.7	1.5 (-9dBm) 3.8 (1dBm)	1.4 (1dBm) 3.55 (10dBm)
Canceller Power (mW)	14.3 (RF x2)	13	<40	44-91/88-182	23-56	7.4/tap	36.2	64 (-9dBm) 154 (1dBm)	135 (1dBm) 207 (10dBm)
Active Area (mm²)	4	0.72	1.12	4.8	1.96	7.2	3.2	6.5	4

wideband SIC at different TX-RX offset frequencies.

The proposed SIC is compared to state of the art in Table 1. The FT-NCSIC architecture demonstrates cancellation at signficantly higher SI power (>8x) compared to state-of-the-art FD/FDD systems demonstrating peak power >10 dBm, with <3.6 dB NF degradation (even with a good stand-alone RX NF of 3 to 4 dB.)

IV. Conclusion

Simultaneous TX and RX is becoming increasingly important in communication and radar systems. The proposed

FT-NCSIC approach breaks the trade-off between SI power handling and noise degradation, by adopting an auxiliary path for noise/distortion cancellation. The N-path based frequency translated scheme enables the dual-path scheme to higher RF frequencies and higher output power. Future work includes increasing bandwidth by emulating SI coupling transfer function in the SIC and NCSIC paths.

Acknowledgment

The authors thank the DARPA Wideband Adaptive RF Protection (WARP) program and the Center for Design of Analog-Digital ICs (CDADIC) for project support, and Dr. Tim Hancock for technical discussions.

References

[1] K. E. Kolodziej, B. T. Perry, and J. S. Herd, "In-band full-duplex technology: Techniques and systems survey," *IEEE TMTT*, July 2019.

[2] J. Zhou *et al.*, "Integrated Wideband SIC in the RF Domain for FDD and Full-Duplex Wireless," *IEEE JSSC*, Dec. 2015.

[3] G. Qi *et al.*, "A SAW-Less Tunable RFFE for FDD and IBFD Combining an EBD and a Switched-LC N-Path LNA," *IEEE JSSC*, May 2018.

[4] K. Chu *et al.*, "A Broadband and Deep-TX SIC Technique for FD and FDD Transceiver Applications," in *IEEE ISSCC*, Feb. 2018, pp. 170–172.

[5] D. van den Broek *et al.*, "An In-Band FD Radio RX With a Passive Vector Modulator Downmixer for SIC," *IEEE JSSC*, Dec. 2015.

[6] A. Nagulu *et al.*, "FD RX with Wideband Multi-Domain FIR Cancellation ..." in *IEEE ISSCC*, Feb. 2021.

[7] E. Zolkov *et al.*, "An Integrated Reconfig. SAW-Less Quad. Balanced N-Path TRX for FD and HD Wireless," in *IEEE RFIC*, June 2022.

[8] H. Alshammary *et al.*, "A Code-Domain RF Signal Proc. Front End With High SI Rejection and Power Handling," *IEEE JSSC*, May 2020.

[9] A. Agrawal *et al.*, "An Interferer-Tolerant CMOS Code-Domain Receiver Based on N-Path Filters," *IEEE JSSC*, 2018.

[10] T. Chi *et al.*, "A mm-Wave Pol.-Division-Duplex TRX FE With an On-Chip Multifeed SIC Antenna," *IEEE JSSC*, Dec. 2018.

[11] J. Zhou *et al.*, "Low-Noise Active Cancellation of TX Leakage and TX Noise in Broadband Wireless RX for FDD," *IEEE JSSC*, 2014.

[12] M. Essawy *et al.*, "A Noise-Cancelling SI Canceller with +7dBm SI Power Handling in 0.18μm CMOS," in *IEEE RFIC*, June 2021.

RTu1C-1

High-Linearity 76-81 GHz Radar Receiver with an Intermodulation Distortion Cancellation and High-Power Limiter

N. Landsberg[#], M. Gordon[#], O. Asaf[#], N. Weisman[#]. K. Ben-Atar[#], S. Levin[#], S. Pellerano[*], W. Shin[*$],
D. Nahmanny[#]

[#]Radar Group, Mobileye, Israel

[*]Intel Labs, USA

[$]now with Apple, Inc., USA

{nlandsbe, gmeir, oasaf1, nweisman, kbenatar, slevin1, dnahmann}@mobileye.com,
stefano.pellerano@intel.com, woorim.shin@gmail.com

Abstract—**A highly linear 76-81 GHz direct conversion receiver has been fabricated in a 16 nm FinFet CMOS process for phased array radar applications. The receiver includes a Low Noise Amplifier (LNA), a semi active mixer, a passive filter and a baseband amplifier. A third order intermodulation distortion cancellation technique is implemented in the LNA and it is used to compensate for distortions in the entire receiver. The peak gain of the receiver is 36 dB with Input IP3 of -8.1 dBm and Noise Figure (NF) of 5.5 dB. The total power consumption of the receiver is 68 mW. To protect the LNA from high input power levels that affect the reliability of the transistors, an innovative limiter circuit is suggested to monitor the operating point of the first stage of the LNA, whereas the mmW signal path is not affected from this monitoring circuitry at tolerable power levels.**

Keywords— **CMOSFET circuits, FinFET, intermodulation distortion cancellation, low noise amplifier, millimeter wave integrated circuits, millimeter wave radar.**

I. INTRODUCTION

The automotive radar is a key element for Advanced Driver Assistance Systems (ADAS) as well as autonomous vehicles. In order to improve spatial resolution of the radar while keeping low form factor of the antenna, the millimeter-wave (mmW) frequencies are used, and mainly the 76-81 GHz band.

Advanced CMOS processes enable implementation of mmW circuits at the above frequency band [1-2] with improved performance and power consumption yet maintaining low-cost solution that is crucial for the automotive industry.

Another aspect of the automotive radar is the harsh environment of the receiver considering interferences caused by nearby radars. To allow the implementation of a sensitive receiver, a high frequency selectivity as well as high chain linearity are required. Typical implementations of high linearity receivers omit the LNA, preventing compression of the most wideband gain block in the design [3]. Yet, at mmW frequencies this results in degraded NF performance, typically exceeding 12 dB for CMOS implementations [4-5]. This work presents an implementation of a low noise and highly linear CMOS receiver for a digital modulation radar that includes an LNA with a third order intermodulation (IM3) cancellation implementation, but unlike traditional implementations the cancellation technique is used for decreasing IM3 of other blocks in the entire chain, including the baseband blocks. This results in a much improved IP3 performance for the entire chain.

The chain is composed of a two-stage LNA, a semi active direct conversion mixer in a complementary topology for enhanced linearity, a passive filter for Out-of-Band (OOB) interference rejection and a Trans Impedance Amplifier (TIA) on the signal path, where a hybrid coupler and Local-Oscillator (LO) driver were implemented for the LO path. The block diagram of the implemented receiver is presented in Fig. 1.

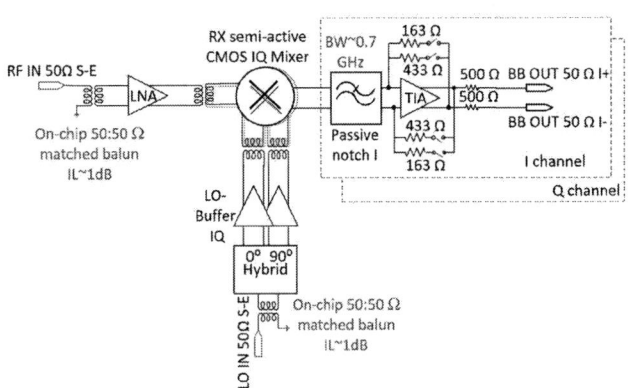

Fig. 1. A block diagram of the suggested receiver. The on-chip balun in front of the LNA is not a part of the design as the LNA input transformer interface is differential, corresponds with the dedicated antenna interface.

II. RECEIVER DESIGN

A. LNA and Power Limiter

The LNA is designed with two stages. The first stage includes a limiter, implemented by a safety circuitry that automatically turns off the stage when the input power is too high, to comply with the transistors reliability requirements. To enable third order distortion cancellation the second stage of the LNA is implemented in a second derivative cancellation topology [6] and its design is presented in [7]. The biases of the main and auxiliary cores are not set to optimize the IP3 of the LNA. Instead, these biases are set to cancel the distortions of the entire chain and improve the overall IP3 of the receiver. This is done by optimizing the IM3 components of the auxiliary stage to be out of phase with those of the baseband blocks, so that when downconverted they destructively interfere and cancel out. A conceptual schematic of the LNA is given in Fig. 2a.

979-8-3503-2123-4/23 $31.00 © 2023 IEEE

A major concern for the LNA of an automotive radar system is the potentially high input power due to transmissions of other radars nearby. Therefore, it is essential to protect the LNA from high power levels to avoid large voltage swings that can harm the transistors. A standard limiter can be easily achieved by using clipping diodes at the input or output of the LNA, or using the ESD diodes already present at the input of the first stage of the LNA. However, these diodes would engage only at extremely high power levels, where the voltage swings on the drain of the transistors is already too high. Moreover, implementing clipping diodes at the drains of the transistors results in increased parasitic seen by the amplifier, hence reducing its gain and degrading NF and linearity. To avoid these degradations, a unique detector was designed, based on sensing the current through the amplifier that is affected by the power level.

A class A amplifier, typical for LNA implementations, has a relatively constant power consumption even at increasing power levels. However, the ESD diodes that are implemented at the input of the first stage of the amplifier cause the input signal to be clipped first at positive values of the voltage swing (due to the gates positive bias voltage), which lowers the average voltage of the signal. Therefore, the average level of the gate voltage is now lower than the desired bias, set by another circuit. By using a sensitive comparator it is possible to detect this offset and turn off the power supply to the first stage of the LNA, avoiding large voltage swings at its output. The limiter and its interface with the first stage of the LNA is shown in Fig. 2b. A series resistor, R_1 (4.3 KΩ), and a shunt capacitor, C_1 (540 fF), are used to further isolate the comparator parasitics from the amplifier, as well as to further filter the high frequency component of the clipped sine wave and allowing its average level to be monitored by the comparator.

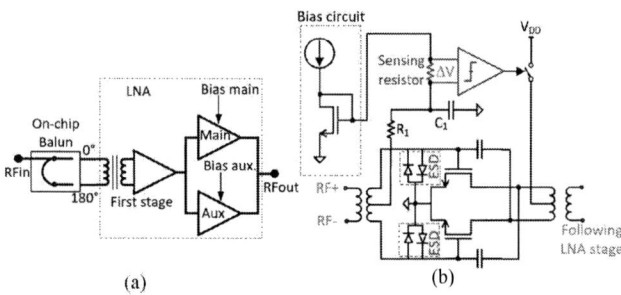

Fig. 2. (a) A block diagram of the implemented LNA: topology is purely differential, including its input transformer, where an on-chip balun was added for measurements purpose; and (b) A schematic of the first stage and the implementation of the limiter.

B. Semi Active Mixer

A new topology of a semi active IQ downconverter is suggested for enhanced linearity, as presented in Fig. 3. Both nmos and pmos transistors are employed for the current switching at the LO frequency. The gate biases of the pmos and nmos transistors is set to allow a quiescent current through the transistors, according to the required common mode level of its following block. Therefore, three-way transformers are required for both the LO input and for the RF input of the mixer to allow the different gate biases as well as the power supply connection to the sources of the pmos and ground connection to the sources of the nmos transistors.

The principle of operation of the suggested mixer is similar to that of a semi active double balanced mixer [8], but in a complementary configuration. This configuration allows the pushing and the pulling of the current from the Base-Band (BB) load to be equivalent to about twice the current of the traditional double balance mixer.

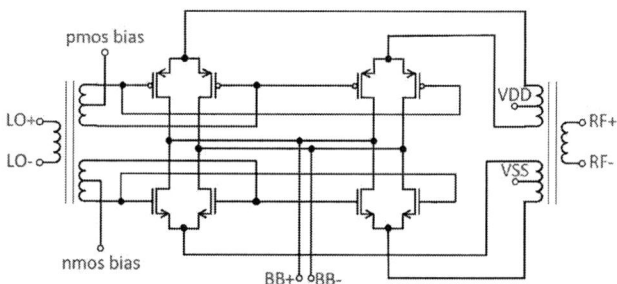

Fig. 3. A block diagram of the implemented downconverting mixer, where a single channel (out of the I/Q channels) is presented.

There are several additional advantages for the presented mixer topology: first, due to the sum of the transconductances of both the nmos and the pmos transistors, the suggested mixer behaves as a robust current source for the next stages of the baseband chain, resulting in a higher gain conversion. In addition, due to the quiescent current, the required LO swings that are needed to switch the transistors are smaller comparing the passive mixer case, leading to overall lower power consumption. Furthermore, the conversion gain has lower dependency on the LO swings. Finally, the complementary topology, similar to a CMOS transmission gate, minimizes the mixer distortions, especially the intermodulation of the third order (IM3), resulting in higher linearity.

C. Passive Filter

The suggested mixer implementation is relatively wide band. It covers well the 76-81 GHz on the RF ports, where the BB outputs can easily cover the 0-3 GHz bandwidth. Therefore, to allow good channel selectivity, the mixer should be followed by a filter, before any additional amplification is implemented. Usually, active filters are used in the BB chain, in order to maintain low form factor and high quality factor implementations. However, to maintain high linearity of such filters, the implemented op-amps gain-bandwidth product must be wider than the required bandwidth of the filter. Therefore, active filter is more sensitive to out-of-band high power level signals. Hence, a low pass *passive* filter should be considered, which also contributes lower noise and in addition saves power compared with the active filter. All the above can justify its larger dimensions.

The filter is required to support a relatively wide bandwidth to enable a digital virtual array, but also an outstanding rejection at a baseband frequency >2 GHz to properly reject aliasing of OOB signals. Therefore, it has been implemented as a notch filter to enhance the rejection closer to its cutoff frequency. Moreover, to address the notch filter lower rejection at higher frequency offsets, several notch filters are cascaded, achieving high rejection up to 5 GHz. The schematic of the filter is presented in Fig. 4a.

To maintain low physical dimensions of the inductors, the differential topology of the filter was leveraged and the inductor of each stage of the filter was implemented in a differential manner: the inductor of the positive chain is wound together with that of the negative chain, as illustrated in Fig. 4b. Thus, the mutual inductance between the negative and positive inductors enhances the overall effective series inductance, resulting in a more compact implementation. The total series resistance of the inductors only slightly affects the gain of the TIA as the impedance seen by the mixer towards the baseband blocks is maintained very low.

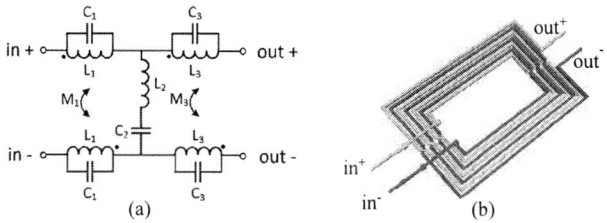

Fig. 4. (a) A schematic of the differential notch filter; (b) a demonstration of an implementation of differential series inductors. Although the layout reassembles to a transformer, this is a set of two differential inductors: the inductor marked in cyan is wound together with its differential pair, marked in magenta. The inductors L_1 and L_3 are implemented in this configuration.

D. TIA

To allow several gain configurations the TIA is implemented with two parallel feedback resistors: 433 Ω and 163 Ω, as shown in Fig. 1. For practical designs, the TIA is loaded by a high impedance block (such as active filter or ADC). However, to allow the measurement of this design with external equipment, a 500 Ω series resistor was implemented at the output of the TIA. This allows measurements of the baseband outputs with either a 50 Ω probe, or with an active, high impedance, probe. Measurements with a 50 Ω probe result in gain loss of 20.8 dB due to the voltage drop on the series resistor (that was de-embedded in measurements), whereas the 500 Ω resistor does not affect the gain when an active probe is used.

III. SIMULATION AND MEASUREMENT RESULTS

The complete receiver is implemented in 16 nm FinFET process. The total area of the chip, including RF, LO, baseband and DC pads is 1580 x 1200 μm². The die photo and power consumption of the receiver is shown in Fig. 5a and 5b respectively.

A. Small Signal Measurements

Small signal gain was measured under initial bias conditions as in the simulations, over different RF and LO frequencies, where the baseband frequency was set constant. A peak gain of 34.5 dB (de-embedded, not including the insertion loss of the

LNA balun, 1 dB, that was measured as a standalone structure) was achieved. Then, the filter rejection was measured by setting the LO frequency fixed and sweeping over the RF frequency to achieve a scan over the baseband frequency of 0-5 GHz.

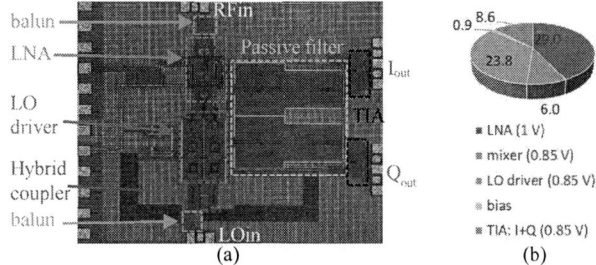

Fig. 5. (a) Die photo of the fabricated chip: RF input is on the top side, LO input is on the bottom side, I and Q outputs are on the right (either I or Q can be measured at a time) and the DC and control pads are on the left; (b) a pie chart of power consumption (in mW) of each block of the receiver.

Next, the NF was measured over the RF frequency band at a 100 MHz baseband frequency, by measuring the gain and output noise floor of the receiver. A NF of 5.5 dB (de-embedded) was demonstrated, thanks to the exceptional NF of the LNA (4.5 dB [7]). Finally, the NF measurement was repeated for different bias configurations of the second stage of the LNA. Since the major contribution to the NF is the first stage of the LNA, and as long as the total gain of the LNA remains constant, the NF had a very small variance over different equal-gain bias configurations. Fig. 6 summarizes the small signal measurement and simulation results.

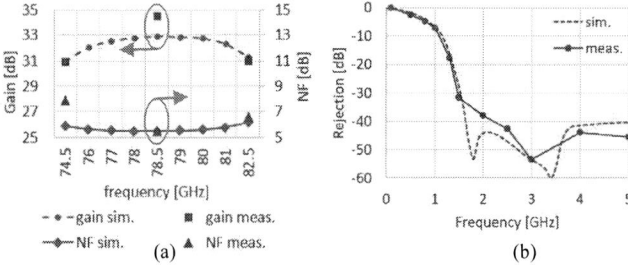

Fig. 6. (a) Small signal gain (left axis) and NF at 100 MHz baseband frequency (right axis) as a function of RF frequency; (b) Chain rejection as a function of baseband frequency.

B. Maximum Input Power

The functionality of the power limiter was tested by an increasing input power into the RF input. Measured power levels of 4 dBm (de-embedded) and above result in triggering the limiter which turns off the power supply to the first stage of the LNA. By that, high voltage levels that can affect the reliability of the LNA transistors are prevented. By increasing the value of the sensing resistor presented in Fig. 2b, even lower threshold levels can be easily set, if required. The simulated response time of the limiter for an input power that turns on the limiter is presented in Fig. 7a. The overall delay of the protection circuit is about 9 nS. The threshold of the limiter is presented in Fig. 7b: measured value of 4 dBm, based on current consumption measurements, and threshold level of 3 dBm in simulations, based on drain voltage.

979-8-3503-2123-4/23 $31.00 © 2023 IEEE

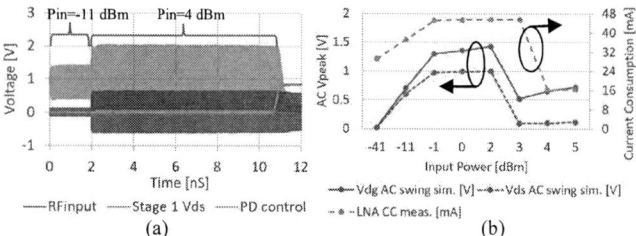

Fig. 7. (a) Transient simulation of an increasing input power level and limiter response; (b) Measured current consumption and simulated AC swings of V_{dg} and V_{ds} of the first stage of the LNA as a function of input power level.

C. IP3 Measurements

Finally, IP3 measurement were done using three sets of tone-pairs: {2, 10} MHz, {2, 100} MHz and {2, 250} MHz offset from the LO frequency, each tone at input power of -50 dBm. The IP3 of the receiver was optimized by sweeping over the biases of the second stage of the LNA. As these biases affect the gain, the latter was also measured. Both gain and IP3 were optimized to achieve maximum output IP3. The optimization point was slightly affected by the chosen two tone pair (where input power does not affect the IP3 level as long as small signal was used). Best OIP3 at all measured baseband offsets was achieved at a bias of 0.43 mA of the auxiliary core and 0.93 mA of the main core of the second stage of the LNA (with a good robustness for IP3 optimization over corners). The bias optimization is described in Fig. 8a. The LNA gain dependency over the different biases of the main and auxiliary cores of the second stage is presented in Fig. 8b.

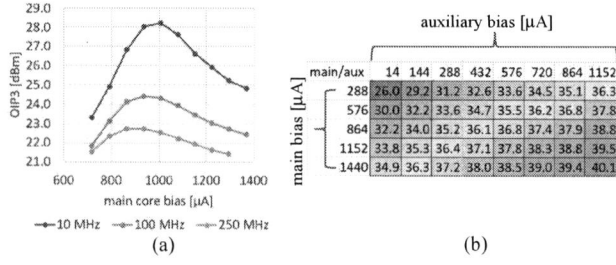

Fig. 8. (a) Receiver OIP3 for intermodulation products at different baseband frequencies for an auxiliary bias of 432µA; and (b) total receiver gain (in dB) at different main and auxiliary biases.

Table 1 summarizes the receiver performance and compares with state-of-the-art radar receivers. The implemented receiver demonstrates state-of-the-art performance for NF and linearity. In addition, this design achieves the lowest silicon area when compared to receivers integrating similar blocks and this is a key enabler for low-cost phased array systems.

IV. CONCLUSION

A receiver for the 76-81 GHz automotive radar systems is demonstrated with state-of-the-art NF and linearity performance. An innovative active mixer in a complementary configuration allows for improved linearity and conversion gain. A passive notch filter, following the mixer, enables out-of-band interference rejection while avoiding the typical high

noise of active filters. A second-derivative cancellation topology is used in the LNA to minimize intermodulation that is generated by baseband components of the chain.

Table 1. Comparison with 76-81 GHz state of the art CMOS receivers.

	This work	RFIC '19 [1]	TMTT '21 [2]	TCAS-I '21 [9]	RFIC '19 [10]
Tech.	16 nm FinFet	40 nm Bulk	28 nm FD-SOI	65 nm Bulk	65 nm Bulk
Gain*[dB]	36.2	30.3	34	38	47
NF [dB]	5.5	9	8.2	17	11
Vdd [V]	1/0.85 RF/BB	1.1/1.8 RF/BB	1		1.2
P_{DC} [mW]	68	143	38	130	-
IIP3 [dBm]	-8.1	-12.3 **	-13.5 **	-18 **	-19 **
Area [mm²] ***	<0.2	0.4	-	1	0.25

* for given IIP3   ** Extrapolated from IP_{1dB} values (IIP3 [dBm] ~ IP_{1dB}+10)   *** Active area only, not including pads, approximated

REFERENCES

[1] T. Murakami, N. Hasegawa, Y. Utagawa, T. Arai and S. Yamaura, "A 9 dB Noise Figure Fully Integrated 79 GHz Automotive Radar Receiver in 40 nm CMOS Technology," *2019 IEEE Radio Frequency Integrated Circuits Symposium (RFIC)*, 2019, pp. 307-310, doi: 10.1109/RFIC.2019.8701748.

[2] G. Papotto *et al.*, "A 27-mW W-Band Radar Receiver With Effective TX Leakage Suppression in 28-nm FD-SOI CMOS," in *IEEE Transactions on Microwave Theory and Techniques*, vol. 69, no. 9, pp. 4132-4141, Sept. 2021, doi: 10.1109/TMTT.2021.3074600.

[3] T. Fujibayashi *et al.*, "A 76- to 81-GHz Multi-Channel Radar Transceiver," in *IEEE Journal of Solid-State Circuits*, vol. 52, no. 9, pp. 2226-2241, Sept. 2017, doi: 10.1109/JSSC.2017.2700359.

[4] S. Krishnamurthy and A. M. Niknejad, "10-35GHz Passive Mixer-First Receiver Achieving +14dBm in-band IIP3 for Digital Beam-forming Arrays," *2020 IEEE Radio Frequency Integrated Circuits Symposium (RFIC)*, 2020, pp. 275-278, doi: 10.1109/RFIC49505.2020.9218301.

[5] A. Paidimarri *et al.*, "A High-Linearity, 24–30 GHz RF, Beamforming and Frequency-Conversion IC for Scalable 5G Phased Arrays," *2021 IEEE Radio Frequency Integrated Circuits Symposium (RFIC)*, 2021, pp. 103-106, doi: 10.1109/RFIC51843.2021.9490413.

[6] W. Gao, Z. Chen, Z. Liu, W. Cui and X. Gui, "A Highly Linear Low Noise Amplifier With Wide Range Derivative Superposition Method," in *IEEE Microwave and Wireless Components Letters*, vol. 25, no. 12, pp. 817-819, Dec. 2015, doi: 10.1109/LMWC.2015.2496793.

[7] N. Landsberg, W. Shin, S. Levin and T. Levinger, "High Linearity 76-81 GHz LNA Using a 16 nm FinFET Technology for Phased Array Radar Applications," *2022 17th European Microwave Integrated Circuits Conference (EuMIC)*, 2022, pp. 123-126, doi: 10.23919/EuMIC54520.2022.9923442.

[8] N. Mazor *et al.*, "Highly Linear 60-GHz SiGe Downconversion/Upconversion Mixers," in *IEEE Microwave and Wireless Components Letters*, vol. 27, no. 4, pp. 401-403, April 2017, doi: 10.1109/LMWC.2017.2678426.

[9] D. Pan *et al.*, "A 76–81-GHz Four-Channel Digitally Controlled CMOS Receiver for Automotive Radars," in *IEEE Transactions on Circuits and Systems I: Regular Papers*, vol. 68, no. 3, pp. 1091-1101, March 2021, doi: 10.1109/TCSI.2020.3042976.

[10] L. Chen, L. Zhang, W. Wu, L. Zhang and Y. Wang, "A Compact 76-81 GHz 3TX/4RX Transceiver for FMCW Radar Applications in 65-nm CMOS Technology," *2019 IEEE Radio Frequency Integrated Circuits Symposium (RFIC)*, 2019, pp. 311-314, doi: 10.1109/RFIC.2019.8701845.

RTu1C-2

Mono/Multistatic Mode-Configurable E-band FMCW Radar Transceiver Module for Drone-Borne Synthetic Aperture Radar

Kangseop Lee, Sirous Bahrami, Kyunghwan Kim, Jiseul Kim, Seung-Uk Choi, Ho-Jin Song

Department of Electrical Engineering, Pohang University of Science and Technology, Republic of Korea

kangseop.lee@postech.ac.kr, hojin.song@postech.ac.kr

Abstract— Drone-borne synthetic aperture radar (SAR) systems are attractive for small and mid-area applications due to easy and temporal deployment capability. In this paper, we present a 77-GHz drone-borne multistatic frequency-modulated continuous-wave radar transceiver (TRX) which enables multiple drones to cooperate for SAR imaging by wirelessly sharing the reference chirp signal between the drones. The TRX can be configured in the monostatic or multistatic (master/slave) mode by integrated RF switches. The antenna module in this work includes microstrip comb-line and planar Yagi-Uda array antennas for SAR signals and wireless synchronization of the reference chirp signal, respectively. The fabricated radar TRX chip size, including PADs, is 2.20 mm². From the on-ground measurement, the SAR module detected a metallic object up to 20 meters away with around 11.2-dB SNR in multi-static mode. The TRX consumes 0.92, 1.3 and 0.4 W for monostatic, multistatic master, and multistatic slave modes, respectively.

Keywords—drone-borne synthetic aperture radar, FMCW radar, monostatic radar, multistatic radar, transceivers, E-band.

I. INTRODUCTION

Drone-borne synthetic aperture radar (SAR) is attracting attention for its cost-effectiveness, accessibility, and ability to be deployed quickly compared to conventional air- and space-borne platforms [1,2]. However, since a drone is not suitable for carrying heavy and large objects, the radar sensor of a drone-borne SAR system must be light and small. Moreover, sensor cost and resolution are important factors. In view of these requirements, a potential solution is to incorporate E-band frequency-modulated continuous-wave (FMCW) transceivers (TRXs) into the sensors [3,4].

In general, an FMCW radar system commonly uses a monostatic mode, in which transmitters (TXs) and receivers (RXs) are co-located, or a multistatic mode in which they are separated. In monostatic drone-borne SAR, there is no need to consider the synchronization problem between the TXs and RXs. In contrast, multistatic radar systems with multiple drones would provide more information from the angular diversity and wide observation areas. But each drone should be synchronized for SAR image reconstruction from the multiple received signals.

For the synchronization of the TX and RX in a multistatic radar system, an independent precision clock can be used [5]. However, this approach is not suitable for drone-borne SAR as the system should be regularly calibrated. In addition, it is obvious that wired means such as a fiber-optic link [6] cannot be employed for synchronizing drones. Passive radar [7] and over-the-air deramping (OTAD) radar [8] use RXs with a fixed position and are thus not suitable for drone-borne systems.

In this paper, we present a 77-GHz drone-borne multistatic FMCW radar TRX and antenna module which enables multiple drones to cooperate for single-scan SAR imaging over a wide range by wirelessly sharing the reference chirp signal between the drones.

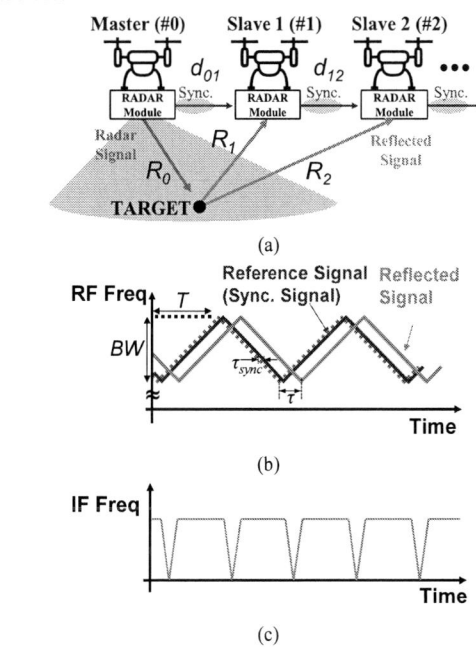

Fig. 1. Drone-borne multistatic radar operation principle. (a) Radar and sync signal path, (b) RF signal diagram, and (c) IF output signal diagram in receivers.

II. MONOSTATIC/MULTISTATIC OPERATION

A. Multistatic SAR with Wireless Synchronization

Fig. 1 illustrates the proposed drone-borne multistatic SAR operation. The master module transmits FMCW signals for SAR (red) and synchronization (blue) in vertical and lateral directions, respectively. Slave 1 then captures the reflected signal from the ground and mixes it with the sync signal shared from the master drone. For scalable multi-drone operation, slave 1 retransmits the sync signal via the other lateral antenna. Assuming the time delay from the TRX circuits is negligible, the propagation distance from the master to slave-n radar module ($R_0 + R_n$) can be expressed as

$$R_0 + R_n = \tau \cdot c = \frac{f_{IF} \cdot T \cdot c}{BW} + \sum_0^{n-1} d_{i,i+1} \quad (1)$$

where τ, c, f_{IF}, T, BW, and $d_{i,i+1}$ are the propagation delay, speed of light, measured IF frequency, time duration of a chirp signal,

979-8-3503-2123-4/23 $31.00 © 2023 IEEE   237   2023 IEEE Radio Frequency Integrated Circuits Symposium

frequency bandwidth of a chirp signal, and the distance between neighboring drones, respectively.

B. Multi-mode SAR Transceiver

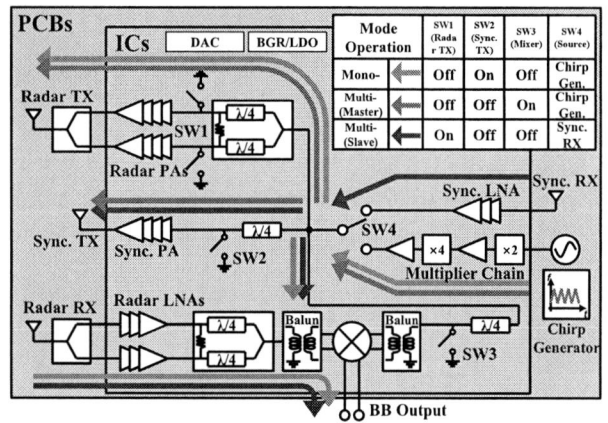

Fig. 2. The block diagram of the mono/multistatic configurable radar module and switch operation.

Fig. 2 shows a block diagram of the proposed radar TRX module. By controlling four RF switches (SW1 - SW4), one can configure the operation mode of the TRX. For each mode, the low-noise amplifiers (LNAs) and/or power amplifiers (PAs) not in use are turned off to reduce DC power consumption. The on-off switches (SW1, SW2, and SW3) with quarter-wavelength transmission lines control the signal path to the radar TX, sync TX, and mixer, respectively. SPDT SW4 selects the sources of the reference chirp signal—either an internally generated one with the ×8 multiplier chain or the synchronization signal received from the synchronization antenna. In the monostatic mode, the three on-off switches should be set to off-on-off, and SW4 selects the chirp generator. Then all amplifiers in the synchronization path are turned off and the TRX works as the conventional FMCW radar does. In the master mode of the multistatic operation, sync TX block should be turned on to broadcast the reference chirp signal for other TRX in the slave mode. In the slave mode, the radar TX block and multiplier chain are turned off. The received radar signal from the radar RX is mixed with the synchronization signal from the sync RX.

III. CIRCUIT, ANTENNA AND MODULE DESIGN

A. Transceiver

Schematics of the functional blocks in the multistatic radar TRX are shown in Fig. 3. The reference chirp signal is generated from the X-band chirp generator (9.4 GHz to 9.65 GHz) and ×8 frequency multiplier, which consists of a push-push doubler, a driver amplifier, a phase-controlled push-push (PCPP) frequency quadrupler, and an output cascode amplifier [Fig. 3a]. Push-push multipliers help to suppress the odd-order harmonic signals. The PCPP ×4 multiplier generates fourth harmonic (75.2 to 77.2 GHz) signals with a single stage compact core and low power consumption [9].

Fig. 3(b) shows a schematic of the cross-coupled capacitive neutralization stacked PA with a driving stage. To enhance the stacking efficiency of the output stage, a shunt inductor L3 between CS and CG transistors was added [10]. It helps to compensate for parasitic capacitance of the huge MOSFET power cell and thus the phase mismatch between CS and CG can be minimized. The simulated gain was larger than 46.5 dB with 1-dB compressed output power of above 11 dBm from 72 to 78 GHz.

Received signals in the radar and synchronization antennas are amplified by a three-stage LNA [Fig. 3c]. The transformer balun at the input provides the noise matching. The simulated gain and noise figure of the LNA were 35 dB or more and 4.8 dB or less, respectively, in the chirp frequency range, 75.2 – 77.2 GHz. The double-balanced passive mixer shown in Fig. 3d provides good linearity and sufficiently large bandwidth.

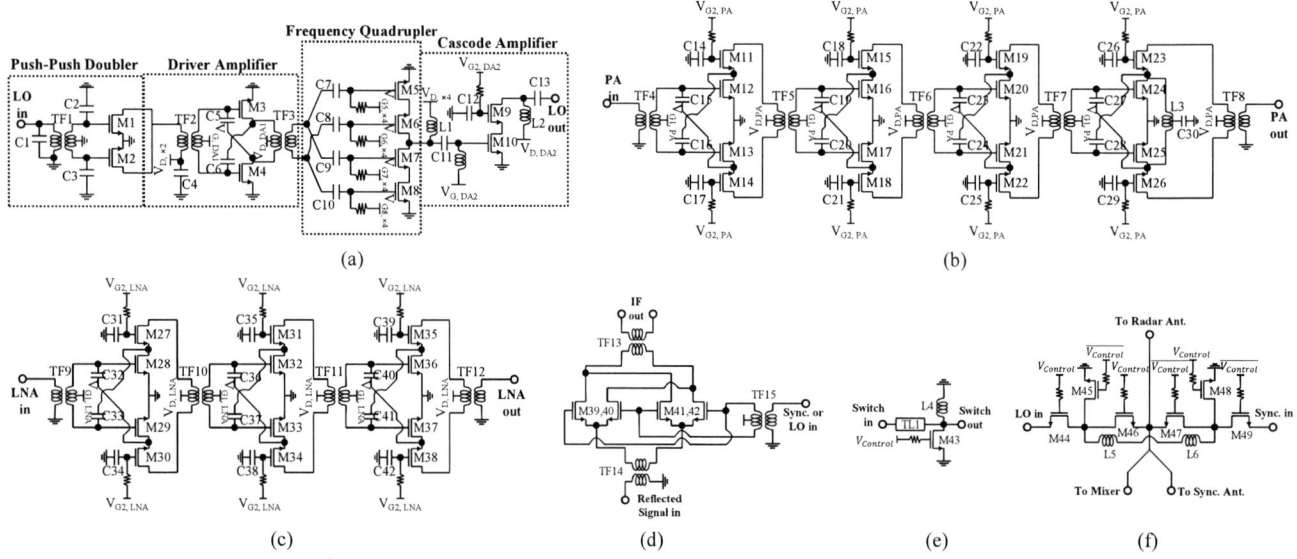

Fig. 3. Schematics of (a) functional blocks in the TRX ×8 frequency multiplier chain, (b) power amplifier with driving stages, (c) low-noise amplifier, (d) mixer, (e) on-off switches for SW1 − SW3, and (f) SPDT switch, SW4.

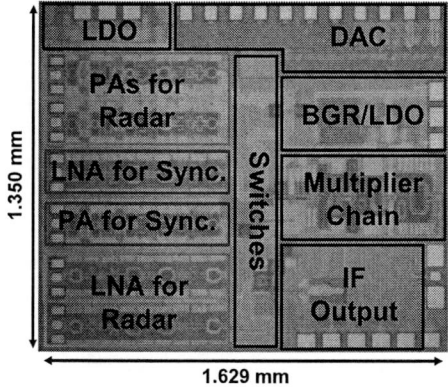

Fig. 4. Microphotograph of the fabricated radar chip.

The on-off switches were implemented with a simple shunt FET and shunt inductor L4 for cancelling out the parasitic capacitance from MOSFET M42. By incorporating with the quarter-wavelength transmission line (TL1), SW1 − SW3 re-route the reference chirp signals. The chirp source selection switch SW4 is based on series-shunt transistor pairs with inductors L5 and L6 for improving off-state isolation.

For stable operation and external control via a microcontroller, SPI, 9b-DACs, BGR, and LDO were also integrated. Fig. 4 shows a microphotograph of the radar chip fabricated in 40-nm bulk CMOS process. The total chip area is 2.20 mm².

B. PCB and Antenna Design

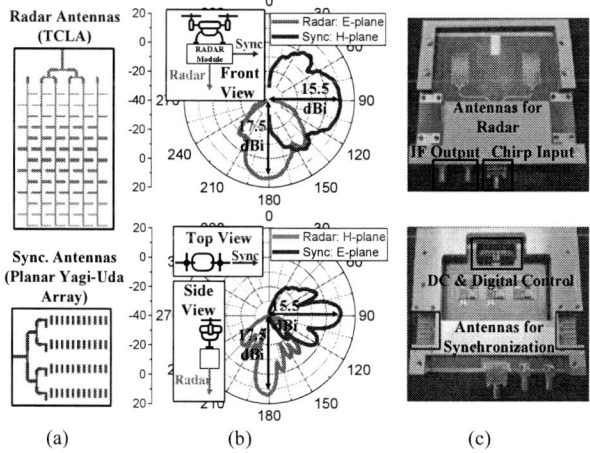

Fig. 5. (a) Antennas layouts for radar and synchronization. (b) Stand-alone antenna measurement results. (c) Front and rear view of the radar module.

As shown in Fig. 5a and Fig. 5b, a microstrip comb-line array (MCLA), providing wide impedance bandwidth and higher gain in the limited area, was used for transmitting and receiving the E-band radar signals, and the MCLA was formed on a 0.25-mm-thick multilayer Taconic TLY-5 substrate. By tapering the radiators, good side lobe level suppression is attainable. The measured peak realized gain of the stand-alone radar antenna was 17.5 dBi. Meanwhile, the planar Yagi-uda antennas, which have an end-fire radiation pattern with wide impedance bandwidth, were used to transmit and receive E-

band synchronization signals. The measured peak realized gain of the stand-alone synchronization antenna was 15.5 dBi. The fabricated radar module is shown in Fig. 5(c). The size and weight of the module are 10.1×8.7 cm² and 197 g, respectively, which is small and light enough to be mounted on a drone.

IV. MEASUREMENT RESULTS

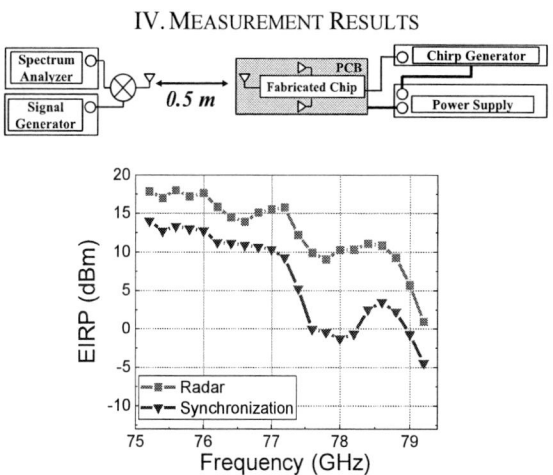

Fig. 6. EIRP measurement setup (top) and measurement results for radar and synchronization (bottom).

Fig. 6 shows the effective isotropic radiated power (EIRP) measurement setup and measurement results. The conversion loss from the down-conversion mixer and reference antenna gain of the setup were calibrated, and the EIRP was calculated using the free-space path loss formula with the distance of 0.5 m. The estimated EIRPs for radar and the sync TX were larger than 13.9 and 9.3 dBm, respectively, in the frequency range of interest.

Fig. 7 shows the measurement setup and results of monostatic and multistatic FMCW radar operation under a static condition. Note that one of the differential ports at the mixer output was terminated with 50 Ω and the IF power P_{IF} was measured 20 times and averaged. The threshold power level $P_{Threshold}$ for recognizing a target was obtained from the measurement results without a target. The chirp period and bandwidth of the triangular waveform were 0.4 ms and 2 GHz, respectively.

For the monostatic and multistatic operation with 0.1-meter module distance d_{multi}, the distance between a 25-dBsm corner reflector target and radar modules was set from 5 to 20 m. As shown in Fig 7(b), in the monostatic mode, the achieved IF output power is more than 11.5-dB higher compared to the threshold power at distances up to 20 meters with 0.92-W DC power consumption. The multistatic case with d_{multi} of 0.1 m is shown in Fig. 7(c). Assuming that traveling distances from the master/slave module to reflectors, R_0 and R_1, are identical, the measured P_{IF} - $P_{Threshold}$ is larger than 11.2 dB at distances up to 20 m. The DC power consumptions of the master and slave modules were 1.3 and 0.4 W, respectively. Fig. 7(d) shows measurement results in the multistatic case with the module distance of 0.5 m. The target was a metal plate, and the distance between the metal plate and radar modules was set from 2 to 12 m. The IF output power is more than 12.9-dB higher compared

979-8-3503-2123-4/23 $31.00 © 2023 IEEE

to the threshold power at within the measured distances. The performance of the proposed radar is summarized in Table 1.

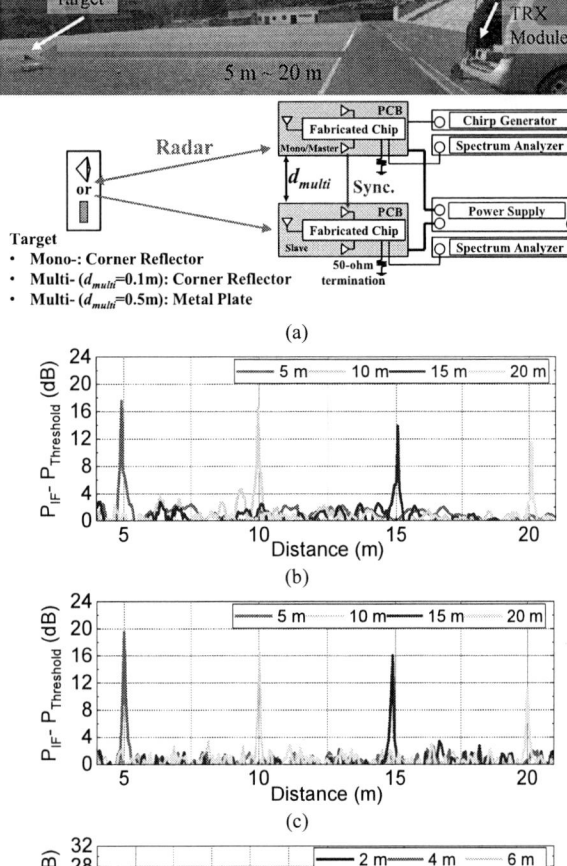

(a)

(b)

(c)

(d)

Fig. 7. (a) Radar measurement setup for monostatic and multistatic cases. (b) Measurement results for monostatic case at distances up to 20 m with corner reflector as target. (c) Measurement results for multistatic case at module distance of 0.1 m with corner reflector as target. (d) Measurement results for multistatic case with module distance of 0.5 m with metal plate target.

V. CONCLUSION

A wireless synchronization method and circuits for multistatic radar are presented. A synchronization chirp signal is directly transmitted from a master radar module to slave radar modules. An SPDT switch and on-off switches with quarter-wavelength transmission lines were used to configure monostatic or multistatic operation by digital control on a single chip. The size and weight of the fabricated radar module are $10.1 \times 8.7 \text{cm}^2$ and 197g, respectively. For the monostatic and multistatic measurements, the module detected a target at a distance of 20 m.

Table 1. Performance Summary

Synchronization Method	Wireless direct connection
Technology	40-nm CMOS
Frequency	75.2 GHz ~ 77.2 GHz
DC Power Consumption	Monostatic: 0.92 W Multistatic (Master): 1.3 W Multistatic (Slave): 0.4 W
Chip Area	2.20 mm²
Chirp BW / Chirp Duration	2 GHz / 400 us
Module Size	10.1×8.7cm²
Module Weight	197 g
EIRP	Radar: 13.9 dBm Synchronization: 9.3 dBm
Measured Distance	up to 20 meters

ACKNOWLEDGMENT

This research was supported by Institute for Information & Communications Technology Planning & Evaluation (IITP) grant funded by the Korea government (MSIT) (2019-0-00060, 'Development of 300 GHz band Tbps beamforming transceiver chip for next generation short range communication', and 2019-0-00762, 'Next-Generation Multistatic Radar Imaging System for Smart Monitoring'). The chip fabrication and EDA tool were supported by the IC Design Education Center(IDEC), Korea.

REFERENCES

[1] A. Bekar, M. Antoniou, and C. J. Baker, "Low-Cost, High-Resolution, Drone-Borne SAR Imaging," *IEEE Transactions on Geoscience and Remote Sensing*, vol. 60, pp. 1-11, 2022.

[2] J. Svedin, A. Bernland, A. Gustafsson, E. Claar, and J. Luong, "Small UAV-based SAR system using low-cost radar, position, and attitude sensors with onboard imaging capability," *International Journal of Microwave and Wireless Technologies*, vol. 13, no. 6, pp. 602-613, 2021.

[3] J. Yan et al., "Full-Coverage Indoor SAR Imaging with a Vehicle-based FMCW Radar System," in 2018 *IEEE/MTT-S International Microwave Symposium - IMS*, 10-15 June 2018, pp. 135-137.

[4] J. Lee, Y. A. Li, M. H. Hung, and S. J. Huang, "A Fully-Integrated 77-GHz FMCW Radar Transceiver in 65-nm CMOS Technology," *IEEE Journal of Solid-State Circuits*, vol. 45, no. 12, pp. 2746-2756, 2010.

[5] P. Lopez-Dekker, J. J. Mallorqui, P. Serra-Morales, and J. Sanz-Marcos, "Phase Synchronization and Doppler Centroid Estimation in Fixed Receiver Bistatic SAR Systems," *IEEE Transactions on Geoscience and Remote Sensing*, vol. 46, no. 11, pp. 3459-3471, 2008.

[6] D. H. Shin, D. H. Jung, D. C. Kim, J. W. Ham, and S. O. Park, "A Distributed FMCW Radar System Based on Fiber-Optic Links for Small Drone Detection," *IEEE Transactions on Instrumentation and Measurement*, vol. 66, no. 2, pp. 340-347, 2017.

[7] M. Antoniou, Z. Zeng, L. Feifeng, and M. Cherniakov, "Experimental Demonstration of Passive BSAR Imaging Using Navigation Satellites and a Fixed Receiver," *IEEE Geoscience and Remote Sensing Letters*, vol. 9, no. 3, pp. 477-481, 2012.

[8] M. Ash, M. Ritchie, K. Chetty, and P. V. Brennan, "A New Multistatic FMCW Radar Architecture by Over-the-Air Deramping," *IEEE Sensors Journal*, vol. 15, no. 12, pp. 7045-7053, 2015.

[9] K. Lee, K. Kim, G. Shin, and H. J. Song, "65.6–75.2-GHz Phase-Controlled Push–Push Frequency Quadrupler With 8.3% DC-to-RF Efficiency in 40-nm CMOS," *IEEE Microwave and Wireless Components Letters*, vol. 31, no. 6, pp. 579-582, 2021.

[10] K. Kim et al., "Analysis and Design of Multi-Stacked FET Power Amplifier With Phase-Compensation Inductors in Millimeter-Wave Band," *IEEE Trans. Micro. Theory Techn.*, early access doi: 10.1109/TMTT.2022.3232167.

RTu1C-3

A W-Band Spillover-Tolerant Mixer-First Receiver for FMCW Radars

Jingzhi Zhang, Sherif S. Ahmed, Amin Arbabian

Department of Electrical Engineering, Stanford University, USA

Abstract—This paper presents a highly linear, low-noise W-band N-path mixer-first receiver for frequency-modulated continuous-wave (FMCW) radar applications. By adopting a high-impedance quarter-wavelength transmission line as the matching network, the input impedance at the fundamental and third-harmonic frequencies has been enhanced, thereby increasing the passive voltage gain and eliminating the reradiation current. Hence, the linearity and noise figure (NF) can be improved. Fabricated in 40 nm CMOS process, the receiver achieves -0.7 dBm out-of-band (OOB) P1dB and 8.0-to-9.5 dB NF simultaneously while consuming 37 mW power.

Keywords—blocker-tolerant receiver, FMCW radar, millimeter-wave, mixer-first receiver, N-path mixer, receiver.

I. INTRODUCTION

Spillover and bumper reflections are the main sources of RF blockers in fully integrated mm-wave automotive radars [1]–[5]. These blockers may compress or saturate the radar receivers, and create nonlinear distortion that results in extra tones corresponding to ghost targets. A spillover of -20 dB at 80 GHz should be considered in the link budget design for margins [5], [6]. Therefore, for a typical 13-to-15-dBm transmitted power [2]–[6], the input P1dB of the receiver should be more than -5 dBm, which is challenging to achieve simultaneously with low noise and low power consumption.

A low-noise amplifier (LNA) is widely adopted in mm-wave radar receivers for low noise figure (NF) [3]–[6]. However, for a high P1dB LNA, higher power and lower gain are necessary to avoid saturation, which consequently decreases the NF of the receiver. As a result, the LNA in radar receivers becomes less effective, while consuming additional power and area. An alternative approach is the passive mixer-first receiver [7], [8] that eliminates the LNA. Therefore, the linearity of the receiver can be significantly improved and the front-end consumes no power. The only problem with the mixer-first receiver is the poor NF. Thus, in this work, we focus on improving the NF of the mixer-first receiver to compete with the LNA-first receiver. The key method in this work is to enhance the input impedance at the fundamental and third-harmonic frequencies by a quarter-wavelength ($\lambda/4$) transmission-line (TL)-based impedance matching network. Consequently, the passive voltage gain [7] is improved, and the reradiation current [9] at the third harmonic is eliminated, thereby improving the NF performance. Fabricated in 40 nm CMOS process, our purposed N-path mixer-first receiver, shown in Fig. 1 [12], can operate from 80 to 90 GHz, and achieve -0.7 dBm out-of-band (OOB) P1dB and 8.0-to-9.5 dB NF with 28-mW local-oscillator (LO) and 9-mW baseband (BB) power consumption.

The rest of this paper is organized as follows. Section II introduces the fundamental issues arising from the mm-wave

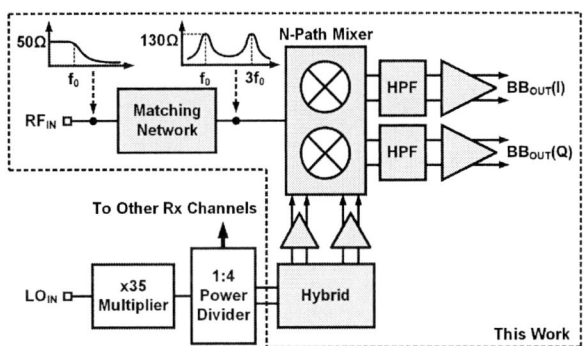

Fig. 1.   Receiver architecture.

N-path mixer architecture. Section III presents the proposed $\lambda/4$ TL-based matching network (MN) and the circuit implementation. Section IV provides the measured results. Finally, we conclude this paper in Section V.

II. INSIGHTS ON MM-WAVE N-PATH MIXER

A. Equivalent Circuit Model

An N-path mixer-first receiver can be simplified to a linear time-variant (LTV) model by modeling the switch transistors to ideal switches in series with ON-resistance R_{SW} [7], [9], as shown in Fig. 2(a). In this model, V_S and R_S are the source voltage and impedance, C_{in} is the input parasitic capacitance, Z_a is the effective source impedance, and R_B is the single-ended input impedance of the baseband stage.

Since the switches are switched ON sequentially, the voltage on the common node of the switches has a stair shape and contains plenty of harmonics. Therefore, it generates current flowing back to the source with harmonics as well. This current is defined as *reradiation current*. As the reradiation current flows back to the source, less energy is delivered to the load. Therefore, the loss of the mixer increases with the existence of the reradiation current, and the NF increases.

According to [9], the LTV model can be further simplified to a linear time-invariant (LTI) model by replacing the switches and the baseband impedance R_B into shunt impedance Z_{sh} in parallel with γR_B, as shown in Fig. 2(b). Here, Z_{sh} models the loss caused by reradiation current, and parameter γ is topology-dependent and equals to $2/\pi^2$ in a four-phase mixer.

B. Gain and NF in N-Path Mixer

The gain and NF of the N-path mixer are determined by the portion of energy delivered to the load. According to the LTI model, a high R_{SW} and a low Z_{sh} result in less gain and high NF. In this work, we focus on two approaches to increase the gain and reduce the NF.

979-8-3503-2123-4/23 $31.00 © 2023 IEEE   241   2023 IEEE Radio Frequency Integrated Circuits Symposium

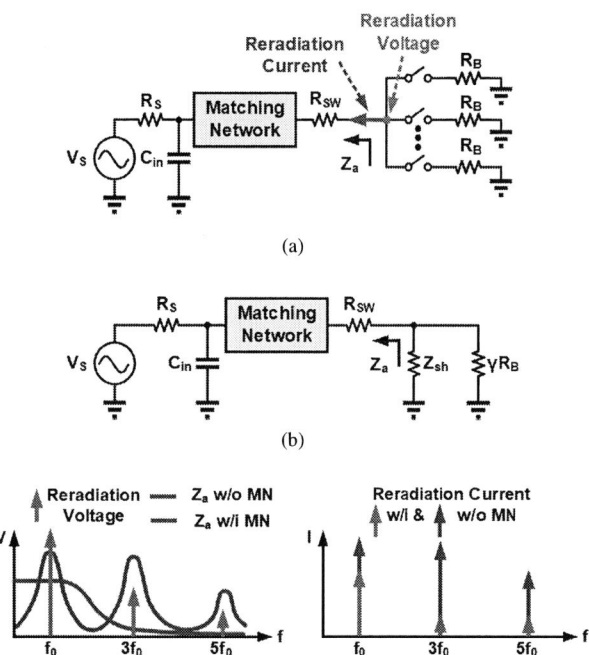

(a)

(b)

(c)

Fig. 2. (a) Equivalent LTV model of the N-path mixer, (b) its LTI equivalent model, and (c) reradiation voltage and current spectrum of four-phase passive mixer-first receivers with and without proposed input matching network.

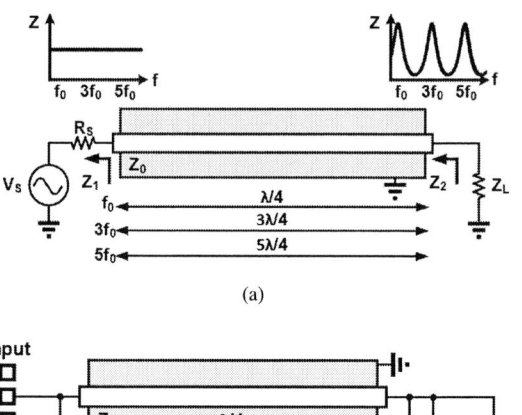

(a)

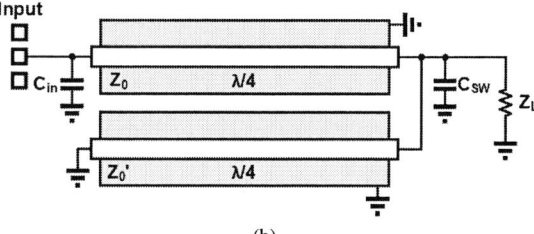

(b)

Fig. 3. (a) $\lambda/4$ TL that boosts the source impedance at odd harmonic frequencies, and (b) shunt $\lambda/4$ TL to resonate out parasitic capacitance.

First, to lower the effect of R_{SW}, we can transform the source impedance to a high level Z_a by the matching network [7]. As a result, the equivalent effect of R_{SW} is reduced in a high-impedance system.

Second, Z_{sh} can be increased by adjusting the harmonic impedance of the source. As illustrated in Fig. 2(c), with the existence of C_{in}, Z_a shows a low-pass filter property, and the generated reradiation current at harmonics becomes higher as the source impedance drops. However, if we increase the source impedance at harmonics, with the same reradiation voltage, the reradiation current can be reduced significantly. Hence, the overall reradiation energy is reduced, and Z_{sh} increases. This effect can be mathematically expressed as [9]

$$Z_{sh} = \left(\sum_{n=1,3,5...}^{\infty} \frac{1}{n^2 Z_a(n\omega_{LO})} \right)^{-1}. \quad (1)$$

Therefore, when we increase the source impedance at harmonics, the total Z_{sh} is increased.

In conclusion, to increase the gain and reduce the NF, we should design a matching network that can transform the source impedance to a high level not only at the fundamental frequency, but also at its odd-harmonic frequencies.

III. CIRCUIT DESIGN

A. Quarter-Wavelength Transmission Line

Compared with the low-frequency implementation, transforming the source impedance at harmonics is challenging at 80 GHz. In this work, we propose a high-impedance $\lambda/4$ TL

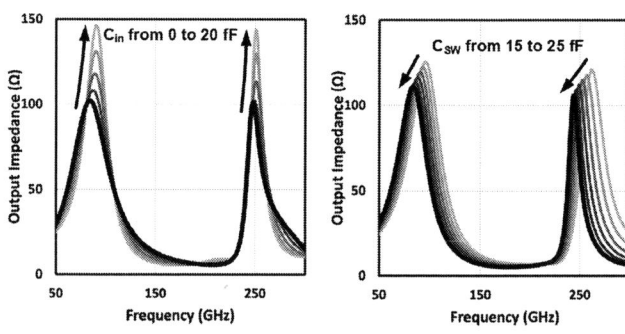

(a)   (b)

Fig. 4. Post-layout simulation results of structure in Fig. 3(b) by varying (a) C_{in} and (b) C_{SW}. Two impedance peaks appear at f_0 and $3f_0$.

matching network to accommodate impedance transformation at fundamental and harmonic frequencies, as well as parasitic resonating.

The source impedance Z_1 can be transformed to Z_2, which can be expressed as

$$Z_2 = Z_0^2/Z_1, \quad (2)$$

where Z_0 is the characteristic impedance of TL, as shown in Fig. 3(a). Therefore, the source impedance at the fundamental frequency is boosted.

Additionally, at the third harmonic of the LO frequency f_0, the TL becomes a $3\lambda/4$ TL. As a result, the impedance transformation at the third harmonic shares the same equation as the fundamental, meaning that the third harmonic impedance is also boosted by the high-impedance TL. This effect applies to all the odd harmonics. Therefore, by adopting a high-impedance $\lambda/4$ TL, the source impedance is transformed to a high impedance at the fundamental frequency

979-8-3503-2123-4/23 $31.00 © 2023 IEEE

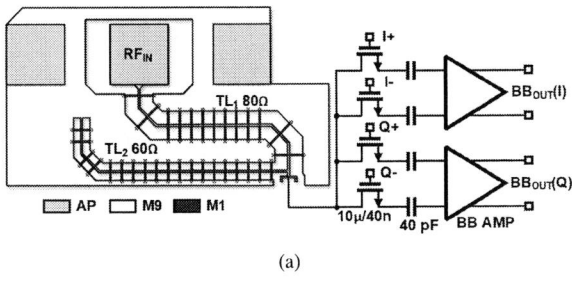

(a)

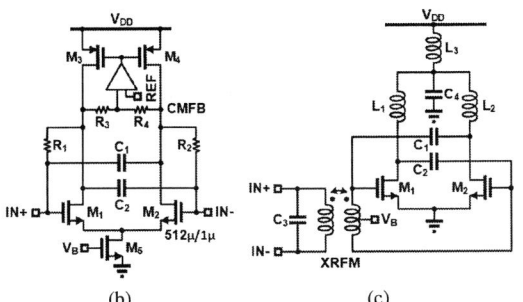

(b)   (c)

Fig. 5. Schematic of (a) the mixer-first receiver, (b) the BB amplifier, and (c) the LO buffer.

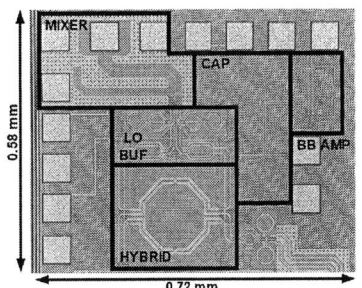

Fig. 6. Die micrograph.

and all the harmonic frequencies, thus increasing Z_{sh} and decreasing the NF.

Besides, the parasitic capacitance from the source and the switches influences the matching property, as the shunt capacitance C_{in} and C_{SW} shown in Fig. 3(b). To reduce the effect of parasitic capacitance, we introduce a shunt $\lambda/4$ TL to resonate out the capacitance. Figure 4 shows the post-layout simulation results of the impedance transformation. Our proposed TL-based matching network can significantly increase the impedance at the fundamental and the third harmonic frequencies with design robustness on the parasitic.

B. Circuit Implementation

The proposed mixer-first receiver is shown in Fig. 5(a). The TL-based matching network is implemented by coplanar waveguides (CPW) with the ground shield. The series and shunt TL have a characteristic impedance of 80 Ω and 60 Ω, respectively. The input 50-Ω source is then transformed to 130 Ω. Four switch transistors are driven by a quadrature LO signal, and the size of the transistors is 10 μm / 40 nm to trade off the parasitic capacitance and the ON-resistance. 40 pF capacitors are connected in series between the switches and the BB amplifier to set the high-pass corner frequency to around 200 kHz for close-in reflection compensation [2].

The schematic of the BB amplifier is shown in Fig. 5(b). The BB amplifier has a fully-differential common source structure with neutralization capacitors C_1 and C_2 to reduce the input capacitance. Thick-oxide devices are used for a higher operating voltage and the amplifier can deliver an around 3.5-V_{pp} differential output voltage swing. Feedback resistors R_1 and R_2 set the gain and the input resistance of the amplifier, and a common-mode feedback (CMFB) loop settles the operation point. The LO buffers

are co-designed with the mixer for higher voltage swing and generate overlapped sinusoidal signals, as shown in Fig. 5(c). We use a capacitance neutralization structure to improve the gain, an input transformer to match input impedance to the 37.5-Ω Hybrid impedance, and an inductive resonant load L_1 and L_2. To further improve the common-mode rejection, an LC-series trap is added at the center tap [7].

IV. MEASUREMENT RESULTS

The prototype chip is fabricated in 40-nm CMOS process. Fig. 6 shows the die micrograph. The die area is 580 $\times$ 720 μm^2. The power consumption of the receiver is 37 mW, where the LO buffers consume 28-mW power and the BB amplifiers consume 9-mW power. The die is wire-bonded to a PCB to provide supply and bias, and we use high-frequency probes for RF and LO signals. A 2.4-GHz LO is generated off-chip, and an on-chip $\times35$ frequency multiplier is used to generate the desired 84-GHz differential LO signals. A Rohde & Schwarz ZVA67 VNA with a ZVA110 frequency extender is used as the RF source for gain and linearity measurement, and a VDI PM4 power meter is used for power and loss calibration. The baseband signal is captured by an oscilloscope. To measure the NF, we use a Quinstar QNSFB15LW 50–110-GHz noise source and a Rohde & Schwarz FSW43 spectrum analyzer, calculating the DSB NF using the Y-factor method.

Figure 7(a) shows the measured voltage gain and NF at 1-MHz IF frequency by changing the LO frequency, with simulation results for comparison. The measured receiver voltage gain is around 20 dB and the NF is 8.0 to 9.5 dB. The bandwidth of the receiver is limited by the on-chip $\times35$ frequency multiplier, which can generate the LO signal from 82 to 88 GHz. However, the simulation results, which matches the measurement results well, with constant LO power over frequencies show a potentially wide bandwidth.

The measurement results over IF frequencies at 84-GHz LO frequency are shown in Fig. 7(b). The IF bandwidth is from 250 kHz to 10 MHz, and the flicker noise corner is below 1 MHz. The OOB IP1dB at 20-kHz IF frequency is -0.7 dBm (Fig. 7(c)), which shows excellent spillover tolerance. Meanwhile, the in-band IP1dB at 1-MHz IF frequency is -5.2 dBm, which indicates good linearity of the mixer-first receiver. Figure 7(d) shows the simulated

Table 1. Performance Summary and Comparison

Reference	ISSCC'21 TI Gen 2 [2]	ISSCC'19 Unhder [3]	JSSC'20 Tsinghua [5]	ISSCC'18 KU Leuven [10]	JSSC'20 UC Berkeley [7]	RFIC'20 IMEC [11]	This Work
Architecture	LNA First				Mixer First		
Application	FMCW Radar	PMCW Radar	FMCW Radar	Communication	Communication	FMCW Radar	FMCW Radar
Frequency (GHz)	76 – 81	76 – 81	76 – 81	75.1	85	61.5	85
Process	45nm CMOS	28nm CMOS	65nm CMOS	28nm CMOS	28nm CMOS	28nm CMOS	40nm CMOS
Phase	Quad.	Quad.	Diff.	Quad.	Quad.	Quad.	Quad.
Gain (dB)	–	–	26.2 – 78.8	28	26	46	20
NF (dB)	9 – 10.8	13	11.8 – 14.3	8.3 – 10	8 – 12.7	10.5	8 – 9.5
In-Band IP1dB (dBm)	–	-6.7	–	-25	-16.8	–	-5.2 [IFN = 4]*
OOB IP1dB (dBm)	-10		-8.5				-0.7 [IFN = 0.08]*
Power per RX (mW)	500**	475**	127**	77.3	12	10.6	37
Area per RX (mm²)	3.14**	3.55**	1.46**	1.09	0.86	0.96**	0.42

*IFN = IF frequency / HPF cut-off frequency (250 kHz). **Estimated by total power/area divided by the number of RF elements.

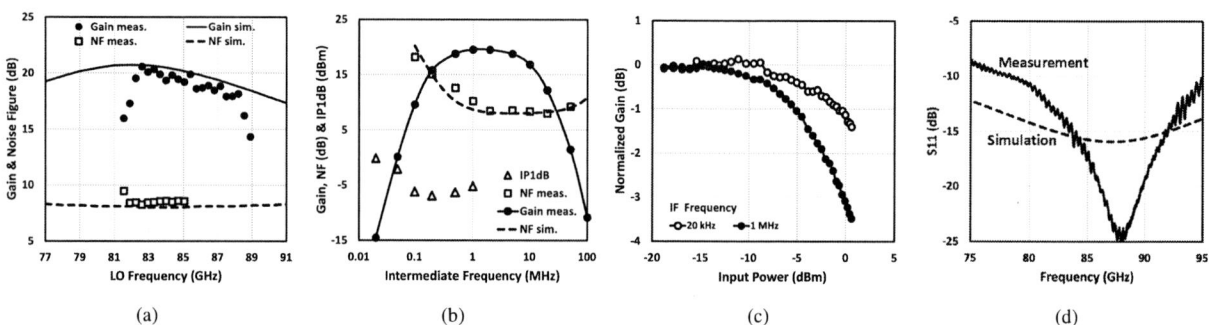

(a)   (b)   (c)   (d)

Fig. 7. Gain, NF, and P1dB at (a) 1 MHz IF frequency, (b) 84 GHz LO frequency, (c) gain compression, and (d) S11.

and measured S11. The -10-dB bandwidth is from 78.4 to 96.0 GHz. Our proposed TL-based matching network shows a wideband matching property.

Measurement results are compared with the state-of-the-art LNA-first and mixer-first receivers in Table 1. Our proposed mixer-first receiver achieves competitive NF and superb in-band as well as OOB IP1dB, indicating excellent spillover tolerance for FMCW radars. Meanwhile, the power and area are much lower than the LNA-first architecture, which shows the potential usage of mixer-first receivers in mm-wave radars.

V. CONCLUSION

In this paper, we present a W-band mixer-first receiver with a $\lambda/4$ TL-based matching network for NF and linearity improvement. By transforming the source impedance to a high value at the fundamental and the third-harmonic frequencies, we achieve a -0.7 dBm OOB IP1dB and an 8.0-to-9.5 dB NF. The improved IP1dB can offer strong spillover tolerance for FMCW radars, and the NF is competitive to the conventional LNA-first receivers with much lower power and area consumption.

ACKNOWLEDGMENT

This work was supported in part by ComSenTer, one of six centers in JUMP, a Semiconductor Research Corporation (SRC) program sponsored by DARPA. The authors acknowledge the TSMC University Shuttle program for chip fabrication.

REFERENCES

[1] A. Kankuppe et al., "A 67-mW D-Band FMCW I/Q radar receiver with an N-path spillover notch filter in 28-nm CMOS," *IEEE J. Solid-State Circuits*, vol. 57, no. 7, pp. 1982-1996, July 2022.

[2] K. Dandu et al., "High-performance and small form-factor mm-wave CMOS radars for automotive and industrial sensing in 76-to-81GHz and 57-to-64GHz bands," in *IEEE Int. Solid-State Circuits Conf. (ISSCC) Dig. Tech. Papers*, Feb. 2021, pp. 39-41.

[3] V. Giannini et al., "A 192-virtual-receiver 77/79GHz GMSK code-domain MIMO radar system-on-chip," in *IEEE Int. Solid-State Circuits Conf. (ISSCC) Dig. Tech. Papers*, Feb. 2019, pp. 164-166.

[4] T. Usugi et al., "A 77 GHz 8RX3TX transceiver for 250-m long range automotive radar in 40-nm CMOS technology," *IEEE J. Solid-State Circuits*, vol. 56, no. 5, pp. 1332-1344, May 2021.

[5] T. Ma et al., "A CMOS 76–81-GHz 2-TX 3-RX FMCW radar transceiver based on mixed-mode PLL chirp generator," *IEEE J. Solid-State Circuits*, vol. 55, no. 2, pp. 233-248, Feb. 2020.

[6] C. M. Hung et al., "Toward automotive surround-view radars," in *IEEE Int. Solid-State Circuits Conf. (ISSCC) Dig. Tech. Papers*, Feb. 2019.

[7] L. Iotti, S. Krishnamurthy, G. LaCaille and A. M. Niknejad, "A low-power 70–100-GHz mixer-first RX leveraging frequency-translational feedback," *IEEE J. Solid-State Circuits*, vol. 55, no. 8, pp. 2043-2054, Aug. 2020.

[8] R. Ying, et al., "A 20–40 GHz high dynamic range HBT N-path receiver with 8.9 dBm OOB B1dB and 8.55 dB NF consuming 130 mW," in *Proc. IEEE Radio Freq. Integr. Circuits Symp. (RFIC)*, 2021, pp. 215-218.

[9] D. Yang, C. Andrews and A. Molnar, "Optimized Design of N-Phase Passive Mixer-First Receivers in Wideband Operation," *IEEE Trans. Circuits Syst. I, Reg. Papers*, vol. 62, no. 11, pp. 2759-2770, Nov. 2015.

[10] M. Vigilante, et al., "A coupled-RTWO-based subharmonic receiver front-end for 5G E-band backhaul links in 28nm bulk CMOS," in *IEEE Int. Solid-State Circuits Conf. (ISSCC) Dig. Tech. Papers*, Feb. 2018.

[11] S. Park et al., "A 62mW 60GHz FMCW radar in 28nm CMOS ," in *Proc. IEEE Radio Freq. Integr. Circuits Symp. (RFIC)*, 2020, pp. 31-34.

[12] J. Zhang et al., "A W-band transceiver array with 2.4GHz LO synchronization enabling full scalability for FMCW radar," in *IEEE Int. Solid-State Circuits Conf. (ISSCC) Dig. Tech. Papers*, Feb. 2023.

979-8-3503-2123-4/23 $31.00 © 2023 IEEE

RTu1C-4

A CMOS 160GHz Integrated Permittivity Sensor with Resolution of $0.05\%\ \Delta\varepsilon_r$

Hai Yu[#1], Xuan Ding[#], Jingjun Chen[*], Sajjad Sabbaghi Saber[#], Qun Jane Gu[#2]

[#]Department of Electrical and Computer Engineering, University of California, Davis, USA

[*]Qualcomm Inc., USA

[1]jetyu@ucdavis.edu, [2]jgu@ucdavis.edu

Abstract— **This paper presents a high-resolution permittivity sensor at 160GHz in 28nm CMOS. It incorporates a THz high Q silicon whispering gallery mode resonator sensor that boosts the system's permittivity sensitivity. Multifold noise reduction techniques are adopted. A novel complementary BPSK signaling suppresses the common mode noise and the low frequency noise from within the system and the environment. Transmitter (TX) local oscillator (LO) feedforward that injection-locks the receiver (RX) mitigates the TX phase noise. The sensing system achieves the best permittivity sensing resolution of $0.05\%\ \Delta\varepsilon_r$ among the state-of-the-art permittivity sensors within 14us of integration time and 54mW power consumption.**

Keywords— **whispering gallery mode resonator, coherent phase noise cancellation, injection locking, sub-terahertz, mm-wave, transceiver, complex permittivity, sensor figure-of-merit.**

I. INTRODUCTION

Permittivity sensing has wide applications in various industries because it can serve as effective indicators of the target for maintenance, treatment, or production. There are essentially two key aspects to improve permittivity sensing resolution. One is to increase the sensor's sensitivity, defined as the sensor's electrical response per unit permittivity change of the material under test (MUT). [1] uses a metamaterial-based sensor to improve sensitivity by utilizing metamaterial structure's intensified electric field. Another common sensing scheme is oscillator-based reactance sensing, which is essentially utilizing the high-Q resonance tank to boost the sensitivity [2]. Although high sensitivity can be achieved, this method does not have a reliable capability to sense the imaginary part of the permittivity. Another direction for high sensing resolution is to reduce system noise. [3] adopts LO feedforward to reduce the TX's noise and correlated double sampling of sensing and reference path to suppress low frequency noise. However, its digitized carrier waveform prevents it from sensing the imaginary part of the permittivity. We present a sub-THz high resolution permittivity sensing system that advances the state-of-the-art permittivity sensing by utilizing a sub-THz high Q whispering gallery mode (WGM) resonator sensor to increase sensitivity and an architectural combination of complementary BPSK chopping scheme with LO feedforward for noise suppression. Working at sub-THz allows the adoption of high-Q WGM disk resonator as the high-sensitivity sensor to enhance resolution.

II. SYSTEM ARCHITECTURE

Fig. 1 illustrates the sensing mechanism and system architecture for material permittivity sensing. The TX outputs

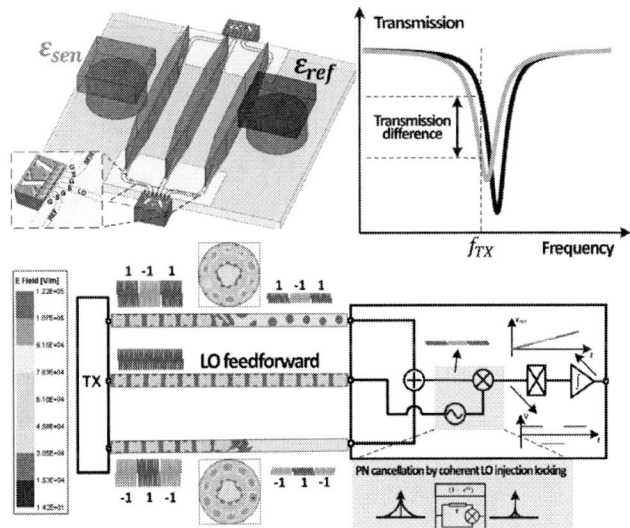

Fig. 1. Proposed sensing system working mechanism and illustration of the integrated sensor architecture.

a complementary pair of BPSK modulated signals through two identical off-chip silicon WGM disk resonator sensors near their band-stop resonant frequency, which are then summed by the combiner of the receiver (RX). The resulting signal reflects the transmission difference (ΔS_{21}) between the sensing and the reference path, which is then down converted by the RX mixer. The proceeding chopper upconverts the low frequency noise and the following integrator filters out the high frequency noise. The complementary BPSK signalling for the sensing and the reference paths suppresses both low frequency noise from the electronics and the impact from the ambient temperature variation. The summation of the complementary signals at the input of the RX reduces signal amplitude at the input of the mixer, which reduces the TX noise impact and mitigate the trade-off between noise and linearity for the RX circuitry. The TX injection-locks the RX LO through a low loss sub-THz LO feedforward path to suppress the TX's close-in phase noise in the down-conversion process at the RX. The architectural combination of balanced sensing and reference path and the LO feedforward overcomes the drift and high close-in phase noise of a THz free running VCO, leading to a high sensing resolution.

III. SENSOR MECHANISM

The working mechanism and the design of the silicon (Si) whispering-gallery mode (WGM) disk resonator sensors are shown in Fig.2. The disk resonator couples and dissipates

979-8-3503-2123-4/23 $31.00 © 2023 IEEE   245   2023 IEEE Radio Frequency Integrated Circuits Symposium

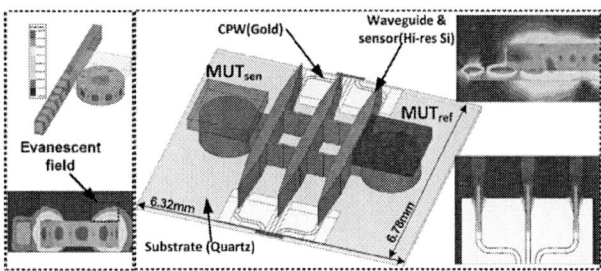

Fig. 2. WGM sensor mechanism and sensor structure

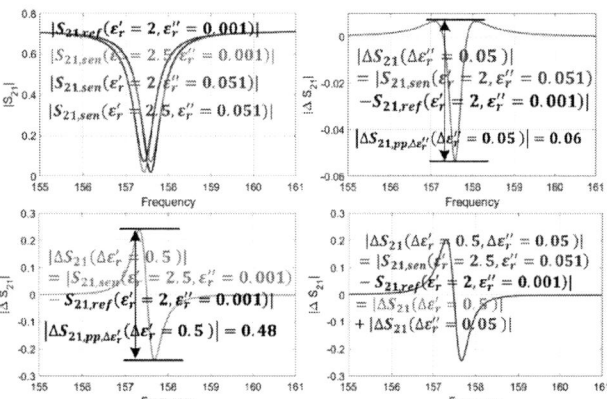

Fig. 3. Simulated sensor's response to complex permittivity difference.

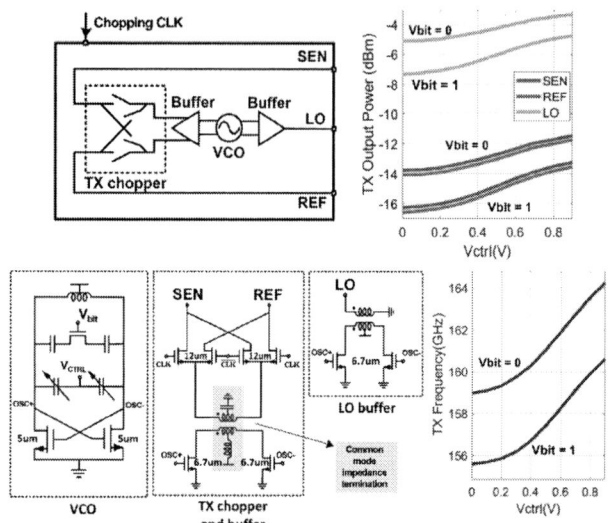

Fig. 4. TX block diagram, schematics, and simulated performance

sub-THz electromagnetic (EM) waves from the Si dielectric waveguide at the resonant frequency, resulting in a transmission notch. The center frequency and the depth of the notch are impacted respectively by the real and imaginary part of the complex relative permittivity $\varepsilon_r', \varepsilon_r''$ of the dielectric material under test (MUT) on top of the disk due to the interaction between the MUT and the evanescent fields of the WGM that extends out of the dielectric resonator. A pair of such identical resonator sensors forms a sensing path and a reference path, so the magnitude of their transmission difference, $|\Delta S_{21}| = |S_{21,ref} - S_{21,sen}|$, can be detected by the RX. As shown in Fig 3, $|\Delta S_{21}|$ near the transmission notch can be decomposed into an odd ($|\Delta S_{21,\Delta\varepsilon_r'}|$) and an even ($|\Delta S_{21,\Delta\varepsilon_r'}|$) waveform, which reflects $\Delta\varepsilon_r'$ and $\Delta\varepsilon_r''$, respectively, between the two MUTs on the sensing and the reference path. The sensor's real and imaginary permittivity sensitivity is defined as

$$Sen_{\varepsilon_r'} \equiv \frac{|\Delta S_{21,pp,\Delta\varepsilon_r'}|}{\Delta\varepsilon_r'}, \; Sen_{\varepsilon_r''} \equiv \frac{|\Delta S_{21,pp,\Delta\varepsilon_r''}|}{\Delta\varepsilon_r''}, \quad (1)$$

where $\Delta S_{21,pp,\Delta\varepsilon_r'}$ and $\Delta S_{21,pp,\Delta\varepsilon_r''}$ are the peak-peak transmission difference between the sensing and the reference path due to $\Delta\varepsilon_r'$ and $\Delta\varepsilon_r''$, respectively. The transmission differences near the notch are larger when the resonators have high Q and are near critical coupling with the feeding waveguide, so that the steep slope and sharp curvature near the resonance produce large response even for small $\Delta\varepsilon_r'$ and $\Delta\varepsilon_r''$, leading to a high sensitivity. From EM simulation in HFSS, the sensor structure shows a Q of 250 and $Sen_{\varepsilon_r'} = 0.96$ and $Sen_{\varepsilon_r''} = 1.2$. The following system analysis will be based on ε_r, which is a representation of either ε_r' and ε_r''. The WGM

sensors and the feeding waveguides are made from high-resistivity Si. The Si waveguides are fed by the co-planar waveguides on a 100um-thick quartz substrate. The sensor is fabricated in the university cleanroom through standard microfabrication processes. The senor fabrication resolution causes mismatches between the sensor and the reference paths, resulting a resonant frequency difference up to 50MHz. However, this static mismatch can be calibrated out during the measurement.

IV. TRANSCEIVER IMPLEMENTATION

A. System signal and noise analysis

The sensing signal strength S at the input of the RX due to a certain permittivity difference $\Delta\varepsilon_r$ between the sensing and the reference paths can be expressed as

$$S = \sqrt{P_{TX}Z_0}Sen_{\varepsilon_r}\Delta\varepsilon_r, \quad (2)$$

where P_{TX} is the TX output power, the system impedance $Z_0 = 50\Omega$, and Sen_{ε_r} is the sensor's permittivity sensitivity defined in the previous section. Given the RX noise $v_{n,RX}$ and the TX noise $v_{n,TX}$, both referred to the input of the RX, the signal-to-noise ratio (SNR) of the sensing system can be expressed as

$$SNR = \frac{S}{N} = \frac{\sqrt{P_{TX}Z_0}Sen_{\varepsilon_r}\Delta\varepsilon_r}{\sqrt{v_{n,RX}^2 + v_{n,TX}^2}}. \quad (3)$$

Then, the minimum detectable permittivity difference, or the permittivity sensing resolution δ_{ε_r}, is given by setting $SNR = 1$:

$$\delta_{\varepsilon_r} = \frac{\sqrt{v_{n,RX}^2 + v_{n,TX}^2}}{\sqrt{P_{TX}Z_0}Sen_{\varepsilon_r}}. \quad (4)$$

According to (4), with $Sen_{\varepsilon_r} = 0.96$, $P_{TX} = -16dBm$ and a total noise of $3uVrms$ at the RX input with a 20kHz noise bandwidth, $\delta_{\varepsilon_r} = 10^{-4}$ can be achieved.

$v_{n,TX}^2$, dominated by the TX phase noise, has the following relationship with the system parameters:

$$v_{n,TX}^2 \propto L_{TX}(f)P_{TX}Z_0|H_c(f)|^2 Sen_{\varepsilon_r}{}^2\Delta\varepsilon_r{}^2. \quad (5)$$

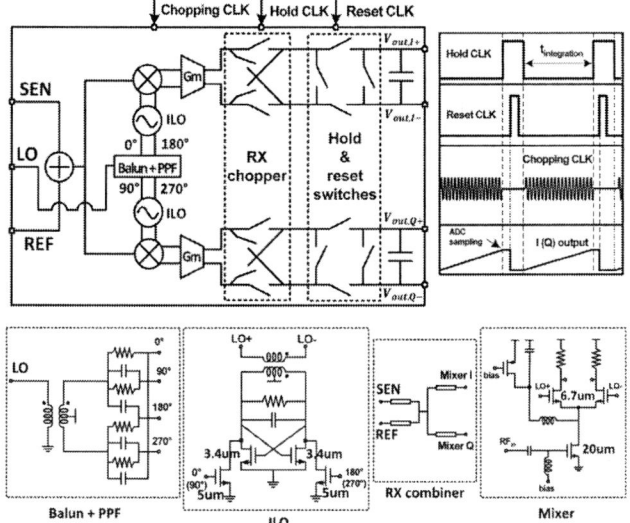

Fig. 5. RX block diagram, schematics, and timing diagram

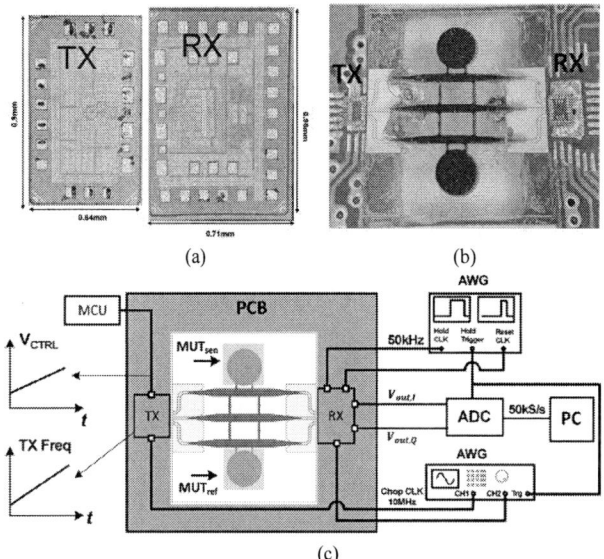

Fig. 6. (a) chip photos; (b) assembled sensing system; (c) permittivity measurement setup.

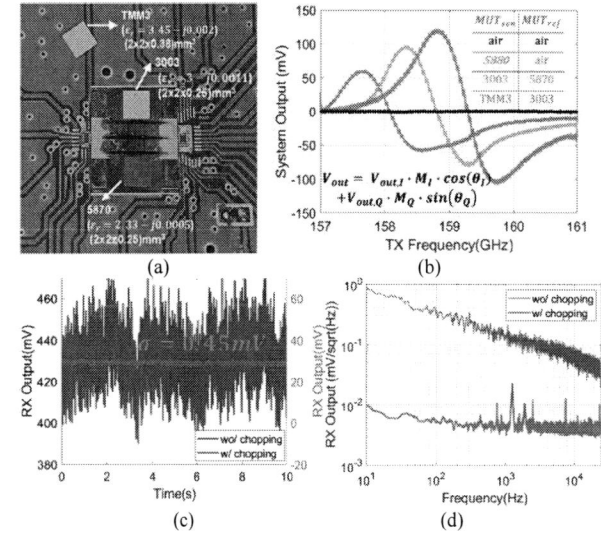

Fig. 7. System measurement results: (a) MUT's on the sensors; (b) system output waveform; (c) system output noise; (d) output noise spectral density.

$L_{TX}(f)$ is the TX phase noise power density in dBc/Hz. $H_c(f) = 1 - e^{j2\pi f t_d}$ is the coherent phase noise cancellation effect, which high-pass filters the close-in phase noise of the TX. t_d, the time delay difference between the LO path and the sensing/reference path, is found to be 1.4ns at its maximum near the sensor's resonant frequencies. This delay gives a higher than 86dB noise suppression below 20kHz. Furthermore, the complementary signaling through the sensing and reference path further suppresses the TX noise through the $Sen_{\varepsilon_r}^2 \Delta\varepsilon_r^2$ term. As a result, the TX noise is suppressed below the RX noise, which is verified in the system measurement process, where the system output noise doesn't change when the TX is turned on and off.

B. TX

The TX's block diagram and circuit schematics are shown in Fig. 4. The L-C voltage-controlled oscillator (VCO)'s output frequency is tuned by the varactors and the discrete capacitor, which enables the system to sweep across an 8-GHz frequency range. The LO buffer delivers the power to the LO feedforward path through a transformer balun, while the signal path buffer drives a double-balanced mixer through a transformer. The transformers are designed to provide a class-A load lines for the buffer amplifiers to maximize the output power. The center tap of the primary and secondary coils of the transformer is terminated by a high and a low impedance respectively to isolate the TX's common mode noise from being transmitted to the RX. The mixer modulates the signals into BPSK at the same chopping frequency of the RX's. The sizes of the mixer transistors are designed to balance the trade-off between the on-off ratio of the modulated signal and the power loss. From simulation, the maximum output power of the LO port is -3.5dBm. Without the 3dB combination of a balun like LO path, and due to the 4dB loss of the mixer and the 1.5dB loss from the additional length of transmission line,

the maximum output power of the sensing/reference port is about -12dBm. The entire TX consumes 24mW DC power, where the oscillator core consumes 6mW and each of the buffer amplifier consumes 9mW.

C. RX

The RX's block diagram and its key circuit schematics are shown in Fig. 5. The input combiner sums the complementary BPSK signals from the sensing path and the reference paths, performs impedance transformation, and delivers the signal voltages to the inputs of the mixers of the in-phase(I) and quadrature(Q) paths. The mixers are implemented with single balanced gilbert cell to serve as an active balun to improve the overall noise figure of the RX. The DC current of the gm cell was bled away through a current bleeding path, improving the

979-8-3503-2123-4/23 $31.00 © 2023 IEEE   247

Table 1. Comparison with the state-of-the-art integrated permittivity sensors

Reference	JSSC2016[2]	JSSC2020[3]	TMTT2018[4]	TMTT2019[5]	CICC2021[6]	**This work**
Technology	65nm CMOS	180nm CMOS	40nm CMOS	65nm CMOS	45nm RFSOI	**28nm CMOS**
Frequency (GHz)	6.5/11/17.5/30	1.8/2.2	0.1-10	3-10	λ=1300nm	**154-160**
Resolution (%)	0.8	0.3	3	0.2	0.08	**0.05**
Integration time(us)	10	20.48	1000	150	10000	**14**
Complex sensing	No	No	Yes	Yes	No	**Yes**
Integrated LO	Yes	Yes	No	Yes	No	**Yes**
Power consump. (mW)	65	77.7	24	64	128	**54**
FoM	63.81	68.44	36.66	64.16	50.87	**87.24**

conversion gain and the noise by reducing the current steering load of the switching pair. A 45pH inductor in the current bleeding path resonates out the parasitic capacitances at the source of the switching pair to further increase the conversion gain and reducing the noise contribution by the switching pair by boosting the impedance at its source node. The combiner and the I/Q mixers form the front end of the RX and are co-designed to have a total voltage gain of 9dB and a noise figure of 21dB and consumes 10mW DC power. In the injection LO path, a balun transformer together with the following poly phase filter split the incoming LO from the TX into 4-phase quadrature signals to drive the two injection-locked oscillators (ILO) for the I/Q mixers. The transistor sizes of the ILO's current injection devices and the cross-coupled pair are designed to balance the trade-off between the ILO's output swing and its lock range. From simulation, with a -10dBm power at the RX LO input, the ILO has a lock range of 20GHz with the output swings larger than 150mV. Each ILO consumes 7.5mW. The gm-C integrator is implemented with a folded telescopic transconductance amplifier and a 52pF integration capacitor to filter out the noise with a 20kHz bandwidth, and the integration time is set by the off-chip hold and reset clocks. Finally, the differential voltage signals from the I/Q outputs are sampled by an off-chip ADC. The RX consumes a total of 30mW power.

V. MEASUREMENT RESULTS

The TX and RX ICs, fabricated in a 28nm bulk CMOS technology, are mounted on the printed circuit board, and connected with the sensor via wire bonding, which are shown in Fig. 6. The permittivity measurement setup is shown in Fig. 6c. The output voltages of the I/Q path of the RX are sampled by the off-chip ADC, which is triggered by the hold clock once an integration cycle of 14us finishes. V_{CTRL} and thus the TX's output frequency is controlled by a micro control unit (MCU). By sweeping V_{CTRL}, the vector waveforms are obtained from the I/Q outputs of the RX. Rogers laminate samples are placed on the sensors of the sensing and the reference paths as MUT's. After calibration, vector waveforms V_{out} can be further analysed to obtain $\Delta\varepsilon_r'$ and $\Delta\varepsilon_r''$. The waveforms and system output noise are shown in Fig.7. From the measurements, this system demonstrates a permittivity sensing resolution of 0.05% within an integration time of 14us. Table 1 compares this work and the state-of-the-art permittivity sensing systems. To have a fair comparison, a sensor figure-of-merit is defined as

$$FoM = \frac{1}{\delta_{\varepsilon_r}\sqrt{P_{DC}\tau_{int}}} \quad , \qquad (6)$$

or, in logarithm scale,

$$FoM = -10\log\left(\delta_{\varepsilon_r}^2 P_{DC}\tau_{int}\right), \qquad (7)$$

where δ_{ε_r} is the permittivity sensing resolution, P_{DC} is the DC power consumption, and τ_{int} is the integration/measurement time.

VI. CONCLUSIONS

This paper presents a high-resolution permittivity sensing system at 160GHz in 28nm CMOS. It advances the state-of-the-art complex permittivity sensing resolution by adopting an off-chip high sensitivity silicon whispering gallery mode resonator sensor, and a novel architectural combination of complementary BPSK chopping and coherent phase noise cancellation with LO feedforward. The sensing system achieves the best permittivity sensing resolution of 0.05% $\Delta\varepsilon_r$ as well as the highest figure-of-merit within 14us of integration time and with 54mW power consumption.

ACKNOWLEDGEMENT

The authors would like to thank National Science Foundation (NSF) for the partial financial support, the UC Davis Center for Nano and Micro Manufacturing (CNM2) and UC Davis FIT of the Physics Department for technical support during the sensor fabrication and packaging, and Dr. Minji Zhu from CHFE of UCLA for the system wire bonding service.

REFERENCES

[1] A. P. Saghati, et al., "A Metamaterial-Inspired Wideband Microwave Interferometry Sensor for Dielectric Spectroscopy of Liquid Chemicals," TMTT, Jul. 2017.

[2] J.-C. Chien and A. M. Niknejad, "Oscillator-Based Reactance Sensors With Injection Locking for High-Throughput Flow Cytometry Using Microwave Dielectric Spectroscopy," JSSC, Feb. 2016.

[3] J.-C. Chien, "A 1.8-GHz Near-Field Dielectric Plethysmography Heart-Rate Sensor With Time-Based Edge Sampling," JSSC, Mar. 2020

[4] G. Vlachogiannakis, et al., "A 40-nm CMOS Complex Permittivity Sensing Pixel for Material Characterization at Microwave Frequencies," TMTT, Mar. 2018

[5] E. Kaya, et al., "A 3–10-GHz CMOS Time-Domain Complex Dielectric Spectroscopy System Using a Contactless Sensor," IEEE TMTT, vol. 67, no. 12, pp. 5202–5217, Dec. 2019.

[6] C. Adamopoulos et al., "Fully Integrated Electronic-Photonic Sensor for Label-Free Refractive Index Sensing in Advanced Zero-Change CMOS-SOI Process," in 2021 IEEE Custom Integrated Circuits Conference (CICC), Apr. 2021, pp. 1–2.

RTu1C-5

A 160-GHz FMCW Radar Transceiver with Slotline-based High Isolation Full-duplexer in 130nm SiGe BiCMOS Process

Xingcun Li[*#$1], Huibo Wu[*1], Shuyang Li[*1], Wenhua Chen[*], Zhenghe Feng[*]

[*]Department of Electronic Engineering, Tsinghua University, China
[#]Analog Design Dept., Sanechips Technology Co., Ltd., China
[$]State Key Lavoratory of Mobile Network and Mobile Multimedia Technology, China
chenwh@tsinghua.edu.cn

Abstract—In this paper, a slotline-based electrical balance duplexer (EBD) with high geometrical symmetry topology and high common mode rejection is proposed to realize high isolation between the transmitter (TX) and receiver (RX). With the proposed EBD structure, a 160-GHz frequency modulated continuous wave (FMCW) radar transceiver with TX/RX antenna sharing architecture is achieved in the 130nm SiGe BiCMOS process. The chirp signal can be generated covering 147 GHz to 165 GHz by adjusting the operating frequency of the push-push voltage-controlled oscillator (VCO) with the external input control voltage. Using spatial power combining slotline-based antennas and a high resistivity silicon lens, the measured effective isotropic radiated power (EIRP) is over 20 dBm. The chip has a die area of 2.21mm² and consumes 0.6 W of DC power.

Keywords—D-band, electrical balance duplexer (EBD), full duplex, frequency modulated continuous wave (FMCW) radar, SiGe BiCMOS.

I. INTRODUCTION

With the rapid development of the silicon-based process, the high-resolution, high output power, and low-cost integrated terahertz FMCW radar transceivers are expected to deliver potential in automotive, security sensing, and non-destructive detection applications. The classical FMCW architecture with separate transmitting (TX) and receiving (RX) antennas is commonly adopted [1]-[2] to realize the high isolation between the TX and RX and avoid the RX blockage caused by leakage signals. However, the configuration with separate TX/RX antennas increases the area of the chip and requires additional phase-shifting units for beamforming and alignment [3], which leads to the high cost of chip fabrication. In addition, due to the short wavelength of terahertz waves, off-chip antennas are difficult to implement, and the power losses in packaging or interconnection are unacceptable. Therefore, the on-chip full-duplex and on-chip antenna are attractive in FMCW radar transceivers.

A high-isolation and low-loss full-duplexer is the major challenge for on-chip antenna sharing FMCW radar. The circulator is one of the commonly used full-duplexers [4]. However, it is impossible to realize the on-chip circulator due to the lack of ferrite material in the silicon-based semiconductor process. M. Porranzl *et al.* proposed a low-loss quasi-circulator based on active devices [5], but the linearity of the quasi-circulator is limited. To achieve high isolation performance, the paper in [6] proposed an impedance tuning network to increase TX-RX isolation to −30 dB, but this

[1] equal contribution

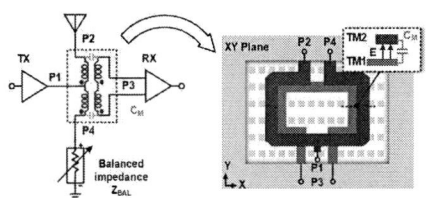

Fig. 1. Conventional EBD based on the hybrid transformer.

technique consumes additional power and reduces the bandwidth. Compared with the quasi-circulator, the passive duplexer has no power consumption and linearity problems, and the reduced passive structure size can be integrated into the chip as the frequency increases to sub-terahertz frequency. X. Chen *et al.* presented a slotline-based duplexing radar transciever without additional 3dB insertion loss both in TX and RX mode, achieving high isolation between TX and RX with adaptive self-interference cancellation technique [11]. Besides that, the passive duplexer can also achieve high isolation at the cost of additional loss, such as the duplex structure based on directional coupler, hybrid transformer, and Wilkinson splitter [7]-[9].

This paper presents a high geometrical symmetry electrical balanced duplexer (EBD) based on CPW and slotline-to-GCPW transition structures. The proposed EBD structure uses slotline to suppress the signal leaking from TX to RX, which can replace the conventional transformer-based EBD, achieving extremely high TX-RX isolation. Based on the proposed full-duplexer, a low-cost and compact D-band FMCW radar transceiver was implemented in the 130nm SiGe BiCMOS process.

II. SLOTLING-BASED ELECTRICAL BALANCE FULL DUPLEXER

The conventional EBD is usually realized by a hybrid transformer with a center tap, as shown in Fig. 1. However, as frequency increases, the parasitic capacitance of the conventional EBD lead to a worse common mode rejection ratio (CMRR) and lower isolation between TX and RX.

To address these problems, a high geometrical symmetry EBD using CPW and slotline-to-GCPW transition structures is proposed, as shown in Fig. 2a, which is called slotline-based EBD. Replacing the traditional transformer with the slotline-to-GCPW transition structure [see Fig.2b] can eliminate the capacitance between windings [10] and improve the common mode rejection. In this work, the center tap of the transition is connected to the transmitter by a CPW structure, as node A in

979-8-3503-2123-4/23 $31.00 © 2023 IEEE 249 2023 IEEE Radio Frequency Integrated Circuits Symposium

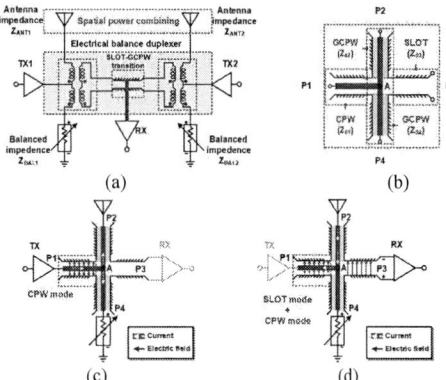

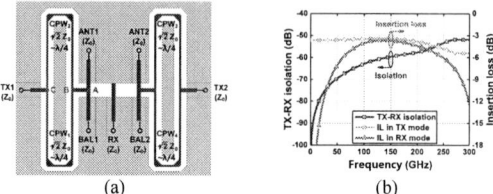

Fig. 3. (a) The proposed layout of the EBD, (b) and the simulation results.

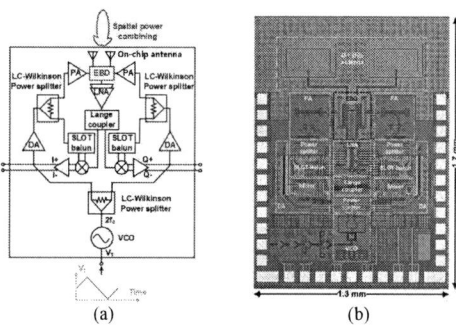

Fig. 2. (a) The proposed high geometrical symmetry EBD, (b) slotline-to-GCPW transition layout replacing the conventional transformer structure, and the basic operating principle of (c) TX and (d) RX mode.

Fig. 2b. Fig. 2c and Fig. 2d present the basic operating principle of the proposed slotline-based EBD. In the TX mode, the signals entering EBD from P1 are transmitted to the slotline-to-GCPW transition by the CPW transmission mode. Then the signals are split into two parts, one for the balanced impedance at P4 and the other for the antenna at P2. When the balanced impedance is equal to the antenna impedance, the signals entering these two ports have no phase difference, and the TM-mode waves in slotline excited by the common-mode signals are suppressed. As a result, no signal is transmitted to the RX, enabling high TX-RX isolation. In the RX mode shown in Fig. 2d, the signals received by the antenna are distributed at node A. One part of the signal flows into the CPW line, but the other part cannot directly reach the balanced impedance because the signal from the antenna to the balanced impedance is truncated by the slotline, which will excite the TE-mode wave in the slot line. There are both CPW transmission mode and slotline transmission mode in the CPW structure. The signals in the slot transmission mode are transmitted by the slotline to P3 and converted into differential signals into the RX, while the common-mode signals cannot be transmitted to the RX by the slotline. Therefore high common rejection is achieved. Assuming the power is distributed equally and the ideal transition is achieved by the slotline-to-GCPW structure, the port impedance matching condition and the characteristic impedance of the transmission line should satisfy the equation as follows:

$$Z_{ANT} = Z_{BAL} = Z_{02} = Z_{04} = Z_0 \quad (1)$$

$$Z_{TX} = Z_{01} = \frac{Z_0}{2} \quad (2)$$

$$Z_{RX} = Z_{03} = 2Z_0 \quad (3)$$

The RX circuit cannot directly connect to the slotline in the actual layout. As a result, a back-to-back slotline-based EBD structure with an additional slotline-to-GCPW transition connecting to RX is proposed, which also greatly improves the symmetry in the layout implementation, and obtains excellent TX-RX isolation.

To verify the prototype, the proposed slotline-based EBD is modeled and simulated in 130nm SiGe BiCMOS process, as shown in Fig. 3a. The $\lambda/4$ folded CPW transmission lines

CPW$_1$ ~ CPW$_4$ are implemented for impedance matching, decreasing loss and making slot line transmission mode equivalent to open in RX mode. Fig. 3b shows the simulation results of the proposed EBD. The simulation results show that the duplexer achieves better than −50 dB of TX-RX isolation over the wideband range. The minimum insertion loss (IL) is 3.4 dB in TX mode and 3.7 dB in RX mode. Over the 150 GHz bandwidth, the IL is less than 5dB in both TX and RX.

III. THE D-BAND FULL-DUPLEX FMCW RADAR TRANSCEIVER DESIGN

The entire block diagram of the proposed D-band full-duplex FMCW radar transceiver is shown in Fig. 4a. The chirp signal covering 147 GHz to 165 GHz can be generated by adjusting the operating frequency of the harmonic VCO. The second harmonic output signal of the VCO is split into TX1 and TX2 by an LC-Wilkinson power splitter and then amplified by the driving amplifiers (DAs). The output signals of the DAs are split into two ways by the LC-Wilkinson power splitter as well. One is amplified by the power amplifier (PA) and the other is used for the local oscillator (LO) signal of the RX. The receiving signals are amplified by LNA and a Lange coupler is utilized to split the RF signals into two quadrature channels. The key implementation details will be introduced in this section.

A. Harmonic VCO

The VCO is one of the core modules in an FMCW radar transceiver system. In the proposed radar, the push-push architecture is adopted to oscillate at a lower frequency because the gain of transistors and the quality factor of the tuning elements are higher. The second harmonic signal is finally extracted for output to obtain wideband tuning.

Fig. 5 shows the schematic of the harmonic VCO. The core of the VCO adopts the Colpitts oscillator structure. The second harmonic signal is extracted and output through the

979-8-3503-2123-4/23 $31.00 © 2023 IEEE   250

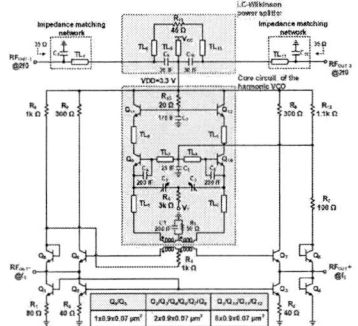

Fig. 5. The schematic of the VCO.

common-mode tap. To monitor the VCO oscillation frequency, a fundamental output port is also configured.

B. Transmitter and Receiver Chain

Fig. 6a shows the schematic of TX. The DA and the PA both use the cascade topology with the linear bias circuit. The DA is single-ended, and the output matching network of the DA is co-designed with the LC-Wilkinson power division network to achieve a compact layout. The PA adopts a two-way power combining topology based on the three-conductor transmission-line-based combiner. The peak small signal gain of the transmitter chain is 16.4 dB, and the 3-dB gain bandwidth covers 138~172 GHz.

The schematics of the LNA and mixer are shown in Fig. 6. To achieve a low noise figure, the LNA adopts the two-stage single-ended cascode amplifier with degenerated inductors TL1 and TL6, and noise cancellation technique using parallel TL2 and TL7. Single-balanced active mixer topology with linear bias is used as a down-converter. The IF signal is filtered and then amplified by the emitter followers. The simulation results show that the noise figure of the receiver chain is better than 12 dB in the range of 10 kHz~10 MHz of the output IF, and the conversion gain is 16 dB at 160 GHz.

C. On-chip antenna

The on-chip full duplexer is the EBD proposed in section II, which can ideally achieve extremely high isolation. However, the antenna impedance and balanced impedance offset will deteriorate the isolation. To avoid the impedance deviation caused by the off-chip antenna and the off-chip interconnection, the on-chip antenna is proposed. As shown in Fig. 7, two folded slot antennas are used for spatial power combining and radiation simultaneously. In addition, a high-resistance silicon lens is used to improve the radiation efficiency and antenna gain. In this work, the width of the slot is 35 μm and the distance between two folded slot antennas is 560 μm. The simulation results show that the gain of the antenna is about 14 dBi in 150~170 GHz.

IV. MEASUREMENT RESULTS

The D-band full-duplex FMCW radar transceiver is implemented in the IHP 130nm SiGe BiCMOS process, with a chip area of 1.3×1.7 mm^2 as shown in Fig. 4b. Fig. 8a shows the measurement setup for TX. The operating frequency is

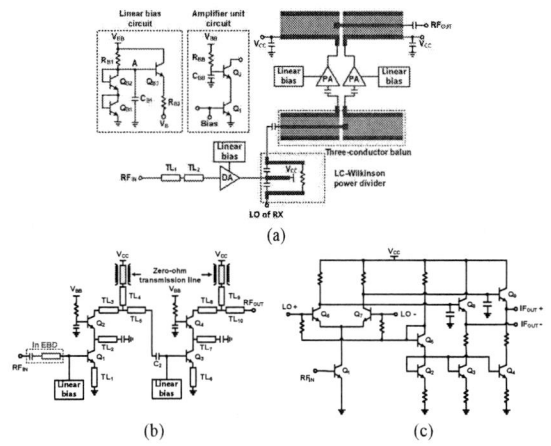

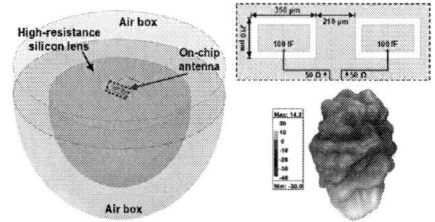

Fig. 6. The schematic of the (a) TX, (b) LNA, (c) mixer.

Fig. 7. Simulation results of the on-chip antenna with a HR-Si lens.

measured through the VDI WR 5.1 (140 - 220 GHz) extender as a down-converter connected to the R&S FSW43 spectrum analyzer. The output frequency results are summarized in Fig 8b. Changing V_T from 0 V to 4 V, the output frequency range is 147~165GHz. Then the output power of the TX is received by a WR 5.1 horn antenna and measured by the Erickson PM5B power meter. The measured received power and the power calculated by the Firrs function are presented in Fig. 8c. The measured EIRP versus distance is also shown in Fig. 8c. It can be seen that the far-field distance needs to be greater than 10 cm for accurate results. Fig. 8d shows the trend of EIRP with V_T at 25 cm. The EIRP is around 22 dBm with a 14-dB gain of the on-chip antenna at an appropriate distance. As a result, the output power of TX can be estimated to be about 8 dBm.

To implement a functional validation of the FMCW radar, a measurement of object detection is performed through the Arbitrary Waveform Generator (AWG) and the R&S FSW43 spectrum analyzer. The measurement setup is presented in Fig. 9a. The triangular wave voltage signal is generated by AWG to control the tuning voltage V_T of the VCO. The sweep frequency f_m of the triangular wave signal is 700 Hz, the sweep period T_m is 1.43ms, the sweep range of voltage is 0~4 V, and the detection target is a cuboid metal plate. The IF spectrum and the measured distance are shown in Fig. 9b when the single object is at 115 cm. Fig. 9c and 9d summarize the results of placing two objects at different locations. The results show the resolution is 10 cm. The balanced and the antenna impedances are same and all port impedances in Fig 3a are 50 Ω for the simulation of the EBD. However, these impedance conditions cannot be fully satisfied in practice and

979-8-3503-2123-4/23 $31.00 © 2023 IEEE

Table 1. Performance comparison of the state-of-the-art FMCW radar.

Reference	Process	TX/RX antenna sharing?	Frequency (GHz)	Antenna type	P_{TX} (dBm)	EIRP (dBm)	Bandwidth (GHz)	DC power (W)	Area (mm^2)
This work	**130nm SiGe**	**Yes**	**148**	**On-chip**	**8**	**22**	**18**	**0.6**	**2.21**
JSSC'17 [1]	130nm SiGe	No	122	On-chip	5	11	10	0.63	10.14
JSSC'21 [7]	130nm SiGe	Yes	168	On-chip	3	8	20	0.86	5.4
JSSC'21 [7]	130nm SiGe	No	168	On-chip	13	18	20	1.05	8.40
JSSC'21 [2]	28nm CMOS	No	145	On-chip	8.6	11.6	13	0.5	6.5
ISSCC'22 [11]	65nm CMOS	Yes	141	On-chip	11.2	25.2	14	0.405	3.1

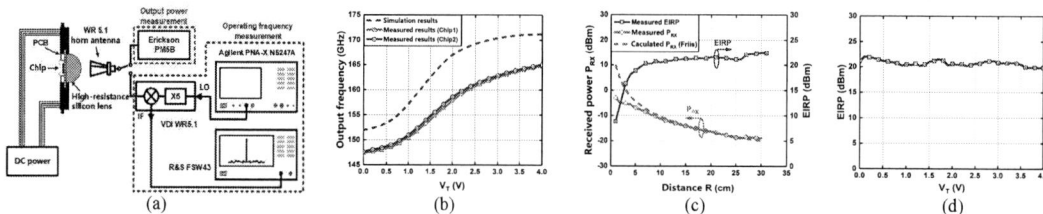

Fig. 8. (a) Measurement setup for TX, the measurement results of (b) operating frequency. (c) ERIP of the TX, received power and the power calculated by the Firrs, and (d) the trend of EIRP with V_T

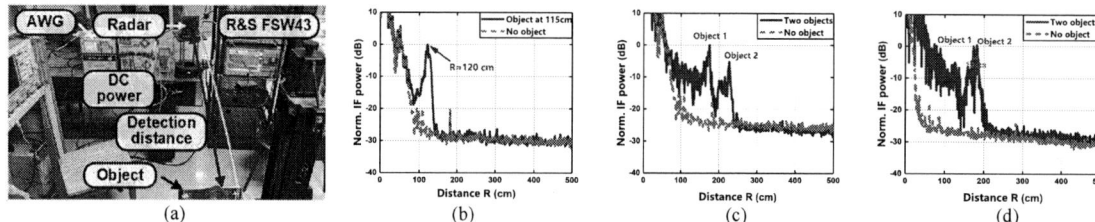

Fig. 9. (a) Measurement setup of object detection. (b) Measured distance of single object. (c) Two objects detection and ΔR = 50 cm. (d) Two objects detection and ΔR = 10 cm.

the tuning range of the balance impedance is limited, which brings the leakage and results in the high IF power below R=1 m in Fig. 9. Besides that, the flicker noise and free running VCO without PLL also deteriorates the IF signals at short distances. The measured performances of the proposed D-band FMCW radar are summarized in Table 1 and compared with the state-of-the-art D-band FMCW radar.

V. CONCLUSION

This paper presents a D-band full-duplex FMCW radar with the TX/RX antenna sharing architecture in 130 nm SiGe BiCMOS. In this radar, a slotline-based electrical balance duplexer with high geometrical symmetry topology is proposed to realize high isolation between the TX and RX. The measured results show the EIRP achieves over 20 dBm with 0.6 W power consumption. The chip area of the radar transceiver is only 2.21 mm^2, which is competitive with other state-of-the-art D-band FMCW radar.

ACKNOWLEDGMENT

This work is supported in part by the National Key Research and Development Program of China under Grant 2019YFB2204701 and in part by the National Science Fund for Distinguished Young Scholars under Grant 62225111.

REFERENCES

[1] H. J. Ng, *et al.*, "Multi-Purpose Fully Differential 61- and 122-GHz Radar Transceivers for Scalable MIMO Sensor Platforms," *IEEE J. Solid-State Circuits*, vol. 52, no. 9, pp. 2242-2255, Sept. 2017.

[2] Visweswaran, *et al.*, "A 28-nm-CMOS Based 145-GHz FMCW Radar: System, Circuits, and Characterization," *IEEE J. Solid-State Circuits*, vol. 56, no. 7, pp. 1975-1993, July 2021.

[3] M. Pauli, *et al.*, "Miniaturized Millimeter-Wave Radar Sensor for High-Accuracy Applications," *IEEE Trans. Microw. Theory Techn.*, vol. 65, no. 5, pp. 1707-1715, May 2017.

[4] A. Tessmann, *et al.*, "Compact single-chip W-band FMCW radar modules for commercial high-resolution sensor applications," *IEEE Tran. Microw. Theory Techn.*, vol. 50, no. 12, Dec. 2002.

[5] M. Porranzl, *et al.*, "An Active Quasi-Circulator for 77 GHz Automotive FMCW Radar Systems in SiGe Technology," *IEEE Microw. Wireless Compon. Lett.*, vol. 25, no. 5, pp. 313-315, May 2015.

[6] M. Porranzl, *et al.*, "77-GHz active quasi-circulator based Doppler radar with phase evaluation for object tracking," in *Proc. IEEE Int. Microw. Symp. (IMS)*, 2017, pp. 60-63.

[7] M. Kucharski, *et al.*, "Monostatic and Bistatic G-Band BiCMOS Radar Transceivers With On-Chip Antennas and Tunable TX-to-RX Leakage Cancellation," *IEEE J. Solid-State Circuits*, vol. 56, no. 3, March 2021.

[8] S. H. Abdelhalem, *et al.*, "Hybrid Transformer-Based Tunable Differential Duplexer in a 90-nm CMOS Process," *IEEE Trans. Microw. Theory and Techn.*, vol. 61, no. 3, pp. 1316-1326, Mar. 2013.

[9] M. Kucharski, *et al.*, "A Monostatic E-Band Radar Transceiver with a Tunable TX-to-RX Leakage Canceler for Automotive Applications," in *Proc. IEEE Int. Microw. Symp. (IMS)*, 2018, pp. 591-594.

[10] X. Li et al., "A 110-to-130 GHz SiGe BiCMOS Doherty Power Amplifier With a Slotline-Based Power Combiner," *IEEE J. Solid-State Circuits*, vol. 57, no. 12, pp. 3567-3581, Dec. 2022.

[11] X. Chen *et al.*, "A 140GHz Transceiver with Integrated Antenna, Inherent-Low-Loss Duplexing and Adaptive Self-Interference Cancellation for FMCW Monostatic Radar," *in IEEE Int. Solid-State Circuits Conf. (ISSCC)*, 2022, pp. 80-82.

979-8-3503-2123-4/23 $31.00 © 2023 IEEE

RTu2B-1

A Diamond Quantum Magnetometer Based on a Chip-Integrated 4-way Transmitter in 130-nm SiGe BiCMOS

Hadi Lotfi[*], Michal Kern[*], Nico Striegler[#], Thomas Unden[#], Jochen Scharpf[#], Patrick Schalberger[*], Ilai Schwartz[#], Philipp Neumann[#], Jens Anders[*]

[*]University of Stuttgart, Germany

[#]NVision Imaging Technologies GmbH, Germany

hadi.lotfi@iis.uni-stuttgart.de, jens.anders@iis.uni-stuttgart.de

Abstract—Solid–state magnetometers based on color centers in diamond are emerging as one of the leading quantum sensors due to their outstanding room-temperature properties, such as high sensitivity and calibration-free long-term stability. However, their integration into compact systems is still an active area of research. To tackle this challenge, in this paper, we present a quantum magnetometer based on negatively charged nitrogen-vacancy (NV) centers using a custom-designed, chip-integrated 4-way transmitter. In combination with a custom-designed microcoil array, the 4-way transmitter delivers microwave magnetic fields up to 226 µT for carrier frequencies around 7 GHz with a conversion gain of ≥32 dB to NV centers. The local oscillator (LO) signal required to drive the on-chip quadrature upconversion mixer is generated by a custom-designed quadrature PLL, which provides a 22% tuning range between 6.4 and 8 GHz, and a low phase noise of -122 dBc/Hz at 1 MHz offset from a 7 GHz carrier, to enable broadband, low-noise magnetometry. To verify the excellent performance of the integrated electronics, we have embedded them into a widefield diamond magnetometer using off-chip scanning optics, achieving a state-of-the-art AC-magnetic field limit of detection of 300 pT/Hz$^{1/2}$.

Keywords—quantum magnetometer, NV centers, ODMR, RF transmitter, power amplifier, SiGe BiCMOS.

I. INTRODUCTION

Due to their excellent room-temperature properties, negatively charged nitrogen-vacancy (NV) centers in diamond are gaining significant attention from both industry and academia as one of the leading physical platforms for quantum sensing. Although the NV center can sense various different physical quantities, including magnetic and electric fields as well as temperature and pressure, it is widely believed that magnetometry will be its first large-scale industrial application [1]. Here, the unique combination of favorable physical properties of NV centers (high sensitivity and stability, optical initialization, and readout) allows for room-temperature sensing of magnetic fields with limits of detection (LODs) below a picotesla [2] with various applications, including the replacement of conventional high-resolution magnetic field sensors [3] but also entirely new domains such as the detection of magnetic nanoparticles in tissue during cancer treatment [4], nanoscale nuclear magnetic resonance (NMR) spectroscopy [5] and single-shot spatially resolved measurements of magnetic fields with submicrometer resolutions. Such widefield NV-based cameras allow for magnetic imaging of bacteria [6], magnetic chemical species [7], or even thermal imaging [8]. Most of the widefield imaging systems in the literature use commercially available high-end signal generators and large

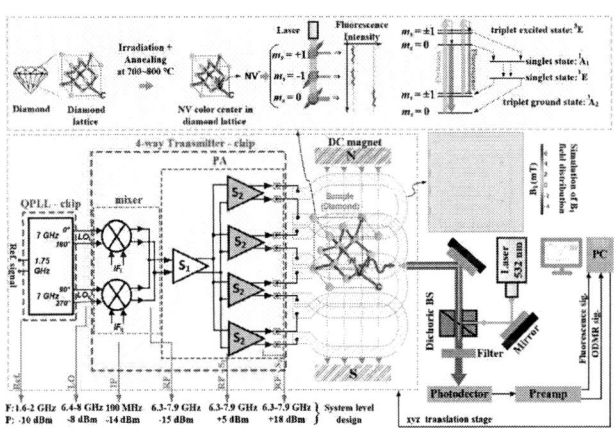

Fig. 1. Architecture of the presented quantum magnetometer.

rack-size microwave (MW) amplifiers connected to simple wires placed over the diamond sample to generate the MW magnetic field (B_1) required for NV-based sensing. This results in relatively poorly defined and strongly inhomogeneous B_1 fields with low conversion efficiency. Moreover, the high-cost and large volume of the employed commercial rack-based electronics present one bottleneck that currently prevents the wide adoption and commercialization of NV-based magnetometers. While chip-integrated NV-magnetometry systems have been introduced in the open literature [9], [10], [11] their focus was the design of compact and scalable magnetometers for general-purpose magnetic field sensing near zero magnetic fields, i.e., operating in the S-band due to the zero field splitting of NV centers of 2.87 GHz. By contrast, in this paper, we present an NV-based quantum sensor that targets nanoscale NMR measurements by directly measuring the magnetic field produced by nuclear spins. Since the spectral resolution of an NMR spectrum scales mostly linearly with the applied magnetic field B_0, the presented NV magnetometer operates at an elevated static B_0 field of approximately 150 mT, corresponding to an operating frequency around 7 GHz.

II. NV MAGNETOMETRY AND SYSTEM ARCHITECTURE

Fig. 1 shows the architecture of the presented NV ensemble sensor. A 532 nm green laser light excites the NV center to a triplet excited state (^{3}E). The electrons then decay either directly to the ground state (^{3}A$_2$) by emitting a red fluorescence signal or, after an intersystem crossing (ISC), via the metastable singlet states (^{1}A$_1$ and ^{1}E) by emitting a photon in the infrared

---

979-8-3503-2123-4/23 $31.00 © 2023 IEEE   253   2023 IEEE Radio Frequency Integrated Circuits Symposium

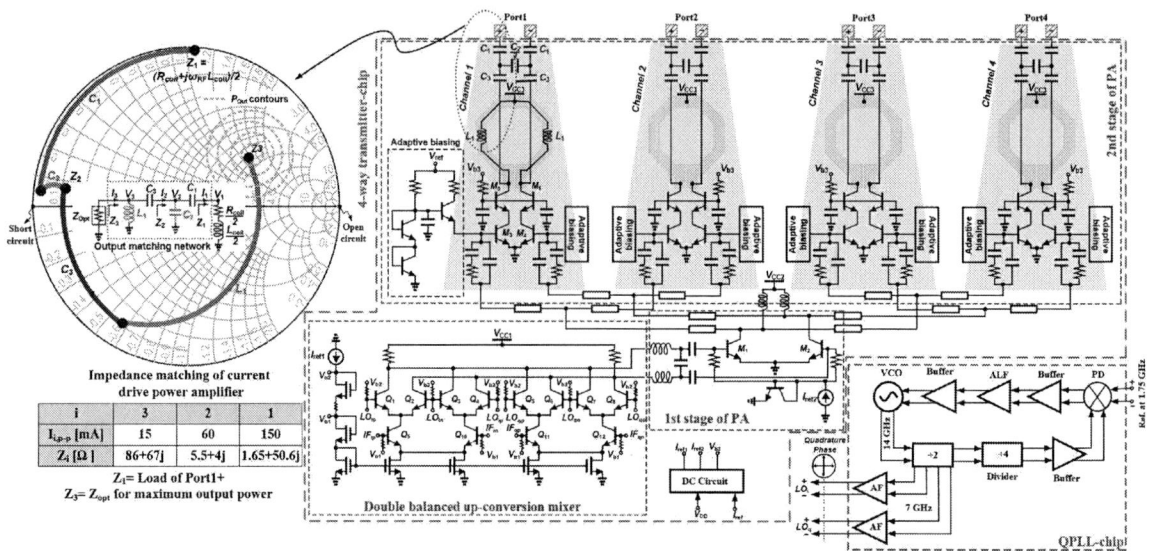

Fig. 2. Left: Illustration of the used output matching for maximum current delivery to the coil array. Right: Detailed schematics of the TX and QPLL chips.

range. Notably, the possibility of decaying through the ISC is higher for the spin states of $|m_s = \pm 1\rangle$ compared to $|m_s = 0\rangle$. Therefore, applying the green laser for a sufficient time initializes almost all NV centers to the state $|m_s = 0\rangle$. Furthermore, by applying an MW magnetic field, B_1 at a frequency of $f_{MW} = 2.87 \text{ GHz} \pm (\gamma/2\pi)B_0$, where $\gamma \approx 2\pi \cdot 28 \text{ GHz/T}$ is the so-called gyromagnetic ratio, the spin states can be excited from $|m_s = 0\rangle$ into $|m_s = \pm 1\rangle$. The resulting change in the population of the $|m_s = 0\rangle$ state results in a detectable change in the red fluorescence signal. This technique is known as optically detected magnetic resonance (ODMR). The ODMR signal of the NV center can be used for DC magnetometry since a change in the magnetic field B_0 leads to a splitting of the resonance lines associated with the $|m_s = \pm 1\rangle$ states by $\Delta f = 2(\gamma/(2\pi))B_0$, which allows for precise extraction of the value of B_0 from the measured Δf. While DC magnetometry is frequently still carried out with a continuous wave MW excitation [10], more advanced pulsed MW excitation schemes also allow for the precise measurements of local AC magnetic fields, including those produced by nuclear spins for nanoscale NMR spectroscopy.

In the presented system, the required MW signal is generated and amplified by a quadrature phase-locked loop (QPLL) and a 4-way transmitter (TX), respectively, both of which are implemented in a 130 nm BiCMOS technology. The 4-way transmitter drives an array of custom-designed microcoils, manufactured on a glass substrate for optical transparency, which generates the B_1 field that excites the NV center ensemble. The QPLL and TX chips are manufactured on two separate dies to maximize experimental freedom.

III. INTEGRATED ELECTRONICS

A. 4-way Transmitter Chip

The schematic of the TX chip is shown as part of Fig. 2. According to the figure, the chip integrates a double-balanced

active quadrature up-conversion mixer, which provides high isolation between the IF, LO, and RF ports. The mixer then connects to a 2-stage, 4-way differential power amplifier (PA), which consists of common-emitter (CE) (M_1, M_2) and cascode (M_3/M_5, M_4/M_6) stages. The differential PA implementation enhances the maximum output power by 3 dB and suppresses supply noise. Moreover, it provides a virtual ground node that helps to avoid gm (and therefore also power gain) reductions due to the RF impedance of bond wire inductances. The output impedance of the mixer is matched to the optimum source-pull impedance of the PA's first stage through custom-designed spiral inductors and shunt MIM capacitors. High-speed heterojunction bipolar transistors (HS-HBTs) are employed as the common-emitter transistors of the PA's first and second stages (M_1-M_4) to provide the required power gain. By contrast, the common-base (CB) transistors (M_5-M_6) of the cascode structures are implemented using high breakdown voltage devices (HV-HBTs) for enhanced output power and reliability. This hybrid cascode structure can partially compensate for the lower breakdown voltage of scaled-down SiGe BiCMOS compared to wide-bandgap III-V technologies. The second stage of the PA is biased with an on-chip adaptive biasing scheme, see Fig. 2, to improve the power-added efficiency (PAE) and linearity while minimizing the dc power consumption in pulsed mode. The cascode configuration of the second stage, including the RC network at its base terminal, is employed to boost the reverse isolation and decrease the undesired low-frequency gain, respectively. Fig. 2 also includes a Smith chart that illustrates the proposed output matching network (OMN), which maximizes the current swing in the coil array to provide the largest possible B_1 field to the NV center ensemble. According to the Smith chart, the OMN turns the inductance of the coil array into a very low impedance by series, close-to-resonant capacitors C_1 before this low impedance is transformed into the optimum impedance for maximum power transfer of the active device. According to

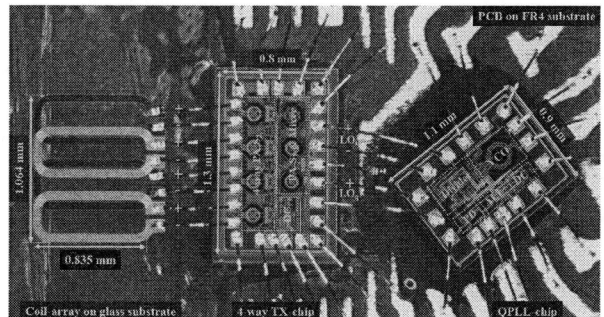

Fig. 3. Micrograph of the probe head containing from left to right: the microcoil array, the TX chip and the QPLL chip.

simulations, in this way, the presented TX chip can deliver more than $200\,\mathrm{mA_{pp}}$ per channel into the custom coil array from 6 to 9 GHz.

B. QPLL Chip

Fig. 2 also contains the detailed block diagram of the presented quadrature PLL. The PLL features a 14 GHz HBT-cross-coupled pair LC VCO, which employs MOS varactors driven by an active loop filter (ALF) for frequency tuning. The VCO output signal is divided by two using a master-slave D-flip-flop to generate the desired quadrature LO signals around 7 GHz for the TX chip. To allow for a large loop bandwidth of 35 MHz, the presented PLL uses a large phase detector (PD) frequency of 1.75 GHz. Here, the large loop bandwidth can be used to produce fast frequency ramps for chirp-like excitation schemes or fast phase and frequency modulations of the carrier for so-called optimal control pulses in pulsed experiments. The PLL requires a reference signal at a power level $\geq -10\,\mathrm{dBm}$ with a frequency between 1.6 GHz and 2 GHz.

C. Custom-designed Coil Array

The microwave B_1 field required to control the spin state of the NV centers is generated by the custom-designed coil array shown in the left part of Fig. 3. The coil array is manufactured on a 700 µm thick glass substrate for optical transparency and directly driven by the TX chip without additional off-chip matching components. Two aluminum metal layers, each with a thickness of 500 nm with an insulating layer of silicon nitride (SiN) in between, are used to realize the four partially overlapping coils. The size of the array coils was selected by considering the following design tradeoffs. First, the signal-to-noise ratio improves with the square root of the number of NV centers in the active volume. Therefore, since the NV center density cannot be arbitrarily increased due to the onset of coupling effects between NV centers for large densities, the sensitive volume of the coil array should be made as large as possible. Then, to ensure proper behavior as an inductor, the maximum coil perimeter is limited by the wavelength λ corresponding to the highest operating frequency. Here, considering the permittivity of the glass substrate of $\varepsilon_r \approx 8$ and constraining the maximum coil perimeter p_{coil} by $p_{\mathrm{coil}} \leq \lambda/10$, we decided on the coil dimensions shown in Fig. 3. The corresponding effective coil inductance seen by each channel of the TX chip – including the mutual coil inductances from the

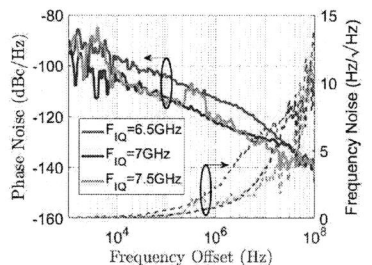

Fig. 4. Measured phase and frequency noise at three different carrier frequencies vs. frequency offset from the carrier.

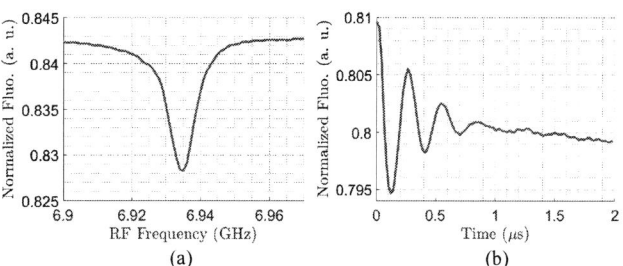

Fig. 5. (a) Measured cw ODMR spectrum of an NV ensemble at a magnetic field of about 150 mT. b) Measured Rabi oscillations in a pulsed ODMR experiment.

neighboring coils – at 7 GHz is 1.3 nH. Therefore, considering the bond wire inductance of around 1 nH, the on-chip matching network discussed in section III.A was designed for a total load inductance per channel of 2.3 nH.

The total resulting array size is $1.064 \times 0.835\,\mathrm{mm^2}$. The overlap between adjacent array coils was optimized by FEM simulations in COMSOL for maximum homogeneity of the produced B_1 field in the region of interest. The electrical parameters of the coil array were extracted from S-parameter simulations performed in ADS.

IV. MEASUREMENT RESULTS

A. Electrical Characterization

The chips described in sections III.A and III.B were manufactured in a 130 nm SiGe BiCMOS technology. Micrographs of the two chips are shown as part of Fig. 3. Before characterizing the overall setup shown in Fig. 3 in its target quantum sensing application, we performed a series of electrical tests. Here, we first measured a PLL tuning range from 12.8 GHz to 16 GHz, corresponding to quadrature LO signals between 6.4 GHz and 8 GHz. Next, we measured the phase and frequency noise performance of the LO signal at three different carrier frequencies of $f_{\mathrm{LO}} = \{6.5, 7, 7.5\}\,\mathrm{GHz}$. The corresponding results are shown in Fig. 4. According to the figure, the phase noise at the target operating frequency of 7 GHz at an offset from the carrier of 1 MHz is $-122\,\mathrm{dBc/Hz}$. Moreover, we verified that by sweeping the supply voltage of the divider, the output level of the QPLL chip can be adjusted between -1 and $-8\,\mathrm{dBm}$, meeting the driving requirement of the TX chip. The measured phase imbalance of the quadrature outputs of the QPLL chip at 7 GHz is $0.8°$.

979-8-3503-2123-4/23 $31.00 © 2023 IEEE   255

Table 1. Performance summary and comparison with the state-of-the-art in chip-integrated NV-based magnetometers.

Reference	Freq. [GHz]	PN @ 3 kHz [dBc/Hz]	Integrated building blocks	Rabi Freq. [MHz]	Sensing Area [μm²]	LOD [nT/√Hz]	Technology
Nature Elec.' 19 [9]	$2.6 - 3.1$	-90 @ 2.87 GHz Offset:1.5 kHz	B_1 generaton Optical detection	N/A	50×50	DC: 3.21×10^4	65 nm CMOS
JSSC' 21 [10]	Around 2.87*	-88 @ 2.87 GHz	B_1 generaton Optical detection	1.2	80×300	DC: 245	65 nm CMOS
This Work	$6.4 - 8$	-102 @ 7 GHz	B_1 generaton	3.66	1064×835	AC: 0.3 DC: 32	130 nm BiCMOS

* The frequency range was not specified.

B. ODMR magnetometry measurements

For the NV experiments, we used a diamond with a natural abundance of ^{13}C (Element Six). The diamond contains a 10 μm thick CVD-grown layer of NV centers with a concentration of 3.5 ppm. To perform the ODMR experiments, we have placed the diamond on top of the coil array shown in Fig. 3, with the NV layer facing the coil array. The ODMR experiments were then performed according to the setup shown in Fig. 1. To this end, the diamond was placed inside an electromagnet that produced the required static magnetic field of $B_0 \approx 150$ mT, corresponding to a resonance frequency of ≈ 7 GHz. The NV centers were then polarized using a green laser pulse with a wavelength of 532 nm, followed by a B_1 magnetic field pulse applied via the setup shown in Fig. 3. Here, the pulse duration τ_π was selected to produce a rotation of the spin magnetization by an angle of $\theta = \pi$ according to $\tau_\pi = \pi/(\gamma \cdot B_1)$, where B_1 is the microwave magnetic field produced by the coil array. The pulsed microwave excitation was then followed by a final laser pulse for readout. By sweeping the frequency of the B_1 field from 6.9 to 7 GH, we have recorded the spectrum shown in Fig. 5a, which clearly shows the resonant absorption of the B_1 field around 6.935 GHz. Next, we recorded the Rabi oscillations shown in Fig. 5b by sweeping the length of an on-resonance ($f = 6.935$ GHz) excitation pulse. From the angular frequency of the Rabi oscillations ω_{Rabi}, we extracted a maximum measured value of the B_1 field of $B_1 = 226$ μT. From the measured B_1 field, we calculated a current of 100 mA$_{\text{pp}}$ in each loop of the microcoil array. Next, we estimated the achievable AC magnetic field LOD using the so-called Qdyne protocol [12]. To this end, we applied an external AC magnetic field with a frequency of 994 kHz by an auxiliary loop coil. From the extracted magnetic field value of 92.1 nT and the measured SNR of 333, we calculated an AC field LOD of 300 pT/√Hz. Finally, we measured the Allan deviation [13] of the presented sensor. The Allan deviation plot displays the expected slope of $-1/\sqrt{T_{\text{av}}}$, reaching a minimum value of 47.6 nT for an averaging time of $T_{\text{av}} \approx 460$ ms, indicating a DC sensitivity of about 32 nT/√Hz. The measured plateau in the Allan deviation occurs due to the drift of the external magnetic field B_0, preventing a further improvement of the LOD by averaging beyond ≈ 460 ms, partially accounting for the difference between the measured AC and DC magnetic field sensitivities. The performance of the proposed magnetometry system is compared against the state-of-the-art in chip-integrated NV magnetometers in Table I, highlighting the excellent performance of the presented design.

V. Conclusion

In this paper, we have presented a platform for magnetic field sensing using NV centers in diamond based on two custom-designed MW ASICs and a custom-designed microcoil array. Thanks to the excellent performance of the presented electronics, the proposed system achieves a performance that rivals high-end rack-based systems at a greatly reduced form factor, power consumption, and price tag. Compared to the state-of-the-art in chip-integrated NV-based magnetometers [9,10], the presented system displays a performance improvement in the achieved DC LOD of more than 7x.

Acknowledgment

This work was sponsored by the German Federal Ministry of Education and Research under contracts no. 13GW0235B, 03SF0565C, 03ZU1110{DC, FE, GA}, and 13N15374.

References

[1] L. Rondin et al., "Magnetometry with nitrogen-vacancy defects in diamond," Rep. Prog. Phys., vol. 77, no. 5, Art. no. 056503, May 2014.

[2] T. Wolf et al., "Subpicotesla Diamond Magnetometry," Phys. Rev. X, vol. 5, no. 4, Art. no. 041001, Oct. 2015.

[3] F. M. Stürner et al., "Integrated and portable magnetometer based on nitrogen-vacancy ensembles in diamond," Adv. Quantum Technol., vol. 4, no. 4, Art. no. 2000111, 2021.

[4] A. Kuwahata et al., "Magnetometer with nitrogen-vacancy center in a bulk diamond for detecting magnetic nanoparticles in biomedical applications," Sci. Rep., vol. 10, no. 1, p. 2483, Feb. 2020.

[5] R. D. Allert et al., "Advances in nano- and microscale NMR spectroscopy using diamond quantum sensors," Chem. Comm., vol. 58, no. 59, pp. 8165–8181, Jul. 2022.

[6] D. Le Sage et al., "Optical magnetic imaging of living cells," Nature, vol. 496, no. 7446, pp. 486–489, 2013.

[7] S. Steinert et al., "Magnetic spin imaging under ambient conditions with sub-cellular resolution," Nature Comm., vol. 4, no. 1, pp. 1–6, Jun. 2013.

[8] Y. Nishimura et al., "Widefield fluorescent nanodiamond spin measurements toward real-time large-area intracellular thermometry," Sci. Rep., vol. 11, no. 1, p. 4248, Feb. 2021.

[9] D. Kim et al., "A CMOS-integrated quantum sensor based on nitrogen–vacancy centres," Nature Electron., vol. 2, no. 7, pp. 284–289, Jul. 2019.

[10] M. I. Ibrahim et al., "High-scalability CMOS quantum magnetometer with spin-state excitation and detection of diamond color centers," IEEE J. Solid-State Circuits, vol. 56, no. 3, pp. 1001–1014, Mar. 2021.

[11] Y. Gao et al., "Diamond NV Centers Based Quantum Sensor Using a VCO Integrated With Filtering Antenna," in IEEE Trans. on Instrum. and Meas., vol. 71, pp. 1-12, Art 2005112, Aug. 2022.

[12] S. Schmitt et al., "Submillihertz magnetic spectroscopy performed with a nanoscale quantum sensor," Science, vol. 356, pp. 832–837, May 2017.

[13] A. A. Wood et al., "dc Quantum Magnetometry below the Ramsey Limit," Phys. Rev. Appl. vol. 18, no. 5, Art. no. 054019, Nov. 2022.

979-8-3503-2123-4/23 $31.00 © 2023 IEEE

RTu2B-2

A Cryo-CMOS DAC-based 40 Gb/s PAM4 Wireline Transmitter for Quantum Computing Applications

Niels Fakkel[1], Mohsen Mortazavi, Ramon Overwater, Fabio Sebastiano, Masoud Babaie

Delft University of Technology, the Netherlands

[1]n.e.fakkel@tudelft.nl

Abstract— State-of-the-art quantum computers already comprise hundreds of cryogenic quantum bits (qubits), and prototypes with over 10k qubits are currently being developed. Such large-scale systems require local cryogenic electronics for qubit control and readout, leaving the digital controllers for algorithm execution and quantum error correction (QEC) at room temperature due to the limited cryogenic cooling budget. The entire process, including qubit readout, data transmission, QEC, and algorithm execution, should be completed well within the qubit decoherence time, thus requiring a low-power high-speed communication link between the cryogenic quantum processor and classical processor located at room temperature. To this end, this paper presents the first cryo-CMOS high-speed 4-level pulse amplitude modulation (PAM4) wireline transmitter. Thanks to a power-efficient serializing architecture driving a 6-bit digital-to-analog converter (DAC), the 40-nm CMOS chip achieves a data rate of 40 Gb/s PAM4 with an efficiency of 2.46 pJ/b and a ratio of level mismatch (RLM) of 97.8% at 4.2 K. While demonstrating an energy efficiency comparable to state-of-the-art transmitters in more advanced CMOS nodes, the extremely wide temperature operating range (4.2 K - 300 K) will enable future large-scale quantum computers.

Keywords— Cryo-CMOS, quantum computing ICs, high-speed DAC, wireline transmitter.

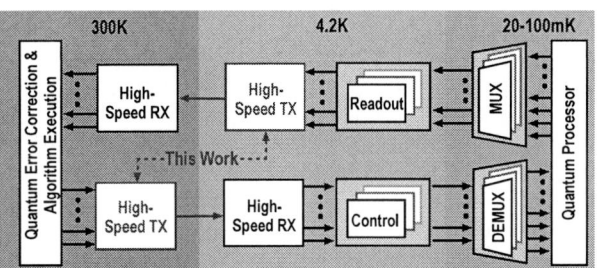

Fig. 1. Simplified block diagram of a scalable quantum computing system incorporating a cryo-CMOS high-speed wireline link.

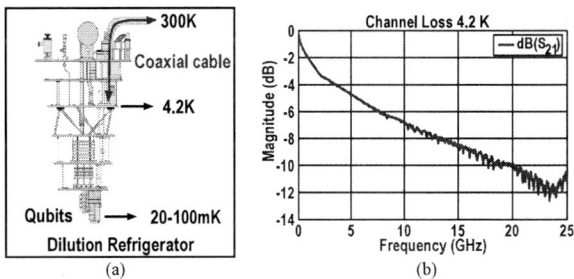

Fig. 2. (a) Dilution refrigerator illustration, and (b) the measured insertion loss (S_{21}) of a typical coaxial cable connecting the fridge 4.2 K stage to its output connector at room temperature.

I. INTRODUCTION

Quantum computers can potentially solve problems intractable by classical computers, with applications ranging from cryptography and pharmaceuticals to artificial intelligence. Yet, the physical implementations of quantum bits (qubits) are too noisy to be used for robust computations. The robustness can be improved by realizing a logical qubit from multiple physical qubits and applying continuous rounds of quantum error correction (QEC). With current technologies, QEC requires reading out each qubit at about 1 Mb/s rate [1], and applying the corresponding corrections in real time. With roadmaps predicting the availability of systems with more than 10k qubits in only four years [2], the amount of data required for QEC grows accordingly. Electronics for qubit control and readout can be integrated within the cryogenic chamber of a quantum computer to tackle the scalability issues. However, the digital controllers for algorithm execution and QEC need to be placed at room temperature (RT) due to the restricted cryogenic cooling budget, as shown in Fig. 1. To prevent any backlog between the QEC and measurement cycles, a large amount of qubit readout data should be serialized and transmitted at >10 Gb/s rate. Hence, a need emerges for a high-speed wireline transmitter (TX) to transfer the data generated by the cryogenic readout circuits to a classical

processor located at RT. Such a cryogenic TX must dissipate low power to comply with the cooling budget of a practical cryogenic refrigerator, while achieving the highest possible speed and linearity despite non-idealities in cryogenic CMOS device behavior, such as higher threshold voltage [3] and worse matching [4].

This paper addresses the imminent need for a cryogenic wireline link in quantum computer applications and tackles the challenges associated with the extremely wide operating temperature range (4.2 K – 300 K). To the authors' best knowledge, this paper presents the first cryogenic CMOS (cryo-CMOS) wireline transmitter. By employing a 4-level pulse amplitude modulation (PAM4) protocol, the TX achieves a 40 Gb/s data rate and 2.46 pJ/b energy efficiency.

II. WIRELINE TRANSMITTER ARCHITECTURE

In a typical dilution refrigerator setup, shown in Fig. 2, a copper coaxial cable connects the 4.2 K stage to its output connector at RT. Considering a channel loss (i.e., S_{21}) of ~10 dB at 20 GHz and the limited TX output bandwidth due to the ESD and pad parasitic capacitance, the PAM4 modulation scheme is chosen over non-return-to-zero (NRZ)

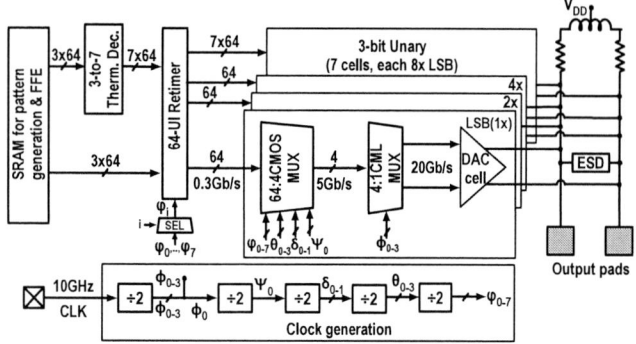

Fig. 3. Simplified block diagram of the proposed DAC-based wireline TX.

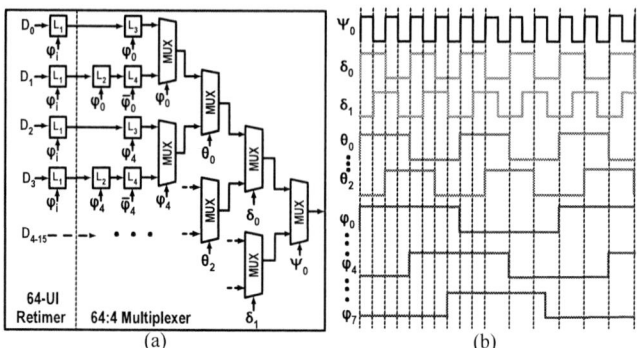

Fig. 4. (a) Schematic of 64:4 MUX, and (b) a timing diagram of its required clock phases to remove latches between the selectors.

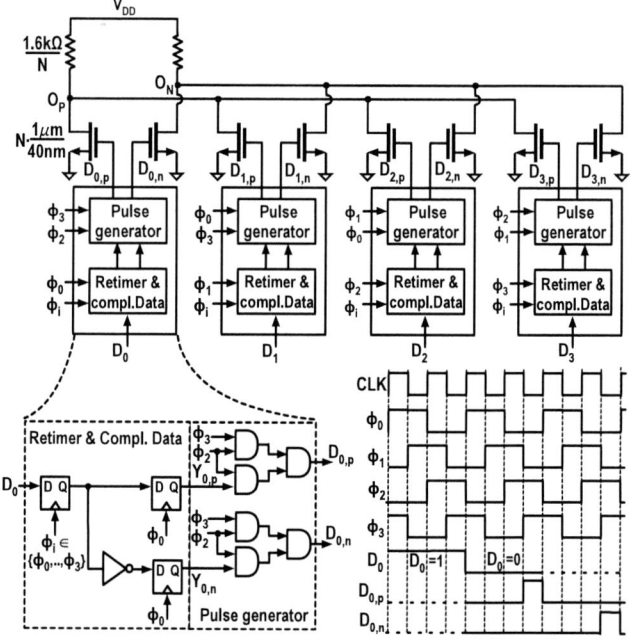

Fig. 5. Implementation of proposed quarter-rate retimer, pulse generator, and 4:1 CML MUX with the corresponding timing diagram.

to avoid extensive loss at higher frequencies and enhance the throughput. A digital-to-analog converter (DAC)-based transmitter architecture is employed since it can support different modulation formats (e.g., NRZ, PAM4), overcome channel imperfections, and transfer data without losing signal integrity. Furthermore, a current-mode logic (CML)-based DAC is adopted instead of a voltage-based output stage, due to its higher bandwidth, better supply rejection, and lower sensitivity to variations of the transistor characteristic at 4.2 K.

The block diagram of the TX architecture is shown in Fig. 3. A programmable $512 \times 64 \times 6$ bit SRAM is implemented to allow for the exploration of different test sequences, data formats, and equalization techniques in measurement. The 64-Unit-Interval (UI) $\times 6$ bit parallel SRAM data is decoded and fed into the 10 DAC slices (3b thermometer + 3b binary coded). In each slice, a 64:4 multiplexer (MUX) structure serializes the data to a high-speed $4UI \times 10$ slice signal. This signal is then retimed by a selectable clock phase to complementary 25% pulses and fed into a 4:1 CML-based MUX. A dedicated CML output driver after the 4:1 MUX is employed to improve TX linearity and bandwidth compared to the case where the 4:1 MUX also acts as the output stage. The output network is designed with two differential $50\,\Omega$ termination resistors, and a center-tapped peaking inductor to compensate for the bandwidth reduction due to the parasitic capacitance of the ESD protection and output pads. A clock generation circuit converts an external clock input into all the divided clock phases necessary for the retimers and multiplexers.

A. 64:4 Multiplexer

Fig. 4 (a) shows the schematic of the 64:4 MUX. All incoming bits from the SRAM are retimed in L_1 latches by one selectable clock phase (i.e., $\varphi_i \in \{\varphi_0, \ldots, \varphi_7\}$) that is digitally calibrated to provide optimal setup- and hold-time margin. To avoid narrow pulses or glitches at the outputs of the first selectors, the time difference between the data transitions at the inputs of each first-rank selector is adjusted to 32UI, using different phases (i.e., φ_{0-7}) of the 64UI clock in L_{2-4} latches. However, this approach is not power-efficient at higher-rank selectors, as it would need 140 more latches operating at higher frequencies. As depicted in

Fig. 4, those latches can be removed if lower-rank selectors use the quadrature clocks generated by dividing the clock frequency of their corresponding higher-rank selector [5].

B. 4:1 Multiplexer

In the quarter-rate retimer and pulse generator, the data is prepared for the 4:1 CML MUX, as shown in Fig. 5. Due to the different clock routing schemes, there is an unknown skew between the 8UI clock domain (i.e., ϕ_0) used in the last stage of 64:4 MUX and 4UI clock domain (i.e., ϕ_{0-3}) employed in the pulse generator. Hence, a retimer is added between these blocks whose clock phase (i.e., $\phi_i \in \{\phi_0, \ldots, \phi_3\}$) is selectable using a multiplexer-based phase rotator. The data is converted to complementary form and again retimed to ensure any delay mismatch from the inverter is compensated. The required 25% duty-cycle pulse (i.e., 1UI) is generated by combining the corresponding 50% overlapping 4UI clock phases and complementary data using two cascaded AND gates. Note that using a single 3-input AND gate would result

979-8-3503-2123-4/23 $31.00 © 2023 IEEE

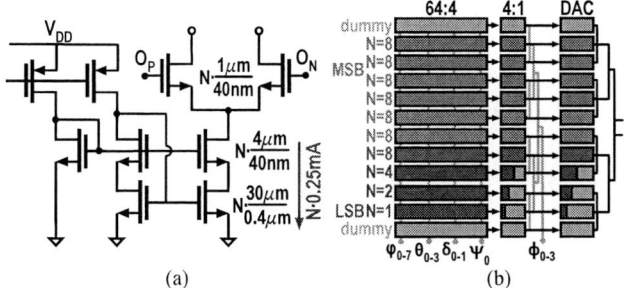

(a)

(b)

Fig. 6. (a) Schematic of LSB DAC cell, and (b) DAC layout floor plan.

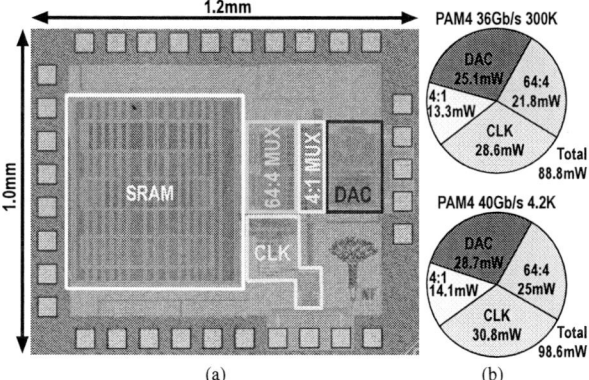

(a)

(b)

Fig. 7. (a) Chip micrograph, and (b) measured power breakdown at 300 K and 4.2 K.

in a much slower rise/fall time at cryogenic temperatures (CT) due to the threshold voltage increase and, therefore, limited voltage overdrive. The pulse generation is done locally, because distributing 25% non-overlapping clock pulses as an alternative would require more power-hungry buffers in the clock path and set tighter skew constraints to the clock distribution. The differential phases of the quadrature clocks use an upper-level metal and are distributed through H-trees to reduce the power consumption of the clock buffer and clock delay mismatches.

The TX maximum speed is determined by the clock-to-q delay and setup time of the high-speed flip-flops as they are in the critical path. Among different flip-flop architectures, a True Single Phase Clock (TSPC) dynamic flip-flop that employs a lower number of stacked devices is adopted, thus maximizing speed by fully exploiting transistors' higher mobility at CT.

The CML-based 4:1 MUX is designed without a tail transistor to counteract the reduced voltage headroom due to the higher threshold at CT and allow for smaller size switches. It combines the interleaved pulses from the quarter-rate retimer to drive the final output stage. Note that the MSB cells have a larger input capacitance than the LSB cells in the DAC structure. Consequently, the load resistance and the transistor size of the corresponding 4:1 MUX as the pre-driver are proportionally scaled to maintain the bandwidth and input voltage amplitude of all DAC slices, thus preventing any systematic delay mismatch.

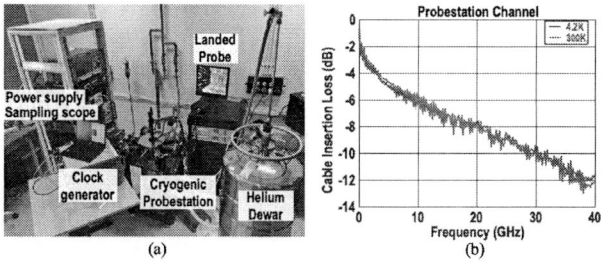

(a)

(b)

Fig. 8. (a) Cryogenic probe station measurement setup, and (b) the measured insertion loss of the probe and cable, realizing the channel between the chip and measurement instrument.

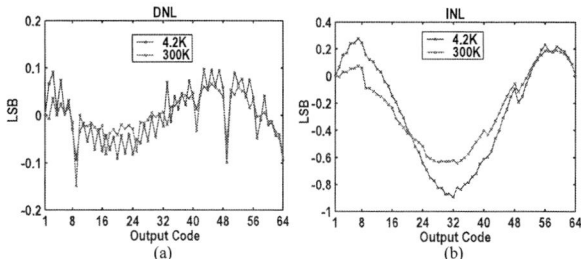

(a)

(b)

Fig. 9. (a) DNL, and (b) INL, measured at 4.2 K and 300 K.

C. DAC

This design employs a 6-bit CML DAC to satisfy the ratio of level mismatch (RLM) of at least 95% defined in wireline standards with sufficient margin. The tail current sources of each DAC cell (Fig. 6 (a)) are sized with a margin to reach the noise and differential non-linearity (DNL) requirements, accounting for the 20% increased device mismatch at CT [4]. To optimize the tradeoff between the integral non-linearity (INL) requirement and the total power consumption due to the number of serializer paths, a 3b binary + 3b unary coded structure is employed. As shown in Fig. 6 (b), the DAC slices, including serializers and retimers, are laid out symmetrically, and dummy rows are added at the top and bottom to minimize mismatch.

III. RESULTS

A TX prototype was fabricated in 40-nm CMOS. As shown in the chip micrograph in Fig. 7 (a), the active area of the transmitter, excluding SRAM, is 0.146 mm². The TX performance was measured with a cryogenic probe station to lower the prototype's ambient temperature to 4.2 K. The measurement setup shown in Fig. 8 introduces a total loss of ∼8 dB at 20 GHz due to probes and cables. Fig. 9 shows that the maximum INL is below 0.6 LSB at RT and, as expected, increases to 0.9 LSB at CT due to increased device mismatch. Fig. 10 displays the measured eye diagrams at RT and CT without using additional feedforward equalization, de-embedding, or scope equalization. To perform the measurement, the on-chip SRAM is loaded with a 2^{15}-length pseudorandom binary sequence (PRBS)-15 for NRZ and a quaternary QPRBS-15 for PAM4. At RT, 20 Gb/s NRZ and 20&36 Gb/s PAM4 signals are measured. At the highest rate, the measured eye heights (widths) of NRZ and PAM4 are 231 mV (0.65UI) and >24.7 mV (0.28UI), respectively, with 96.5% RLM. Due

979-8-3503-2123-4/23 $31.00 © 2023 IEEE

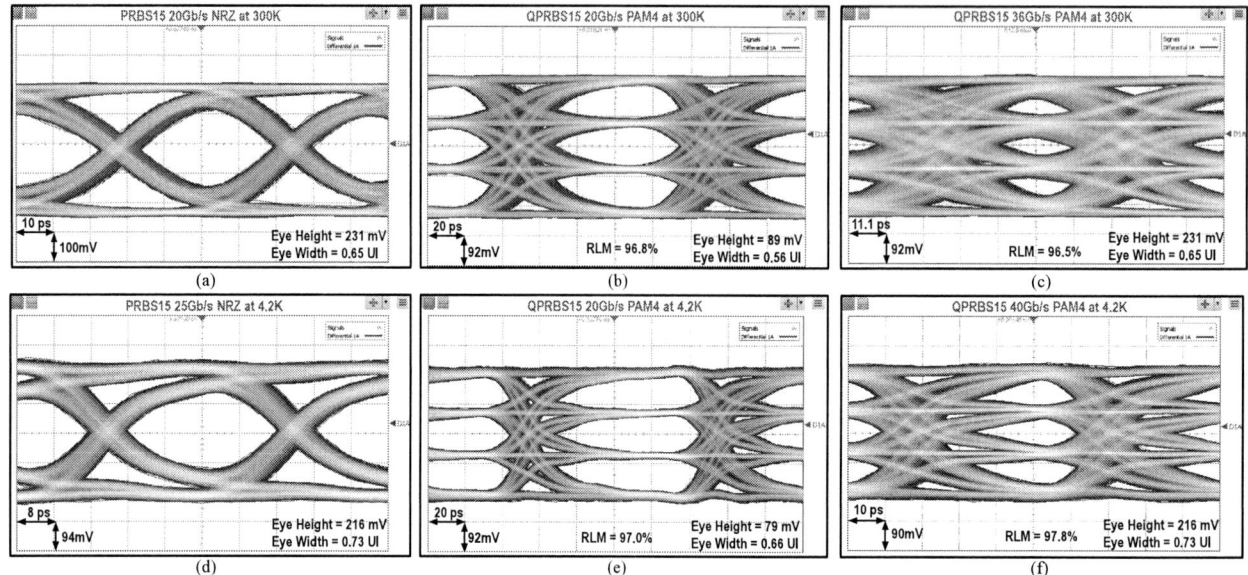

Fig. 10. Measured eye diagrams: (a) 20 Gb/s NRZ at 300 K, (b) 20 Gb/s PAM4 at 300 K, (c) 36 Gb/s PAM4 at 300 K, (d) 25 Gb/s NRZ at 4.2 K, (e) 20 Gb/s PAM4 at 4.2 K, and (f) 40 Gb/s PAM4 at 4.2 K.

Table 1. Comparison table with prior DAC-based and multi-tap wireline transmitters.

	This work		[6] ISSCC'17	[7] ISSCC'18	[8] ISSCC'19
Temperature [K]	300	4.2	300	300	300
Data-rate [Gb/s]	36	40	36	45	112
Power [mW]	88.8	98.6	84	120	175*
Efficiency [pJ/b]	2.47	2.46	2.33	2.67	1.56*
RLM [%]	96.5	98.8	-	92	94
Max. Vpp	0.8		0.8	1.3	1
Signalling	PAM4		PAM4	PAM4	PAM4
Output driver	CML		CML	SST	H-bridge
FFE technique	DAC		4-taps	DAC	DAC
Supply [V]	1.1		1/1.5	1	0.9/1.2
Technology [nm]	40		28	28	7
Active area [mm²]	0.146		0.05	0.28	0.193

*including PLL

to the digital speed improvement at CT, the baud rate could be increased, and consequently, 25 Gb/s NRZ and 20&40 Gb/s PAM4 are measured. At the highest rate, the measured eye heights (widths) of NRZ and PAM4 are 216 mV (0.73UI) and >38.5 mV (0.47UI), respectively, with 97.8% RLM. The power breakdown chart is shown in Fig. 7 (b). At RT (CT), the TX can achieve a maximum configurable swing of 0.8 Vpp and consumes 88.8 mW (98.6 mW) at 36 Gb/s (40 Gb/s), thus achieving 2.47 pJ/b (2.46 pJ/b) energy efficiency. Measurements above those baud rates violate the timing constraints in the retimers before the 4:1 MUX, and therefore result in incorrect behavior.

Table 1 benchmarks the performance of this work with prior art. At RT, even by using a less advanced technology node, the proposed transmitter can reach similar energy efficiency as state-of-the-art TXs while achieving the highest RLM. Moreover, by demonstrating, for the first time, both full functionality and high efficiency over the wide temperature range down to CT, the proposed TX enables the required

high-speed wireline link for quantum computing applications.

IV. Conclusion

This paper demonstrates the first cryogenic wireline transmitter. At CT (RT), the prototype achieves 40 Gb/s (36 Gb/s) PAM-4 transmission with 2.46 pJ/b (2.47 pJ/b) efficiency and 97.8%(96.5%) RLM. By circumventing the disadvantages of cryo-CMOS devices (higher threshold, larger mismatch) and exploiting their higher speed in the design of the serializer and DAC, the transmitter maintains high power efficiency, linearity, and data rate down to CT. This result enables high-speed data communication between classical and quantum processors, which is essential in the scale-up of future quantum computers.

References

[1] R. W. J. Overwater *et al.*, "Neural-network decoders for quantum error correction using surface codes: A space exploration of the hardware cost-performance tradeoffs," *IEEE Transactions on Quantum Engineering*, vol. 3, pp. 1–19, 2022.

[2] J. Gambetta. (2022, May) Expanding the IBM Quantum roadmap to anticipate the future of quantum-centric supercomputing. [Online]. Available: https://research.ibm.com/blog/ibm-quantum-roadmap-2025

[3] A. Beckers *et al.*, "Physical model of low-temperature to cryogenic threshold voltage in MOSFETs," *IEEE Journal of the Electron Devices Society*, vol. 8, pp. 780–788, 2020.

[4] P. A. T. Hart *et al.*, "Subthreshold mismatch in nanometer CMOS at cryogenic temperatures," *IEEE Journal of the Electron Devices Society*, vol. 8, pp. 797–806, 2020.

[5] Y. Chang *et al.*, "An 80-Gb/s 44-mW wireline PAM4 transmitter," *IEEE Journal of Solid-State Circuits*, vol. 53, no. 8, pp. 2214–2226, 2018.

[6] A. Nazemi *et al.*, "A 36Gb/s PAM4 transmitter using an 8b 18GS/S DAC in 28nm CMOS," in *IEEE International Solid-State Circuits Conference*, San Francisco, USA, Feb. 2015, pp. 58–59.

[7] M. Bassi *et al.*, "A 45Gb/s PAM-4 transmitter delivering 1.3Vppd output swing with 1V supply in 28nm CMOS FDSOI," in *IEEE International Solid-State Circuits Conference*, San Francisco, USA, Feb. 2016, pp. 66–67.

[8] E. Groen *et al.*, "A 10-to-112Gb/s DSP-DAC-Based transmitter with 1.2Vppd output swing in 7nm FinFET," in *IEEE International Solid-State Circuits Conference*, San Francisco, USA, Feb. 2020, pp. 120–121.

RTu2B-3

A mm-wave CMOS/Si-Photonics Hybrid-Integrated Software-Defined Radio Receiver Achieving > 80-dB Blocker Rejection of < -10 dBm In-Band Blockers

Ramy Rady, Yu-Lun Luo, Christi Madsen, Samuel Palermo, Kamran Entesari

Analog and Mixed-Signal Center, Texas A&M University, USA

ramyrady@tamu.edu

Abstract — This paper presents a hybrid-integrated mm-wave software-defined radio (SDR) receiver front-end implemented with silicon photonics and CMOS chips. The proposed SDR leverages a programmable silicon photonics IC (PIC) with high-Q filters to perform re-configurable channel-selection/image-rejection and jammer-rejection with tunable center frequency of 30-45 GHz, and 3-5 GHz bandwidth. Up to four out-of-band blockers are automatically detected and rejected simultaneously. Also, the desired mm-wave signal is mixed with a tunable local oscillator (LO) carrier and down-converted to a 2.5-GHz IF center frequency. Subsequently, the CMOS IC converts the current signal into an amplified voltage signal, thus compensating for PIC losses. The PIC is fabricated using a silicon-over-insulator (SOI) process, and the CMOS is fabricated using 28nm process. The receiver achieves > 80-dB rejection for two blockers and > 65-dB rejection for four blockers. The EVM measures -30-dB using a 100-MSymbol/s 64-QAM signal at the presence of a 10-dBc out-of-band blocker.

Keywords — Millimeter-Wave radio receivers, programmable silicon photonics filters, software-defined-radio, RFICs.

I. INTRODUCTION

Next-Generation wireless communication systems demand mm-wave wideband receivers to meet the growing wireless access and the required data throughput. A major challenge for these receivers is to maintain a high tolerance for both in-band and out-of-band (OOB) blockers. This motivates employing a re-configurable software-defined radio (SDR) receiver that aggregates multiple contiguous frequency bands while sufficiently tolerating in-band and filtering OOB blockers, thus providing multi-channel wideband operation. Traditional electronic SDRs face significant challenges at mm-wave frequencies due to lack of programmable electronic filters and the limited quality factor of passives on silicon substrate which makes it hard to implement high-order filtering. There has been increasing interest in mixer-first receiver architectures for addressing in-band blockers such as ([1], [2], [3]) but achieving high OOB blocker rejection and high spurious free dynamic range (SFDR) simultaneously, is still a challenge. This limits their maximum instantaneous BW, especially for mm-wave operation. Recently, there has been a growing interest in other technologies to realize SDR such as photonic-based multi-band transceiver [4], but it is a complex and bulky implementation. RF Silicon photonics technology can realize integrated band-pass and notch filters to perform channel-selection and interference cancellation at mm-wave frequencies [5], and has the potential to assist

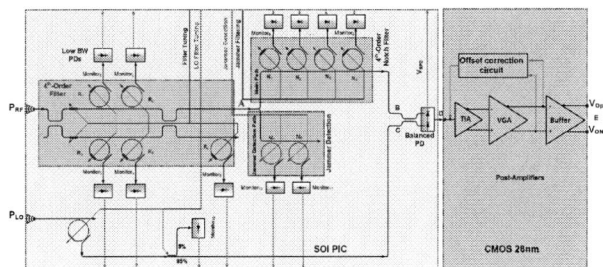

Fig. 1. The proposed photonically-assisted hybrid SDR architecture.

CMOS amplifying stages in the receiver by providing the necessary filtering solutions for OOB blockers.

This work presents a photonically-assisted SDR receiver featuring high-Q factor silicon photonics filters that are highly programmable and capable of rejecting strong OOB blockers and a CMOS post-amplifying stage. A proof-of-concept receiver prototype in SOI Photonics and 28nm CMOS process demonstrates the reconfigurability to cover multiple operating bands within 30-45 GHz frequency range and > 35-dB interference rejection using the channel-select band-pass filter (BPF) at adjacent and alternate channels, and an overall rejection > 80-dB using both bandpass and notch filters.

II. PROPOSED PHOTONICALLY-ASSISTED SDR

Fig. 1 illustrates the proposed photonically-assisted hybrid SDR consisting of a PIC and CMOS amplifiers. Two input grating couplers are available to the PIC for optically-modulated mm-wave signal, P_{RF}, and mm-wave LO, P_{LO}, respectively. The top path consists of a tunable wideband BPF for channel-selection/image-rejection and jammer-rejection and a tunable notch filter with spectrum monitoring and jammer rejection capabilities. The bottom path comprises of LO image-reject filter. Both of these paths are combined through a directional coupler and fed into a balanced PD. The output current signal by the PD is centered at the frequency difference between the two paths $f_{IF} = f_{RF} - f_{LO}$ which enters the CMOS unit. The CMOS unit provides signal conditioning by three amplifying stages and an offset correction circuit. The trans-impedance amplifier (TIA) converts the current signal into a differential amplified voltage signal, then the variable gain amplifier (VGA) provides more tunable gain and the buffer provides an output to drive

979-8-3503-2123-4/23 $31.00 © 2023 IEEE   261   2023 IEEE Radio Frequency Integrated Circuits Symposium

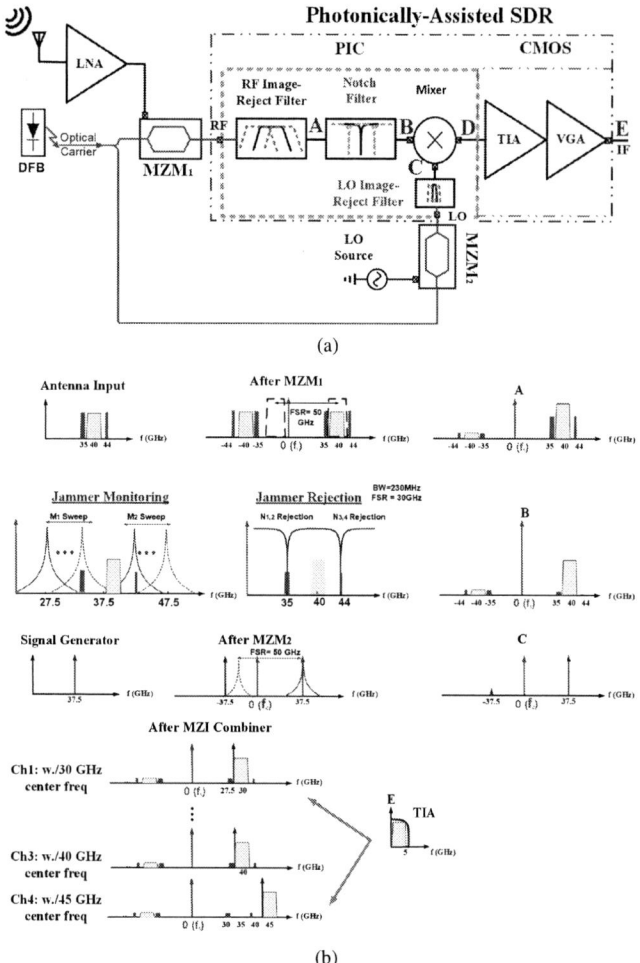

Fig. 3. (a) The full receiver system of the proposed photonically-assisted SDR architecture, and (b) frequency planning of the SDR.

50-Ω load. An analog control signal re-programs PIC filter responses, while a local monitor PD for each ring returns a signal to indicate pole/zero locations. Moreover, the monitor path signal is able to locate the interferes.

Fig. 3a shows the full SDR receiver system including photonically-assisted SDR, two external Mach-Zehnder Modulators (MZM$_{1,2}$) and low noise amplifier (LNA). Tunable

RF image-rejection filter, notch filter, LO image-rejection filter, and mixer are all encompassed in the PIC part and gain stages are built into the CMOS unit of the proposed photonically-assisted SDR. The system is driven by two inputs; a mm-wave received signal by the antenna is amplified by the LNA to optimally drive MZM$_1$ and improve the overall system gain, SFDR and reduce the overall NF. Then, MZM$_1$ converts it to optical domain going to P_{RF} input, also MZM$_2$ converts the LO tone into optical domain coupled to P_{LO} port. A frequency plan for the proposed photonically-assisted SDR receiver system is shown in Fig. 3b. The desired mm-wave signal (27.5-32.5,...,42.5-47.5 GHz) and OOB blockers both are received through the antenna. MZM$_1$ is biased at Q-point for maximum linearity while modulating the light input from a distributed feedback laser (DFB) at 1550-nm. The BPF is configured for one of the channels and its corresponding BW, and thus image-rejection and partial inference rejection of > 35-dB is performed at point A. Two ring monitors detect the location of two interferers and automatically configure the notch filters to their proper locations. The notch filter provides an extra 30-dB rejection to interference frequencies at Point B. The LO signal (27.5,...,42.5 GHz) drives MZM$_2$ and the LO image-reject filter selects the upper side-band of the modulated signal, thus rejecting LO image at point C. The mixer combines both signals at each of the four channels and its corresponding LO to provide a 2.5-GHz output from the PIC at point D. Finally, the CMOS amplifies these signal to drive the loading instrument at point E.

III. CIRCUIT IMPLEMENTATION

A. Silicon Photonics Circuits

Fig. 2 illustrates the circuit schematic of the photonically-assisted SDR. The RF photonics 4th-order BPF (Ring$_{1-5}$) is designed similar to [5] to provide a band-pass response with a free spectral range (FSR) of 50-GHz, a re-configurable 3~5-GHz BW, and center frequency to cover 30~45 GHz range at point A. The simulated filter response in MATLAB shows an insertion loss of -4.8-dB and an OOB rejection > 35-dB ($\approx$ 40-dB at 12.7 GHz offset from center), which agrees with the measurement results of Fig. 5a. The notch filter starts with a tunable coupler which divides the signal between main and monitor paths by 90% coupling ratio to the main path. The main path consists of four rings

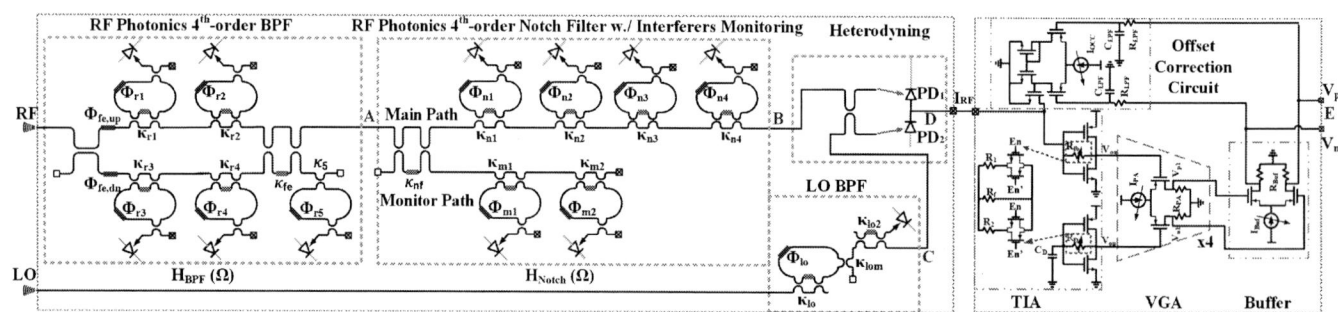

Fig. 2. The circuit implementation of the photonically-assisted SDR.

979-8-3503-2123-4/23 $31.00 © 2023 IEEE

acting as four notch filters (N_{1-4}), while the monitor path consists of two rings (M_{1-2}) acting as BPF from input to each local monitor. The unit element of the notch filter is a ring resonator with FSR of 30-GHz, two heaters to control resonance frequency and coupling ratio which determine the order of the filter and the level of rejection, respectively. For the interferer monitoring rings, two rings are chosen to divide the search range evenly and thus reduce the sweeping time required by half. Each monitor ring provides a peak at the location of any interferers in its search range, which is used to tune the notch filter resonance frequency. The LO BPF comprises of a ring resonator with FSR of 50-GHz with output from the band-pass response going to an MZI coupler tuned to provide 90% coupling ratio to the main path at point C and 10% to a local monitor. The resonance frequency and rejection level of all the ring resonators are controlled by $N+$ resistive heaters using the thermo-optic effect, where applying a DC voltage/current shifts the pole/zero magnitude level and frequency location. A local PD monitor is used with each ring to indicate the location of the pole/zero of each ring, which can be automatically-controlled by an external controller [6]. Heterodyning is done through a 50% coupler and a balanced PD where the coupler combines the top and bottom paths into two different outputs. The balanced PD, consisting of two PDs back-to-back ($PD_{1,2}$), boosts the electrical IF output current I_{RF} by 3 dB over single PD at the expense of doubling the parasitic capacitance.

The CMOS TIA is an inverter amplifier with three switchable feedback resistances R_{fb} (as either R_f, R_1, or R_2) for gain control and a dummy amplifier stage with dummy capacitor C_D. The balanced PD output is directly connected to the TIA input so it is dc-biased at $V_{DD}/2$. The anode of PD_2 is connected to ground, and the cathode of PD_1 is connected to 2-V_{DC}. Four cascaded VGA amplifiers as current steering resistive-loaded stages and a current steering resistive-loaded buffer are used to provide a tunable gain and sufficient fan-out to the load impedance of 50-Ω. Also, VGA and buffer gain control is provided through the tail current (I_{PA} and I_{BUF}, respectively) control. The offset correction circuit of C_{LPF} R_{LPF} low pass filter and differential to single ended current steering operational trans-impedance amplifier is used to compensate for offset at the input node. The full CMOS stage provides a gain of 60-dBΩ with 20-dB gain control and 3-dB BW of 5-GHz, while consuming 18~75.6 mW from 0.7~1.2 V_{DD}.

IV. MEASUREMENT RESULTS

The micro-graph of the hybrid-integrated SDR receiver prototype is shown in Fig. 4. Since the PIC is thinned to 75-μm to reduce thermal coupling [6], the PIC and CMOS chips do not have similar heights, hence; the CMOS chip is placed inside a cavity and the two chips are directly connected through a bond-wire. The PIC is fabricated in SOI Si-photonics process occupying an area of 9.42 mm^2, while the CMOS is fabricated in 28nm occupying an area of 0.0736 mm^2. The PIC response is measured by optical vector analyzer (OVA)

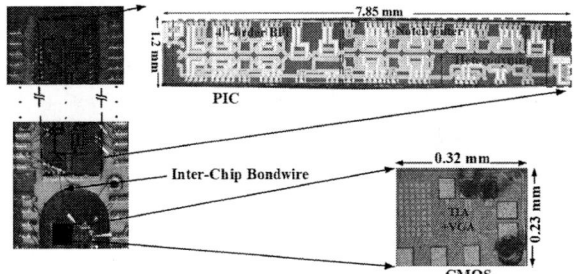

Fig. 4. The two chips micro-graphs and their inter-chip bondwire.

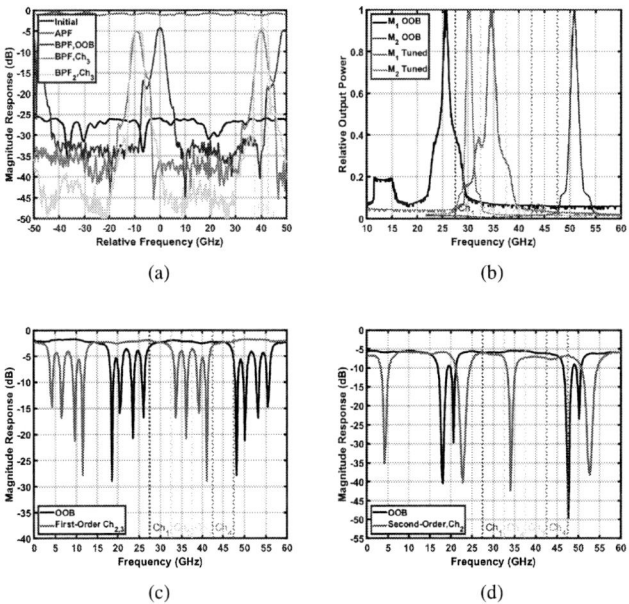

Fig. 5. Measured PIC optical response at different programming options; (a) initial, APF and BPF at different center frequencies and BWs, (b) spectrum monitoring, (c) first-order notch filter, and (d) second-order notch filter.

and the initial response is shown in Fig. 5a, which is then automatically calibrated to display an all-pass filter (APF) and two different BPF responses with either BPF_1 (placed at OOB or at Ch_3) and -3dB-BW of 3-GHz or BPF_2 at Ch_3 and -3dB-BW of 5-GHz. The spectrum monitoring is done through M_{1-2} and the spectrum for detecting two interferers is shown in Fig. 5b, where the monitor is first placed at the location of OOB and then tuned to search for the band of operation. The first-order notch filter demonstrates quadruple interferer rejection capability with maximum of 30-dB as shown in Fig. 5c, which is shown for OOB and then tuned in a way that two notches are located in Ch_2 and the other two are located in Ch_3 (i.e. dual-band). Furthermore, the notch filter is tuned as a second-order filter to show up to two interferers rejection with > 45-dB initially at OOB and then at Ch_2 as shown in Fig. 5d. The full top path response is then a combination of Fig. 5a and Fig. 5d that shows a maximum rejection of > 80-dB (> 35-dB from BPF and 45-dB from notch filter).

For measuring the RF performance, the PIC is placed

Table 1. Performance summary of the presented hybrid-integrated SDR and comparison with the state-of-the-art.

Specification	This Work	RFIC 2022 [1]	RFIC 2021 [2]	RFIC 2020 [3]
Blocker Rejection Technique	PIC BPF and Notch Filter	Time-Approximation Filter	N-Path Filter	N-Path Filter
Frequency Range (GHz)	30 ~ 45	31 ~ 37	6 ~ 31	10 ~ 35
Signal Bandwidth (MHz)	3000 ~ 5000	550 ~ 660	320 ~ 470	400
OOB Blocker Rejection (dB)	First-Order: 60 Second-Order: 80	45	25	9
RF Gain (dB)	25	4 ~ 6	-6.6 ~ -4.5	11 ~ 15
NF (dB)	9.9 ~ 12.9	12 ~ 17	5 ~ 20	12.5 ~ 15.7
IIP_3 (dBm)	3.1	NA	1.4 ~ 6.3	+10 ~ +14.1
P_{1dB} (dBm)	-6.4	-6	-7.4 ~ -2	-2.5 ~ 0
SFDR (dB)	50	NA	NA	NA
Total Chip Areas (mm^2)	9.42	0.46	0.69	NA
Technologies	SOI PIC and 28nm CMOS	28nm CMOS	45nm SOI	28nm CMOS
External Components	$MZM_{1,2}$, LNA and Post-Amplifier	—	—	—

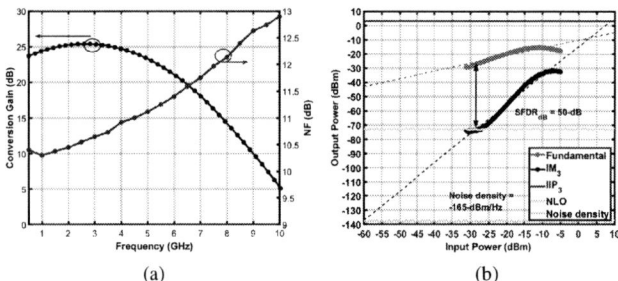

Fig. 6. Measured RF responses of the SDR receiver; (a) CG and NF. vs frequency, and (b) linearity metrics vs input power at 2.5-GHz.

Fig. 7. Modulation measurement w./o (left) and w./ (right) interferer filtering.

in an APF mode with two inputs provided to $MZM_{1,2}$ mm-wave input ports and output from the buffer bond-wired to a transmission line terminated by SMA. Both pre-amplifier (83051A) and post-amplifier (ZX60-6013E) serve as LNAs and base-band amplifiers, improving the NF and overall gain. The measured link conversion gain (CG) and noise figure (NF) are shown in Fig. 6a, with a max. CG (CG_{max}) of 25-dB and a min. NF (NF_{min}) of 9.9-dB. Fig. 6b illustrates the measured linearity metrics vs input power, where the fundamental output power, IM_3, noise level at the output (NLO) and noise density are shown. The link achieves IIP_3 of 0-dBm, P_{1dB} of -9.5-dBm and SFDR of 50-dB at IF frequency of 2.5 GHz.

The measured constellations from a blocker test with and without notch filtering is presented in Fig. 7. Si-PIC is supplied with a 100-MSymbol/s 64-QAM signal at -25 dBm and two mm-wave blockers at -10 dBm. The modulated signal carrier frequency is 30-GHz, while blocker frequencies are (35 and 40)-GHz. Fig. 7 confirms that the received signal without

filtering cannot be reconstructed due to the large in-band folded blocker. Then, automatic-detection of this blocker starts and the location information is provided to the notch filter. The notch filter is then reconfigured as a second-order notch filter centered at 35-GHz. Therefore, the demodulated signal has improved constellation as shown in Fig. 7 and EVM has improved from -23.5 to -30.0-dB.

V. Conclusion

A CMOS/Si-PIC hybrid-integrated SDR is investigated in this paper. Channel selection is demonstrated for four different bands and OOB interference is properly rejected. This work demonstrates an automatically calibrated, programmable receiver with OOB blocker rejection capability. A comparison between the presented integrated SDR receiver performance and other recent mm-wave integrated CMOS SDRs is shown in table 1. Future works should consider the integration of the two MZMs within the PIC and the LNA/post-amplifier within the CMOS chip.

References

[1] C. Yang, S. Su, and M. S.-W. Chen, "A millimeter-wave mixer-first receiver with non-uniform time-approximation filter achieving > 45-dB blocker rejection," in *2022 IEEE Radio Frequency Integrated Circuits Symposium (RFIC)*. IEEE, jun 2022.

[2] S. Hari, C. J. Ellington, and B. A. Floyd, "A 6-31 GHz tunable reflection-mode n-path filter," in *2021 IEEE Radio Frequency Integrated Circuits Symposium (RFIC)*. IEEE.

[3] S. Krishnamurthy and A. M. Niknejad, "10-35ghz passive mixer-first receiver achieving +14dbm in-band IIP3 for digital beam-forming arrays," in *2020 IEEE Radio Frequency Integrated Circuits Symposium (RFIC)*. IEEE.

[4] F. Scotti, F. Laghezza, P. Ghelfi, and A. Bogoni, "Multi-band software-defined coherent radar based on a single photonic transceiver," *IEEE Transactions on Microwave Theory and Techniques*, vol. 63, no. 2, pp. 546–552, feb 2015.

[5] R. Rady, C. Madsen, S. Palermo, and K. Entesari, "A 20-43.5-GHz wideband tunable silicon photonic receiver front-end for mm-wave channel selection/jammer rejection," *Journal of Lightwave Technology*, pp. 1–14, 2022.

[6] R. Rady, G. Choo, C. Madsen, S. Palermo, and K. Entesari, "External modulator-based automatic tuning of reconfigurable silicon photonic 4th-order apf-based pole/zero filters," in *2021 Optical Fiber Communications Conference and Exhibition (OFC)*, 2021, pp. 1–3.

RTu2B-4

Mixer-Free Phase and Amplitude Comparison Method for Built-in Self-Test of Multiple Channel Beamforming IC

Seonjeong Park, Eun-Taek Sung, Seunghun Wang, Songcheol Hong
Dept. of Electrical Engineering, KAIST, Republic of Korea
{seonjeong95, schong1234}@kaist.ac.kr

Abstract—This paper presents a built-in self-test (BIST) method of a multiple channel beamforming IC, based on successive comparisons of phase and amplitude differences between a pair of signals in beamforming channels. This is implemented with a simple difference detection circuit without a mixer and a reference LO. The phase and gain differences of the channels are obtained independently using the proposed simple detector. The phase difference detection accuracy is improved by removing the DC offset caused by the amplitude difference of the signals. The detection circuit is integrated into a 4-channel beamforming IC and the BIST concept verified for both the transmitter and receiver paths. The errors of phase and gain differences between the pair of channels measured through the proposed circuits are as low as -18.1° to 9.8° and -0.29dB to 0.38dB for the transmitter and -8.8° to 15.3° and -0.31dB to 0.41dB for the receiver, respectively.

Keywords— built-in self-test, phased array, 5G, beamforming, mm-wave transceiver.

I. INTRODUCTION

Beamforming technology using a phased array antenna has been intensively studied in millimeter-wave frequency bands to support high data rate communications. Recently, large scale arrays have been announced for base-stations or for backhaul for high SNR. However, in multi-channel beamforming IC, phase and gain variations between channels caused by fabrication mismatches, asymmetric layouts and inter-channel couplings [1] degrade the beamforming quality. The phase variations increase the sidelobes and lower the directivity of the array antenna [2]. In addition, the gain variations can deteriorate the EVM by different amplitude and phase distortions of the power amplifiers [3]. Various BIST methods have been presented to solve the inter-channel variations [4]-[7]. Reducing channel variations via BISTs can significantly reduce overall beamformer calibration times, especially in the large-scale arrays. As shown in Fig 1. (a), the characteristics of channels are measured one by one using a common IQ mixer [4]; consequently, it takes a long time to test these and the I/Q mismatch can reduce the test accuracy. As shown in Fig. 1 (b), the loopback method [5], sends a test signal from Tx to Rx. It is difficult to make the identical loopback test paths simply as the number of channels increases. The proposed mixer-free comparison method uses a simple detection circuit to figure out the phase and gain difference between channels (Fig. 1 (c)). Since an additional mixer and a reference LO for self-testing are not required, the proposed method is convenient to use and it has small area and low power consumption. The proposed BIST circuits are integrated into 4-channel beamforming IC to verify its accuracy.

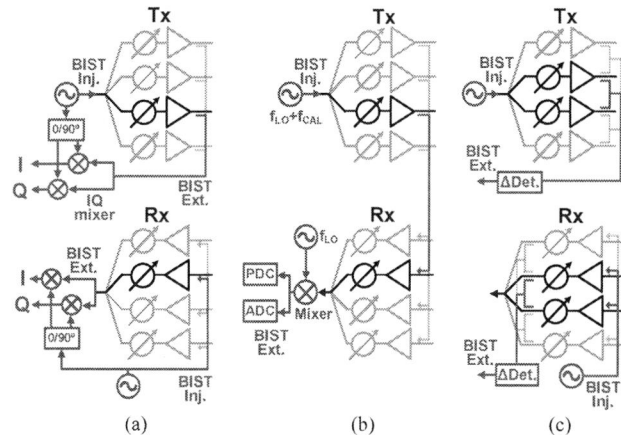

Fig. 1. Various BIST methods: (a) channel test using I/Q mixer and reference LO, (b) loopback test of Tx and Rx channel combinations, (c) the proposed mixer-free comparison test of Tx and Rx channels, respectively.

II. PROPOSED MIXER-FREE PHASE AND AMPLITUDE COMPARISON METHOD

A. Proposed BIST method and building blocks

Fig. 2 shows a block diagram of a 4-channel beamforming IC with the proposed BIST circuits. The black, red and blue lines represent RF beamforming circuits, Tx and Rx BIST circuits, respectively. These compare phase and amplitude between channels using a simple difference detection circuit without a mixer and a reference LO. For comparison between Tx channels, a BIST signal is applied to P5. The test signals containing phase and gain information of Tx beamformer channels are extracted through switched capacitive couplers. A pair of test signals are directed to a difference detector through on-chip waveguides, which are coplanar waveguides with ground (GCPW) lines indicated by ⓕ. The difference detector figures out their phase and gain differences. In order to compare the receiving channels, the BIST signal is input into the P6. It is distributed uniformly through the Wilkinson divider and the GCPW line (shown by ⓒ and ⓓ). Similarly, the test signals are sent to two LNAs by turning on the switched capacitive couplers of the two comparing channels. The BIST signals passing through each Rx beamformer are extracted by differential switched couplers indicated by ⓑ, and these are compared in a difference detector. Fig. 3 (a) is a detailed block diagram of the BIST circuit operation assuming a comparison between CH1 and CH2. The amplitude and phase of the test signals passing through each channel of Tx or Rx beamformer are expressed as $A_N \angle \theta_N$. When the self-test circuits are not

979-8-3503-2123-4/23 $31.00 © 2023 IEEE

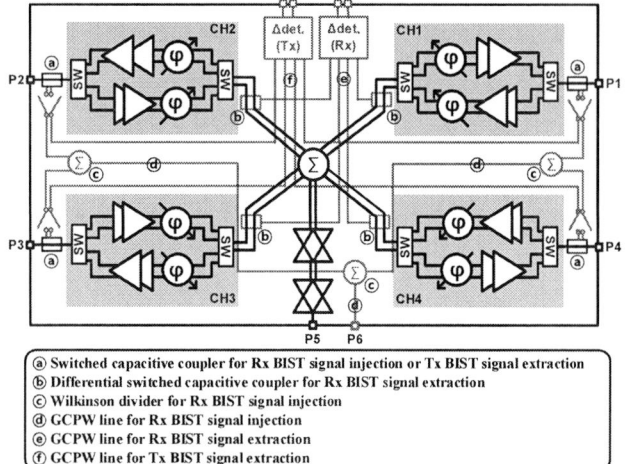

Fig. 2. Block diagram of the proposed 4-channel beamforming IC with the proposed BIST circuit blocks for successive comparison tests between channels.

ⓐ Switched capacitive coupler for Rx BIST signal injection or Tx BIST signal extraction
ⓑ Differential switched capacitive coupler for Rx BIST signal extraction
ⓒ Wilkinson divider for Rx BIST signal injection
ⓓ GCPW line for Rx BIST signal injection
ⓔ GCPW line for Rx BIST signal extraction
ⓕ GCPW line for Tx BIST signal extraction

used, there should be no degradations of the noise figure and linear output power. Therefore, a switched capacitive coupler with very small insertion loss must be used. When the switch is turned on, the coupling factor is -15dB at 24GHz. The extracted BIST signals are transmitted to the channel selection switch and finally to the difference detector located in the middle of the chip though the GCPW lines, which have to have identical electrical lengths. Compared to the loopback calibration method, it has a simple routing of the test. The channel selection switches are used to increase isolation between the BIST signal and the leakage signal. The simulated isolation between output ports of the channel selection switch is -30dB. The phase and gain difference detector consists of a switched RC-CR circuit and a power detector. Fig. 3 (b) is a microphotograph of the phase and gain difference detector. It was fabricated using a 65-nm RF CMOS process and the core size is as small as $70\times70um^2$. Section B details the operating principles of the difference detector.

B. Phase and amplitude difference detection circuit

A magnitude of the BIST signal applied to the switched RC-CR circuit is $\alpha\beta A_N$, which is multiplied by amplitude factors of the coupler, GCPW lines and selection switch as shown in Fig. 3 (a). To compare the phases of the two channels, SW1 which is integrated in the RC-CR circuits is turned on. The current amplitudes of the BIST signals injected to the power detector are

$$|i_{L1}| = |i_1 + i_4| = \alpha\beta\sqrt{\omega^2 C^2 A_1^2 + \frac{A_2^2}{R^2} - \frac{2\omega C A_1 A_2}{R}\sin\Delta\theta} \quad (1)$$

$$|i_{L2}| = |i_2 + i_3| = \alpha\beta\sqrt{\frac{A_1^2}{R^2} + \omega^2 C^2 A_2^2 + \frac{2\omega C A_1 A_2}{R}\sin\Delta\theta} \quad (2)$$

Since the current flowing in the capacitor leads by 90° compared to that of resistance, the amplitude of i_{L1} and i_{L2} can be expressed as a function of $\Delta\theta$, i.e., the phase difference between the two channels ($\theta_1 - \theta_2$). The output voltage of power detector, V_{out}, is obtained as

$$V_{out,SWon} = \gamma(i_{L1}^2 - i_{L2}^2)$$
$$= (\alpha\beta)^2\gamma\left\{\left(\omega^2 C^2 - \frac{1}{R^2}\right)(A_1^2 - A_2^2) - \frac{4\omega C A_1 A_2}{R}\sin\Delta\theta\right\}$$

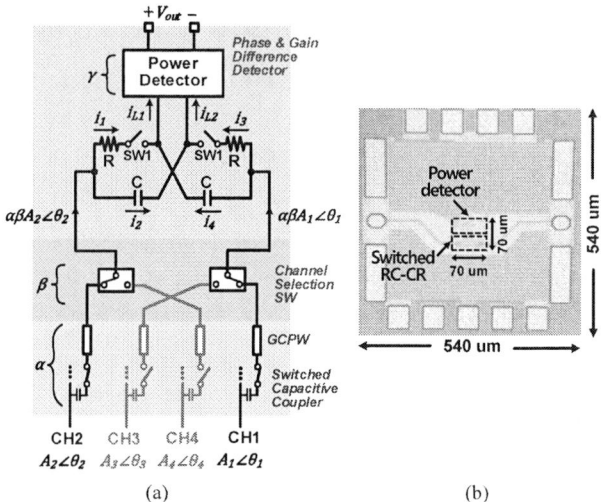

Fig. 3. (a) Schematics of the BIST circuit including capacitive couplers, channel combination selection switches, and phase and gain difference detector. (b) Microphotograph of the phase and gain difference detector.

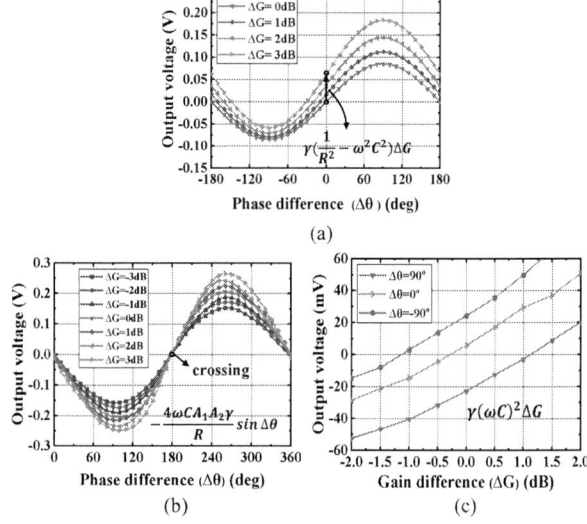

Fig. 4. (a) Simulated output voltages of the power detector with DC offsets which are due to the gain differences between channels. (b) The measured output voltage of the power detector versus phase difference in various amplitude differences with the switch on and (c) the measured output voltage of the power detector versus amplitude difference in various phase differences with the switch off.

$$= (\alpha\beta)^2\gamma\left\{(\omega^2 C^2 - \frac{1}{R^2})\Delta G - \frac{4\omega C A_1 A_2}{R}\sin\Delta\theta\right\} \quad (3)$$

γ is a conversion coefficient of the power detector. $V_{out,SWon}$ is given as the sum of the DC offset term due to gain difference (ΔG) and the sin function term of phase difference. Fig. 4 (a) shows the simulated V_{out} with respect to the phase difference in various gain differences. Note that an offset occurred in the sine function due to the gain difference. In order to achieve phase difference regardless of the gain difference, the term with ΔG in equation (3) should be removed. If $\omega = 1/RC$ satisfies, (3) becomes

$$V_{out,SWon} = -(\alpha\beta)^2\gamma\frac{4\omega C A_1 A_2}{R}\sin\Delta\theta \quad (4)$$

The phase difference can be easily estimated without offset

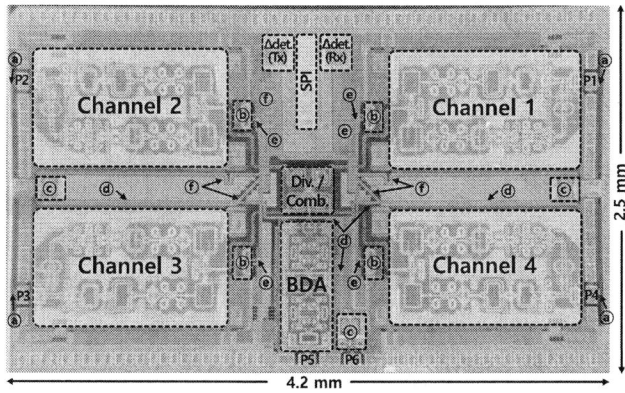

Fig. 5. Microphotograph of the proposed 4-channel beamforming IC with the BIST circuit blocks.

by determining the R and C values according to the test frequency. Considering the parasitic components of the SW1, R and C are set to 28Ω and 30fF, respectively. Fig. 4 (b) shows the measured V_{out} of the detector versus phase difference in various amplitude differences with the switch on. Even if there is an amplitude difference, it was measured in the sine function of $\Delta\theta$ without offset as shown in equation (4). There is no phase difference between two channels regardless of the gain difference if V_{out} is zero. Since the phase difference between channels in a phased array antenna is typically set between ±90°, it can be estimated by calculating $sin^{-1}(V_{out}/V_{out,max})$.

To compare the gains of two channels, the SW1 is turned off. The amplitudes of currents flow into the power detector, i_{L1} and i_{L2}, and $V_{out,SWoff}$ are

$$|i_{L1}| = |i_4| = \alpha\beta\omega CA_1 \quad (5)$$
$$|i_{L2}| = |i_2| = \alpha\beta\omega CA_2 \quad (6)$$
$$V_{out,SWoff} = \gamma(i_{L1}^2 - i_{L2}^2) = \gamma(\alpha\beta\omega C)^2\Delta G \quad (7)$$

Fig 4. (c) shows the measured V_{out} versus ΔG in various phase differences with the switch off. As shown in equation (7), V_{out} is proportional to the gain difference in the ±2dB range. It can be deduced that there is no gain difference between channels when V_{out} is crossing to zero in the case of $\Delta\theta = 0°$. If a phase difference exists, leakage signals are added by the off-capacitance of SW1, resulting in an offset. To eliminate the offset, one can obtain the extrapolated line of V_{out} with respect to the gain difference by pre-measurement. The offset can be compensated using the y-intercept point. For example, if the slope and the y-intercept point are a and b, respectively, the ΔG can be estimated by calculating $(V_{out} - b)/a$.

III. MEASUREMENT RESULTS OF THE 4-CH BIST CIRCUIT

The proposed BIST circuit is integrated into a 4-channel beamforming IC, for which concept is verified for both transmitter and receiver paths. Fig. 5 is the microphotograph of the proposed 4-channel beamforming IC with the BIST circuits. It was fabricated using a 65-nm RF CMOS process. The chip size is 4.2×2.5mm². Fig. 6 shows measurement set-ups for comparison between CH1 and CH2 at the transmitter and CH2 and CH3 at the receiver using the proposed difference detection circuit. The phase and gain differences between channels are measured by fixing the phase and gain bits of a certain channel

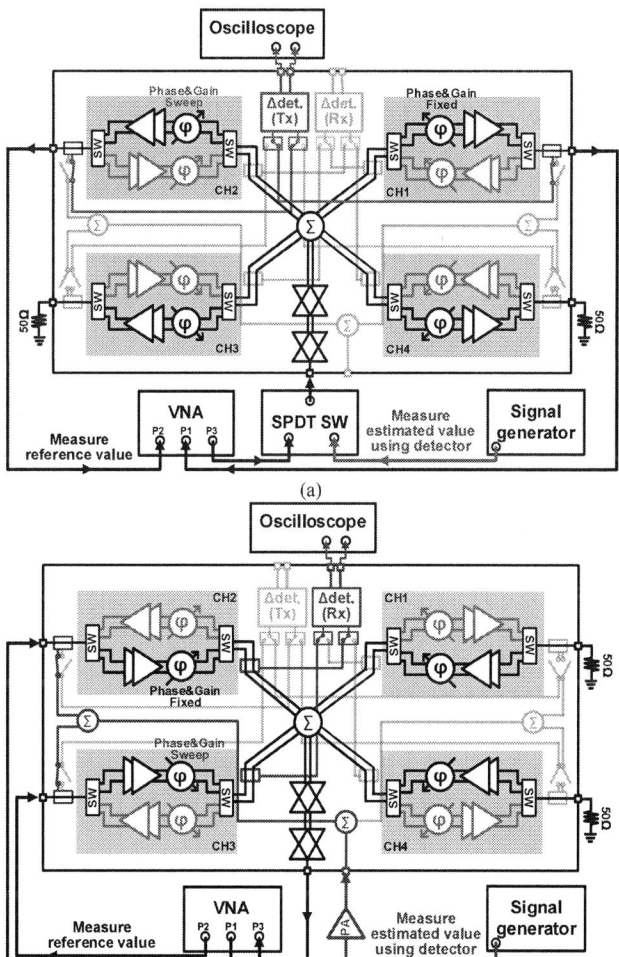

Fig. 6. Measurement setups to test the BIST circuits to measure the phase and gain differences of (a) the transmitting beamforming parts and (b) the receiving beamforming parts.

as a reference channel and sweeping the phase and gain bits of the other channel under comparison. With the same set-up, the reference channel values can be measured by VNA to evaluate the error of $\Delta\theta$ and ΔG, which are also estimated by the BIST circuits.

Fig. 7 (a) and (b) show the measured V_{out} versus $\Delta\theta$ and ΔG between CH1 and CH2 of the transmitter path by the BIST circuits. As previously described, the measured V_{out} appears in the form of a sine function of $\Delta\theta$ when the switch is turned on and in the form of a linear function of ΔG when the switch is turned off. In Fig. 7 (c) and (d), the black and red lines are reference values measured by a VNA and estimated values measured by the detector, respectively. It is found that they show similar results. For various gain differences, the phase error was measured to be as low as -18.1° to 9.8° (Fig. 7 (e)). The measured gain errors for several phase differences are low with in the -0.29dB to 0.38dB range as shown in Fig. 7 (d). The comparison results between CH2 and CH3 of the receiving path are shown in Fig. 8 (a) and (b). When compared with the values measured by a VNA, similar results are achieved (Fig. 8 (c) and (d)). The measured phase and gain errors for several gain

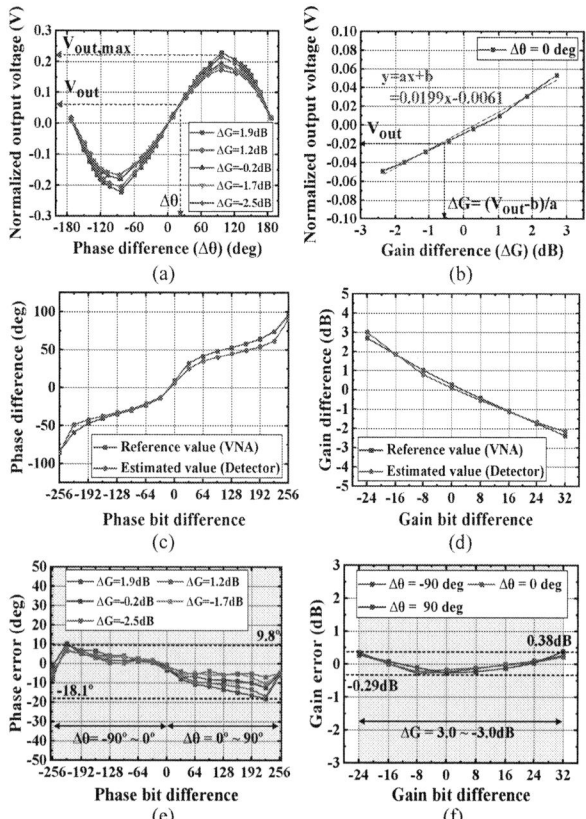

Fig. 7. The measured phase and gain differences between CH1 and CH2 of the transmitting beamforming parts. (a) The power detector output voltage with various gain differences with respect to phase difference, (b) The power detector output voltage versus gain difference. (c) The phase difference and (d) gain differences measured by a reference VNA and the proposed detector. (e) The phase errors and (f) gain errors between the reference and estimated values.

and phase states, respectively, are as low as within -15.3° to 8.8 ° and -0.31dB to 0.41dB.

IV. CONCLUSION

The BIST circuits that can detect the gain and phase differences between channels without a mixer and a LO are proposed. The DC offset caused by the gain difference during phase difference extraction are effectively removed. The proposed difference detection circuits are integrated into the Tx and Rx of the 4-channel beamformer IC to verify the self-test concept. The errors of the phase and gain difference are as low as -18.1° to 9.8° and -0.29dB to 0.38dB for the Tx and -8.8° to 15.3° and -0.31dB to 0.41dB for the Rx, respectively.

ACKNOWLEDGMENT

This work was supported by the Institute of Information & Communications Technology Planning & Evaluation (IITP) grant funded by the Korea Government (MSIT) (2019-0-00933, Development on multi-beam antenna technology). Dr. Eun-Taek Sung is currently working at Samsung Electronics, Republic of Korea. Dr. Seunghun Wang is currently working at the Electronics and Telecommunications Research Institute (ETRI), Republic of Korea.

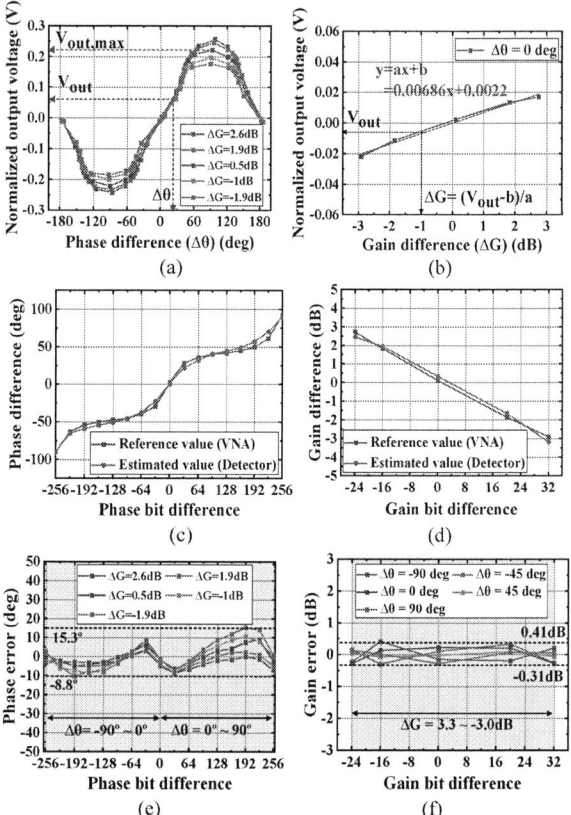

Fig. 8. Measurement results of the phase and gain differences between CH2 and CH3 of the receiving beamformer parts. (a) The power detector output voltage with various gain differences with respect to phase difference, (b) The power detector output voltage versus gain difference. (c) The phase difference and (d) gain differences measured by a reference VNA and the proposed detector. (e) The phase errors and (f) gain errors between the reference and estimated values.

REFERENCES

[1] T. Yu and G. M. Rebeiz, "A 22–24 GHz 4-element CMOS phased array with on-chip coupling characterization," in IEEE Journal of Solid-State Circuits, vol. 43, no. 9, pp. 2134-2143, Sept. 2008.

[2] S. Lee and H. -J. Song, "Accurate statistical model of radiation patterns in analog beamforming including random error, quantization error, and mutual coupling," in IEEE Transactions on Antennas and Propagation, vol. 69, no. 7, pp. 3886-3898, July 2021.

[3] Y. Yin et al., "A 37–42-GHz 8 × 8 phased-array with 48–51-dBm EIRP, 64–QAM 30-Gb/s data rates, and EVM analysis versus channel RMS errors," in IEEE Transactions on Microwave Theory and Techniques, vol. 68, no. 11, pp. 4753-4764, Nov. 2020.

[4] S. Y. Kim, O. Inac, C. -Y. Kim, D. Shin and G. M. Rebeiz, "A 76–84-GHz 16-element phased-array receiver with a chip-level built-in self-test system," in IEEE Transactions on Microwave Theory and Techniques, vol. 61, no. 8, pp. 3083-3098, Aug. 2013.

[5] Y. Wang et al., "A 39-GHz 64-element phased-array transceiver with built-in phase and amplitude calibrations for large-array 5G NR in 65-nm CMOS," in IEEE Journal of Solid-State Circuits, vol. 55, no. 5, pp. 1249-1269, May 2020.

[6] S. Choi, Y. Aoki, H. -C. Park, S. -G. Yang and H. -J. Song, "Sequential loopback built-in self-test algorithm for dual-polarization millimeter-wave phased-array transceivers," 2021 IEEE Radio Frequency Integrated Circuits Symposium (RFIC), 2021, pp. 55-58.

[7] J. Park and S. Hong, "Calibration method of multiple channel beamforming transceiver IC with successive channel signal comparisons," U.S. Patent Application 17919623, Oct. 18, 2022.

979-8-3503-2123-4/23 $31.00 © 2023 IEEE

RTu2C-1

A 24-30 GHz 4-Stream CMOS Transceiver Based on Dual-LO Phase-Shifting Fully Connected Architecture

Qingfeng Zhang[1], Yiming Yu[1], Dongming Duan[1], Xin Xie[1], Shaoyu Meng[1], Haoran Wang[1],
Chenxi Zhao[1], Huihua Liu[1], Yunqiu Wu[1], Wenquan Che[2], Quan Xue[2], Kai Kang[1*]
[1]University of Electronic Science and Technology of China, China
[2]South China University of Technology, China
*kangkai@uestc.edu.cn

Abstract—This paper presents a wideband low-complexity multi-stream beamforming transceiver (TRX) based on multiple local-oscillator (multi-LO) phase-shifting fully-connected (FC) architecture for MIMO communication. In supporting the same beams, the proposed architecture halves the complexity of the FC combiners as well as the ADCs/DACs overhead. Meanwhile, the FC combining network of each stream is implemented at IF-domain, which benefits from avoiding the mutual overlap between the combiners in the RF-domain, but also reducing combiner loss. To verify the proposed architecture, a 24-30 GHz 4-stream dual-LO phase-shifting FC TRX chip is fabricated in 65-nm CMOS process. Each single-stream path integrates a TX and RX channel with independent up/down-mixer, 0.4-dB gain and 6-bit phase control, where the whole TRX IC only has one PA and LNA to amplify 4-stream incoherent signals saving chip area and power consumption. The measured peak conversion gain (CG) of the TX is 31.4 dB, with a 3-dB bandwidth of 24.4-30.1 GHz. In the RX mode, the measured CG versus IF is 18.3 dB, with a 3-dB bandwidth of 3.3-11.8 GHz (112.6% fractional bandwidth). Besides, the TRX IC achieves a maximum OP$_{1dB}$ of 15.4 dBm and minimum NF of 4.6 dB including T/R switch. Additionally, it demonstrates a 26 dB gain control and 360° tuning range, where the maximum gain variation of only ±0.1 dB during 360° phase tuning at 27.5 GHz without any calibration.

Keywords—CMOS, fully-connected, multi-stream, multiple local-oscillator phase-shifting, transceiver.

I. INTRODUCTION

Phased-array beamformers with multi-stream capability at millimeter-wave (mm-wave) frequencies have emerged as the preferred to further boost the data rates and channel capacity. Compared with the multi-stream partially-connected (PC) beamforming architectures, the fully-connected (FC) ones have better spectral efficiency, superior spatial filtering and higher energy efficiency due to each stream FC to all available antenna elements [1]-[3]. However, FC architectures increase hardware design complexity significantly in massive arrays, which is difficult to scale to more streams.

To tackle this challenge, an emerging solution to mitigate the complexity is proposed in this paper, which demonstrates a multi-stream FC beamforming based on multiple local-oscillator (multi-LO) phase-shifting (PS) architecture for the first time. As shown in Fig. 1, conventional RF PS FC architecture suffers from severe complexity, while the proposed beamforming architecture has several innovative attributes: (i) a multi-LO PS allows multi-stream signals to be distinguished by frequency, hence multi-stream signals can

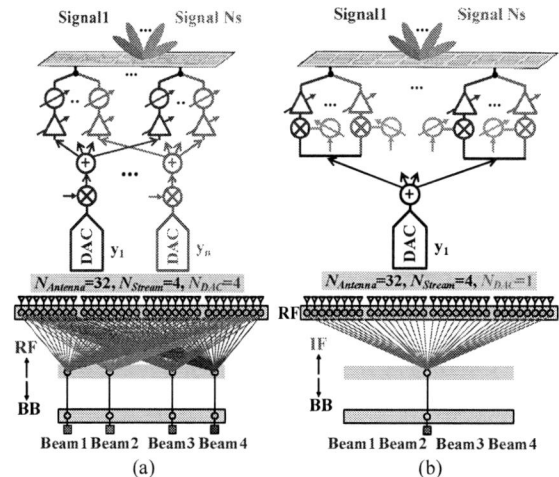

Fig. 1. Multi-stream beamforming architectures: (a) conventional RF PS fully-connected and (b) proposed multi-local oscillator PS fully-connected.

share same combiner and ADC/DAC. In supporting the same number of beams [1], the system complexity and baseband processor overhead can be significantly mitigated; (ii) the FC combining network of each stream is implemented at IF-domain, thereby avoiding the mutual overlap between the combiners in the RF-domain, but also reducing combiner loss contrast to [1], [2]; (iii) the LO power distribution in massive arrays is provided by chip-to-chip through a single-wire, which simplifies the LO power splitters and reduces the required power compared to [4], [5]. With the help of the above three key contributions, the FC complexity of more beams in massive arrays has been greatly alleviated. In addition, it is worth mentioning that the LO PS architecture can prominently mitigate gain variation during phase tuning, which will alleviate the distortion of the radiation pattern and a high side-lobe level.

II. MULTIPLE LOCAL-OSCILLATOR PHASE-SHIFTING FULLY-CONNECTED ARCHITECTURE

Fig. 2 illustrates the system diagram of a phased-array transmitter employing the proposed multi-LO PS FC architecture. With multiple incoherent RF beams of the same frequency, by adopting different LO PS assignments in each channel, the n-stream IF signals can be distinguished by frequency and then separated through a specific filter. Thus, such n-stream different frequency signals can be concurrently input to the system sharing DACs and combiners. In contrast,

979-8-3503-2123-4/23 $31.00 © 2023 IEEE
269
2023 IEEE Radio Frequency Integrated Circuits Symposium

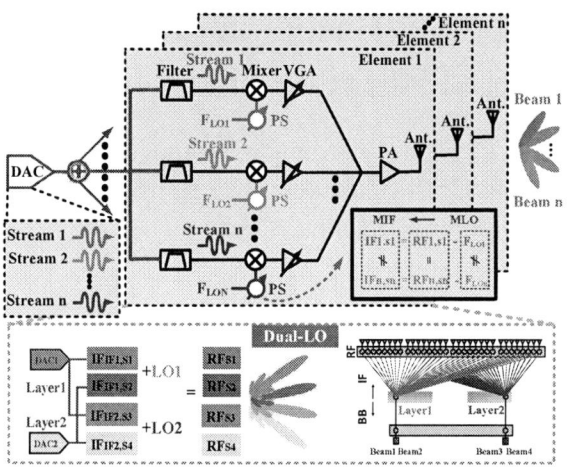

Fig. 2. Block diagram of a phased-array transmitter employing the proposed multi-LO phase-shifting fully-connected beamforming architecture.

the conventional RF or single-LO PS FC architecture is unable to distinguish n-stream IF signals by frequency, resulting in the inability to share DACs and combiners, and thus suffering from severe complexity in massive arrays. Specifically, for four incoherent streams with the same frequency (RF_{S1}, RF_{S2}, RF_{S3} and RF_{S4}), the application of a dual-LO ($F_{LO1} \neq F_{LO2}$) PS will produce two different frequencies of IF signals (IF_{IF1} and IF_{IF2}). Then, such signals ($IF_{IF1,S1}$ and $IF_{IF2,S3}$, $IF_{IF1,S2}$ and $IF_{IF2,S4}$) can recognize by specific filters respectively, which then can be concurrently input to the system sharing corresponding DAC and combiner. Therefore, relying on dual-LO, the complexity of the FC combining network of each stream and overhead of the DACs can be reduced by 50%. Additionally, those combiners are moved to IF-domain and the FC combining of each stream can be realized by each layer of a two-layer board, as shown in Fig. 2. Unlike tree-based crossbar combiners in RF-domain [1], [2], this method lessens the overlap and loss between the multi-stream FC combiners.

To validate the proposed multi-LO PS FC architecture, a 16-element 2-stream beamforming transceiver (TRX) based on dual-LO PS FC is firstly built by discrete modules, as shown in Fig. 3. Compared with the conventional FC architecture in two-streams [1], [2], the proposed prototype only uses one FC combiner and one ADC/DAC. Fig. 3 also shows the measured beam patterns in the TX mode. It can be observed that no matter how LO1 and LO2 beams vary

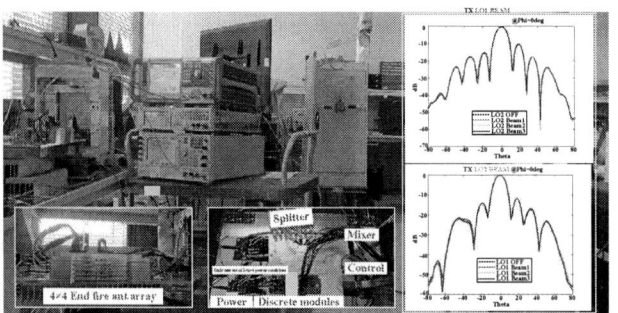

Fig. 3. 16-element two-stream transceiver assembled with discrete modules and measured beam patterns.

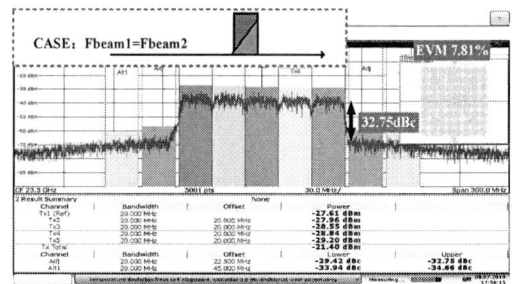

Fig. 4. Measured spectrum and constellation.

relative to each other, the beam patterns basically keep constant, which indicates that there is no mutual interference between the multi-LO signals. Furthermore, the spectrum and constellation are measured with incoherent same frequency (Fbeam1=Fbeam2) under 64-QAM modulation, as presented in Fig. 4. The TRX achieves an ACLR of -32.75 dBc and EVM of 7.81% with 6 dBm average output power at 23.5 GHz. Taken together, the proposed FC architecture based on multi-LO PS is feasible and can satisfy the desired specifications.

III. 4-Stream Dual-LO Phase-Shifting Fully-Connected Transceiver IC Design

The multi-LO PS FC prototype assembled with discrete modules has low-complexity and high performance, but suffers from larger area. For that, a 24-30 GHz 4-stream dual-LO PS FC beamforming TRX chip is designed using 65-nm CMOS process, as shown in Fig. 5. Each single-stream path integrates a TX and RX channel, mainly consisted of variable gain amplifiers (VGAs), up/down-mixers, drive amplifiers and LO phase shifters. Each two-stream path shares one LO power and frequency doubler. While the whole TRX IC only has one PA and LNA to simultaneously amplify 4-stream incoherent signals saving chip area and power consumption. Except for the switches, differential structures are adopted for all other blocks to robust the CMOS imperfect ground.

Furthermore, an additional active power divider is innovatively integrated in each LO input port of the chip. As shown in Fig. 5 (top left), one port distributes LO power to the chip itself, and another one is connected to the additional integrated signal pad (LO*) facilitating driving the LO of the next chip by single-wire. Compared with the conventional method using the Wilkinson power distribution [4], [5], it saves off-chip bulky complex power dividers and required LO power in massive arrays. In addition, the frequency plan of this TRX IC is that the RF implements 24-30 GHz to simultaneously cover the 5G NR bands (n257, n258 and n261); to form a certain guard-band which eases filter design and enhance the scalability of the system, the IF and LO operate at 3-10 GHz (108% fractional bandwidth (FBW)) and 8-13 GHz (47.6% FBW), respectively. To extend the operating bandwidth while considering the gain, noise, etc, a system pole analysis theory and a module pole tuning method are introduced in this work.

Besides the complexity and bandwidth consideration, the phase and gain errors of the TRX is critical to beamforming quality. As a result, 0.4 dB fine gain step control and 6-bit LO

979-8-3503-2123-4/23 $31.00 © 2023 IEEE   270

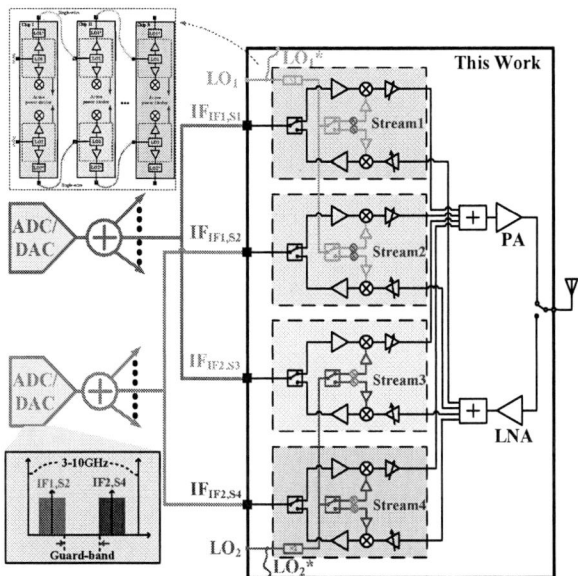

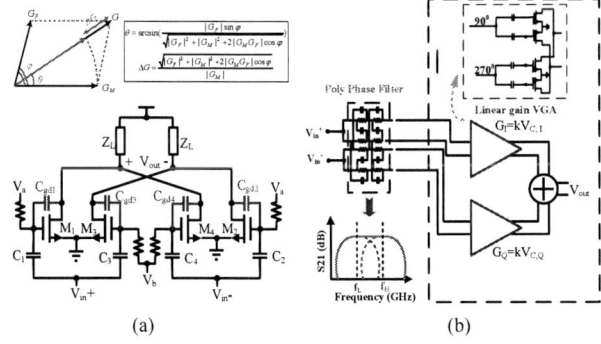

Fig. 5. Block diagram of the proposed 4-stream dual-LO phase-shifting fully-connected beamforming transceiver IC.

PS are chosen. Base on theoretical derivation, the feedforward signal generated by parasitic capacitor C_{gd} is one of the key factors that result in phase variation during gain tuning of CS VGA. To achieve phase compensation and robust against the PVT, an active cross-coupling unstacked topology is proposed, as plotted in Fig. 6(a). For the wideband LO PS design, a variable gain PS architecture with linear phase control technique is employed, as shown in Fig. 6(b). Meanwhile, a two-stage poly-phase filter is adopted to expand the bandwidth.

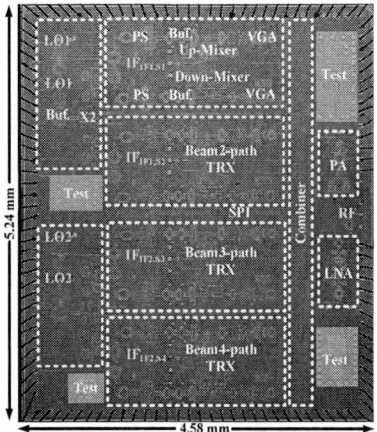

Fig. 6. (a) Core circuit schematic of digitally controlled RF VGA and (b) wideband LO phase shifter.

IV. MEASUREMENT RESULTS

The proposed 4-stream dual-LO PS FC beamforming TRX is implemented in 65-nm CMOS process and the overall chip size including pads is 4.58×5.24 mm², as shown in Fig. 7.

Fig. 8 presents the measured small-signal performance of single-stream path of the TRX. With a fixed LO of 10.5 GHz, the measured maximum conversion gain (CG) of the TX is 31.4 dB, with a 3-dB bandwidth of 24.4-30.1 GHz. In the RX mode, the measured CG versus IF is 18.3 dB, with a 3-dB bandwidth of 3.3-11.8 GHz (112.6% FBW). The CG of the TRX is also measured at different fixed LO (8-13 GHz, 0.5

Fig. 7. Die micrograph fabricated in 65-nm CMOS.

GHz step) and RF frequencies (22-30 GHz, 1 GHz step), respectively. The results demonstrate this work can operate at wideband and satisfy various frequency combination scenarios.

Fig. 9 shows the measured gain and phase responses of the TRX, which achieves 26 dB gain control with 0.4 dB fine gain step and 360° tuning range. Without any calibration, it also achieves RMS phase error of 1.7° at 24 GHz and gain error of 0.13 dB at 26 GHz, as shown in Fig. 9(c). Besides, the maximum gain variation is only ±0.1 dB during 360° range at 27.5 GHz, as plotted in Fig. 9(d). Additionally, in Fig. 10, the TRX achieves a maximum output 1-dB compression point (OP$_{1dB}$) of 15.4 dBm and minimum noise figure (NF) of 4.6 dB including T/R switch.

The measured performance is summarized and compared to state-of-the-art mm-wave beamformer ICs in Table 1. This work applying dual-LO PS architecture achieves the lowest system complexity and the minimum overhead of ADC/DACs in supporting the same number of beams. The TRX IC also realizes the widest bandwidth at IF and LO. Without any calibration, it demonstrates ultra-low gain and phase errors as well as invariant gain during phase shifting.

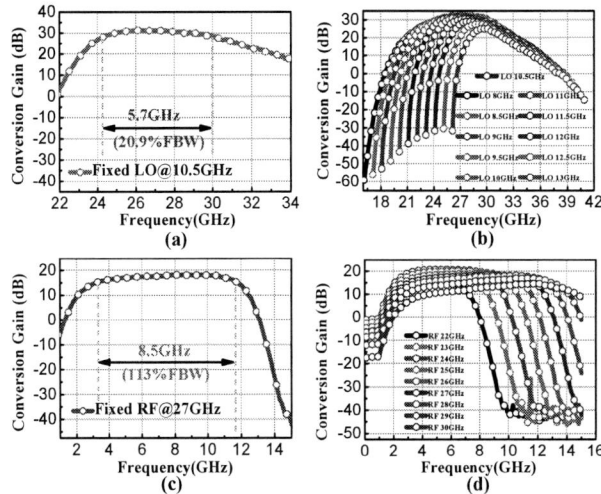

Fig. 8. Measured transmitter CG versus RF frequency (a) at fixed LO 10.5 GHz and (b) at different LO frequencies. Measured receiver CG versus IF frequency (c) at fixed RF 27 GHz and (d) at different RF frequencies.

979-8-3503-2123-4/23 $31.00 © 2023 IEEE   271

Table 1. Performance summary and comparison

	This Work	**Carnegie Mellon JSSC,2018[1]**	**Zhejiang Univ. RFIC,2020[2]**	**Tokyo Tech JSSC,2019[4]**	**IBM ISSCC,2022[6]**	**Samsung ISSCC,2020[5]**
Technology	**65nm CMOS**	65nm CMOS	65nm CMOS	65nm CMOS	130nm SiGe BiCMOS	28nm CMOS
Architecture	**Dual-LO Beamf. FC**	HB Beamf. FC	RF Beamf. FC	Single-LO Beamf. PC	RF Beamf. PC	RF Beamf. PC
No. of Streams /Element	**4**	2	8	1	1	2
Average No. of ADC or DAC or Combiner/Beam	**0.5(50% savings)**	1	1	1	1	1
RF Frequency (GHz)	**24-30**	25-30	7.5-9	26.5-29.5	24-30	37-40
IF Frequency (GHz)	**3-10(108% FBW)**	N/A	—	4	3	10-11.2
5G NR band Support	**Yes(n257,n258,n261)**	Yes(n257,n261)	—	Yes(n257,n261)	Yes(n257,n258,n261)	Yes(n260)
TX Gain (dB)	**31.4**	—	—	10	31	60
OP1dB (dBm)	**15.4**	—	—	15.7*	16	16.5*(Psat)
RX Gain (dB)	**18.3**	34	20	12	30	59
Nf_{min} (dB)	**4.6**	7.3*	3.6*	4.1*	3	4.2*
Gain Control (dB)	**26**	34	15.75	10	20(TX), 27(RX)	30(TX), 43(RX)
Phase Control	**360^0, 6 bit**	360^0	360^0, 6 bit	360^0, 2+3+10 bit	360^0	360^0, 4 bit
RMS Gain Error (dB)	**0.13@26GHz**	N/A	<0.3	<0.04**	±0.5	<0.33
RMS Phase Error (^{0})	**1.7@24GHz**	N/A	<2	0.3**	1.2	<3.3
Gain variation (dB)	**±0.1@27.5GHz**	N/A	N/A	0.2**	N/A	N/A
Integration	**TRX, IF, LO (w/o PLL)**	RX, IF, LO	RX	TRX, IF, LO (w/o PLL)	BFIC+FCIC	TRX, IF, LO (w/o PLL)
Power/Element /Beam(mw)	**237.5(TX+LO) 203.5(RX+LO)**	27.5	27	299(TX+LO) 148(RX+LO)	180(TX) 90(RX)	105(TX) 39(RX)
Die Area(mm²)	**4.58*5.24**	6.16	5.7*3.9	12	37	30

*T/R Switch excluded.   **With 2+3+10 bit control.

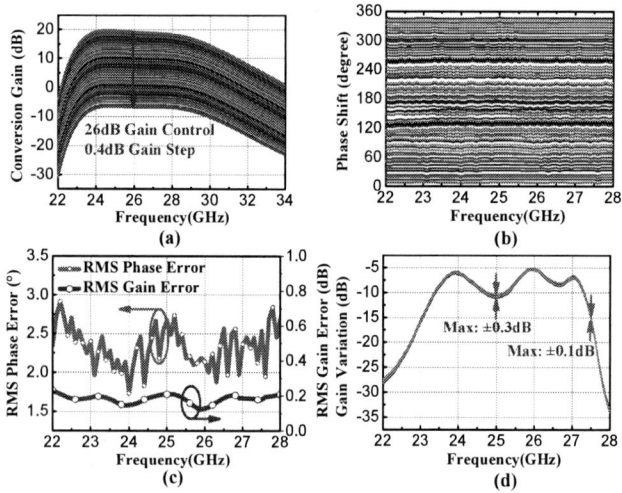

Fig. 9. Measured (a) gain and (b) phase responses. (c) Measured RMS gain and phase errors. (d) Gain variation during the 360° phase shifting.

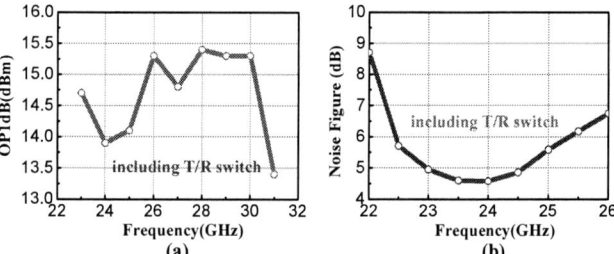

Fig. 10. Measured (a) OP_{1dB} and (b) NF including T/R switch.

V. Conclusion

This paper demonstrates a 24-30 GHz 4-stream FC beamforming TRX based on multi-LO PS architecture for the first time. Relying on dual-LO, the complexity of the FC combining network of each stream is reduced by 50%. The TRX IC also achieves the widest bandwidth in IF (>112%) and LO (>47%). Additionally, it has the characteristics of ultra-low gain and phase errors as well as gain invariant phase tuning. The proposed prototype can be deployed in massive arrays and fully supports to scale to more streams.

Acknowledgment

This work is supported by the National Key R&D Program of China (Grant No.2020YFB1805003), and the National Natural Science Foundation of China (Grant No. 62025106, 61931007).

References

[1] S. Mondal, R. Singh, A. I. Hussein and J. Paramesh, "A 25–30 GHz Fully-Connected Hybrid Beamforming Receiver for MIMO Communication," in *IEEE Journal of Solid-State Circuits*, vol. 53, no. 5, pp. 1275-1287, May 2018.

[2] N. Li *et al.*, "A 4-Element 7.5-9 GHz Phased Array Receiver with 8 Simultaneously Reconfigurable Beams in 65 nm CMOS Technology," *2020 IEEE Radio Frequency Integrated Circuits Symposium (RFIC)*, Los Angeles, CA, USA, 2020.

[3] E. Naviasky *et al.*, "14.1 A 71-to-86GHz Packaged 16-Element by 16-Beam Multi-User Beamforming Integrated Receiver in 28nm CMOS," *2021 IEEE International Solid- State Circuits Conference (ISSCC)*, San Francisco, CA, USA, 2021.

[4] J. Pang *et al.*, "A 28-GHz CMOS Phased-Array Transceiver Based on LO Phase-Shifting Architecture With Gain Invariant Phase Tuning for 5G New Radio," in *IEEE Journal of Solid-State Circuits*, vol. 54, no. 5, pp. 1228-1242, May 2019.

[5] H. . -C. Park *et al.*, "A 39GHz-Band CMOS 16-Channel Phased-Array Transceiver IC with a Companion Dual-Stream IF Transceiver IC for 5G NR Base-Station Applications," *2020 IEEE International Solid-State Circuits Conference-(ISSCC)*, San Francisco, CA, USA, 2020.

[6] B. Sadhu *et al.*, "A 24-to-30GHz 256-Element Dual-Polarized 5G Phased Array with Fast Beam-Switching Support for >30,000 Beams," *2022 IEEE International Solid- State Circuits Conference (ISSCC)*, San Francisco, CA, USA, 2022.

RTu2C-2

A 39 GHz 2×16-Channel Phased-Array Transceiver IC With Compact, High-Efficiency Doherty Power Amplifiers

Joonho Jung[1], Jooseok Lee[2], Daehyun Kang, Jinhyun Kim, Woojae Lee, Hansik Oh, Jae-hong Park, Kihyun Kim, Dong-hyun Lee, Sangho Lee, Jeong Ho Lee, Ji Hoon Kim, Younghwan Kim, Taewan Kim, Sangyong Park, Seungwon Park, Seungjae Baek, Bohee Suh, Soyoung Oh, Dongsoo Lee, Juho Son, Sung-gi Yang

Samsung Electronics Co., Ltd., S. Korea
[1]joonho5.jung@samsung.com, [2]joos85.lee@samsung.com

Abstract—This paper presents a 39 GHz 2×16-channel phased-array transceiver IC with compact, high-efficiency Doherty power amplifiers. The IC was implemented using a 28-nm bulk CMOS process with flip-chip packaging and evaluated using 100 MHz 8-carrier component (CC) fifth-generation (5G) new radio (NR) signal (total 800 MHz). With single transformer-based compact Doherty power amplifiers in transmitter (TX) path, the phased-array IC shows excellent TX performance, achieving high average output power of >8 / >9.3 dBm/ch. at error-vector-magnitude (EVM) of -32 / -28 dB while consuming low DC power of <109 / <118 mW/ch. Furthermore, the receiver (RX) path demonstrates low system noise figure (NF) of 5.8 – 6.5 dB, and an EVM of -29.8 dB with low DC power consumption of <42.3 mW/ch.

Keywords—39 GHz, 5G, millimeter-wave (mm-Wave), phased-array, bulk, CMOS, RFICs, transceiver, transmitter, receiver, Doherty power amplifiers.

I. INTRODUCTION

Growing demands for wireless communication systems with multi-gigabit-per-second speeds and ultra-low milliseconds latency are accelerating the development of the fifth-generation new radio (5G NR) cellular network technology with data rates surpassing 1-10 Gb/s. To meet the requirements / specifications for 5G communication systems, millimeter-wave (mm-Wave) bands such as 28 or 39 GHz are being utilized for the implementation of 5G NR systems. Recent efforts to deploy 5G NR service using mm-Wave technologies have led to notable transceiver IC designs based on bulk/SOI CMOS for 28 or 39 GHz [1]-[8].

In order to achieve high transmitter (TX) equivalent isotropic radiated power (EIRP) with array gain factor of 20log(N), the RF phased-array transceivers with N multiple channels are employed. However, there remains a difficulty in implementing a high-efficiency mm-Wave power amplifier (PA) which is necessary to operate at a large power back-off (PBO) condition (PAPR reaching ~10dB) to satisfy the strict linearity specifications in 5G systems (<3.5% EVM for 256QAM OFDM signals) [10]. A recent work of 28 GHz phased-array IC has focused on adopting a single TF-based parallel-combined Doherty topology for mm-Wave PA, enhancing PBO efficiencies within a compact die area [8]. Due to this approach, the work shows noteworthy performances, especially in terms of TX efficiency and linearity.

This work presents a new and successful example of a 39 GHz 2×16-channel (2-stream, 16-channel) phased-array

transceiver IC implemented in a 28-nm bulk RF CMOS process with superior TX performance and a lower power consumption compared to other recent state-of-the-art transceivers. PAs inside the transceiver adopt an improved version of the Doherty topology used in [9]. In addition, outstanding receiver (RX) performance with low noise figure (NF) and good error-vector modulation (EVM) results is demonstrated, owing to well-designed RX path inside the phased-array transceiver.

II. PHASED ARRAY TRANSCEIVER ARCHITECTURE

Fig. 1 shows the 39 GHz phased-array transceiver IC architecture of this work. The RF phased-array IC allows TX/RX operation with signal BW up to 800 MHz within the RF frequency band from 37 to 40 GHz (n260). The RFIC sends/receives IF signals to/from a baseband transceiver chip, which operates from 4.5 to 7.5 GHz. A local oscillator (LO) input is externally provided and multiplied by 3 for up/down conversion with up/down-mixers (UMIX/DMIX). In this paper, only the RF phased-array IC will be discussed.

The 2×16-channel RF phased-array IC is comprised of 2-stream beam-former arrays, with each stream containing 16 channels. Each channel in the RF phased-array IC includes a power amplifier (PA), a low noise amplifier (LNA), a passive phase shifter (PS) and a single-pole, double-throw (SPDT) type antenna switch (ASW) as a front-end circuit.

With the SPDT ASW, dual-polarization multi-input, multi-output (2-MIMO) operation becomes possible, reducing the number of RFICs and antenna arrays required for 5G NR FR2 handset / base-station applications [11], compared to previous well-known mm-Wave RFIC works [1], [6]-[7].

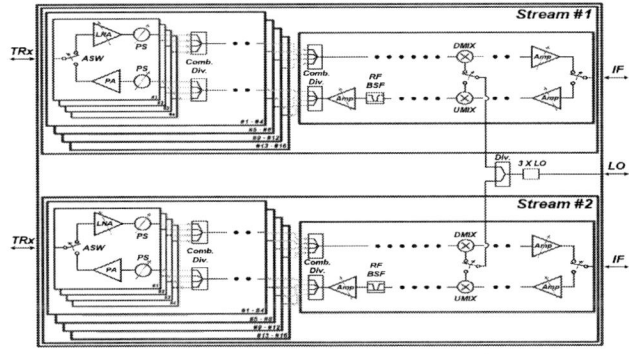

Fig. 1. 39 GHz phased-array transceiver IC architecture.

979-8-3503-2123-4/23 $31.00 © 2023 IEEE   273   2023 IEEE Radio Frequency Integrated Circuits Symposium

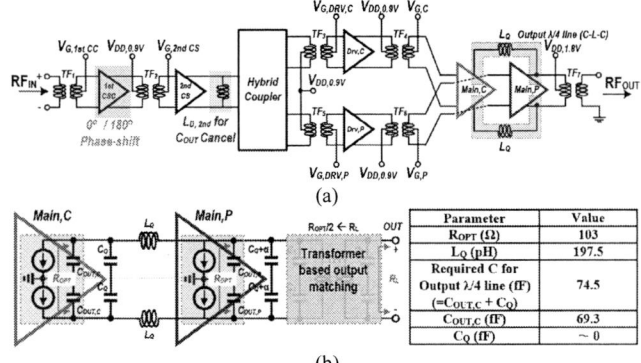

(a)

(b)

Fig. 2. Single transformer-based compact Doherty power amplifier: (a) A top-level schematic and (b) a detailed schematic and design parameters of output matching network.

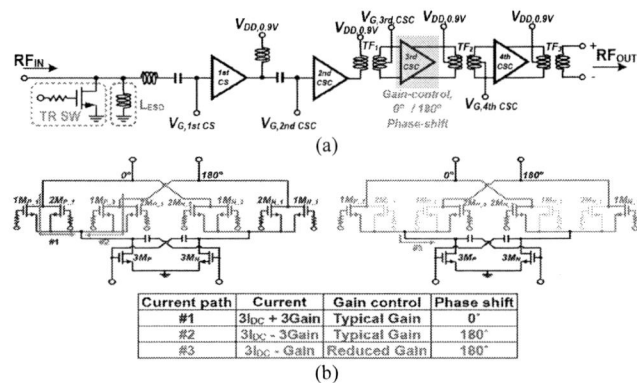

(a)

Current path	Current	Gain control	Phase shift
#1	$3I_{DC} + 3Gain$	Typical Gain	0°
#2	$3I_{DC} - 3Gain$	Typical Gain	180°
#3	$3I_{DC} - 3Gain$	Reduced Gain	180°

(b)

Fig. 3. Four-stage single-ended to differential type LNA in RX path: (a) A top-level schematic and (b) Gain-control / Phase-shift mechanism by the 3rd stage of LNA.

The PA adopts an improved version of the single transformer-based compact Doherty topology in [9], leading to superior TX performance in terms of good linearity and high efficiency. Detailed information regarding the PA will be discussed in Section III–(A).

The single-ended to differential type LNA is designed in 4-stage cascaded structure for a high gain and low NF, contributing to the good performance of RX path. Section III–(B) describes the designed LNA in detail.

The differential 6-bit PS is a cascade of L-C filter-type transition providing 5.625-degree phase resolution with an insertion loss less than 5.7 dB for all phase controls, except for 0°/180° phase –shift controls, which are included in PA/LNA.

The TX/RX paths show wide dynamic gain ranges, due to several RF / IF amplifiers and RF / IF attenuators in TX/RX paths. In addition, RF band-stop filter (BSF) is introduced between UMIX and an RF amplifier to eliminate LO feed-through (LOFT) in TX path.

III. CIRCUIT DESIGNS

A. Design of Transmitter Path

Fig. 2 (a) shows a top-level schematic of the 4-stage Doherty PA in TX path. A single TF-based parallel-combined Doherty PA output matching network has been adopted to achieve effective load modulation with a compact die area [9]. The Doherty PA includes a 2-stage pre-power amplifier (pre-PA), which adopts a cascode (CSC) topology for the 1st stage and a common-source (CS) topology for the 2nd stage. The 1st stage also implements 0°/180° phase shift, which eliminates the need for 180° phase shift stage in PS, reducing the size and insertion loss of the PS in TX path. The remaining 3rd and 4th stages of the Doherty PA consist of a hybrid coupler, common-source driver stages (Drv,C and Drv,P), and stacked main stages (Main,C and Main,P). All differential PA cells utilize a cross-coupled neutralization capacitor scheme to improve both gain and stability. The Doherty PA use 1.8 V DC supply voltage for the stacked main stages, which is perfectly separated from the 0.9 V DC supply voltage used by the remaining 1st, 2nd, and 3rd stages. This DC supply isolation of the final stage from the preceding stages at the envelope frequency improves the linearity of the Doherty PA.

Compared to a recent work of the 28 GHz phased-array IC based on the single transformer-based compact Doherty topology [8], this work has an additional inductor component ($L_{D,2nd}$) for resonating out intrinsic drain capacitance (C_{out}) of the 2nd stage, which is effective in improving efficiency and linearity of the TX path.

Fig. 2 (b) shows a detailed schematic and design parameters of output matching network. Since the intrinsic output capacitance of Main,C ($C_{out,C}$) is large enough for the implementation of Doherty load modulation in 39 GHz band (74.5 fF of required C for output $\lambda/4$ line ≈ 69.3 fF of $C_{out,C}$), additional passive output capacitor in the output of Main,C (C_Q) is no longer necessary, and therefore removed in this work.

In addition, the output matching network of PA is co-designed with SPDT ASW and input matching network of LNA for optimized TX performance. Due to these approaches, the proposed phased-array transceiver IC achieves outstanding TX performance compared to other mm-Wave FR2 (n260) phased-array transceiver IC works [1]-[5].

B. Design of Receiver Path

The RX path includes the 4-stage LNA, which is designed in single-ended to differential type for interface with SPDT ASW and following inner circuit blocks (PS, 4-way Wilkinson Combiners, etc.).

Fig. 3 (a) shows a top-level schematic of the 4-stage LNA. For the optimum NF of LNA, the input of the first stage has been co-designed with a transistor switch (TR SW) for RX ON/OFF operation and an ESD inductor (L_{ESD}). The first stage of LNA adopts a single-ended common-source (CS) topology, while 2nd, 3rd and 4th stages of LNA are implemented in cascode (CSC) topology for low leakage current in case of RX OFF by turning off the common-gate (CG) transistors.

Fig. 3 (b) shows the gain-control / phase-shift mechanism by the 3rd stage of LNA. In addition to main current paths by CG transistors in the 3rd stage ($1M_{P\_1}$, $2M_{P\_1}$, $2M_{N\_1}$, $1M_{N\_1}$), additional current paths for 180° phase shift ($1M_{P\_2}$, $2M_{P\_2}$, $2M_{N\_2}$, $1M_{N\_2}$) have been cross-connected between the source and the drain terminals of CG transistors. Under the condition of typical gain mode / 0° phase shift, only the main current paths are on, leading to 0° phase shift by the 3rd stage (current path #1 in Fig.3. (b)).

979-8-3503-2123-4/23 $31.00 © 2023 IEEE

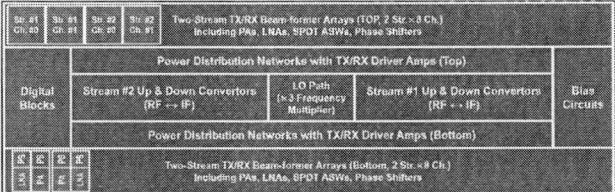

Fig. 4. Die micrograph of the 2×16-channel (2-stream, 16-channel) phased-array transceiver IC (die area: 9.8 mm × 6.3 mm).

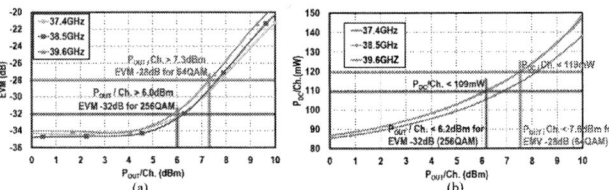

Fig. 5. Measured TX path (conduction) results for 16-channels (one-stream only) under 100 MHz 8-CC (total 800 MHz) 5G NR 64/256-QAM OFDM signals. (a) EVM and (b) P_{DC}/Ch. vs. P_{OUT}/Ch. for the frequencies of 37.4, 38.5, and 39.6 GHz.

In case of typical gain mode / 180° phase shift, only the additional current paths for 180° phase shift are on, leading to 180° phase shift by the 3rd stage (current path #2 in Fig.3. (b)). This implementation of 180° phase shift by the 3rd stage eliminates the need for 180° phase shift stage in PS, reducing the size and insertion loss of the PS in RX path. In case of reduced gain mode / 180° phase shift, two transistors in each of main/additional current paths are turned on, leading to gain reduction due to decreased current gain by the 3rd stage (current path #3 in Fig.3. (b)). From this gain-control / phase-shift mechanism, the LNA supports a gain range of 24 dB with a 12 dB gain step, increasing the dynamic range of the RX system. Along with LNA, an IF amplifier and RF / IF attenuators also contribute to large dynamic range of RX path, supporting 28 to 78 dB total RX gain range with 1 dB gain step.

C. Design of Blocks for LO Feed-Through Rejection

To satisfy the spectrum emission mask, the unwanted LO feed-through (LOFT) has been dealt with by adopting the following design approaches. The UMIX performs LOFT calibration by tuning the bias current of the double balanced mixer. To compensate both magnitude and phase of the LOFT, I/Q quadrature LO signal which is coming from the hybrid coupler in LO path drives UMIX. In addition, the RF BSF, which is a 3rd order Chebyshev type, has been included in the TX path for further LOFT suppression. Cap banks are employed in the RF BSF to cover the LO frequency range achieving greater than 30dBc out-of-band rejection.

IV. IMPLEMENTATION AND MEASUREMENT

Fig. 4 shows the die micrograph of the proposed 39 GHz phased-array transceiver IC. The 39 GHz transceiver with 2×16 channels (2-stream, 16-channel) is fabricated using a 28-nm bulk CMOS technology with flip-chip packaging, within the die area of 9.8 × 6.3 mm². The implemented IC is measured in a conduction mode test using a 16-way combiner on a customized test board and all input/output losses due to combiner/package are de-embedded up to the reference points.

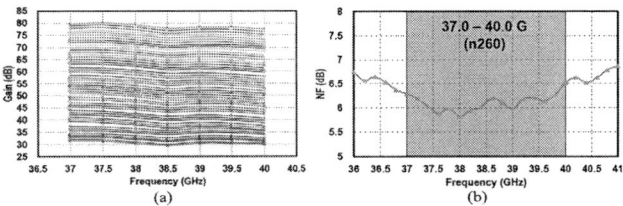

Fig. 6. Frequency response of the measured RX path. (a) Gain range with 1-dB gain step (b) NF at RX gain mode = 60 dB

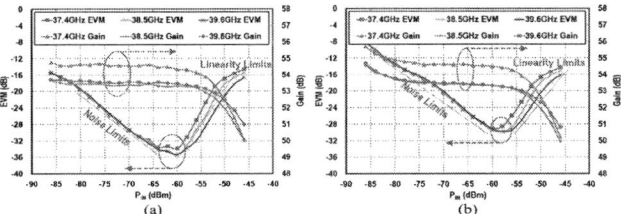

Fig. 7. Measured RX path (conduction) results for 16 channels (one-stream only) for the frequencies of 37.4, 38.5, and 39.6 GHz under 5G NR 64-QAM OFDM signals of (a) 100 MHz 1-CC and (b) 100 MHz 8-CC (total 800 MHz).

The one carrier component (CC) of the NR signal has 120 kHz subcarrier spacing, 64 QAM, 10 dB PAPR, and 100 MHz BW. The eight CCs of the NR signal are aggregated for the 800 MHz BW test.

A. TX Measurement

Figs. 5 (a) and (b) show the full 16-element (one-stream only) TX measurement results of the proposed transceiver under the modulation signal of the 100 MHz 8-CC (total 800 MHz) 5G NR 64/256-QAM OFDM. According to 5G NR standard by 3GPP Release 17 [10], the required EVM level of NR carrier for 64-QAM / 256-QAM inside FR2 Base Station is -28 dB (8%) / -32 dB (3.5%) respectively. In this work, the target TX EVM of -28/-32 dB for 64-/256-QAM is determined by assigning extra margins of 6/3 dB to the 5G NR FR2 specifications. At the frequencies from 37.4 to 39.6 GHz for fully occupied 800 MHz BWs, the TX path shows remarkable performance, exhibiting P_{OUT} per channel (P_{OUT}/Ch.) of >9.3 dBm and DC power consumption per channel (P_{DC}/Ch.) of <118 mW at EVM of -28 dB for 64-QAM OFDM. This P_{OUT} is calculated by de-embedding a loss of 2.0 dB from TX/RX switch and package. Furthermore, the transceiver IC shows superb TX performances by supporting the 256-QAM with P_{OUT}/Ch. >8.0 dBm at EVM of -32dB, consuming P_{DC}/Ch. <109 mW.

B. RX Measurement

Fig. 6 (a) shows the gain step response of the measured RX path across the RF frequency from 37 to 40 GHz, supporting 28 to 78 dB total RX gain range with 1 dB gain step. The system NF performance of 5.8 to 6.5 dB (3.8 to 4.5 dB when de-embedding a loss of 2.0 dB from TX/RX switch and package) over the frequency range of 37.0-40.0 GHz is shown in Fig. 6 (b) at RX gain = 60 dB. Figs. 7 (a) and (b) show the 16-channel RX Gain/EVM performances over the input power range for the frequencies of 37.4, 38.5, and 39.6 GHz under 100/800 MHz 5G NR 64-QAM OFDM signals. Different gain mode settings are used for different frequencies and signal BWs.

Table 1. Comparison with the state-of-the-art mm-wave (FR2) phased-array transceivers.

	This work	[1] ISSCC2020	[2] JSSC2020	[3] RFIC2019	[4] JSSC2022		[5] ISSCC2022	
Technology	**28-nm CMOS**	28-nm CMOS	65-nm CMOS	28-nm CMOS	65-nm CMOS		28-nm SOI CMOS	
RF Freq. (GHz)	**37.0 - 40.0**	37.0 - 40.0	39	37.0 - 40.0	26.5 - 29.0	35.25 - 38.0	24.25 - 29.5	37.0 - 40.0
IF Freq. (GHz)	**4.5 - 7.5**	10.8 - 11.2	3.9	4 - 6.8, 10.56	0.1		8.0 - 10.0	
Modulation & BW (MHz)	**64QAM OFDM Up to 800**	64QAM OFDM Up to 800	64QAM OFDM Up to 400	64QAM OFDM Up to 100	64 QAM 1.5 G/s		DFT-s-OFDM (TX), CP-OFDM (RX)	
# of Arrays	**32 (Dual streams)**	16	4	8	8		16	
# of PS Bit (Resolution)	**6 (5.625 deg.)**	4 (22.5 deg.)	3+10 (0.05 deg.)	5 (11.25 deg.)	-		-	
TX Gain (dB)	**25 - 70**	30 - 60	-	45 - 48	43.5	40	-	
PA P_{sat} (dBm)	**> 17**	>16.5	15.5	-	15.5	15.6	-	
TX P_{out}/Ch. (dBm)	**6 (8*) / 7.3 (9.3*)**	6* / 8.8*	2.5+	3.6	1.2 - 3.3 @ P1dB	-	4.5+	3.0+
TX EVM (dB)	**-32 / -28**	-34 / -27	-27+	-24.6	-32.57	-27.7	-27.1	-26
TX P_{dc}/Ch. (mW)	**109 / 118**	105 / 115	375	339 @ P1dB	241.3		265	275
TX Eff./Ch. (%)	**3.7 (5.8*) / 4.6 (7.2*)**	3.8* / 6.6*	-	-	-		1.1+	0.7+
RX Gain (dB)	**28 - 78**	16 - 59	33**	36.9 - 41.9	44	37	-	
RX NF (dB)	**5.8 - 6.5 (3.8* - 4.5*)**	4.2* - 4.6*	7**	6 - 7.6	7.9	8.8	4.3 - 6.4	5.3 - 6.8
RX EVM (dB)	**-29.8 / -35.4 (800/100 MHz)**	-34.8 / -37.9 (800/100 MHz)	-	-29.8	-27.9	-26.3	-	
RX P_{DC}/Ch. (mW)	**42.3**	39.0	125.0	78.5	98.8		43	51
Die Area (mm^2)	**62.0**	30.0	12	17.2	12.7		25.1	

*ASW excluded, **LNA only results, +Graphically estimated.

The best EVM levels for 100/800 MHz are -33.9 / -29.1 dB (at 37.4 GHz), -35.4 / -29.8 dB (at 38.5 GHz), and -35.4 / -29.8 dB (at 39.6 dB), respectively. As the signal BW decreases from 800 to 100 MHz, the EVM level improves due to better output SNR of 100 MHz signal BW. This EVM improvement with 100 MHz signal BW becomes more evident at low input power condition, showing 9 dB lower EVM compared to 800 MHz signal BW, which is consistent with the theoretical expectation.

C. Comparison with the other state-of-the-art works

Table 1 summarizes this work and shows comparisons with other state-of-the-art mm-Wave FR2 (n260) phased-array transceivers with number of channels greater than 4. This work outperforms other mm-Wave phased-array ICs in terms of TX characteristics, delivering the highest TX efficiency of >7.2/>5.8% at EVM of -28/-32 dB under the modulation signal of the 100 MHz 8-CC (total 800 MHz) 5G NR 64/256-QAM OFDM. Furthermore, this work shows outstanding RX performance, exhibiting the lowest NF of 3.8 to 4.5 dB (de-embedding a loss of 2.0 dB from TX/RX switch and package) over the frequency range of 37.0-40.0 GHz among the state-of-the-art mm-Wave FR2 (n260) phased-array transceivers for 5G NR applications.

V. CONCLUSION

A 39 GHz 2×16-channel phased-array IC for 5G application is presented in this paper. The transceiver architecture of the RFIC adopts a single transformer-based Doherty PA in its TX path, demonstrating excellent TX performance in terms of high average output power per channel of >8/>9.3 dBm at EVM of -32/-28 dB under low DC power consumption per channel of <109/<118 mW. In addition, good RX performance is observed, exhibiting the low system NF of 5.8 – 6.5 dB, and an EVM of -29.8 dB under low DC power consumption per channel of <42.3 mW. With these outstanding TX/RX performances, the proposed phased-array IC becomes a critical component for 5G NR FR2 (n260) communication systems.

ACKNOWLEDGMENT

The authors would like to thank all RF Part members of Modem Group in Samsung Electronics for their valuable technical discussions and supports.

REFERENCES

[1] H. -C. Park et al., "A 39GHz-band CMOS 16-channel phased-array transceiver IC with a companion dual-stream IF transceiver IC for 5G NR base-station applications," IEEE ISSCC, pp. 76-78, 2020.

[2] Y. Wang et al., "A 39-GHz 64-element phased-array transceiver with built-in phase and amplitude calibrations for large-array 5G NR in 65-nm CMOS," in IEEE JSSC, vol. 55, no. 5, pp. 1249-1269, May 2020.

[3] A. G. Roy et al., "A 37-40 GHz phased array front-end with dual polarization for 5G MIMO beamforming applications," IEEE RFIC, pp. 251-254, 2019.

[4] S. Mondal, L. R. Carley and J. Paramesh, "Dual-band, two-layer millimeter-wave transceiver for hybrid MIMO systems," in IEEE JSSC, vol. 57, no. 2, pp. 339-355, Feb. 2022.

[5] A. Verma et al., "A 16-channel, 28/39GHz dual-polarized 5G FR2 phased-array transceiver IC with a quad-stream IF transceiver supporting non-contiguous carrier aggregation up to 1.6GHz BW," IEEE ISSCC, pp. 1-3, 2022.

[6] Y. Cho et al., "A 16-element phased-array CMOS transmitter with variable gain controlled linear power amplifier for 5G new radio," IEEE RFIC, pp. 247-250, 2019.

[7] Y. Yoon et al., "A highly linear 28GHz 16-element phased-array receiver with wide gain control for 5G NR application," IEEE RFIC, pp. 287-290, 2019.

[8] H. -C. Park et al., "Single transformer-based compact Doherty power amplifiers for 5G RF phased-array ICs," in IEEE JSSC, vol. 57, no. 5, pp. 1267-1279, May 2022.

[9] S. Kim et al., "A 24.5–29.5GHz Broadband Parallel-to-Series Combined Compact Doherty Power Amplifier in 28-nm Bulk CMOS for 5G Applications," IEEE RFIC, pp. 171-174, 2021.

[10] 3GPP 5G NR, "Base Station (BS) radio transmission and reception," 3GPP TS 38.104 version 17.7.0 Release 17, Oct 2022.

[11] B. Sadhu et al., "7.2 A 28GHz 32-element phased-array transceiver IC with concurrent dual polarized beams and 1.4 degree beam-steering resolution for 5G communication," IEEE ISSCC, pp. 128-129, 2017.

RTu2C-3

A 14-nm Low-Cost IF Transceiver IC with Low-Jitter LO and Flexible Calibration Architecture for 5G FR2 Mobile Applications

Wanghua Wu[#], Jeiyoung Lee[*], Pak-Kim Lau[#], Taeyoung Kang[#], Kim Kiu Lau[#], Si-Wook Yoo[#], Xingliang Zhao[#], Ashutosh Verma[#], Ivan Siu-Chuang Lu[#], Chih-Wei Yao[#], Hou-Shin Chen[#], Gennady Feygin[#], Pranav Dayal[#], Kee-Bong Song[#], Sangwon Son[#]

[#]Samsung Semiconductor, Inc., USA
[*]Samsung Electronics Co., Ltd., Korea
wanghua.wu@samsung.com, jeiyoung.lee@samsung.com

Abstract—We present a low-cost dual-stream IF transceiver IC (IFIC) for 5G mm-wave mobile applications. It up/down-converts the baseband signal to an intermediate frequency of 8.4–10 GHz and forms a heterodyne transceiver system together with beamforming ICs to support all FR2 bands. The IFIC features a compact transceiver RF circuitry, low-jitter reconfigurable LO suitable for 256-QAM and non-contiguous carrier aggregation, and an integrated MCU in the digital baseband for flexible calibration and control of both transceiver ICs. The IFIC is implemented in 14-nm FinFET and occupies 16.2 mm². The overall chain IPN measured at 39-GHz band is as low as 114 fs. Thanks to the flexible calibration architecture, digital-pre-distortion (DPD) is demonstrated in TX, which allows for >1 dB of increase in EIRP for both DFT-s- and CP-OFDM signals at EVM of 5.5% at 39-GHz band.

Keywords— intermediate frequency (IF), transceiver, 5G mm-wave, low cost, low jitter, local oscillator (LO), phase noise, digital-pre-distortion (DPD), calibration, testing.

I. INTRODUCTION

The time is ripe for many commercialized products and services to ride the wave of unparalleled high-speed connectivity and minimal latency enabled by millimetre-wave (mmW) 5G, i.e., frequency range 2 (FR2). However, making 5G FR2 affordable for every smart phone remains a challenge. The die size and the power consumption of the transceiver chipset have to be minimized. Meanwhile, the implementation needs to be scalable to support higher data rate via wider aggregated signal bandwidth (BW), higher-order quadrature amplitude modulation (QAM), and multiple-input multiple-output (MIMO) operation.

Fig. 1 shows a mobile 5G mmW system, employing an intermediate frequency (IF) interface for low power and better scalability. It consists of a modem, an IF transceiver IC (IFIC), and multiple phased-array beamforming ICs (BFIC), integrated with antenna module, placed on the side or the back of the phone for enhanced spherical coverage. While previous publications [1]–[3] have demonstrated 5G mmW front-ends, a key part of the system, the IF transceiver, including baseband (BB), PLLs, and mixed signal components, has been under addressed. The combination of wide aggregated signal BW (e.g., >1 GHz) and high-order modulation such as 256-QAM presents a difficult challenge in both design and test.

This paper focuses on the IFIC to address the aforementioned challenges. For power and performance

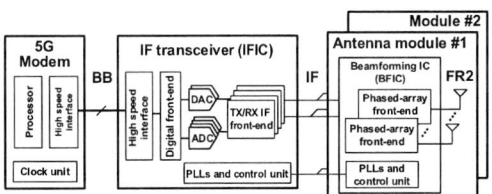

Fig. 1. Block diagram of a mobile 5G mmW system.

scalability, a reconfigurable dual-PLL local oscillator (LO) is presented, providing a low-power mode for 64-QAM of up to 800-MHz signal BW. It also offers a high performance mode either for intra-band non-contiguous carrier aggregation (NC-CA) of 1.4-GHz BW or to reduce the overall chain's integrated phase noise (IPN) by ~1.5 dB. To ease the transceiver testing and calibration, a flexible architecture utilizing an integrated micro-controller unit (MCU) is incorporated in this 14-nm IFIC, realizing TX digital-pre-distortion (DPD) together with a companion phased-array module. With DPD, the 39-GHz TX equivalent isotropically radiated power (EIRP) is increased by more than 1 dB for both discrete Fourier transform spread OFDM (DFT-s-OFDM) and cyclic prefix-OFDM (CP-OFDM) signals at error vector magnitude (EVM) of 5.5%.

II. IFIC ARCHITECTURE

Fig. 2 shows the system diagram of the IF transceiver. It can interface to two phased-array modules with concurrent operation in low-band (24.25–29.5 GHz) and high-band (37–43.5 GHz). For each module, there are two transceiver chains, i.e., H- and V-path, providing IF_H and IF_V for dual polarization MIMO. Transmitters (TX) are shared between two modules to halve the TX front-end area at the expense of 0.3-dB extra loss at IF due to the additional loading of the driver amplifier (DA). The receiver (RX) has four signal chains, each of 800-MHz maximum signal BW, which can operate concurrently with two dual-polarized antenna modules to improve SNR. IF-PLL1 is shared between TX and RX, as 5G New Radio (NR) is a time division duplex (TDD) system. Moreover, an IF-PLL2 that is identical as IF-PLL1 is added to support a low-jitter mode for the upcoming 256-QAM FR2 downlink (DL), which is elaborated in Section III.

As cellular providers do not have 800-MHz contiguous spectrum available, dual-LO receive mixers are employed to support intra-band NC-CA up to 1.4 GHz. This approach halves

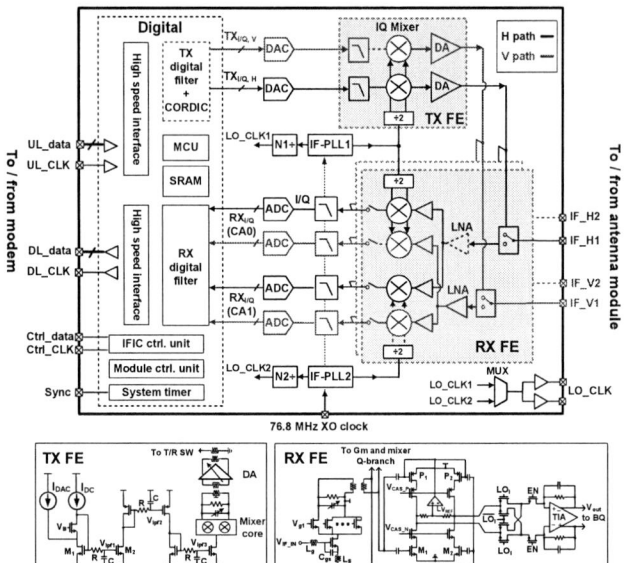

Fig. 2. IFIC Architecture and TX/RX front-end (FE) circuits.

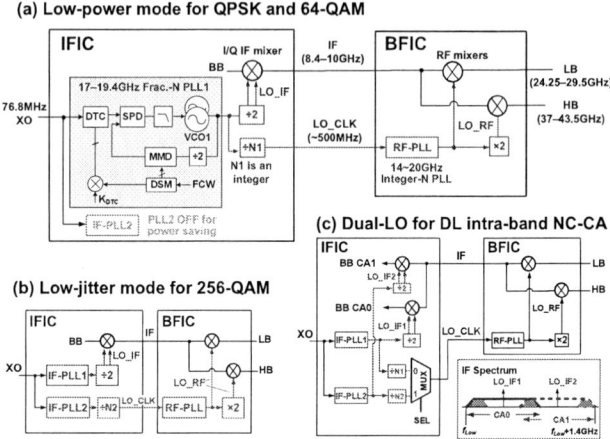

Fig. 3. LO configurations for various operational scenarios.

the sampling rate requirement of the ADC and BW of the BB filters at a fractional cost of one extra RX mixer in each RX chain. Existing IF-PLL2, BB filters, and ADCs are reconfigured to support this mode.

In contrast to [3], where analog I/Q interfaces requiring burdensome PCB-matching are used, the presented IFIC interfaces to the modem via a fully digital interface for data and controls. To ease the IFIC characterization, an MCU-based control and calibration architecture is integrated, which also enables faster and simultaneous control of both IFIC and BFICs, as well as flexible transceiver calibration.

The TX and RX front-end circuits are also shown in Fig. 2. In TX, a 10-bit current DAC drives a third-order current mode low-pass-filter (LPF) for image suppression. LPF also provides DC bias for the active mixer, which is then followed by a class-AB driver amplifier. In each RX, after one shared gain stage (i.e., LNA), push-pull Gm buffers are allocated for each I- and Q-path of the down-conversion mixer with 50% duty-cycle LOs. Similar to what is presented in [2], a 3rd-order butterworth LPF is realized using transimpedance amplifier (TIA) and Tow-Thomas Biquad with a configurable 3-dB bandwidth of ~450 MHz.

III. RECONFIGURABLE LO SCHEME

Fig. 3 depicts the complete LO topology for the heterodyne transceiver system. The IFIC and BFIC interface at an IF of 8.4–10 GHz for all the FR2 bands so that it is high enough to avoid interference with other radios in a mobile phone, including cellular sub-6 GHz bands, UBW (mandatory channel 9: 7.7–8.3 GHz), and Wi-Fi, and low enough to reduce the cost of the flexible cable connecting to the two chips. The IFIC also generates the reference clock for PLLs in the BFIC, i.e., LO_CLK in Fig. 3. The LO_CLK is ~500 MHz so that its strong 2nd and 3rd harmonics will not desensitize the global navigation satellite system (GNSS) receiver. Such a high reference clock rate allows the use of a wider PLL BW (e.g., 10 MHz) to suppress VCO noise, thereby effectively addressing

the stringent design tradeoff between VCO tuning range and phase noise (PN). Thus, a low-cost LO_RF generation can be realized in the BFIC using only one integer-N PLL (i.e., RF-PLL in Fig.3) with a single VCO tunable from 14 to 20 GHz to support both HB and LB. For HB, an additional frequency doubler multiplies the RF-PLL output to obtain LO_RF at ~30 GHz for the RF mixers. The chain IPN is the sum of the PN of LO_IF (used for IF mixers) and LO_RF, both of which are dominated by the IF-PLLs in the IFIC.

Fig. 3 shows three different LO configurations that can be supported. Low-power mode in Fig. 3a is the default mode for 64-QAM DL/UL up to 800-MHz signal BW. IF-PLL1 is used for the generation of both LO_CLK and quadrature LO_IF via divide-by-32 and divide-by-2, respectively. IF-PLL2 is powered down for power saving. Both PLLs use a digital-to-time converter (DTC) based fractional-N PLL topology for low jitter and high figure-of-merit [4]. To reduce the LO power consumption and die size, the VCO oscillates directly at ~18 GHz, eliminating the frequency tripler used in [1] and [4]. Consequently, the required DTC delay range is reduced by 33% compared to a 6-GHz PLL, lowering fractional spurs and in-band PN. To ease the low-PN VCO design at 18 GHz, the required VCO tuning range is kept <15% in the proposed LO scheme.

For 256-QAM, a lower chain IPN is needed. To reduce the rms jitter of IF-PLL1, VCO1 supports a dual-core mode by employing two identical NMOS cross-coupled LC-VCO in parallel, which reduces VCO noise by 3 dB at the expense of doubled VCO current. Further increase in the number of VCO cores is not cost-effective, as VCO1 only contributes ~30% of the total chain IPN, and the rest are from the crystal oscillator (XO: ~30%), IF-PLL1 in-band circuitry (~25%), and the RF-PLL (~15%). Besides, it is difficult to make a multi-core VCO reconfigurable to reduce power consumption when a lower-order QAM modulation is used. Alternatively, we propose to use two PLLs to generate LO_IF and LO_CLK, respectively, as shown in Fig. 3b. Consequently, the PN of the LO_IF and LO_RF are mostly uncorrelated (except the noise from the XO), leading to a maximum 3-dB lower chain IPN.

In intra-band NC-CA scenario, dual-LO receive mixers extend the RF BW from 800 MHz to 1.4 GHz, as 200-MHz

979-8-3503-2123-4/23 $31.00 © 2023 IEEE   278

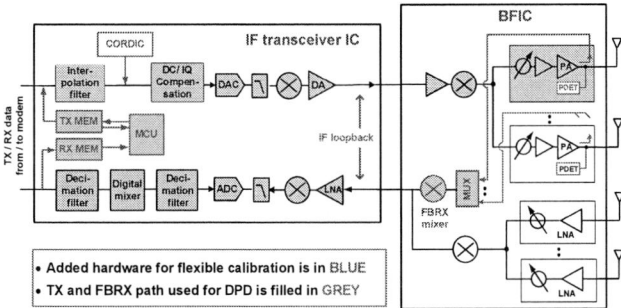

Fig. 4. MCU-based flexible calibration architecture.

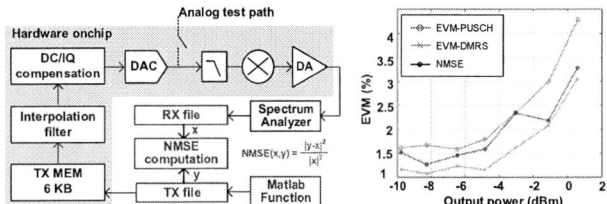

Fig. 5. Memory-based TX NMSE measurement.

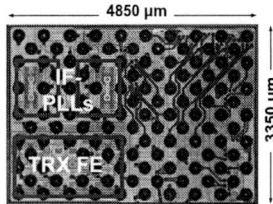

Fig. 6. Chip micrograph.

overlap is reserved between CA0 and CA1 in order to support arbitrary non-contiguous spectrum allocation, as shown in Fig. 3c. LO_IF1 and 2 are set to the center of the CA0 and CA1, respectively, to reduce the required VCO and IF range. LO_CLK is configured from PLL1 if uplink (UL) spectrum is allocated within CA0; otherwise PLL2 path is selected. This topology simplifies the TX LO distribution, as only LO_IF1 is needed for the TX mixer.

IV. FLEXIBLE CALIBRATION ARCHITECTURE

The low-cost flexible calibration architecture in the IFIC is shown in Fig. 4. It consists of a RISC-V MCU with floating-point unit and dedicated RX and TX memory (MEM) of 6 KB, each with a maximum access clock rate of 245.76 MHz. Its chip area is negligible compared to the digital BB filter. For DL, quadrature sample data can be captured from various digital BB filter stages and stored into the RX MEM, i.e., RX capture. The captured data can be read out by the MCU for onchip processing or transferred to external for post-processing. Thus, RX chain performance can be characterized by applying a modulated signal at RF input and analyzing RX MEM data. For UL, a desired data pattern generated by test software can be written into the TX MEM, and injected into the TX digital BB filter input, i.e., TX playback. Moreover, both TX playback and RX capture can operate simultaneously, forming a feedback receiver (FBRX) to support various applications, such as DPD and transceiver performance built-in test by looping back TX output to RX.

A. DPD calibration using FBRX

DPD is a well-known technique used to compensate for power amplifier (PA) non-linearity. AM/AM and AM/PM lookup tables (LUTs) are constructed and applied to the digital baseband signals to pre-compensate for the PA non-linearity. The aforementioned FBRX architecture is used to implement DPD calibration for a companion dual-polarized antenna module. Fig. 4 shows the two-chip FBRX architecture. In the IFIC, one IF path (either H or V) is used to transmit and the other is used as a feedback receiver. In the BFIC, each PA is loaded by a power detector (PDET) and a directional coupler to sample the PA signal to a dedicated feedback receiver path, which is shared among all the PA paths. The main purpose of DPD calibration is to learn the PA characteristic by sending a known training signal via the TX MEM playback. By comparing the training waveform and the combined the PA characteristics loop-backed to the RX MEM, PA characteristic can be assessed and used to generate DPD LUTs. A 100-MHz CP-OFDM QPSK signal is used for DPD training, which has a peak-to-average power ratio (PAPR) of ~10 dB to cover a wide range of PA responses.

For each polarization, one antenna port is turned on at a time to get PA loopback signal for that particular path. It is fair to assume that the signals from all paths are coherently combined over the air, and hence we correct the timing and the phase of each path and scale them according to the individual PDET read. The combined FBRX signals are used to construct polynomial models for the PA and generate AM/AM and AM/PM LUTs. These post processing and DPD LUT generation operations are all done by the MCU. The total DPD calibration time is less than one second and can be done during the factory calibration and then stored in non-volatile memory.

B. Memory-based TX NMSE measurement

To characterize the TX of the IFIC, conventionally, an external signal generator sends a 5G-NR signal in voltage domain to TX BB via an analog test path, and the EVM is measured at the IF output. However, the actual signal path is in current mode generated by DAC, as shown in Fig. 2. To improve test coverage, TX MEM playback can be used, which exercises the complete TX signal path. However, it is not possible to measure the actual NR waveform based EVM, as it requires a much larger sized MEM to store the 5G NR waveform. Alternatively, a close approximation can be obtained using time-domain normalized-mean-squared-error (NMSE). The test setup and results are illustrated in Fig. 5. Matlab generates the specified non-standard compliant signal (one symbol of 100-MHz, QPSK-OFDM without cyclic prefix is used), which is loaded to the TX MEM and transmitted repeatedly. A spectrum analyzer captures the time-domain waveform at IF output and NMSE is computed in Matlab, which involves time/frequency/phase correction and a multi-tap equalizer.

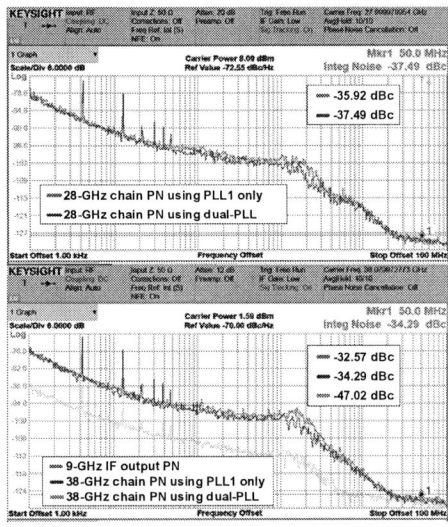

Fig. 7. 28- and 38-GHz chain PN w/wo dual-PLL.

To verify the NMSE measurement, a 3GPP compliant CP-OFDM QPSK signal is sent to the analog test path, and the EVM of 5G NR physical uplink shared channel (PUSCH) and the demodulation reference signal (DMRS) are measured by the test equipment and plotted in Fig. 5. For fair comparison, a repeated one symbol non-standard compliant signal is loaded via the same test path and NMSE is computed, which falls between the two EVM curves. As good correlation between NMSE and EVM is demonstrated, the former can be used for TX testing to improve test coverage.

V. MEASUREMENT RESULTS

The IF transceiver is manufactured in a 14-nm FinFET process and occupies 16.2 mm². The chip micrograph is shown in Fig. 6. When connected with the front-end module and the modem, the system achieves a maximum throughput of 5 Gb/s at 39-GHz band in the configuration of 2×2 MIMO (8CC DL, 2CC UL, DL/UL ratio is 80/20).

The IPN is measured at the IF output as well as at the cascaded BFIC output, and plotted in Fig. 7. In low-power mode, the overall IPN values on 28- and 38-GHz chains are -35.9 dBc and -32.6 dBc, respectively, integrated from 1 kHz to 100 MHz, corresponding to an EVM floor of 2.3% and 3.3%, respectively. The measured IPN at 9-GHz IF is -47 dBc. At low-jitter mode, IPN is improved by ~1.6 dB for both LB and HB at the expense of 38 mW more power consumed by IF-PLL2. The chain rms jitter of 114 fs (i.e., -34.3 dBc at 38 GHz) is sufficiently low for 256-QAM.

Fig. 8 shows the transmitter EVM results measured on a phone prototype in non-signaling mode. DPD calibration firmware is implemented in the MCU using the FBRX calibration path. TX operates in dual-polarization and TX EVM is measured at either polarization (only H-pol is shown in Fig. 8) via an over-the-air setup. With DPD, the EIRP can be increased by at least 1 dB for both DFT-s- and CP-OFDM signals, while keeping EVM ≤5.5% at 39.95-GHz channel. The performance parameters of the IFIC are summarized in Table 1. Compared to the recently reported IF transceivers, this work achieves a

Table 1. Performance summary and comparison to prior arts.

	[2] A. Verma ISSCC 2022	[3] H.C. Park ISSCC 2020	[5] B. Jann ISSCC 2019	This work
Technology	14 nm FinFET	65 nm CMOS	28 nm CMOS	14nm FinFET
Chains	4 TX / 4 RX	2 TX / 2 RX	1 TX / 1 RX	4 TX / 4 RX
CA	Intra-band 2-CA	Not supported	2	Intra-band 2-CA
ADC BW/ENOB	400 MHz / 8	Off-chip ADC	200 MHz / 7.7	400 MHz / 8
Aggregated BW	1.6 GHz	800 MHz	800 MHz	1.4 GHz
RMS jitter at IF / chain	NA / 205 fs	Off-chip PLL	336 fs / NA	112 fs / 139 fs →114 fs with dual-PLL scheme
TX Linearity	-37.3 dB EVM @-9.8 dBm (64-QAM OFDM)	-40.4 dB EVM @-10 dBm (64-QAM OFDM)	-31dB EVM @5 dBm (64-QAM OFDM)	-40 dB EVM @-9 dBm (64-QAM OFDM)
RX noise figure	10 dB	NA	15.5 dB@10.56 GHz	9.5 dB
RX IP1dB	NA	NA	NA	-27 dBm @ max. gain
TX Power consumption	291 mW (1×1)	325 mW (SISO)	596 mW	215 mW (1×1)
	438 mW (2×2)	490 mW (MIMO)	NA	373 mW (2×2)
RX Power consumption	381 mW (1×1, 2-CA)	455 mW (SISO)	446 mW	302 mW (1×1, 2-CA)
	542 mW (2×2, 2-CA)	700 mW (MIMO)	NA	450 mW (2×2, 2-CA)
Die size (mm²)	20.1	33.4	NA	16.2

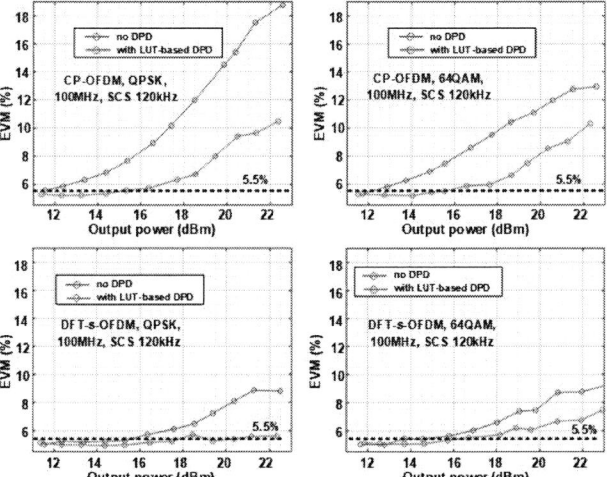

Fig. 8. TX EVM of H-pol. at 39.95 GHz w/wo DPD.

smaller die size, lower IPN (>3 dB lower), while consuming ~25% less power in both transmit and receive modes (ADC/DAC and LO power is included).

VI. CONCLUSION

A 14-nm low-cost dual-stream IF transceiver for 5G FR2 is presented. The IFIC achieves low power consumption and low jitter thanks to the reconfigurable LO scheme implemented in it. Moreover, a built-in flexible calibration architecture facilities low-cost testing and calibration.

REFERENCES

[1] J.D. Dunworth et al., "A 28GHz Bulk-CMOS Dual-Polarization Phased-Array Transceiver with 24 Channels for 5G User and Base-station Equipment", *ISSCC*, pp. 70–72, Feb. 2018.

[2] A. Verma et.al. "A 16-channel, 28/39 GHz dual-polarized 5G FR2 phased-array transceiver IC with a quad-stream IF transceiver supporting aggregate carrier aggregation upto 1.6 GHz BW", *ISSCC*, pp. 1–3, Feb. 2022.

[3] H. C. Park et.al. "A 39GHz-band CMOS 16-channel phased-array transceiver IC with a companion dual-stream IF transceiver IC for 5G NR base-station applications", *ISSCC*, pp. 76–78, Feb. 2020.

[4] W. Wu et al., "A 5.5–7.3GHz Analog Fractional-N Sampling PLL in 28-nm CMOS with 75fsrms Jitter and -249.7dB FoM", in *IEEE RFIC Symp.*, pp. 52–55, June 2018.

[5] B. Jann et.al. "A 5G Sub-6GHz Zero-IF and mm-Wave IF Transceiver with MIMO and Carrier Aggregation", *ISSCC*, pp. 352–354, Feb. 2019.

RTu2C-4

A Quad-Band RX Phased-Array Receive Beamformer with Two Simultaneous Beams, Polarization Diversity, and 2.1–2.3 dB NF for C/X/Ku/Ka-Band SATCOM

Zhaoxin Hu, Oguz Kazan, Gabriel M. Rebeiz

University of California San Diego, USA

{z1hu,okazan,grebeiz}@eng.ucsd.edu

Abstract — This paper presents a wideband 16-channel dual-beam receive (RX) phased-array beamformer in a 90-nm SiGe BiCMOS process. Radio-frequency (RF) beamforming is implemented, and each channel has 5-bit phase control and 25 dB gain control. The dual-beam channel outputs are combined with two wideband Wilkinson networks. The chip has 27.3 dB electronic gain with a 3.4–28.7 GHz 3-dB bandwidth and a 2.1–2.3 dB noise figure (NF) up to 21 GHz. To the authors' knowledge, this work achieves the widest operating bandwidth among RX beamformers for satellite communication (SATCOM). Application areas are in phased arrays for C/X/Ku/Ka-band SATCOM ground terminals.

Keywords — beamformer, low-noise amplifier (LNA), phased array, receiver, satellite communication (SATCOM), SiGe BiCMOS, wideband.

I. INTRODUCTION

Low-earth orbit (LEO) and medium-earth orbit (MEO) satellite constellations provide internet coverage to under-resourced areas when the cellular infrastructure is too expensive to implement. Electronically-scanned arrays (ESA) equipped with silicon beamformer chips are the workhorse of the next-generation SATCOM ground terminals as they can track the satellite with low latency and with no mechanical movement. Commercial SATCOM systems utilize C, X, Ku, and Ka bands (Fig. 1), and a terminal supporting all these bands in a single ESA offers a low-cost solution to the end users. Prior work has demonstrated SATCOM RX beamformers with competitive performance [1]–[6], but with frequency coverage of only one band.

This paper introduces a 16-channel 3.4–28.7 GHz dual-beam RX phased-array beamformer in 90-nm SiGe BiCMOS. The chip offers complete reconfigurability, allowing users to synthesize two independent beams at C, X, Ku, and Ka bands, or any combination of bands (e.g. C/X and Ku/Ka). Each beam can be in any polarization (linear, rotated-linear, LHCP, or RHCP). Both beams can also operate at the same band for make-before-break handover between two satellites. The beamformer demonstrates very low NF (<2.3 dB up to 21 GHz) which is important to realize a high gain-to-noise-temperature (G/T) performance.

II. DESIGN

Fig. 2(a) presents the beamformer block diagram. The chip has 8 inputs intended for a 2×2 quad-antenna unit cell with dual polarizations (Fig. 2(b)). An LNA and a

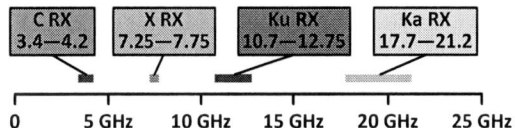

Fig. 1. C, X, Ku, and Ka SATCOM frequency bands.

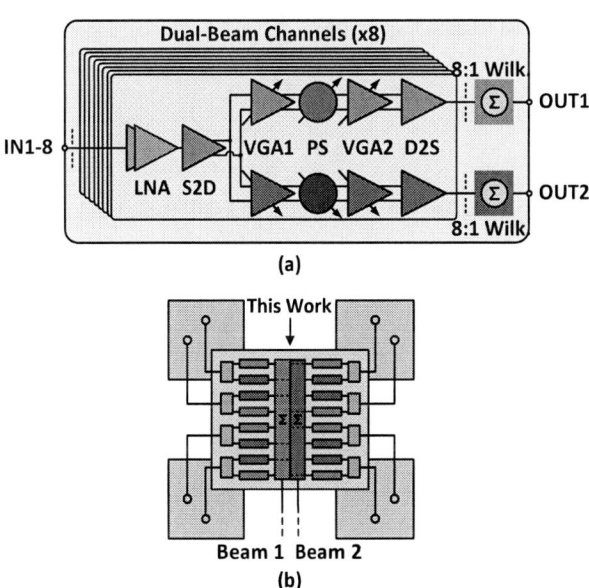

Fig. 2. (a) Beamformer block diagram. (b) 2×2 quad-antenna unit cell diagram.

single-ended-to-differential active balun (S2D) are placed at each antenna input and shared between two channels. Each channel consists of two VGAs (VGA1 and VGA2), one PS, and a differential-to-single-ended active balun (D2S). The two VGAs and the PS adopt differential topologies to reduce channel-to-channel coupling and improve stability. The channel outputs are summed using two independent 8:1 Wilkinson combining networks for dual-beam outputs.

A. Technology

The beamformer is implemented in Tower Semiconductor's 90-nm SiGe BiCMOS process [7]. The process features high-performance heterojunction bipolar transistors (HBTs) with peak f_t/f_{max} of 285/310 GHz and NF_{min} of <0.5 dB

979-8-3503-2123-4/23 $31.00 © 2023 IEEE

281

2023 IEEE Radio Frequency Integrated Circuits Symposium

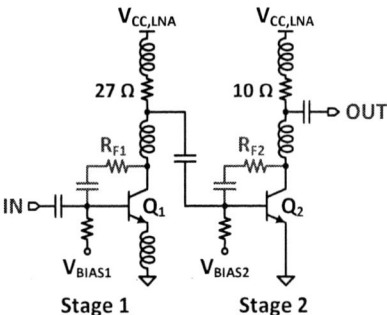

Fig. 3. Wideband 2–24 GHz LNA.

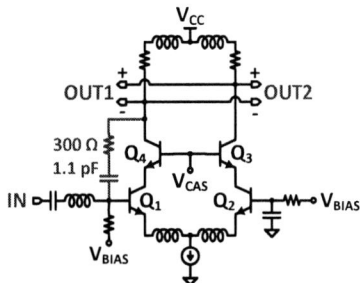

Fig. 4. S2D current splitter (with resistive feedback).

at a current density of 0.1–0.5 mA/μm[1] at 10 GHz. Two thick metal layers are included at the top of the stack-up for high-Q inductors.

B. LNA

The LNA employs two common-emitter (CE) stages with resistive feedback (Fig. 3). The feedback resistors ($R_{F1,2}$) are designed to be high to improve input and inter-stage matching while not compromising input-to-output isolation and NF. $Q_{1,2}$ have large device sizes ($L_e = 4 \times 5$ μm) and are biased at $J_c = 0.5$ mA/μm for small base resistance and good noise matching.

The LNA does not employ an on-chip inductor at the input, as the finite Q-factor (<15) degrades NF_{min} though better noise matching is achieved [8]. As seen in Section III, the input matching is still acceptable at 10–25 GHz, but a small input inductor should be used for matching below 10 GHz. It can be implemented using transmission lines (with Q of 40) on a printed circuit board (PCB) without degrading the NF when the beamformer is flipped on a PCB.

Load networks consisting of series resistors and peaking inductors are employed for gain flatness, as a purely inductive load has insufficient gain at <5 GHz. This is applied to other circuit blocks and improves output matching at the same time.

The LNA has a simulated gain of 22 dB with a 3-dB bandwidth of 2–24 GHz. The NF is 1.8–2.2 dB at 2–21 GHz and the input 1-dB compression point (IP_{1dB}) is -17±3 dBm. It consumes 29 mW from a 1.2 V supply.

---

[1]$J_c = I_c/L_e$, where J_c is the current density, I_c is the collector current, and L_e is the emitter length

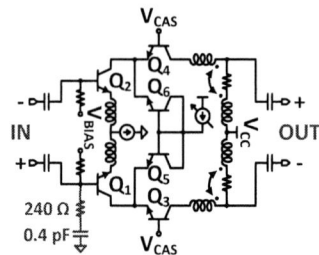

Fig. 5. VGA1 (with RC compensation).

C. S2D

A differential cascode stage with one input AC grounded is used as the S2D and a current splitter to drive two channels (Fig. 4). For wideband input matching, a feedback resistor (R_F) is connected between the input and one of the output branches. The resulting mismatch between the output branches can be compensated by an RC network at VGA1 input as described later. The S2D has a simulated gain of 12 dB, an NF of 5.5 dB, and an IP_{1dB} of -22 dBm, with 20 mW power consumption.

D. VGAs

Each channel has two current-steering VGAs. VGA1 also compensates for the mismatch between the S2D output branches (Fig. 5). The slightly capacitive input impedance and the RC network on one side of the inputs can be designed to result in a lower mismatch than a 50-Ω interface. VGA2 has a similar topology (without the compensation RC) with resistors across the positive and negative input terminals for better matching. Both VGAs use T-coil load networks with series resistors for gain flatness and better output matching.

When used in phased arrays, the gain of VGA2 is lowered first to preserve the channel NF. VGA1 gain can be reduced to improve channel IP_{1dB} which is limited by later stages, or if further gain reduction is needed.

VGA1 has an NF of 5 dB and an IP_{1dB} of -13±2 dBm, while VGA2 has an IP_{1dB} of -16±1 dBm. The linearity is high enough to handle the ~30-dB gain in front of the VGAs and results in an acceptable channel IP_{1dB}. The VGAs consume 16 and 12 mW, respectively.

E. PS and D2S

The vector-modulation-based PS includes a quadrature all-pass filter (QAF) to generate the I/Q signals and two VGAs for weighted summation (Fig 6). In the QAF, series resistors are used to lower the network Q-factor and increase the bandwidth, despite higher loss and NF [9]. This is acceptable since the preceding blocks have >30 dB gain. The VGA input impedance introduces I/Q imbalance, and this is compensated by the series LC network, resulting in a gain error of <0.5 dB and a phase error of $<8°$. The I/Q signals are weighted by the

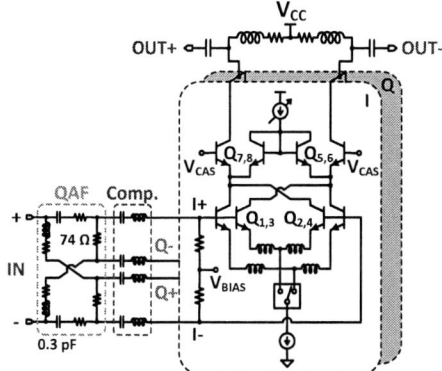

Fig. 6. PS with wideband QAF and LC compensation network.

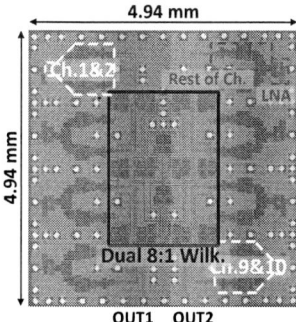

Fig. 7. Wideband 16-channel beamformer chip micrograph.

current-steering VGA and summed at the RL network at the output.

The PS has a simulated gain of -0.5±1.5 dB, an NF of 11.5±1.5 dB, an IP_{1dB} of -10±2 dBm, and consumes 41 mW. The simulated root-mean-square (RMS) gain and phase errors have maximum values of 0.6 dB and 6.5° for all 64 states (6-bit). The RMS phase error is < 5.6° when 32 out of the 64 states are selected, resulting in a true 5-bit response.

The last stage of the channel is the D2S which employs stacked HBTs to combine the differential inputs in phase. The D2S has a simulated gain of 1 dB and an IP_{1dB} of -7 dBm with 8.1 mW power consumption.

F. Channel and Wilkinson Network

The entire channel achieves ∼37 dB simulated gain, 1.9–2.3 dB NF, and -50 dBm IP_{1dB} at 3–28 GHz. In SATCOM, the received power at the ground terminal is typically low, and an IP_{1dB} of -50 dBm is sufficient. Except for the LNA, the channel operates from a 2.4 V supply.

Two-section Wilkinson combiners are used in the 8:1 Wilkinson networks for a wideband response. Each combiner has an ohmic loss of ∼1.3 dB and an isolation of > 13.7 dB at 7.5–28.5 GHz. The 8:1 Wilkinson network results in an ohmic loss of 5 dB including transmission lines. The entire chip thus has a simulated electronic gain of ∼32 dB.

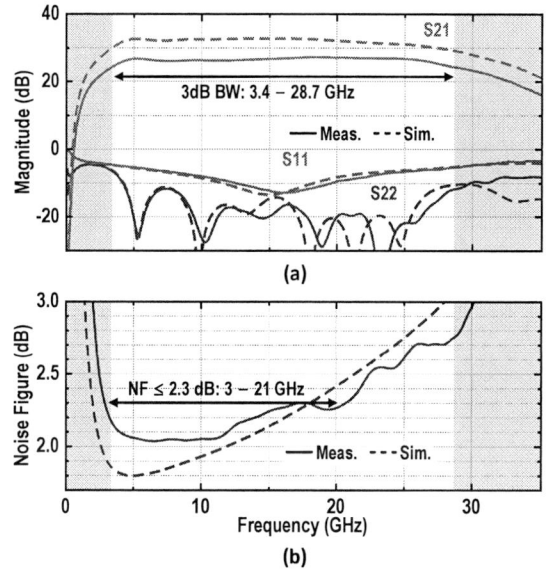

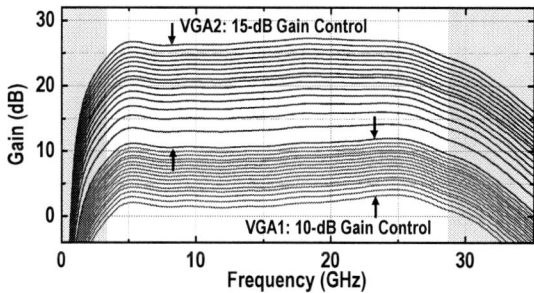

Fig. 8. Measured and simulated beamformer (a) S-parameters and (b) NF.

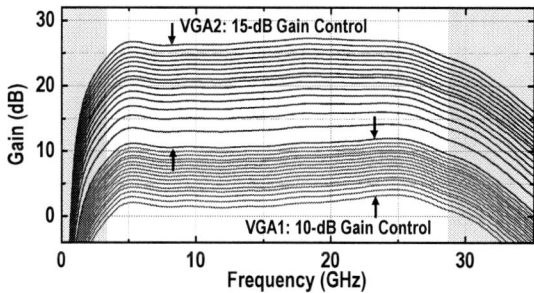

Fig. 9. Measured gain response at different VGA1 and VGA2 gain states.

III. MEASUREMENTS

Fig. 7 presents the beamformer, which is measured using RF probes and a Keysight N5247A PNA-X network analyzer. The chip consumes 1.8 W in total (113 mW per channel). The beamformer can be chip-scale packaged and assembled on a PCB and it has a size of 4.9×4.9 mm², mostly determined by the 0.4-mm bump-pad spacing.

The measured beamformer input and output matching (S_{11} and S_{22}) and electronic gain (S_{21}) are shown in Fig. 8(a). The gain is 27.3 dB with a 3-dB bandwidth of 3.4–28.7 GHz. The gain is lower than the simulated value by 4–6 dB which is typical in large complex chips.

The measured NF is < 2.1 dB at C, X, and Ku SATCOM frequency bands and < 2.3 dB at Ka band (Fig. 8(b)). The measured IP_{1dB} is -47 dBm, mostly limited by VGA2. The reverse isolation is > 60 dB.

The gain responses at different VGA gain states are shown in Fig. 9. VGA1 and VGA2 have 10 and 15 dB gain control, respectively. The phase variation is < 5° for either VGA at all gain states. The phase responses at all 64 phase states are shown in Fig. 10(a). The RMS phase error is < 5.6° at 5.5–9.3 and 15.6–22.5 GHz (Fig. 10(b)). When trimmed to 32 states at

979-8-3503-2123-4/23 $31.00 © 2023 IEEE   283

Table 1. Performance Comparison of C/X/Ku/Ka-Band Beamformers

Reference	This Work	TMTT'23 [1]	TMTT'21 [2]	TMTT'18 [3]	TMTT'16 [4]	IMS'16 [10][†]	TMTT'10 [11][†]
Technology	90-nm SiGe BiCMOS	65-nm CMOS 0.1-μm GaAs	65-nm CMOS	250-nm SiGe BiCMOS	130-nm SiGe BiCMOS	130-nm SiGe BiCMOS	200-nm SiGe BiCMOS
Frequency (GHz)	3.4–28.7	17.7–20.2[a]	7.5–9[a]	19.7–21[a]	9–11[a]	2–16	13–15[a]
Function	RX	RX	RX	RX	TRX	RX[b]	RX
# Channels	16	8	8	2	1	8	8
# Beams	2	1	8	1	1	4	4
Gain (dB)	27.3	20.3	20	21.6	25	11.5	6
NF (dB)	2.1–2.3	1.7–2.1	3.6	5	3	11.5	10–11
IP_{1dB} (dBm)	-47	-	-	-34	-18	-15	-3.3
Gain Control (dB)	25	31.5	34	-	-	8	17
Phase Res. (bit)[*]	5	6	6	3	5	4	4
P_{DC}/Ch. (mW)	113	30.2	108	120	352	265	227.5

[*] Effective resolution   [†] To be used with external LNA   [a] Single-band   [b] Downconverter included

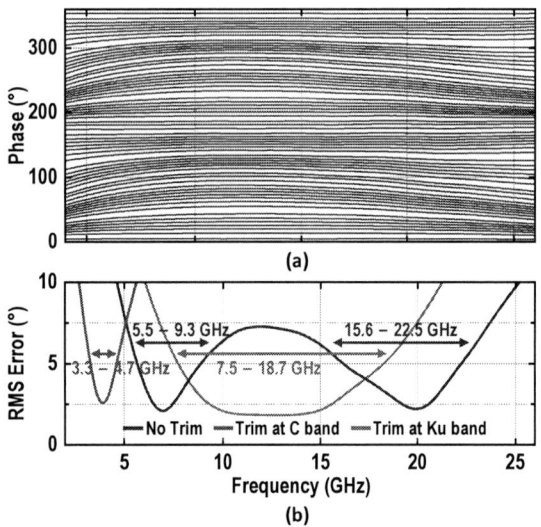

Fig. 10. Measured (a) (relative) phase response of all 64 phase states and (b) RMS phase error before and after trimming at mid-band.

C and Ku bands, the RMS phase errors are $< 5.6°$ at 3.3–4.7 and 7.5–18.7 GHz. Similar trimming can be done at other frequencies for a lower RMS phase error. The RMS gain error is < 1 dB at 4.2–24.6 GHz.

The electronic gain is within ± 2 dB of the average over all 16 channels, and all channels have similar gain, phase, NF, and IP_{1dB} responses. The channel-to-channel coupling, as defined in [11], is < -36 dB. The beamformer is compared with previous work at C, X, Ku, and Ka bands in Table 1. The beamformer covers all 4 bands with high gain, low NF, and low power consumption.

IV. CONCLUSION

An RX beamformer that integrates 16 channels and supports dual-beam operation is presented in this paper. The beamformer chip achieves a very wide bandwidth by using active circuits with resistive feedback, resistive loading, and two-section Wilkinson combiners. Excellent performance is achieved and the beamformer is suitable for C/X/Ku/Ka-band RX phased arrays for SATCOM.

ACKNOWLEDGMENT

This work is supported by the MITRE Corporation. The authors would like to thank Tower Semiconductor for chip fabrication and Keysight Technologies for the measurement instruments.

REFERENCES

[1] D. Zhao et al., "A K-band hybrid-packaged temperature-compensated phased-array receiver and integrated antenna array," IEEE Trans. Microw. Theory Techn., vol. 71, no. 1, pp. 409–423, Jan. 2023.

[2] N. Li et al., "A four-element 7.5–9-GHz phased-array receiver with 1–8 simultaneously reconfigurable beams in 65-nm CMOS," IEEE Trans. Microw. Theory Techn., vol. 69, no. 1, pp. 1114–1126, Jan. 2021.

[3] F. Tabarani et al., "0.25-μm BiCMOS system-on-chip for K-/Ka-band satellite communication transmit–receive active phased arrays," IEEE Trans. Microw. Theory Techn., vol. 66, no. 5, pp. 2325–2339, May 2018.

[4] C. Liu et al., "A fully integrated X-band phased-array transceiver in 0.13-μm SiGe BiCMOS technology," IEEE Trans. Microw. Theory Techn., vol. 64, no. 2, pp. 575–584, Feb. 2016.

[5] M. Li et al., "14.7 an adaptive analog temperature-healing low-power 17.7-to-19.2GHz RX front-end with ±0.005dB/°C gain variation, <1.6dB NF variation, and <2.2dB IP1dB variation across -15 to 85°C for phased-array receiver," in IEEE Int. Solid- State Circuits Conf. (ISSCC) Dig. Tech. Papers, vol. 64, Feb. 2021, pp. 230–232.

[6] "F6121 and F6122 dual-beam Rx active beamforming IC," Renesas, USA. [Online]. Available: https://www.renesas.com/us/en/products/ rf-products/phased-array-beamformers#parametric_options

[7] S. Phillips et al., "Advances in foundry SiGe HBT BiCMOS processes through modeling and device scaling for ultra-high speed applications," in Proc. IEEE BiCMOS Compound Semiconductor Integr. Circuits Technol. Symp. (BCICTS), 2021, pp. 1–5.

[8] T. Kanar and G. M. Rebeiz, "X- and K-Band SiGe HBT LNAs with 1.2- and 2.2-dB mean noise figures," IEEE Trans. Microw. Theory Techn., vol. 62, no. 10, pp. 2381–2389, Oct. 2014.

[9] S. Y. Kim, D.-W. Kang, K.-J. Koh, and G. M. Rebeiz, "An improved wideband all-pass I/Q network for millimeter-wave phase shifters," IEEE Trans. Microw. Theory Techn., vol. 60, no. 11, pp. 3431–3439, Nov. 2012.

[10] M. Sayginer and G. M. Rebeiz, "An 8-element 2–16 GHz phased array receiver with reconfigurable number of beams in SiGe BiCMOS," in Proc. IEEE Int. Microw. Symp. (IMS), May 2016, pp. 1–3.

[11] D.-W. Kang, K.-J. Koh, and G. M. Rebeiz, "A Ku-band two-antenna four-simultaneous beams SiGe BiCMOS phased array receiver," IEEE Trans. Microw. Theory Techn., vol. 58, no. 4, pp. 771–780, Apr. 2010.

RTu3B-1

An Ultra-Wideband and Compact Active Quasi-Circulator With Phase Alternated Differential Amplifier

Dongho Yoo, Jun Hwang, Byung-Wook Min

School of Electrical & Electronic Engineering, Yonsei University, Korea

{dongho.yoo, qkdxo159, bmin}@yonsei.ac.kr

Abstract— This paper presents a new design for an active quasi-circulator that utilizes a phase alternated differential amplifier to achieve ultra-wideband and compact size in a 28-nm CMOS process. The proposed active quasi-circulator is based on a differential two-stage distributed amplifier, where the differential outputs of the phase alternated amplifier in second stage are cross-connected to the differential outputs of the amplifier in first stage. The use of the phase alternated differential amplifier allows for wideband isolation regardless of frequency. Furthermore, interstage inductor, which is located between two transistors of amplifier makes the circulator more wideband by maintaining frequency response of two transmission lines identically. The measured transmitter (TX) to receiver (RX) isolation is >21 dB, -3 dB bandwidth of TX to antenna (ANT) is from 20 to 38.5 GHz with 4.5 dB of peak gain at 29.8 GHz, minimum insertion loss of ANT to RX is 4.3 dB. The measured TX to ANT output power 1 dB compression point is 7.3 dBm at 28 GHz with DC power consumption of 107 mW. The circulator occupies only 0.07 mm^2, thanks to coupled inductors which contribute to compact size.

Keywords— Active quasi-circulator, wideband, distributed amplifier, non-magnetic circulator, full-duplex, CMOS.

I. INTRODUCTION

Conventional magnetic circulators have several disadvantages such as high cost and bulky size, and incompatibility with on-chip integration. To overcome these disadvantages, non-magnetic circulators have been developed [1]-[6]. Resonators or gyrator based circulators realize non-reciprocal characteristic by modulating switches with multiple clock phases [1]-[2]. However, clock generation using frequency divider is hard to realize in millimeter-wave (mmW) frequency band. Although clock generation using poly-phase filter is possible in mmW, both methods of clock generation are power hungry and bulky [2]. Therefore, as an alternative of clock based circulators, active quasi-circulator (QC) using unilateral characteristic of amplifier are introduced [3]-[6]. In [3], an isolation is achieved by using 180° phase difference between common-source and source-follower. Nevertheless, it suffers from high insertion loss and narrow bandwidth. In order to achieve high gain, distributed amplifier (DA) is modified by using 50Ω termination of the output line as a receiver (RX) port and incorporating tunable phase offset lines [4]. However, this design is only operational at $\lambda/4$ transmission line, which results in narrow bandwidth. [5]-[6] propose multiple DA or combination of couplers to achieve wideband operation. Still, wideband active QCs are bulky due to a number of amplifiers and passive elements.

In this paper, an ultra-wideband and compact active QC is proposed. The circulator is designed differentially with

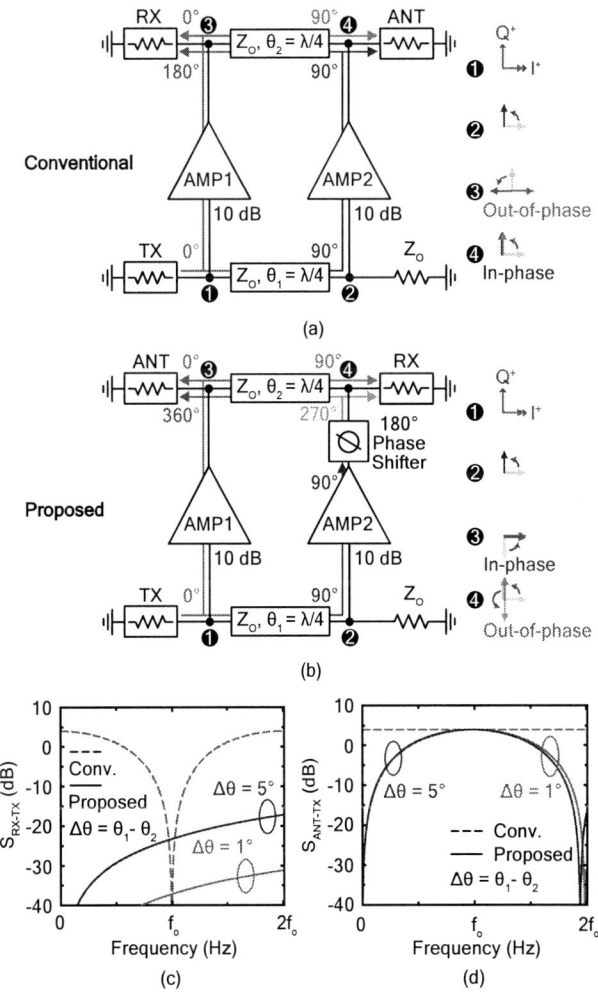

Fig. 1. (a) A configuration of conventional active QC based on DA and (b) proposed active QC based on DA with 180° phase shifter. Simulated results of (c) TX to RX and (d) TX to ANT paths of two configurations depending on phase difference between input and output transmission line when the gain of amplifier is 10 dB.

a two-stage DA configuration, utilizing a phase alternated amplifier in the second stage. The outputs of the second stage amplifier are cross-connected with the outputs of the first-stage amplifier, resulting in the cancellation of transmitter (TX) to RX leakage signal across a wide frequency range. Furthermore, to achieve compact die area, coupled inductor which occupies less space than a single-ended inductor with same inductance is employed. Additionally, interstage inductor inserted between

979-8-3503-2123-4/23 $31.00 © 2023 IEEE   285   2023 IEEE Radio Frequency Integrated Circuits Symposium

the transistors of amplifier, makes the difference between the input and the output capacitance of amplifier constant. This maintains frequency response of two transmission lines, in which the parasitic capacitance of the amplifier is absorbed, identically over a wide frequency range.

II. WIDEBAND ACTIVE QUASI-CIRCULATOR

Fig. 1(a) represents a basic configuration of conventional active QC based on a single-ended two-stage DA. The circulator is composed of two ideal amplifiers and transmission lines where the input and output port of the DA are used as TX and antenna (ANT), respectively. This establishes the gain of TX to ANT path. A Z_o termination of output line is used as the RX port. The transmission lines are quarter-wavelength at operating frequency f_o, and located between the two amplifiers. A TX to RX signal passing through second stage amplifier experiences two transmission lines more than a signal passing through first stage amplifier. Therefore, the two signals meet out-of-phase at the RX port and destructive interference occurs. As the destructive interference at the RX port strongly relies on the $\lambda/4$ transmission line, TX to RX isolation is sharply deteriorated as the frequency deviates from the f_o.

Fig. 1(b) represents a basic configuration of the proposed active QC. Unlike Fig. 1(a), the position of the ANT port and the RX port are changed. Furthermore, 180° phase shifter is located at the output of the second stage amplifier. Therefore, although both of two TX to RX signals experience equal number of amplifier and transmission line, the outputs of the two amplifier stages meet out-of-phase at the RX port due to additional 180° phase shift in the second stage. Moreover, as the destructive interference does not rely on $\lambda/4$ transmission line, the proposed QC has infinite TX to RX isolation bandwidth. A TX to ANT signal passing through the second amplifier stage experiences two transmission lines and a 180° phase shifter more than a signal passing through the first stage amplifier. Therefore the two TX to ANT signals meet in-phase at the ANT port and constructive interference occurs. In addition, the TX to ANT transmission is not deteriorated

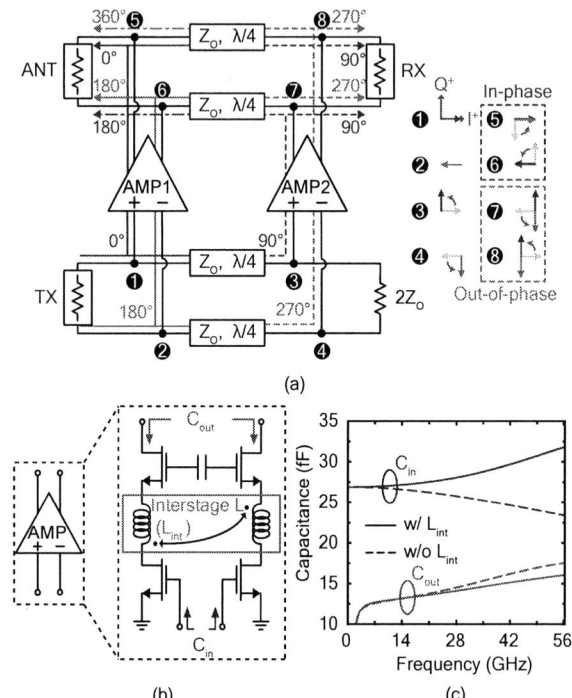

Fig. 2. (a) The configuration of proposed differential active QC with phased alternated amplifier. (b) The schematic of differential amplifier core and (c) simulated input and output capacitance of differential amplifier depending on interstage inductor.

as much as the TX to RX isolation of conventional QC, as the frequency deviates from the f_o.

Fig. 1(c) and Fig. 1(d) show simulated results of TX to RX and TX to ANT paths of two configurations with gain of amplifier 10 dB and phase difference between the two transmission lines. As seen in Fig. 1(c), the bandwidth of TX to RX isolation of the proposed QC is broader than conventional QC. Even with the phase difference of 5°, the isolation of the proposed QC maintains its wideband characteristic, while the isolation of conventional QC is still narrow and the peak

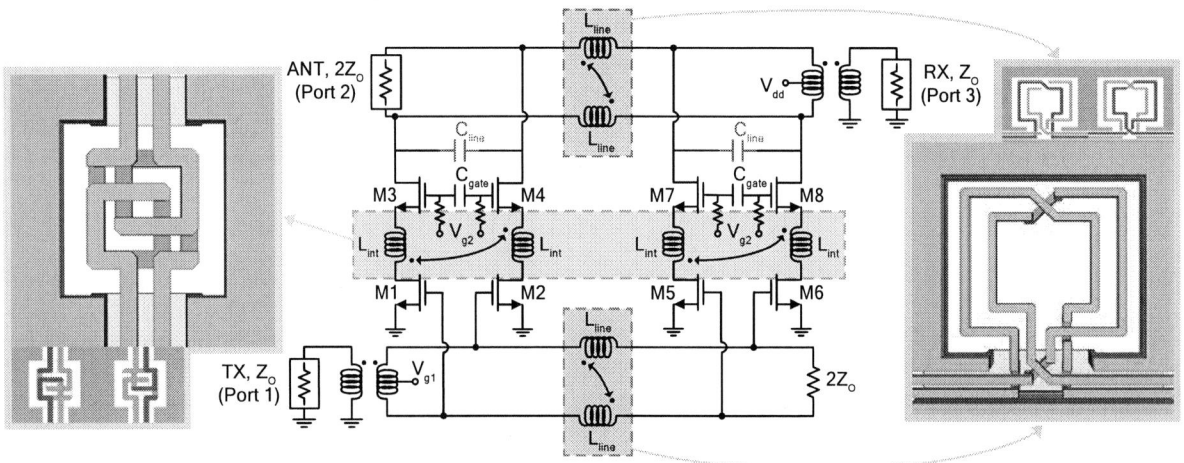

Fig. 3. Full schematic of the proposed active QC with phase alternated differential amplifier and layout of coupled inductors.

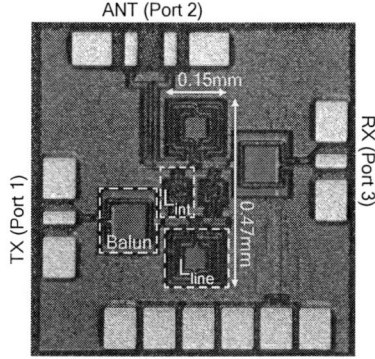

Fig. 4. Die photograph of the proposed active QC.

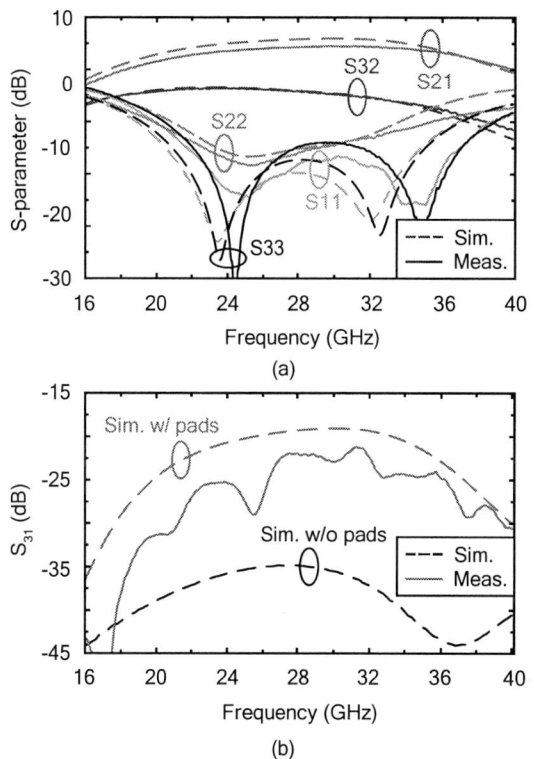

Fig. 5. Measured and simulated S-parameters of (a) TX to ANT, ANT to RX and return losses. (b) Simulated S-parameters of TX to RX depend on presence of pads and measured S-parameters of TX to RX.

is shifted from the f_o. In terms of TX to ANT gain, the bandwidth of the proposed QC is narrower than conventional QC as seen in Fig. 1(d). However, unlike the isolation of conventional QC, the gain of proposed QC differs gradually as the frequency deviates from the f_o. Thus, considering both the bandwidth of TX to RX isolation and TX to ANT gain, the proposed QC has wide bandwidth than conventional QC.

III. DIFFERENTIAL ACTIVE QUASI-CIRCULATOR DESIGN

A configuration of the proposed differential active QC with phase alternated amplifier is shown in Fig. 2(a). The 180° phase shifter in Fig. 1(b) is implemented by using phase alternated differential amplifier, in which positive (+) and negative (-) node of the second stage amplifier are connected to - and + node of the first stage amplifier, respectively. This configuration eliminates the need for additional phase shifter, which would require insertion loss to be compensated using a different amplifier at the second stage, making the design more complex and difficult to achieve wideband operation.

A. Differential Amplifier Core Design

Fig. 2(b) shows schematic of the differential amplifier core in QC. Two-stack amplifier is utilized so as to increase power handling, allowing nominal DC voltage of 2 V at drain. Gate capacitance of stacked field-effect-transistor (FET) is chosen for equal drain-source voltage distribution. The finite value of gate capacitance ensures stability of the amplifier. The width of the transistor is selected according to the parasitic capacitances at the input (C_{in}) and the output (C_{out}) of the amplifier which are absorbed in transmission lines.

The wideband characteristic of TX to RX isolation is maintained regardless of the electrical length of transmission lines, if the two transmission lines are equal over the frequency. However, the C_{in} and C_{out} are varied in the opposite direction due to the miller effect which weakens gain of the amplifier as frequency goes up as seen in Fig. 2(c). Inserting interstage inductor (L_{int}) between the drain of common source stage and source of stacked FET solves this problem by increasing gain of the amplifier and therefore maintaining the difference of C_{in} and C_{out} over the frequency.

B. Synthetic Transmission Line Design

The characteristic impedance of transmission lines is 50Ω to match the impedance at all ports. $\lambda/4$ transmission line is too long to be integrated, therefore synthetic line is used instead and capacitor-inductor-capacitor low pass π structure is adopted to absorb C_{in} and C_{out}. The inductors in synthetic line (L_{line}) are implemented using coupled inductor to obtain large inductance occupying same area as much as single-ended inductor.

C. Active Quasi-Circulator Integration

The full schematic of the proposed active QC and layout of coupled inductors are shown in Fig. 3. The width and gate capacitor (C_{gate}) of stacked FETs are 2×25 μm and 180 fF, respectively. The common-source gate bias voltage (V_{g1}) and common-gate gate bias voltage (V_{g2}) are 0.7 V and 1.7 V with drain bias voltage (V_{dd}) of 2 V. An additional metal-oxide-metal (MOM) capacitor (C_{line}) of 11 fF is attached at the drain of the amplifier to compensate the mismatch between C_{in} and C_{out}. The coupled inductors are considered to design in terms of symmetry which is important in differential circuit. Transformer baluns are inserted at TX and RX ports for a single-ended to differential transition and differential feeding lines are used in the ANT port. In addition, the gate and drain bias of the amplifier is biased through the center tap of the transformer balun.

Table 1. Performance Comparison of State-of-the-Art Circulators

	This work	[1]	[2]	[3]	[4]	[5]	[6]
Technology	28-nm CMOS	65-nm CMOS	45-nm SOI CMOS	0.18-μm CMOS	45-nm CMOS	90-nm CMOS	90-nm CMOS
Topology	Phase Alternated Differential DA	N-Path-Filter based Circulator	Poly-Phase-Filter based Circulator	Phase Cancellation	DA with Tunable Line	Cascading DA	DA with Coupler
Frequency (GHz)	$20 \sim 38.5$	0.75	$22.7 \sim 27.3$	24	$5.3 \sim 7.3$	$10 \sim 67$	$14 \sim 67$
TX to ANT (S_{21}) (dB)	$1.5 \sim 4.5$	-1.7	$-4.2 \sim -3.3$	-9	$9 \sim 10.5$	$0.5 \sim 4.8$	$-3.5 \sim 4.3$
ANT to RX (S_{32}) (dB)	$-10 \sim -4.3$	-1.7	$-4.1 \sim -3.2$	-8.5	$-8 \sim -5$	$1.1 \sim 4.3$	$-9 \sim 1.2$
TX to RX $\lvert S_{31}$ (dB)$\rvert$	>21	>20	>18.5	>30	>30	$23.4 \sim 31.3$	$12.9 \sim 22.8$
ANT to RX NF (dB)	$13.7 \sim 18.7^*$	4.3	$3.3 \sim 4.4$	N/R	20	$7.9 \sim 12.8$	$7.1 \sim 12.6$
TX to ANT OP$_{1dB}$ (dBm)	7.3	N/R	>17.2	-22	14	7	-3
P$_{DC}$ (mW)	107	59	78.4	9.12	415	67.8	41.8
Area (mm^2)	$0.07^\dagger$	$0.64^{\dagger\dagger}$	2.16	0.35	1.57	0.51	0.93

*Simulated result. †Excluding baluns and ANT feeding lines. ††Excluding pads. N/R : Not reported.

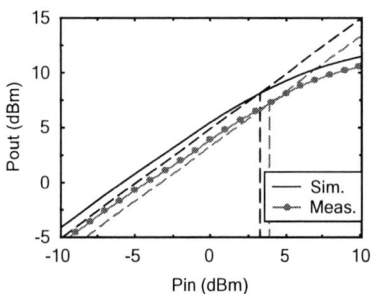

Fig. 6. Measured and simulated results of large signal performance of TX to ANT path.

IV. MEASUREMENT RESULTS

Fig. 4 is a die photograph of the chip. Thanks to the coupled inductors, the circulator is integrated in a compact area. Therefore, the size of the circulator in which transformer baluns, feeding lines, and pads are not included is 0.07 mm^2.

The S-parameters of proposed active QC are measured by 4-port vector network analyzer. Fig. 5 shows the measured and simulated S-parameters of TX to ANT (S_{21}), ANT to RX (S_{32}), TX to RX (S_{31}), and return losses (S_{xx}) at all ports. The -3 dB bandwidth of S_{21} is from 20 to 38.5 GHz with peak gain of 4.5 dB at 29.8 GHz. The measured S_{32} is -4.3 to -10 dB. The S_{11}, S_{22} and S_{33} are better than -10.2 dB, -7.8 dB and -9.1 dB respectively over the same bandwidth. The S_{31} is measured by 2-port measurement with the two ANT ports terminated with DC block capacitors and 50Ω terminations due to calibration complexity. The parasitic capacitance of pads causes impedance mismatch at the ANT port and therefore S_{31} is degraded. Nevertheless, the measured S_{31} is better than -21 dB as shown in Fig. 5(b). Practically, the isolation of the circulator, which is defined as the ratio of RX leakage power to ANT output power, is S_{31}-S_{21} and better than -25.7 dB.

Large signal performance of TX to ANT path is shown in Fig. 6. The measured output power 1 dB compression point (OP$_{1dB}$) is 7.3 dBm at 28 GHz with DC power consumption (P$_{DC}$) of 107 mW. Noise figure (NF) of ANT to RX path is higher than the insertion loss of ANT to RX path due to the noise generated by the 50Ω termination of the input line. The simulated NF is 13.7 to 18.7 dB from 20 to 38.5 GHz.

Table 1 compares the performance of proposed active QC with other state-of-the-arts. The proposed active QC represents the ultra-wideband operation in the smallest die area.

V. CONCLUSION

The 28-nm CMOS active QC with phase alternated differential amplifier is presented. Wideband operation is realized by connecting the output of the second stage alternately to the first stage and using interstage inductor between the transistors of amplifier. Thanks to coupled inductor, the size of active QC is only 0.07 mm^2. The TX to ANT peak gain is 4.5 dB and ANT to RX minimum insertion loss is 4.3 dB. The TX to RX isolation is >21 dB and TX to ANT OP$_{1dB}$ is 7.3 dBm at 28 GHz with P$_{DC}$ of 107 mW.

ACKNOWLEDGMENT

This work was supported by Samsung Electronics Co., LTD (IO201209-07875-01), and IITP grant funded by the Korea government (MSIT) (No. 2020000218). The EDA tool was supported by the IC Design Education Center (IDEC), Korea. D. Yoo and J. Hwang contributed equally to this paper.

REFERENCES

[1] N. Reiskarimian, J. Zhou, and H. Krishnaswamy, "A CMOS Passive LPTV Nonmagnetic Circulator and Its Application in a Full-Duplex Receiver," in *IEEE Journal of Solid-State Circuits*, vol. 52, no. 5, pp. 1358-1372, Mar. 2017.

[2] T. Dinc, A. Nagulu, and H. Krishnaswamy, "A Millimeter-Wave Non-Magnetic Passive SOI CMOS Circulator Based on Spatio-Temporal Conductivity Modulation," in *IEEE Journal of Solid-State Circuits*, vol. 52, no. 12, pp. 3276–3292, Dec. 2017.

[3] D. J. Huang, J. L. Kuo, and H. Wang, "A 24 GHz Low Power and High Isolation Active Quasi-Circulator," in *2012 IEEE MTT-S International Microwave Symposium Digest*, Montreal, QC, Canada, 2012, pp. 1–3.

[4] K. Fang and J. F. Buckwalter, "A Tunable 5–7 GHz Distributed Active Quasi-Circulator With 18-dBm Output Power in CMOS SOI," in *IEEE Microwave Wireless Components Letters*, vol. 27, no. 11, pp. 998–1000, Nov. 2017.

[5] S. D. Tang, C. M. Lin, S. H. Hung, K. W. Cheng, and Y. H. Wang, "Ultra-Wideband Quasi-Circulator Implemented by Cascading Distributed Balun With Phase Cancelation Technique," in *IEEE Transactions on Microwave Theory and Techniques*, vol. 64, no. 7, pp. 2104–2112, Jul. 2016.

[6] S. H. Hung, K. W. Cheng, and Y. H. Wang, "An Ultra-Wideband Quasi-Circulator With Distributed Amplifiers Using 90 nm CMOS Technology," in *IEEE Microwave Wireless Components Letters*, vol. 23, no. 12, pp. 656–658, Dec. 2013.

RTu3B-2

A D-band Calibration-Free Passive 360° Phase Shifter With 1.2° RMS Phase Error in 45 nm RFSOI

Mohammadreza Abbasi[1], Wooram Lee

The Pennsylvania State University, USA

[1]mka6056@psu.edu

Abstract — **This paper presents a new concept of passive phase shifters based on manipulating propagation delay through two parallel transmission lines periodically connected via digitally controlled switch networks. The proposed approach enables precise phase control and flat amplitude response across different phase settings. The prototype IC is fabricated in a 45 nm RFSOI process and occupies only 0.033 mm^2. The phase control operates with 11.25° steps over 360° at 140 GHz while maintaining an RMS phase error of 1.2°. The insertion loss is 11.5 dB with $< \pm 0.8$ dB variation. Among published D-band phase shifters, this work achieves the lowest RMS phase error and reports bi-directional phase control over 360° and calibration-free operation.**

Keywords — **Calibration-free, bi-directional, D-band, mm-wave, passive, phase shifter, phased array.**

I. INTRODUCTION

D-band frequency spectrum (110-170 GHz) has attracted considerable interest to accommodate the explosive growth in wireless data capacity and emerging sensing applications [1]. To overcome severe path loss and limited transistor performance at such high frequencies, the use of a large-scale phased array transceiver is essential, thereby requiring the development of a D-band front-end phase shifter. Several D-band phase shifters have been proposed using different topologies including switched transmission line [2], a vector modulator [3], [4], and a reflection-type phase shifter (RTPS) combined with an active phase inverter [5], [6]. However, the following design requirements remain unaddressed. First, bi-directional phase control is necessary so that TX and RX front-ends can share a single phase shifter to reduce the die area of beamformer ICs. The area of antenna-in-package (AiP) is scaled down with frequency to maintain $\lambda/2$ antenna spacing. The resulting small AiP dimensions require compact beamformer ICs to fit into the available space. Second, the challenge of modeling transistors and electromagnetic (EM) structures above 100 GHz makes precise phase control over broad bandwidth difficult. Finally, calibration-free, direct digital phase control with constant insertion loss is necessary for simple array calibration and fast beam switching.

To address the challenges above, we propose a new concept of passive phase shifters based on manipulating propagation delay through two parallel transmission lines periodically connected via transistor switch networks. We implement the prototype phase shifter, which operates with 11.25° steps over 360° at 140 GHz, in a 45 nm RFSOI process. The measured RMS phase error is 1.2° without calibration, which is the lowest among published phase shifters in similar

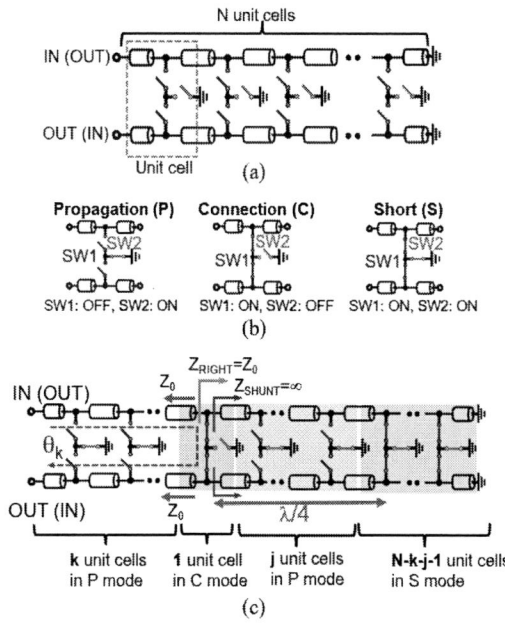

Fig. 1. (a) Proposed phase shifter concept consisting of cascaded N unit cells, (b) different configurations of the unit cell, and (c) operation principle of the proposed phase shifter implemented with ideal switches.

frequency ranges. The measured insertion loss is 11.5 dB with $< \pm 0.8$ dB variation. The proposed phase shifter also reports calibration-free, bi-directional phase control over 360° at D-band.

II. PROPOSED PHASE SHIFTER

Fig. 1 describes the concept of the proposed phase shifter. It consists of two parallel transmission lines periodically connected via switch networks, which can be considered N cascaded unit cells. Each unit cell is formed with two series switches (SW1) that connect the two parallel transmission lines and one shunt switch (SW2) that connects the middle node to the ground. The unit cell can be configured in three different modes for the proposed phase shifter operation: *Propagation* mode, *Connection* mode, and *Short* mode as shown in Fig. 1(b). Fig. 1(c) illustrates the operation principle of the proposed phase shifter when the switches are ideal ($r_{ON}=0$ and $C_{OFF}=0$). An input signal propagates along the upper transmission line of the unit cells in *Propagation* mode until it reaches the one in *Connection* mode. Then, the signal is redirected and propagates along the lower transmission line toward the output port in the opposite direction. The total

979-8-3503-2123-4/23 $31.00 © 2023 IEEE   289   2023 IEEE Radio Frequency Integrated Circuits Symposium

phase shift between the input and output ports is given by $\theta_k = (2k + 1)\tau_0 \cdot \omega$, where k is the number of the unit cells in *Propagation* mode before the connection and τ_0 is the propagation delay per unit cell. The phase shift between the input and output ports is programmable by selecting different k with phase steps of $2\tau_0\omega$. To ensure that an input signal travels towards the output port without disruption, the impedance seen after the connection Z_{SHUNT} should be ∞ and the signal sees only the characteristic impedance Z_0 continuously ($Z_{RIGHT}=Z_0$) from the lower transmission line. To this end, a quarter-wave impedance transformer terminated by a short is formed by 1) having j unit cells in *Propagation* mode after the connection for 90° phase shift at the operation frequency and 2) having the rest of the unit cells in *Short* mode.

The proposed phase shifter has only a single unit cell in *Connection* mode for any phase state; the overall insertion loss remains constant across different phase settings as it is dominated by the loss of two series switches (SW1) of the unit cell in *Connection* mode. This feature addresses the trade-off between insertion loss and phase resolution/tuning range as well as large loss variation across different phase states, as exhibited by prior switch-based phase shifters [2].

III. CIRCUIT-LEVEL IMPLEMENTATION IN 45-NM RFSOI

The 140-GHz prototype phase shifter is designed using GlobalFoundries' 45-nm RFSOI process. Fig. 2 shows the schematic of the unit cell with the 3-D layout view. The transistor switch layout is RC-extracted using PVS-QRC, and the surrounding EM structure is modeled by EMX. The switch network that connects the two coplanar waveguides is implemented with two transistors (M_{SW1}) in series and one shunt transistor (M_{SW2}) in the middle. M_{SW2} is turned on for *Propagation* mode as well as *Short* mode to minimize capacitive coupling between the forward and backward signal paths through $C_{off,SW1}$. This is critical for precise phase control, which comes at the cost of larger $r_{on} \times C_{off}$ than a single transistor-based switch. The selected size of M_{SW2} is 91μm/40nm to provide a small r_{on} enough for isolation between the forward and backward signal path as well as short-circuit termination. The ground shield is also placed between the two signal lines for isolation as shown in Fig. 2. The phase shift per unit cell in *Propagation* mode is given by $\theta_0 \simeq \omega\sqrt{L_0 \cdot (C_0 + C_{off,SW1})}$, where L_0 and C_0 are the inductance and capacitance of the transmission line per unit cell, respectively. The selected θ_0 is 22.5° at 140 GHz for 45° phase steps. The size of M_{SW1} is chosen from the trade-off between $r_{on,SW1}$ and $C_{off,SW1}$. A larger size of M_{SW1} reduces $r_{on,SW1}$ for lower loss in *Connection* and *Short* modes while increasing $C_{off,SW1}$. Larger $C_{off,SW1}$ requires a shorter transmission line (smaller L_0) for θ_0 of 22.5°, posing an EM modeling challenge in layout for a reliable inductance design ($L_0<10$ pH). The selected Z_0 is 22 Ω from the optimum size of M_{SW1} of 25μm/40nm. It is noted that the increase in $C_{off,SW1}$ also reduces the characteristic impedance in *Propagation* mode, given by $Z_0 = \sqrt{L_0/(C_0 + C_{off,SW1})}$. This leads to higher insertion loss

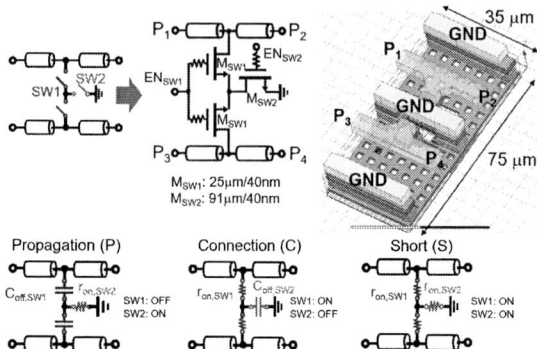

Fig. 2. Schematic and 3-D layout view of the unit cell design implemented with NMOS transistors and coplanar waveguides. Different configurations of the unit cell are shown with the RC models of on/off state transistors.

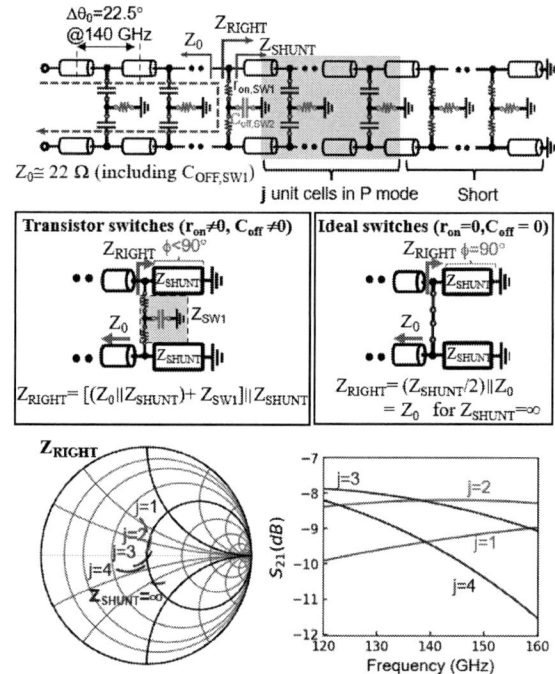

Fig. 3. Impedance matching condition ($Z_{RIGHT}=Z_0$) at the connection node by selecting the number of the unit cells, **j**, between *Connection* and Short.

and narrower bandwidth for a matching network to 50 Ω at the input and output ports. To minimize signal reflection at the connection node for precise phase control, Z_{RIGHT} must be matched to Z_0 as shown in Fig. 3. Z_{RIGHT} is tuned via Z_{SHUNT} by changing the number of unit cells in *Propagation* mode after the connection node, **j**. j=2 is chosen based on the simulated matching condition on the Smith chart, achieving the lowest insertion loss over a wide bandwidth. It is noted that in the ideal-switch-based implementation (r_{on}=0 and C_{off}=0), **j** should be chosen for the quarter-wave impedance transformation (ϕ=90°, $Z_{SHUNT}=\infty$), resulting in $Z_{RIGHT}=Z_0$ as discussed in Section II.

Fig. 4 shows the configuration table of 12 cascaded unit cells to control phase shift with 45° phase steps when θ_0 is 22.5°. Two dummy cells in *Short* mode are placed at the end to

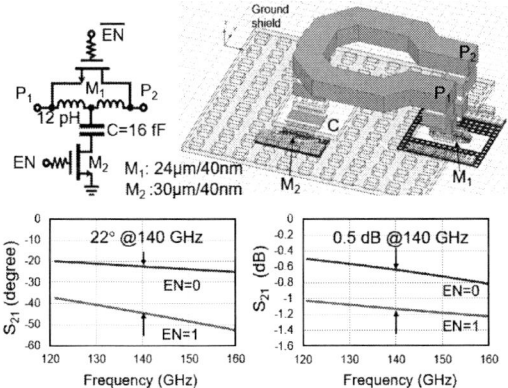

Fig. 4. Configuration table of 12 cascaded unit cells to control $\Delta\theta$ with 45° phase steps, where $\Delta\theta$ is a relative phase shift to the minimum phase shift.

Fig. 5. Schematic and 3-D layout view of 22.5° 1-bit fine-tuning phase shifter with the simulated phase shift and insertion loss.

provide consistent Z_{SHUNT} for larger phase shift settings. The minimum phase step achievable in the proposed phase shifter is limited by $r_{on} \times C_{off}$ of transistor switches in a 45-nm RFSOI and the minimum inductance that can be reliably designed. To operate with phase steps of 11.25° over 360°, two 1-bit fine-tuning phase shifters (0/22.5° and 0/11.25°) are designed. The schematic of the 1-bit 22.5° phase shifter with the 3-D layout view is shown in Fig. 5. The phase shift occurs when M_1 is off and M_2 is on to switch to larger inductance and capacitance in the signal path. The simulated phase shift and insertion loss variation between the two states are 22° and 0.5 dB at 140 GHz, respectively. The 1-bit 11.25° phase shifter is designed using the same topology. Fig. 6 shows the full schematic of the prototype 140-GHz phase shifter integrated with two fine-tuning phase shifters and an on-chip decoder. The pad capacitance and 90-μm transmission line form an L-match network to transform Z_0=22 Ω to 50 Ω.

IV. MEASUREMENT

The prototype phase shifter was fabricated in GlobalFoundries' 45-nm RFSOI process. The chip photograph is shown in Fig. 6; the chip area is only 0.033 mm². The fabricated IC is characterized on a wafer for a VDD of 1.0 V by measuring S-parameters with Keysight N5242B PNA-X, N5292A mmWave test controller, and N5262BW06 D-band VNAX modules. Fig. 7 shows the measured phase shift for

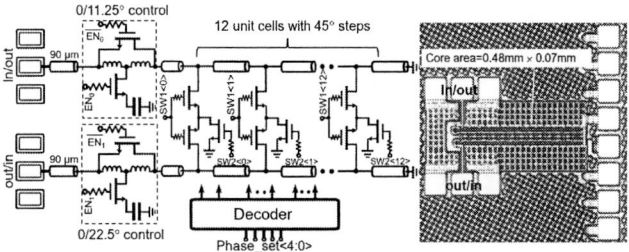

Fig. 6. Schematic and die photo of the proposed phase shifter.

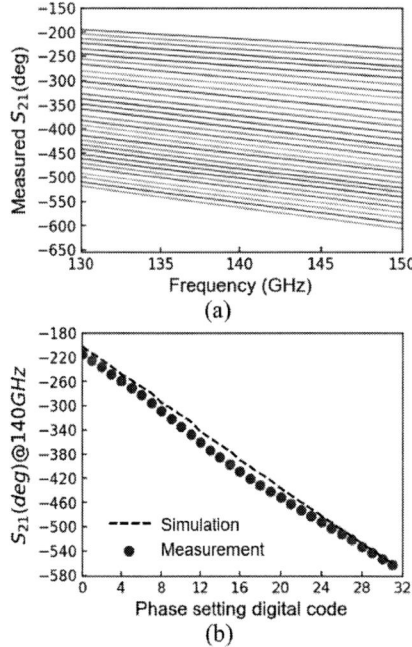

Fig. 7. (a) Measured phase shift for all phase states (b) measured phase shift at 140 GHz with respect to phase setting in comparison to simulation.

all phase states with respect to frequency and the measured phase shift at 140 GHz across phase settings in comparison to the simulation. The proposed phase shifter presents a phase tuning range greater than 360° at frequencies higher than 140 GHz with uniform phase steps. The phase tuning range and step increase with frequency since the proposed phase tuning approach is based on the manipulation of true-time delay. Fig. 8 shows the measured insertion loss and input return loss for all phase states in comparison to the simulation results for the phase states of 0 and 31. The measured insertion loss ranges from 10.5 to 12.3 dB at 140 GHz including the measured pad loss (~0.6 dB). The insertion loss variation across different phase settings is less than ±0.8 dB without calibration. It is noted that the measured insertion loss is reduced to 9.7~11.6 dB due to smaller r_{ON} for a VDD of 1.2 V. The measured input return loss is higher than 10 dB at frequencies up to 154 GHz for all phase states.

Fig. 9 shows the measured phase shift and insertion loss in polar plots at 140 GHz, 145 GHz, and 150 GHz. The RMS phase errors are 1.2° with 11.25° phase steps, 1.5° with 11.6° phase steps, and 2.0° with 12° phase steps for

979-8-3503-2123-4/23 $31.00 © 2023 IEEE

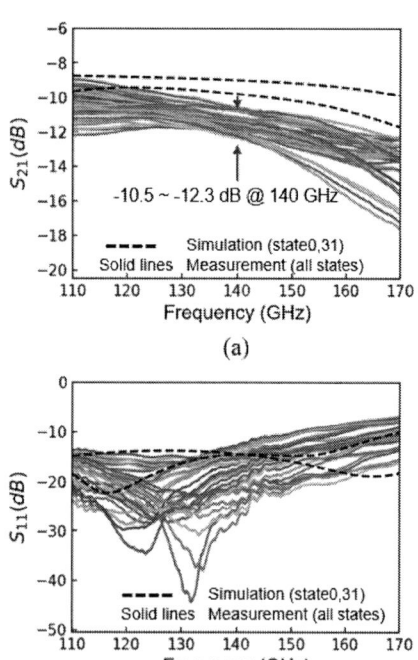

(a)

(b)

Fig. 8. Measured insertion loss and input return loss for all phase states.

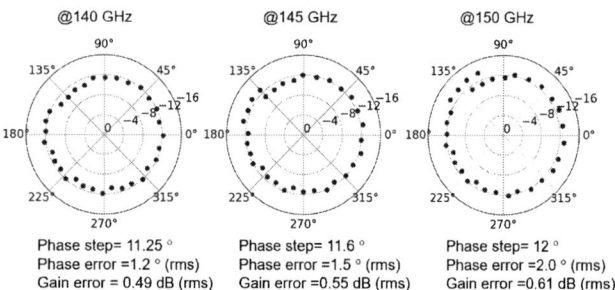

@140 GHz
Phase step= 11.25 °
Phase error =1.2 ° (rms)
Gain error = 0.49 dB (rms)

@145 GHz
Phase step= 11.6 °
Phase error =1.5 ° (rms)
Gain error = 0.55 dB (rms)

@150 GHz
Phase step= 12 °
Phase error =2.0 ° (rms)
Gain error =0.61 dB (rms)

Fig. 9. Measured phase and insertion loss for all phase states on a polar plot at 140 GHz, 145 GHz, and 150 GHz.

140 GHz, 145 GHz, and 150 GHz, respectively. The RMS gain error ranges from 0.5 dB to 0.6 dB across frequencies. The achieved phase error at 140 GHz is the lowest among published phase shifters in similar frequency ranges. It is also noteworthy that the phase and gain errors are achieved without calibration. Fig. 10 is the measured phase shift and insertion loss across different phase settings at 140 GHz for different IC samples. The phase and gain variations are less than ±1.9° and ±0.5 dB, respectively. This feature is desirable to reduce the complexity of the front-end calibration in a large-scale phase array. Table 1 summarizes the performance of the proposed phase shifter in comparison to other published phase shifters in similar frequency ranges. The proposed phase shifter is the first bi-directional 360° tuning, calibration-free phase shifter at D-band with superior phase accuracy and loss variation.

V. Conclusion

This work demonstrates a novel 140-GHz passive phase shifter to enable precise phase control and flat amplitude

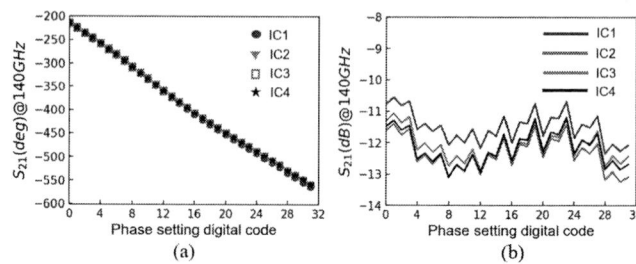

(a)

(b)

Fig. 10. Measured phase and insertion loss for different samples.

Table 1. Performance summary in comparison with other published phase shifter above 100 GHz.

	This work	[2]	[5]	[6]	[3]	[4]
Frequency(GHz)	**140**	110-170	116-128	110-145	115	140-160
Type	**Passive**	Passive	Active+ Passive	Active+ Passive	Active	Active
Phase range(°)	**360**	TTD***	360	360	360	360
S$_{21}$ (dB)	**-10.6~-12.3*** **-9.7~-11.6****	-20~-22	-5.8	<-5.5	0.5	-4.5
RMS Gain error(dB)	**0.5**	1.4	0.1~0.5	<1.25	1.6	1.4
RMS Phase error(°)	**1.2**	N.A.	2.2	<13	5.5	7.5
Number of states	**32**	16	32	16	16	32
P$_{dc}$(mw)	**0**	6.22	30	21	33	50
Bidirectional	**Yes**	Yes	No	No	No	No
Calibration-free	**Yes**	N.A.	No	No	No	No
Area(mm^2)	**0.033**	1.16	0.405	0.572	NA	0.05
Process	**45RFSOI**	130-nm SiGe	0.12um SiGe	90-nm SiGe	130-nm SiGe	55-nm SiGe

* VDD=1V, ** VDD=1.2V (including the pad loss of 0.6 dB)
*** TTD (True-time delay) with the maximum delay of 6.64 ps.

responses across different phase settings without calibration. Fabricated in a 45 nm RFSOI process, the prototype IC demonstrates the lowest RMS phase error (1.2°) and the first bi-directional 360° tuning, calibration-free operation among state-of-the-art D-band phase shifters.

Acknowledgment

The authors would like to thank GlobalFoundries for chip fabrication and L. Zhong and S.A. Uddin for technical discussions.

References

[1] T. S. Rappaport *et al.*, "Wireless Communications and Applications Above 100 GHz: Opportunities and Challenges for 6G and Beyond," *IEEE Access*, vol. 7, pp. 78 729–78 757, 2019.

[2] A. Karakuzulu *et al.*, "Broadband 110 - 170 GHz True Time Delay Circuit in a 130-nm SiGe BiCMOS Technology," in *IEEE MTT-S Int. Microw. Symp. Dig.*, 2020, pp. 775–778.

[3] S. Afroz and K.-J. Koh, "A *D*-Band Two-Element Phased-Array Receiver Front End With Quadrature-Hybrid-Based Vector Modulator," *IEEE Microwave and Wireless Components Letters*, vol. 28, no. 2, pp. 180–182, 2018.

[4] D. d. Rio *et al.*, "A Compact and High-Linearity 140–160 GHz Active Phase Shifter in 55 nm BiCMOS," *IEEE Microwave and Wireless Components Letters*, vol. 31, no. 2, pp. 157–160, 2021.

[5] R. B. Yishay and D. Elad, "D-Band 360° Phase Shifter with Uniform Insertion Loss," in *IEEE MTT-S Int. Microw. Symp. Dig.*, 2018, pp. 868–870.

[6] S. G. Rao and J. D. Cressler, "A D-Band Reflective-Type Phase Shifter Using a SiGe PIN Diode Resonant Load," *IEEE Microwave and Wireless Components Letters*, vol. 32, no. 10, pp. 1191–1194, 2022.

RTu3B-3

A 140GHz RF Beamforming Phased-Array Receiver Supporting >20dB IRR with 8GHz Channel Bandwidth at Low IF in 22nm FDSOI CMOS

Shenggang Dong[#*], Navneet Sharma[#*], Sensen Li[#], Michael Chen[#], Xiaohan Zhang[$], Yaolong Hu[$], Jiantong Li[#], Yong Su[#], Xinguang Xu[#], Vitali Loseu[#], Eunyoung Seok[#], Taiyun Chi[$], Won-Suk Choi[#], Gary Xu[#]

[#]Samsung Research America, USA
[$]Rice University, USA
s.dong@samsung.com

Abstract— **A 140GHz 4-element RF beamforming phased-array receiver (RX) has been demonstrated in 22nm FDSOI CMOS. The proposed single-side-band architecture provides >25dB and >20dB measured image rejection ratio (IRR) across 4GHz and 8GHz channel bandwidth centered at 7GHz intermediate frequency (IF). Each front-end element consists of a wideband low-noise amplifier (LNA) and a vector-modulator phase shifter. The 4 elements are combined on chip through power combiners and driver amplifiers before the double-balanced mixer, which is driven by an on-chip multiplier (×9). The receiver consumes 480mW DC power and provides <10dB noise figure from 135 to 147 GHz. The RX is measured up to 32 and 24Gb/s in the probe and over-the-air test. To the authors' knowledge, this CMOS RF beamforming RX presents the largest channel bandwidth (8GHz) with 20dB IRR$_{min}$ at low IF consuming the lowest DC power per element (120mW) among the published phased-array RX in the 140GHz band.**

Keywords—**phased array, receiver, 6G, D-band, RF-beamforming, wireless, communication**

I. INTRODUCTION

The sub-terahertz (THz) portion of spectrum can provide a large bandwidth for 6G wireless communication, therefore allowing unprecedented services such as truly immersive mixed-reality [1]. To achieve sufficient signal-to-noise ratio and a large scan range, a 2D scalable phased-array receiver is needed on both base station and mobile devices. The digital/IF beamforming phased array receivers at 140GHz band have been demonstrated [2], [3]. Because the digital/IF beamforming requires a separate mixer and LO driver for each front-end element, it increases the layout complexity and DC power consumption, making it difficult to implement a scalable 2D array where footprint and thermal management capacity are limited. A 2D array using the RF beamforming is implemented in the SiGe BiCMOS process in [4]. Until now, there is still no CMOS RF beamforming phased-array receiver at 140 GHz.

A high IRR is preferred for the sub-THz RX to attenuate the noise and potential interferences from image bands. A high IF (11.5 GHz) is adopted to move images out of the RF front-end (FE) bandwidth in [3]. However, for a multi-chip large-array integration, the high IF scheme suffers from large IF routing trace loss and degraded direct IF sampling performance, therefore entailing a 2nd wideband down-converter. With RF FE filtering effect only, as the LO and image move closer to the passband, the IRR will also become worse. To mitigate these

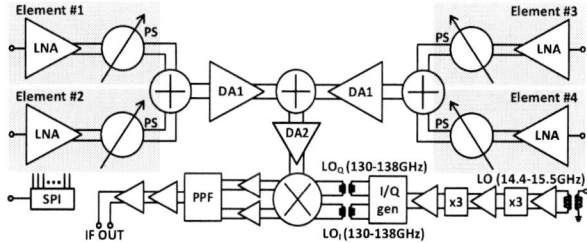

Fig. 1. Frequency plan.

Fig. 2. Block diagram of the 140-GHz RF beamforming RX

challenges, this article proposes an RF beamforming single-side-band (SSB) phased-array receiver in 22nm FDSOI CMOS. 4 elements are integrated on chip to demonstrate the scalability and provide flexibility on the system design to cover different application scenarios. Fig. 1 shows the RX frequency plan. With different LO frequencies, the RX can switch among 3 channels of 4GHz bandwidth. The center of IF frequency can be fixed at 7 GHz or below to ease the large-array integration and allow direct IF sampling using high speed ADCs. Section II discusses the RX architecture and its implementation. Measurement results are presented in Section III. Finally, this article is concluded in Section IV.

II. RECEIVER ARCHITECTURE AND BUILDING BLOCK

Fig. 2 shows the receiver block diagram. Each front-end element consists of a wideband LNA and an active phase shifter (PS). 4 elements are combined on chip through power combiners and driver amplifiers (DA) before the double-balanced mixer, which is driven by a quadrature LO generated from an on-chip multiplier (×9). The mixer outputs are amplified and then recombined through a polyphaser filter (PPF) to realize the SSB down-conversion [5]. Thanks to low IF, no

---

* equal contribution

979-8-3503-2123-4/23 $31.00 © 2023 IEEE   293   2023 IEEE Radio Frequency Integrated Circuits Symposium

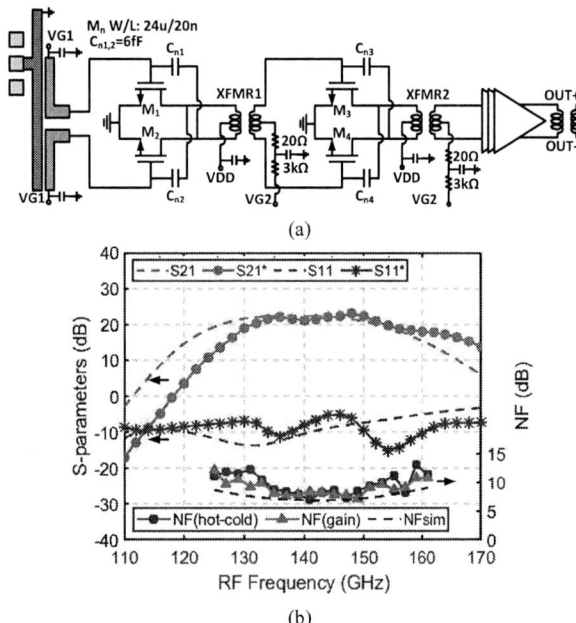

Fig. 3. (a) LNA schematic (b) test structure measurement results (dashed lines: simulated results)

bulky inductor or transformer is needed in the wideband IF amplifier chain. A serial-to-parallel interface (SPI) is integrated to control the phase shifters.

A. LNA

Fig. 3(a) shows the 5-stage LNA schematic. Each stage is made of a differential amplifier with neutralization capacitors (C_n). The value of C_n is chosen to increase the maximum available gain of the transistor core and achieve the highest stability factor [3]. The transistor layout is optimized to minimize the gate resistance and EM modelling mismatch. Its parasitic including C_n is extracted using HFSS. Low-K transformers are adopted for LNA inter-stage matching to extend the bandwidth [6].

To verify the LNA performance, an LNA test structure is taped-out and measurement results are shown in Fig. 3(b). The last transformer in the LNA is changed to a balun for a 50Ω probe load. The S-parameter test shows the LNA has a 3dB bandwidth of 20 GHz (133-153 GHz) with a peak gain of 22dB. The noise figure (NF) of the LNA is measured using both hot-cold (Y factor) and gain methods achieving 7-10dB from 133 to 153 GHz. The 1dB worse NF than simulation can be caused by the extra input impedance mismatch and process variation.

B. Phase shifter

To achieve the full field-of-view RF beamforming with high-resolution, a D-band 5-bit digitally-controlled vector-modulator phase rotator is designed for a 360° phase-shifting range with the resolution of 3°. The RF signals are converted to differential I/Q phase through the differential quadrature coupled-line coupler and fed to the I/Q variable gain amplifiers (VGA) (Fig. 4(a)). Each VGA has a 5-bit control on the binary-weighted cell implemented as differential amplifiers and

controlled by tail switches similar to [7]. The stack transistor is removed for larger headroom as shown in Fig. 4(b). The PS consumes only 25mW DC power under 1V supply with -5dBm output P_{1dB} in simulation. 1024 points in the phase constellation diagram are formed after combining the I and Q vectors. The simulated RMS phase error is 1.2° with 3° phase resolution. At 140 GHz, the simulated PS NF and conversion loss are 16.5dB and 9.5dB at 65 °C. With the LNA, the noise impact of PS is reduced to less than 1dB additional NF at RX input.

C. Power combining path

Transformer-based Wilkinson power combiners are used to combine the 4 front-end elements. 3-stage differential driver amplifiers (DA1+DA2) with neutralization capacitors are inserted in the power combining path to compensate the trace loss and reduce mixer noise impact. The DAs can provide 13dB total gain at 65°C in the post-layout simulation at 140 GHz.

D. SSB frequency downconverter

The SSB frequency downconverter consists of a Gilbert-cell based double balanced IQ mixer, a wideband IF amplifier chain and a frequency multiplier (×9), which generates the 130-to-138GHz I/Q local oscillator (LO) signals. The quadrature phase of the LO is generated by the same differential I/Q generation block in the phase shifter. The available power at the mixer gate is 0dBm in simulation to compensate process and temperature variation. The LO chain consumes 100mW DC power in total. Since 4 FE elements share the same LO chain, the DC power contribution due to LO chain is only 25mW/element.

The outputs of the mixer are fed to the trans-impedance amplifiers (TIA). The low input impedance of TIA reduces the impact of mixer IF trace mismatch and extends the IF bandwidth [8]. Another differential amplifier is inserted between TIA and PPF to provide common mode rejection ratio (CMRR) and compensate the signal loss from PPF. The high CMRR helps to provide a clean I/Q signal to PPF. The PPF is based on a 3-stage type-II RC-CR architecture designed

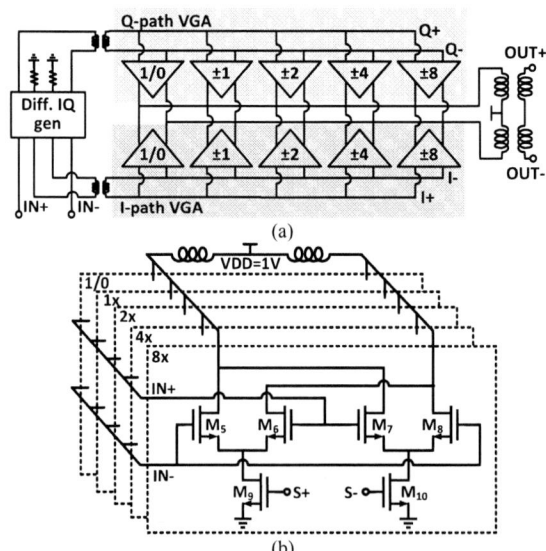

Fig. 4. (a) phase shifter block diagram and (b) I-path VGA schematic

979-8-3503-2123-4/23 $31.00 © 2023 IEEE

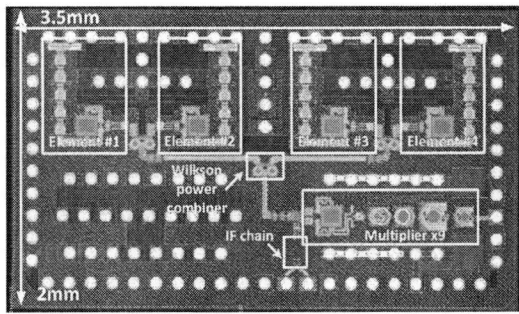

Fig. 5. Die photograph of the 140GHz RX

for >30dB IRR from 5 to 9 GHz in simulation. After the PPF, a wideband Cherry-Hooper amplifier is added to boost the gain and broaden the bandwidth [9]. Finally, the output is fed to the digital backend through a 50Ω data driver. To block the DC offset and create a high frequency gain peaking, AC-coupling capacitors are inserted after mixer and PPF. Post-layout simulation including the LO, mixer and IF chain shows the SSB down-converter itself provides 12GHz 3dB gain bandwidth at 134GHz LO with >30dB IRR from 5 to 9GHz IF.

III. Measurement Results

The SSB RF beamforming RX is fabricated in the GF 22nm FDSOI CMOS process, occupying an area of 3.5 mm × 2 mm (Fig. 5). The pad frame is designed to be compatible with the antenna-in-package (AiP). The chip is measured using both on-chip probing and the over-the-air link. All the DC biases are provided through bonding wires for the initial verification.

A D-band multiplier (WR6.5-VNAX) is used as the TX for the continuous wave measurement. The TX output is calibrated with the power meter (VDI PM5) and then attenuated by a tunable attenuator (STA-60-06-D1) before being fed to the RX input through a GSG probe (GGB, 110-170GHz). Fig. 6 shows the single-element small-signal conversion gain with a fixed IF frequency (5 GHz) when RF and LO frequencies are swept. The RX is measured with a 3dB gain bandwidth of 12 GHz (131-143 GHz). Compared to the simulation, the FE center frequency is shifted down by about 5 GHz and 3dB bandwidth is reduced by about 2 GHz. This can be caused by the extra dummy fill and EM modelling mismatch in the PS and DA.

The single-element NF is characterized using hot-cold method. The IF is fixed at 5 GHz when LO frequencies are swept with a D-band noise source (Eravant STZ-06-L1) replacing the VDI TX and its attenuator. During the test, phase shifters in the other 3 elements are set to provide the minimum gain. Fig. 6 shows the NF of the full RX chain is below 10dB from 135 to 147 GHz. Since the equalizer is included in the digital baseband to account for the channel response, the 10dB smooth gain drop (140-147 GHz) can be easily compensated. As long as the NF is flat, the gain droop will not influence the demodulated SNR.

With RX system bandwidth of 12 GHz (135-147 GHz), small-signal conversion gain is measured at 130, 134 and 138GHz LO frequencies, which match the three 4GHz channels in Fig. 1. Fig. 7(a) and (b) present the measured conversion gain and its corresponding IRR. The IRR for all 3 LO frequencies is

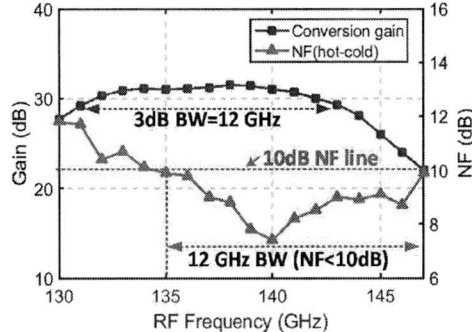

Fig. 6. RX conversion gain and noise figure with fixed IF

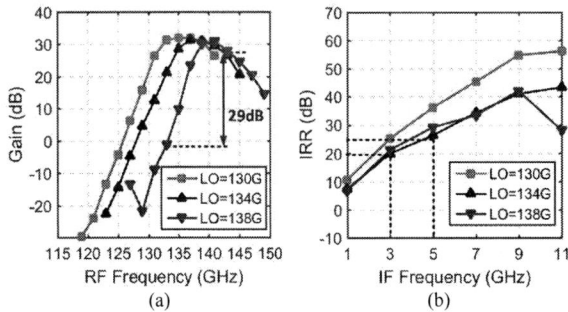

Fig. 7. (a) RX conversion gain at 3 LO frequencies (b) its IRR

maintained >25dB from 5 to 9 GHz IF and >20dB from 3 to 11 GHz IF. This high IRR is the combined results of both the SSB down-conversion and RF front-end filtering. To decouple these two, the conversion gain at 133 and 143 GHz is compared with 138 GHz LO, due to the similar FE gain, the 29dB IRR is mainly from the SSB down-converter.

RF phase shifter is measured based on a similar setup as conversion gain. The RX output is fed to a signal analyzer (DSA-X 91604A). An arbitrary waveform generator (Keysight M8195) is used to improve the instrument synchronization. The phase shifter control bits are swept through on-chip SPI. Within 1dB gain variation, the measured RMS phase error is 1.4° with phase resolution of 3°.

To verify the demodulation performance, the RX is measured at first with on-chip probing with 16-QAM at 4 GBaud/s for all three 4GHz channels. Measured RMS error vector magnitude (EVM_{RMS}) shows only about 1dB variation. Fig. 8 shows the EVM_{RMS} vs RX available input power for the 135-139GHz channel. The lowest EVM_{RMS} for 4GBaud/s 64-QAM (24Gb/s) and 16-QAM (16Gb/s) demodulation are 5.8% and 6.7% respectively. The gain of DA and mixer is tuned down by 12dB in total to increase the RX input P_{1dB} for 1-element probe characterization at the expense of 1-2dB NF penalty. Only one element is probed, therefore each power combiner introduces 3dB extra loss due to the asymmetry. Nominal bias is applied for the over-the-air test. Thanks to the high IRR at low IF, 8GBaud/s 16-QAM (32Gb/s) signals are also demodulated successfully using 135-143GHz RF bandwidth at 3-11GHz IF frequency (Fig. 8). This provides sufficient flexibility for various IF choices and modulation schemes.

A 64-element RX module with 16 RX chips is built to perform an over-the-air test. The input RF GSG pads of each

979-8-3503-2123-4/23 $31.00 © 2023 IEEE   295

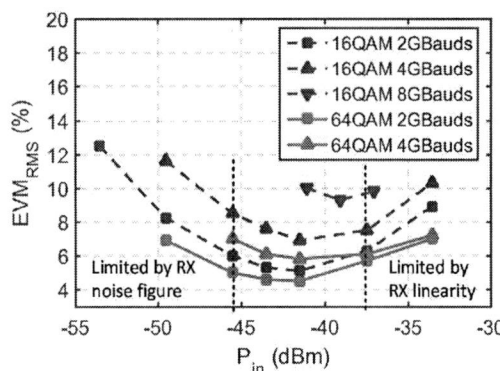

Fig. 8. EVM$_{RMS}$ versus RX available input power

Table 1. Performance comparison of D-band phased array receivers.

	This work	[2]	[3]	[4]
Element per chip	4	4	8	4$^{\&}$
Beamforming	**RF**	IF	IF	RF
RF 3dB gain BW (GHz)	**131-143**	125-145	137-151	122-145$^{@}$
NF (dB) (Freq. (GHz))	**7.5-10 (135-147)**	8.5*	6.4-7.5 (134-149)	7.5-10 (125-145)
Gain (dB)	**32**	27	26.5	30
Mod. BW (GHz)	**4-8/carrier**	20	4-5/carrier	-$^{\&}$
IP1dB (dBm) (@Freq. (GHz))	**-45/-33$^{\#}$ @140**	-30 @135	-28.5 @139	-27$^{\&}$
IF Freq. (GHz) (Min. IRR)	**5-9 (25dB) / 3-11 (20dB)**	-	9-14 (20dB$^{\wedge}$)	-$^{\&}$
SSB	**Yes**	No	No	No
Max. data rate (probe/air) (Gb/s)	**32$^{\#}$/24**	-/1.92	-/10	30$^{\&}$/-
PS resolution /RMS error (°) (Freq. (GHz))	**3°/1.4° (140)**	-	11.25°/<6° (143-147)	-
Technology	**22nm CMOS SOI**	22nm CMOS SOI	45nm CMOS SOI	130nm SiGe BiCMOS
DC power (mW) /element	**120/105$^{\#}$**	198	145	200$^{\&}$

*simulated, $^{@}$estimated from S21, $^{\&}$requires a separate chip for down-conversion. $^{\#}$Low bias for high P$_{1dB}$, $^{\wedge}$: Averaged IRR.

element are bonded to the 1×3 sub-array PCB patch antenna and IF signals are combined on board before being fed to the signal analyzer. Compared to other assembly options at D-band [2], PCB provides a lower cost and shorter turn-around time at the expense of RF performance. Fig. 9 shows the fully assembled RX and the measurement setup. The beam-aligned 64-element RX successfully demodulates the 138GHz 4GBaud/s 64-QAM (24Gb/s) waveforms at 6.4% EVM$_{RMS}$ with 18dBm EIRP VDI TX at 1-meter distance. More tests about beam scanning are ongoing when this paper is submitted.

Table 1 compares the performance of the published phased-array RX at D-band. To the authors' knowledge, this CMOS RF beamforming receiver consumes the lowest DC power per element (120mW) to achieve the largest channel bandwidth (8 GHz) with 20dB minimum IRR at low IF in 140GHz phased

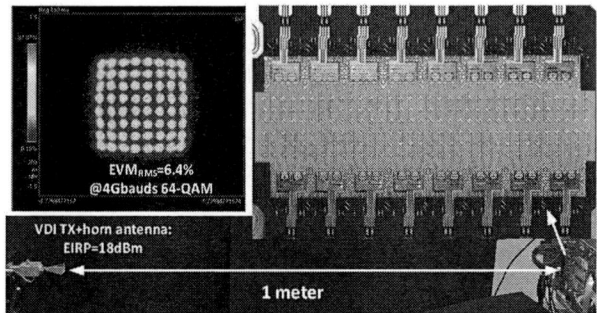

Fig. 9. 64-element RF beamforming RX array over-the-air test

array receivers. More importantly, the low-IF architecture allows direct IF sampling, low-loss and low-cost package, greatly simplifying the multi-chip large-array integration.

IV. CONCLUSION

A 140GHz 4-element RF beamforming phased-array receiver has been demonstrated in 22nm FDSOI CMOS. The proposed RX provides >25dB and >20dB IRR across 4GHz and 8GHz channel bandwidth at low IF (7GHz). The lowest DC power per element, high IRR at low IF and compact footprint are ideal for future integration in a large-scale 2D array for 6G communication at D band.

ACKNOWLEDGMENT

The authors would like to thank GlobalFoundries for chip fabrication and Professor Ho-Jin Song from Pohang University of Science and Technology for helpful technical discussions.

REFERENCES

[1] Samsung Research, "6G: The Next Hyper Connected Experience for All.," White Paper, pp. 1–44, 2020.

[2] A. A. Farid et al., "135GHz CMOS / LTCC MIMO Receiver Array Tile Modules," *2021 IEEE BiCMOS and Compound Semiconductor Integrated Circuits and Technology Symposium (BCICTS)*, Monterey, CA, USA, 2021, pp. 1-4

[3] S. Li, Z. Zhang, B. Rupakula and G. M. Rebeiz, "An Eight-Element 140-GHz Wafer-Scale IF Beamforming Phased-Array Receiver with 64-QAM Operation in CMOS RFSOI," in *IEEE Journal of Solid-State Circuits*, vol. 57, no. 2, pp. 385-399, Feb. 2022.

[4] M. Elkhouly et al., "Fully Integrated 2D Scalable TX/RX Chipset for D-Band Phased-Array-on-Glass Modules," *2022 IEEE International Solid-State Circuits Conference (ISSCC)*, 2022, pp. 76-78

[5] S. F. Behbahani et al., "CMOS mixers and polyphase filters for large image rejection," in *IEEE Journal of Solid-State Circuits*, vol. 36, no. 6, pp. 873-887, June 2001.

[6] S. V. Thyagarajan, A. M. Niknejad and C. D. Hull, "A 60 GHz Drain-Source Neutralized Wideband Linear Power Amplifier in 28 nm CMOS," in *IEEE Transactions on Circuits and Systems I: Regular Papers*, vol. 61, no. 8, pp. 2253-2262, Aug. 2014

[7] Y. Hu, X. Zhang and T. Chi, "A 28GHz Hybrid-Beamforming Transmitter Array Supporting Concurrent Dual Data Steams and Spatial Notch Steering for 5G MIMO," *2021 IEEE Custom Integrated Circuits Conference (CICC)*, Austin, TX, USA, 2021, pp. 1-2.

[8] S. Dong et al., "A 10-Gb/s 180-GHz phase-locked-loop minimum shift keying receiver," *IEEE Journal of Solid-State Circuits*, vol. 56, no. 3, pp. 681–693, March 2021.

[9] A. Townley et al., "A Fully Integrated, Dual Channel, Flip Chip Packaged 113 GHz Transceiver in 28nm CMOS supporting an 80 Gb/s Wireless Link," *2020 IEEE Custom Integrated Circuits Conference (CICC)*, Boston, MA, USA, 2020, pp. 1-4.

RTu3B-4

A mm-Wave Blocker-Tolerant Receiver Achieving <4 dB NF and -3.5 dBm B1dB in 65-nm CMOS

Erez Zolkov, Nimrod Ginzberg, Emanuel Cohen

Faculty of Electrical Engineering, Technion, Israel

Abstract—**Digital beam-forming requires highly linear receivers (RXs), as null steering is performed only in digital baseband (BB). This paper presents a highly linear RX, with a high out-of-band (OOB) blocker tolerance without sacrificing performance or power, utilizing a highly linear inverter low-noise amplifier (LNA), followed by an N-path mixer with tunable filtering properties and a BB transimpedance (TIA) amplifier for linearity enhancement. The N-path mixer design trade-offs are discussed, and several linear LNA topologies are presented and compared. A chip prototype was manufactured in a TSMC 65 nm CMOS process. In our implementation, a <4 dB noise figure (NF) is achieved, with an RX gain of 40 dB, in-band (IB) IIP3 of -20 dBm and -3.5 dBm B1dB at a 500 MHz offset, while occupying an active area of 1.62 mm^2 and drawing a total power of 76.8 mW, at a frequency range of 22-31 GHz.**

Keywords — **CMOS, digital beam-forming, mm-wave receiver, N-path mixer.**

I. INTRODUCTION

Mm-Wave phased array and multi-input multi-output (MIMO) systems are used in to improve channel capacity and system reliability. Analog beamforming at RF or IF allows for a robust and power efficient operation, but is limited to a single-beam only [1]. Digital beamforming, on the other hand, has become increasingly popular due to its ability to perform multi-beam operation through digital signal processing (DSP).

The implementation of digital beamforming requires highly linear receivers (RXs), as in-band (IB) and out-of-band (OOB) interferers are only cancelled in digital baseband (BB) [1]. The low-noise amplifier (LNA)-first with an active mixer RX [2], as shown in Fig. 1a, achieves a low noise figure (NF) but has limited linearity in terms of blocker-induced compression (B1dB), as it does not consist of any blocker rejection mechanisms. Mm-Wave mixer-first RXs (MFRXs) [3]–[5], shown in Fig. 1b, incorporate N-path mixers (Fig. 1c) instead of conventional active mixers to achieve blocker rejection by creating a frequency tunable band-pass shaped impedance at RF, resulting in a high B1dB at the cost of a higher NF.

In this work, we propose the combination of a linear single-stage LNA followed by an N-path mixer to achieve both low NF and good B1dB. Shown in Fig. 1d, the BB chain includes a transimpedance amplifier (TIA) and a shunt capacitor [6], [7]. Since the LNA comprises only a single amplification stage, all RF voltage nodes except the LNA's input experience virtual ground in-band (IB) and filtering out-of-band (OOB), resulting in both high IB linearity and B1dB.

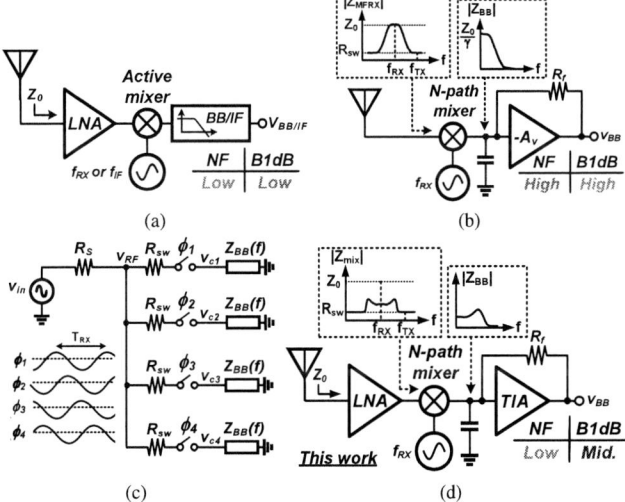

Fig. 1. (a) LNA-first RX with an active mixer, (b) mixer-first RX, (c) N-path mixer (N=4) with its clocks timing diagrams, and (d) LNA-first RX with an N-path mixer and a BB TIA.

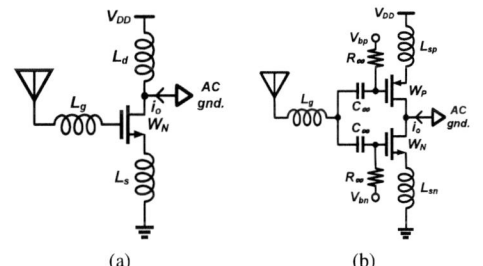

Fig. 2. Inductive degenerated LNA topologies drawings, of: (a) common-source amplifier, and (b) inverter.

II. CIRCUIT DESIGN

A. Highly Linear LNAs

For the following analysis, we assume an ac ground at the LNA's output, as in our architecture the LNA is followed by an N-path mixer with BB TIA and shunt capacitance, resulting in a low input impedance both IB and OOB. For simplicity, we also assume that the ratio of c_{gd}/c_{gs} is constant for all device widths. The common source (CS) amplifier with inductive source degeneration (SD), as shown in Fig. 2a, is a common mm-wave LNA topology due to its implementation simplicity and its capability to achieve low NF and input matching simultaneously. The input impedance is given by [8]:

$$Z_{in} \approx sL_g + \left(\frac{1}{sc_{gd}} \, \| \, \left(\frac{1}{sc_{gs}} + sL_s + \omega_T L_s \right) \right) \quad (1)$$

979-8-3503-2123-4/23 $31.00 © 2023 IEEE   297   2023 IEEE Radio Frequency Integrated Circuits Symposium

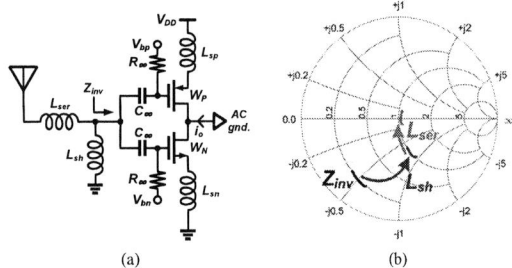

(a)   (b)

Fig. 3. (a) Inductive degenerated inverter LNA with the proposed MN, and (b) MN principle of operation drawn on a smith chart.

where $\omega_T = g_m/c_{gs}$ and is assumed to be constant for all device widths [8] at a given bias. Impedance matching is achieved by setting L_s so that the real part of the input impedance is matched, and L_g is set to achieve resonance at the frequency of interest. Note that $\omega_T L_s$ has to be set slightly higher than the source impedance due to the shunt c_{gd}, which causes scaling down of the real part of the input impedance. The transconductance at the resonance frequency is then given by:

$$ G_m = \frac{v_{gs}}{v_{ant}} g_m \approx \frac{(2\omega_0 Z_0 c_{gs})^{-1}}{\left(1 + \frac{c_{gd}}{c_{gs}}\right)\left(1 + j\omega_0\omega_T L_s \frac{c_{gd}c_{gs}}{c_{gd}+c_{gs}}\right)} g_m \quad (2) $$

with v_{ant} as the source voltage. The NF is given by:

$$ F \approx 1 + \frac{R_g}{Z_0} + \gamma g_m Z_0 $$
$$ \times \left(\frac{\omega_0}{\omega_T}\right)^2 \left(1 + \frac{c_{gd}}{c_{gs}}\right)^2 \left| 1 + j\omega_0\omega_T L_s \frac{c_{gd}c_{gs}}{c_{gd}+c_{gs}} \right|^2 \quad (3) $$

where R_g accounts for the MOS gate induced noise and is approximately equal to $R_{poly} + 1/(5g_m)$, and γ is the MOS channel's noise coefficient. Since the LNA's output is ac ground, the RX linearity is mainly determined by the compression at the LNA input, i.e. the v_{gs} voltage swing. Thus, it is desirable to lower v_{gs}/v_{ant} while increasing g_m by the same factor to maintain a constant G_m. By examining (2), we see that the above can be achieved by increasing the device's width, leading to the increase of both c_{gs} and g_m by the same factor. However, such increment results in higher bias currents, leading to a trade-off between power and linearity.

To mitigate this trade-off, the LNA can be implemented as an inductive source-degenerated inverter, as shown in Fig. 2b. CMOS based inverter amplifiers are known to have superior linearity compared to CS stages [9] due to their ability to maintain a constant G_m over wider input swings (given that the NMOS and PMOS devices were designed for relatively equal transconductance), and provide higher G_m values for a given bias current due to current reuse. However, to achieve matching, the proposed technique dictates the choice of much higher source inductors compared to the CS stage, as the generated real input impedance of the NMOS and PMOS branches appear in parallel, along with the addition of the higher equivalent c_{gd} and c_{gs} values. The above constraints lead in turn to severe G_m and NF degradation, especially for large device widths, according to equations (2,3).

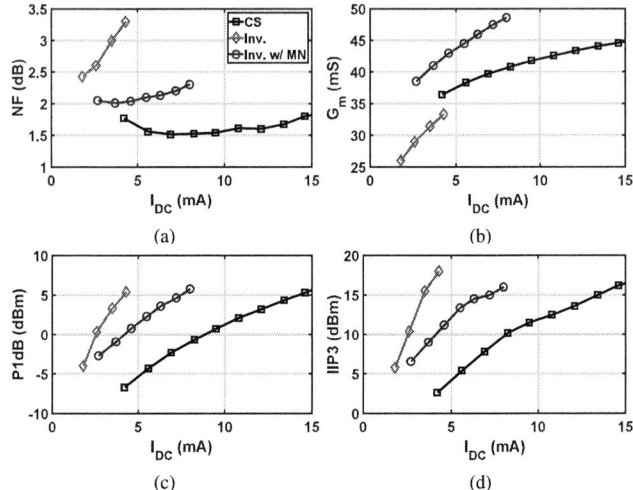

(a)   (b)

(c)   (d)

Fig. 4. Comparison of simulation results between the 3 presented topologies, of :(a) NF, (b) G_m (c) P1dB and (d) IIP3. Simulations were conducted at 28 GHz, with ideal output AC ground and ideal inductors with $Q_L = 20$.

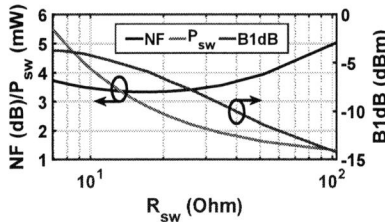

Fig. 5. Simulation results of NF, B1dB and estimated LO power vs. the NMOS switches' R_{sw}, for an inverter LNA with MN (I_{DC}=3.7 mA) followed by a 4-phase N-path mixer at 28 GHz.

We propose to mitigate the inverter LNA's NF and G_m degradation by implementing an inverter LNA with an alternative matching network (MN) instead of the series inductor technique. The proposed topology is shown at Fig. 3a, with a MN comprised of a shunt inductor L_{sh} and a series inductor L_{ser}. The MN's principle of operation is depicted in Fig. 3b and is as follows. The input impedance of the inverter (Z_{inv}) is comprised of a small real part due to the low source inductance values, and a negative imaginary part due to the c_{gs} and c_{gd} of the devices. L_{sh} shifts Z_{inv} to the $Re\{Z\} = Z_0$ circle at the smith chart, and L_{ser} resonates the remaining imaginary part. The advantage of this MN over the standard series inductor is that inductors with lower values can be picked for SD (down to the value of a CS stage source inductor), leading to lower NF and higher G_m values according to (2,3). The proposed MN can be implemented compactly by using coupled inductors.

A comparison of the simulated performance at 28 GHz of all three presented LNA topologies is presented in Fig. 4. A constant bias-width ratio was maintained for all devices, and for each bias current point the input inductors were adjusted to achieve matching, while the devices' SD inductors values were kept constant. We see that besides a 0.25-0.5 dB NF penalty, the inverter with a MN is favorable over the CS stage, in terms of achievable G_m, P1dB and IIP3 for low bias currents.

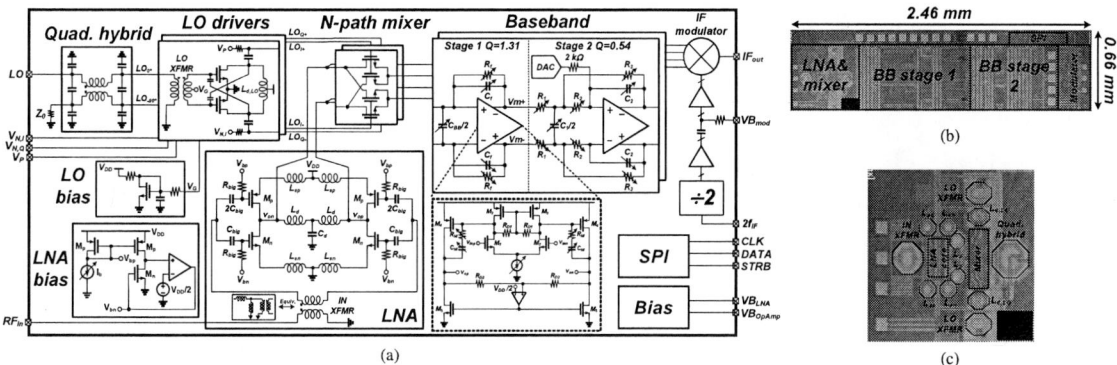

Fig. 6. (a) Chip schematic of the implemented RX, and micrograph pictures of: (b) entire RX chain, and (c) LNA, mixer and LO phases generation circuits zoomed in.

B. N-path Mixer

The performance of mm-wave N-path mixers is mainly determined by the switches' resistance, namely R_{sw} [5]. Widening the switch results in lower R_{sw}, which reduces the NF and increases B_{1dB} [6] at the cost of higher local oscillator (LO) power consumption and higher phases overlapping losses [5]. Thus, a trade-off exists between power, B1dB and NF. The above claims are verified in Fig. 5, which plots the above merits at 28 GHz vs. an NMOS LVT device's R_{sw}, for an RX with an inverter-based LNA with MN (I_{dc}=3.7 mA) followed by an N-path mixer fed by sine wave LOs with an optimized crossover point for power and noise reduction, and an ideal virtual ground at the BB node. We see that the NF reaches a "sweet spot" at about 13-14 ohms, where R_{sw} is low but the overlapping losses are still negligible. The B1dB point rises as R_{sw} decreases, until it saturates at -3.5 dBm due to the LNA's B1dB limit, which is approximately 3 dB lower than the P1dB shown in Fig. 4c.

III. IMPLEMENTATION

We have implemented the RX in TSMC's 65nm CMOS process. The chip schematic is shown in Fig. 6a, and its micrograph is shown in Fig .6b, with a zoom-in of the LNA, mixer and LO phases generation circuits at Fig. 6c. The RX occupies an active area of $1.62\ mm^2$ and consumes 76.8 mW at the frequency range of 22-31 GHz.

The LNA is implemented as a differential inductive SD inverter, with an input transformer that acts both as a matching network and a balun, and an output inductor load to resonate the LNA's output capacitance and the N-path mixer's parasitic capacitance. The NMOS and PMOS gate bias voltages are generated by a dedicated bias block. The PMOS gate voltage is set according to the variable bias current, and the NMOS gate voltage is set accordingly by the OpAmp negative feedback path so that an output common mode voltage of $V_{DD}/2$ is obtained. The LNA's outputs are fed directly to a differential 4-phase N-path mixer with special layout attention so that a minimal distance is obtained between the LNA and the mixer, to achieve a virtual ground at the LNA's outputs.

To generate the LO phases, an external LO is fed to a quadrature hybrid (QH), whose two quadrature outputs are

followed by two LO drivers. Each LO driver converts the QH's output to two differential signals by a balun, which are then amplified by differential cross-coupled inductive loaded CS stages. Finally, the LO drivers' outputs are followed by high-pass stages with controllable mixer gate bias voltages. Simulations show that a -165 dBc/Hz phase noise (PN) is achieved in our design. To compensate for any mixer or LO phases mismatches that lead to DC offsets and IIP2 degradation, the positive phases share the same external bias voltage (V_P), but each negative IQ LO phases has its independent external bias voltage ($V_{N,I}$ and $V_{N,Q}$).

The BB portion consists of two stages to obtain a 4th-order Butterworth low-pass filter. Each stage incorporates a two-stage Miller-compensated OpAmp with a common mode feedback (CMFB) circuitry. The first stage achieves a high Q by utilizing a shunt-feedback TIA technique [10] to enable a 2-pole blocker rejection while causing minimal noise degradation. This stage consumes a relatively high bias current to achieve low noise and high bandwidth. The second stage is designed as a Rauch filter due to its simplicity in achieving a high gain for the required Q of 0.54, with a single OpAmp. An R2R digital-to-analog converter (DAC), whose output is placed at the second OpAmp positive input, is used to compensate for any DC offset. Finally, to ease measurements, an intermediate frequency (IF) modulator converts the four IQ BB outputs to an RF frequency of 1 GHz. The IF modulator is designed as a passive IQ up-converter with digital clocks generated by an on-chip digital divider, and controllable pulse widths set externally by VB_{mod}.

IV. MEASUREMENT RESULTS

The fabricated chip was measured using a probe station. Note that in the following measurement results we have de-embedded the modulator's loss. Fig. 7a shows input matching for LO of 28 GHz. A <-10 dB S11 is achieved at the frequency range of 24-40 GHz. The maximum RX gain was measured for LO frequencies between 22 to 31 GHz and is shown at Fig. 7b, along with the simulated LNA gain normalized to the maximum gain achieved at 28 GHz LO. We see that the achievable RFBW is 215 MHz with a maximum gain of 40 dB at 28 GHz, and that the RX gain plots match the

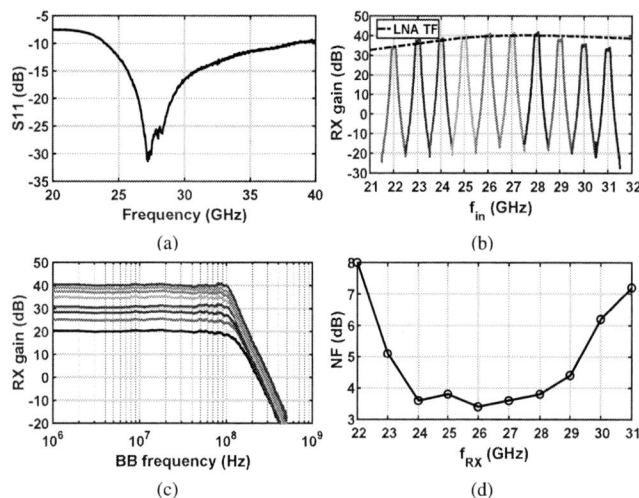

Fig. 7. Small signal measurements of: (a) S11 as f_{RX}=28 GHz, (b) RX maximum gain for various LO frequencies, (c) RX gain for all gain steps at 28 GHz LO, and (d) NF at maximum gain for various LO frequencies.

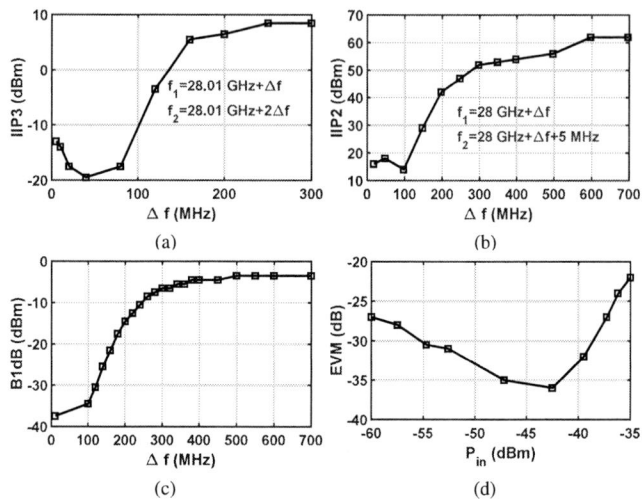

Fig. 8. Large signal measurements, as f_{RX}= 28 GHz and maximum RX gain, of: (a) IB and OOB IIP3, (b) IB and OOB IIP2, (c) B1dB, and (d) EVM vs. input power for a 100 MHz 64QAM OFDM modulated input signal.

LNA gain up to 28 GHz while deviating from it at higher LO frequencies due to the LO drivers network limited frequency of operation. The RX gain at 28 GHz LO can be configured to provide 20-40 dB DC gain with >100 MHz corner frequency and 80 dB/dec low pass filtering (Fig. 7c). The measured NF at maximum gain is presented at Fig. 7d, showing an NF between 3.4 to 3.8 at LO frequencies between 24-28 GHz.

Large signal measurements at f_{RX}=28 GHz and maximum RX gain are presented in Fig. 8. The IIP3 measurements (Fig. 8a) were obtained by injecting two tones at $f_1 = 28.01GHz + \Delta f$ and $f_2 = 28.01GHz + 2\Delta f$, with Δf swept, so that the IM3 product would always fall at 28.01 GHz. We see that the IB IIP3 is bigger than -20 dBm and that at far OOB frequencies a 9 dBm OOB IIP3 is obtained. Similarly, the IIP2 measurements (Fig. 8b) were obtained by injecting two tones at $f_1 = 28GHz + \Delta f$ and $f_2 = 28GHz + \Delta f + 5$ MHz so that the IM2 would fall at 5 MHz for every Δf. The resulting IB IIP2 and OOB IIP2 were measured to be 20 dBm and 62 dBm, respectively. B1dB measurements (Fig. 8c) reveal an IB P1dB of -34.5 dBm (B1dB is 3 dB lower than P1dB) and a B1dB of -3.5 dBm at >500 MHz offset. Fig. 8d shows the EVM vs. input power plot for a 100 MHz RFBW 64QAM OFDM signal. We see that the minimum achievable EVM is -36 dB, a limit which is set by our measurement equipment.

V. CONCLUSION

In this work we have presented a blocker-tolerant mm-wave RX with low NF and high IB and OOB linearity. Several highly-linear LNA topologies have been presented and compared and N-path mixer design considerations were discussed. Implementation in TSMC 65-nm CMOS process is shown. Measurement results show 40 dB RX gain with 4th order BB low-pass filtering, low RX NF and high B1dB, OOB IIP2 and IB IIP3 while drawing relatively low power, comparing to prior art (Table 1).

Table 1. Comparison to prior published mm-wave RXs

Features	GOMAC 19' [3]	JSSC 19' [2]	RFIC 20' [4]^a	JSSC 20' [5]^b	This work
Architecture	MFRX	LNA w/ active mixer	MFRX	MFRX	LNA w/ N-path mixer
Technology	SOI CMOS 45 nm	SOI CMOS 45 nm	CMOS 28 nm	CMOS 65 nm	CMOS 65 nm
f_{RF} (GHz)	5-30	24.5-43.5	10-35	21-29	22-31
RFBW (MHz)	200	19000	500	1000	215
Max gain (dB)	21-23	35	11-15	3-6	40
NF (dB)	5-9.8	3.2-6.1	12.5-15.7	12-14.5	3.4-3.8^c
IB IIP3@ max gain (dBm)	N/R	-17.3	+10-+14.1	N/R	-20
B1dB@ max gain (dBm)	-10	-28.5^d	-5.5 to -3^d	6.5 ($\frac{\Delta f}{BW}$=6)	-3.5 ($\frac{\Delta f}{BW}$=2.5)
Power (mW)	164-300	60	LO:19-37 BB:22.8	LO:13.2 BB:9.6	LNA:9.6 LO:24 BB:43.2
Active area (mm²)	N/R	0.66	N/R	0.63	1.62

^{a} Nominal settings ^{b} BB Amp ON ^{c} At 24-28 GHz ^{d} Est. from IP1dB

REFERENCES

[1] C. Fulton et al., "Digital Phased Arrays: Challenges and Opportunities," in Proceedings of the IEEE, Mar. 2016.

[2] M. -Y. Huang et al., "A 24.5–43.5-GHz Ultra-Compact CMOS Receiver Front End With Calibration-Free Instantaneous Full-Band Image Rejection for Multiband 5G Massive MIMO," in JSSC, May 2020.

[3] S. Hari, "A 5 to 31 GHz four-phase mixer-first receiver," GOMAC., Mar. 2019.

[4] S. Krishnamurthy and A. M. Niknejad, "10-35GHz Passive Mixer-First Receiver Achieving +14dBm in-band IIP3 for Digital Beam-forming Arrays," RFIC, 2020.

[5] P. Song and H. Hashemi, "mm-Wave Mixer-First Receiver With Selective Passive Wideband Low-Pass Filtering," in JSSC, May 2021.

[6] E. Zolkov and E. Cohen, "A Quadrature Hybrid Transimpedance-Amplifier-Based Mixer-First Receiver," ISCAS, 2022.

[7] J. Jin et al., "An FDD Auxiliary Receiver with a Highly Linear Low Noise Amplifier," ESSCIRC, 2022.

[8] D. K. Shaeffer and T. H. Lee, "A 1.5-V, 1.5-GHz CMOS low noise amplifier," in JSSC, May 1997.

[9] H. Zhang and E. Sánchez-Sinencio, "Linearization Techniques for CMOS Low Noise Amplifiers: A Tutorial," in TCAS I, Jan. 2011.

[10] E. Säckinger, "The Transimpedance Limit," in TCAS I, Aug. 2010.

RTu3C-1

A Reactive Passive Mixer for 16-QAM Cartesian IoT Transmitters in 22 nm FD-SOI CMOS

Lorenzo Tomasin[#1], Daniele Vogrig[#], Andrea Neviani[#], Andrea Bevilacqua[#]

[#]University of Padova, Italy

[1]lorenzo.tomasin@studenti.unipd.it

Abstract — The use of a reactive passive mixer is proposed to implement an efficient Cartesian transmitter for IoT applications, capable of supporting high-order modulations, and high data rates. Prototypes in a 22 nm FD-SOI CMOS technology show a 5.5 dBm output-referred 1 dB compression point with 34.1% system efficiency in CW operation. Under a 2.4 Mbaud, 16-QAM modulation at 2.7 dBm average output power, they achieve 9.6 Mb/s data rate, EVM = -24.5 dB, ACLR = -32 dBc, and $P_{alt} = -36$ dBm with 22% system efficiency.

Keywords — Reactive passive mixer, IoT, transmitter, FD-SOI

I. INTRODUCTION

In Internet-of-Things (IoT) wireless systems, maximizing the efficiency is a primary goal. This is achieved by leveraging simple modulation formats, enabling the use of leaner transmitter (TX) architectures, e.g., polar modulators [1], [2]. With this approach, however, the achievable data rates are somewhat limited, precluding the possibility to support important emerging IoT applications requiring high data rates such as, multimedia IoT systems, video streaming, disaster monitoring, etc. When high-order modulations are employed to increase the data rate, Cartesian (I/Q) architectures are used to achieve sufficient linearity at low power, without incurring in penalties in terms of bandwidth expansion and need for calibration and pre-distortion [2], [3], [4]. Compared to polar modulators, I/Q modulators require, however, a more complex baseband interface, which adds power consumption and ultimately limits the system efficiency.

This work explores the use of a reactive passive mixer to implement an efficient I/Q transmitter for IoT applications. Offering a high impedance at the baseband port, the reactive mixer is compatible with the use of a simple, low power, resistive digital-to-analog converter (DAC), greatly simplifying the baseband digital-to-analog interface. The proposed IoT TX, developed as a test vehichle to show the potential of the reactive passive mixer, operates in the 2.4 GHz ISM band, supports high-order modulations (16-QAM), and achieves up to 9.6 Mb/s data rate at 22% system efficiency.

A block diagram of the proposed Cartesian IoT TX is shown in Fig. 1. Baseband in-phase and quadrature DACs directly drive the I/Q passive reactive mixer without the need of any additional active stage. The upconverted modulated signal is then amplified by a 2-stage power amplifier (PA), that drives the antenna port. Because of the power savings due to the use of the reactive passive mixer, most of the power in the TX (>90%) is dissipated by the power amplifier, as it should be in an efficient transmitter.

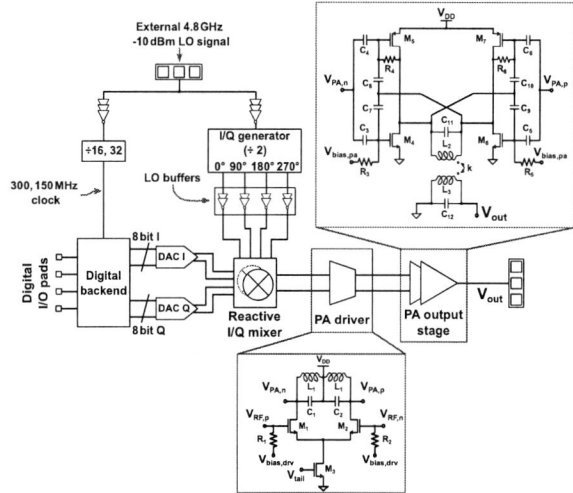

Fig. 1. Block diagram of the proposed low power transmitter.

II. ANALYSIS OF THE REACTIVE PASSIVE MIXER

Upconversion reactive mixers based on varactors are known to provide large gains at output frequencies in excess of 100 GHz [5]. The key feature used here is, however, that, in contrast to commonly used voltage-mode resistive passive mixers, capacitive reactive mixers can be driven, at baseband, by a high source impedance without compromising the in-band conversion gain. On the contrary, the filtering effect resulting from the large source resistance can be exploited as a reconstruction filter for the DAC output, without the need of an explicit circuit, and the related power consumption.

An equivalent circuit (Fig. 2) allows to analytically assess the in-band conversion gain, $A_{v,mix}$, of the reactive passive mixer:

$$A_{v,mix} = \frac{2}{\pi} \frac{\frac{C_{v,max}}{C_{v,min}} - 1}{\left(\frac{C_{v,max}}{C_{v,min}} + 1\right)\left(\frac{C_{drv}}{C_c} + 1\right) + 2\frac{C_{drv}}{C_{v,min}}} \quad (1)$$

The gain increases as the ratio between the maximum ($C_{v,max}$) and minimum ($C_{v,min}$) capacitance of the varactor increases: the upper bound for $A_{v,mix}$ is $2/\pi$. The input capacitance of the PA driver, C_{drv}, decreases $A_{v,mix}$ as it loads the varactor, effectively reducing the capacitance variation. The varactor sizing hence stems, for a target $A_{v,mix}$, from a given C_{drv}.

979-8-3503-2123-4/23 $31.00 © 2023 IEEE   301   2023 IEEE Radio Frequency Integrated Circuits Symposium

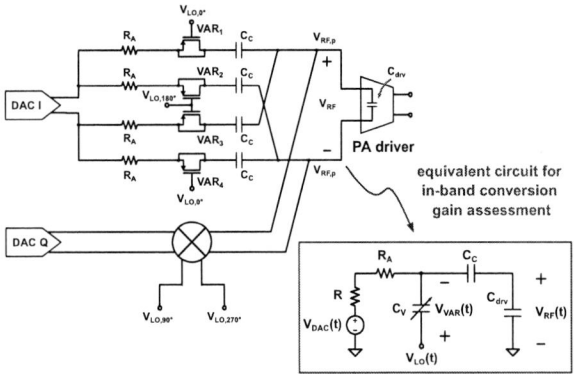

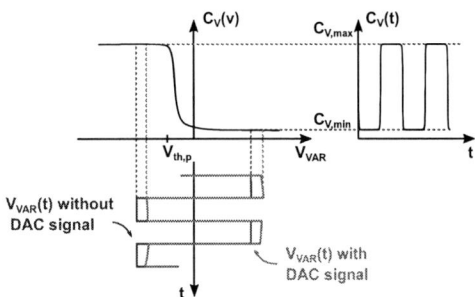

Fig. 2. Schematic of the reactive mixer and equivalent circuit for the assessment of the in-band conversion gain.

Fig. 3. Sketch of the voltage across the mixer varactors and varactor capacitance waveform.

A varactor with a sharp transition between $C_{v,min}$ and $C_{v,max}$ in the C-V characteristic is beneficial. In this condition, the varactor capacitance waveform, $C_v(t)$, and, consequently, the mixer gain, is basically independent on the input baseband signal as long as the amplitude of the latter is smaller than the swing of the local oscillator (Fig. 3). This largely improves the mixer linearity. The proposed reactive mixer is based on inversion-mode pMOS varactors for sharper transition between $C_{v,min}$ and $C_{v,max}$. Due to the use of a FD-SOI technology in this design, the accumulation region is avoided in the used varactor (the bulk is isolated), and the C-V characteristic is monotonic.

The mixer is double-balanced: $V_{RF,p}$ and $V_{RF,n}$ in Fig. 2, are ac-grounds for the odd harmonics of the local oscillator (LO), while the 2nd harmonics of the in-phase and quadrature LO signals cancel each other at nodes $V_{RF,p}$ and $V_{RF,n}$.

III. DESIGN OF THE CARTESIAN TX FOR IoT

A. TX Building Blocks

The baseband 8-bit pseudo-differential DACs are implemented with a simple R-2R topology. The choice of R = 10 kΩ limits the power consumption of each DAC to about 100 μW. Resistances $R_A = 90$ kΩ are added to set the DAC reconstruction filter bandwidth to 4 MHz.

The power amplifier is a 2-stage design (Fig. 1). The driver stage is a class-A tuned amplifier. The output stage is a class AB pseudo-differential design: a complementary topology is used to maximally exploit the reduced supply

(0.8 V in this design) without incurring in any reliability issue related to the voltage stress of the transistors. The output stage makes use of neutralization capacitors (C_7-C_{10}). It is loaded by a doubly-tuned transformer network, that also implements differential-to-single-ended conversion. The target peak output power is about 8 dBm.

B. Optimization of the PA Driver-Reactive Mixer Cascade

The PA driver and the reactive mixer are sized with the goal of minimizing the power consumption of the cascade, while ensuring that a signal large enough is provided to the PA output stage, such that the linearity of the overall TX chain is limited by the PA output stage.

Both the gain and the power consumption of the PA driver are approximately proportional to its input capacitance, C_{drv}. The reactive mixer is passive. However, dynamic power is dissipated at the LO port. It can be shown that the equivalent capacitance at the LO port increases with both the varactor capacitance $C_v(t)$, and the coupling capacitance C_c. Hence, for a given mixer gain, (1) dictates that the capacitance at the LO port also increases with C_{drv}. The power consumption of the PA driver-reactive mixer cascade is consequently related to C_{drv}. Decreasing C_{drv} results in a lower power consumption, but also lower gain, for the PA driver. For a target overall gain of the cascade, the gain of the mixer must be consequently increased, which calls for larger $C_v(t)$ and C_c, and, hence, larger LO dynamic power. Hence, for a given gain of the cascade there is an optimal C_{drv} that minimizes the overall power consumption. In the presented design, the minimum power consumption of the PA driver-mixer cascade is achieved setting $A_{v,mix}$ roughly 1 dB lower than its maximum value, i.e., the one for nil C_{drv}, limited by the attainable varactor capacitance ratio $C_{v,max}/C_{v,min}$, as shown by (1). Trying to further increase the mixer gain beyond the optimal value requires $C_v(t)$ and C_c, and hence the LO dynamic power, to skyrocket. In this design, the optimal gain partition is achieved setting the varactor capacitance to $C_{v,max} = 100$ fF with $C_{v,max}/C_{v,min} = 5.8$, while $C_c = 280$ fF and $C_{drv} = 13$ fF.

C. Comparison with the Use of a Resistive Mixer

To show the advantage of using a reactive mixer, a comparison is carried out with a conventional passive resistive mixer in the TX. The only required modification in the TX chain is that the source and drain terminals of VAR$_1$-VAR$_4$ in Fig. 2 are not shorted.

The simulated TX output power vs. the input code of the DACs is shown in Fig. 4 for the TX with the reactive mixer (blue curve) and the one with the conventional resistive mixer (red curve). Clearly, the use of a conventional resistive mixer is not compatible with a high impedance signal source, such as the used R-2R DACs. The interaction of the (large) source resistance with the load capacitance C_{drv} introduces low-pass filtering. While in the proposed reactive mixer such a filtering operates on the baseband signal before upconversion takes place, in the conventional resistive mixer the signal is first

979-8-3503-2123-4/23 $31.00 © 2023 IEEE 302

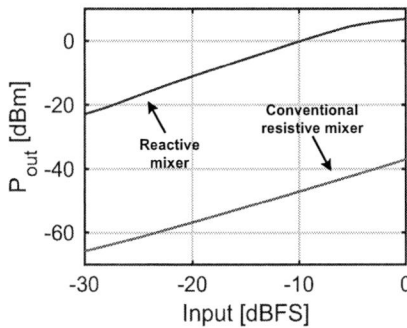

Fig. 4. Simulated TX output power vs. input code: TX with reactive mixer (blue curve); TX with the conventional resistive mixer (red curve).

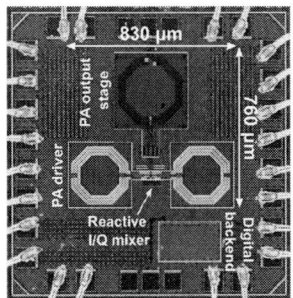

Fig. 5. Die microphotograph. Active area is $830 \times 760\,\mu m^2$

upconverted and then filtered, which results in a much lower conversion gain (Fig. 4).

IV. MEASUREMENT RESULTS

A proof-of-concept prototype of the proposed TX is implemented in a 22 nm FD-SOI CMOS technology for operation in the 2.4 GHz ISM band. The TX active area is $830 \times 760\,\mu m^2$ (Fig. 5).

A LO at twice the carrier frequency is externally fed to the chip, and processed by an on-chip frequency divider by 2 to generate quadrature phases and drive the reactive mixer. The LO is also divided by a programmable divider to clock the baseband DACs at 150 or 300 MHz (Fig. 1). A simple digital backend is integrated for testing purposes. A 1k-symbol 8-bit shift register is integrated, giving the flexibility to test various modulation formats. The complex symbols are loaded and then looped. The symbol rate can be set to 1.2 or 2.4 Mbaud. Two integrated digital pulse-shaping interpolation filters (interpolation ratio equal to 128) process the symbols and drive the in-phase and quadrature DACs.

Prototypes are first tested in continuous wave operation (Fig. 6). The TX shows a measured saturated output power $P_{sat} = 6.6$ dBm, and an output-referred 1 dB compression point OP1dB = 5.5 dBm. The measured system efficiency at OP1dB is 34.1%. The TX 3 dB bandwidth spans from 2.27 to 2.55 GHz.

Next, measurements with modulated signals are carried out. The saturated average output power in this case is about 6.2 dBm, with little dependence on the used modulation

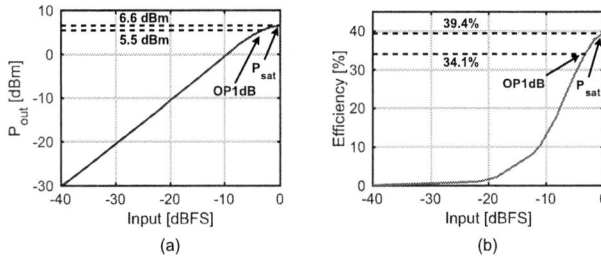

Fig. 6. Measured CW TX performance: (a) output power; (b) system efficiency.

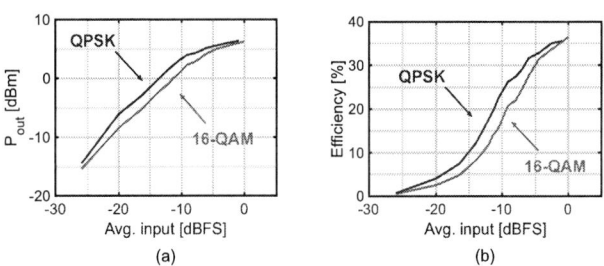

Fig. 7. Measured TX performance at 2.4 Mbaud symbol rate: (a) average output power; (b) system efficiency.

format, QPSK or 16-QAM (Fig. 7). A wideband spectrum measurement with 16-QAM modulation and 2.4 Mbaud symbol rate is reported in Fig. 8. The average output power is 2.7 dBm. All spurious spectral content is below -55 dBc, with most relevant elements due to limited attenuation of the digital interpolation filter, or to FM radio interference with the measurement setup. The lack of strong TX signal replicas at ±300 MHz offsets emphasizes the effectiveness of the baseband filtering embedded into the operation of the proposed capacitive reactive mixer.

The linearity of the proposed TX is evaluated using both EVM and figures of merit borrowed from the Bluetooth v5.1 standard [8], namely adjacent channel leakage ratio (ACLR) and alternate channel power (P_{alt}). No predistortion has been used. QPSK modulated signals at 2.4 Mbaud symbol rate and 5.1 dBm average output power show EVM = -27.9 dB, ACLR = -29 dBc, and $P_{alt} = -35$ dBm (Fig. 9). The

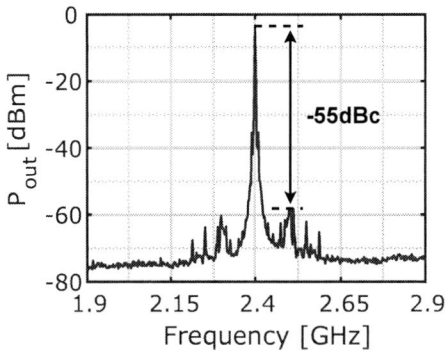

Fig. 8. Wideband TX output spectrum with 16-QAM, 2.4 Mbaud modulation at 2.7 dBm average output power. Spurious signals at 87 to 106 MHz offsets are likely due to FM radio interference with the measurement setup.

979-8-3503-2123-4/23 $31.00 © 2023 IEEE

Table 1. Performance summary and comparison with the state-of-the-art.

	This work	JSSC'20 [3]	VLSI'17 [2]	SSCL'20 [4]	JSSC'16 [1]	JSSC'08 [6]	JSSC'17 [7]
Technology	**22 nm FD-SOI**	65 nm	40 nm	65 nm	28 nm	130 nm	28 nm
Active area [mm²]	**0.63**	0.49	0.95*	0.5	0.65	3*	1.9*
Supply voltage [V]	**0.8**	1.2/1.8	1.2	1.2	0.5/1	1.2/3.3	0.5/1
Carrier frequency [GHz]	**2.4**	2.4	2.4	1	2.4	2.4	2.5
Symbol rate [Mbaud]	**2.4**	1	1	16	1	1	1
Modulation	**QPSK/16-QAM**	GFSK/8-DQPSK	GFSK/8-DQPSK	16-QAM	GFSK	GFSK/8-DQPSK	GFSK
Architecture	**Cartesian**	Cartesian	Polar/Cartesian	Cartesian	Polar	Polar	Polar
Data rate [Mb/s]	**4.8/9.6**	1/3	1/3	64	1	1/3	1
Average P_{out} [dBm]	**5.1/2.7**	6.4/4	0/0	2.5	0/3	2	0
EVM [dB]	**-27.9/-24.5**	NA/-21.9	NA/-25.2	-31	NA	NA/-24.4	NA
ACLR [dBc]	**-28/-32**	-29.3/-30.3	NA#	-41	NA#	NA#	NA#
P_{alt} [dBm]	**-37/-36**	-34/-33	NA#	NA	NA#	NA#	NA#
DC power [mW]	**10.2/8.5**	77.2/64.6	10**/15**	45	4.4**/6.3**	89.2**/99.3**	3.7**
System efficiency [%]	**31.6/22**	5.7/3.9	10**/6.7**	4	23**/32**	1.8**/1.6**	27**

* includes both TX and RX ** includes LO generation # ACLR and P_{alt} not reported but TX compliant with Bluetooth spectral mask

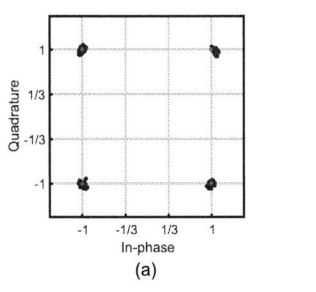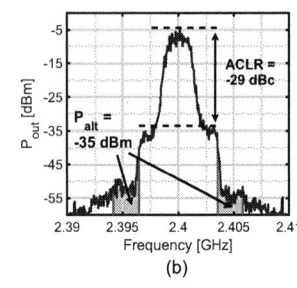

Fig. 9. Measured QPSK modulated signal at 2.4 Mbaud symbol rate and 5.1 dBm output power: (a) normalized constellation diagrams; (b) output spectrum.

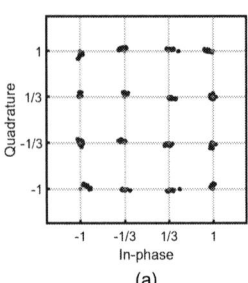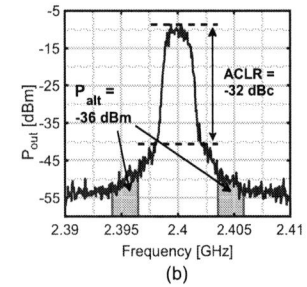

Fig. 10. Measured 16-QAM modulated signal at 2.4 Mbaud symbol rate and 2.7 dBm output power: (a) normalized constellation diagrams; (b) output spectrum.

corresponding system efficiency is 31.6% (Fig. 7). Changing the modulation format to 16-QAM, to increase the data rate to 9.6 Mb/s, yields EVM $= -24.5$ dB, ACLR $= -32$ dBc, and $P_{alt} = -36$ dBm at 2.7 dBm average output power (Fig. 10), with 22% system efficiency. Performing the measurements at 1.2 Mbaud symbol rate gives similar results.

The performance of the measured prototypes is summarized in Table 1. Compared to the state-of-the-art, the system efficiency of the proposed TX is superior to other Cartesian modulators, and almost in line with polar transmitters.

V. CONCLUSION

The proposed voltage-mode capacitive reactive mixer can be driven by a high impedance source, which allows to simplify the baseband circuitry of a Cartesian TX, saving power and increasing the system efficiency. A system with high linearity, able to transmit signals with high-order modulations, emerges. Achieving high system efficiency is intrinsically difficult in a low power, high data rate system because of the relatively low power delivered to the antenna: the use of the reactive mixer helps solving this issue.

ACKNOWLEDGMENT

The authors wish to thank the HiSilicon Sponsorship Program for funding the MPW prototyping and EUROPRACTICE MPW and design tools support.

REFERENCES

[1] M. Babaie et al., "A Fully Integrated Bluetooth Low-Energy Transmitter in 28 nm CMOS With 36% System Efficiency at 3 dBm," *IEEE Journal of Solid-State Circuits*, vol. 51, no. 7, pp. 1547–1565, 2016.

[2] A. Zolfaghari et al., "A multi-mode WPAN (Bluetooth, BLE, IEEE 802.15.4) SoC for low-power and IoT applications," in *2017 Symposium on VLSI Circuits*, 2017, pp. C74–C75.

[3] M. V. Praveen and N. Krishnapura, "High Linearity Transmit Power Mixers Using Baseband Current Feedback," *IEEE Journal of Solid-State Circuits*, vol. 55, no. 2, pp. 272–281, 2020.

[4] K. Vasilakopoulos and A. Liscidini, "A Reconfigurable Passive Switched-Capacitor TX RF Front End With -57 dB ACLR2," *IEEE Solid-State Circuits Letters*, vol. 3, pp. 294–297, 2020.

[5] Z. Chen, W. Choi, and O. Kenneth, "270-to-300GHz Double-Balanced Parametric Upconverter Using Asymmetric MOS Varactors and a Power-Splitting-Transformer Hybrid in 65nm CMOS," in *2021 IEEE International Solid- State Circuits Conference (ISSCC)*, vol. 64, 2021, pp. 324–326.

[6] W. W. Si et al., "A Single-Chip CMOS Bluetooth v2.1 Radio SoC," *IEEE Journal of Solid-State Circuits*, vol. 43, no. 12, pp. 2896–2904, 2008.

[7] F.-W. Kuo et al., "A Bluetooth Low-Energy Transceiver With 3.7-mW All-Digital Transmitter, 2.75-mW High-IF Discrete-Time Receiver, and TX/RX Switchable On-Chip Matching Network," *IEEE Journal of Solid-State Circuits*, vol. 52, no. 4, pp. 1144–1162, 2017.

[8] "Bluetooth core specification v5.1," Bluetooth SIG, Tech. Rep., January 2019.

RTu3C-2

A 110-170 GHz Phase-Invariant Variable-Gain Power Amplifier Module with 20-22 dBm P$_{sat}$ and 30 dBm OIP3 Utilizing SiGe HBT RFICs

Mustafa Sayginer, Michael Holyoak, Mike Zierdt,
Mohamed Elkhouly, Joe Weiner, Yves Baeyens, Shahriar Shahramian

Nokia Bell Labs, USA

mustafa.sayginer@nokia-bell-labs.com

Abstract— A phase-invariant variable-gain PA RFIC in 130-nm SiGe BiCMOS supporting multi-QAM waveforms over the entire D-band (110-170 GHz) is presented. The chip has a very low phase variation of ±2° over a 15 dB gain control range as well as self-testing and fault-detection features (power detectors, temperature sensors, ADC, SPI control) to ease the implementation and testing of multi-chip PA modules. A WR-6 interface packaged module combining four RFICs on a glass substrate achieves an average P$_{sat}$, OIP3 and gain of 21 dBm, 30 dBm and 14 dB, respectively over 110-170 GHz while the return-losses are better than 10 dB. TX constellations of 256-QAM (16-Gb/s with 3.6% EVM at 11 dBm P$_{out}$) and 64-QAM (36-Gb/s with 8.6% EVM at 15.5 dBm P$_{out}$) are demonstrated at 140 GHz. The output power×bandwidth performance of the packaged all-silicon module is better than the state-of-the-art commercially available III-V parts.

Keywords— D-band, power amplifiers, RFIC, SiGe HBT, VGA, 5G, 6G, phased arrays, III-V, glass substrate.

I. INTRODUCTION

Developers of next-generation wireless communications (e.g. 5G, 6G, backhaul) are looking for opportunities in the mm-wave bands including >100 GHz where the III-Vs are natural candidates for such hardware with their high-power, high-efficiency and low-noise performance. However, the III-Vs lack high integration level and have costlier development and operation cycles than their silicon counterparts. [1]–[8].

This work (Fig.1) presents a packaged all-silicon power amplifier (PA) module with WR-6 interface operating at full D-band (110-170 GHz). The module employs eight-way power combining on a low-loss glass interposer using four dual-output RFICs. The PA design is optimized for best power×bandwidth performance metric which outperforms reported III-V chipsets and modules to date.

The architecture allows for scalable and flexible use-cases where the output power can be determined by the number of RFICs used and without sacrificing the bandwidth (BW). Substrate integrated waveguide (SIW) power splitting (4-way) and combining (8-way) provide a low-loss and flat pass-band response. The RFIC design trades off between the BW and efficiency. The die-size and output power is optimized to relax package thermal design for high-power targets (e.g. >20 dBm). The module is highly-integrated with features like phase-invariant gain control, power detectors, temperature sensors, ADC and SPI. These features not only allow for flexible use-cases during normal operation but also help with self-testing and fault detection in manufacturing.

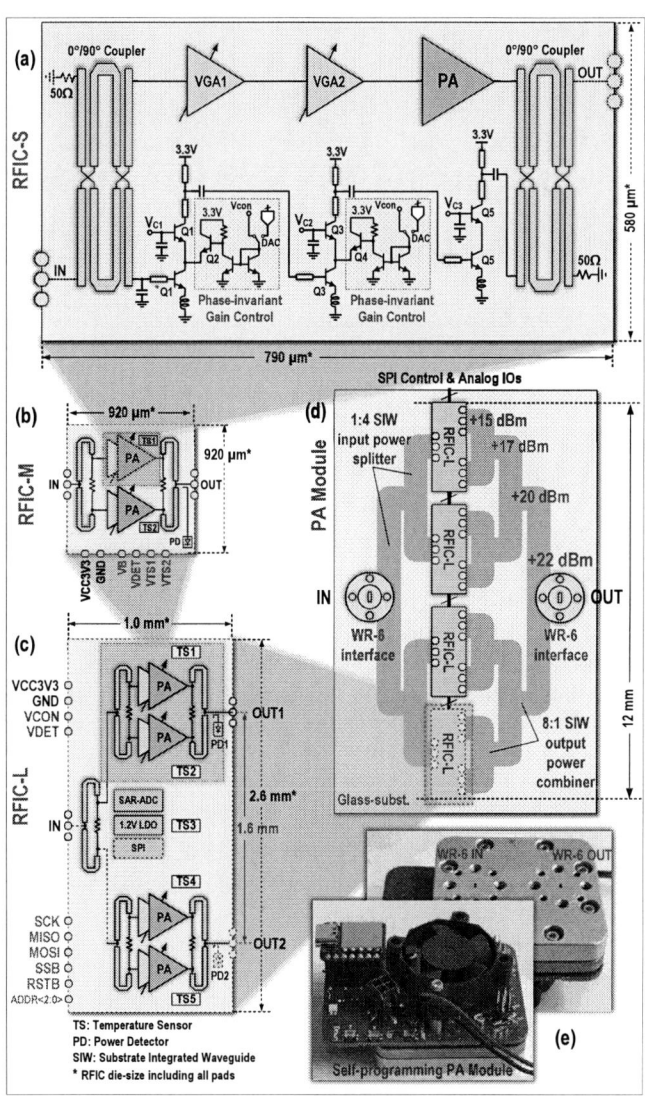

Fig. 1. Circuits and blocks to develop scalable and flexible packaged all-silicon PA modules with WR-6 interface at D-band (110-170 GHz). (a) RFIC-S: the small-size PA design exposing the phase-invariant gain control circuitry in a balanced topology. (b) RFIC-M: The medium-size RFIC combining two RFIC-Ss using integrated Wilkinson combiners and with integrated two temperature sensors (TS) and a power detector (PD). (c) RFIC-L: The large-size RFIC made of two RFIC-Ms but without combining their outputs. This chip adds five-TSs and two-PDs as well as other integrated features. (d) A power combining module with four RFIC-L dies on a glass interposer to output >20 dBm power. (e) A stand-alone and self-programmable packaged all-silicon PA module of (d) with WR-6 interface.

979-8-3503-2123-4/23 $31.00 © 2023 IEEE   305   2023 IEEE Radio Frequency Integrated Circuits Symposium

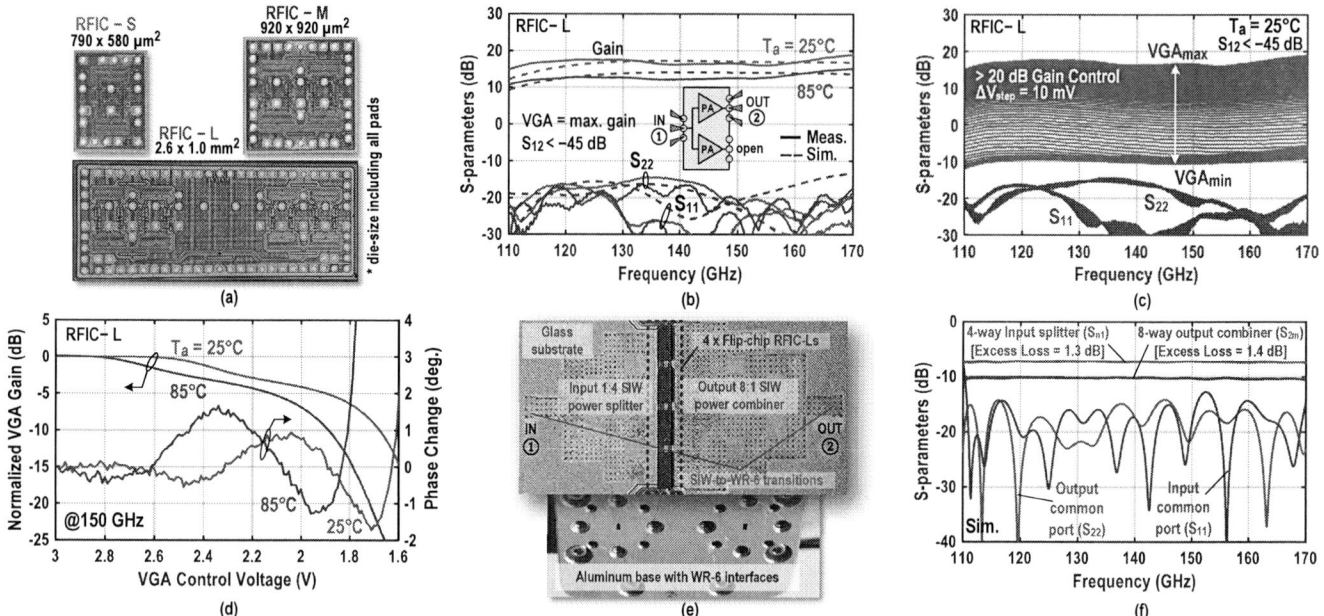

Fig. 2. (a) Fabricated flip-chip compatible RFIC-S, -M and -L. On-wafer small-signal measurements for RFIC-L: (b) 25°C and 85°C S-parameters at max. gain (c) Gain control vs frequency (d) Normalized VGA gain and phase change vs gain control voltage at 25°C and 85°C (e) PA module with four RFIC-Ls are assembled on a glass interposer and employing 1:4 SIW input power splitting and 8:1 SIW output power combining networks and WR-6 interface. (f) Simulated S-parameters for 1:4 SIW input power splitting and 8:1 SIW output power combining networks including WR-6 interface and glass-to-chip transition.

II. RFIC DESIGN

A balanced topology is used as the core PA with two 0°/90° couplers (~1-dB loss) shown as RFIC-S (i.e. RFIC-Small) in Fig.1(a). Medium (RFIC-M) and large (RFIC-L) versions are derived using the same core (Fig.1(b-c)). All chips are flip-chip compatible, operating at entire D-band and have incremental die area, output power and integrated features. In this design, the overall efficiency is traded off for wideband and high-linearity operation. The core active circuitry in RFIC-S consists of two single-ended phase-invariant variable-gain cascode stages followed by a fixed-gain final power amplifier stage. RFIC-M combines two of the RFIC-S blocks using integrated Wilkinson combiners (~1-dB loss) and adds a power detector (PD) and two temperature sensors (TS).

RFIC-L further integrates two RFIC-Ms and additional features to ease the multi-chip power combining (Fig.1(c)). The two outputs of RFIC-L are taken off-chip to interface with a low-loss power combiner in glass. The spacing (1.6-mm) of these outputs and the chip size are carefully optimized for both thermal considerations and the physical SIW implementation in the glass. However, the input is split on-chip to save area on the interposer (Fig.1(c-d)). Up to eight RFIC-Ls can be controlled through a single external SPI controller via 3-bit chip addressing. A programmable analog and/or digital gain control, 8-bit SAR-ADC, five TSs over the die and two output PDs all make a flexible chip platform for constructing high-power modules with self-test and fault-detection capability for various applications from communication to instrumentation. Fig.2(a) shows the

fabricated RFICs in 130-nm IHP SiGe BiCMOS technology.

III. PA MODULE

A packaged PA module with four flip-chip RFIC-Ls on a glass interposer is constructed for >20 dBm P_{sat} over 110-170 GHz. In glass 1:4 input and 8:1 output SIW networks with very low simulated excess path losses and including the WR-6 and glass-to-chip transitions are shown in Fig.2(e-f). An aluminum support is designed to house the glass substrate assembly and incorporate two WR-6 waveguide interfaces. In addition, a copper heat-sink with fan is used to thermally manage the PA module (Fig.1(e)). The module operates from a single 3.3V supply and draws 2.7A total dc current during normal operation. An integrated 1.2V LDO in each RFIC-L provides the supply voltage for SPI and other digital blocks.

IV. MEASUREMENT RESULTS

Extensive on-wafer RFIC and PA module measurements were performed. The S-parameters of RFIC-L at 25°/85°C are given in Fig.2(b) for one output path. The measured average RFIC gain is 16 dB at 25°C and shows excellent return losses (RL) of better than 15 dB across the entire D-band. RFIC-S and -M have 22 dB and 20 dB of gain, respectively, while preserving better than 15 dB of RLs, not shown here. The analog gain control function is also measured over 110-170 GHz with >20 dB gain control and resulting in very good gain flatness (Fig.2(c)). An excellent measured phase-invariance of <±2° is observed over 15 dB gain control and measured at 150 GHz under both 25°C and 85°C conditions (Fig.2(d)).

A packaged PA module with four RFIC-Ls (Fig.1(e)) is also characterized at room temperature (25°C). S-parameters

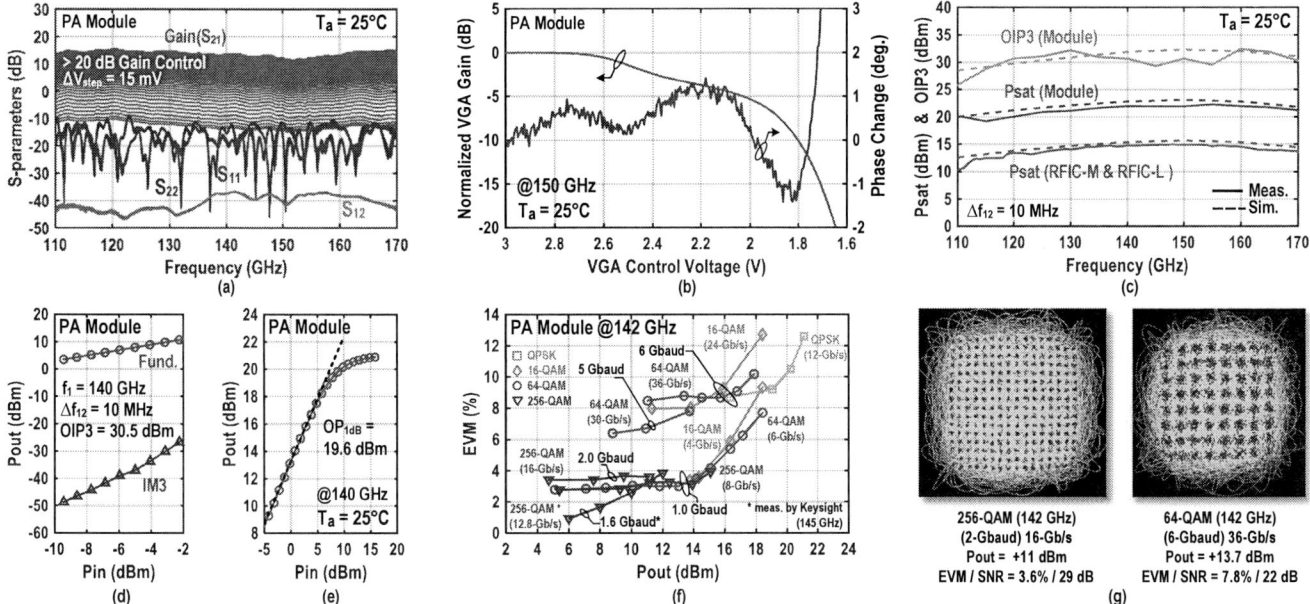

Fig. 3. Measured PA module performance: (a) S-parameters vs frequency with analog gain control (b) Normalized VGA gain and phase change vs gain control voltage (c) Psat (RFIC-L and PA module) and OIP3 (PA module) vs frequency (d) IM3 term with two-tone test at 140 GHz (e) Pin vs Pout showing a sharp-compression curve (140 GHz) (f) EVM(rms) vs Pout at various QAM modulations (g) Selected constellations for 256-/64-QAMs

are shown in Fig.3(a) where the four flip-chip dies share the same analog gain control voltage. The module has input and output RL better than 10 dB across the entire 110-170 GHz and reverse isolation is more than 37 dB. The module's measured phase-invariant response exceeds that of a single RFIC to be better than $\pm 1°$. This is due to an averaging effect with eight-way power combining scheme (Fig.3(b)).

The measured P_{sat} of RFIC-L is as high as 15 dBm while the PA module provides a maximum P_{sat} of 22 dBm at 150 GHz with an average P_{sat} of 21 dBm across the entire D-Band frequency range. A 30 dBm average OIP3 is also measured with the module across the D-band (Fig.3(c)). A two-tone test with 10 MHz Δf_{12} and IM3 term extrapolates to an OIP3 of 30.5 dBm at 140 GHz (Fig.3(d)). A P_{in} vs P_{out} curve is shown in Fig.3(e) which demonstrates a sharp compression characteristic. This is unlike the typical compression curves of III-V PAs which consequently require stronger levels of predistortion for PA linearization [5], [6]. All the measured

results are in good agreement with the simulations.

The module is also tested across a wide range of modulation formats at low (1.0-/1.6-/2.0-Gbaud) and at high (5.0-/6.0-Gbaud) data-rates as well as at different back-off powers. The resulting rms EVMs are summarized in Fig.3(f). The PA module demonstrates 140 GHz TX constellations up to 256-QAM (16-Gb/s with 3.6% EVM at 11 dBm P_{out}) and 64-QAM (36-Gb/s with 7.8-/8.6-% EVM at 13.7-/15.5-dBm P_{out}) as shown in Fig.3(g). The setup in Fig.4(a) is used to perform various data transmission measurements. It is important to note that a second identical PA module is also used to compensate for the setup losses and insufficient overall system gain. The high data-rate measurements are primarily limited by the wide-band frequency response of the measurement setup which requires heavy equalization. To the best of the authors' knowledge, the demonstrated data-rates and EVMs for the given module output power represent state-of-the-art performance for packaged PA at D-band.

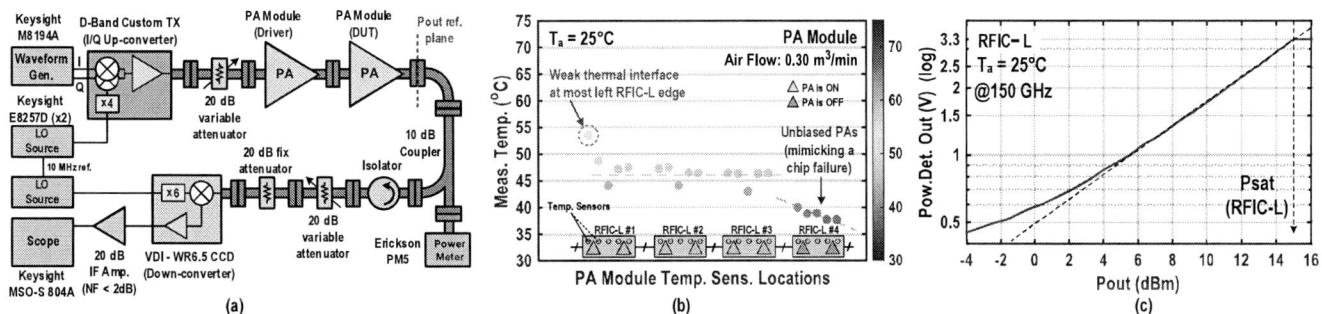

Fig. 4. (a) Test setup for modulation measurements (b) Measured temperature data over 25-sensors in the PA module (across four RFIC-Ls each with five-sensors) (c) Measured power detector output voltage in RFIC-L via analog voltage output pin (ADC reads back the same curve as well).

979-8-3503-2123-4/23 $31.00 © 2023 IEEE

Table 1. Measured performance of prior published D-band PAs in Silicon and III-V technologies.

	This Work — RFIC-S	RFIC-M	RFIC-L	PA Module	[1] JSSC'22	[2] ISSCC'20	[3]# ESSCIRC'19	[4] JSSC'22	[5] EuMIC'21	[6] TMTT'19	[7] VDI	[8] Eravant
Technology	130-nm SiGe-HBT			130-nm SiGe-HBT + Glass Subst.	45-nm CMOS-SOI	16-nm FinFET	130-nm SiGe-HBT	130-nm SiGe-HBT	250-nm InP HBT	100-nm GaN HEMT	250-nm InP WG-Amplifier	Waveguide Amplifier
Supply (V)	3.3			3.3	2.4	1.0	1.5 / 3.3	4	2.43 / 2.65	15	9	8
Topology	Balanced	Balanced + 2-way pow. comb.	Balanced + 2-way pow. comb. (dual)	4 x RFIC-L (8-way) pow. comb. on glass	8-way pow. comb.	2-way pow. comb.	2-way pow. comb.	Doherty + 8-way slot-line pow. comb.	8-way pow. comb.	4-way pow. comb.	-	-
In/Out Interface	Wafer probing $\mathbb{W}$			WR-6 ⊖	$\mathbb{W}$	$\mathbb{W}$	$\mathbb{W}$	$\mathbb{W}$	$\mathbb{W}$	$\mathbb{W}$	⊖	⊖
Frequency (GHz)	>110 – 170 *			110 – 170	133 – 148	110 – 128	112 – 142	107 - 135	124 – 166	107-148	110 – 170	110 – 150
BW (GHz)	>60 *			60	15	18	30	28	42	41	60	40
P_{sat} (dBm)	14	15	15 (dual)	20-22	18.5	15	15	22.5	23	23 ⊖	17	13
OIP3 (dBm)	-	-	-	30 ⊖	-	-	-	-	-	-	-	-
Peak PAE (%)	4	2.8	2.5	2	11	12.8	8	17	17.8	6 ⊖	2	1
Gain (dB)	22	20	16 **	14.5	24.8	20.5	20	20	20	30	20	25
In & Out RL (dB)	Both >15			Both > 10	Both >10	>8 & >3	>8 & >5	>10 & >7	>5 & >3	>5 & >5	>7 & >10	6
P_{DC} (W)	0.62	1.2	2.25	9.0	0.64	0.207	0.37	1.0§	1.2§	3.3§	5.4	3.2
Die Area (mm²)	0.46 (incl. pads)	0.85 (incl. pads)	2.6 (incl. pads)	4 x RFIC-L	0.46 (excl. pads)	0.041 (excl. pads)	1.06 (incl. pads)	1.1 (incl. pads)	1.34 (incl. pads)	7.5 (incl. pads)	-	-
Integrated Features	(a)	(b)	(c)	(c) + Self programming	-	-	-	-	-	-	-	-

TX Modulation (PA Module @ 142 GHz: Rate / EVM / Pout; [5] @ 131.5 GHz: Rate / EVM / Pout)

	PA Module (142 GHz)	[5] (131.5 GHz)
QPSK	12-Gb/s, 12.6%, 21-dBm	-
16-QAM	24-Gb/s, 12.7%, 8.4-dBm / 4-Gb/s, 3.5%, 14-dBm	8-Gb/s, 11.6%, 13.7-dBm
64-QAM	36-Gb/s, 8.6%, 15.5-dBm / 4.8-Gb/s, 3.5%, 12-dBm	4.8-Gb/s, 10.9%, 3.8-dBm
256-QAM	16-Gb/s§, 3.6%, 11-dBm / 12.8-Gb/s§, 2.6%, 10-dBm	-

(a) Flip-chip compatible pads & pinout   **(b)** Everything in *(a)* + Pow. Detector, Temp. Sens.   **(c)** Everything in *(b)* + Dual-RF out, Analog & digital VGA, 8-bit SAR-ADC, SPI, 3-bit hard-wired chip addr., LDO (1.2V), dual & symmetrical pinout
* Actual BW is limited by WR-6 setup   ** Input to one output gain (add 3-dB when both output combined)   # Balun loss included   § Estimated / calculated   ⊖ Average   & Measured by Keysight at 145 GHz   WG: Waveguide

To demonstrate some of the fault-detection features, all temperature sensors (i.e. 20 of them from 4-dies) are read back using the integrated ADCs/SPIs and a temperature gradient across the four-dies is observed for a faulty PA module (Fig.4(b)). The temperature profile indicates a weak thermal interface at the edge of the first die as well as internal biasing faults for the fourth die in the module. The integrated sensor information aids in factory testing and reduces production time and cost. The integrated power detector is also used to measure a wide range of output power during normal operation (Fig.4(c)) and can indicate an issue of RF bump failure or a non-working RF core in case of any unexpected low-power read-back.

Table 1 compares the prior published D-band PAs as well as commercially available III-V modules.

V. CONCLUSION

A fully packaged all-silicon variable-gain PA module with WR-6 interface is demonstrated at D-band with an output power (20-22 dBm) and OIP3 (30 dBm) vs frequency (110-170 GHz) better than commercially available PAs. The presented RFIC and PA module show state-of-the-art phase-invariant ($\pm 1°$) gain control (>15 dB) with a flat frequency response over the full D-band. All the other integrated features make the presented RFIC a suitable platform for designing multi-chip PA modules with a comprehensive self-testing and fault detection capability.

To the best of authors' knowledge, the presented work is the first highly integrated and packaged all-silicon PA operating at entire D-band with the highest output power×bandwidth performance and showing state-of-the-art multi-QAM data-rates with high power and low EVM.

ACKNOWLEDGMENT

The authors would like to thank Terry Xian, Alex Sun, Victor Yin and Derek Tattersall for their design expertise, Hernan Castro for his layout support, Osamu Kusano (Keysight), Ahmed Ibrahim and Waleed Mansha for their measurement support, Maurizio Moretto and Pierre Lopez for their project and technical support, Pascal Roux, Akshay Visweswaran, Jaegeun Ha and Philipp Thomas for their technical discussions.

REFERENCES

[1] S. Li and G. M. Rebeiz, "High Efficiency D-Band Multiway Power Combined Amplifiers With 17.5–19-dBm Psat and 14.2–12.1% Peak PAE in 45-nm CMOS RFSOI," *IEEE Journal of Solid-State Circuits*, vol. 57, no. 5, pp. 1332–1343, 2022.

[2] B. Philippe and P. Reynaert, "A 15dBm 12.8%-PAE Compact D-Band Power Amplifier with Two-Way Power Combining in 16nm FinFET CMOS," in *2020 IEEE International Solid- State Circuits Conference - (ISSCC)*, 2020, pp. 374–376.

[3] A. Visweswaran, B. Vignon, X. Tang, S. Brebels, B. Debaillie, and P. Wambacq, "A 112-142GHz Power Amplifier with Regenerative Reactive Feedback achieving 17dBm peak Psat at 13% PAE," in *ESSCIRC 2019 - IEEE 45th European Solid State Circuits Conference (ESSCIRC)*, 2019, pp. 337–340.

[4] X. Li, W. Chen, H. Wu, S. Li, X. Yi, R. Han, and Z. Feng, "A 110-to-130 GHz SiGe BiCMOS Doherty Power Amplifier With a Slotline-Based Power Combiner," *IEEE Journal of Solid-State Circuits*, vol. 57, no. 12, pp. 3567–3581, 2022.

[5] A. S. Ahmed, M. Seo, A. A. Farid, M. Urteaga, J. F. Buckwalter, and M. J. Rodwell, "A 200mW D-band Power Amplifier with 17.8% PAE in 250 nm InP HBT Technology," in *2020 15th European Microwave Integrated Circuits Conference (EuMIC)*, 2021, pp. 1–4.

[6] M. Ćwikliński, P. Brückner, S. Leone, C. Friesicke, H. Maßler, R. Lozar, S. Wagner, R. Quay, and O. Ambacher, "D-Band and G-Band High-Performance GaN Power Amplifier MMICs," *IEEE Transactions on Microwave Theory and Techniques*, vol. 67, no. 12, pp. 5080–5089, 2019.

[7] Virginia Diodes Inc. (2023, Jan.) D-band Amplifier (WR-6.5 Amp). [Online]. Available: https://www.vadiodes.com/en/products/amplifier

[8] Eravant. (2023, Jan.) D-band Amplifier (SBP-1141543010-0606-E1). [Online]. Available: https://www.eravant.com/products/amplifiers

RTu3C-3

A D-band 20.4 dBm OP$_{1dB}$ Transformer-Based Power Amplifier With 23.6% PAE In A 250-nm InP HBT Technology

Senne Gielen[$][#], Yang Zhang[#], Mark Ingels[#], Patrick Reynaert[$]

[$]MICAS, KU Leuven, Belgium

[#]imec, Belgium

{senne.gielen, patrick.reynaert}@kuleuven.be

Abstract — This paper presents a high-efficiency transformer-based D-band power amplifier (PA) in 250-nm InP HBT. The PA has a saturated output power of 21 dBm and peak power-added efficiency (PAE) of 23.6 %. The small-signal gain and bandwidth are 19.8 dB and 24.4 GHz respectively. Careful design of the biasing networks results in a record OP$_{1dB}$ and associated PAE of 20.4 dBm and 23% respectively. To the authors best knowledge this is the highest PAE ever reported at P$_{1dB}$ for D-band power amplifiers, resulting in record output power and efficiency during modulated measurements up to 20 Gb/s.

Keywords — D-band, InP, III-V, millimeter-Wave, high-efficiency, power amplifiers, transformers

I. INTRODUCTION

The ever-increasing demand for higher data-rates keeps pushing the operating frequency of communication circuits in order to obtain more bandwidth. While historically CMOS-based technologies were preferred due to their high level of integration and low cost, the performance of CMOS technologies at millimeter-wave frequencies is generally subpar due to a combination low f$_{max}$, limiting the gain, and the low supply voltage, limiting the output power of circuits in these technologies.

III-V based technologies have shown higher gain and better power handling capabilities than CMOS-based technologies at millimeter-wave frequencies due to their higher f$_{max}$ and breakdown voltage respectively. Especially InP-based technologies have shown the highest f$_{max}$ while also maintaining decent power handling capabilities, making it an excellent candidate for millimeter-wave power amplifier design [1].

Typical III-V-based circuits are generally single-ended transmission line based due to their model accuracy the available metal-stack in these technologies [2]. In mm-wave CMOS on the other hand, differential transformer-based designs are very common [3], [4].

This paper presents a power amplifier (PA) that combines the design principles of millimeter-wave CMOS passives with InP HBT active devices. While previous transformer-based designs in InP HBT utilized the common-emitter topology at lower frequencies [5], in this work they are used at D-band in combination with the common-base topology. The result is a PA with a combination of both record PAE and record OP$_{1dB}$ of 23.6 % and 20.4 dBm respectively. Additionally the PA achieves record PAE$_{avg}$ and P$_{avg}$ with modulated measurements up to 20 Gb/s

II. CIRCUIT IMPLEMENTATION

The power amplifier is designed in a 250-nm InP DHBT technology that offers four Au metal layers, MIM capacitors with a density of $0.3\,\text{fF}/\mu\text{m}^2$ and thin film resistors of $50\,\Omega/\text{sq}$. The HBTs have a BV$_{CEO}$ of $4.5\,\text{V}$, a peak f$_{max}$ of $650\,\text{GHz}$ and a maximum current density of $12\,\text{mA}/\mu\text{m}^2$.

A. PA Topology

The schematic of the PA is shown in figure 1. The common-base topology is chosen instead of the typical common-emitter topology as this provides a higher G$_{max}$ without any neutralization [2]. Common-base amplifiers also show better power handling capabilities due to their increased breakdown voltage as the low base impedance allows avalanche current to flow freely out of the base [6]. The drawback of the common-base stage is that it requires accurate modelling of the base inductance for stability and gain. An additional benefit is low-Q input impedance of the common-base stage as it allows for a broadband input match of the PA.

A differential topology is chosen for this design as this naturally doubles the output power of the amplifier. The differential topology also allows the use of the virtual ground to connect to the base of the common-base stages. The layout of the driver stages and the PA core is shown in figure 2. Positioning the base terminals in a differential pair as close as possible to each other lowers the base inductance and therefore increases the gain and stabilizes the amplifier. At the output stage, multiple transistors are used in parallel, increasing the minimum distance between the base terminals. In order to stabilize the output stage, a small base resistor is added in shunt with a bypass capacitor to cancel out the resulting base inductance.

Transformers are used to implement area-efficient and low loss matching networks. This is especially beneficial for differential designs where the virtual ground at the center tap can be used to connect the power supply and provide a second harmonic short. The high resistivity of the InP substrate, the availability of multiple thick metal layers and the low dielectric constant of interlayer dielectric allow for high-Q transformers with a high self-resonant frequency. In order to increase the low input impedance of a common-base stage, a LC-network is placed after the transformer. A ground plane in the lowest metal layer is used to provide a well modelled return path for common-mode currents and for easy distribution of the

979-8-3503-2123-4/23 $31.00 © 2023 IEEE

Device	Q_1	Q_2	Q_3	Q_4	Q_5
Size (μm^2)	4x0.25	8x0.25	4x5x0.25	2x0.25	5x0.25

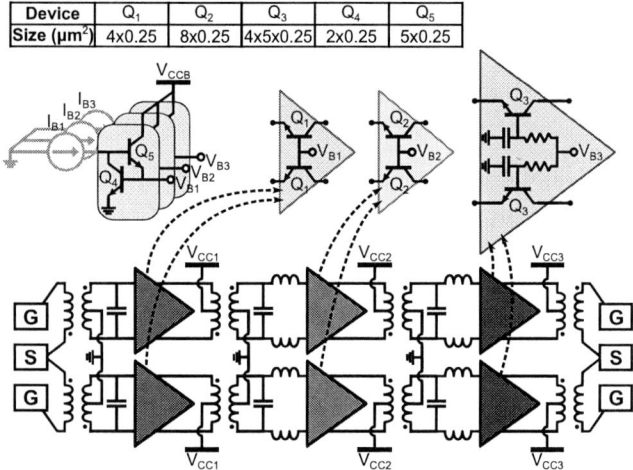

Fig. 1. Schematic of the power amplifier

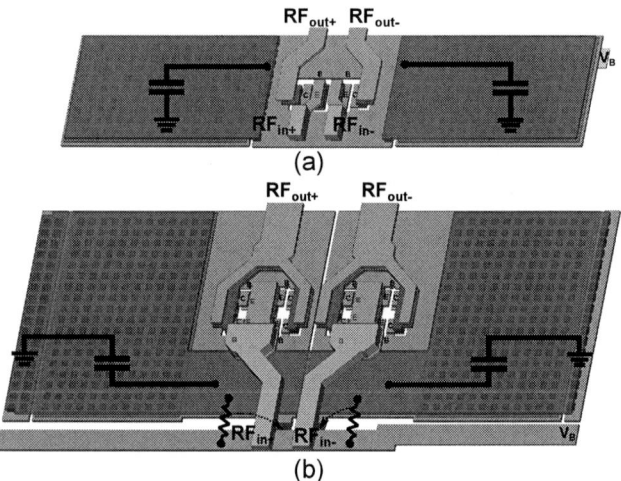

Fig. 2. Core layout of (a) the driver stages and (b) the PA stage

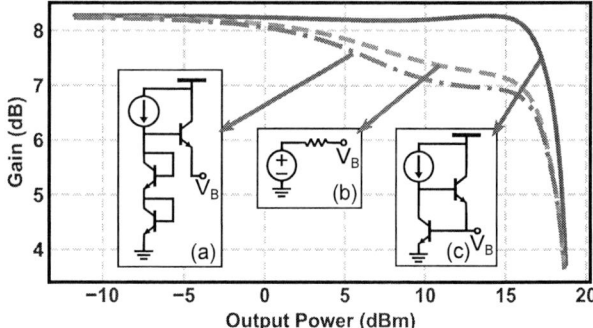

Fig. 3. Simulated gain compression of a PA stage when biased with (a) a current mirror with emitter follower , (b) a voltage source with resistor and (c) a buffered current mirror biasing network.

DC ground. Through-substrate vias and ground plane slots are used to suppress substrate modes.

The input stage is biased in moderate class AB and is sized big enough to avoid compression while providing a broadband input match for the PA. The second stage is biased in deep class AB and sized such that the resulting gain expansion increases the OP_{1dB} of the PA.

B. Biasing Network

In order to achieve higher PAE at power backoff (PBO), the PA and driver stages are biased in deep class AB. Therefore the collector current increases with the input power. At the same time, the low β (~ 25) in this technology induces a considerable amount of base current that also increases with the input power. As the base should be able to pull this current from its basing network in order to prevent early gain compression of the PA, it should provide a low output impedance [7]. Figure 3 shows the simulated gain compression of the PA stage when it is biased with either a current mirror with emitter follower, a voltage source and a $50\,\Omega$ resistor

or a buffered current mirror. The buffered current mirror outperforms the other presented biasing networks as it provides the lowest gain compression for the PA, and thus it is used for all stages in the design. Due to the linearity enhancement of the biasing network, the driver stages can be sized more aggressively to increase the efficiency without compromising the linearity of the PA. The PA further benefits from the differential design as it allows this biasing network to be easily connected to the virtual ground nodes formed between the base terminals of the common-base stages.

The supply voltage of these buffers can be much lower than the rest of the PA in order to minimize the impact on efficiency. The base biasing network only consumes $5.5\,\%$ of the total power consumption of the PA at P_{1dB}.

C. Power combiner-divider

Current combining naturally increases the load-impedance seen by the power amplifier, making it an excellent match for the high load-line impedance originating from the high supply voltage in this technology. The designed PA uses a two-way transformer-based power combiner. The power combiner and GSG pads have a combined simulated insertion loss of a maximum $0.8\,dB$ across the entire operating bandwidth of the PA. At the input, a similar transformer-based parallel power divider is used with an additional MIM capacitor to tune out the reactive part of the impedance. Combined with the common-base topology, this results in a very broadband input match, as seen in the measurement results.

III. MEASUREMENT RESULTS

The amplifier is fabricated in a 250-nm InP HBT process, and figure 4 shows the die photo. The dimensions of the chip are $682\,\mu m \times 900\,\mu m$. The chip is mounted on a standard FR4 PCB and is wire bonded to provide the supplies and bias currents. Wire bond capacitors are used on the supply lines to handle the high dynamic current consumption caused by the deep class AB operation of the PA. The input and output of the PA are probed using GSG probes with a pitch of $50\,\mu m$. The PA and second driver stage supplies are set to $2.9\,V$ for maximum linearity, while the input stage has a supply of $2.65\,V$ for better overall efficiency. The bias buffer devices have a supply $2.0\,V$.

979-8-3503-2123-4/23 $31.00 © 2023 IEEE

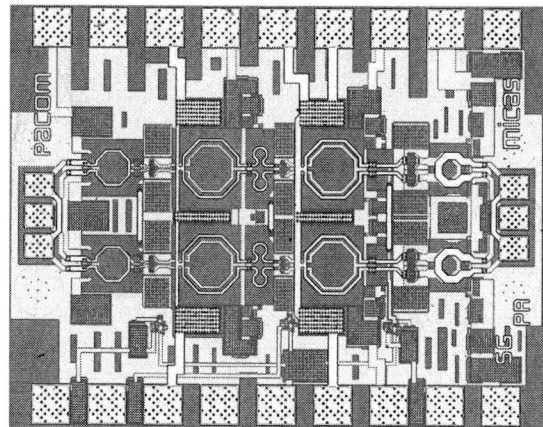

Fig. 4. Die photo of the power amplifier

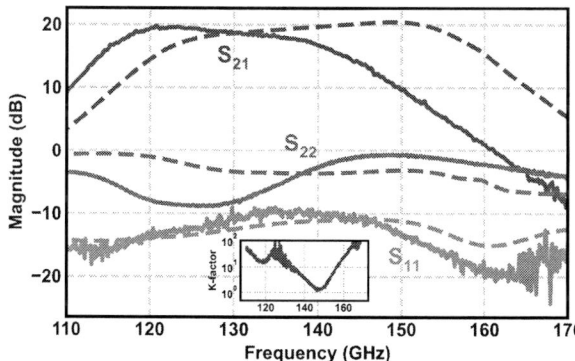

Fig. 5. Measured (solid) and simulated (dashed) S-parameters of the PA

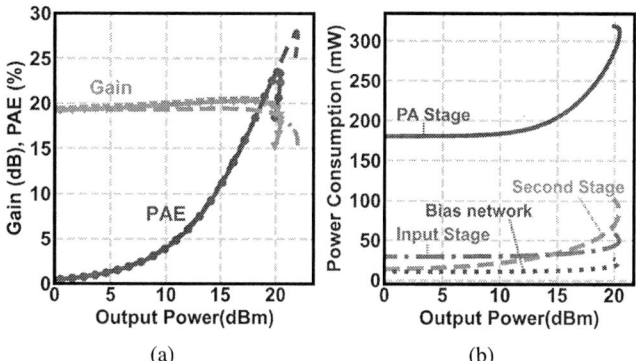

Fig. 6. (a) Measured (solid) and simulated (dashed) CW performance of the PA at 125 GHz (simulated at 140 GHz). (b) Measured DC power consumption of the PA stages at 125 GHz

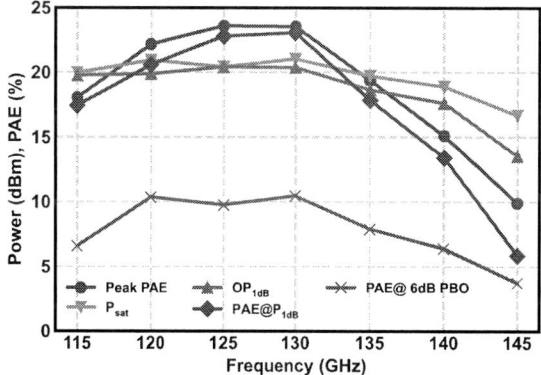

Fig. 7. Measured CW performance of the PA across frequencies

A. Small-signal

This small-signal S-parameters are measured using a R&S VNA in combination with D-band R&S VNA extenders. The VNA is calibrated using a CS-15 calibration substrate and LRRM calibration. The measured and simulated S-parameters are presented in figure 5. As shown in the figure, there is a frequency shift which can be attributed to errors in the model of the device parasitics. The PA has a peak gain of 19.8 dB and a 3-dB bandwidth of 24.4 GHz. The input match of the PA is better than -9.9 dB across the entire D-band. The K-factor of the PA is always greater than 1, indicating that the PA is unconditionally stable.

B. Large signal

For large signal measurement, the GSG probe loss is measured and the R&S D-band extenders are power calibrated using an Erickson PM5B power meter with the reference plane at the GSG probe mounting points. The measured probe loss and power calibration of the extenders allow the VNA to sweep the input power of the PA by varying the input power for the extenders. The measured large-signal S-parameters are used to calculate the output power of the PA. A waveguide attenuator is used at the output to prevent the PA from damaging the frequency extenders. The DC power consumption is measured using a combination of the R&S and Keysight PSU's built-in current measurements and a Keithley

multimeter. The measured and simulated gain and PAE at 125 GHz is shown in figure 6a. In order to account for the shift, the simulated frequency is 140 GHz. The PA achieves a OP_{sat} of 20.4 dBm and a peak PAE of 23.6 %. The power consumption of the different PA stages is shown in figure 6b, showing that both the PA and second driver stage are operating in deep class AB. The large-signal performance over frequency is shown in figure 7. The PA maintains a OP_{1dB} and associated PAE of greater than 19.5 dBm and 16 % respectively from 115 GHz to 135 GHz.

C. Modulated Signal Measurements

The modulated signal measurement setup is shown in figure 8. The PA's average output power and input power are measured with a PM5B at the S-bend flanges. The EVM is measured without any predistortion and is normalized to the RMS value of the constellation point magnitudes. The average output power of the PA is set to the value that results in an EVM corresponding to a BER of $\sim 10^{-3}$ for the selected modulation type. The resulting constellation diagrams for 16-QAM, 32-QAM and 64-QAM modulated data are shown in figure 9. For 10 Gb/s 16-QAM, the PA achieves a record P_{avg} and PAE_{avg} of 16.3 dBm and 11.9 % respectively. Using frequency domain equalization, the PA achieves 24 Gb/s with -16 dB EVM for an average output power and PAE of 15.7 dBm and 10.6 % respectively. Thanks

979-8-3503-2123-4/23 $31.00 © 2023 IEEE

Table 1. Comparison between state-of-the-art D-band PAs

Reference	This work		IMS'20 [2]	ISSCC'20 [3]	JSSC'22 [4]	JSSC'22 [8]	
Technology	**250-nm InP HBT**		250-nm InP HBT	16-nm FinFet	45-nm CMOS SOI	130-nm SiGe	
Frequency (GHz)	**115 - 140**		125 - 150	114 - 136	130 - 151	107 - 135	
Gain(dB)	**19.8**		20.3	20.5	24.8	21.8	
P_{sat}/OP_{1dB} (dBm)	**21/20.4**		20.5/17	15/9.2	18.5/13.5	22.6/17.1*	
PAE @ OP_{1dB} (%)	**23**		14*	5*	1.1*	12.7*	
PAE @ 6dB PBO (%)	**10.4**		2.4*	1.2*	<1*	11.7	
Modulation scheme	**16-QAM**	**64-QAM**	-	-	-	16-QAM	64-QAM
Data rate(Gb/s)	**10 (20§)**	**6**	-	-	-	8	4.8
EVM (%)	**15.7(15.9§)**	**7.95‡**	-	-	-	11.6‡	10.9‡
P_{avg} (dBm)	**16.3(15.7§)**	**12.9**	-	-	-	13.7	13.8
PAE_{avg} (%)	**11.9(10.6§)**	**6.5**	-	-	-	7.8	7.9
Area (mm²)	**0.61(0.286†)**		0.68	0.062(0.041†)	0.46†	1.11(0.58†)	

*= Graphically estimated. †= Core area not including DC and RF pads. ‡= Measured using postdistortion.§= Measured using equalization

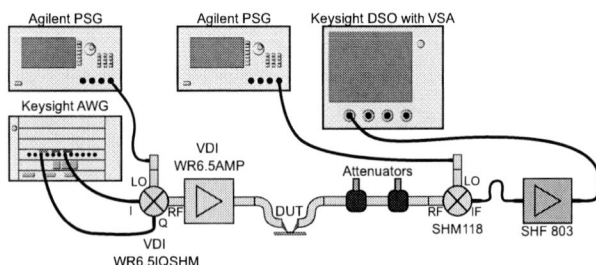

Fig. 8. EVM measurement setup

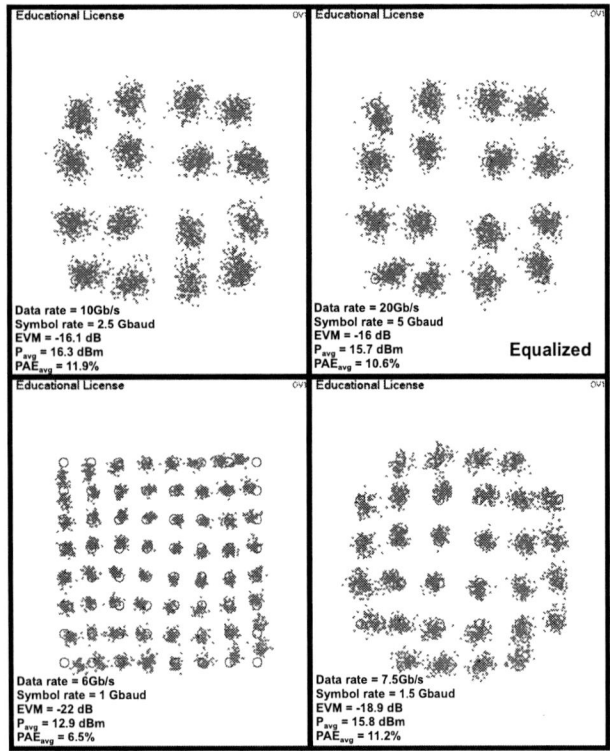

Fig. 9. Constellation diagram, EVM, P_{avg}, PAE_{avg} and data rate for 16-QAM, 32-QAM and 64-QAM modulation

to the high OP_{1dB} of the PA, it is able to operate very close to P_{sat} and still achieve reasonable EVM performance. For 64-QAM modulation, the PA has to operate very deep into PBO, mainly due to the AM-PM distortion that can be seen by the rotation of the outer points in the constellation diagram.

IV. CONCLUSION

This paper presented a high-power, high-efficiency D-band amplifier in a 250-nm InP HBT technology. The use of transformers in combination with the common-base stages results in high efficiency and high output power. Moreover, the OP_{1dB} is very close to the P_{sat} thanks to the biasing network and deep class AB operation of the PA, allowing the PA to operate very close P_{sat} while still maintaining reasonable EVM. In comparison with other state-of-the-art D-band PAs in table 1, this PA provides the highest OP_{1dB} and the highest associated PAE. In modulated data measurements, the PA shows record P_{avg} and PAE_{avg}, making it suitable for D-band communication systems.

REFERENCES

[1] M. J. W. Rodwell, J. Rode, H. W. Chiang, P. Choudhary, T. Reed, E. Bloch, S. Danesgar, H.-C. Park, A. C. Gossard, B. J. Thibeault, W. Mitchell, M. Urteaga, Z. Griffith, J. Hacker, M. Seo, and B. Brar, "THz Indium Phosphide Bipolar Transistor Technology," in *2012 IEEE Compd. Semicond. Integr. Circuit Symp. CSICS*, Oct. 2012, pp. 1–4.

[2] A. S. H. Ahmed, M. Seo, A. A. Farid, M. Urteaga, J. F. Buckwalter, and M. J. W. Rodwell, "A 140GHz power amplifier with 20.5dBm output power and 20.8% PAE in 250-nm InP HBT technology," in *2020 IEEEMTT- Int. Microw. Symp. IMS*, Aug. 2020, pp. 492–495.

[3] B. Philippe and P. Reynaert, "24.7 A 15dBm 12.8%-PAE Compact D-Band Power Amplifier with Two-Way Power Combining in 16nm FinFET CMOS," in *2020 IEEE Int. Solid- State Circuits Conf. - ISSCC*, Feb. 2020, pp. 374–376.

[4] S. Li and G. M. Rebeiz, "High Efficiency D-Band Multiway Power Combined Amplifiers With 17.5–19-dBm Psat and 14.2–12.1% Peak PAE in 45-nm CMOS RFSOI," *IEEE J. Solid-State Circuits*, vol. 57, no. 5, pp. 1332–1343, May 2022.

[5] Z. Liu, T. Sharma, C. R. Chappidi, S. Venkatesh, and K. Sengupta, "Transformer-based Broadband mm-Wave InP PA across 42-62 GHz with Enhanced Linearity and Second Harmonic Engineering," in *2020 IEEEMTT- Int. Microw. Symp. IMS*, Aug. 2020, pp. 1295–1298.

[6] C. M. Grens, J. D. Cressler, J. M. Andrews, Q. Liang, and A. J. Joseph, "The Effects of Scaling and Bias Configuration on Operating-Voltage Constraints in SiGe HBTs for Mixed-Signal Circuits," *IEEE Trans. Electron Devices*, vol. 54, no. 7, pp. 1605–1616, Jul. 2007.

[7] Y. Yang, K. Choi, and K. Weller, "DC boosting effect of active bias circuits and its optimization for class-AB InGaP-GaAs HBT power amplifiers," *IEEE Trans. Microw. Theory Tech.*, vol. 52, no. 5, pp. 1455–1463, May 2004.

[8] X. Li, W. Chen, H. Wu, S. Li, X. Yi, R. Han, and Z. Feng, "A 110-to-130 GHz SiGe BiCMOS Doherty Power Amplifier With a Slotline-Based Power Combiner," *IEEE J. Solid-State Circuits*, vol. 57, no. 12, pp. 3567–3581, Dec. 2022.

RTu3C-4

305-GHz Cascode Power Amplifier Using Capacitive Feedback Fabricated Using SiGe HBT's with f_{max} of 450 GHz

Suprovo Ghosh, Frank Zhang, Haidong Guo, Kenneth K. O

TxACE & Department of ECE, The University of Texas at Dallas, USA

suprovo.ghosh@utdallas.edu

Abstract—A 305-GHz power amplifier (PA) fabricated in a 130-nm SiGe HBT BiCMOS technology with HBT f_t / f_{max} = 350/450 GHz is presented. The PA employs 4 cascode amplification stages with capacitive feedback between the collector of common base stage and the base of common emitter stage that increases power gain of each stage by ~4 dB and a 4-way power combiner at the output. The PA achieves a measured P_{sat} of 7.5 dBm and OP_{1dB} of 4.5 dBm at 290 GHz. The design reaches a peak small signal gain of 14.5 dB at 305 GHz. The circuit consumes 1008 mW DC power from a 4-V supply and achieves a PAE_{max} of 0.39%. The PA exhibits the highest P_{sat} and OP_{1dB} at 290 GHz, and the highest small signal gain at 305 GHz among the PA's fabricated using SiGe HBT's with f_{max} less than 500 GHz.

Keywords—Power amplifier (PA), feedback capacitance, gain-boosting, power combiner, SiGe HBT, cascode, millimeter wave.

I. INTRODUCTION

Utilizing the frequency bands located above 100 GHz has attracted interests due to their potential to provide a large communication bandwidth. With the development of SiGe-HBT BiCMOS processes with f_{max} > 450 GHz [1], amplifications at frequencies over 300 GHz has become possible. Gain and P_{sat} of power amplifiers (PA) become limited as the operating frequency approaches ½ f_{max} due to the rapid decrease of maximum available/stable gain (MAG/MSG) of transistors. Numerous design techniques for increasing the gain and output power of PA's operating near ½ f_{max} have been reported over the years [2]-[8]. These include both higher order power combining and circuit techniques to increase the intrinsic gain of the amplifier cores. However, it remains a challenge to achieve high gain and output power for SiGe HBT PA's operating at frequencies close to 300 GHz.

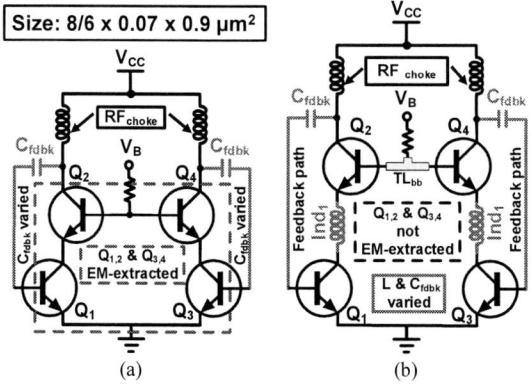

Fig. 1. (a) Schematic of an EM-extracted cascode amplifier with feedback capacitance. (b) Schematic of a cascode amplifier (not extracted) with feedback capacitance and inter-stage inductance.

In this paper, a 4-stage cascode PA with capacitive feedback employing a 4-way power combining technique is presented. The PA achieves a measured peak gain of 14.5 dB at 305 GHz, measured P_{sat} of 7.5 dBm and OP_{1dB} of 4.5 dBm at 290 GHz, which are the highest at the specified frequencies for PA's implemented using SiGe HBT's with f_{max} less than 500 GHz. The P_{sat} and OP_{1dB} at 305 GHz could not be measured because of the instrumentation limitation. Simulations suggest that P_{sat} and OP_{1dB} at 305 GHz should be ~1 dB higher than that at 290 GHz. The capacitive feedback increases the power gain and output power capability especially at frequencies close to 300 GHz. The PA is fabricated in the IHP 130-nm SiGe BiCMOS process with 7-layer Aluminium metallization which supports HBT f_t / f_{max} of 350/450 GHz.

II. 300-GHz PA DESIGN

A. PA Unit Design

The simulated maximum available gain (without including the interconnect parasitics) of a common-emitter amplifier in this technology is 3.2 dB, while that of a cascode amplifier is 8 dB at 280 GHz. Hence, a pseudo-differential cascode topology is chosen. The transistor sizes are 6 x 0.07 x 0.9 μm² for the 1st and 2nd stages and 8 x 0.07 x 0.9 μm² for the 3rd and 4th stages.

Fig. 1(a) shows a differential cascode pair with a feedback capacitance (C_{fdbk}) between the input of common-emitter stage and output of the common-base stage. The 3-D model of the circuit is shown in Fig. 2. The bases of the CB stages are shorted together to provide a virtual ground in the middle and is connected in series to a 1-kΩ resistor. This boosts the small signal gain and improves the common-mode stability of the circuit [2]-[3]. The passives and interconnects are simulated/extracted using EM simulation of a 21-port structure in Ansys HFSS (Fig. 2).

B. Gain-boosting using Feedback Capacitance

The maximum available gain/maximum stable gain MAG/MSG of the circuit in Fig. 1(a) at V_{cc} = 4 V is simulated at varying frequencies and plotted versus feedback capacitance in Fig. 3. Negative values of capacitance represent cross-coupled feedback capacitance [9] or inductive feedback between the input and output [10] and are often used to boost the gain at lower frequencies. However, above 240 GHz, the feedback capacitance is more effective for increasing the gain. At 280 and 300 GHz, the feedback capacitance can boost the gain by 4 dB. This is attributed to the impact of parasitic inductance of the interconnect between the output of CE stage

979-8-3503-2123-4/23 $31.00 © 2023 IEEE   313   2023 IEEE Radio Frequency Integrated Circuits Symposium

and input of CB stage at frequencies greater than ~240 GHz. Also, it is observed that the feedback capacitance to peak the gain increases with frequency.

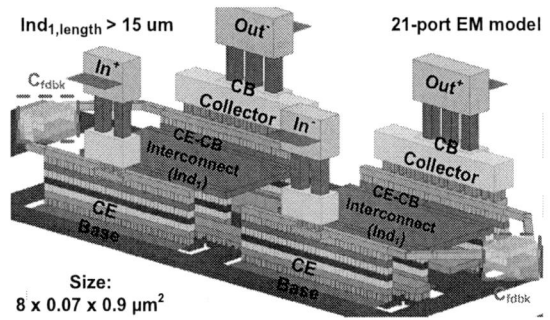

Fig. 2. EM-extracted layout of a cascode amplifier with feedback capacitance.

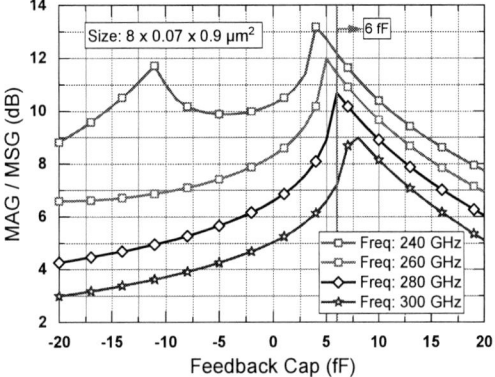

Fig. 3. Simulated MAG/MSG of an EM-extracted cascode amplifier versus feedback capacitance at varying frequencies.

To better understand this, the circuit in Fig. 1(a) is further examined using the native model for the transistors without extraction and inductors (Ind$_1$) with Q of 10 at 280 GHz as shown in Fig. 1(b). 'Ind$_1$' represents the inductance of the interconnect between the CE and CB stages. Furthermore, the base interconnect in the CB stage that adds series inductance and increases the gain [2],[5],[11] is included using an extracted 3-port network (TL$_{bb}$) from EM simulations. In Fig. 4, the MAG/MSG is plotted at 280 GHz versus feedback capacitance for different values of inductance. For C$_{fdbk}$ = 0 fF, it is seen as the inductance increases, the gain decreases and can be degraded by ~3 dB when Ind$_1$ = 12 pH. This inductance in the layout is unavoidable because of the design-rule limitations in placing the CB and CE transistors closer as shown in Fig. 2. For Ind$_1$ = 0 & 4 pH, the gain peaks higher for negative capacitances, but as the inductance increases, the gain peaks at positive capacitances. The gain reaches its maximum for Ind$_1$ = 8 pH. Fig. 5 shows the frequency dependence of MAG/MSG for varying values of C$_{fdbk}$ at Ind$_1$ = 8 pH. As the frequency increases, the feedback capacitance to maximize MAG/MSG also increases. At 280 GHz, the gain reaches its maximum for C$_{fdbk}$ of 9.4fF.

After fixing the feedback capacitance (stages 1,2 = 4.5 fF and stages 3,4 = 6.5 fF), load-pull and source-pull simulations are used to determine the optimal termination impedances for each of the four stages. The interstage matching is implemented using a T-shaped transmission line network [2] with ~2-dB loss at 280 GHz. The 4-way combiner/divider is based on a three-conductor transmission line balun [5]. The overall loss of the combiner/divider is ~2-2.5 dB at 280 GHz.

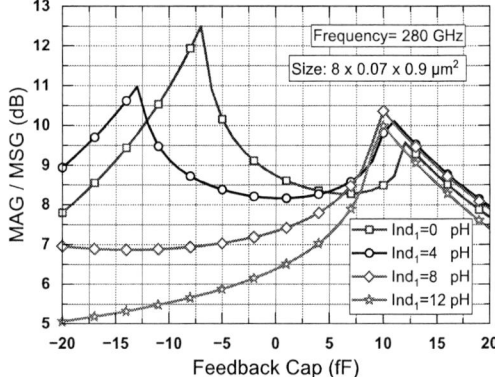

Fig. 4. Simulated MAG/MSG of a cascode amplifier (not extracted) versus feedback capacitance for different inductances at 280 GHz.

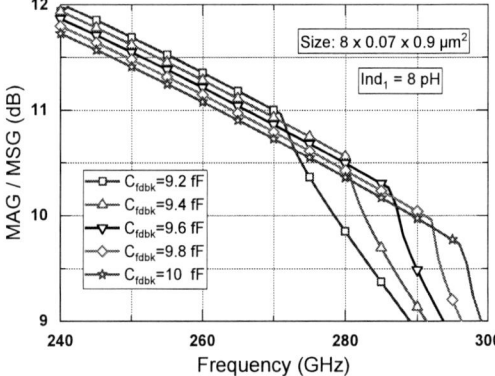

Fig. 5. Simulated MSG/MAG of a cascode amplifier (not extracted) versus frequency at varying capacitances at Ind$_1$ = 8 pH.

C. 4-Stage PA with 4-way power combining

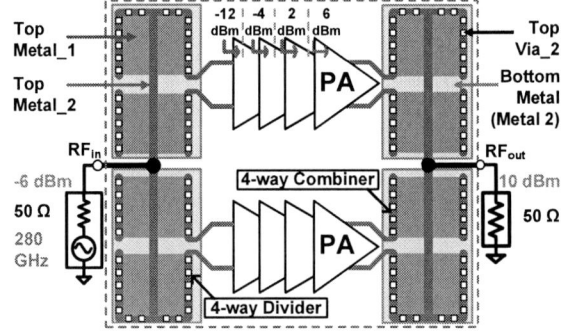

Fig. 6. Block diagram of the 4-stage PA with 4-way output combining.

Fig. 6 shows the overall block diagram of the 4-stage PA with 4-way output combining. The input power for each stage is shown through the chain. Fig. 7 shows the simulated output power and the |S$_{21}$| of the overall chain versus frequency. The 3-dB P$_{out}$ bandwidth of the chain for P$_{in}$ = 2 dBm or P$_{sat}$ is approximately 60 GHz while the simulated |S$_{21}$| 3-dB bandwidth is 13 GHz. The small signal 3-dB bandwidth is

979-8-3503-2123-4/23 $31.00 © 2023 IEEE   314

reduced by the frequency response of gain enhancement technique using capacitive feedback. As shown in Fig. 3, for a feedback capacitance of 6 fF, the peak gain of an amplification stage is 10.5 dB at 280 GHz while it is ~7.0 dB at 300 GHz.

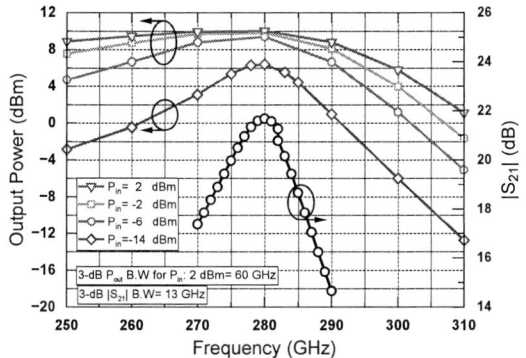

Fig. 7. Simulated P_{out} for different values of P_{in} and $|S_{21}|$ versus frequency.

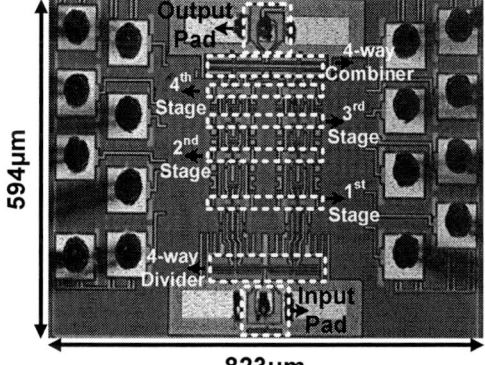

Fig. 8. Die photograph of 4-stage PA with 4-way combining at the output.

III. MEASUREMENT RESULTS

Fig. 8 shows a die photograph of the fabricated PA. Including bond pads, the circuit occupies an area of 594 μm x 823 μm. The 1st and 2nd stages are biased at 49 mA current and 77 mA for the 3rd and 4th stages from a 4-V supply. The PA is measured using two GSG Formfactor infinity waveguide probes (WR-3 band), while the DC connections are made on a PCB using bond wires. The S-parameters are measured using a Keysight PNA-X along with OML WR-3 VNA frequency extenders. A SOLT CS-15 substrate is used for the calibration. Fig. 10 compares the simulated and measured S-parameters of the PA. The measured peak $|S_{21}|$ is 14.5 dB at 305 GHz while the simulated peak $|S_{21}|$ is 21 dB at 280 GHz. The measured 3-dB bandwidth is approximately 13 GHz which is the same as the simulated. The measured k-factor in Fig. 11 shows that the amplifier is unconditionally stable throughout the entire WR-3 band. The measured centre frequency is shifted up by ~10% from the simulated frequency. This discrepancy is currently being investigated. Fig. 9 shows the setup for large signal measurements. An input signal generated by a Keysight PSG drives a 270-290 GHz VDI AMC-S221 chain to provide the RF input signal for the PA. The output power is measured by a PM4 Erickson power meter. The losses of WR-3 waveguide S-bends, WR-3 to WR-10 tapers and probes are de-embedded from the measured input and output power.

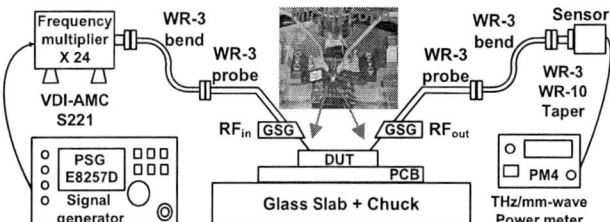

Fig. 9. Setup for the large-signal measurements.

The measured and simulated output power, power gain and PAE are plotted as a function of input power in Fig. 12. The PA measures a saturated output power (P_{sat} : P_{out} at 5-dB compression point) of 7.5 dBm when P_{in} is 2.4 dBm, output 1-dB compression power (OP_{1dB}) of 4.5 dBm, PAE_{max} of 0.39% and small signal power gain of 9 dB at 290 GHz. When P_{in} is lowered to 1.6 dBm, P_{out} is reduced to 7.3 dBm. The output power is also plotted as a function of frequency between 282-292 GHz when P_{in} is ~1.6 dBm and shown in Fig. 13. Over the 10 GHz, output power varies less than 1.5 dB, indicating that the bandwidth when operating near P_{sat} is substantially higher than the 3-dB bandwidth from the small signal frequency response. The output power of 7.5 dBm at 290 GHz is the

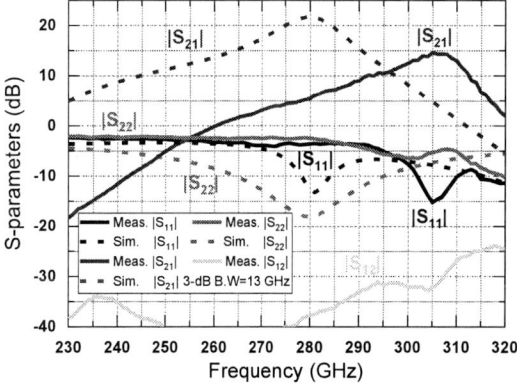

Fig. 10. Comparison of measured and simulated S-parameters of the PA.

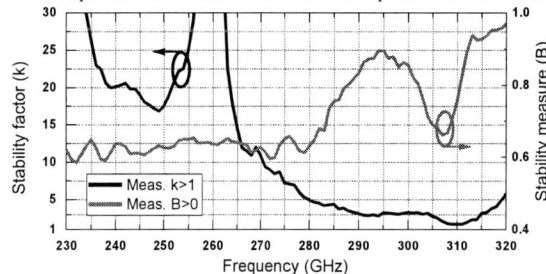

Fig. 11. Measured k-factor and B of the PA.

highest among the power amplifiers fabricated using SiGe HBT's with f_{max} less than 500 GHz. The power measurements are currently limited by the VDI-AMC source which works till 290 GHz. From the simulated P_{out} versus frequency for $P_{in} = 2$ dBm in Fig. 7, P_{out} is ~1 dB lower when measured at the frequency ~15 GHz lower than the peak gain frequency. This means if the PA is measured at 305 GHz with P_{in} of 2 dBm its P_{out} should be ~ 8.5 dBm and PAE should also be higher. A frequency multiplier for the measurements at 305 GHz is on order.

979-8-3503-2123-4/23 $31.00 © 2023 IEEE   315

Table 1. Comparison with the state-of-the-art SiGe HBT power amplifiers operating between 250 and 300 GHz.

Ref.	Tech.	f_{max} (GHz)	Freq. (GHz)	Topology	V_{cc} (V)	P_{sat} (dBm)	OP_{1dB} (dBm)	PAE (%)	$\|S_{21}\|$ (dB)	3-dB BW $\|S_{21}\|$ (GHz)	P_{dc} (mW)	Area (mm²)
[4]	130-nm SiGe	550	275	Diff. 6-stages CB	1.77	>-8	-10	0.07	10	7	123	0.38
[2]	130-nm SiGe	500	252	Diff 3-stages Casc.	3.3	>0	-3.7	0.3 (at P_{1dB})	21.5	11	149	0.17
[5]	130-nm SiGe	450	290	4-way Comb. 4-stages Casc.	2.5	5	3.3	1.19	15 (297 GHz) 13.5 (305 GHz)	67 (247-314)	268	0.58
[6]	130-nm SiGe	650	290 259	Diff. 3-stages Casc.	4	7.4 8.2(9.7*)	5 5.2(6.7*)	1.25 1.38	20.1 (254 GHz) 16 (305 GHz)	63 (239-302)	360	0.26
[7]	130-nm SiGe	650	290 275	8-way Comb. 3-stages Casc.	3	10.1 11.5(13.5*)	8 7.5(9.5*)	- 1.26* (at P_{1dB})	15 (250 GHz) 10 (305 GHz)	100 (220-320)	710	0.8
[8]	130-nm SiGe	650	290 267	Diff. 3-stages Casc.	-	6 6.4(8.1*)	- 5.2	- 0.92	17.9 (267 GHz) 13 (305 GHz)	59 (239-298)	417	0.26
This work	130-nm SiGe	450	290	4-way Comb. 4-stages Casc.	4	7.5	4.5	0.39	14.5 (305 GHz)	13	1008	0.49

* With de-embedding the losses of input and output baluns and pads.

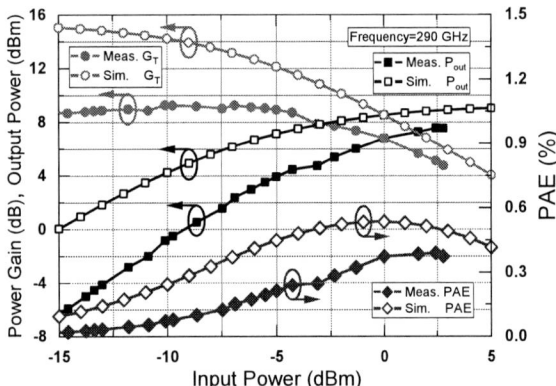

Fig. 12. Comparison of measured and simulated large signal parameters.

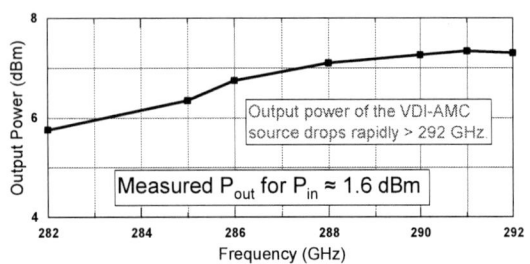

Fig. 13. Output power of the PA versus frequency for a P_{in} of 1.6 dBm.

IV. CONCLUSION

Capacitive feedback in a cascode SiGe HBT differential pair increases the MAG/MSG of PA's more than inductive feedback [10] at frequencies above ~240 GHz. A 4-stage cascode PA with capacitive feedback employing a 4-way power combiner at the output achieves P_{sat} of 7.5 dBm and OP_{1dB} of 4.5 dBm at 290 GHz, and a peak $\|S_{21}\|$ of 14.5 dB at 305 GHz. At 305 GHz, P_{sat} is expected to be ~1dB higher which should also increase the PAE. The PA achieves the highest P_{sat} and OP_{1dB} at 290 GHz and the highest $\|S_{21}\|$ at 305 GHz among the PA's fabricated using SiGe HBT's with f_{max} less than 500 GHz.

ACKNOWLEDGMENT

This work was supported in part by the SRC Joint University Microelectronic Program (JUMP) and DARPA, and in part by the Regents of the University of California. The authors would also like to thank Aniello Franzese of Infineon (IHP) for the technical discussions.

REFERENCES

[1] P. Chevalier et al., "SiGe BiCMOS Current Status and Future Trends in Europe," IEEE BiCMOS and Compound Semiconductor Integrated Circuits and Technology Symposium, pp. 64-71, 2018, San Diego, CA.

[2] H. Li et al., "A 250-GHz Differential SiGe Amplifier With 21.5-dB Gain for Sub-THz Transmitters," in IEEE Transactions on Terahertz Science and Technology, vol. 10, no. 6, pp. 624-633, Nov. 2020.

[3] M. H. Eissa et al., "A 13.5-dBm 200–255-GHz 4-Way Power Amplifier and Frequency Source in 130-nm BiCMOS," in IEEE Solid-State Circuits Letters, vol. 2, no. 11, pp. 268-271, Nov. 2019.

[4] S. Malz et al., "A 275 GHz amplifier in 0.3 μm SiGe," 2016 11th European Microwave Integrated Circuits Conference (EuMIC), London, UK, 2016, pp. 185-188.

[5] X. Li et al., "A 250–310 GHz PA With 15-dB Peak Gain in 130-nm SiGe BiCMOS Process for Terahertz Wireless System," in IEEE Transactions on Terahertz Science and Technology, vol. 12, no. 1, pp. 1-12, Jan. 2022.

[6] T. Bücher et al., "A Broadband 300 GHz Power Amplifier in a 130 nm SiGe BiCMOS Technology for Communication Applications," in IEEE Journal of Solid-State Circuits, vol. 57, no. 7, pp. 2024-2034, July 2022.

[7] E. Mohamed et al., "220–320-GHz J-Band 4-Way Power Amplifier in Advanced 130-nm BiCMOS Technology," in IEEE Microwave and Wireless Components Letters, vol. 32, no. 11, pp. 1335-1338, Nov. 2022.

[8] T. Bücher et al., "A 239–298 GHz Power Amplifier in an Advanced 130 nm SiGe BiCMOS Technology for Communications Applications," ESSCIRC 2021 - IEEE 47th European Solid State Circuits Conference (ESSCIRC), Grenoble, France, 2021, pp. 369-372.

[9] W. L. Chan and J. R. Long, "A 58–65 GHz Neutralized CMOS Power Amplifier With PAE Above 10% at 1-V Supply," in IEEE Journal of Solid-State Circuits, vol. 45, no. 3, pp. 554-564, March 2010.

[10] H. Bameri and O. Momeni, "A 200-GHz Power Amplifier With a Wideband Balanced Slot Power Combiner and 9.4-dBm Psat in 65-nm CMOS: Embedded Power Amplification," in IEEE Journal of Solid-State Circuits, vol. 56, no. 11, pp. 3318-3330, Nov. 2021.

[11] Y. Su and K. K. O, "An 800-μW 26-GHz CMOS tuned amplifier," IEEE Radio Frequency Integrated Circuits Symposium, pp. 151-154, June 2006, San Francisco, CA.

A

Abbasi, Mohammadreza 289
Aghighi, Amin 161, 229
Ahmed, Amr 93
Ahmed, Sherif S. 241
Alavi, Morteza S. 189
Alizadeh, Amirreza 85
Anders, Jens 253
Andersen, H. ....NA
Andree, Marcel 137
Arbabian, Amin 241
Arias-Purdue, Andrea 65
Arrunategui, Viviana 5
Asaf, O. 233
Avitabile, Gianfranco 169
Azevedo-Goncalves, J. 61

B

Babaie, Masoud 257
Babakhani, Aydin 153, 213
Baek, Min-Seok 125
Baek, Seungjae 273
Baeyens, Yves 305
Bahr, Bichoy 117
Bahrami, Sirous 237
Bai, Zhanjun 201
Ben-Atar, K. 233
Bevilacqua, Andrea 301
Bhagavatula, Venumadhav 53
Bilato, Andrea 133
Bluemm, Christian 13
Boon, Chirn Chye 105
Boyer, Chris ....NA
Buckwalter, James F. 5, 65

C

Carnu, Ovidiu 217
Cassiau, Nicolas 97
Chan, Lye Hock Kelvin ....NA
Chang, Mau-Chung Frank 3, 89
Chang, Ming-Cheng ....NA
Chang, Po-Yu 181
Chang, Tienyu 53
Chao, Wei-Pang 181
Chattopadhyay, Goutam 89
Che, Wenquan 269
Chen, Chien-Wei 113, 181
Chen, Hou-Shin 277
Chen, Jingjun 245
Chen, Long 81
Chen, Michael 293
Chen, Qian 105
Chen, Tsung-Ming 181
Chen, Wenhua 81, 249
Chen, Xiaofan 81
Chen, Ying 53
Chen, Zhilin 193
Chen, Zhiyu 201
Chevalier, P. 61
Chew, Koi Wai ....NA
Chi, Baoyong 29, 145, 165
Chi, Taiyun 293
Chiang, Pei-Yuan 201

Choi, Han-Woong 125
Choi, Hongseok 49
Choi, Kyujong 49
Choi, Seung-Uk 237
Choi, Sunkyu 125
Choi, Won-Suk 293
Chou, Yu-Ting 181
Chow, Wai Heng ....NA
Chowdhury, Srabanti 69
Chun, Dohoon 49
Chung, Yuan-Hung 181
Clemente, Antonio 97
Cohen, Emanuel 221, 297
Courouve, Pierre 97
Crémer, S. 61

D

Davidson, Alfred 225
Dayal, Pranav 277
De Filippi, Guglielmo 133
Dehos, Cedric 97
del Rio, David .... NA
Deng, Lei 197
Deng, Mingxing 145
Deng, Wei 29, 145, 165
Dens, Kristof 13
de Vreede, Leo C.N. 189
Dey, Samrat 53
Ding, Xuan 245
Dolt, David 109
Dong, Shenggang 293
Duan, Dongming 269
Duan, Zongming 197
Duriez, B. 61
Dyck, Alexander 13

E

Elkhouly, Mohamed 305
Entesari, Kamran 121, 261
Ermolov, Vladimir .... NA
Ershengoren, Natan 33
Essawy, Mostafa 161, 229

F

Fache, T. 61
Fakkel, Niels 257
Fang, Ran 145
Feng, Zhenghe 81, 249
Feygin, Gennady 277
Foglia-Manzillo, Francesco 97
François, B. .... NA
Frecassetti, Mario Giovanni .... NA
Fu, Jierui 121
Fu, Tao 17

G

Gaillard, F. 61
Gao, Hao 197
Gao, Huiyan 177
Gao, Yang 205
Garikapati, Sasank 225
Garimella, Sastry 225
Ghaedi Bardeh, Mohammad 121
Ghosh, Suprovo 313
Gielen, Senne 309
Ginzberg, Nimrod 221, 297
Gonzalez-Jimenez, Jose Luis 97
Gordon, M. 233
Grzyb, Janusz 137
Gu, Junjie 197
Gu, Qun Jane 177, 245
Guan, Pingda 145
Gui, Ping 17
Guimaraes, Gabriel 13
Gungor, Berke 13
Guo, Benqing 21
Guo, Haidong 313

H

Hamani, Abdelaziz 97
Han, Kefeng 197
Harjani, Ramesh 57
Haroun, Baher 117
Hartmann, J.-M. 61
He, Fei 149
Heinemann, Bernd 137
Heo, Yun Jung 101
Hoentschel, Jan .... NA
Holyoak, Michael 305
Hong, Songcheol 41, 45, 265
Hong, Wei-Kai 181
Hsu, Min-Shun 181
Hsueh, Yu-Li 113, 181
Hu, Jie 193
Hu, Suoping 201
Hu, Yaolong 293
Hu, Zhaoxin 281
Huang, Chang-Cheng 181
Huang, Xiangrong 165
Huang, Zhihong 9
Hung, Chao-Ching 113, 181
Hwang, Jun 49, 285

I

Ingels, Mark 309
Issakov, V. 209
Iwamoto, Masaya 69

J

Jafari Nokandi, Mostafa 37
Javorka, Peter ....NA
Jia, Haikun 29, 145, 165
Jin, Hyoungkyu 45
Jung, Joonho 273
Jung, Minjae 93

K

Kadota, Igor 225
Kalia, Sachin 117
Kambale, Akash 217
Kammar Nagaraja, Raghavendra ....NA
Kang, Daehyun 273
Kang, Hyungryul 9
Kang, Kai 269
Kang, Taeyoung 277
Kazan, Oguz 281
Kerdiles, S. 61
Kern, Michal 253
Kim, Choul-Young 125
Kim, Inhyun 9

Kim, Ji Hoon 273
Kim, Jinhyun 273
Kim, Jiseul 237
Kim, Joon-Hyung 125
Kim, Juwon 49
Kim, Kihyun 273
Kim, Kyunghwan 237
Kim, Sanghoek 101
Kim, Taein 101
Kim, Taewan 273
Kim, Yanghyo 89
Kim, Younghwan 273
Knorr, Andreas ....NA
Kooshkaki, Hossein Rahmanian 173
Krishnaswamy, Harish 225
Kuan, Yen-Cheng 177
Kumar, Ankur 9
Kumaran, Anil Kumar 189
Kuo, Chechun 53
Kwon, Kuduck 25
Kwon, Sungwon 49

L

Lam, Eythan 65
Lammert, V. 209
Landsberg, N. 233
Lau, Kim Kiu 277
Lau, Pak-Kim 277

Lee, Chong-Min 125
Lee, Donggu 25
Lee, Dong-hyun 273
Lee, Dongsoo 273
Lee, Eun-Gyu 125
Lee, Gyuha 45
Lee, Hyeonkeon 101
Lee, Jae-Eun 125
Lee, Jaehun 45
Lee, Jeiyoung 277
Lee, Jeong Ho 273
Lee, Jooseok 273
Lee, Kangseop 237
Lee, Sangho 273
Lee, Woojae 273
Lee, Wooram 289
Lee, Youngjoo 49
Leinonen, Marko E. 37
Levin, S. 233
Leyrer, M.L. 209
Li, Jiantong 293
Li, Lei 21
Li, Linjie 93
Li, Nayu 177
Li, Sensen 293
Li, Shuyang 249
Li, Xingcun 249
Liang, Ting-Wei 181
Liang, Xin 145

Liang, Yuan 105
Lim, Jung-Taek 125
Lin, Yui 217
Lipp, Dieter ....NA
Liu, Andrew 217
Liu, Edward 185
Liu, Huihua 269
Liu, Junqian 5
Liu, Ming-Chung 181
Liu, Ruida 9
Liu, Sen-You 181
Liu, Weitian 197
Long, Zhijun 193
Loo, Timothy 217
Loseu, Vitali 293
Lotfi, Hadi 253
Lu, Eric 181
Lu, Hang 177
Lu, Ivan Siu-Chuang 53, 277
Lu, Min 193
Lucci, L. 61
Luo, Yu-Lun 261
Luong, Howard C. 205
Lv, Guansheng 81

M

Ma, Xiaoxiao 193
Madsen, Christi 261
Maharry, Aaron 5
Maier, M. 209
Malka, Omer 221
Mangiavillano, Christoph 141
Manstretta, Danilo 169
Mazzanti, Andrea ....133, NA
Melamed, Itamar 221
Meng, Shaoyu 269
Mercier, Patrick P. 173
Meyer, A. 209
Min, Byung-Wook 49, 129, 285
Mitta, Rohish Kumar Reddy 141
Montaseri, Mohammad Hassan 37
Moody, Jesse 157
Morand, Y. 61
Moretto, Maurizio ....NA
Mortazavi, Mohsen 257
Moss, Stephen ....NA
Movaghar, Ghazal 5
Myeong, Jonghoon 129

N

Nagulu, Aravind 225
Nahmanny, D. 233
Nam, Mi Song 101
Narravula, Sridhar 217
Naseh, Navid 121
Natarajan, Arun 161, 229
Nemati, Hossein Mashad 189
Neumann, Philipp 253
Neviani, Andrea 301
Ni, Menghu 149

O

O, Kenneth K. 313
Oh, Hansik 273
Oh, Soyoung 273
O'Malley, Everett 65
Ong, Shih Ni ....NA
Ouyang, Keqing 193
Overwater, Ramon 257

P

Pahl, Philipp 69
Palermo, Samuel 9, 109, 261
Pallotta, Andrea ....NA
Paramesh, Jeyanandh 121
Park, Honghyeon 101
Park, Jae-hong 273
Park, Kyutae 49
Park, Sangyong 273
Park, Seonjeong 265
Park, Seungwon 273
Pärssinen, Aarno 37
Pashaeifar, Masoud 189
Pellerano, S. 233
Peng, Qiuyu 145
Perez Martinez, Rafael 69
Pfeiffer, Ullrich 137
Phan, Khoi T. 205
Piotto, Lorenzo 133
Pirbazari, Mahmoud ....NA
Poojary, Jitesh 57
Potier, C. ....NA
Pretl, Harald 141

Q

Qin, Haoqi 197

R

Rady, Ramy 261
Rahkonen, Timo 37
Rashed, Kareem 229
Razavian, Sam 153, 213
Razzaghi, Alireza 217
Rebeiz, Gabriel M. 93, 281
Reynaert, Patrick 13, 309
Ricco, Paolo 169
Rodwell, Mark J.W. 85
Roux, Pascal ....NA

S

S., Ramprasath 57
Sabbaghi Saber, Sajjad 245
Sahoo, Amit Kumar .... NA
Säily, Jussi ....NA
Sankaran, Swaminathan 117
Sapatnekar, Sachin S. 57
Sayginer, Mustafa 305
Schalberger, Patrick 253
Scharpf, Jochen 253
Schow, Clint 5
Schumacher, Tim 141
Schwartz, Ilai 253
Sebastiano, Fabio 257
Seo, Munkyo 85
Seok, Eunyoung 293
Sevillano, Juan Francisco ....NA
Shahramian, Shahriar 305
Sharma, Navneet 293
Shen, Xiaoliang 197
Shin, W. 233
Siligaris, Alexandre 97
Singh, Sumit Pratap 37
So, Cheol 41
Socher, Eran 33
Son, Jeong-Taek 125
Son, Juho 273
Son, Sangwon 53, 201, 277

Song, Chunyi 177
Song, Ho-Jin 237
Song, Jae-Hyeok 125
Song, Kee-Bong 277
Soylu, Utku 85
Stelzer, Andreas 141
Striegler, Nico 253
Su, Yong 293
Suh, Bohee 273
Suh, Bosung 49
Sung, Eun-Taek 41, 45, 265
Syed, Shafiullah ....NA

T

Tam, Sai-Wang 217
Tan, Kirby Kheng Seong ....NA
Tang, Adrian 89
Tang, Xiaochen 73, 77
Tao, Jonathan 65
Thomas, Sidharth 153, 213
Thota, B. ....NA
Ting, Hsiu-Hsien 181
Tomasin, Lorenzo 301
Tsai, S.H. ....NA
Tsang, Randy 217

U

Unden, Thomas 253

V

Vaes, Joren 13
Verma, Ashutosh 277
Vogrig, Daniele 301

W

Wan, Chee Wai ....NA
Wan, Ruichen 29
Wang, Haishi 21
Wang, Haoran 269
Wang, Hua 185
Wang, Seunghun 265
Wang, Shaogang 177
Wang, Xiyu 193
Wang, Yao-Chi 181
Wang, Yicheng 73, 77
Wang, Yong 73, 77
Wang, Zhaowu 73, 77
Wang, Zheng 149
Wang, Zhenyu 73, 77
Wang, Zhihua 165
Weiner, Joe 305
Weisman, N. 233
Wen, Xianshan 17
Wong, Alden 217
Wong, Jen Shuang ....NA
Wu, Huibo 249
Wu, Pi-An 181
Wu, Wanghua 201, 277
Wu, Yunqiu 269

X

Xie, Changsong 13
Xie, Qian 149
Xie, Xin 269
Xu, Gary 293
Xu, Hao 197
Xu, Jianjun 69
Xu, Weiwei 217
Xu, Xinguang 293
Xu, Zhiwei 177
Xue, Jiamin 145
Xue, Quan 269

Y

Yan, Na 197
Yang, Qiao .... NA
Yang, Sung-gi 273
Yao, Chih-Wei 201, 277
Ye, Wenjing 29
Yeh, C.S. .... NA
Yeo, Sung-Ku 125
Yi, Il-Min 9
Yin, Rui 197
Yoo, Dongho 49, 285
Yoo, Si-Wook 277
Younkin, Todd 1
Yu, Hai 245
Yu, Xiaohua 53
Yu, Yiming 269
Yuan, Yuan 9
Yuan, Yuexiaozhou 177
Yun, Sukju 25

Z

Zachl, Georg 141
Zhan, Jing-Hong Conan 181
Zhang, Frank 313
Zhang, Jingzhi 241
Zhang, Qingfeng 269
Zhang, Xiaohan 293
Zhang, Yang 309
Zhao, Chenxi 269
Zhao, Fuyuan 29
Zhao, Xingliang 277
Zhao, Zhixing .... NA
Zhou, Jingying 177
Zhou, Wanting 21
Ziegler, C. 209
Zierdt, Mike 305
Zolkov, Erez 297
Zussman, Gil 225

IEEE
445 Hoes Lane
Piscataway, NJ 08854-4141

ISBN 979-8-3503-2123-4